Geologic and Tectonic Development of the Caribbean Plate Boundary in Southern Central America

Edited by

Paul Mann
Institute for Geophysics
University of Texas at Austin
8701 Mopac Boulevard
Austin, Texas 78759

THE GEOLOGICAL SOCIETY OF AMERICA · 1888 ·

SPECIAL PAPER

295

1995

Published by The Geological Society of America, Inc.
3300 Penrose Place, P.O. Box 9140, Boulder, Colorado 80301

Printed in U.S.A.

GSA Books Science Editor Richard A. Hoppin

Library of Congress Cataloging-in-Publication Data

Geologic and tectonic development of the Caribbean Plate boundary in southern Central America / edited by Paul Mann.
p. cm. — (Special paper ; 295)
Includes bibliographical references and index.
ISBN 0-8137-2295-0
1. Geology—Central America. 2. Plate tectonics—Central America. I. Mann, Paul, 1956– . II. Series: Special paper (Geological Society of America) ; 295.
QE210.G46 1995
557.2—dc20 94-41170
CIP

Cover: Free-air gravity anomaly of the Caribbean Sea and Pacific Ocean adjacent to southern Central America. The gridded data file (world_grav.img.4) used to make this image is available from Sandwell et al. (1994). The image was prepared by L. Gahagan of the University of Texas Institute for Geophysics using the Generic Mapping Tools (GMT) software package described by Wessel and Smith (1991).

Sandwell, D. T., Yale, M. M., and Smith, W.H.F., 1994, ERS-1 geodetic mission reveals detailed tectonic structures, 1994 Fall Meeting Abstract Supplement: EOS, Transactions of the American Geophysical Union, v. 75, p. 155.

Wessel, P., and Smith, W.H.F., 1991, Free software helps map and display data: EOS, Transactions of the American Geophysical Union, v. 72, p. 445–446.

10 9 8 7 6 5 4 3 2 1

Dedication

This volume on the geology and tectonics of southern Central America is dedicated to Professor Richard Weyl (1912–1988), a pioneer in geologic and tectonic studies of the circum-Caribbean region.

* * * * *

Early years. Richard Weyl was born on August 10, 1912, in the city of Kiel in the province of Schleswig-Holstein on the north German coastal plain. His interest in geology began in early childhood in Kiel, and as an adolescent he pursued his interest in geology at the universities of Kiel, Innsbruck, Freiburg, and Heidelberg. He obtained his Ph.D. in 1934 at the University of Heidelberg under the supervision of Professor J. L. Wilser. Weyl's dissertation topic on the Rhine graben fault system along the Black Forest rift shoulder of southern Germany introduced him to the study of igneous rocks and structural geology. Following graduation from Heidelberg, he returned to the University of Kiel to assume a teaching position. At Kiel, he continued research on Tertiary stratig-

---

Contributed by Wolfgang Franke and Fritz Stibane with minor modifications by Paul Mann and Christoph Heubeck.

raphy, sedimentary facies, sedimentary petrography, and paleontology of the north German coastal plain. One specialty he developed during this time was the use of heavy minerals from sedimentary rocks as a tool for stratigraphic correlation. He supervised several Ph.D. projects at Kiel on heavy mineral studies and would later apply heavy minerals studies to rocks in Central America.

Greater Antilles. In 1938 and 1939 Weyl had the opportunity to make an extended field excursion to the Dominican Republic in the Greater Antilles in the northeastern Caribbean. Funding and organization for the trip were provided by the German-Dominican Tropical Research Institute, a research organization founded in 1937 and disbanded in 1939 at the outbreak of World War II. His written descriptions and photographs made during field excursions provide some of the first geologic descriptions of the Enriquillo Valley and the Cordillera Central of the Dominican Republic. Weyl would continue this pattern of publishing descriptions of high-quality photographs from his geological reconnaissance trips for the rest of his career. His detailed study of the structure and development of deeply eroded Cretaceous-Eocene island arc rocks of the Cordillera Central earned him tenure on the faculty at the University of Kiel. Weyl continued work in the Dominican Republic until the early 1950s and made a reconnaissance trip to the active Lesser Antilles island arc in the early 1960s. As did his contemporaries Harry H. Hess and Walter H. Bucher in the United States, Weyl struggled with pre–plate tectonic concepts of how extinct island arcs evolve into Alpine-type mountain belts.

During World War II, Weyl remained as a professor at the University of Kiel, where he developed an interest in the history of geology and was especially fascinated by the precise observations and deductions of Leonardo da Vinci and the wide-ranging travels and scientific observations of Alexander von Humboldt. During Weyl's career, he adapted the view of da Vinci and von Humboldt that geology is a subfield encompassed within the larger discipline of science and philosophy. He wrote four articles in the period of 1949 to 1958 on the geological observations of da Vinci.

Weyl's heavy mineral studies in Germany became a major focus of his work in the late 1940s and 1950s. He also wrote several articles during this period on the mining of metals in North America.

Central America. Shortly after World War II, Weyl initiated a program of regional geologic studies in northern and southern Central America that would occupy him for the next 35 years. Original fieldwork in El Salvador, Guatemala, Nicaragua, and Costa Rica focused on several areas: petrology and geochemistry of the Middle America volcanic arc (mostly in Costa Rica and El Salvador); stratigraphic, crustal, paleogeographic, and geomorphic development of Central America; and economic geology of Central America, with original studies of magnetite sands and metallogenesis. In 1956, Weyl left the University of Kiel to become chairman of the geology department at the University of Giessen. Collaborators on many of his Central American studies after leaving Kiel included Professor Hans Pichler (now at the University of Tübingen) and Professor Fritz Stibane (now at the University of Giessen).

Weyl was fascinated by the volcanic and tectonic evolution of the Central American land bridge through time and tried to find explanations at a time when plate tectonic paradigms were not yet available. His method was to make careful observations, compile these observations into detailed maps, and work on a regional scale without regard to political boundaries or the present-day land-sea interface. Many of his studies of the older section of arc crust exposed in Central America concentrated on the rapidly uplifting and deeply eroded Sierra de la Talamanca (see de Boer et al., this volume). Key discoveries by Weyl and colleagues included the realization that the southern Middle America island arc was built on oceanic crust (see Di Marco et al., this volume), the large amount of late Tertiary mountain building along the length of the land bridge (see Collins et al., this volume), and geomorphic evidence for Pleistocene glaciation in the Cordillera de Talamanca.

He expressed his multidisciplinary approach in two classic review volumes put out by the publishing house Gebrüder Borntraeger in their series on regional geology of the

Earth. The first volume, entitled *Geologie der Antillen,* was published only in German in 1966 and was never translated in English. The second volume, *Geology of Central America,* was originally published in German by Borntraeger in 1961, but published again by Borntraeger in a revised English version in 1980. The 1980 version reflects the impact of plate tectonics on models for the geologic and tectonic development of the region. In both the Greater Antilles and Central America volumes, Weyl displays his extensive knowledge of Caribbean regional geology and his concern for fostering future geologic studies in both regions. Both volumes would later serve novices and veterans of Caribbean geology as valuable source books for carefully compiled and accurate geologic information essential for planning further studies.

Later years. In 1956, Weyl became chairman of and reorganized the geology department at the University of Giessen that had been severely damaged during World War II. He remained at Giessen until his retirement in 1980, despite offers of chaired professorships from other universities. During this period, he was a popular figure among his colleagues and was elected to the dean of the faculty of science in 1960–1961 and to the rectorship of the university in 1968–1969. In the 1970s, Weyl, Pichler, and Stibane initiated regional studies of igneous rocks in northwestern South America and in Mexico. During this period, Weyl supervised one Ph.D. student, Axel von Erffa, in a study of sedimentation and erosion along the north coast of Colombia. He also became interested in the topic of soil conservation and environmental protection in Central and South America. In the late 1970s, he devoted himself to the updating of the English language revision of his volume *Geology of Central America,* that was published in 1980. Weyl passed away on December 15, 1988, after a prolonged illness. He is survived by his wife, Herta, and three children. Selections from his scientific legacy of 176 articles and books are categorized and listed at the back of this volume.

Printed in U.S.A.

Contents

Plates
(in pockets)

Preface

The large number of published plate tectonic models for the Jurassic to Recent tectonic development of the Caribbean region attests to its complex geologic and tectonic development. Southern Central America has long been recognized as a critical area for the study of Caribbean and eastern Pacific tectonics because it currently forms a segment of the circum-Pacific "ring of fire" of volcanically and seismically active volcanic arc systems; forms a Neogene land bridge between the North and South American continents; forms an oceanographic barrier or, alternatively, an oceanic corridor or "gateway" between the Caribbean Sea and the Pacific Ocean during the Cenozoic; and forms a seismically and tectonically active focal point where four large lithospheric plates converge in a complex manner (Fig. 1).

Paleoplate reconstructions, such as those by Bullard et al. (1965) and Pindell and Barrett (1990), leave no space for Central America in its present position when South America is closed against North America in pre–Late Jurassic time (Fig. 2). Faulting and deformed belts between diverse continental, arc, and oceanic fragments in the Central American region further suggests that plate mobility and mountain building events have occurred at the site of the present-day Central American land bridge over the last 165 m.y. (Fig. 3).

For the purpose of this volume, I define "southern Central America" as the combined land areas of the countries of Costa Rica and Panama and their adjacent offshore areas in the Pacific Ocean and Caribbean Sea (Fig. 1). For comparative purposes, the combined land areas of Panama (75,517 km^2) and Costa Rica (51,100 km^2) occupy about one-third the area of California (411,049 km^2) along the Pacific margin of the United States (Fig. 4). The discrepancy in geologic understanding of the two regions is considerable, as the combined population of southern Central America (about 5.5 million with perhaps two hundred professional geologists) is about 20% of that of California (about 30 million, including several thousand professional geologists).

Molnar and Sykes (1969) first placed southern Central America into a plate tectonic framework by pointing out that the region forms a tectonically active junction between four major lithospheric plates: Caribbean, South America, Cocos, and Nazca (Fig. 1). This recent plate tectonic activity has imposed a confusing overprint of late faults, large-scale rotations,

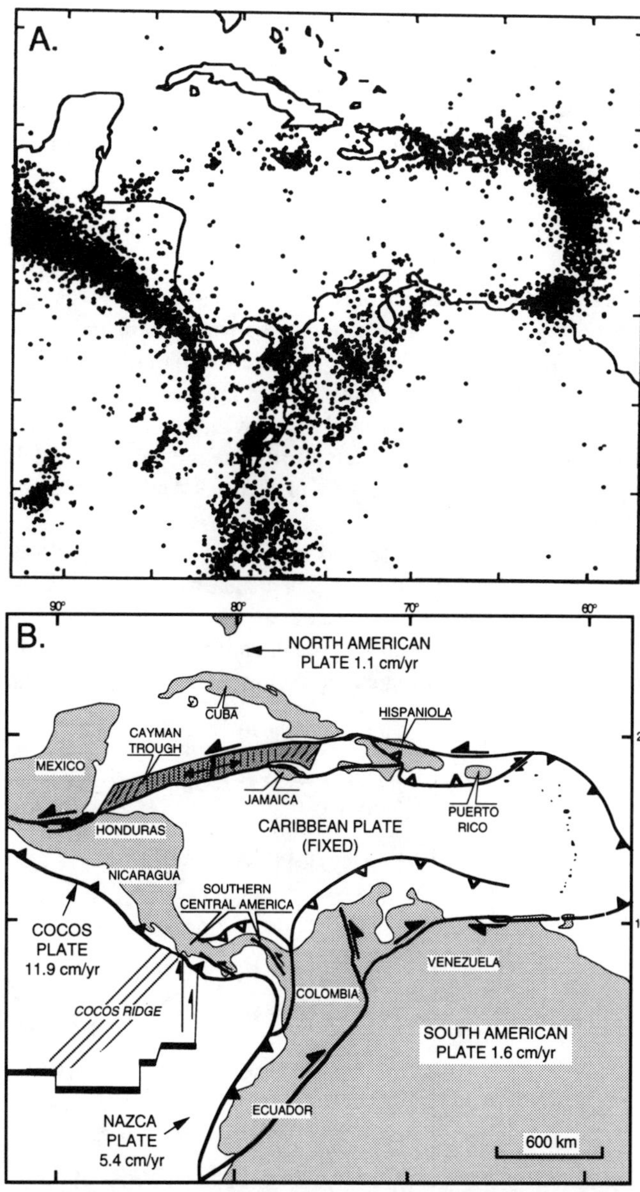

Figure 1. A, Earthquake epicenters of the Caribbean region from 1962 to 1992. B, Plate structure of the Caribbean. Note the diffuse seismicity and structural complexities in the area of southern Central America. Open thrust barbs indicate zones of overthrusting not associated with active volcanic arcs.

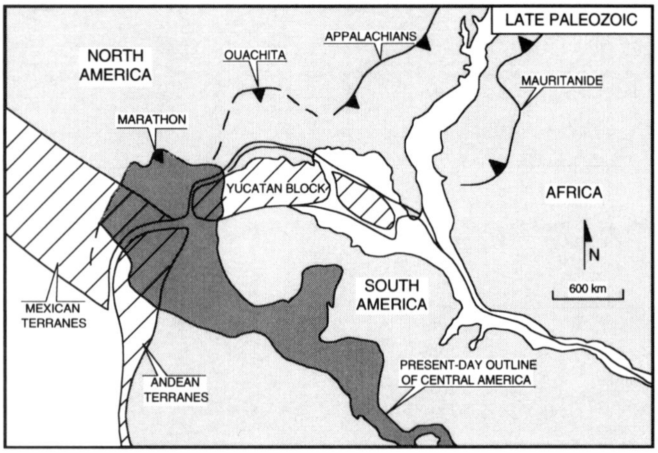

Figure 2. Late Paleozoic reconstruction of North and South America and Africa (modified from Pindell and Barrett, 1990). The present-day outline of Central America shows its Paleozoic overlap position with the various continental fragments.

and uplift but has also allowed exposure of older basement "blocks" recording earlier Mesozoic geological events and the history of plate motions in the region between North and South America (Fig. 3). The Jurassic to Recent geologic record exposed in southern Central America and inferred from offshore geophysical studies forms the basis for this volume and provides a rich source of new data for evaluating various models for the plate tectonic evolution of Central America and the Caribbean as a whole.

This volume is the second in what it is hoped will develop into a series of volumes that I plan to organize on the geology of the circum-Caribbean plate margins. The first volume, upon which the goals and organization of this volume are closely based, is "Geologic and Tectonic Development of the North America–Caribbean Plate Boundary in Hispaniola" (Mann et al., 1991).

PURPOSE

The goals of this volume, as outlined by me to invited volume participants in 1991, included the following:

• Continue the efforts of Professor Richard Weyl (see Dedication and Weyl, 1980) by establishing a *systematic geological data base and coherent stratigraphic nomenclature* for onland southern Central America in Panama and Costa Rica (Fig. 5). Such efforts are particularly needed in Panama, where there has been less academic research and consequently a smaller geological literature. Moreover, there has been little effort in correlating the onland stratigraphy and structure of Panama and Costa Rica.

• Begin efforts at better *correlation between onshore geology and marine geology.* There have been few efforts at such correlations in either Panama or Costa Rica because few workers attempt both marine and land studies as part of a single study.

• *Test models for the tectonic evolution* of southern Central America in Panama, Costa Rica, and offshore areas using various types of geologic and geophysical data.

• Attempt a *better integration of the disciplines of the geologic sciences* to solve regional geologic problems (for example, sedimentation and paleontology, marine and land geology, petrology and tectonics, basin analysis and tectonics, paleomagnetism and tectonics, seismology and tectonics, and geodetic studies and tectonics).

• Establish southern Central America and its offshore areas as an *important area in which to study a variety of plate boundary processes at all scales.* Global tectonic processes

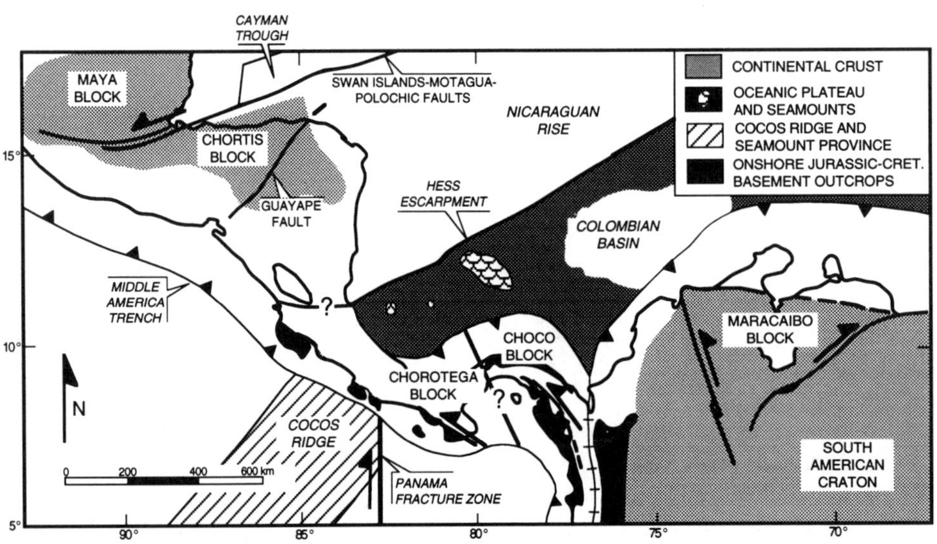

Figure 3. Basement types of Central America (from Case et al., 1990). Note the small size and faulted character of the diverse basement blocks.

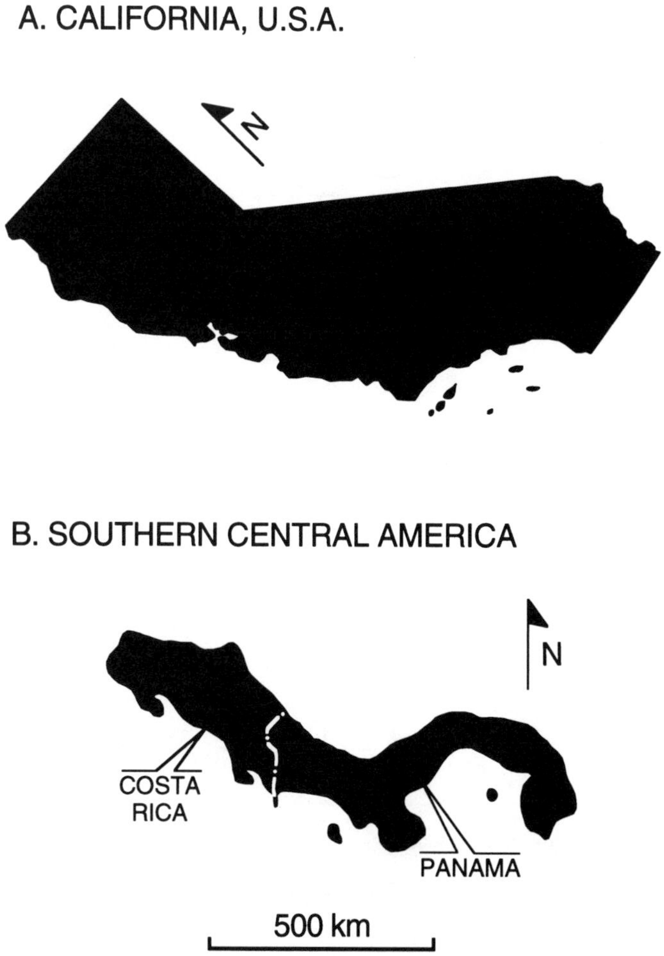

Figure 4. Comparison of the land areas of the state of California and southern Central America (Panama and Costa Rica). The areas shown are Albers equal-area projections that avoid size distortions related to latitude.

addressed in this volume include: (1) Accretion/subduction of oceanic and island arc terranes in arc settings and their effects on the growth of arc crust, (2) Origin and evolution of intra-oceanic island arcs, (3) Effects of oblique and shallow subduction on active and extinct island arcs, and (4) Growth of accretionary margins.

To achieve these goals, I have assembled 17 chapters on the geology and tectonics of Panama, Costa Rica, and offshore areas. Five of the chapters describe onshore geology, three describe a combination of onshore geology and offshore marine geophysical data and attempt land-sea correlations, six describe marine geophysical data from offshore areas, and three rely entirely on either seismological or geodetic data.

ORGANIZATION

I have organized the chapters of this volume into three sections. The latter half of this preface provides information that helps place the results reported in this volume into a better historical and geographic context. Part 1, of four chapters, describes and interprets the composition, structure, and paleomagnetism of the Late Cretaceous to Recent metamorphic, plutonic, volcanic, and related sedimentary rocks of the Middle America volcanic arc and proto-arc and of the oceanic rocks that form the "backbone" of basement rocks beneath western Panama and Costa Rica (see area of "Chorotega block" in Fig. 3, and Case et al. [1990] and Di Marco et al. [this volume] for definitions of crustal blocks shown). Part 2, of eight chapters, describes and interprets the Neogene tectonics and sedimentation of the Panama microplate. Five of these chapters rely on offshore geophysical data, much of which is published here for the first time; two attempt correlations between offshore geophysical data and onshore geologic studies; one relies on the results of a five-year campaign of Global Positioning System (GPS) measurements of plate and microplate motions interpreted together with gravity measurements. Part 3, of five chapters, focuses on the effects of shallow subduction of the Cocos Ridge on the structure of the Middle America arc in Costa Rica. One of these chapters relies on onshore geologic data, two rely on correlations between on- and offshore data, on relies entirely on offshore geophysical data, and two use seismology to infer the structure and deformation of the area beneath and within the crust of Costa Rica.

HISTORY OF GEOLOGIC INVESTIGATIONS IN PANAMA AND COSTA RICA

I subdivide the history of geology in this region in a manner similar to Weyl (1980) and Draper and Dengo (1990). The reader is inferred to these articles as well as Terry (1956) for additional information on the history of geology in southern Central America.

Spanish exploration and settlement

As in many areas of the New World, gold exploration was the major incentive for initial Spanish exploration and colonization in southern Central America. The Caribbean coasts of Costa Rica and Panama were discovered by Europeans during the voyage of Rodrigo de Bastidas in 1501 and of the fourth and final voyage to the Caribbean in 1502–1504 of Christopher Columbus (Fig. 6). Columbus had sailed from the area of northern Honduras to the Isthmus of Panama and paused long enough in Costa Rica to optimistically name the area "Rich Coast" on the basis of gold trinkets brought to him by the Indians. Vasco Núñez de Balboa walked along Indian trails across the narrowest part of the Isthmus of Panama in eastern Panama or Darién in 1513 to discover the "South Sea" or Pacific Ocean. Following the execution of Balboa in 1517, the new Spanish governor sponsored expeditions in 1522–1523 that explored from the present area of the Panama Canal along the Pacific coasts of Panama and Costa Rica to as far north as El Salvador and Honduras.

Although some gold and pearls were bartered for or taken

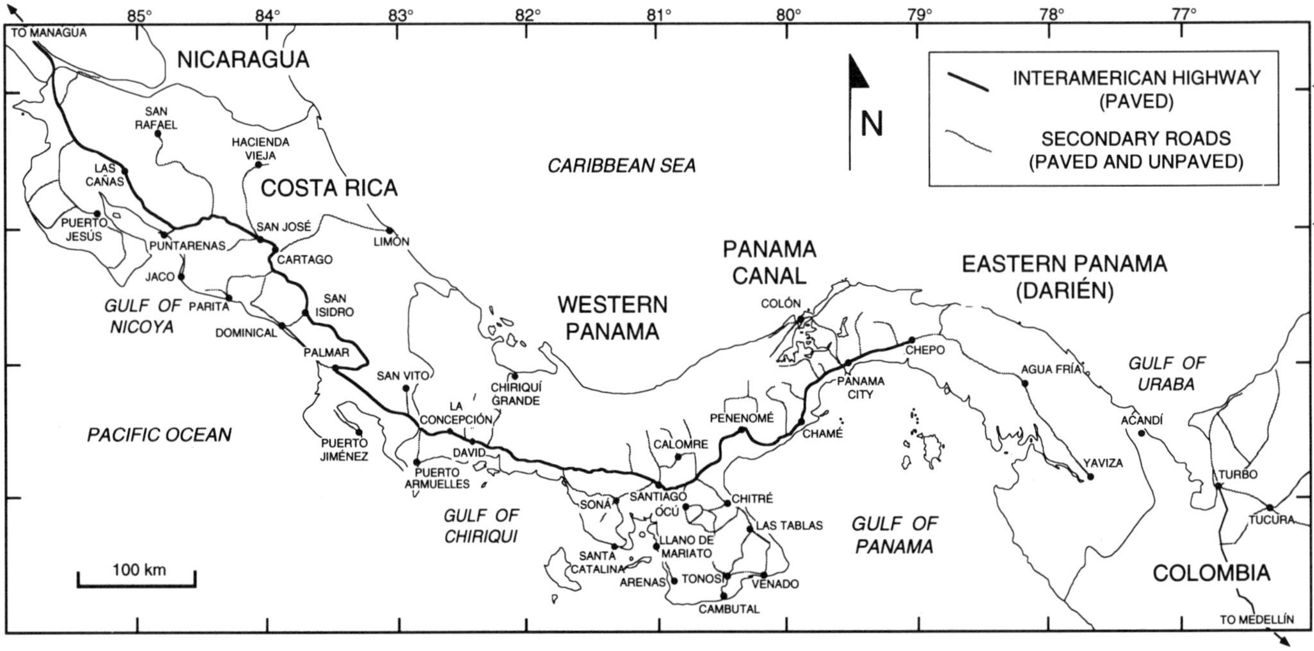

Figure 5. Location and road map of southern Central America.

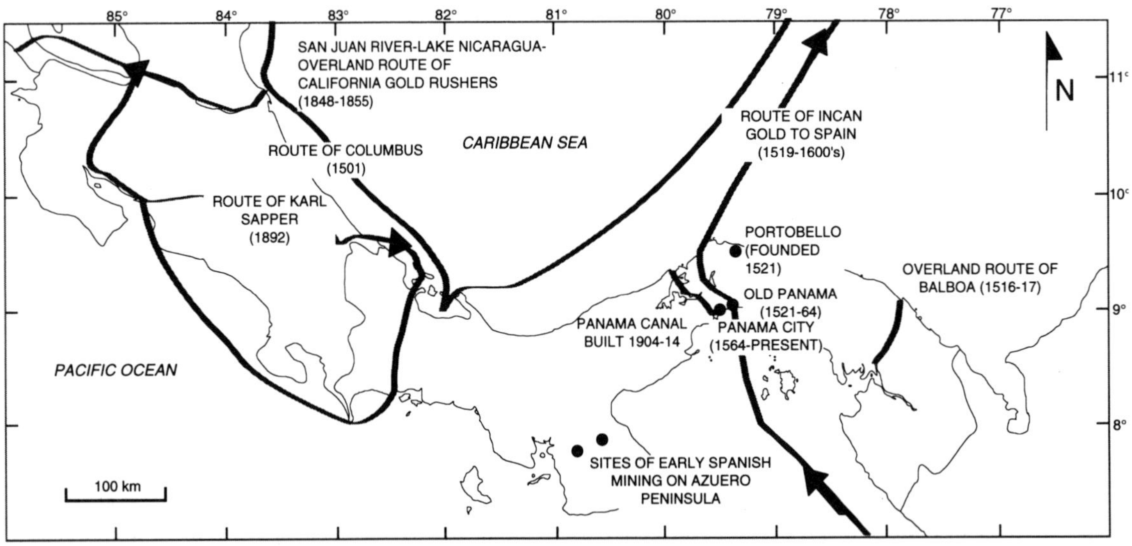

Figure 6. Areas of historical interest in southern Central America (see text for discussion).

from the Indians of Panama and Costa Rica, greater wealth was found to the south in Peru, where Francisco Pizarro used the Spanish colonial settlements in Panama as a staging area for his conquest of the Inca empire. Inca gold was transported by ship to Panamá Nuevo near the site of the present-day Panama City and from there by mules to the ports of Nombre de Dios and Portobelo near the site of the present-day Colón for shipment back to Spain (Fig. 6). Southern Central America therefore achieved its early importance in the Spanish empire as a transshipment and trade center, and exploitation of its native mineral resources was largely ignored.

Few economic mines were discovered in Panama or Costa Rica, although alluvial gold was panned by the Indians and colonists in rivers heading in plutonic and volcanic rocks of the arc rocks forming the continental divide. According to Terry (1956), the Spanish worked and exhausted a few areas of shallow gold mineralization in central and eastern Panama. The most famous mine in Panama was the Espíritu Santo gold mine of eastern Panama that was worked intermittently during the seventeenth and early eighteenth centuries by a succession of French, English, and Spanish adventurers and pirates (Woakes, 1923). The mine was rediscovered in the late nineteenth century, and production figures for the period 1899 to 1907 are given by Terry (1956).

Age of foreign explorer-geologists

Little geologic investigation of southern Central America seems to have occurred between the time of the early Spanish explorers and the late nineteenth century, when the area was visited by several U.S. and European explorer-geologists. U.S. geologist R. C. Taylor made an early foray along the central Caribbean coast of Panama to investigate historical gold panning localities (Taylor, 1852). German geographer and geologist M. Wagner visited several coastal and river localities along the Caribbean coast of Panama and noted the presence of plutonic rocks of the arc basement (1861, 1862). German geologist Karl von Seebach made the first regional study of volcanoes of Central America in 1864–1965, but this work was not published until after his death in 1892 (von Seebach, 1892). In 1873–1774, U.S. geologist William M. Gabb carried out stratigraphic studies on Tertiary rocks in the Limon belt along the Caribbean coast of Costa Rica (Gabb, 1875, 1881) and explored remote regions of the Cordillera de Talamanca of Costa Rica, where he recognized the post-Miocene age of the plutonic rocks (Gabb, 1895). The tireless German geologist Karl T. Sapper was dubbed "the father of geological research in Central America" by Charles Schuchert (1935) to commemorate Sapper's regional mapping efforts in both northern and southern Central America. Although most of Sapper's work was in northern Central America, he traveled through the area (Fig. 6) and published an early paper on mountain building in southern Central America (Sapper, 1905) and a later review of the regional geology of southern Central America (Sapper and Staub, 1937).

Overland and canal routes between the Caribbean and Pacific

A major impetus for the geologic exploration of Central America was travel along overland routes between the Caribbean and Pacific and the eventual construction of the Panama Canal by the United States between 1903 and 1914 (Fig. 6). Because the Central American isthmus is narrowest in Nicaragua, Costa Rica, and Panama, over 400,000 gold rushers passed overland in the years following the 1848 California gold rush. One popular route was to sail from the Caribbean Sea up the San Juan River, forming the border between Nicaragua and Costa Rica, and onto Lake Nicaragua. Ports on the Pacific Ocean lay less than 50 km west of the western shore of the lake. A transcontinental railroad across central Panama was completed by U.S. investors near the present site of the Panama Canal in 1855. The railroad assisted in transporting gold rushers to California and helped with planning canal construction. In 1902, the United States purchased the rights for a canal from the French consortium whose canal-building efforts between 1880 and 1889 were unsuccessful because of epidemics and inadequate equipment. The United States could not obtain permission from the Colombian government that controlled the isthmus, so helped sponsor a separatist revolution in 1903 that created a sovereign Panama state. In 1903 the United States obtained sovereignty over a 16-km-wide Canal Zone and completed construction of the canal in 1914.

Choice of the central Panama Isthmus as the site of the canal followed reconnaissance geologic studies by several teams of U.S. geologists in Nicaragua, Costa Rica, and Panama. U.S. reports on the geologic feasibility of a canal through Nicaragua included Hayes (1899) and through Panama included Carson (1874) and Maack (1874); later reports that accompanied or closely preceded excavation of the canal across the U.S.-controlled Canal Zone included Hill (1898), Hershey (1901), and Howe (1907). The accessibility and excellent exposures afforded by the construction and maintenance of both the Panama Canal and the railway parallel to the canal led to a large number of stratigraphic and paleontologic studies of the Miocene and younger marine sedimentary rocks in central Panama. Many type localities of the rich tropical micro- and macrofauna were described from the Canal Zone outcrops. Paleontologic studies included work by the eminent U.S. paleontologists and stratigraphers A. P. Brown and H. A. Pilsbry (Brown and Pilsbry, 1911–1913) and Thomas Wayland Vaughan (Vaughan, 1919a, b). Other paleontologic and stratigraphic studies carried out by French and German workers included Lemoine and Douvillé (1904), Toula (1909–1911), Dall (1912), and Douvillé (1915).

Canal maintenance and widening programs along with seismic hazard assessment led to continued geologic research by U.S. geologists employed by the U.S. Geological Survey and the Canal Zone. Included among their reports were MacDonald, 1913, 1915, 1919, and 1937; Kirkpatrick, 1939; Keen and Thompson, 1946; Jones, 1950. Distinguished North American regional geologist and paleontologist Wendell P. Woodring produced numerous reports (Woodring, 1949, 1954, 1955, 1966, 1974; Woodring and Thompson, 1949).

Oil exploration

As in other areas of the circum-Caribbean, oil replaced gold as the major impetus for geologic exploration in Panama and Costa Rica during the first half of the twentieth century. Despite geologic mapping, seismic profiling of offshore areas, and wildcat wells on- and offshore over a period of about 70 years, there have been no significant discoveries of oil or gas (Fig. 7). The first wildcat wells in Panama were drilled in the 1920s by U.S. companies on surface anticlines nested within the onshore, synclinal Bayano-Chucanauqué basin of eastern Panama (Johnson and Headington, 1971; Mann and Kolarsky, this volume). Small discoveries of oil were found but did not prove to be economic. More sophisticated submarine exploration using multichannel seismic data was carried out during the 1970s by Mobil and Exxon along the Pacific shelves of Panama. Several offshore wells were drilled, but oil was not discovered. Exploration in Costa Rica has mainly been carried out by a national oil company, RECOPE, which has also not been successful in discovering major oil or gas deposits despite numerous wildcat wells in the vicinity of surface oil and

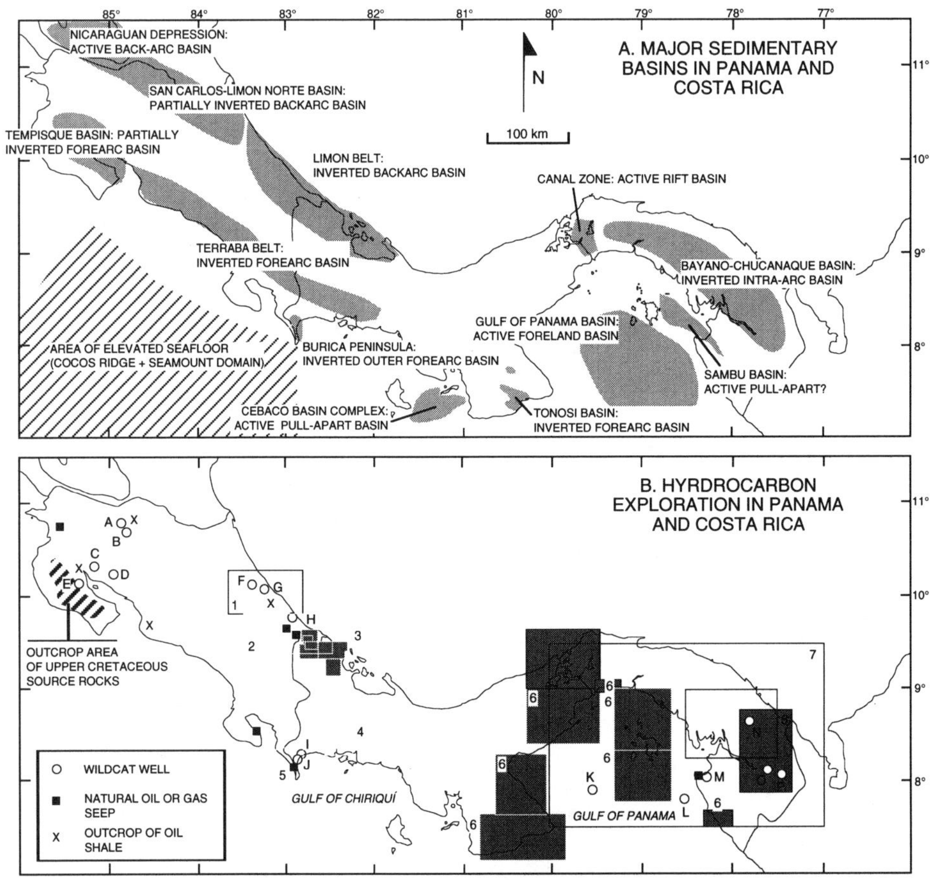

Figure 7. A, Major sedimentary basins in Panama and Costa Rica. The types of basins are classified according to the results presented in this volume. B, Hydrocarbon exploration in Panama and Costa Rica. Numbers indicate study areas of various companies: 1, 2 = RECOPE, 3 = Union, 4 = Ball Associates, 5 = Sinclair, 6 = Mobil, and 7 = Exxon. Letters indicate wildcat wells discussed by Sheehan et al. (1990).

gas seeps and high-quality oil source rocks (Sheehan et al., 1990) (Fig. 7). Several of the chapters in this volume may benefit future exploration efforts by placing the major basins into a better tectonic and stratigraphic framework (current interpretations of basin origins based on chapters in this volume are given in Fig. 7A).

Mineral exploration

The Mesozoic-Cenozoic geology of southern Central America was largely formed or influenced by arc volcanism and plutonism (Carr and Stoiber, 1990) (see chapters in Part 1 of this volume). Arc activity present over most of Costa Rica and in parts of Panama has led to a variety of mineral occurrences that have been catalogued by Kesler et al. (1990) (Fig. 8). Many of the mineralized areas are concentrated along the axis of the late Neogene volcanic arc where heat flow related to volcanism and plutonism is highest. High rates of tectonic uplift have produced exposed deeper levels of mineralization in both the volcanic arc and Pacific forearc areas of Panama and Costa Rica (de Boer et al., this volume). Although several rich deposits have been found in Panama and Costa Rica, falling mineral prices have delayed or closed most mining operations.

Beginning in the late 1960s, the United Nations Development Programme (UNDP) undertook a major program of geologic mapping and mineral exploration in Panama. This program has produced a series of four publicly available geologic map transects across the strike of the volcanic arc at a scale of 1:250,000 (Fig. 10). These maps established the basic stratigraphy and continue to form the basis for much of our understanding of the arc rocks of Panama.

Topographic mapping

From the 1950s to the present, the Interamerican Geodetic Survey has worked closely with government surveys in Panama and Costa Rica to produce and update a series of topographic maps at scales of 1:250,000 and 1:50,000. To increase the usefulness of this volume as a field guide, I have included an index map of available topographic sheets at both scales in Panama and Costa Rica (Figs. 9, 10). These maps are available for purchase at the national geographic institutes in Panama City and San José, Costa Rica, if a letter of introduction is obtained from a collaborating Panamanian or Costa Rican government agency.

Academic research

Academic research by both foreign and native geologists has made steady progress over the past 40 years in both Panama and Costa Rica. Professor Richard Weyl of the University of Giessen in Germany played a major role in promoting academic research in southern Central America through his original research and the 1980 publication of his outstanding geologic synthesis, *Geology of Central America* (see Dedication). This volume in English helped bridge the gap between geologic literature in German, Spanish, and English and remains the best compendium of geologic information on Central America.

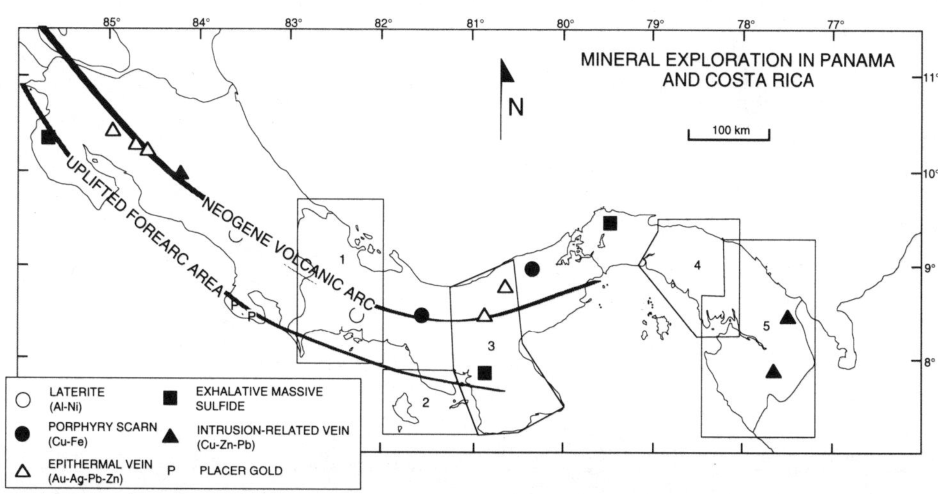

Figure 8. Map showing major mineral deposits of Panama and Costa Rica (modified from Kesler et al., 1990). Maps 1–5 are colored geologic maps made at a scale of 1:250,000 by geologists from the United Nations Development Programme.

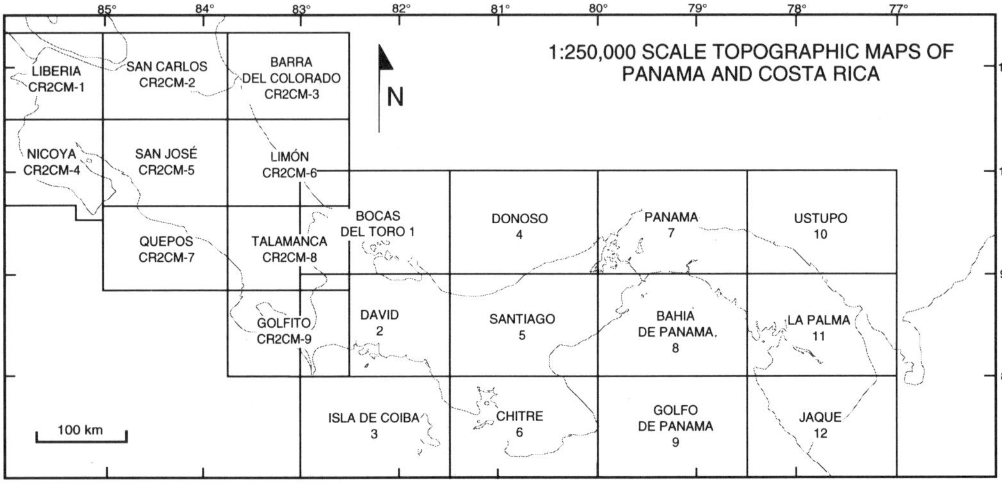

Figure 9. Key for 1:250,000 scale topographic maps in Panama and Costa Rica.

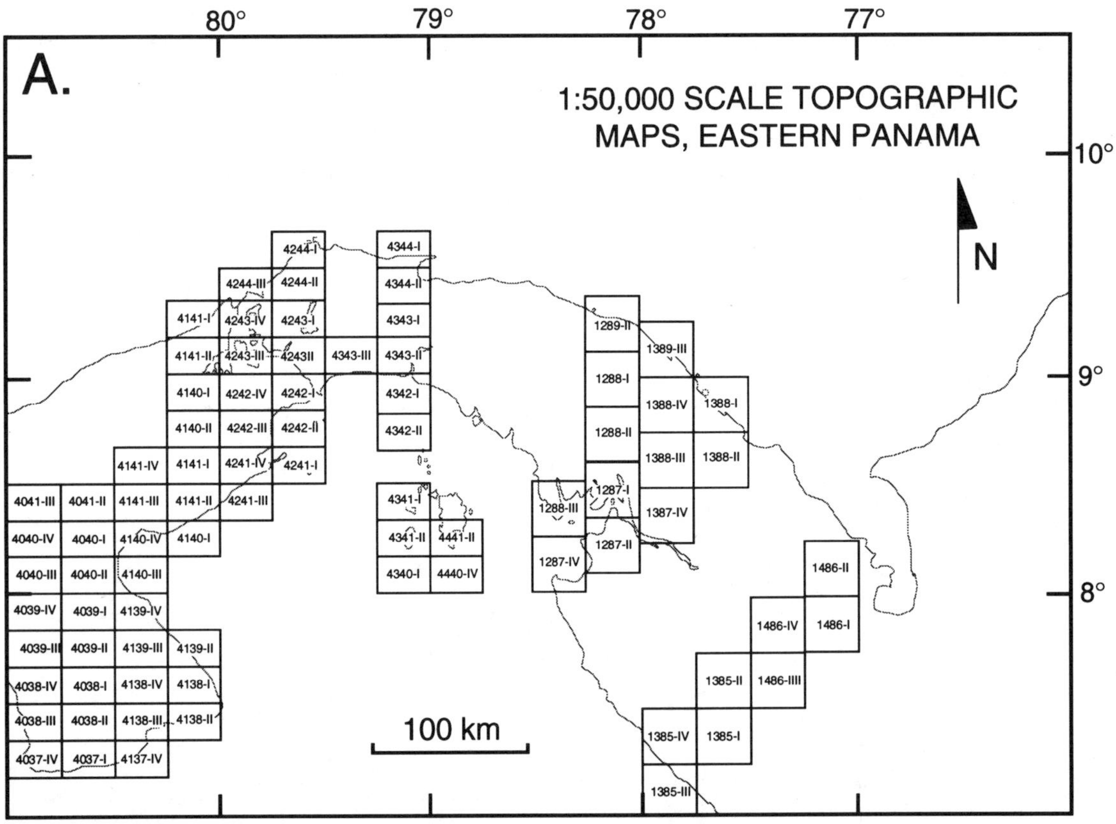

Figure 10 (on this and facing page). A, Key for 1:50,000 scale topographic maps in eastern Panama (above). B, Key for 1:50,000 scale topographic maps in western Panama and Costa Rica (facing page).

A list of 55 Master's and Ph.D. theses completed in the period 1952 to 1992 at U.S., German, and French universities attests to the long-term interest of academic geologists in the geology of southern Central America (Table 1; Fig. 11). Several universities, including Louisiana State University, the University of California at Santa Cruz, Oregon State University, and the University of Texas at Austin in the United States and Giessen and Marburg in Germany, have developed long-term research programs that have produced a series of related thesis projects in southern Central America. The results of several recent thesis studies appear in the chapters of this volume.

The Central American School of Geology was established in 1971 at the University of Costa Rica as a regional training center for geologists from all Central American countries. Funding is provided by Costa Rica and foreign development agencies in Germany and Switzerland. The geology faculty includes Costa Ricans and visiting geology professors from Germany and Switzerland who live, teach, and supervise student research for several years in Costa Rica before returning to their countries. The school has produced about 86 undergraduate geology majors who are required to complete a field-based senior thesis project. These senior thesis projects have examined a variety of problems in Costa Rica and other Central American countries (Table 2; Fig. 12). Professors at the school periodically synthesize the student thesis work into the regional reports that are compiled in Table 3. The University of Costa Rica–Heredía also has a geology department that specializes in earthquake seismology.

The student projects emphasize practical applications of geology and geophysics to earthquakes, hydrology, environmental studies, and geotechnical topics because most of these geologists are later employed by government or private agencies requiring these types of skills (Table 2). In recent years there has been an increase in the number of graduates from the Central American School of Geology who continue their post-graduate geology education abroad in the United States, Germany, Switzerland, or France.

Since 1984, the Central American School of Geology has

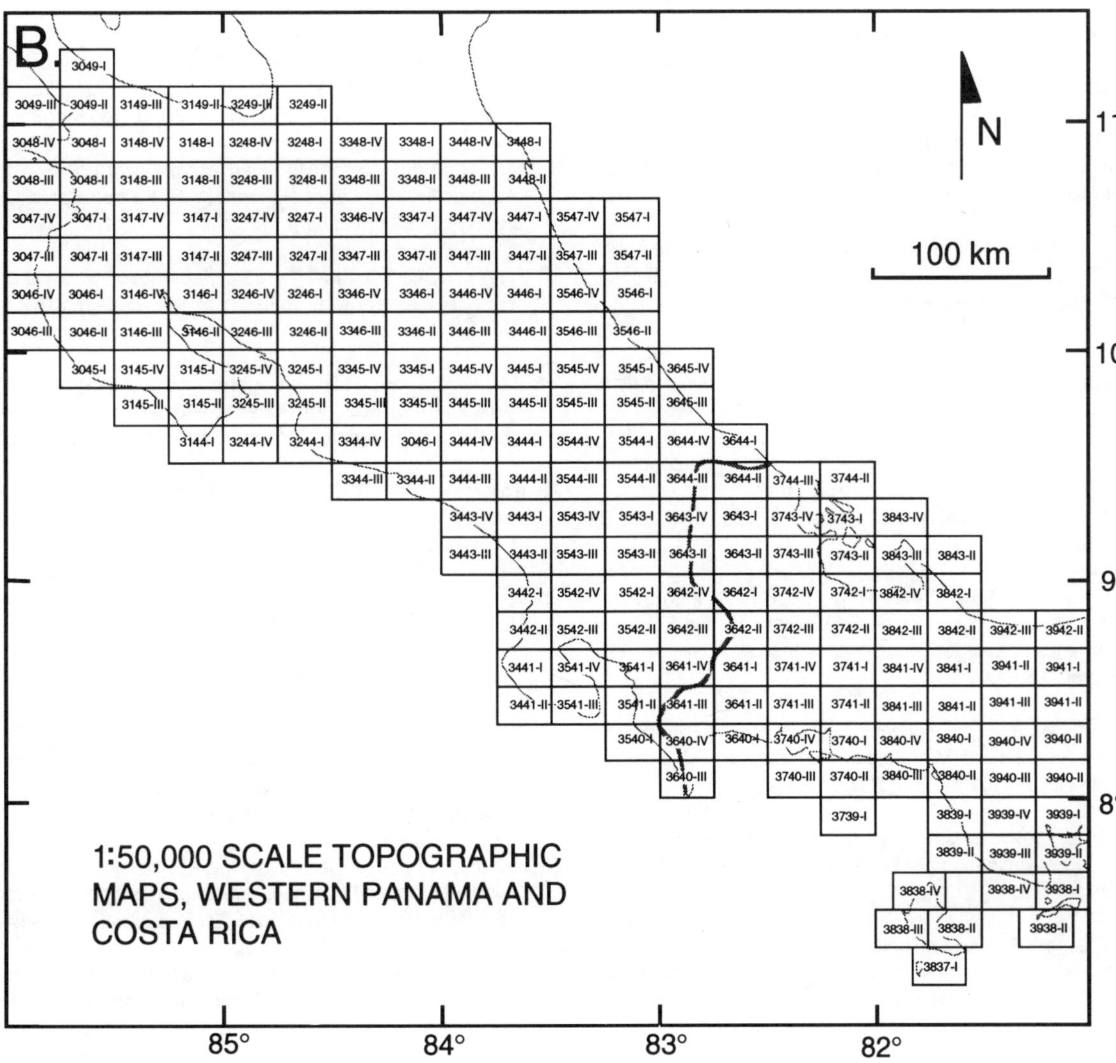

TABLE 1. LIST OF UNPUBLISHED THESES CONCERNING THE GEOLOGY AND TECTONIC EVOLUTION OF COSTA RICA, PANAMA, AND OFFSHORE AREAS (1952-1992)*

1. Adamek, S., 1986, Earthquake studies in the Panamá-Costa Rica region [M.S. thesis]: Austin, University of Texas, 211 p.
2. Barday, R. J., 1974, Structure of the Panamá basin from marine gravity data [M.S. thesis]: Corvallis, Oregon State University, 99 p.
3. Bentley, L. 1974, Crustal structure of the Carnegie Ridge, Panamá basin, and Cocos Ridge [M.S. thesis]: Manoa, Univeristy of Hawaii, 49 p.
4. Bowland, C. L., 1984, Seismic stratigraphy and structure of the western Colombian basin, Caribbean Sea [M.A. thesis]: Austin, Univeristy of Texas, 248 p.
5. Briceño-Guarupe, L. A., 1978, The crustal structure and tectonic framework of the Gulf of Panamá [M.S. thesis]: Corvallis, Oregon State University, 71 p.
6. Burbach, G. V., 1985, Intermediate and deep seismicity and lateral structure of subducted lithosphere in the circum-Pacific region [Ph.D. thesis]: Austin, University of Texas, 115 p.
7. Cassell, D. T., 1986, Neogene foraminifera of the Limon basin of Costa Rica [Ph.D. thesis]: Baton Rouge, Louisiana State University, 323 p.
8. Castillo, P. R., 1984, Geology and geochemistry of Cocos Island; Implications for the evolution of the aseismic Cocos Ridge [Ph. D. thesis]: St. Louis, Missouri, Washington University, 310 p.
9. Chinn, E. W., 1986, Diagenesis of Tertiary volcanogenic sandstones of the Térraba belt, southwest Costa Rica [M.S. thesis]: Baton Rouge, Louisiana State University, 72 p.
10. Clark, L. F., 1989, The geology, geochemistry, and petrogenesis of El Valle volcano, Panamá [M.S. thesis]: Tampa, University of South Florida, 130 p.
11. Coltrin, D. G., Jr., 1987, Seismic reflection imaging problems resulting from a rough surface at the top of the accretionary prism at convergent margins [M.A. thesis]: Austin, University of Texas, 116 p.
12. Corrigan, J. D., 1986, Geology of the Burica Peninsula, Panamá–Costa Rica; Neotectonic implications for the Southern Middle America convergent margin [M.A. thesis]: Austin, University of Texas, 152 p.
13. Dean, B. W., 1976, Focal mechanism solutions and tectonics of the Middle America arc [M.A. thesis]: Hanover, New Hampshire, Dartmouth College, 81 p.
14. Dewey, J. W., 1971, Seismicity studies with the method of joint hypocenter determination [Ph.D. thesis]: Berkeley, University of California, 173 p.
15. **Di Marco, G.,** 1993, Evolution sédimentaire et tectonique d'un prisme d'accretion tertiare et mise en place des terrains du sud du Costa Rica (Amerique Central) [Ph.D. thesis]: Lausanne, Switzerland, University of Lausanne, 245 p.
16. Drake, P. G., 1989, Quaternary geology and tectonic geomorphology of the coastal piedmont and range, Puerto Quepos area, Costa Rica [M.S. thesis]: Albuquerque, University of New Mexico, 183 p.
17. **Güendel, F. D.,** 1978, On the relationship between earth tides and volcanic activity of Arenal volcano, Coast Rica [M.A. thesis]: Austin, University of Texas, 32 p.
18. **Güendel, F. D.,** 1986, Seismotectonics of Costa Rica; An analytical view of the southern terminus of the Middle America trench [Ph.D. thesis]: Santa Cruz, University of California, 142 p.
19. Gursky, H.-J., 1984, Die sedimentgesteine im ophiolithischen Nicoya-Komplex (Ober-Jura bis Alt-Tertiär von Costa Rica); Ihre verbreitung, fazies, und geologische geschichte mit besonderer berücksichtiung der radiolarite [Ph.D. thesis]: Marburg, Germany, University of Marburg, 394 p.
20. Gurksy, M. M., 1986, Tektonische und thermische deformationen im ophiolithischen Nicoya-Komplex und seinem sedimentären auflager (Nicoya-Halbinsel, Costa Rica) und ihre bedeutung für die geodynamische entwicklung im südlichen Zentralamerika [Ph.D. thesis]: Marburg, Germany, University of Marburg, 230 p.
21. **Hardy, N. C.,** 1992, A marine geophysical study of the Pacific margins of Colombia and southwest Panamá [Ph. D. thesis]: Birmingham, University of Birmingham, United Kingdom, 253 p.
22. Heil, D., 1988, Response of an accretionary prism to transform ridge collision south of Panamá [M.S. thesis]: Santa Cruz, University of California, 88 p.
23. Heywood, C. E., 1984, Forearc deformation in southern Costa Rica: A sequence of collision of the aseismic Cocos Ridge [M.S. thesis]: Santa Cruz, University of California, 104 p.
24. Jackson, T. E., 1991, Neogene geochemistry of the Central American arc, western Panamá and southeastern Costa Rica, [M.S. thesis]: Tampa, University of South Florida, 104 p.
25. **Kolarsky, R. A.,** 1992, Late Cretaceous–Cenozoic tectonics and sedimentation in southern Central America [M.A. thesis]: Austin, University of Texas, 351 p.
26. Lew, L. R., 1983, The geology of the Osa Peninsula, Costa Rica; Observations and speculations about the evolution of part of the outer arc of the southern Central American orogen [M.S. thesis]: State College, Pennsylvania State University, 128 p.
27. Liaw, H. B., 1982, Modified regularization method applied to joint determination of hypocenters and velocity structure and its application to earthquake data from northern Costa Rica [Ph.D. thesis]: Dallas, University of Texas, 135 p.
28. Lowery, B. J., 1979, Sedimentology and tectonic implications of the Middle to Upper Miocene Curre Formation, southwestern Costa Rica [M.S. thesis]: Baton Rouge, Louisiana State University, 100 p.
29. Lundberg, N. S., 1982, Evolution of the forearc landward of the Middle America trench, Nicoya Peninsula, Costa Rica, and southern Mexico [Ph.D. thesis]: Santa Cruz, University of California, 256 p.

TABLE 1. LIST OF UNPUBLISHED THESES CONCERNING THE GEOLOGY AND TECTONIC EVOLUTION OF COSTA RICA, PANAMA, AND OFFSHORE AREAS (1952-1992)* (continued)

30. MacKay, M., 1988, Accretionary response to trench sediment variation along the south Panamá margin [M.S. thesis]: Tulsa, Oklahoma, University of Tulsa, 44 p.

31. Malavassi, E., 1979, Geology and petrology of Arenal volcano, Costa Rica [M.S. thesis]: Manoa, University of Hawaii, 111 p.

32. Malavassi, E., 1991, Magma sources and crustal processes at the southern terminus of the Central American volcanic arc [Ph.D. thesis]: Santa Cruz, University of California, 192 p.

33. **McIntosh, K. D.,** 1992, Geologic structure and processes of the eastern Pacific margin: California and Costa Rica [Ph.D. thesis]: Santa Cruz, University of California, 171 p.

34. McKee, W. H., 1985, Tertiary foraminifera of Puntarenas Province, Costa Rica [M.A. thesis]: Baton Rouge, Louisiana State University, 276 p.

35. Ogujiofor, I. J., 1985, Active convergent margins of northwest South America [M.S. thesis]: Honolulu, University of Hawaii, 100 p.

36. Pennington, W. D., 1980, Subduction of the eastern Panamá basin and seismotectonics of northwestern South America [Ph. D. thesis]: Madison, University of Wisconsin, 175 p.

37. Phillips, P. J., 1983, Stratigraphy, sedimentology, and petrologic evolution of Tertiary sediments in southwestern Costa Rica [M.S. thesis]: Baton Rouge, Louisiana State University, 153 p.

38. Pinter, N., 1988, Isostasy and eustasy on the Osa Peninsula, Costa Rica: Unraveling of late Quaternary geologic history through uplift and sea level change [M.S. thesis]: University Park, Pennsylvania State University, 144 p.

39. Prosser, J. T., 1983, The geology of Poás volcano, Costa Rica [M.S. thesis]: Hanover, New Hampshire, Dartmouth College, 165 p.

40. **Protti-Quesada, J. M.,** 1991, Correlation between the age of the subducting Cocos plate and the geometry of the Wadi-Benioff zone under Nicaragua and Costa Rica [M.S. thesis]: Santa Cruz, University of California, 66 p.

41. Restrepo, J. F., 1987, A geochemical investigation of Pleistocene to Recent calc-alkaline volcanism in western Panamá [M.S. thesis]: Tampa, University of South Florida, 103 p.

42. Richerson, P. M., 1990, Petrogenesis of La Yeguada volcanic complex, western Panamá, via both differentiation and partial melting [M.S. thesis]: Tampa, University of South Florida, 133 p.

43. Schuster, G. T., 1977, Seismic studies of crustal structure in Nicaragua and Costa Rica [M.S. thesis]: Houston, Texas, University of Houston, 79 p.

44. Shelton, B. J., 1952, Geology and petroleum prospects of Darien, southeastern Panamá [M.S. thesis]: Corvallis, Oregon State University, 59 p.

45. Tagudin, J., 1988, A correlation between fault vergence, fault spacing, and sediment type in the North Panama thrust belt [M.S. thesis]: Santa Cruz, University of California, 32 p.

46. Taylor, G. D., 1975, The geology of the Limon area of Costa Rica [Ph.D. thesis]: Baton Rouge, Louisiana State University, 116 p.

47. Tournon, J., 1984, Magmatismes du Mesozoique a l'Actuel en Amerique Central: L'exemple de Costa Rica, des ophiolites aux andesites [Ph.D. thesis]: Paris, France, Marie et Pierre Curie University, 335 p.

48. Von Breymann, A., 1989, Marine geophysical studies of the Atlantic margin of Costa Rica [Ph.D. thesis]: College Station, Texas A&M University, 245 p.

49. Wildberg, H.G.H., 1980, Die Magmatite des Nicoya-Komplexes, Costa Rica [Ph.D. thesis]: Münster, Germany, University of Münster, 122 p.

50. Winsemann, J., 1992, Tiefwasser-sedimentationsprozesse und -produkte in den forearc-becken des mittleamerikanischen insel-bogensystems: Eine sequenzstratigraphische analyse [Ph.D. thesis]: Stuttgart, Germany, University of Stuttgart, 321 p.

51. Wolters, B., 1983, Seismicity and earthquake focal mechanisms and their tectonic implications in southern Central America and northwstern South America [M.S. thesis]: Stanford, California, Stanford University, 89 p.

52. Yuan, P. B., 1984, Stratigraphy, sedimentology, and geologic evolution of eastern Terraba trough, southwestern Costa Rica [Ph.D. thesis]: Baton Rouge, Louisiana State University, 110 p.

53. Verdonck, D., 1990, Subduction of the Cocos Ridge and vertical movement of western Costa Rica [M.S. thesis]: University Park, Pennsylvania State University, 52 p.

54. Victor, L., 1976, Structures of the continental margin of Central America from northern Nicaragua to northern Panamá [M.S. thesis]: Corvallis, Oregon State University, 76 p.

*Note: Authors contributing to this volume are in bold face type.

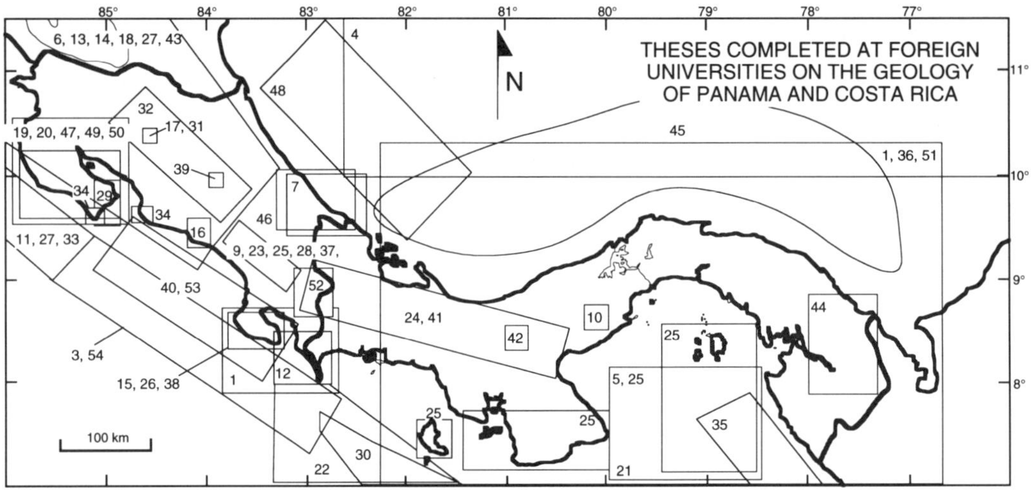

Figure 11. Map showing locations of subjects of theses completed at foreign universities on the geology of Panama and Costa Rica. Numbers are keyed to Table 1.

published a high-quality biannual geology journal: the *Geological Magazine of Central America.* The contents of this journal up to October 1992 are translated into English and compiled as Table 4. The journal contains articles by professors at the Central American School of Geology and the University of Costa Rica–Heredía, summaries of student thesis work, work by foreign geologists, summaries of earthquakes and recent volcanic activity, and announcements.

UNRESOLVED GEOLOGIC CONTROVERSIES IN CENTRAL AMERICA FOR THE READER TO CONSIDER

I suggest several unresolved controversies for readers to keep in mind. The first controversy can be summed up as the question: Did the Caribbean form *in situ* or is it a trapped piece of exotic Pacific seafloor? The *in situ* school of thought, as most recently put forward by Frisch et al. (1992), proposes that the Caribbean region formed during the period 130 to 80 Ma as South America moved to the southeast away from North America (Fig. 13A). Igneous upwelling in the space that formed between the two continents is thought to have produced the anomalously thick oceanic plateau crust of the Caribbean and perhaps Central America (Bowland and Rosencrantz, 1988; Burke, 1988) (Fig. 3).

An alternative school of thought, first formulated by Wilson (1966) and later elaborated by Pindell and Barrett (1990), proposes that the Caribbean is a piece of Pacific ocean floor and oceanic plateau that has been rafted behind an eastward-moving arc system that entered the oceanic corridor between North and South America (Fig. 13B). Normal Pacific ocean crust may have been modified into an oceanic plateau as the Caribbean region passed over the Galapagos hot spot in the eastern Pacific (Duncan and Hargraves, 1984).

A second, related controversy revolves around the origin of the Mesozoic basement of southern Central America. Was the basement formed by the accretion of fragments of the subducting Pacific crust to the arc, was the basement formed by volcanic and plutonic arc growth, or is the basement the uplifted edge of the Caribbean oceanic plateau? Bowland and Rosencrantz (1988) take the view that the basement of southern Central America is the upturned edge of the Caribbean Sea and that the Caribbean Sea is an oceanic plateau derived from the eastern Pacific. In this volume, DiMarco et al. support the idea that much of the basement is the upturned Caribbean but also show evidence for accretion of Pacific material.

A third major controversy relates to the definition and formation of late Neogene plate boundaries. Early workers like Jordan (1975) treated southern Central America as part of the stable Caribbean plate. Work reported by several groups in this volume suggests that some or all of southern Central America may be a detached microplate that moves independently of both the Caribbean and the Pacific. Much of the late Neogene deformation may be related to the collision between the Panama arc and the South American margin and to the uplift of the present-day land bridge (Wadge and Burke, 1983) (Fig. 14).

A fourth controversy surrounds the cause of the topographic and structural anomalies at the southern terminus of the Middle America trench and volcanic arc in Costa Rica and western Panama. Are these anomalies produced by shallow subduction of the Cocos Ridge (e.g., Kolarsky et al., this volume) (Fig. 13B) or are they produced by more subtle effects of variations in the age of the crust entering the trench (e.g., Protti et al., this volume)?

TABLE 2. LIST OF SENIOR THESES OF GEOLOGY MAJORS AT THE CENTRAL AMERICAN SCHOOL OF GEOLOGY, UNIVERSITY OF COSTA RICA, SAN JOSE*

15[§]. Aguilar-Alvarez, A. T., 1978 Fauna from a section of the Térraba Formation, 84 p.

39. Aguilar-Díaz, A., 1984, Seismicity from September 3-9, 1980, and its relationship with the geology of the headwaters of Río Navarro, Cartago, 168 p.

35. Alán-Mora, M. A., 1983, Geology of laterites in the extreme northeast of the Valle de El General, Costa Rica, 166 p.

28. Alvarado-Campos, M. E., 1982, Sedimentological studies of the Pacacua Formation (Miocene), Costa Rica, 185 p.

40. Alvarado-Induni, G. E., 1984, Petrologic-geologic aspects of the volcanoes and lava units of the upper Cenozoic of Costa Rica, 183 p.

31. Alvarado-Valverde, R., 1983, Geologic reconnaissance of the middle drainage of the Río Usumacinta with investigation of a possible dam site named "La Línea," 85 p.

38. Alvarado-Villalón, F., 1984, Structural geology and tectonics of the southern Valle Central of Costa Rica (Tarbaca), 108 p.

29. Alvarez-Echeverría, F., 1982, Geology of Bustamante and surrounding areas, Province of San José, Costa Rica, 53 p.

84. Arias-Molina, O., 1992, Computerized methcds of structural analysis and their application to structural geology in Costa Rica, 171 p. and 5 diskettes.

57. Astorga-Gättgens, A., 1987, The Upper Cretaceous and Paleogene of the Pacific slope of **southern Nicaragua**[†] and northern Costa Rica: Origin, evolution, and dynamics of deep basins in relation to the convergent margin of Central America, 250 p.

42. Avila-Ballar, M., 1984, Geology and microseismicity of San Ignacio de Acosta and the surrounding area (southwestern Valle Central of Costa Rica), 44 p.

74. Avila-Romero, G. A., 1989, Program to analyze seismologic data from Costa Rica (ANASIS program, v. 1.0, User's Manual), 105 p.

79. Barquero-Picado, R., 1990, Seismicity and tectonics of the region northeast of Costa Rica with emphasis on the area of the Miravalles geothermal project, 117 p.

3. Baxter-Gandara, S., 1976, Geologic study of the Mata Limón and Punta Carballo Formations, Costa Rica, 66 p.

36. Bolanos-Irigaray, K., 1983, Geologic evaluation of coal deposits of the Lower Talamanca for a pre-feasibility study, 2 volumes.

70. Boschini-López, I. M., 1989, Incidence of strong earthquakes in the Caribbean region of Costa Rica, 97 p.

58. Calvo-Vargas, C., 1987, Neritic limestone along the Pacific margin of nothern Costa Rica and **southern Nicaragua**[†]; Epochs and systems associated with the opening and evolution of the convergent margin of southern Central America, 165 p.

60. Campos-Bejarano, L., 1987, Geology of the Fila Asunción and neighboring zones, Atlantic coast, Costa Rica, 83 p.

13. Carballo-Hernández, M. A., 1978, The San Miguel Formation, Lower Miocene, Costa Rica, 114 p.

46. Castro-Muñoz, J. F., 1985, Geochemical survey of part of the Montes del Aguacate (Desmonte area), 2 volumes.

85. Castro-Zuñiga, S., 1993, Petrography and alteration mineralogy of wells in the Tenorio geothermal field, Guanacaste (Costa Rica), 86 p.

81. Cerdas-Salas, A., 1991, Analysis of slope stability along the margin of the Río Pacuare, Siquirres hydroelectric project, Limón Province, Costa Rica, 120 p.

72. Cervantes-Loaiza, J. F., 1989, Geological framework for the geotechnical analysis of the Siquirres hydroelectic project, Limón Province, Costa Rica, 117 p.

71. Cosillo-Pinto, A. G., 1989, Geochemistry and structural geology of the Bellavista gold mine, Miramar, Puntarenas, Costa Rica, 169 p.

21. Chávez-Cernas, J., 1980, Geologic aspects of the Río Nimoyores basin, Guanacaste, 41 p.

66. Chinchilla-Chavez, A. L., 1989, Geologic study of the Montezuma Formation (Plio-Pleistocene), Nicoya Peninsula, Costa Rica, 148 p.

25. Darce-Rivera, M., 1980, Site location and evaluation of magnetite sand deposits along the Pacific coast of **Nicaragua**, 75 p.

10. Denyer-Chavarría, P., 1977, Notes about manganese mineralization of the Nicoya ophiolite complex, Guanacaste Province, Costa Rica, 65 p.

TABLE 2. LIST OF SENIOR THESES OF GEOLOGY MAJORS AT THE CENTRAL AMERICAN SCHOOL OF GEOLOGY, UNIVERSITY OF COSTA RICA, SAN JOSE* (continued)

17. Domínguez-Alvarez, R. A., 1978, Preliminary study of the Fortuna project dam site and geotechnical observations, 77 p.

26. Echandi-Echeverría, E., 1981, Volcanic units from the northern slope of the Río Virilla basin, 123 p.

18. Elizondo-Méndez, J. A., 1979, Preliminary hydrogeological study of the Río Navarro basin, Cartago Province, 111 p.

47. Estrada-Barrios, F. J., 1985, Geologic-geotechnical study for the construction of the Santa Rosa dam, Tilarán, Guanacaste Province, 108 p.

56. Estrada del Llano, A., 1986, Geologic-geotechnical study of the San Blas landslip, Río Reventado, Cartago Province, Costa Rica, 2 volumes.

78. Fernández-Acre, M., 1990, Activity of Poás volcano, Costa Rica: Seismic analysis during the period 1980-89, 185 p.

41. Fernández-Aragonés, E., 1984, Geology and hydrothermal alteration at the Miravalles geothermal field, Guanacaste Province, 92 p.

54. Fernández-Solórzano, J. A., 1987, Geology of the Tucurrique topographic sheet (scale 1:50,000, no. 3445-I), 205 p.

55. Flores-Rivera, J. A., 1986, Hydrogeologic investigation of the recharge zone in karsted limestone of the El Cajón hydroelectric project, **Honduras,** Central America, 92 p.

4. Flores-Rodríguez, W. Z., 1976, Geologic study of sulfur mineralization, Punta Gorda, Nicoya Peninsula, Costa Rica, 35 p.

14. Franco-Austin, J. C., 1978, The Coris Formation (Miocene), Valle Central, Costa Rica, 87 p.

61. Gómez-Cruz, A. L., 1987, Evaluation of aquifer potential and design for containment of subterranean waters in the Puente Mulas zone, Heredia Province, Costa Rica, 66 p.

83. Gómez-Venegas, M. A., 1992, Evaluation of the diatomite mine in Quebrada Grande Liberia, Guanacaste (Costa Rica), 101 p.

75. Herrera-Ibañez, I. R., 1989, Geochemistry and pedology of soils in the western Valle Central, Alajuela Province, Costa Rica, 303 p.

22. Herrera-Ocampo, J. A., 1980, Hydrogeology of the Río Andamojo basin, Santa Cruz, Guanacaste, 81 p.

9. Jager-Contreras, G., 1977, Geology of chromite mineralization of the eastern Santa Elena Peninsula, Costa Rica, 136 p.

11. Laguna-Morales, J., 1977, Geology and sedimentology of three Paleocene units near Sámara, Nicoya Peninsula, Guanacaste Province, Costa Rica, 79 p.

64. Laurito-Mora, C. A., 1988, Sedimentological analysis of the El Fraile Formation on the southwest, Pacific coast of **Nicaragua** near Carmen and Punta Carballo, 117 p.

62. Leandro-Montoya, C. E., 1987, Geophysical-geotechnical study for the feasibility of a tunnel construction near Pejibaye, Cartago, Costa Rica, 76 p.

33. León-Bonilla, C. G., 1983, Geologic-geotechnical study for the feasibility of the Sandilla-Cañas hydroelectric project, 2 volumes.

30. Lezama-Fernández, G., 1983, Geophysical study for the Savegre hydroelectric project, 147 p.

8. Losilla-Penón, M., 1977, Subterranean waters in the Río Virilla basin and a proposal for their exploitation, 45 p.

24. Madrigal-Castro, O., 1980, Geology and potential of part of the Abangares mineral district, Guanacaste, Costa Rica, 87 p.

43. Madrigal-Morales, C., 1985, Geologic-geotechnical study for the feasibility of the Guayabo hydroelectric project, Turrialba, Costa Rica, 2 volumes.

52. Malavassi-Rojas, L. R., 1985, General geology of the area south of Baja Talamanca in relation to coal deposits, Limón Province, San José, Costa Rica, 89 p.

37. Marín-Guzmán, F., 1983, Regional geochemistry of the central part of Guanacaste Province, Costa Rica, 236 p.

6. Miranda-Gómez, R., 1976, Geology of the area of Loma Espino, Tonosí area, Los Santos Province, **Panamá**, 87 p.

80. Molina-Zúñiga, F., 1990, The Alto de Tapezco landslip near Santa Ana, San José Province, Costa Rica, 44 p.

77. Moore-Coleman, F. A., 1990, Analysis of the dynamic behavior of soils during earthquakes in the area of Managua, Nicaragua, 102 p.

16. Mora-Castro, D., 1978, Hydrogeologic aspects of the Nicoya Peninsula, 83 p.

19. Mora-Castro, S., 1979, Geologic study of the southeastern part of the Valle de El General, Puntarenas, Costa Rica, 3 volumes.

63. Mora-Protti, O., 1988, Geologic-petrologic study of the pyroclastic rocks near Bagaces, Guanacaste Province, Costa Rica, 61 p.

50. Mora-Rojas, C., 1985, Sedimentology and geomorphology of the southern Nicoya Peninsula (Puntarenas Province, Costa Rica), 148 p.

TABLE 2. LIST OF SENIOR THESES OF GEOLOGY MAJORS AT THE CENTRAL AMERICAN SCHOOL OF GEOLOGY, UNIVERSITY OF COSTA RICA, SAN JOSE* (continued)

68. Morán-Orellana, G. N., 1989, Evaluation of seismic risk around the Berlin geothermal area, **El Salvador**, Central America, 73 p.

32. Obando-Acuña, L. G., 1983, Stratigraphy and petrography of outcrops south of the Valle Central–Tarbarca area, 136 p.

53. Obando-Rodríguez, J. A., 1986, Sedimentology and tectonics of the Cretaceous and Paleogene near Golfito and the Burica and Osa Peninsulas, Puntarenas Province, Costa Rica, 211 p.

2. Paniagua-Pérez, S., 1975, Geology and hydrogeological aspects of the first field of wells for supplementation of water for the city of San José, Costa Rica, 152 p.

5. Paredes-Paz, J. R., 1976, Preliminary geochemical reconnaissance of an area on part of the Abra and Río Grande sheets, Costa Rica, 55 p.

51. Peña-Ruiz, O., 1985, Geologic-geotechnical study of the design and construction of the impermeable base of the El Cajón dam, **Honduras**, Central America, 83 p.

65. Pérez-Arias, C. L., 1989, Preliminary geologic-geotechnical characterization of the rocky core at the site of the Siquirres hydroelectric site, Limón Province, Costa Rica, 77 p.

82. Pinilla-Bonilla, G. A., 1992, Regional geochemistry of the Abra and Barba topographic sheets, Valle Central, Costa Rica, 2 volumes.

34. Pinzón-Jeréz, O. E., 1983, Investigation and potential of basic metals, Canaan-Managua project, Izabel y Zacapa Department, **Guatemala**, 108 p.

48. Pizarro-Araya, D. M., 1985, Biostratigraphy of the Uscari Formation (Middle to Upper Miocene) using planktic foraminifera, 34 p.

27. Protti-Quesada, R., 1981, Geology and geophysics applied to hydrogeology in the Jicara–San Rita coastal plain, Nicoya Peninsula, 47 p.

67. Quezada-Muñoz, A. A., 1989, Geologic and geophysical study of the Las Pailas geothermal field, Rincón de la Vieja National Park, Guanacaste, Costa Rica, 67 p.

1. Ramírez-Escribano, O., 1973, Geologic and geotechnical study of a strip of land between Tilarán and Fortuna, Costa Rica, 87 p.

73. Rodríguez-Altamirano, D., 1989, Geologic-structural study of hydrothermal alteration in the area of the Mina de Agua–Rincón, Managua, **Nicaragua**, 100 p.

23. Rodríguez-Estrada, H. V., 1980, Recharge and discharge conditions for water wells near La Valencia, Heredia, Costa Rica, 29 p.

20. Rojas-Esquivel, J. F., 1979, Hydrogeochemistry of the Barranca aquifer, Puntarenas Province, Costa Rica, 74 p.

86. Rojas-Quesada, W., 1993, Catalogue of historical and recent seismicity in Central America: Development and analysis, 91 p.

44. Sáenz-Sánchez, L. F., 1985, Geophysical study for the feasibility of the Angostura hydroelectric project, 2 volumes.

49. Salazar-Cyrman, A. J., 1985, Feasibility study for improving the Azul de Turrialba quarry, Cartago, Costa Rica, 132 p.

45. Sánchez-González, F., 1985, Applied field techniques in the geophysical method of seismic reflection: Their interpretation and analysis, 91 p.

7. Santana-Bendix, A., 1977, Geology of a part of the Miravalles and Tierras Morenas sheets, Costa Rica, 89 p.

12. Ulloa-Chaverri, F., 1977, Aspects of the eastern part of the Santa Elena Peninsula (from Cuajiniquil to Santa Rosa from the Guanacaste intersection), 76 p.

59. Valdez-Carrillo, M. R., 1987, Preliminary geologic-geotechnical study of the Savegre hydroelectric project along the Pacific margin, Costa Rica, 2 volumes.

69. Valverde-Guillén, R., 1989, Geologic-environmental investigation of the Pejibaye tunnel project, Cartago, Costa Rica, 2 volumes.

76. Vásquez-Rojas, R., 1989, Statistical analysis of geochemical data from three fluvial sedimentary deposits of epithermal gold, Cordillera de Tilarán–Montes del Aguacate, Costa Rica, 111 p.

*All theses in Spanish; titles translated from Spanish to English by P. Mann; copies of theses at reproduction and mailing costs available through Ms. Sonia Castillo, Geology Librarian, Central American School of Geology, Box 35, San Pedro, San José, Costa Rica (telephone: 506-25-7941; fax: 506-34-2347).

†Study areas in Guatemala, Honduras, Nicaragua, and Panamá that are not shown in Figure 12 are shown in bold in this list.

§Number corresponds to position in total number of theses completed at the Central American School of Geology. For example, Ramirez-Escribano completed the first thesis in 1973 and Rojas-Quesada completed the 86th thesis in 1992.

**Acronym RECOPE = Refinadora Costarricense de Petróleo, S. A. (Costa Rican National Oil Company).

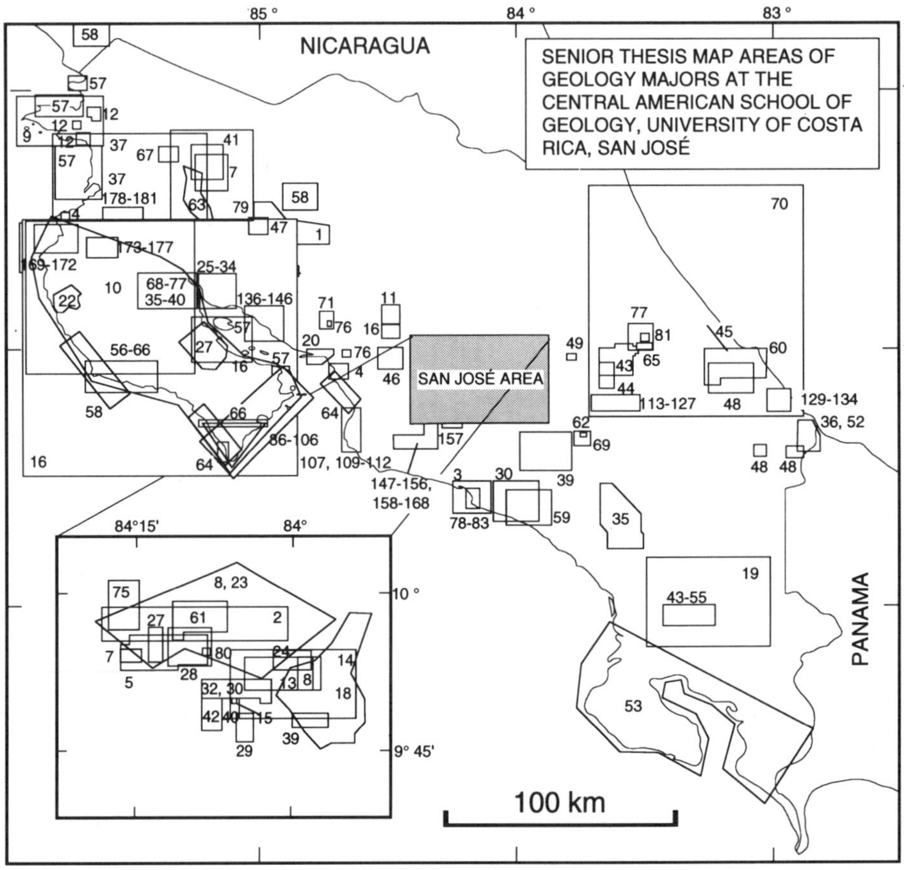

Figure 12. Map showing locations of senior thesis map areas of geology majors at the Central American School of Geology, University of Costa Rica, San José. Numbers are keyed to Table 2.

TABLE 3. SYNTHESES OF FIELDWORK BY PROFESSORS AT THE CENTRAL AMERICAN SCHOOL OF GEOLOGY*

1. Denyer-Chavarría, P., and Soto-Bonilla, G. J., eds., 1988, Synthesis of field geology, 1988; Geology of the southwest Valle Central, Costa Rica, 34 p.
2. Denyer-Chavarría, P., ed., 1989 field synthesis, 21 p.
3. Denyer-Chavarría, P., and Arias, O., eds., Geology of part of the Matapalo and Belén sheets, Guanacaste, Costa, Rica, field geology synthesis, February 1990, 19 p.
4. Rivier, F., 1983, Geology synthesis and geologic map of the area of Tempisque Bay, Guanacaste, Costa Rica, January-June, 1983, p. 7-30.

*All syntheses in Spanish; titles translated from Spanish to English by P. Mann.

TABLE 4. LIST OF ARTICLES PUBLISHED IN THE "GEOLOGICAL MAGAZINE OF CENTRAL AMERICA" PUBLISHED BY THE CENTRAL AMERICAN SCHOOL OF GEOLOGY*

Volume 1, October 1984

Laguna, J., Effects of hydrothermal alteration and weathering on volcanic rocks from the Aguacate Group, Costa Rica.

Alguilar, T., Presence of sponges (Porifera) in the Oligocene of Costa Rica (Valle Central).

Morales, L. D., and Montero, W., Felt earthquakes in Costa Rica from 1973-1983 and their relation to the seismicity of the country.

Baumgartner, P. O., Mora, C. R., Butterlin, J., Sigal, J., Glacon, G., Azéma, J., and Bourgois, J., Sedimentation and paleogeography of the Cretaceous and Cenozoic of the Pacific margin of Costa Rica.

Anonymous, Technical note; Petroleum exploration in Costa Rica.

Volume 2, April 1985

Marín, F., Regional geochemical profile of the 1:200,000 scale Liberia sheet (Costa Rica).

River, F., Geologic cross section from the Pacific across Costa Rica.

Paniagua, S., Geochemistry of trace elements in Upper Cenozoic volcanic rocks of the central region of Costa Rica.

Seyfried, H., and Sprechmann, P., On the formation of the southern Central American land bridge with emphasis on its development from Campanian to Eocene.

Oldemar, R., Technical note: Coal in Costa Rica.

Volume 3, October 1985

Seyfried, H., Sprechmann, P., and Aguilar, T., Sedimentology and paleoecology of an estuary along the Pacific coast of the primordial Central American landbridge (Middle Miocene), Costa Rica.

Morales, L. D., Seismic zones of Costa Rica and its surroundings.

Alvarado, G., Petrologic considerations of the stratovolcanoes of Costa Rica.

Mora, S., Unstable slopes of Costa Rica.

Oldemar, R., Technical note: Peat in Costa Rica.

Alvaro, D., Alavarado, F., and Protti, M., Technical note: Computerized data base for tectonics and structural geology of Central America and the Caribbean (Tectobase).

National Seismological Network, UCR-ICE, Technical note: A decade of continuous seismic recording in Costa Rica by UCR-ICE (1975-85).

Volume 4, April 1986

Alvarado, G., Evidence for large mammal fossils in Costa Rica.

Cohen, A. D., Raymond, R., Mora, S., Alvarado, A., and Malavassi, L., Geologic characteristics of peat deposits in Costa Rica (preliminary study).

Oldemar, R., Rocks that are weak or of low resistance: Chalk as an analytical case study.

Paniagua, S., Technical note: Considerations about volcanic risk in Costa Rica.

Losilla, M., Technical note: Hydrogeologic map of the Valle Central, Costa Rica.

Volume 5, October 1986

Lucas, S. G., Pyrthere systematics and a Caribbean route for land mammal dispersal during the Paleocene.

Montero, W., Recurrence periods and types of seismic sequences of inter- and intraplate earthquakes in the region of Costa Rica.

Obando, L., Stratigraphy of the Venado Formation and overlying rocks (Miocene-Recent), Alajuela Province, Costa Rica.

National Seismological Network, Technical note: Summary of seismic and volcanic activity in Costa Rica.

Villata, C., Exploration of gold in Costa Rica.

Volume 6, April 1987

Calvo, C., and Bolz, A., The Venado sequence: A tropical, lagoonal estuary of the Middle Miocene, San Carlos, Costa Rica.

Chiesa, S., The largest Plinian eruption of Arenal volcano, Costa Rica.

Aguilar, T., Comparison between the malacologic fauna of the principal formations of the marine Pliocene of Costa Rica, Central America.

Rymer, H., Morales, L., and Brown, G. C., Microgravity monitoring at Poas volcano (1983-86).

Berrangé, J. P., Technical note: Gas seeps on the margins of the Golfo Dulce pull-apart basin, southern Costa Rica.

Bolivar, S., Castillo, R., Page, N., and Azuola, H., Technical note: Mineral resource assessment program for Costa Rica.

National Seismological Network, Technical Note: Summary of seismic and volcanic activity of Costa Rica.

Volume 7, October 1987

Pizarro, D., Biostratigraphy of the Uscari Formation based on planktic foraminifera.

Darce, M., Geology of the La Libertad mineral district, Nicaragua.

Kussmaul, S. Petrology of the Neogene intrusive igneous rocks of Costa Rica.

Denyer, P., Feoli, S., Murillo, G., and Rodríguez, C., Geologic mapping of a sector near the high basin of the Río Niñey, Limón, Costa Rica.

TABLE 4. LIST OF ARTICLES PUBLISHED IN THE "GEOLOGICAL MAGAZINE OF CENTRAL AMERICA" PUBLISHED BY THE CENTRAL AMERICAN SCHOOL OF GEOLOGY (continued)

Amalberto, S., Chiné, B., and Trogolo, D., Technical note: Research on gold mining in Costa Rica.

Vergara, A., Technical note: Importance of prevention.

Zaragoza, M., Technical note: Pemex and Petrocanada form partnership.

National Seismological Network, Technical note: Summary of seismic and volcanic activity of Costa Rica (January-June, 1987).

Volume 8, April 1988

Meschede, M., Frisch, W., and Sick, M., Geodynamic interpretation of the ophiolite complexes of Costa Rica.

Gursky, M., Tectonic analysis of the Nicoya Peninsula (Costa Rica) and its significance for the structural-geodynamic development of southern Central America.

Nyström, J., Levi, B., Troëng, B., Ehrenborg, J., and Carranza, G., Geochemistry of volcanic rocks in a traverse through Nicaragua.

Boschini, I., Alvarado, I., and Rojas, W., The Buenavista de Pérez Zeledón earthquake (July 3, 1983): Evidence for a previously unknown, strong intraplate earthquake in Costa Rica.

Anonymous, Technical note: RECOPE, 1988, year of its 25th anniversary.

Anonymous, Technical note: Annual report of the Costa Rican committee of the International Geological Correlation Project (ICGP).

Paniagua, S., Technical note: Workshop on the prevention and prediction of natural disasters in Central America.

National Seismological Network, Summary of seismic and volcanic activity in Costa Rica.

Volume 9, November 1988

Astorga, A., Geodynamics of the Upper Cretaceous–Paleocene basins in the forearc region of southern Nicaragua and northern Costa Rica.

Laurito, C., Ichthyology of marine sediments of the Carmen Formation, Cóbano District, Puntarenas, Costa Rica.

Rivier, F., and Calvo, C., Tertiary of the southern Valle Central:Stratigraphic section of the Cerro Caraigres, San José Province, Costa Rica.

Alvarado, G., Morales, L., Montero, W., Climent, A., and Rojas, W., Seismologic and morphotectonic aspects of the extreme western part of the volcanic Cordillera Central of Costa Rica.

Anonymous, Technical note: RECOPE is concerned about conserving the environment.

National Seismological Network, Technical note: Summary of seismic and volcanic activity for Costa Rica.

Morales, L., and Alvarado, G., Technical note: 100 years of seismograph studies.

Volume 10, October 1989

Mathers, S., Costa Rican diatomite: A review of the existing knowledge and future potential.

Gursky, H-J., Presence and origin of sedimentary rocks in the ophiolitic basement of Costa Rica.

Berrangé, J., The Osa Group: An auriferous Pliocene sedimentary unit from the Osa Peninsula, southern Costa Rica.

Aguilar, A., and Fernández, J., Technical note: Convention of technical assistance of RECOPE-PCIAC.

National Seismological Network, Technical note: Summary of seismic and volcanic activity in Costa Rica (July-December, 1988).

Aguilar, T., and Soto, G., Annual report for the Costa Rican committee for the International Geological Correlation Project (IGCP).

Soto, G., and Alvarado, G., Conference report: Costa Rican volcanism workshop, Washington, D.C.

Paniagua, S., Conference report: Symposium on the energy and mineral potential of Central America and the Caribbean, San José, Costa Rica.

Volume 11, October 1990

Singer, D. A., Page, N. J., Bagby, W. C., Cox, D. P., and Ludington, S., Evaluation of mineral resources of Costa Rica.

Kriz, S., Tectonic evolution and origin of the Golfo Dulce gold placers in southern Costa Rica.

Laurito, C., Study of a juvenile example of *Cuvieronius hyodon,* Río Nacamoe, Guanacaste, Costa Rica.

Alguilar, T., Evolution of mollusc communities and their relation with the geologic development of southern Central America.

TABLE 4. LIST OF ARTICLES PUBLISHED IN THE "GEOLOGICAL MAGAZINE OF CENTRAL AMERICA" PUBLISHED BY THE CENTRAL AMERICAN SCHOOL OF GEOLOGY (continued)

Mora, R., Cerdas, A., Molina, F., and Vega, E., Geologic characterization of the Chiz canal (Turrialba, Cartago Province, Costa Rica).

Montero, W., and Morales, L., Deformation and neotectonic forces in Costa Rica.

Berrangé, J., and Mathers, S., Technical note: Industrial, non-metallic minerals in Costa Rica.

Denyer, P., Technical note: Details on the geology of Punta Morales–Coyolito-Manzanillo.

Aguilar, T., and Soto, G., Annual report of the Costa Rican committee for the International Geological Correlation Project (IGCP).

Volume 12, April 1991

Denyer, P., and Arias, O., Stratigraphy of the central region of Costa Rica.

Arias, O., and Denyer, P., Structural geology of the region including the topographic sheets, Abra, Caraigres, Candelaria, and Río Grande, Costa Rica.

Barrantes, M., Petrographic comments on some rocks cropping out in the central region of Costa Rica.

Arias, O., and Denyer, P., Neotectonic and geologic aspects of Puriscal and surroundings, Costa Rica.

RECOPE, Technical Note: Technical cooperative project between Noruega, Costa Rica, and Nicaragua.

Aguilar, T., and Soto, G., Technical note: 1990 annual report of the Costa Rican committee of the International Geologic Correlation Project (IGCP).

ICE-UCR National Seismic Network, Technical Note: Summary of seismic and volcanic activity in Costa Rica in 1990.

Alvarado, M. E., Modern fan deltas in Golfito Bay, Costa Rica.

Lopez, A., Biographical note, John Graham Ramsay (60th anniversary).

Volume 13, October 1991

Camacho, E., The Puerto Armuelles earthquake (southwestern Panamá) on July 18, 1934.

Cortés, J., The coral reefs of Golfo Dulce, Costa Rica: Geological aspects.

Astorga, A., Fernández, J., Barboza, G., Campos, L., Obando, J., Aguilar, A., and Obando, L., Sedimentary basins of Costa Rica; Geodynamic evolution and hydrocarbon potential.

Obando, L., Bottazzi, G., and Alvarado, F., Sedimentology of some facies of the Coris Formation (Middle-Upper Miocene), Valle Central, Costa Rica, Central America.

Chiesa, S., A biotite pumice flow in the Río Liberia (Guanacaste), Costa Rica, Central America.

Lucas, S., and Avarado, G., The most southern occurrence of a *Mammut americanum:* The case of the mastodon of San Pedro Sula, Honduras.

Calvo, C., and Bolz, A., Technical note: The Espíritu Santo Formation (Costa Rica): Systems of autochthonous carbonate platforms in the Upper Paleocene–Lower Eocene.

Lucas, S., and Alvarado, G., Technical note: Comment on the mastodon of the Barra Honda (Río Nacaome), Guanacaste, Costa Rica.

Volume 14, October 1992

Montero, W., Paniagua, S., Kussmaul S., and Rivier, F., Internal geodynamics of Costa Rica.

Berrangé, J. P., Gold from the Golfo Dulce placer province, southern Costa Rica.

Nelson, C. E., and Arauz, A., Deposits of large tonnage precious metal deposits in the Caribbean region.

Chiesa, S., Civelli, G., Gillot, P.-Y., Mora, O., and Alvarado, G. E., Pyroclastic rocks associated with the formation of the Guayabo caldera, Cordillera de Guanacaste, Costa Rica.

Mora, R., Probablistic analysis of peak water flow for the El Humo station, Reventazón-Parismina, Costa Rica.

Salazar, L. G., Obando, L., and Mora, R., Metropolitan aqueduct in the section damming the El Llano–Río Navarro (Costa Rica): A low risk area.

National Seismic Network, Technical Note: ICE-UCR summary of seismic and volcanic activity in Costa Rica in 1991.

Aguilar, T., and Soto, G., Technical note: 1991 annual report of the Costa Rica committee for the International Geologic Correlation Program (IGCP).

Vargas, F., and Alfaro, A., Technical note: Presence of serpentinites and alkaline basalts in the northern Atlantic zone of Costa Rica.

Astorga, A., Technical note: Discovery of Mesozoic oceanic crust in northern Costa Rica and southern Nicaragua.

*Papers in English shown in bold face type; titles of Spanish papers translated to English by P. Mann; copies of articles at reproduction and mailing costs available through Ms. Sonia Castillo, Geology Librarian, Central American School of Geology, Box 35, San Pedro, San José, Costa Rica (telephone: 506-25-7941; fax: 506-34-2347).

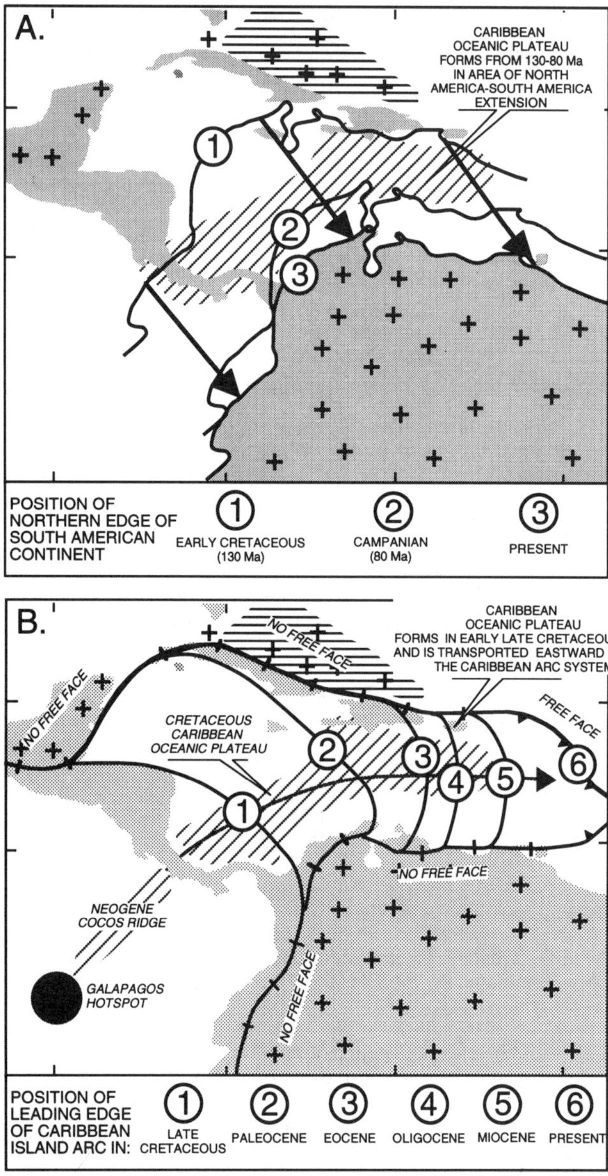

Figure 13. Two possible origins for the Caribbean: A, The Caribbean oceanic plateau forms by the separation of North and South America during the period 130 to 80 Ma (Frisch et al., 1992). Crosses indicate areas of continental crust. The numbers give positions of the northern margin of South America according to Pindell and Barrett (1990). B, The Caribbean oceanic plateau forms as normal Pacific oceanic crust drifts over the Galapagos hot spot, is thickened in the mid- to late Cretaceous, and passes into the gap between North and South America. The numbers give positions of the leading edge of the Caribbean arc system and oceanic plateau through time, according to Pindell and Barrett (1990). Note that the positions of continental masses allow "no free face" for arc migration. The "free face" of the Atlantic Ocean acts to channel the arc in an eastward direction.

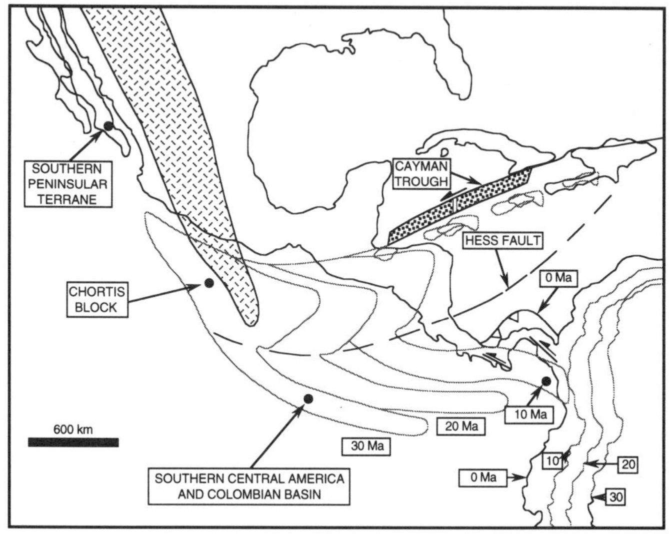

Figure 14. Reconstruction of the position and shape of Central America according to Wadge and Burke (1983). Note that the rotation and translation of Central America involve significant internal deformation in southern Central America during the period from 10 to 0 Ma. Stippled pattern indicates extent of original arc and continuous arc along the western margin of Mexico and the north Caribbean according to Wadge and Burke (1983). Dotted pattern indicates 1,400-km length of thinned or oceanic crust produced by left-lateral offset along the north margin of the Caribbean.

ACKNOWLEDGMENTS

My work in southern Central America began in 1984 with small grants from the National Geographic Society and Amoco that supported fieldwork by myself and a University of Texas Master's degree candidate, Jeff Corrigan. Work continued with institutional support from the University of Texas Institute for Geophysics and the Department of Geological Sciences and a grant from the Petroleum Research Fund of the American Chemical Society that supported a second University of Texas Master's degree candidate, Radim Kolarsky. Our fieldwork in Panama and Costa Rica was greatly facilitated by the cooperation of government geologists in both countries and by contributions of marine seismic reflection data by Mobil, Shell International, and Texaco. I would like to thank the following colleagues for volunteering their time to review the papers presented in this volume: W. Gose, J. Hagstrum, W. Frisch, S. May, F. McDowell, G. Alvarado, J. Walker, E. Malavassi, M. Perfit, T. Donnelly, M. Lagoe, D. Reed, J. Corrigan, R. Stern, D. Smith, J. Case, G. Westbrook, P. Lundgren, A. Kudrna, G. Moore, J. de Boer, K. McIntosh, M. Gordon, D. Orange, K. McMillen, D. Davis, J. Dolan, T. Shipley, C. Bowland, T. Gardiner, G. Keller, N. Lundberg, F. Tajima, C. Frohlich, J. Dewey, and L. Astiz. I would also like to give special thanks to Richard A. Hoppin (GSA Books Editor) for his expert help in getting this volume published.

REFERENCES CITED

Bowland, C. L., and Rosencrantz, E., 1988, Upper crustal structure of the western Colombian basin, Caribbean Sea: Geological Society of America Bulletin, v. 100, p. 534–546.

Brown, A. P., and Pilsbry, H. A., 1911–1913, Fauna of the Gatún Formation, Isthmus of Panama: Proceedings of the Academy of Natural Sciences of Philadelphia, v. 63, p. 36–420; v. 64, p. 500–519.

Bullard, E. C., Everett, J. E., and Smith, A. G., 1965, The fit of the continents around the Atlantic: A symposium on continental drift: Philosophical Transactions of the Royal Society of London, Series A, v. 258, p. 41–51.

Burke, K., 1988, Tectonic evolution of the Caribbean: Annual Review of Earth and Planetary Sciences, v. 16, p. 201–230.

Carson, J. P., 1874, Geological report on the Darién route and Nercalagua River of San Blas, *in* Selfridge, T. O., ed., Reports of explorations and surveys to ascertain the practicability of a ship canal between the Atlantic and Pacific oceans by way of the Isthmus of Panama: Washington, D.C.

Carr, M. J., and Stoiber, R. E., 1990, Volcanism, *in* Dengo, G., and Case, J. E., eds., The Caribbean region: Geological Society of America, The Geology of North America, v. H, p. 375–392.

Case, J. E., MacDonald, W. D., and Fox, P. J., 1990, Caribbean crustal provinces: Seismic and gravity evidence, *in* Dengo, G., and Case, J. E., eds., The Caribbean Region: Geological Society of America, The Geology of North America, v. H, p. 15–36.

Dall, W. H., 1912, New species of fossil shells from Panama and Costa Rica: Smithsonian Miscellaneous Collections, v. 59, p. 1–10.

Douvillé, H., 1915, Les couches a orbitoïdes de l'isthme de Panama: Comptes Rendus Sommaires de la Societe Geologique de France, no. 16, p. 129–131.

Draper, G., and Dengo, G., 1990, History of geological investigation in the Caribbean region, *in* Dengo, G., and Case, J. E., eds., The Caribbean Region: Geological Society of America, The Geology of North America, v. H, p. 1–14.

Duncan, R. A., and Hargraves, R. B., 1984, Plate tectonic evolution of the Caribbean region in the mantle reference frame, *in* Bonini, W. E., Hargraves, R. B., and Shagam, R., eds., The Caribbean–South American plate boundary and regional tectonics: Geological Society of America Memoir 162, p. 81–93.

Frisch, W., Meschede, M., and Sick, M., 1992, Origin of the Central American ophiolites: Evidence from paleomagnetic results: Geological Society of America Bulletin, v. 104, p. 1301–1314.

Gabb, W. M., 1875, Note on the geology of Costa Rica: American Journal of Science, v. 9, p. 198–204.

Gabb, W. M., 1881, Descriptions of new species of fossils from the Pliocene clay beds between Limon and Moen, Costa Rica, together with notes on previously known species from there and elsewhere in the Caribbean area: Journal of the Academy of Natural Sciences of Philadelphia, v. 8, p. 349–380.

Gabb, W. M., 1895, Informe sobre la exploración de Talamanca verificada durante los años 1873–1874: Anales del Instituto Fisico-geografico Nacional de Costa Rica, v. 5, p. 67–90.

Hayes, C. W., 1899, Physigraphy and geology of regions adjacent to the Nicaragua canal route: Geological Society of America Bulletin, v. 10, p. 285–488.

Hershey, O. H., 1901, The geology of the central portion of the Isthmus of Panama: Bulletin of the Department of Geology, University of California, v. 2, p. 231–267.

Hill, R. T., 1898, The geological history of the Isthmus of Panama and portions of Costa Rica: Bulletin of the Museum of Comparative Zoology, v. 28, p. 154–281.

Howe, E., 1907, Geology and the Panama Canal: Economic Geology, v. 2, p. 639–658.

Johnson, M. S., and Headington, E., 1971, Panama: Exploration history and petroleum potential: Oil and Gas Journal, v. 69, p. 96–100.

Jones, S. M., 1950, Geology of Gatún Lake and vicinity: Geological Society of America Bulletin, v. 61, p. 893–920.

Jordan, T. H., 1975, The present-day motions of the Caribbean plate: Journal of Geophysical Research, v. 80, p. 4433–4439.

Keen, A. M., and Thompson, T. F., 1946, Notes on the Gatún Formation (Miocene), Panama: Bulletin of the Geological Society of America, v. 57, p. 1260.

Kesler, S. E., Levy, E., and Martín F., C., 1990, Metallogenic evolution of the Caribbean, *in* Dengo, G., and Case, J. E., eds., The Caribbean Region: Geological Society of America, The Geology of North America, v. H, p. 459–482.

Kirkpatrick, R. Z., 1939, Trigger forces—Canal Zone earthquakes: Bulletin of the Seismological Society of America, v. 28, p. 15–22.

Lemoine, P., and Douvillé, R., 1904, Sur le genre *Lepidocyclina* Gümbel: Memoire de la Societe Geologique de France, Paleontologie, no. 32.

Maack, G. A., 1874, Report on the geology and natural history of the Isthmus of Choco, of Darién, and of Panama, *in* Selfridge, T. O., ed., Reports of explorations and surveys to ascertain the practicability of a ship canal between the Atlantic and Pacific Oceans by way of the Isthmus of Darién: Washington, D.C.

MacDonald, D. F., 1913, Isthmian earthquakes: Canal Record (December 10), p. 144–149.

MacDonald, D. F., 1915, Some engineering problems of the Panama Canal: Bulletin of the United States Bureau of Mines, no. 86, p. 88.

MacDonald, D. F., 1919, Sedimentary formations of the Panama Canal Zone with specific reference to the stratigraphic relations of the fossiliferous beds: Bulletin of the United States National Museum, no. 103.

MacDonald, D. F., 1937, Contributions to Panama geology: Journal of Geology, v. 45, p. 655–662.

Mann, P., Draper, G., and Lewis, J. F., 1991, Geologic and tectonic development of the North America–Caribbean plate boundary in Hispaniola: Geological Society of America Special Paper 262, 401 p.

Molnar, P., and Sykes, L. R., 1969, Tectonics of the Caribbean and Middle America regions from focal mechanisms and seismicity: Geological Society of America Bulletin, v. 80, p. 1639–1684.

Pindell, J. L., and Barrett, S. F., 1990, Geological evolution of the Caribbean region: A plate tectonic perspective, *in* Dengo, G., and Case, J. E., eds., The Caribbean Region: Geological Society of America, v. H, p. 405–432.

Sapper, K., 1905, Ueber gebirgsbau und boden des südlichen mittelamerika: Gotha, Germany Petermann's Geographische Mitteilungen, Ergänzungsheft, v. 32, 82 p.

Sapper, K., and Staub, W., 1937, Mittelamerika, *in* Steinmann and Wilckens, eds., Handbuch der regionalen geologie: Heidelberg, Carl Winter, 160 p.

Schuchert, C., 1935, Historical geology of the Antillean-Caribbean region, or the lands bordering the Gulf of Mexico and the Caribbean Sea: New York, John Wiley and Sons, 811 p.

Sheehan, C. A., Penfield, G. T., and Morales, E., 1990, Costa Rica geologic basins lure wildcatters: Oil and Gas Journal (April 30), p. 73–79.

Taylor, R. C., 1852, Substance of notes made during a geological reconnaissance in the auriferous porphyry region next to the Caribbean Sea in the province of Veraguas and Isthmus of Panama: Journal of the Academy of Natural Sciences of Philadelphia, v. 2, p. 81–88.

Terry, R. A., 1956, A geological reconnaissance of Panama: Occasional Papers of the California Academy of Sciences, no. 23, 91 p.

Toula, F., 1909–1911, Eine jungetertiäre fauna von Gatún am Panama-Kanal: Kaiserlichköniglichen Geologischen Reichsandstalt, Jahrbuch, v. 58, p. 673–760; v. 59, p. 487–530.

Vaughan, T. W., 1919a, The stratigraphic horizon of the beds containing *Lepidocyclina chaperi* on Haut Chagres, Panama: Proceedings of the National Academy of Sciences, v. 12, p. 519–522.

Vaughan, T. W., 1919b, Contributions to the geology and paleontology of the Canal Zone, Panama, and geologically related areas in Central America and the West Indies: Bulletin of the United States National Museum, v. 103, p. 547–612.

von Seebach, K., 1892, Uber die Vulkane Zentralamerikas: Göttingen, Abhandlungen Kgl. Gesellschaft Wissenschaft, 251 p.

Wadge, G., and Burke, K., 1983, Neogene Caribbean plate rotation and associated Central America tectonic evolution: Tectonics, v. 2, p. 633–643.

Wagner, M., 1861, Beiträge zu einer physisch-geographischen skizze des isthmus von Panama: Gotha, Germany Petermann's Geographische Mittheilungen, Ergänzungsheft, p. 1–25.

Wagner, M., 1862, Eine reise in das innere der landenge von San Blas und der cordillere von Chepo in der Provinz Panama: Gotha, Germany Petermann's Geographische Mittheilungen, Ergänzungsheft, p. 128–141.

Weyl, R., 1980, Geology of Central America: Berlin, Gebrüder Borntraeger, 371 p.

Wilson, J. T., 1966, Are the structures of the Caribbean and Scotia arcs analogous to ice rafting? Earth and Planetary Science Letters, v. 1, p. 335–338.

Woakes, E. R., 1923, Darién gold mines: Mining Magazine (London) (November), p. 270–278.

Woodring, W. P., 1949, The Panama landbridge: Science, v. 109, p. 437.

Woodring, W. P., 1954, Caribbean land and sea through the ages: Geological Society of America Bulletin, v. 65, p. 719–732.

Woodring, W. P., 1955, Geologic map of the Canal Zone and adjoining parts of Panama: U.S. Geological Survey Miscellaneous Geologic Investigations Map I-1, scale 1:100,000.

Woodring, W. P., 1966, The Panama land bridge as a sea barrier: Proceedings of the American Philosophical Society, v. 110, p. 425–433.

Woodring, W. P., 1974, The Miocene Caribbean faunal province and its subprovinces, *in* Contributions to the Geology and Paleobiology of the Caribbean and Adjacent Areas: Naturforschende Gesellschaft in Basel, Verhandlungen, v. 84, p. 209–213.

Woodring, W. P., and Thompson, T. F., 1949, Tertiary formations of the Panama Canal Zone and adjoining parts of Panama: Bulletin of the American Association of Petroleum Geologists, v. 33, p. 223–247.

Printed in U.S.A.

Geological Society of America
Special Paper 295
1995

Late Cretaceous–early Tertiary paleomagnetic data and a revised tectonostratigraphic subdivision of Costa Rica and western Panama

Gianni Di Marco and Peter O. Baumgartner
Institut de Géologie et Paléontologie BFSH2, Université de Lausanne, CH-1015 Lausanne, Switzerland
James E.T. Channell
Department of Geology, University of Florida, Gainesville, Florida 32611

ABSTRACT

New paleomagnetic data are presented in the context of a revised tectonostratigraphic subdivision of the Mesozoic-Tertiary oceanic basement of Costa Rica. We present the tectonostratigraphic characteristics and areal extent of four terranes (Chorotega, Nicoya, Golfito, and Burica) bordered toward the Middle America Trench by the Tertiary Osa-Caño Accretionary Complex. The paleomagnetic data are derived from Upper Cretaceous and Paleocene pelagic limestones.

The Chorotega Terrane constitutes most of the southern Middle American Landbridge and was the western edge of the Caribbean plate during the Late Cretaceous–Paleocene. The Nicoya Terrane comprises the Santa Elena Peninsula and most of the outer Nicoya Peninsula. The Nicoya Terrane includes the Nicoya Complex *(sensu stricto)* and should therefore probably be regarded as a composite terrane. The Golfito Terrane forms the Golfito region and extends into Panama to the Azuero Peninsula. The Late Cretaceous basement of the terrane is thought to have formed a marginal piece of the Caribbean oceanic plateau, transported northward by strike slip along the rim of the Caribbean plate. The Burica Terrane forms the Burica Peninsula. The terrane is thought to represent an accreted, structurally high piece of a primitive island arc. The inner Osa Peninsula is formed by a thick pile of oceanic basalts including Late Cretaceous to Eocene sediments. The outer Osa Peninsula and the Caño Island are built by the Osa-Caño Accretionary Complex, a mélange-type complex characterized by strongly deformed turbidites and hemipelagic and pelagic sediments that range in age from Late Cretaceous to Miocene. The exotic terranes are thought to have originated outboard in the Paleopacific, been brought into contact with the Caribbean plate boundary by plate convergence, and then been moved farther north by strike-slip motion along the margin.

The paleomagnetic data for the Chorotega Terrane indicate an origin close to its present latitude and no significant rotation relative to South America since Late Cretaceous time. The paleomagnetic data obtained from the Nicoya Terrane imply a low southerly Late Cretaceous paleolatitude and almost no rotation relative to the Chorotega Terrane. The Nicoya Terrane was about 16° of latitude south relative to

Di Marco, G., Baumgartner, P. O., and Channell, J.E.T., 1995, Late Cretaceous–early Tertiary paleomagnetic data and a revised tectonostratigraphic subdivision of Costa Rica and western Panama, *in* Mann, P., ed., Geologic and Tectonic Development of the Caribbean Plate Boundary in Southern Central America: Boulder, Colorado, Geological Society of America Special Paper 295.

the Chorotega Terrane in Late Cretaceous times. The paleomagnetic data from the Golfito Terrane indicate a Late Cretaceous equatorial paleolatitude and counterclockwise rotation of about 60° relative to the Chorotega Terrane. Similar paleomagnetic data were obtained from the Azuero Peninsula in southwestern Panama. The paleomagnetic data from the Burica Terrane indicate a low northerly latitude in the Paleocene and a counterclockwise rotation of nearly 90° relative to the Chorotega Terrane.

INTRODUCTION

Oceanic basement rocks and associated sediments crop out along the Pacific Coast of Costa Rica, Panama, and northwestern Colombia. The Santa Elena and Nicoya Peninsulas, in northwestern Costa Rica, are the most extensively studied parts of this coastal outcrop. In the Nicoya Peninsula, Dengo (1962) defined the Nicoya Complex as a pre-Senonian tectonized body of basaltic basement, associated with oceanic sediments and unconformably overlain by Upper Cretaceous deep-water sediments. A refined analysis and definition of the Nicoya Complex was provided by Kuijpers (1980), who was the first to recognize nappe structures in the Nicoya Complex. Bourgois et al. (1984) and Azéma et al. (1985) documented the emplacement history and established the tectonic relationship between the Nicoya Complex, composed of the Esperanza and Matapalo units, and the overriding Santa Elena Ultramafic Unit. Ages ranging from late Liassic to Santonian have been documented for parts of the Nicoya Complex (Schmidt-Effing, 1979; Baumgartner, 1984; de Wever et al., 1985).

Many authors have included the bulk of the basaltic basement and associated oceanic sediments outcropping in Costa Rica and Panama with the Nicoya Complex (Henningsen and West, 1967; Pichler and Weyl, 1973; Schmidt-Effing, 1979; Berrangé and Thorpe, 1988; Bowland and Rosencrantz, 1988; Escalante, 1990; Frisch et al., 1992). However, it has been recognized that there are fundamental differences in age, lithology, tectonic structure, geochemical characteristics, and paleomagnetic signature in different areas (Wildberg, 1984; Baumgartner, 1984; Baumgartner et al., 1984; Baumgartner, 1987; Frisch et al., 1992). Baumgartner (1987) considered the basaltic basement south of the Nicoya Peninsula as independent units distinct from the Nicoya Complex (*sensu stricto*). Baumgartner et al. (1989) and Baumgartner (1990) subdivided the Costa Rican Pacific coast into three tectonostratigraphic suspect terranes (Nicoya Complex, Golfito and Burica Terranes) and defined the Caño Accretionary Complex (an accretionary mélange) formed as a result of Tertiary accretion in the Middle America Trench.

Few paleomagnetic surveys have been undertaken in Costa Rica. De Boer (1979) discussed two mean paleomagnetic directions obtained from igneous rocks on the Nicoya Peninsula. These data are not assigned to a precise location and are therefore of no use in further studies. The first useful data are given by Gose (1983), who initiated comprehensive paleomagnetic studies in Costa Rica and Nicaragua. According to these data, the Nicaraguan sites have been close to their present latitude since the Cretaceous, whereas the Costa Rican sites show a Southern Hemisphere origin. Gose's (1983) sites are included in our dataset, although sites 5 and 6 are excluded because of probable remagnetization and poor within-site precision, respectively. Frisch et al. (1992) presented geochemical and paleomagnetic data from Costa Rica and western Panama (see also Sick, 1989). Their results imply a broadly defined equatorial origin for the whole region, with variable amounts of local block rotation.

This chapter presents a synthesis of an on-going study of the tectonostratigraphic subdivision of Costa Rica and western Panama being carried out by P. O. Baumgartner and associates. The Costa Rican isthmus is subdivided into four tectonostratigraphic terranes and the Osa-Caño Accretionary Complex (Fig. 1) (Di Marco, 1994). The tectonostratigraphic characteristics of these units are described. We also present paleomagnetic data and discuss the degree of allochthony of the tectonostratigraphic units with respect to the Caribbean plate.

TECTONOSTRATIGRAPHIC SUBDIVISION

Introduction

The Mesozoic and early Tertiary oceanic basement and associated deep-water sediments of Costa Rica and Panama continue to be included with the Nicoya Complex. Over 30 years ago, Dengo (1962) expressed doubts as to whether basaltic basement complexes outcropping in southern Costa Rica should be included with the Nicoya Complex, principally because of the absence of the pre-Campanian deformational phase characteristic of the Nicoya Complex (*sensu stricto*). Baumgartner et al. (1984) documented fundamental differences in magmatism, sedimentary environment, age, and tectonic history between units of the northern and the southern Pacific coast of Costa Rica (see also Baumgartner, 1987; Seyfried et al., 1991).

We propose a revised subdivision of the Costa Rican basement and overlying Mesozoic–early Tertiary sediments into several tectonostratigraphic units, each of which is characterized by a distinct tectono-sedimentary history. This subdivision (Fig. 1) is based on the detailed stratigraphic work carried out in the last 12 years by members of the Escuela Centroamericana de Geologia, by various foreign research groups, and by the authors. Our own fieldwork in the last few

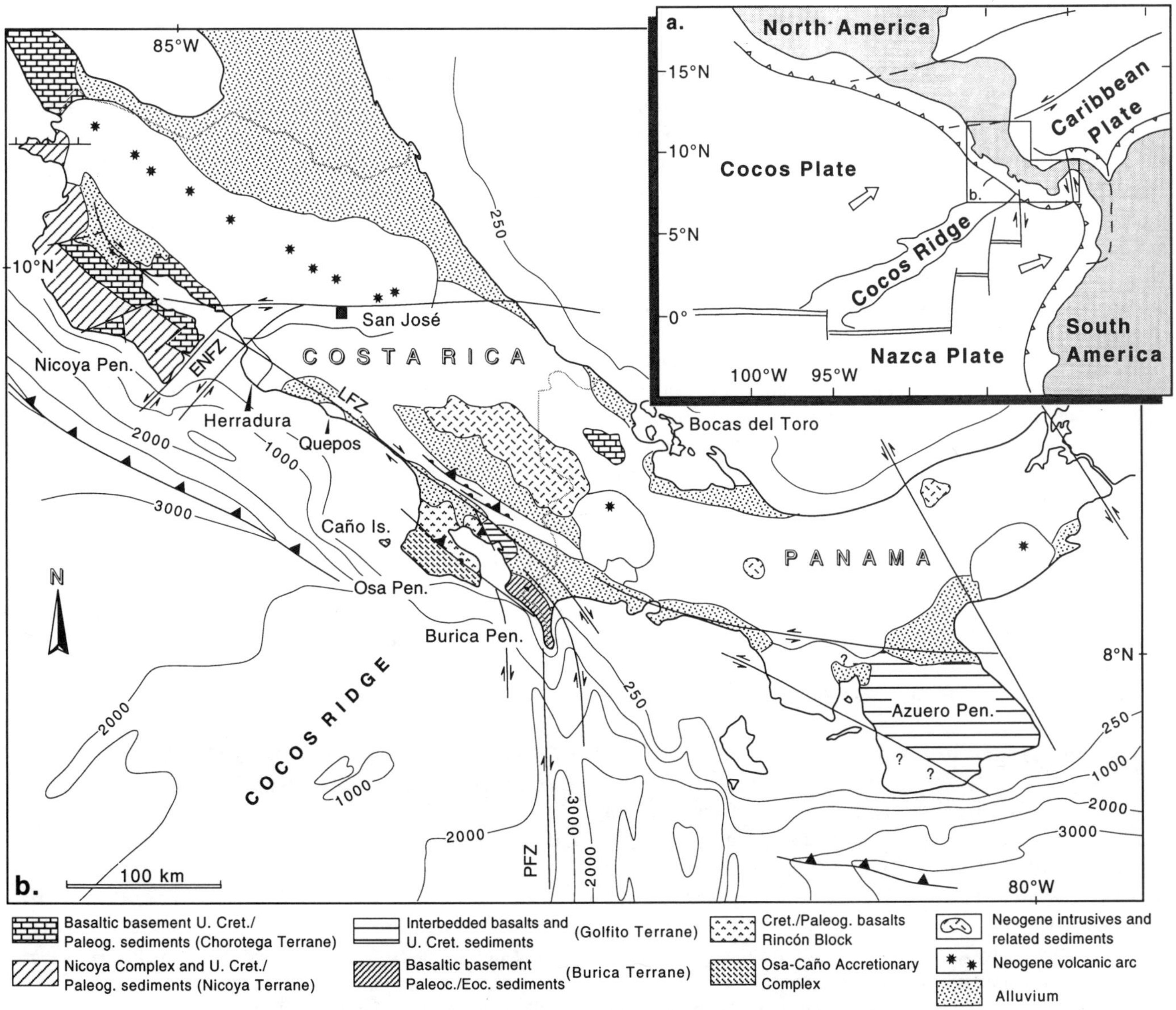

Figure 1. a, tectonic setting of Central America. Modified from Case and Holcombe (1980). b, geological map showing the main tectonic features and distribution of the tectonostratigraphic terranes of Costa Rica and western Panama (overlap sequences have been removed). ENFZ = East Nicoya Fracture Zone. LFZ = Longitudinal Fault Zone. PFZ = Panama Fracture Zone.

years has been focused on the southern Pacific area (Quepos, Osa, Golfito, Burica). Results of this work will be reported in detail in future publications (see Di Marco, 1994). This chapter summarizes the igneous and sedimentary characteristics leading to the proposed tectono-stratigraphic subdivision and documents important stratigraphic events in each area that constrain emplacement and accretion history.

We adopt the terrane nomenclature (Howell et al., 1985) because the outcrops along the Pacific coast of Costa Rica reveal a number of fault-bounded basement blocks, each of which is covered by a more or less unique stratigraphic sequence. Except for the outer Osa Peninsula and parts of the Nicoya Complex (*sensu stricto*), the Costa Rican Pacific coast can be regarded as a collage of stratigraphic terranes that represent fragments of ocean basins, oceanic seamounts, and/or island arcs. The Osa-Caño Accretionary Complex outcropping in the outer Osa Peninsula should be regarded as a disrupted terrane (Howell et al., 1985) that includes blocks of ophiolitic rocks but shows no evidence of an igneous basement (see below).

Many authors have attempted to organize the sedimentary basins of northern Costa Rica into a single paleogeography (Dengo, 1962; Lundberg, 1982; Baumgartner et al., 1984; Seyfried and Sprechmann, 1985; Seyfried et al., 1991). Although the sequences in the Tempisque area and southern Nicaragua show common features (Sprechmann et al., 1987; Winsemann,

1992), correlation between this "inner forearc" and the "outer forearc" in the outer Nicoya Peninsula is difficult. These areas are separated by a number of faults, and where the sedimentary cover on the basaltic basement is preserved, completely different sedimentary sequences are juxtaposed along these faults (see below). The igneous and sedimentary sequences recorded in the Burica, Osa, and Golfito areas have little in common with each other and even less in common with those of northern Costa Rica. These areas are separated from each other by major faults or fault zones and are designated as terranes (see Howell et al., 1985). In most cases, these fault zones do not represent the original suture that resulted from the accretion of the terrane but are Neogene to Recent (active) faults. The present-day juxtaposition of terranes is the result of a multiphase emplacement, involving both initial plate convergence and successive phases of strike-slip motion. Amalgamation and accretion of terranes can be documented by local and regional sedimentary and tectonic events such as the onset of coarse-clastic sediments and deformational phases. The exotic terranes are thought to have originated somewhat outboard in the Paleopacific, been brought into contact with the Caribbean plate boundary by plate convergence, and then been moved farther north by strike-slip motion along the margin. The collage of terranes is bordered toward the Middle America Trench by the Tertiary Osa-Caño Accretionary Complex.

Areal extent and boundaries of terranes

The Chorotega Terrane constitutes most of the southern Middle American Landbridge. It includes the southern Nicaragua (Rivas) Basin and the Tempisque Basin, including the inner part of the Nicoya Peninsula (Fig. 2). It is the original basement of the Neogene to Recent volcanic arc and extends into Panama at least as far as the Bocas del Toro area (Fig. 1), where late Campanian and younger sediments crop out in the Changuinola River. The Chorotega Terrane was the western edge of the Caribbean plate during the Late Cretaceous–Paleocene. The other terranes proposed in this chapter are allochthonous relative to the Chorotega Terrane.

The boundary between the Chorotega and the Nicoya Terranes is at present poorly understood and is therefore drawn somewhat tentatively. The boundary is represented by northeast-southwest–trending faults (Figs. 2 and 3) that probably had an important strike-slip component. These faults are offset in the southern part of the Nicoya Peninsula by younger southwest-northeast–trending strike-slip faults that could be related to the East Nicoya Fracture Zone (ENFZ) (Corrigan et al., 1990), which trends perpendicular to the mainland coast in the southeastern part of the Nicoya Peninsula (see Baumgartner et al., 1984; Burbach et al., 1984) (Fig. 1). Multiphase faulting has caused the area of Puerto Carrillo (Fig. 2) to become tectonically isolated from the main Chorotega Terrane.

The Nicoya Terrane comprises the Santa Elena Peninsula and most of the outer Nicoya Peninsula facing the Pacific. The Nicoya Terrane includes the Nicoya Complex (*sensu stricto*) and should therefore probably be regarded as a composite terrane (Howell et al., 1985). South of the ENFZ, a heterogeneous collage of fault-bounded promontories and peninsulas borders the Chorotega Terrane. The Herradura promontory lies at the junction of the Longitudinal Fault Zone (LFZ) and the ENFZ and is affected by both fault systems (Fig. 1). This zone is tentatively included in the Chorotega Terrane and awaits more detailed analysis. The Quepos promontory (Fig. 1) is separated from the Chorotega Terrane by the LFZ and displays a unique sedimentary sequence (Baumgartner et al., 1984; Winsemann, 1992). The interpretation of the Quepos promontory as a separate terrane will be addressed elsewhere.

The Osa Peninsula, the Golfito Terrane, and the Burica Terrane are separated from each other by high-angle reversed faults that are either observed in the field (Fig. 4) or inferred from unpublished seismic data. These units are separated from the Chorotega Terrane by the LFZ. The Osa Peninsula comprises two fundamentally different units: (1) the Rincón Block, a tectonically thickened pile of basaltic basement with rare occurrences of Late Cretaceous to Eocene sediments, and (2) the Osa-Caño Accretionary Complex, a pile of accreted oceanic and trench-fill sediments including some ophiolitic blocks. The two units are separated by a trench-parallel high-angle reverse fault that is the trace of an earlier strike slip fault.

STRATIGRAPHY AND SEDIMENTARY EVOLUTION OF EACH TERRANE

Chorotega Terrane

Basement and associated Cretaceous-Paleocene sediments. The Chorotega basement comprises oceanic basalts of middle-Late Cretaceous or older age. Few data are available on their geochemistry, since most recent work has been focused on the outer Nicoya Peninsula (Nicoya Terrane). We believe, however, that this basement was part of the Caribbean seafloor, thickened by the Caribbean sill event (Burke et al., 1978). The oldest associated sediments range in age from Albian to Santonian and are likely interbedded in the volcanic rocks of the Caribbean sill event. These sediments are more than 600 m thick and include bituminous shales, siliceous mudstones, and tuffs in the Tempisque area (Azéma et al., 1979; Astorga, 1987). This facies association is unknown in the Nicoya Terrane, where relatively thin radiolarites prevail during the same time interval. Widespread basaltic breccias and conglomerates of Campanian age interfinger with or are overlain by Campanian deep-water sequences, and in the Bajo Tempisque area (Colorado, Fig. 2), deep-water breccias contain reworked shallow-water fossils (Rivier, 1983; Seyfried and Sprechmann, 1986; Winemann, 1992). In the area of Tambor-Paquera (Fig. 3), the basal breccias are more than 100 m thick and contain several-meter-sized blocks of radiolarite, basalt, and basaltic breccia (Punta Pochote, Playa Curú, Islas Tortugas). These coarse rockfall and debris-flow deposits document important post-Santonian erosion due to submarine relief. The late Campanian–early Maastrichtian is usually represented by more than 50 m of 0.5- to 1-m-thick beds of white micritic limestones rich in *Globotruncana* sp, radiolarians, with some fine-grained basaltic detritus (Bahía Murcielago, Río Changuinola). Interbeds of dark shale may occur. Although no shallow-water fossils were found in these sequences, we suspect the thick bedding to be the result of periplatform ooze input into a hemipelagic environment of sedimentation. Sediments of this unit contrast with the usually condensed Senonian sediments of the Nicoya Terrane. Arc-derived siliciclastic sedimentation prevailed in the Chorotega Terrane from late Maastrichtian probably until Paleocene–middle Eocene time. Up to 3,000 m of turbiditic forearc sequences were deposited in this time interval. In contrast, hemipelagic siliceous limestones were deposited during the upper Paleocene–middle Eocene in the Nicoya Terrane.

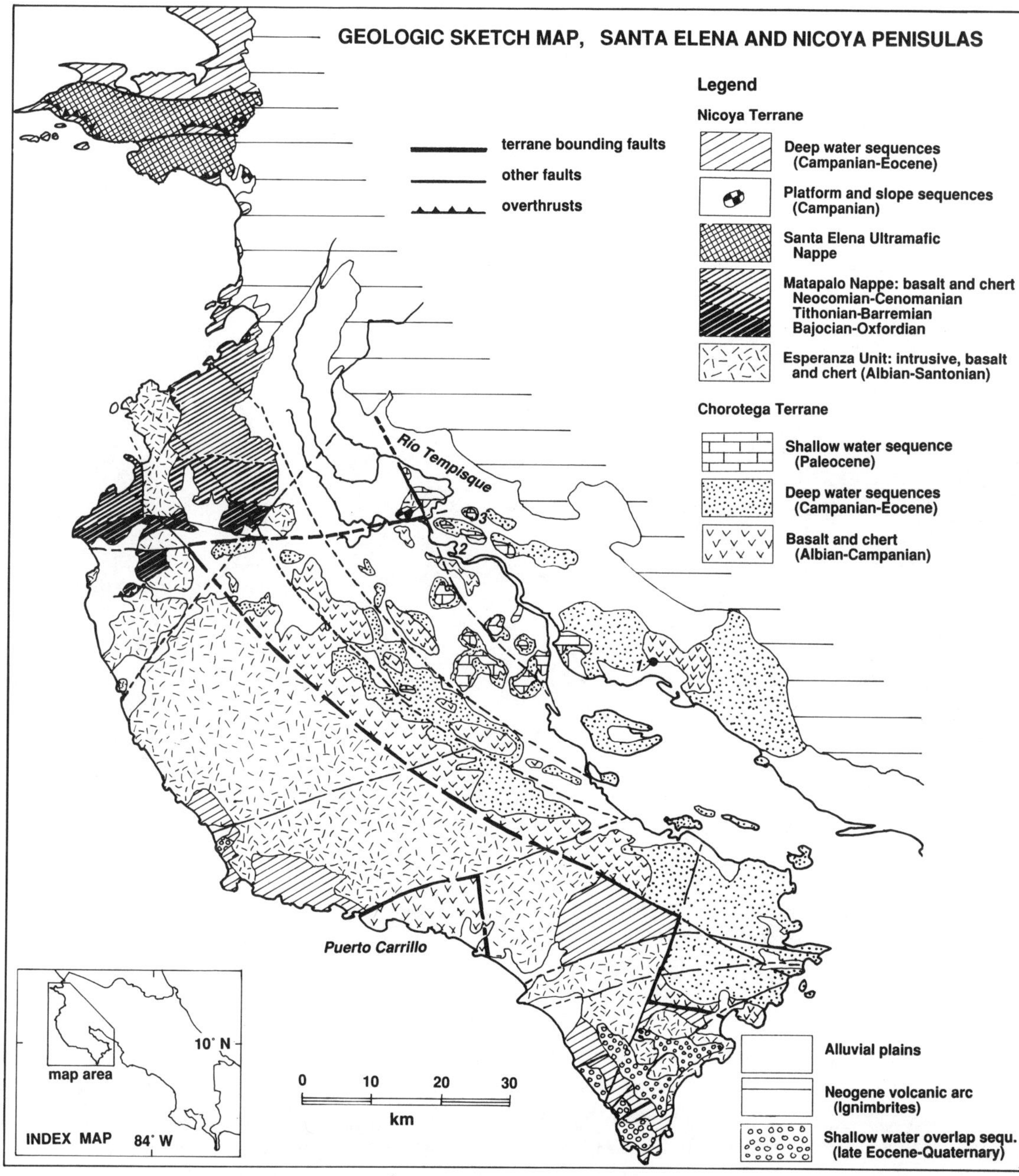

Figure 2. Map of the Santa Elena and Nicoya Peninsulas showing a tentative tectonostratigraphic subdivision. Data from Rivier (1983), Baumgartner (1984), Azéma et al. (1985), Arias and Denyer (1992a, b), Denyer and Arias (1992), Gursky (1988), and unpublished maps by P. Denyer, E. P. Kuijpers, and the authors. 1: Colorado; 2: Cerros Guayacán; 3: Cerro Espiritu Santo.

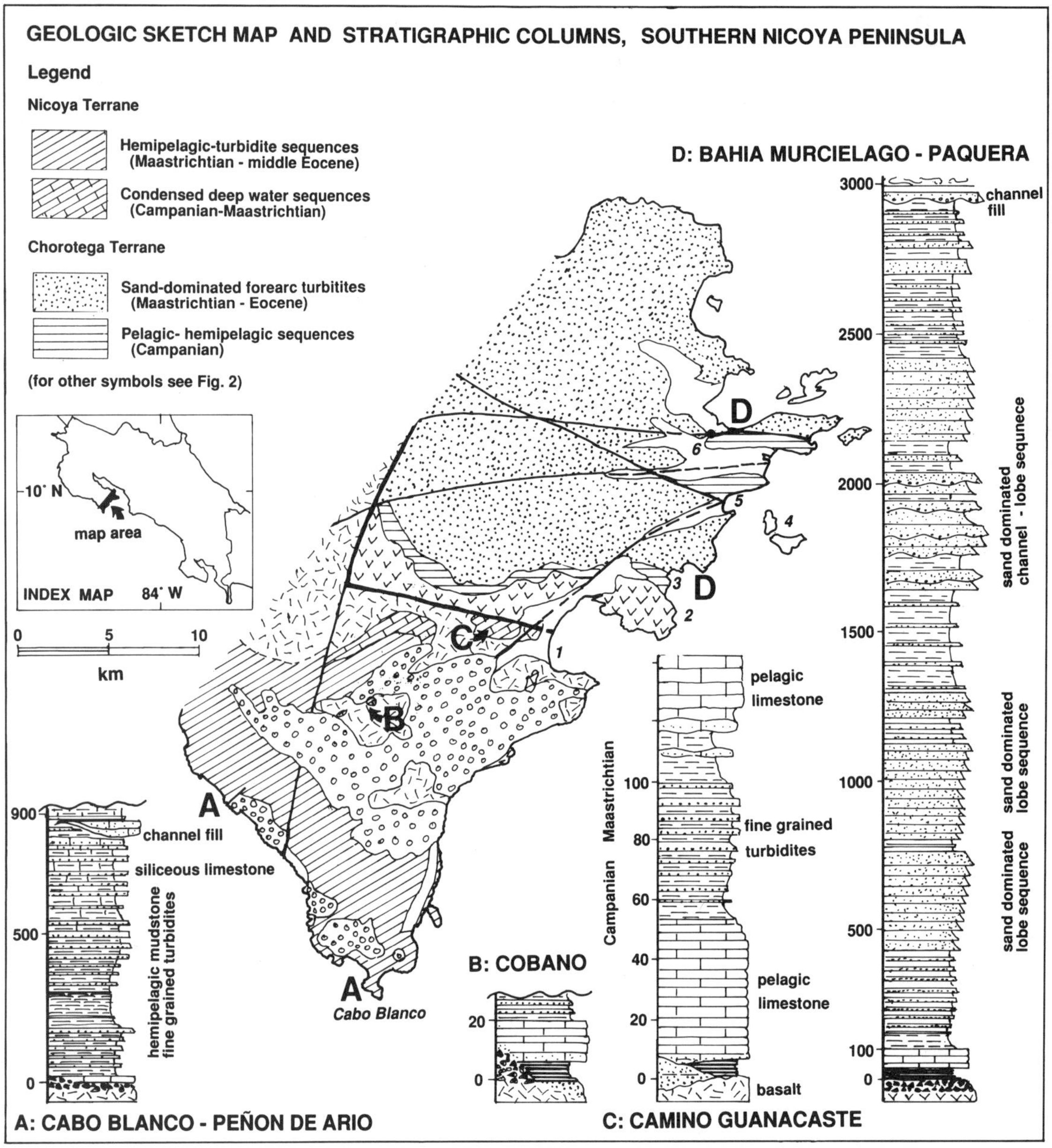

Figure 3. Geologic map and representative stratigraphic sections of the Southern Nicoya Peninsula: evidence for the boundary between the Chorotega and Nicoya Terranes. Data from Mora (1985), Baumgartner et al. (1984), Winsemann (1992), and unpublished data by the authors. 1: Tambor; 2: Punta Pochote; 3: Bahía Murcielago; 4: Islas Tortugas; 5: Playa Curú; 6: Paquera. Sections A to C are included in the Nicoya Terrane. Section D is included in the Chorotega Terrane. Note different scales of sections.

Barra Honda Platform. The Barra Honda limestones, micritic limestones of predominantly restricted platform origin (Mora, 1981), unconformably overlie Upper Cretaceous–Eocene deep-water sequences in the Tempisque area. Rivier (1983) suggested a middle Eocene or younger flooding of a slightly deformed area to explain the unconformable onset of deposition of these limestones. Sprechmann et al. (1987) correlated the Barra Honda limestones with the rudist-bearing limestones of Santa Elena (El Viejo Formation; Schmidt-Effing, 1975) on the basis of early Maastrichtian globotruncanids that they recovered from sheared mudstones below the base of the limestones. They also mentioned Campanian larger foraminifers recovered from unspecified limestones. Neither Sprechmann et al. (1987) nor Calvo (1987) provided exact sample locations or measured sections that would show how the reported Late Cretaceous fossils relate to the micritic Barra Honda facies, which is generally de-

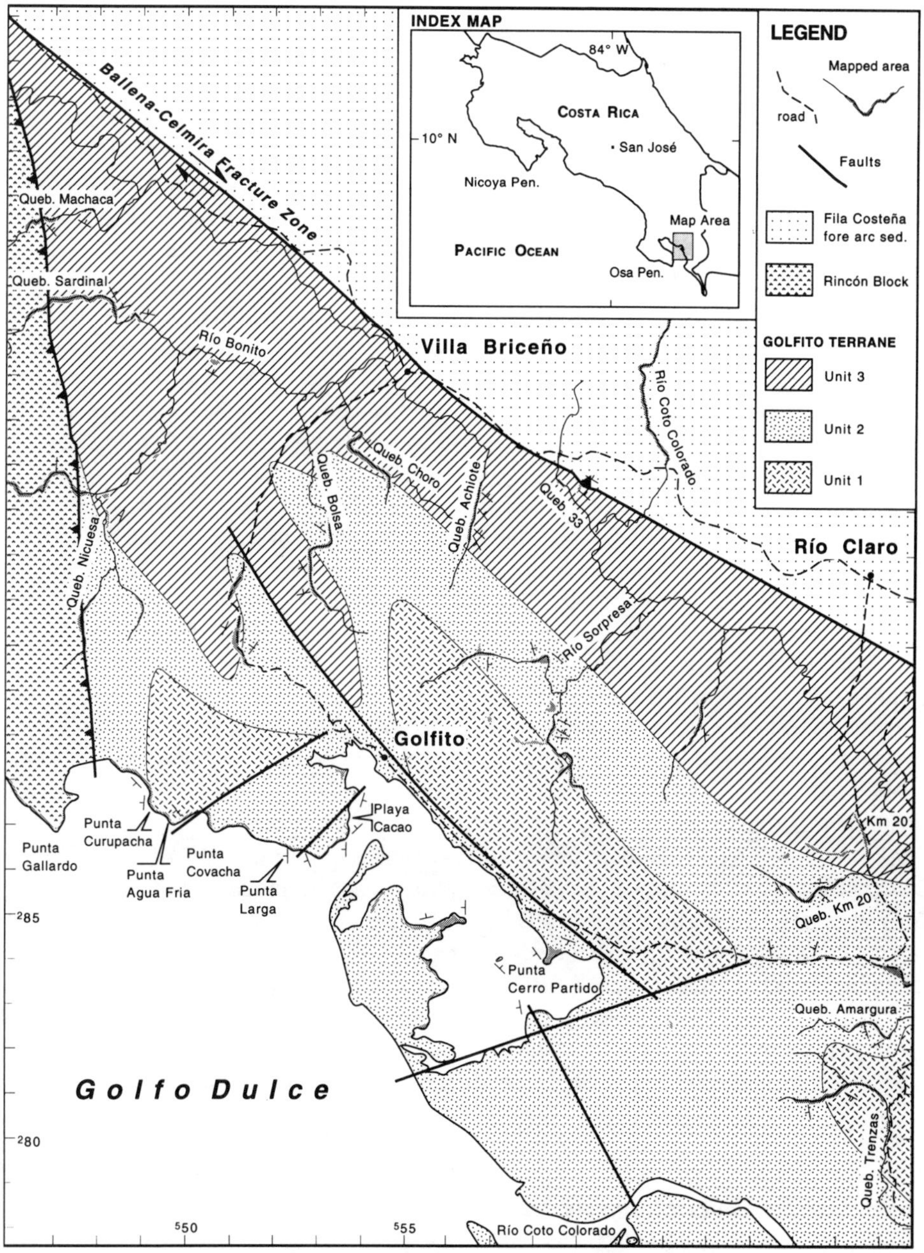

Figure 4. Interpretive geologic sketch map of the pre-Neogene geologic units of the Golfito Terrane. Mapped areas are shaded. Lithostratigraphic sections are shown in Figure 5.

void of age-diagnostic fossils. Late Cretaceous reworked shallow- and deep-water material is, however, quite common in the Paleocene-Eocene sequences that underlie the Barra Honda limestones (Seyfried and Sprechmann, 1985; Winsemann, 1992). The presence of reworked Campanian larger foraminifers and rudistids in Paleocene-Eocene mass flows demonstrates the erosion of Campanian carbonates produced in an open marine, high-energy environment that is distinct from the mud-dominated Barra Honda environment. Calvo (1987) assumed that the rudist-foraminifer facies is marginal to the restricted Barra Honda facies but provided no evidence of lateral continuity between the two formations. In view of the lack of precise stratigraphic data, we doubt this interpretation and consider that a late Cretaceous age of the Barra Honda limestones has not been established.

A number of observations are the basis of our interpretation of the Barra Honda limestones. The occurrence of the late Campanian rudist and larger foraminifer–bearing platform and slope facies (El Viejo Formation) is restricted to the Santa Elena and northwestern Nicoya Peninsula (Fig. 2) and is therefore a feature of the Nicoya Terrane. In the Bajo Tempisque area (Colorado, Fig. 2), rudistids and Campanian larger foraminifers are reworked into coarse ophioclastic deposits that directly overlie the basaltic basement (Barbudal Formation of Rivier, 1983; Seyfried and Sprechmann, 1985). These deposits probably formed during the late Campanian but do not contain clasts of Barra Honda–type carbonates. The first appearance of Barra Honda–type clasts in coarse mass-flow and channel-fill deposits associated with deep-water formations is systematically middle-late Paleocene in age (Sprechmann et al., 1987; Calvo, 1987; Winsemann, 1992). The Barra Honda–type micrites are therefore unlikely to be much older than middle-late Paleocene.

To the east of the Tempisque River, Barra Honda Limestones extend at least as far north as Cerros Guayacán (Fig. 2), where we found small *Discocyclina,* characteristic of a Paleocene-Eocene age, associated with the micritic facies. A few kilometers to the north, in the Cerro Espiritu Santo (Fig. 2), a more-than-50-m-thick sequence of platform limestones rich in upper Paleocene *Ranikothalia* rests with a well-exposed transgressive unconformity on siliceous pelagic mudstones (not on "Nicoya Complex" as illustrated by Calvo and Bolz, 1991). These mudstones, assignable to the Upper Cretaceous–Paleocene Sabana Grande Group (Dengo, 1962), have a thickness of about 100 m and rest unconformably on basalt. The transgression of the limestones is preceded by 5 m of reddish altered siliceous mudstones reminiscent of a paleosoil and by 4 m of carbonate conglomerate, including well-rounded and bored cobbles of Barra Honda–type algal micrites, rhodoliths, and *Ranikothalia*-bearing clasts, passing upward into bioclastic grain and packstones. To us, this sequence represents a marginal facies to the restricted Barra Honda platform. As a consequence, we conclude that this platform certainly existed during the late Paleocene and may have originated during the late early to middle Paleocene. Calvo and Bolz (1991) reported late Paleocene–early Eocene platform limestones from another locality of the Chorotega Terrane: Fila Chonta, about 20 km north of Quepos in the Fila Costeña.

Nicoya Terrane

Pre-Campanian units of the Santa Elena and Nicoya Peninsulas. The Nicoya Complex and the Santa Elena Ultramafic Unit are among the best-studied units in southern Central America. Nevertheless, the structural interpretation of these mafic and ultramafic units remains controversial. The presence of nappe structures was first proposed by Kuijpers (1980) for the Nicoya Peninsula and by Azéma and Tournon (1980) for the Santa Elena Peninsula. It was Kuijpers (1980) who interpreted the northeastern Nicoya Peninsula as a nappe structure, resulting from pre-Campanian compressive tectonics. Radiolarian biostratigraphic data confirmed these ideas, leading to the acceptance of a megastructure comprising the Santa Elena Ultramafic Unit and the northwestern Nicoya Peninsula (Bourgois et al., 1984; Azéma et al., 1985). Four units (Fig. 2) are recognized in this megastructure (see Baumgartner, 1987): (1) The Esperanza Unit can be considered as the relatively autochthonous unit forming the major part of the outcrops in the Nicoya Terrane. (2) The Matapalo Unit overlies the Esperanza Unit along a thrust contact. The Matapalo Unit is preserved as isolated klippes in the northwestern Nicoya Peninsula. (3) A volcano-sedimentary unit underlies the Santa Elena Unit in small outcrops and tectonic windows. (4) The Santa Elena Unit is the structurally highest unit in the nappe pile. Both in Santa Elena and Nicoya, there is clear evidence for compressive deformation with a southward vergence (Strebin, 1982; Azéma et al., 1985; Meschede et al., 1988).

The Matapalo Unit and the volcano-sedimentary unit of the Santa Elena Peninsula represent remnants of accretionary prisms that after their formation became emplaced on the Esperanza Unit and are overthrust by the Santa Elena Unit. The accretionary history preserved in these two units indicates Middle and Late Cretaceous subduction followed by Santonian–early Campanian obduction. The age distribution within the Matapalo Unit indicates a southward dip during its accretion. The predominantly primitive island arc origin postulated for the Esperanza Unit (Wildberg, 1984) may well be associated with this intraoceanic subduction, which probably began in pre-Albian times, although it is clearly documented only from the Cenomanian until late Santonian–early Campanian, when all units became obducted (Azéma et al., 1985). This obduction has again a southern vergence (Azéma et al., 1985; Meschede et al., 1988) and could well be related to relative motion along an intraoceanic transform fault system. The Middle America Trench system became established after this obduction and is independent of the pre-Campanian events described above.

Late Senonian–early Tertiary mesoautochthonous sequences. The late Santonian–early Campanian tectonic phase created an important (mostly submarine) relief that was subject to intense erosion. This relief produced marked facies changes in the late Senonian sedimentary cover. The basal unconformity that separates the highly deformed Nicoya Complex from the much less deformed late Senonian sedimentary cover is, however, only evident in the Santa Elena Peninsula and northwestern-central Nicoya Peninsula, where klippes of the Matapalo Unit are present and the deformation of the underlying Esperanza Unit is intense. The nappe front probably was never much farther south than the location indicated by the Matapalo klippes. In southwestern Nicoya Peninsula the late Senonian cover rests almost conformably on the basalts of the Esperanza Unit.

The structurally most elevated parts (Santa Elena Peninsula and northwestern Nicoya) reached into shallow waters and became partially subaerial. This is suggested by well-rounded boulder conglomerates that characterize the unconformity at the base of rudistid bioherms. In the remaining parts of the Nicoya Terrane, both rounded and angular boulder breccias are always associated with siliceous and/or calcareous pelagic sediments and suggest submarine relief, causing rockfall and debris flows in a deep-water environment during the Campanian-Maastrichtian.

Siliceous mudstones occasionally overlie the basal breccias and represent the oldest sediments of the sedimentary cover of the Nicoya Complex. Pelagic limestone sedimentation started everywhere in the late Campanian, both overlying the siliceous sequence described above as well as directly in contact with basalts or basaltic conglomerates. No intrusive contacts with Campanian limestones have been observed in the Nicoya Terrane. All contacts of sediments with basalts are cold, sedimentary contacts. In contrast with coeval facies of the Chorotega Terrane, the upper Campanian–lower Maastrichtian sediments are usually condensed and pelagic. During this time, sedimentation became more uniform over the entire Nicoya Terrane. Former neritic areas subsided to subphotic depths and sedimentation became pelagic. During the Maastrichtian, distal turbidites began to appear. Both in the Santa Elena and the Garza-Samara area, these early detrital facies are dominated by basaltic material and indicate local erosion rather than far-traveled turbidite systems.

Middle or upper Paleocene coarse channel-fill and overbank deposits (Winsemann, 1992), including boulders of andesites and platform limestones, indicate the initial juxtaposition of the Nicoya Terrane with the Chorotega island arc. These deposits may be indicative of the amalgamation of the two terranes. The source area for these beach-derived andesites and platform limestone boulders may have been farther to the southeast, owing to Paleocene-Eocene strike-slip motion of the Nicoya Terrane after initial amalgamation.

Golfito Terrane

The Golfito Terrane crops out along the northeastern shores of the Golfo Dulce (Fig. 4) and apparently extends into Panama to the Azuero Peninsula. The terrane is characterized by oceanic basalts and dolerites and by stratigraphically overlying Campanian-Maastrichtian pelagic limestones, argillites, basaltic turbidites, and tuffs. In contrast to the Chorotega Terrane, no thick Maastrichtian-Paleogene sand-dominated forearc sequences exist. Except for the youngest (late Maastrichtian–Paleocene ?) sediments, which contain more acidic volcanic glass, the volcaniclastic material appears to be derived from local basaltic basement. Pelagic sediments are interbedded with volcanic breccias and flows and crosscut by dikes, indicating basaltic igneous activity until the middle Maastrichtian.

Detailed mapping allowed us to define a three-fold subdivision of the volcano-sedimentary sequences observed in the Golfito Terrane (Di Marco, 1994): (1) the basement composed of dolerites and basalts, (2) a sequence of interbedded volcanic flows and hemipelagic sediments, and (3) a sequence of volcaniclastic sediments (Fig. 5).

The basement (Unit 1). The central part of the Golfito region is floored by massive dolerites and oceanic basalts (Fig. 4). The minimum age of this unit can be determined as latest Campanian, based on planktonic foraminifers recovered from the base of the stratigraphically overlying sediments.

Interbedded volcanic flows and sediments (Unit 2). A suite of pelagic limestones bearing planktonic foraminifers and radiolarians, calcareous to siliceous mudstones, volcanic siltstones and sandstones, and basaltic breccia (Fig. 5) rests with a stratigraphic contact on the basement. This very variable sequence is intruded by basaltic dikes and sills and intercalated with basaltic flow units. Pelagic limestones from Unit 2 generally include Maastrichtian *sensu largo* foraminiferal assemblages. At Río Sorpresa (Fig. 4), a section of sedimentary rocks with thickness greater than 50 m contains several massive doleritic/basaltic flows (Fig. 5). In Quebrada Chorro, a middle Maastrichtian age was determined in pelagic limestones stratigraphically overlying a massive volcanic flow. Intermittent basaltic volcanic activity occurred during hemipelagic/pelagic sedimentation since the Campanian-Maastrichtian and lasted at least until the middle Maastrichtian.

The volcaniclastic sequence (Unit 3). Unit 3, characterized by the absence of volcanic flows and by the occurrence of coarse breccias and boulder beds, is well exposed in the northern and northeastern parts of the Golfito Terrane (Fig. 4). In the Quebrada Chorro and Quebrada Bolsa (Fig. 5), we observed olistostromes with blocks measuring up to several meters in diameter composed of pelagic limestone, basalt, and siliceous mudstone set in a volcaniclastic matrix. Some of the blocks were included in the matrix as semilithified blocks. These coarse-clastic rocks are locally (Quebrada Bolsa, Quebrada Nicuesa, and on the Golfito–Villa Briceño road, Figs. 4 and 5) capped by an unconformable sequence of fine-grained, massive green siliceous volcanic tuffs and tuffites. The presence of acidic volcanic glass shards suggests influence of an explosive volcanism. These rocks may be considered as an overlap sequence. Unit 3 is only dated at its base as middle Maastrichtian (Fig. 5); its top could range into the Paleocene.

The coarse clastic sediments with olistostromes clearly record a tectonic event that could be related to the initial collision of the Golfito Terrane with the Chorotega margin. This event is actually poorly dated. It must have happened after the cessation of middle Maastrichtian basaltic submarine volcanism, probably during the Paleocene. Arc-derived volcanic detritus is not present in the above-mentioned overlap sequence, as is the case for the Tertiary sediments of the Chorotega Terrane. We speculate that the Golfito Terrane, once in contact with the Chorotega margin, became an elevated area that was bypassed by arc-derived clastics. The stratigraphic sequence recorded in the Golfito terrane—mainly Unit 2 with its basalt flows interbedded with Late Cretaceous volcaniclastic and pelagic sediments—is typical of that from oceanic plateaus of Cretaceous age in the western Pacific (Bowland and Rosencrantz, 1988). Therefore, the terrane is thought to have formed a marginal piece of the Caribbean oceanic plateau, transported northward by strike slip along the rim of the Caribbean plate.

The Osa-Caño Accretionary Complex

Detailed mapping of coastal and river outcrops in the Osa Peninsula revealed two fundamentally different units. In the Osa isthmus and the inner Osa Peninsula (Fig. 6), a thick pile of oce-

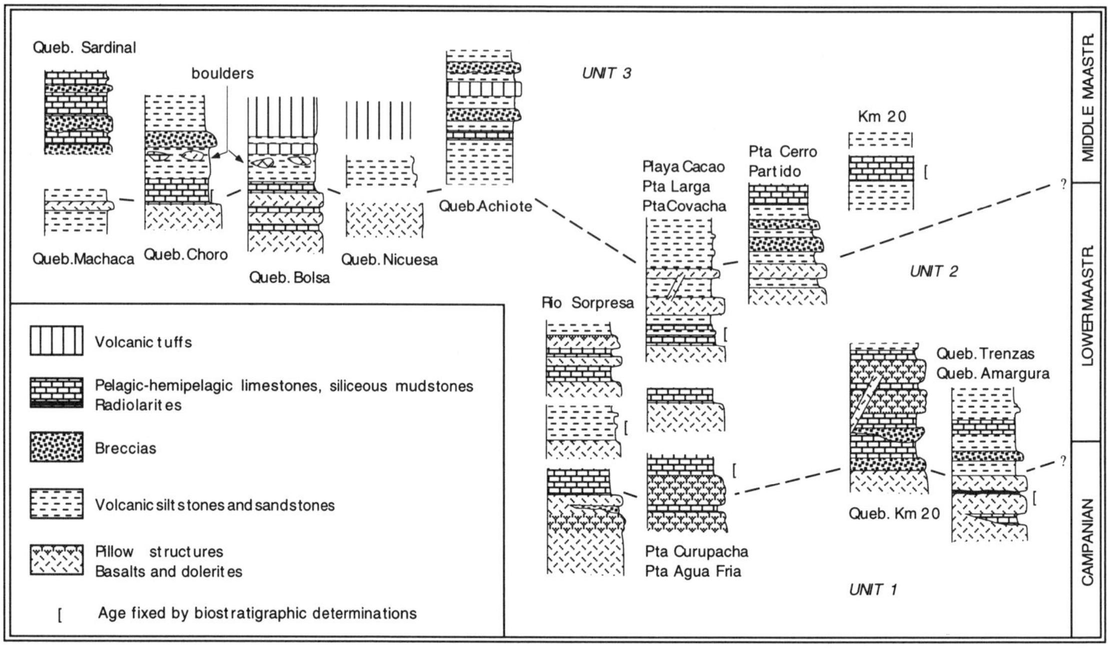

Figure 5. Lithostratigraphic sections in the Golfito Terrane (for localities see Fig. 4).

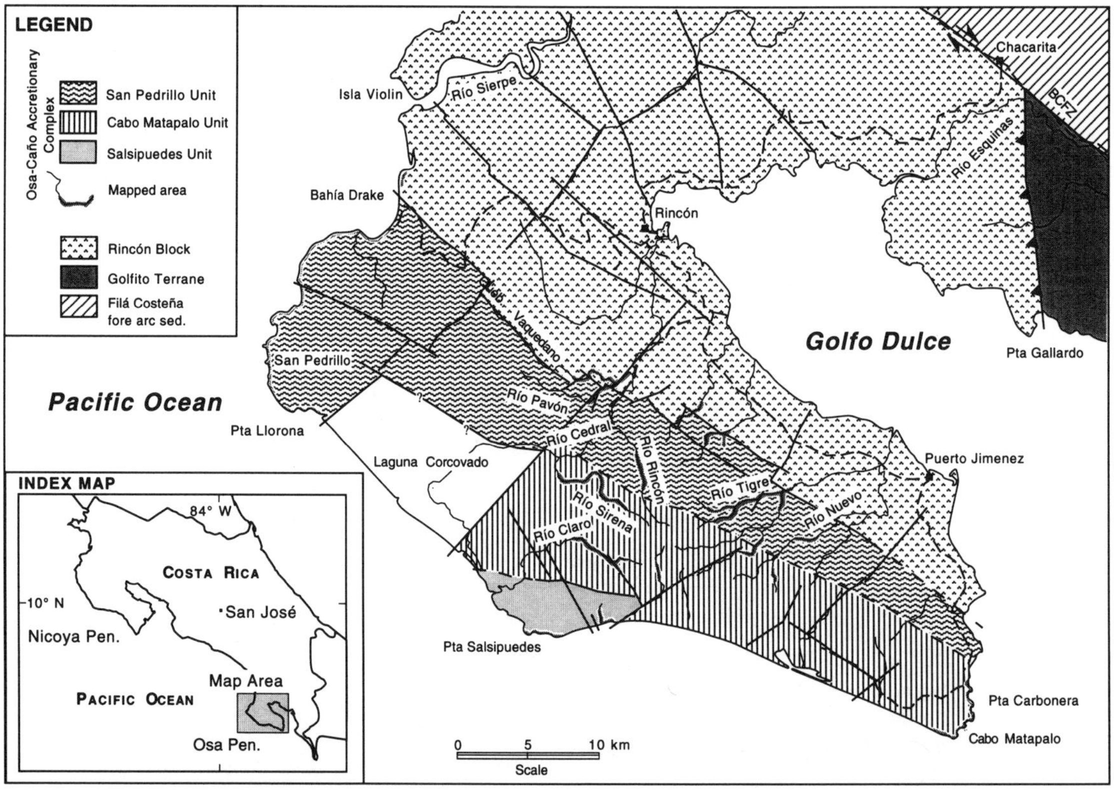

Figure 6. Interpretive sketch of the pre-Neogene geology of the Osa Peninsula. Mapped areas are shaded. BCFZ = Ballena-Celmira Fracture Zone (Corrigan et al., 1990). Note that the BCFZ is part of the LFZ (Longitudinal Fault Zone).

anic basalts, including small amounts of sediments of Late Cretaceous to Eocene age, crops out. We refer to this unit as the Rincón Block (Fig. 6). This unit is dealt with in more detail by Di Marco (1994). The outer Osa Peninsula and the Caño Island are built by the Osa-Caño Accretionary Complex characterized by strongly deformed turbidites and hemipelagic and pelagic sediments (over 95% of volume) that range in age from Late Cretaceous to Miocene. In some places (San Pedrillo–Marenco) 10- to 100-m-sized blocks of oceanic basalt, dolerite, and minor amounts of gabbro and acidic cumulates occur in a hemipelagic sedimentary matrix. In contrast to published work on this area, we cannot confirm the existence of any basement underlying sediments in the outer Osa Peninsula.

The Osa-Caño Accretionary Complex forms the most external geological unit of the Costa Rican Pacific coast (Fig. 1). It comprises an accretionary mélange formed during the Paleogene–early Neogene that is unconformably overlain by late Neogene to Quaternary subaeric and upper-slope sediments (Corrigan et al., 1990). Lew (1983) and Tournon (1984) documented biostratigraphic ages ranging from Late Cretaceous to Eocene for the accretionary complex. Berrangé and Thorpe (1988) and Berrangé et al. (1989) proposed a backarc origin for igneous rocks of the Osa Peninsula. These authors misinterpreted the outcrops of dark-colored, strongly deformed and diagenetically altered mudstones along the outer Osa coast as basaltic basement. Thick Neogene–Quaternary polymict conglomerates, including a suite of fresh ophiolitic rocks, with pyroxenites, gabbros, dolerites, and even foliated amphibolites (see Tournon, 1984), have been mapped in the upper reaches of most rivers on the peninsula. The Osa Peninsula can be excluded as a source area for these conglomerates, since none of these rock types crop out in the interior. The presence of sedimentary mélange in this area was first documented on Caño Island (Baumgartner, 1986) and then described from the outer Osa Peninsula (Obando, 1986; Baumgartner et al., 1989).

Mélangelike rock bodies are often associated with convergent plate margins. Raymond's (1984) definition of the term mélange fits well with field observations from central and outer Osa. The genesis of a mélange involves tectonic (e.g., offscraping, dismembered strata, etc.) and sedimentary (e.g., olistostromes) processes.

The Osa-Caño Accretionary Complex is subdivided into three units, characterized by matrix and associated blocks. These are: (1) the San Pedrillo unit, (2) the Cabo Matapalo unit, and (3) the Salsipuedes unit (Fig. 6). Detailed description and interpretation of the Osa-Caño Accretionary Complex are given in Di Marco (1994).

San Pedrillo Unit. The San Pedrillo Unit is well exposed along the northwestern tip of the Osa Peninsula, from Bahía Drake to Punta Llorona (Fig. 6). The Río Rincón, Río Cedral, and Río Tigre also have good exposures of this unit. The matrix includes blocks of dark graywacke (often siliceous), red radiolarian chert and mudstone (Bahía Drake, Río Cedral), and shallow-water redeposited limestones (Bahía Drake). The radiolarian chert and mudstone occur in sequences a few meters thick (Bahía Drake, Río Cedral) to hundreds of meters thick (Río Rincón) as dismembered interbeds in a matrix of volcaniclastic graywacke. Shallow-water carbonates are often included in the graywacke matrix as centimetric to metric clasts or blocks or as calciturbidites, supposed to originate from the inner wall of the trench.

Blocks of oceanic basalt and associated sediments (basaltic breccias, red radiolarian cherts/mudstones, red siliceous limestones) are found in variable amounts in the San Pedrillo Unit. These blocks are interpreted to have formed part of the subducting oceanic plate and to have been sporadically incorporated into the trench-fill sediments by offscraping and dismembering of the top of the incoming oceanic crust. Between Bahía Drake and San Pedrillo (Fig. 6), the ophiolitic component is locally very important and represents up to 80 to 90% of the volume of rocks. However, the basaltic bodies are always within a volcaniclastic sedimentary matrix. The age of the matrix is middle Eocene (Azéma et al., 1983). Lew (1983) proposed a late Paleocene to middle Eocene (most probably early Eocene) age for red radiolarian cherts included in basalt blocks near San Pedrillo.

Cabo Matapalo Unit. Typical sections for the Cabo Matapalo Unit crop out from Punta Carbonera to Cabo Matapalo on the southeastern tip of the Osa Peninsula (Fig. 6). Examples of this unit are also found in the Río Nuevo, Quebrada Piedras Blancas (a tributary to the Río Tigre), and the Río Sirena. The matrix of the mélange is generally a dark-brownish volcanic graywacke including, in minor amounts, bedded volcanic siltstones, sandstones, and breccias. Radiolarian cherts and redeposited shallow-water carbonates are absent. The age of the matrix has not been directly determined; however, an Eocene to Miocene age is probable according to the ages of the included blocks. The blocks are mainly light-brownish siliceous pelagic limestones that occur as one-meter to several-meters-sized bodies within the graywacke matrix. The limestones include planktonic foraminifers, calcified radiolarians, and sponge spicules. The siliceous fraction sometimes resulted in chert beds alternating with the micritic limestones. The chert beds are usually 10 cm thick and can represent more tan 60 to 70% of the rock volume. The limestones at Punta Carbonera include decimetric blocks of basalts. This observation suggests that these limestones represent the base of a sedimentary section overlying an oceanic basaltic basement. Blocks of basalts up to hundreds of meters across are found in the Río Sirena and in the Quebrada Piedras Blancas. The assemblages of planktonic foraminifers determined in the limestone blocks indicate ages ranging from late middle Eocene to middle Miocene. The outcrops between Punta Carbonera and Cabo Matapalo and the biostratigraphic control suggest that the limestone blocks originally formed a continuous pelagic sequence overlying a basaltic basement. Large bodies of this sequence were supposedly accreted by offscraping of the sedimentary cover of the incoming plate and subsequent tectonic dismembering and mélange formation.

Salsipuedes Unit. Prominent massive limestone bodies characterize the coastal promontory of Punta Salsipuedes (Fig. 6). These limestone bodies are up to 200 m across and occur in a dark-brown fine-grained graywacke matrix that also includes numerous smaller limestone blocks. In the coastal outcrops around Punta Salsipuedes, bedding is preserved in a large limestone block, and we can observe a steeply landward-dipping sequence of alternating meter-thick beds of pelagic limestone and 1- to 5-m-thick beds of shale and graywacke. Unfortunately, no well-preserved fossils could be extracted from this sequence. Phantoms of planktonic foraminifers suggest, however, a Paleocene-Eocene age. Lew (1983) and Tournon (1984) mention basaltic rock in this area. Our mapping cannot confirm this observation; however, dark-colored mudstone dikes crosscutting the white limestones are quite numerous. Azéma et al. (1981) reported a Paleocene age from outcrops near Quebrada Hedionda that are included with this unit. The Salsipuedes Unit probably represents an offscraped oceanic sequence that had little contact with trench-fill sediments. The nature of the limestone bodies and the origin of the interbedded fine graywackes require further investigation.

Burica Terrane

The Burica Terrane is represented by Late Cretaceous–Paleogene basement and associated sediments of the Burica Peninsula. The Burica Peninsula is a northwest-southeast–oriented promontory shaped by the Panama Fracture Zone (Fig. 1, Fig. 7). Structurally, the Burica Peninsula is a tilted fault block of oceanic basement and overlying Paleogene sediments on which thick Neogene (Pliocene-Quaternary) forearc sequences onlap with angular unconformity.

The Neogene sequences and the recent structure have been studied by Corrigan et al. (1990). Only one (unpublished) report exists on the Paleogene sequences (Obando, 1986). The outcrops of oceanic basalts and dolerites along the western coast of the Burica Peninsula (Fig. 7) contain some interflow sediments that have recently yielded Campanian radiolarians. At Playa Mangle, *Ranikothalia*-bearing shallow-water carbonate material occurs as infill in fissures penetrating several meters into an altered surface of oceanic basalt (Fig. 7). A discontinuous limestone bed, rich in resedimented shallow-water clasts mixed with planktonic foraminifers, covers the basalt surface (Fig. 7). Upsection, we measured a 30-m-thick sequence of siliceous pelagic limestone interbedded with mudstones and with limestone turbidites in the upper part. *Morozovella velascoensis* and other planktonic foraminifers that characterize the late Paleocene were identified both in the basal resedimented limestones and in the bedded pelagic limestones. Several Paleogene sections in rivers inland from Playa Mangle show similar sections, with a general increase in reworked shallow-water material toward the east. In the Quebrada Piedra Azul (Fig. 7), we measured a section of over 60 m comprising coarsening-upward limestone turbidites and breccias, overlain by a boulder bed, at least 20 m thick, containing middle Eocene larger foraminifers and rhodoliths. This limestone sequence is underlain by the same siliceous pelagic facies that form the top of the Playa Mangle sequence. However, the basaltic base of the section does not crop out in Quebrada Piedra Azul.

In the headwaters of Río Palo Blanco, Panama (Fig. 7), we observed large blocks of *Ranikothalia*-rich shallow-water limestones that occur near the base of the unconformable Neogene sediments. Even in the absence of good outcrop, these blocks suggest the presence of a late Paleocene carbonate platform in the immediate vicinity. In the Río Palo Blanco, we also discovered a more-than-50-m-thick sequence of siliceous limestones with interbedded turbidite and mass-flow deposits containing late Eocene shallow-marine material. The upper Paleocene–upper Eocene section described above is unique to the Burica Peninsula. The nearest other area that yielded late Paleocene–early Eocene pelagic lithologies with reworked *Ranikothalia*-bearing carbonates is the Quepos area.

The described outcrops suggest emergence of a basaltic area and the formation of an insolated carbonate platform since the late Paleocene, which lasted (probably with interruptions) until the late Eocene. Very proximal slope breccias (Quebrada Piedra Azul) and more distal carbonate turbidites suggest resedimentation of the platform carbonates into deep-water environments on the nearby slopes of an oceanic "seamount" (Obando, 1986). However, Frisch et al. (1992) reported an Island Arc Tholeiite (IAT) affinity of basaltic rocks exposed along the Costa Rican coast of Burica. Overturned pillow basalts found along the coast near Playa Mangle indicate that the basaltic basement underwent important tectonization prior to the deposition of the upper Paleocene–upper Eocene sequence. We therefore interpret the emergence of the Burica basement as the result of a tectonic event that affected a portion of a Late Cretaceous primitive island arc near the western edge of the thickened Caribbean-type crust. Post-Eocene plate convergence and strike-slip transport along the western edge of the Caribbean plate brought the Burica Terrane into its present position.

PALEOMAGNETIC DATA

We focused our paleomagnetic sampling on Upper Cretaceous and Paleocene sedimentary rocks outcropping in southwestern Costa Rica and western Panama. The sampled area includes the Chorotega, Nicoya, Golfito, and Burica Terranes. We drilled 34 sites along the Pacific coast and three sites near the Caribbean northwestern coast of Panama (Fig. 8; Table 1), with an average of 10 samples per site. We preferably sampled fine micritic limestones with biostratigraphic age control. From a total of 37 sites, five are from Paleogene limestone blocks included in the Osa-Caño Accretionary Complex and are not presented in this study. Thirteen sites sampled in

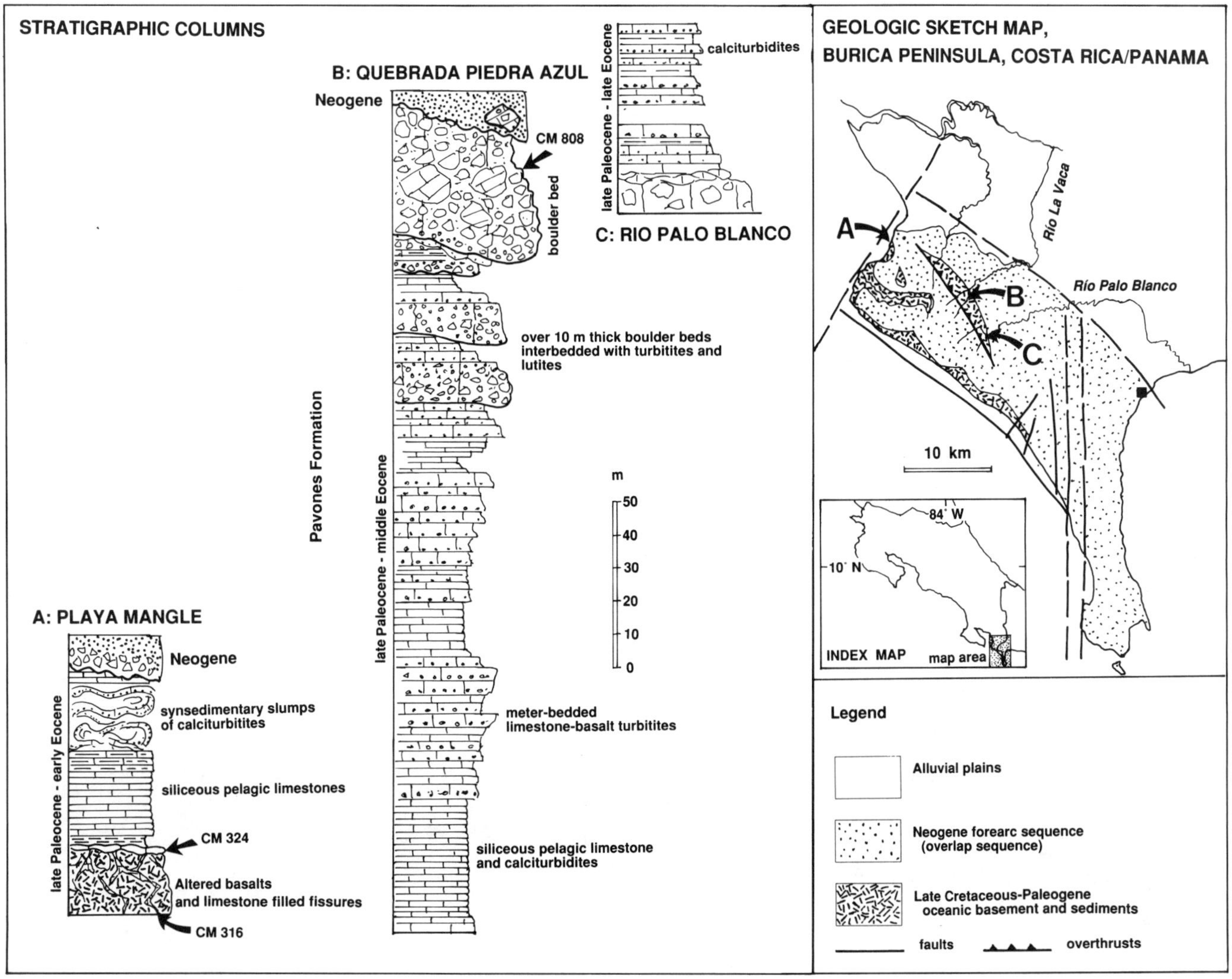

Figure 7. Sketch map of the Burica Peninsula and representative stratigraphic columns of the pre-Neogene stratigraphy. Sections include data from Obando (1986).

Upper Cretaceous pelagic limestones were discarded as a result of either unstable demagnetization behavior or poor within-site grouping of magnetization components (α_{95} >16). A total of 19 sites yielded magnetization components with sufficiently good within-site grouping.

For all samples collected in this study, progressive thermal demagnetization in 50°, 25°, or 15°C steps was carried out until the magnetization intensity dropped close to magnetometer noise level. The magnetization vector during progressive demagnetization was plotted in orthogonal projection. The magnetization component directions were derived by visually picking straight-line segments from the orthogonal projections and using the standard principal component analysis (Kirschvink, 1980) to compute the component direction. For some sites both a high unblocking temperature (HBT) and a low unblocking temperature (LBT) component could be resolved. The site mean magnetization components were computed by combining sample magnetization components using Fisher (1953) statistics (Table 1).

Chorotega Terrane

Changuinola River (Panama). A section of Upper Cretaceous limestones crops out in the upper part of the Changuinola River, near the western Caribbean coast of Panama (Changuinola Formation; Fisher and Pessagno, 1965). Sites 20 and 21 were drilled in light-gray to white micritic bioturbated limestones, close to the confluence of the Río Culubre with the Changuinola River (Peña Blanca locality; Fig. 8). At site 20, the limestones include poorly preserved planktonic foraminifers. The presence of *Globotruncanita stuarti* and *Globotruncana ventricosa* confirms the late Campanian–early Maastrichtian age given by Fisher and Pessagno (1965).

These limestones showed variable behavior during ther-

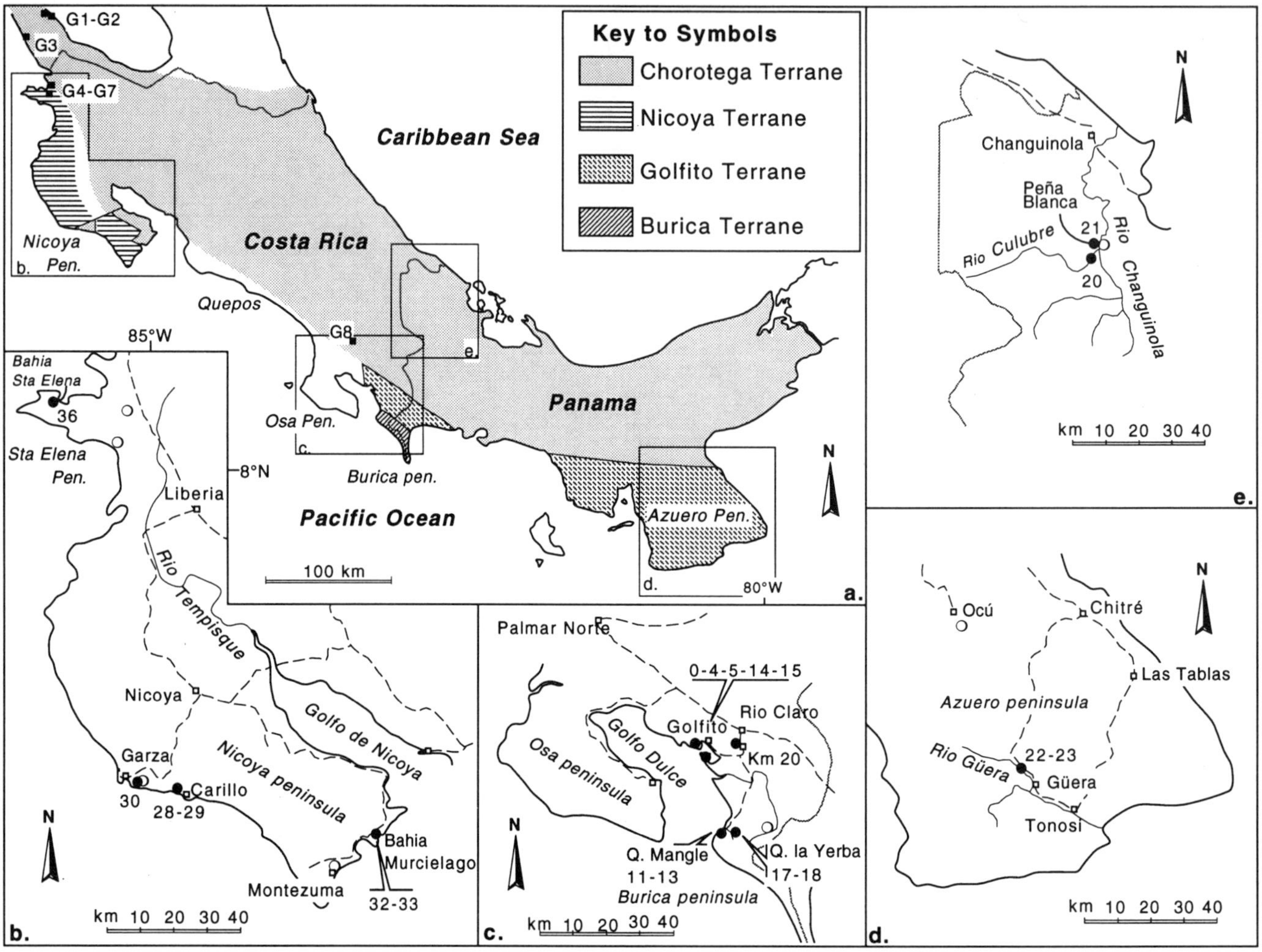

Figure 8. a, Distribution of the Late Cretaceous–Paleogene tectonostratigraphic terranes of Costa Rica and western Panama with location of Gose's (1983) paleomagnetic sites (solid squares). b through e, Location of paleomagnetic sites of this study (solid circles); open circles represent rejected sites.

mal demagnetization (Fig. 9). Samples from site 20 exhibited a discrete blocking temperature spectrum in the 500 to 600°C temperature range. Blocking temperatures are more distributed for site 21, and a magnetization component was usually defined in the 400 to 500°C temperature range. The HBT components of site 20 cluster around a south and shallow up direction after structural tilt correction (Fig. 10). Magnetization component directions resolved from site 21 cluster in the antiparallel direction, except for one sample that provides a component almost parallel to the HBT components from site 20 (Fig. 10). The presence of dual polarity magnetization components and the improved antiparallelism of the component directions after tilt correction (Fig. 10) suggest the presence of a primary magnetization component at these two sites. The precision (K) of the mean of the two site means improves markedly after structural tilt correction (Table 2).

Nicoya Peninsula. At Puerto Carrillo (sites 28, 29; Fig. 8), basaltic breccias are overlain by several tens of meters of upper Campanian to Maastrichtian calciturbidites with thin interbeds of fine gray-black calcareous mudstones (Baumgartner et al., 1984). The section at Bahía Murcielago in southeastern Nicoya Peninsula (Fig. 8) includes well-exposed light-gray to light-green limestones, changing to red marly limestones in the upper part of section. We sampled sites 32 and 33 just below this contact: site 32 at Punta Murcielago and site 33, 150 m to the south. The two sites are separated by a minor fault, but site 33 clearly lies above site 32 in view of their position relative to the contact with the marly limestones. At both sites, we identified a middle Maastrichtian foraminiferal assemblage (Table 3; Plate I on p. 25), which can be correlated to the gansseri Zone of Robaszynski et al. (1984).

At Puerto Carrillo and Bahía Murcielago (sites 28, 29, 32, 33), an LBT component was defined below the 350°C demagnetization step, and a major component (HBT) was resolved in most of the samples in the 350 to 625°C temperature range. For some samples from site 28, the HBT component was re-

TABLE 1. PALEOMAGNETIC DATA FROM COSTA RICA AND PANAMA

Site ID and Location	Age	Component Designator	N	Before Tilt Correction R	Decl.	Incl.	K	a95	After Tilt Correction R	Decl.	Incl.	K	a95
						Golfito Region							
(0) Pta Curupacha	U. Camp.	HBT	6N	5.7	300.5	15.9	18.1	16.2	5.8	297.5	2.4	29.1	12.6
(4) Km 20	Mid. Maast.	HBT	9N	8.5	315.1	-11.4	17.4	12.7	8.6	315.3	13.8	17.9	12.5
		LBT	8	7.5	8.5	23.4	14.0	15.4	7.6	20.6	29.5	16.1	14.2
(5) Playa Cacao	Camp./Maast.	HBT	8N	7.8	306.2	18.6	40.0	8.9	7.8	306.9	6.1	34.5	9.6
		LBT	7	4.8	6.2	30.7	2.8	44.6	4.8	2.3	19.7	2.7	45.5
(14) Pta Curupacha	U. Camp.	HBT	13N	12.8	311.6	25.1	53.9	5.7	12.8	307.5	6.4	55.8	5.6
		LBT	6	5.9	0.2	27.1	83.5	7.4	5.9	348.2	20.3	62.7	8.5
(15) Pta El Cabro	Camp./Maast.	HBT	15N	14.7	292.1	38.0	52.8	5.3	14.7	293.1	-3.0	52.7	5.3
		LBT	10	9.7	9.2	39.7	29.0	9.1	9.7	347.8	19.2	28.8	9.2
						Burica							
(11) Up Q. Mangle	U. Paleoc.	HBT	5R	4.9	94.4	11.3	75.6	8.9	4.9	98.4	-19.4	71.2	9.1
		LBT	9	8.8	13.5	21.4	39.7	8.3	8.8	11.1	-32.2	33.4	9.0
(13) Down Q. Mangle	U. Paleoc.	HBT	8R	7.9	124.3	-30.1	63.7	7.0	7.9	119.0	-13.7	63.7	7.0
		LBT	8	7.7	9.4	19.6	27.3	10.8	7.7	1.0	24.0	27.2	10.8
(17) Q. La Yerba	U. Paleoc./Eoc.	HBT	12R	11.9	82.2	-0.3	74.8	5.1	11.9	84.7	-8.6	75.1	5.0
(18) Q. La Yerba	U. Paleoc./Eoc.	HBT	9R	8.6	94.4	-4.5	22.4	11.1	8.6	99.3	-13.0	22.4	11.1
		LBT	8	7.6	2.0	34.7	28.1	10.6	7.8	5.6	4.2	28.2	10.6
						Panama							
(20) Rio Changuinola	Camp./Maast.	HBT	12R	11.9	162.7	3.8	179.1	3.3	11.9	161.9	-7.7	190.4	3.2
(21) Rio Changuinola	Camp./Maast.		10N/1R	10.6	341.3	11.5	28.1	8.8	10.6	339.8	14.1	28.2	8.7
(22) Rio Güera	U. Cret.	HBT	11R	10.8	97.4	13.1	58.3	6.0	10.8	100.1	0.9	58.1	6.0
(23) Rio Güera	U. Cret.	HBT	7R	6.9	99.3	33.5	40.6	9.6	6.9	103.3	3.2	58.1	8.0
						Nicoya–Sta Elena							
(28) Playa Carrillo	Camp./Maast.	HBT	11N	10.6	351.9	27.8	26.9	9.0	10.7	347.3	11.3	29.3	8.6
(29) Playa Carrillo	Camp./Maast.	HBT	9R	8.9	169.1	-43.6	64.3	6.5	8.9	156.6	-19.4	64.0	6.5
(32) Bahia Murcielago	Mid. Maast.	HBT	11R	10.7	164.6	-43.8	30.3	8.4	10.7	177.8	-22.2	30.2	8.4
(33) Bahia Murcielago	Mid. Maast.	HBT	9N	8.7	8.1	49.8	28.0	9.9	8.8	5.1	12.4	35.7	8.7
(30) Playa Garza	Mid. Maast.	HBT	14N/R	12.8	338.0	-10.4	10.6	12.8	12.8	324.4	-17.0	10.9	12.6
		LBT	16	14.4	15.0	3.4	8.7	13.2	14.3	15.0	3.4	8.7	13.2
(36) Bahia Sta Elena	Camp./Maast.	HBT	10N/R	9.7	342.3	13.5	26.0	9.6	9.8	340.2	-21.3	27.8	9.3
						Previous Workers							
(G1) Nicaragua	U. Cret.		18		356.4	24.3	71.9	4.1		352.8	32.4	72.3	4.1
(G2) Nicaragua	U. Cret.		11		3.9	15.8	264.6	2.8		1.6	21.7	262.4	2.8
(G3) Nicaragua	Paleoc.		20		178.9	-4.2	56.3	4.4		176.3	-16.0	56.6	4.0
(G4) Sta Elena	U. Cret.		14		2.4	29.0	87.6	4.3		1.7	-08.6	63.6	5.0
(G7) Sta Elena	Eoc.(?)		27		0.3	22.8	65.3	3.5		1.0	-14.1	44.3	4.2

solved after demagnetization at temperatures above 500°C. The HBT magnetization components for sites 28, 29, 32, and 33 cluster into two groups of directions, which are either north-northwest and down or south-southeast and up after tilt correction (Fig. 11). According to the middle Maastrichtian age and stratigraphic superposition of sites 32 and 33, we conclude that the negative inclinations of site 32 were acquired during chron C31R and the positive inclinations of site 33 during chron C31N (Fig. 12).

Nicoya Terrane

Site 30 is located at Playa Garza, on the southwestern coast of the Nicoya Peninsula (Fig. 8). Near the eastern end of Playa Garza, intercalations of light-red marly limestone are observed within a thick sequence of gently folded basaltic conglomerate overlying a basaltic basement (Baumgartner et al., 1984). We sampled two lenses of the red marly limestones assigned to the middle Maastrichtian, gansseri Zone of Robaszynski et al. (1984) (Plate I on p. 25, Table 3; see also Schmit-Effing, 1979).

On the Santa Elena peninsula, along the northern coast, Upper Cretaceous to Eocene sediments crop out. We sampled site 36 in the Bahía Santa Elena, toward the western end of the beach (Fig. 8). The section is floored by shallow-water limestones bearing rudist fragments and orbitoids and overlain by basaltic breccia. Some ten m of red to gray pelagic limestones

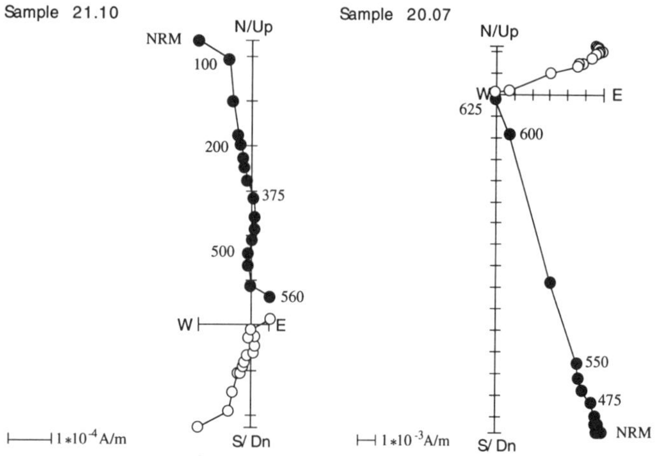

Figure 9. Orthogonal projections of thermal demagnetization data from Changuinola Formation (sites 20 and 21), after tilt correction. Open and solid circles represent projection on the vertical and horizontal planes, respectively. The temperatures in degrees Celsius are indicated for some of the demagnetization steps.

overlying the breccias were sampled for paleomagnetism, avoiding turbidites present in the upper part. A sample collected within the first 0.5 m of pelagic limestones provided a planktonic foraminifer assemblage that places the base of the pelagic sequence into the uppermost Campanian (calcarata Zone, Robaszynski et al., 1984) (Table 3, Plate I on p. 25). The sampled section is overlain by a coarsening-upward turbiditic sequence.

The red marly limestones from Garza and Santa Elena (sites 30, 36) yielded two magnetization components (Fig. 13). Below 250/350°C, an LBT direction was resolved. Above 300°C, the HBT component can be defined. Above 650°C, the directions became scattered. The LBT components of site 30 at Playa Garza cluster around the present field direction before tilt correction (Fig. 14). They are interpreted to be of Recent origin. The HBT components are poorly grouped and orientated northwest and up or southeast and down (Fig. 14). The HBT components with northwest declinations and negative inclinations are observed in the top 0.5 m of the sampled section and overlie the samples with southeast declinations. In view of the middle Maastrichtian age (gansseri Zone) of the section, the northwest declinations and negative inclinations are correlated to chron C31N (Fig. 12). This implies that the site originated in the Southern Hemisphere.

At Bahía Santa Elena, the HBT directions from site 36 lie along two antiparallel directions, which are either northwest and up or southeast and down. The negative inclinations appeared at the base of the sampled section (between 0 and 0.55 m) and at the top (above 1.65 m). Four samples collected between 0.80 and 1.20 m yielded positive inclinations (Fig. 15). The assemblage of planktonic foraminifers observed in samples collected at the base of the section (0.20 m) indicated a late Campanian age (calcarata Zone) (see above). Since this biozone correlates to a time of normal polarity (chron 33N) (Fig. 15), the basal negative inclinations must have been recorded in a normal polarity field. This implies a south equatorial paleolatitude for site 36, as for the Playa Garza site (site 30).

Golfito Terrane

Golfito region. Sites 0 and 14 were sampled at Punta Curupacha, a coastal outcrop on the northeastern side of the Golfo Dulce (Fig. 8). At this locality, fine white micritic limestones stratigraphically overlie pillow basalts. Planktonic foraminifers

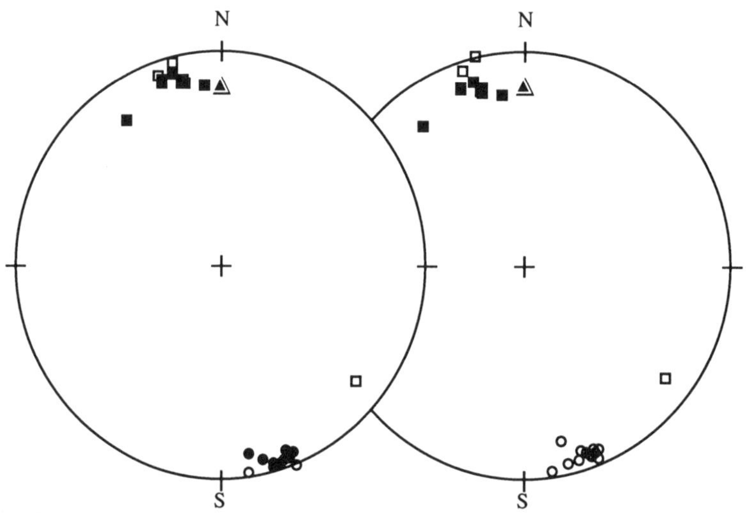

Figure 10. Equal area projection of high unblocking temperature magnetization components before (left) and after (right) tilt correction for sites 20 (circles) and 21 (squares) at Changuinola. Solid and open symbols represent positive (downward) and negative (upward) inclinations, respectively. The triangle represents the Present field direction.

TABLE 2. GROUPING OF SITE MEAN DIRECTIONS

Group ID	Group of Sites	N	Before Tilt Correction R	Decl.	Incl.	K	a95	After Tilt Correction R	Decl.	Incl.	K	a95
Chorotega Block	20,21,28,29,32,33,G1, G2,G3	9	8.5	352.7	24.4	15.6	13.4	8.8	350.9	17.7	44.4	7.8
	28,29,32,33,G1,G2,G3	7	6.7	356.3	30.1	20.5	13.6	6.9	354.0	19.6	49.4	8.7
	28,29,32,33	4	3.9	172.9	-41.6	46.7	13.6	3.9	351.8	16.6	38.8	14.9
	20,21	2	2.0	342.0	3.9	55.7	34.1	2.0	340.9	10.9	290.5	14.7
Nicoya Terrane	30,36,G4,G7	4	3.8	350.3	14.1	15.1	24.4	3.8	347.1	-15.8	20.1	21.0
	30,36	2	1.9	340.1	1.6	22.4	55.6	2.0	332.2	-19.3	56.4	34.5
Golfito Terrane	0,4,5,14,15,22,23	7	6.2	297.9	6.0	7.9	23.0	6.8	297.6	3.1	31.8	10.9
	0,4,5,14,15	5	4.8	306.4	18.1	17.4	18.9	4.9	304.8	5.4	46.1	11.4
	22,23	2	2.0	98.3	23.3	31.4	46.2	2.0	101.7	2.1	846.4	8.6
Burica Terrane	11,13,17,18	4	3.7	98.1	-6.0	11.4	28.5	3.9	100.3	-14.0	31.6	16.6

TABLE 3. BIOSTRATIGRAPHY AT SAMPLING SITES

Tertiary					Upper Cretaceous																											
Paleocene					Middle Maastricht.	Lower Maastricht.	Upper Campanian																									
P5	P4	P3	P2	P1	G. gansseri Zone	Gtr. falso-stuarti Zone	Gta. calca-rata Zone	Sites	*A. mckannai*	*A. primitiva*	*M. conicotruncata*	*M. aequa*	*M. velascoensis*	*G. gansseri*	*Glla. citae*	*Glla. havanensis/citae*	*Glla. havanensis*	*Rugoglobigerina sp.*	*Rugotruncana sp.*	*R. contusa*	*Gta. angulata*	*R. fornicata*	*Rosita sp.*	*Gta. stuarti*	*Gta. stuartiformis*	*Gta. stuarti/stuartiformis*	*Gtr. falsostuarti*	*Gtr. linneiana*	*Gtr. bulloides*	*Gtr. ventricosa*	*Gta. subspinosa*	*Gta. calcarata*
					■			32, 33	|	|	|	|	|	*	*	|	|	|	|	*	*	|	|	*	*	|	*	*	*	*	|	|
					■			30	|	|	|	|	|	*	|	|	|	*	|	*	*	*	|	*	*	|	*	*	|	*	|	|
							■	36 (base)	|	|	|	|	|	|	|	|	|	|	|	|	|	*	|	|	|	|	|	*	|	*	|	*
							■	0, 14	|	|	|	|	|	|	|	|	|	|	|	|	|	|	|	|	*	|	|	|	|	*	*	*
						■		5	|	|	|	|	|	|	|	*	|	|	|	|	|	|	*	|	*	*	|	*	*	*	|	|
					■			4	|	|	|	|	|	|	|	|	*	|	*	|	*	*	*	*	|	|	|	*	*	*	|	|
■	■							11, 13	*	*	*	*	*	|	|	|	|	|	|	|	|	|	|	|	|	|	|	|	|	|	|	|

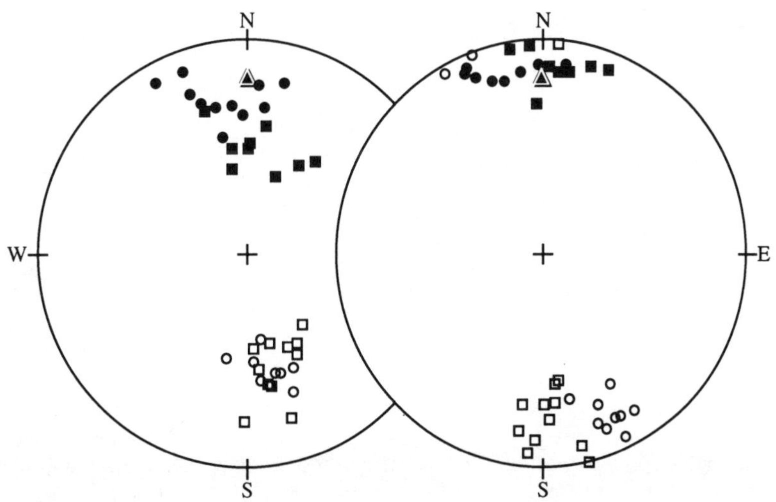

Figure 11. Equal area projection of high unblocking temperature components before (left) and after (right) tilt correction for sites 28 and 29 at Puerto Carrillo (circles) and 32 and 33 at Bahía Murcielago (squares). Symbols as for Figure 10.

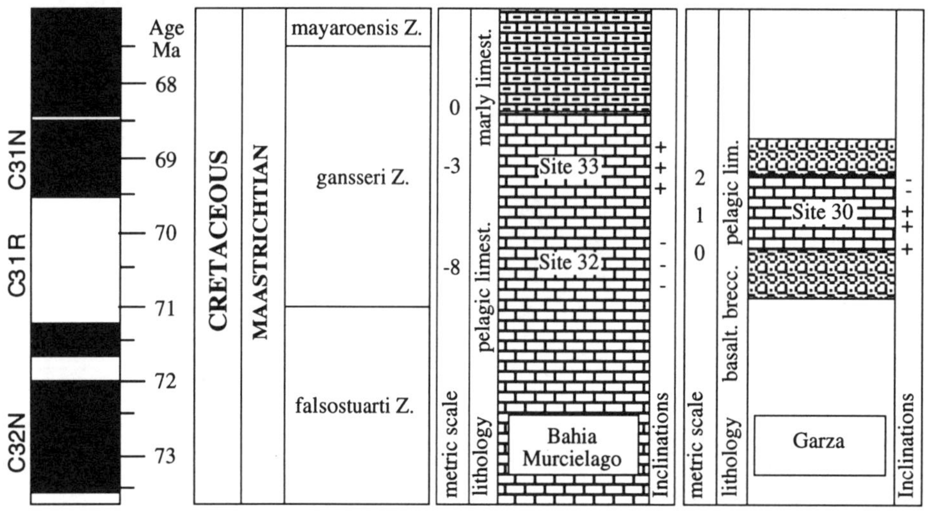

Figure 12. Correlation of the lithostratigraphy at Bahía Murcielago (sites 32, 33) and Garza (site 30) to the geomagnetic polarity timescale (GPTS) using the foraminiferal biozonation. Biozones refer to Robaszynski et al. (1984) and correlation to the GPTS follows Kent and Gradstein (1985).

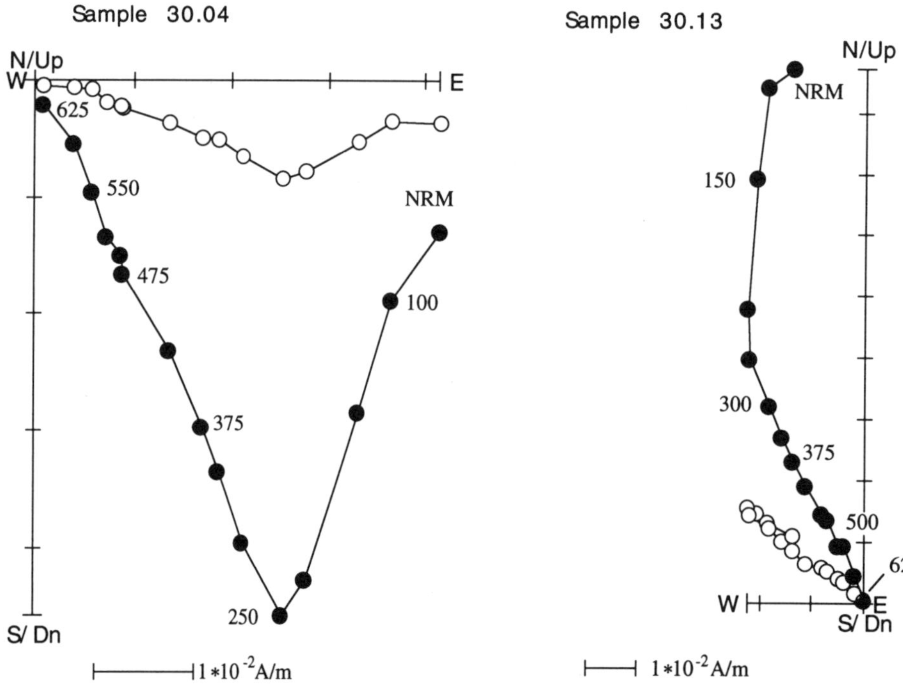

Figure 13. Orthogonal projections of thermal demagnetization data from Garza limestones (site 30), after tilt correction. Symbols as for Figure 9.

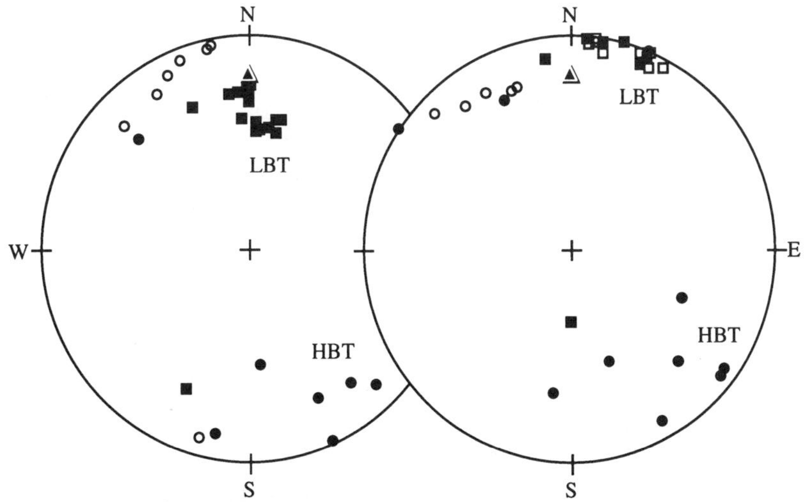

Figure 14. Equal area projection of HBT (circles) and LBT (squares) components before (left) and after (right) tilt correction for Garza (site 30). Symbols as for Figure 10.

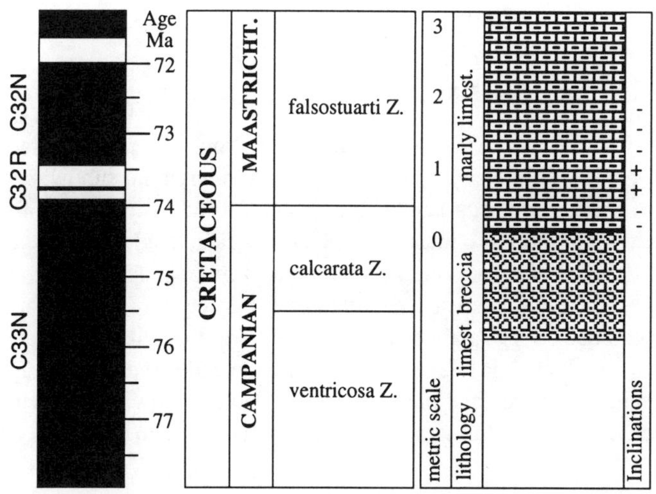

Figure 15. Correlation of the lithostratigraphy at Bahía Santa Elena (site 36) to the GPTS using the foraminiferal biozonation. References as for Figure 12.

are abundant in the limestones. We determined a faunal assemblage indicating a latest Campanian age (calcarata Zone; Robaszynski et al., 1984; Table 3). Site 5 was sampled at Playa Cacao (Fig. 8) in white to gray laminated micritic limestones interbedded with calcilutites. The planktonic foraminifers observed (Table 2) tend to place the section in the falsostuarti Zone of the early Maastrichtian (Robaszynski et al., 1984). Site 15 was sampled at Punta el Cabro, south of Golfito (Fig. 8), in white micritic limestones in fault contact with massive dolerites. Site 4 is located at the locality "km 20" on the Golfito–Río Claro road (Fig 8). To the west side of the road, white micritic limestones crop out. They include centimetric levels rich in organic matter that form a regular black lamination. The faunal assemblage fixes the age to the middle Maastrichtian (gansseri Zone; Robaszynski et al., 1984; Table 3).

The Golfito sites (sites 0, 4, 5, 14, 15) exhibited multicomponent magnetization during thermal cleaning (Fig. 16). Below 350°C, LBT magnetization directions were defined in nearly all the samples. Above 350°C, an HBT component was removed. Between 570 to 580°C, the major part of the intensity has been removed and the directions become scattered. The LBT components from the Golfito sites give directions almost parallel to the present Earth's field before tilt correction (e.g., site 14, Fig. 17) and are therefore interpreted to be of recent origin. The HBT components cluster around a northwest and shallow direction after tilt correction (e.g., site 14, Fig. 17). The overall mean of HBT components from all five Golfito site means passes a regional fold test at the 95% confidence level (McElhinny, 1964) (Table 2). The HBT components therefore predate the (probably) Paleogene folding (see above).

Azuero Peninsula (Panama). Few outcrops of Upper Cretaceous sediments are known from the Azuero Peninsula. We sampled thin calcareous interbeds intercalated in volcanic arenites from the so-called Güera Formation exposed on the banks of Río Güera (sites 22 and 23), approximately 10 km north of Güera village (Fig. 8). The sampled lithologies include calcified radiolaria and debris of planktonic foraminifers, but the age could not be precisely determined.

For the Güera Formation (sites 22, 23), an LBT component was defined below demagnetization temperatures of 200°C and an HBT in the 200 to 580°C range (Fig. 18). The HBT component does not always decay to the origin of the projection, suggesting the presence of an additional magnetization component, but as magnetization directions became scattered above 580°C, this final component could not be resolved. The HBT components from the Güera sites (22, 23) cluster after tilt correction around an east and shallow direc-

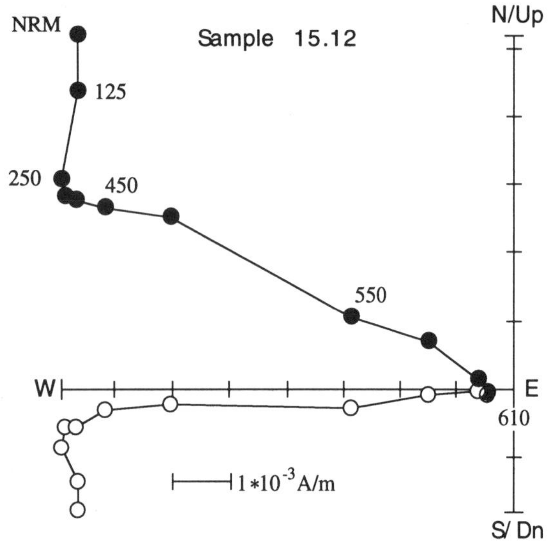

Figure 16. Orthogonal projection of thermal demagnetization data from Golfito limestones (site 15), after tilt correction. Symbols as for Figure 9.

tion. The two site mean directions are very similar after structural tilt correction (Table 1), and the precision of the mean of the two site means improves after tilt correlation (Table 2), giving a positive fold test at the 95% confidence level using the criteria of McElhinny (1964).

Burica Terrane

The sampled sections on the Burica Peninsula include pelagic siliceous limestones interbedded with redeposited shallow-water sediments overlying oceanic basalts and dolerites. At Playa Mangle and in the Mangle and La Yerba rivers, upper Paleocene to middle Eocene limestones crop out. Sites 11 and 13 were sampled in micritic limestones and calcilutites from the Mangle river (Fig. 8). The planktonic foraminiferal assemblage indicates that the sediments are late Paleocene in age (Zone P4-P5 from Toumarkine and Luterbacher, 1985; Table 3). Sites 17 and 18 are located in the La Yerba river (Fig. 8), where fine calcilutites rich in calcified radiolaria were sampled. These sediments are thought to be of late Paleocene–middle Eocene age (Obando, 1986).

The Burica samples (sites 11, 13, 17, 18) showed multicomponent magnetization, and two component directions could generally be defined (Fig. 19). During the first steps of demagnetization (100 to 200°C) an LBT component could be defined, and an HBT component was resolved in the 200 to 500°C temperature range. The observable LBT components from sites 11, 13, and 18 all more or less correspond before tilt correction to the present field direction (e.g., site 11, Fig. 20) and are thought to be of Recent origin. The HBT directions cluster around an east and shallow up direction after tilt correction. The precision (K) of the overall mean from the four Burica sites improves after structural tilt correction (Table 2); however, the improvement is insufficient to constitute a positive regional fold test.

DISCUSSION OF THE PALEOMAGNETIC RESULTS

The site mean magnetization directions were grouped according to the newly proposed tectonostratigraphic subdivision of the Costa Rican landbridge (Fig. 21). The overall mean for each group of site mean directions was calculated using Fisher (1953) statistics (Table 2). The data from Gose (1983) are included in the analysis. The data of Frisch et al. (1992) are not included in the analysis as these data are difficult to evaluate. In Frisch et al. (1992) indeed, the site mean declinations and locality mean declinations are variable, with shallow scattered inclinations. As the data are based on blanket demagnetizations

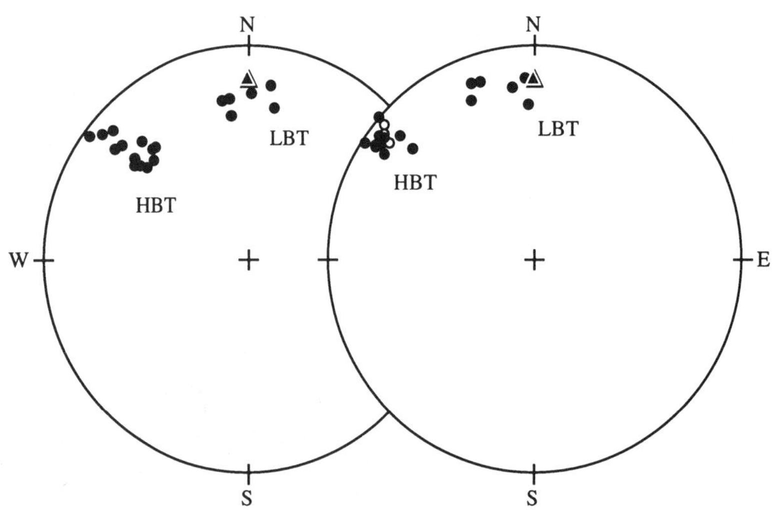

Figure 17. Equal area projection of HBT and LBT components before (left) and after (right) tilt correction for one Golfito site (site 14). Symbols as for Figure 10.

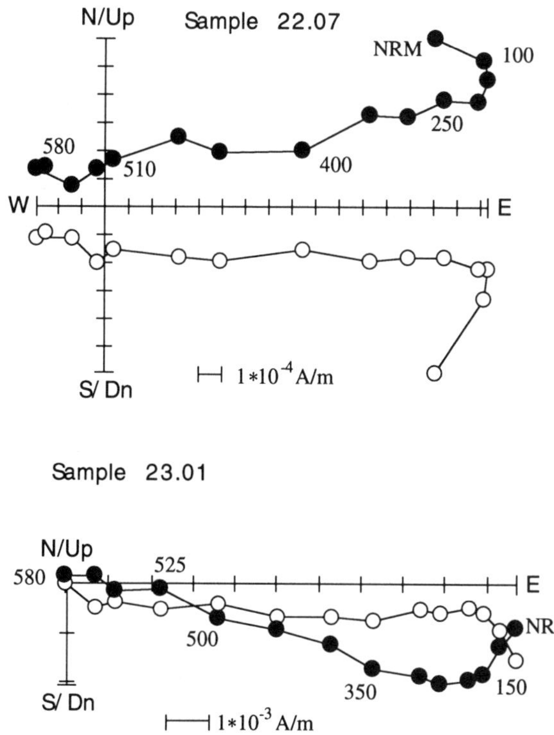

Figure 18. Orthogonal projections of thermal demagnetization data from Güera sediments (sites 22 and 23), after tilt correction. Symbols as for Figure 9.

(at peak alternating fields of 15mT or at temperatures in the 350 to 400°C range), part of the within-site and between-site scatter may be due to single magnetization components' not having been resolved. Furthermore, the structural correction for the sites derived from basalts (more than 60% of the sites included in the study) was made using the anisotropy of magnetic susceptibility (AMS), assuming that the short axes of the AMS coincides with the pole of bedding of the nearby sedimentary rocks. Figure 7 of Frisch et al. (1992) demonstrated, on the contrary, that these two directions are different (over 10°); therefore, their structural corrections are inexact for the igneous lithologies. Blanket demagnetizations and inexact structural corrections may be the reason for the scatter of the data.

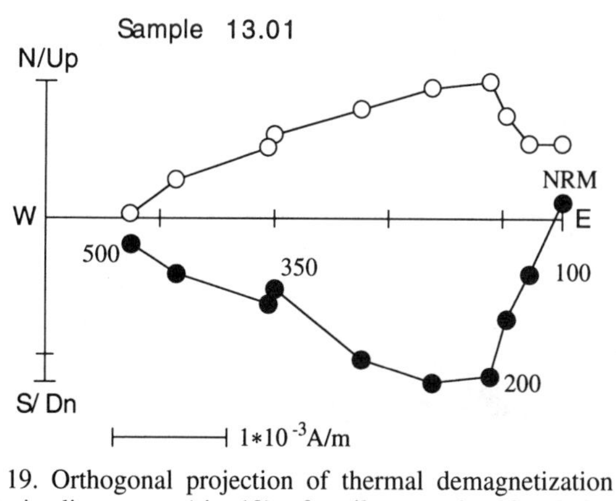

Figure 19. Orthogonal projection of thermal demagnetization data from Burica limestones (site 13), after tilt correction. Symbols as for Figure 9.

Chorotega Terrane

The first group of paleomagnetic data includes sites 28, 29, 32, 33 located in the Nicoya Peninsula (Puerto Carrillo, Bahía Murcielago); sites G1, G2, G3 (Gose, 1983) from southern Nicaragua; and sites 20, 21 sampled in northwestern Panama (Changuinola river). Site G8 of Gose (1983) was not included in the data set since it was sampled from sediments that probably are Oligocene in age. The Chorotega Terrane fold test is positive at the 95% level of confidence (N = 9; k2/k1 = 2.85; Table 2) using the McElhinny (1964) criteria. This implies that the site mean directions comprise magnetization components that predate the folding. This result mainly depends on the Changuinola data, which are well grouped after tilt correction (sites 20, 21; Table 2). Without sites 20 and 21, the site mean directions do not pass the fold test at the 95% confidence level, although the grouping is improved (N = 7; k2/k1 = 2.41). Furthermore, reversals were observed at each of the localities. Although reversals are no guarantee that a rock unit is not remagnetized, they add reliability to the result (Van der Voo, 1990).

The Chorotega Terrane group of data is characterized by mean paleomagnetic directions (Decl = 350.9; Incl = 17.7; α_{95} = 7.8°) approximately parallel, after tilt correction, to present-day Earth's field (Table 2; Fig. 21). The paleopole of this mean direction lies at 81.1°N, 180.4°E. This pole is not distinct at the 95% confidence level from the Late Cretaceous (70 Ma) mean paleopole for South America (Irving and Irving, 1982). Therefore, we conclude that the Chorotega Terrane underwent no significant rotation relative to South America and no significant northward shift. This result implies that the sum of latitudinal displacement of the Caribbean plate, relative to South America, between the Late Cretaceous and the Present is not significant (Fig. 22). Site G8 (Gose, 1983) suggests a southward displacement of the Chorotega Terrane between Late Cretaceous (70 Ma) and Oligocene (30 Ma). More data are needed to confirm this trend.

Nicoya Terrane

Sampling sites associated with the Nicoya Terrane include the Santa Elena sites and some sites on the Nicoya Peninsula (sites 30, 36, G4, G7). The mean paleomagnetic direction after tilt correction for this group of sites is northward with negative inclination (Decl = 347.1° and Incl = –15.8°, α_{95} = 21.0°) (Table 2, Fig. 21). Reversals observed at sites 30 and 36 improve the reliability of the result, although we have no field test to constrain the age of these site mean magnetizations. The detailed biostratigraphy of the sampled sites leads us to conclude that the northerly overall mean di-

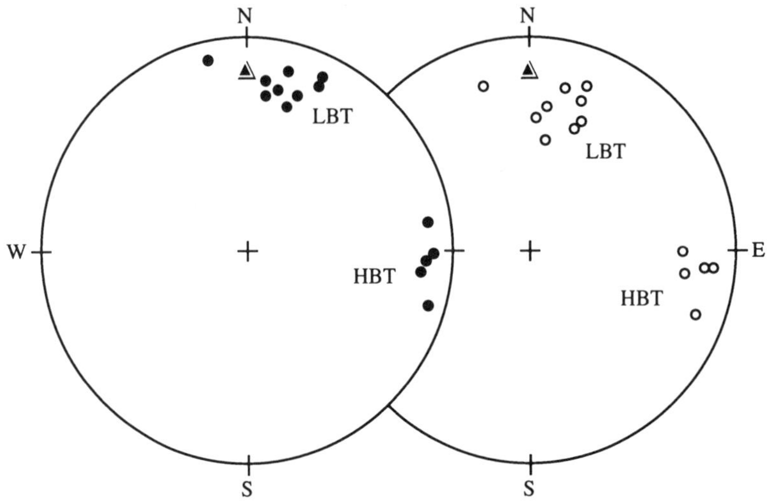

Figure 20. Equal area projection of HBT and LBT components, before (left) and after (right) tilt correction for one Burica site (site 11). Symbols as for Figure 10.

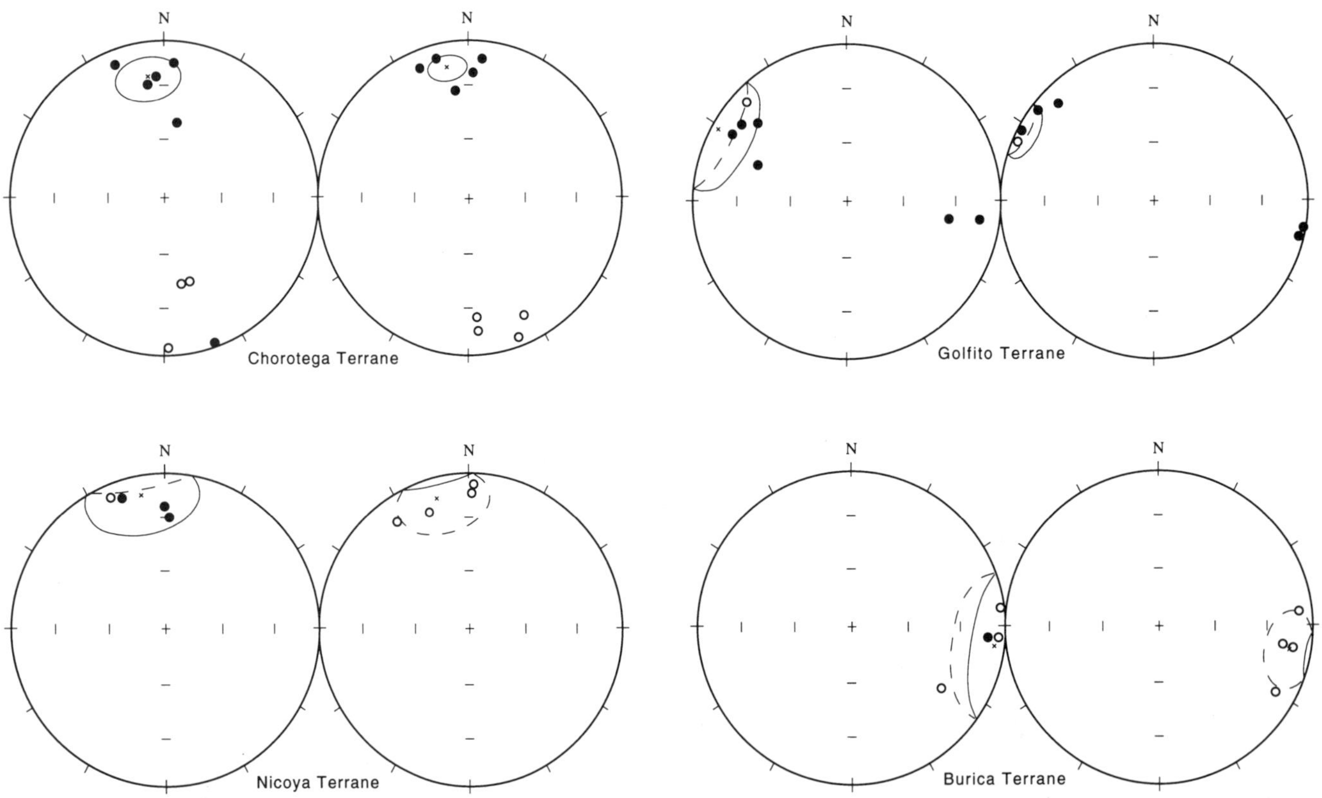

Figure 21. Equal area projections of site mean directions before (left) and after (right) tilt correction for each tectonostratigraphic unit defined in this chapter. Symbols as for Figure 10.

rection represents a normal polarity direction, indicating a small counterclockwise rotation and a low south-equatorial paleolatitude in the Late Cretaceous (Table 2, Fig. 22). In spite of the large error associated with this mean, the data are significantly different at the 95% confidence level from the Chorotega Terrane data (e.g., the 95% confidence ellipses do not overlap; Table 2, Fig. 21). The poor precision of the overall mean is due to scattered site mean declinations, which are probably a result of local tectonic rotations.

Golfito Terrane

The sites associated with the Golfito Terrane (sites 0, 4, 5, 14, 15, 22, and 23) are from the Golfito and Azuero areas. The overall mean direction after tilt correction (Decl = 297.6°, Incl = 3.1°, α_{95} = 10.9°) (Table 2, Fig. 21) is characterized by very shallow inclinations. The precision of the overall mean improves substantially after structural tilt correction, constituting a positive fold test at the 95% significance level (N = 7; k2/k1 = 4.03; Table 2) using the McElhinny (1964) criteria. The Golfito sites alone (sites 0,, 4, 5, 14, 15), in the absence of the Güera sites (22, 23), do not pass the fold test, but the precision value does increase on structural tilt correction (N = 5; k2/k1 = 2.65).

The polarity determinations based on foraminiferal biostratigraphy correlated with the magnetic polarity timescale imply about 60° counterclockwise rotation of the Golfito Terrane relative to the Chorotega Terrane, combined with a small northward shift, since Late Cretaceous time.

Burica Terrane

The paleomagnetic data from the Burica Peninsula are associated with the Burica Terrane. The paleomagnetic mean direction of the Burica group of sites (Decl = 100.3° and Incl = –14.0°, α_{95} = 16.6°) is not distinct at the 95% confidence level from either the Golfito or the Güera groups of sites (Fig. 21, Table 2). However, as documented above, the stratigraphy of the Burica Peninsula is completely different from that from Golfito Terrane. We consider that the terrane underwent since the late Paleocene a very small northward migration from low northerly latitudes, with a counterclockwise rotation of about 90° relative to the Chorotega Terrane. The polarity uncertainty implies an alternative solution with a south equatorial origin of the Burica Terrane. This would imply a northward shift of about 15° between the late Paleocene and the Eocene.

CONCLUSIONS

The Chorotega Terrane constitutes most of the southern Middle American Landbridge and was the western edge of the Caribbean Plate during the Late Cretaceous–Paleocene. Paleomagnetic data for the Chorotega Terrane indicate an origin close to its present latitude and no significant rotation relative to South America since the Late Cretaceous.

The Nicoya Terrane includes the Nicoya Complex (*sensu stricto*), a thrust complex including three units of oceanic origin, unconformably overlain by Upper Cretaceous and Tertiary sedimentary rocks. Lower Cretaceous to middle Paleocene detrital facies are dominated by basaltic material derived from

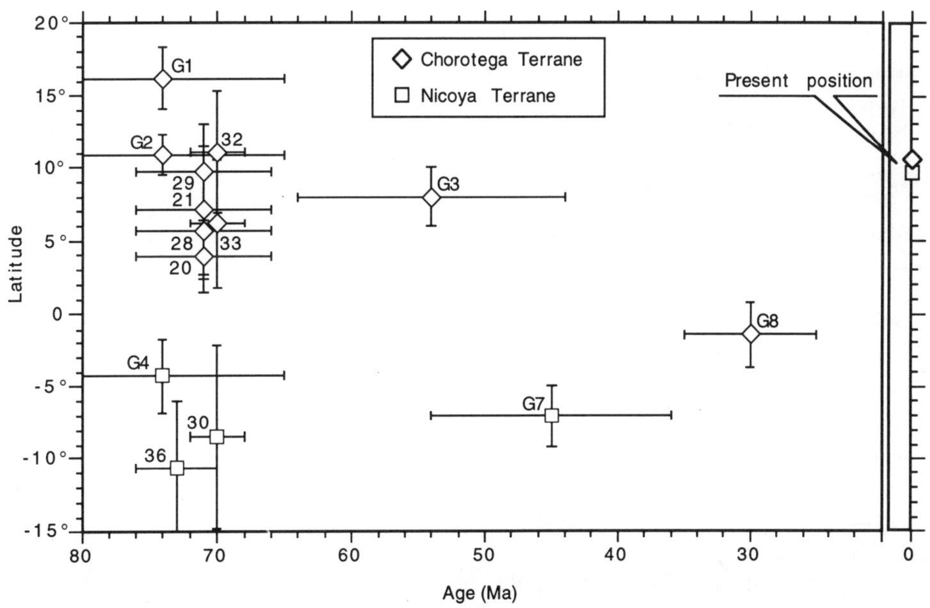

Figure 22. Paleolatitude of the Chorotega Terrane and Nicoya Terrane data sets. Numbers refer to the site numbers; G1 to G8 refer to sites from Gose (1983) (Table 1). Vertical error bars are the paleolatitude uncertainties calculated on the base of the Fisher (1953) α_{95} values for each site. Horizontal lines show ranges in age.

local erosion within the terrane. Late Paleocene channel-fill deposits contain large andesitic boulders marking the initial amalgamation of the Nicoya Terrane with the Chorotega Terrane. The paleomagnetic data obtained from the Nicoya Terrane indicate a low southerly Late Cretaceous paleolatitude with almost no rotation relative to the Chorotega Terrane. The Nicoya Terrane was positioned about 16° of latitude south of the position of the Chorotega Terrane, in Late Cretaceous times. The pre-Campanian Nicoya Complex (*sensu stricto*), therefore, must have originated in the Pacific, at quite some distance from its present position. Consequently, the term Nicoya Complex (*sensu stricto*) should no longer be applied to oceanic basement areas of Central America, distinct from the Nicoya Terrane.

The Golfito Terrane is composed of a sequence of volcano-sedimentary rocks including: (1) a basaltic basement, (2) a sequence of interbedded volcanic flows and sediments of Campanian–middle Maastrichtian age, and (3) a volcaniclastic sequence. The paleomagnetic data from the Golfito Terrane indicate a Late Cretaceous equatorial paleolatitude and a counterclockwise rotation of about 60° relative to the Chorotega Terrane. The basement of the Golfito area is thought to have formed a marginal piece of the Caribbean oceanic plateau, transported northward by strike slip along the rim of the Caribbean plate.

The volcanic basement of the Burica Terrane (Burica Peninsula) is overlain by proximal shallow carbonate debris flows that suggest a near carbonate platform mounted on an emergent piece of a primitive island arc. The paleomagnetic data from the Burica Terrane indicate a low northerly paleolatitude in the Paleocene, slightly south of its present position, and a counterclockwise rotation of nearly 90° relative to the Chorotega Terrane. A south equatorial origin of the Burica Terrane and, thus, a northward shift of about 15° between the late Paleocene and the middle Eocene are not excluded, although these are less likely considering the reduced northward shift of the Farallon and Caribbean plates during this time.

ACKNOWLEDGMENTS

This project was funded by the Swiss National Science Foundation, grants 21-26564.89 and 20-30852.91. The Foundation Herbette (UNIL, Switzerland) funded the field excursions of 1987 and 1989. We are very grateful to M. Murillo and CIMAR (Costa Rica) for their logistic support and hospitality in Golfito. We thank the Servicio de Parques Nacionales for working permits and facilities in the national parks. We thank F. Bujan, A. Steck, C. Mora R., and P. Denyer for letting us include their unpublished data. Thanks to the forementioned and to R. Marchant, B. Reymond, F. Philippossian, M. Richards, D. Simic, A. Isely, T. Carbonara, and G. Venturini for their friendship and help in the field and for inspiring discussions. We want to express our gratitude to M. Caron and J. B. Saunders for checking the determinations of the planktonic foraminifers. We also express our thanks to E. Kuijpers for taking care of the sample shipments to Europe.

Plate I. Selected species of Late Cretaceous planktonic foraminifers from sampled sections at Bahía Murcielago (1 through 7), Playa Garza (8 through 11), and Bahía Santa Elena (12 through 15). Species: (1) *Globotruncanella citae;* (2, 3) *Gansserina gansseri;* (4, 8) *Globotruncanita angulata;* (5, 9) *Rosita fornicata;* (6, 10) *Rosita contusa;* (7) *Globotruncana linneiana;* (11) *Globotruncanita stuarti;* (12, 13, 14) *Globotruncanita calcarata;* (15) *Globotruncana ventricosa.* Scale A for all figures except figures 10 and 11, which are Scale B.

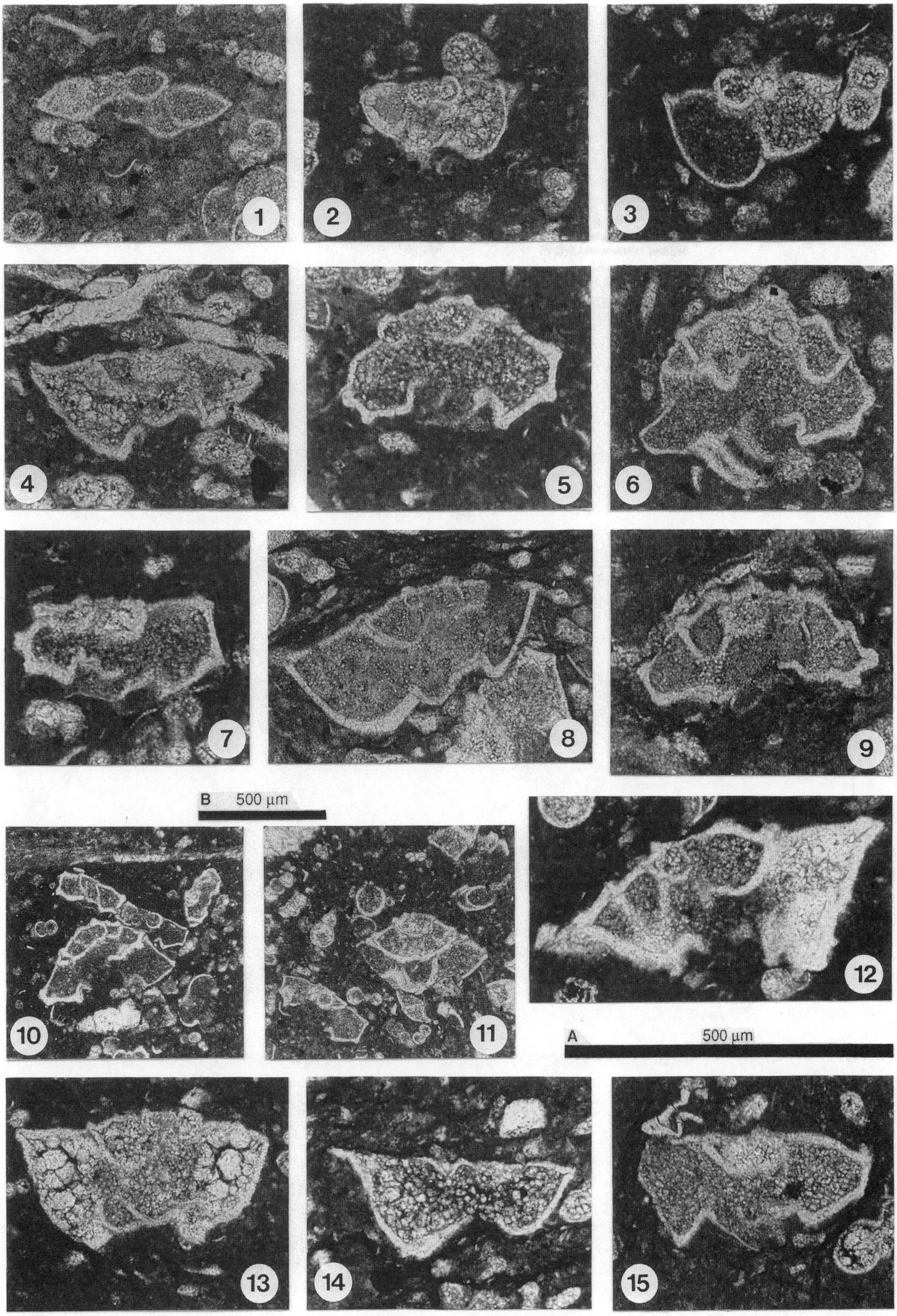
1
2
3
4
5
6
7
8
9
B 500 μm
10
11
12
A 500 μm
13
14
15

REFERENCES CITED

Arias, O., and Denyer, P., 1992a, Mapa geológico de la hoja Carrillo Norte, Guanacaste, Costa Rica: Instituto Geográfico Nacional de Costa Rica, Hoja 3047 I, scale 1:50,000.

Arias, O., and Denyer, P., 1992b, Mapa geológico de las hojas Matapalo y Punta Gorda, Guanacaste, Costa Rica: Instituto Geográfico Nacional de Costa Rica, Hoja 3047 II y IV, scale 1:50,000.

Astorga, A., 1987, El Cretácico Superior y el Paleógeno de la vertiente Pacífica de Nicaragua meridional y Costa Rica septentrional: Orígen, evolución y dinámica de las cuencas profundas relacionadas con el margen convergente de Centroamérica [Tesis de Licenciatura]: San José, Universidad de Costa Rica, 250 p.

Azéma, J., and Tournon, J., 1980, La Péninsule de Santa Elena, Costa Rica: Un massif ultrabasique charrié en marge pacifique de l'Amérique centrale: Comptes rendus de l'Académie des Sciences de Paris, v. 290, p. 9–12.

Azéma, J., Sornay, J., and Tournon, J., 1979, Découverte d'Albien supérieur à ammonites dans le matériel volcano-sédimentaire du "Complexe de Nicoya" (Province de Guanacaste, Costa Rica): Comptes rendus sommaires de la Société géologique de France, v. 3, p. 71–76.

Azéma, J., Glaçon, G., and Tournon, J., 1981, Nouvelles données sur le Paléocène à foraminifères planctoniques de la bordure pacifique du Costa Rica (Amérique Centrale): Comptes rendus sommaires de la Société géologique de France, v. 3, p. 85–88.

Azéma, J., Butterlin, J., Tournon, J., and de Wever, P., 1983, Presencia de material volcano-sedimentario de edad Eoceno medio en la Península de Osa (provincia de Puntarenas, Costa Rica): Conferencia Geológica del Caribe, 10a, Cartagena, Colombia, Abstract.

Azéma, J., Bourgois, J., Baumgartner, P. O., Tournon, J., Desmet, A. and Auboin, J., 1985, A tectonic cross-section of the Costa Rican Pacific littoral as a key to the structure of the landward slope of the Middle America Trench off Guatemala, *in* von Huene, R., and Auboin, J., eds., Initial reports of the Deep Sea Drilling Project, Volume 84: Washington, D.C., U.S. Government Printing Office, p. 831–850.

Baumgartner, P. O., 1984, El complejo ofiolitico de Nicoya (Costa Rica): Modelos estructurales analizados en función de las edades de los radiolarios (Calloviense a Santoniense), *in* Sprechmann, P., Manual de Geología de Costa Rica: San José, Editorial Universidad de Costa Rica, p. 115–123.

Baumgartner, P. O., 1986, Discovery of subduction-related mélanges on Caño Island and Osa Peninsula (Pacific, Costa Rica, Central America): Réunion annuelle des Sciences de la Terre, 11ème, Clermont-Ferrand, France, Abstract, p. 12.

Baumgartner, P. O., 1987, Tectónica y sedimentación del Cretácico superior en la zona pacífica de Costa Rica (América Central): Actas de la Facultad de Ciencias de la Tierra U.A.N.L. Linares, v. 2, p. 251–260.

Baumgartner, P. O., 1990, Mesozoic and Tertiary arcs, seamount and accretionary terranes of Costa Rica (Central America): Geological Society of America Abstracts with Programs, v. 22, p. A338–A339.

Baumgartner, P. O., and six others, 1984, Sedimentación y paleogeografía del Cretácico y Cenozoico del litoral pacífico de Costa Rica: Revista geológica de América Central, v. 1, p. 57–136.

Baumgartner, P. O., Obando, J. A., Mora, C., Channell, J.E.T., and Steck, A., 1989, Paleogene accretion and suspect terranes in southern Costa Rica (Osa, Burica, Central America), *in* Larue, D. K., and Draper, G.: Transactions, 12th Caribbean Geological Conference, St. Croix, U.S. Virgin Islands: Miami, Florida, Miami Geological Society, p. 529.

Berrangé, J. P., and Thorpe, R. S., 1988, The geology, geochemistry and emplacement of the Cretaceous-Tertiary ophiolitic Nicoya Complex of the Osa Peninsula, southern Costa Rica: Tectonophysics, v. 147, p. 193–220.

Berrangé, J. P., Bradley, D. R., and Snelling, N. J., 1989, K/Ar age dating of the ophiolitic Nicoya Complex of the Osa Peninsula, southern Costa Rica: Journal of South American Earth Sciences, v. 2, p. 49–59.

Bourgois, J., Azéma, J., Baumgartner, P. O., Tournon, J., Desmet, A., and Auboin, J., 1984, The geologic history of the Caribbean-Cocos plate boundary with special reference to the Nicoya Ophiolite Complex (Costa Rica) and D.S.D.P. results (Leg 67 and 84 off Guatemala): A synthesis: Tectonophysics, v. 108, p. 1–32.

Bowland, C. L., and Rosencrantz, E., 1988, Upper crustal structure of the western Colombian Basin, Caribbean Sea: Geological Society of America Bulletin, v. 100, p. 534–546.

Burbach, G., Frohlich, C., Pennington, W. D., and Matumoto, T., 1984, Seismicity and tectonics of the subducted Cocos Plate: Journal of Geophysical Research, v. 89, B9, p. 7719–7735.

Burke, K., Fox, P. J., and Sengör, M. C., 1978, Buoyant ocean floor and the evolution of the Caribbean: Journal of Geophysical Research, v. 83, p. 3949–3954.

Calvo, C., 1987, Las calizas neríticas de la vertiente Pacífica del norte de Costa Rica y sur de Nicaragua: Épocas y sistemas, asociados con la apertura y evolución del margen convergente de la América Central Meridional [Tesis de Licenciatura]: San José, Universidad de Costa Rica, 165 p.

Calvo, C., and Bolz, A., 1991, La Formacion Espíritu Santo (Costa Rica): Sistemas de plataforma carbonatada autótona del Paleoceno superior–Eoceno inferior: Revista geológica de América Central, v. 13, p. 91–95.

Case, J. E., and Holcombe, T. L., 1980, Geologic-tectonic map of the Caribbean region: U.S. Geological Survey Miscellaneous Investigation Series Map I-100, scale 1:2,500,000.

Corrigan, J., Mann, P., and Ingle, J. C., Jr., 1990, Forearc response to subduction of the Cocos Ridge, Panama–Costa Rica: Geological Society of America Bulletin, v. 102, p. 628–652.

De Boer, J., 1979, The outer arc of the Costa Rican Orogen (oceanic basement complexes of the Nicoya and Santa Elena Peninsulas): Tectonophysics, v. 56, p. 221–259.

Dengo, G., 1962, Tectonic-igneous sequence in Costa Rica, *in* Engel, A.E.J., James, H. M., and Leonard, B. F., eds., Petrologic studies: A volume in honor of A. F. Buddington: Boulder, Colorado, Geological Society of America, p. 133–161.

Denyer, P., and Arias, O., 1992, Mapa geológico de la hoja Belén, Guanacaste, Costa Rica: Instituto Geográfico Nacional de Costa Rica, Hoja 3047 II, scale 1:50,000.

De Wever, P., Azéma, J., Tournon, J., and Desmet, A., 1985, Découverte de matériel océanique du Lias-Dogger inférieur dans la péninsule de Santa Elena (Costa Rica, Amérique Centrale): Paris, France, Comptes rendus de l'Académie des Sciences de Paris, v. 300, p. 759–764.

Di Marco, G., 1994, Les terrains accrétés du sud du Costa Rica: Evolution tectonostratigraphique de la marge occidentale de la plaque Caraïbe: Mémoires de Géologie (Lausanne), v. 20, 185 p.

Escalante, G., 1990, The geology of southern Central America and western Colombia, *in* Dengo, G., and Case, J. E., eds., The Caribbean Region: Geological Society of America, The Geology of North America, v. H, p. 201–230.

Fisher, R. A., 1953, Dispersion on a sphere: Proceedings of the Royal Society of London, Series A, v. 217, p. 295–305.

Fisher, S. P., and Pessagno, E. A., Jr., 1965, Upper Cretaceous strata of Northwestern Panama: American Association of Petroleum Geologist Bulletin, v. 49, p. 433–444.

Frisch, W., Meschede, M., and Sick, M., 1992, Origin of the Central American ophiolites: Evidence from paleomagnetic results: Geological Society of America Bulletin, v. 104, p. 1301–1314.

Gose, W. A., 1983, Late Cretaceous–Early Tertiary tectonic history of Southern Central America: Journal of Geophysical Research, v. 88, p. 10585–10592.

Gursky, M., 1988, Analisis tectónico de la Peninsula de Nicoya (Costa Rica) y su significado para el desarrollo estructural-geodinamico de América Central Meridional: Revista geológica de América Central, v. 8, p. 19–75.

Henningsen, V. D., and Weyl, R., 1967, Ozeanische Kruste im Nicoya-

Komplex von Costa Rica (Mittelamerika): Geologische Rundschau, v. 57, p. 33–47.

Howell, D. G., Jones, D. L., and Schermer, E. R., 1985, Tectonostratigraphic terranes of the Circum-Pacific region, *in* Howell, D. G., ed., Tectonostratigraphic terranes of the Circum-Pacific region: Houston, Texas, Circum-Pacific Council for Energy and Mineral Resources, p. 3–30.

Irwing, E., and Irwing, G. A., 1982, Apparent polar wander paths Carboniferous through Cenozoic and the assembly of Gondwana: Geophysical Survey, v. 5, p. 141–188.

Kent, D. V., and Gradstein, F. M., 1985, Cretaceous and Jurassic geochronology: Geological Society of America Bulletin, v. 96, p. 1419–1427.

Kirschvink, J. L., 1980, The least squares line and plane and the analysis of paleomagnetic data: Geophysical Journal of the Royal Astronomical Society, v. 62, p. 699–718.

Kuijpers, E. P., 1980, The geologic history of the Nicoya ophiolite Complex, Costa Rica, and its geotectonic significance: Tectonophysics, v. 68, p. 233–255.

Lew, L., 1983, The geology of the Osa Peninsula, Costa Rica: Observations and speculations of the outer arc of the Southern Central American orogen [M.S. thesis]: University Park, Pennsylvania State University, 128 p.

Lundberg, N., 1982, Evolution of the slope landward of the Middle America Trench, Nicoya Peninsula, Costa Rica, *in* Leggett, J. K., ed., Trench-forearc geology: Geological Society of London Special Publication, v. 10, p. 131–147.

McElhinny, M. W., 1964, Statistical significance of the fold test in paleomagnetism: Geophysical Journal of the Royal Astronomical Society, v. 8, p. 338–340.

Meschede, M., Frisch, W., and Sick, M., 1988, Interpretación geodinámica de los complejos ofioliticos de Costa Rica: Revista geológica de América Central, v. 8, p. 1–17.

Mora, C. R., 1985, Sedimentología y geomorphología del sur de la Peninsula de Nicoya (Provincia de Puntarenas, Costa Rica [Tesis de Licenciatura]: San José, Universidad de Costa Rica, 148 p.

Mora, S., 1981, Barra Honda: San José, Editorial Universidad Estatal a Distancia, Serie Educación Ambiental, v. 5, 94 p.

Obando, J. A., 1986, Sedimentología y tectónica del Cretácico y Paleógeno de la región de Golfito, Península de Burica y Península de Osa, Provincia de Puntarenas, Costa Rica [Tesis de Licenciatura]: San José, Universidad de Costa Rica, 211 p.

Pichler, H., and Weyl, R., 1973, Petrochemical aspects of Central American magmatism: Geologische Rundschau, v. 62, p. 357–396.

Raymond, L. A., 1984, Classification of mélanges: Their nature, origin and significance: Geological Society of America Special Paper 198, p. 7–20.

Rivier, F., 1983, Síntesis geológica y mapa geológica del área del Bajo Tempisque Guanacaste, Costa Rica: San José, Ministerio Obras Publicas y Transporte, Instituto Geográfico Nacional, Informe semestral, p. 7–30.

Robaszynski, F., Caron, M., Gonzales Donoso, J. M., and Wonders, A.A.H., E.W.G.P.F., 1984, Atlas of Late Cretaceous Globotruncanids: Revue de Micropaléontologie, v. 26, p. 145–305.

Schmidt-Effing, R., 1975, El primer hallazgo de amonites en América Central Meridional y notas sobre las facies cretácicas en dicha región: San José, Instituto Geográfico Nacional, Informe semestral, p. 53–61.

Schmidt-Effing, R., 1979, Alter und Genese des Nicoya-Komplexes, einer ozeanischen Paläokruste (Oberjura bis Eozän) im südlichen Zentralamerika: Geologische Rundschau, v. 68, p. 457–494.

Seyfried, H., and Sprechmann, P., 1985, Acerca de la formación del Puente-Istmo Centroamericano Meridional con énfasis en el desarollo acaecido desde el Campaniense al Eoceno: Revista geológica de América Central, v. 2, p. 63–87.

Seyfried, H., and Sprechmann, P., 1986, Über die Frühgeschichte (Campan bis Eozän) des südlichen mittelamerikanischen Landbrücke: Neues Jahrbuch Geologie und Palaeontologie Monatsch, 1986, p. 38–55.

Seyfried, H., and six others, 1991, Anatomy of an evolving island arc: Tectonic and eustatic control in the south Central American fore-arc area, *in* MacDonald, D.I.M., ed., Sea-level changes at active plate margin: processes and products: Special Publication of the International Association of Sedimentologists, v. 12, p. 217–240.

Sick, M., 1989, Paleomagnetism of the ophiolite complexes from the Southern Middle America Landbridge (Costa Rica and western Panama): Tübinger Geowissenschaftliche Abhandlungen, Reihe A, Geologie, Paläontologie, Stratigraphie, v. 4, 108 p.

Sprechmann, P., Astorga, A., Bolz, A., and Calvo, C., 1987, Estratigrafía del Cretácico de Costa Rica: Actas de la Facultad de Ciencias de la Tierra U.A.N.L. Linares, v. 2, p. 69–83.

Strebin, M., 1982, Zur Deformationgeschichte des Ophiolith-Komplexes der Halbinsel Nicoya (Costa Rica) und seines sedimentären Auflagers: Geowissenschaftliches Lateinamerikanisches Kolloquium, Göttingen (BRD), v. 8, 94 p.

Tourmarkine, M., and Luterbacher, H., 1985, Paleocene and Eocene planktic foraminifers, *in* Bolli, H. M., Saunders, J. B., and Perch-Nielsen, K., eds., Plankton stratigraphy, v. 1: Cambridge, Cambridge University Press, p. 87–154.

Tournon, J., 1984, Magmatismes du Mésozoïque à l'Actuel en Amérique Centrale: l'exemple du Costa Rica, des ophiolites aux andésites: Mémoires des Sciences de la Terre, Université Pierre et Marie Curie, v. 84, 335 p.

Van der Voo, R., 1990, The reliability of paleomagnetic data: Tectonophysics, v. 184, p. 1–9.

Wildberg, H.G.H., 1984, Der Nicoya-Komplex, Costa Rica, Zentralamerika: Magmatismus und Genese eines polymagmatischen Ophiolith-Komplexes: Münstersche Forschungen zur Geologie und Paläontologie, v. 62, 123 p.

Winsemann, J., 1992, Tiefwasser-Sedimentationsprozesse und-produkte in den Forearc-Becken des mittleamerikanischen Inselbogensystems: Eine sequenzstratigraphische Analyse: Dissertation an der Universität Stuttgart, Profil, v. 2, 218 p.

Manuscript Accepted by the Society February 7, 1994

Geological Society of America
Special Paper 295
1995

Early Tertiary arc volcanics from eastern Panama

René C. Maury
Département des Sciences de la Terre et UA1278, Université de Bretagne Occidentale, 29275 Brest, France
Marc J. Defant
Department of Geology, University of South Florida, Tampa, Florida 33620
Hervé Bellon
Département des Sciences de la Terre et UA1278, Université de Bretagne Occidentale, 29275 Brest, France
Jelle Zeilinga de Boer
Department of Earth and Environmental Sciences, Wesleyan University, Middletown, Connecticut 06459-0139
Robert H. Stewart
Department of Geology, University of South Florida, Tampa, Florida 33620
Joseph Cotten
Département des Sciences de la Terre et UA1278, Université de Bretagne Occidentale, 29275 Brest, France

ABSTRACT

Early Tertiary volcanic clasts were collected from the streambed of the Rio Morti near the village of Morti, eastern Panama, as part of a reconnaissance study. The samples range from basalts to rhyolites. K-Ar dates cluster around 58 Ma. The phenocryst mineralogy of the samples is typical of that found in arc-related volcanics: plagioclase (the dominant phase), clinopyroxene, titanomagnetite, and minor orthopyroxene. The clinopyroxenes are augites that plot in the field of orogenic lavas. The geochemistry of the rocks—high ratios of large-ion lithophile elements to high-field strength elements, negative Nb and Ta anomalies, positive Ba anomalies, and relatively low Th to U values—confirms that they are arc related (specifically the calc-alkaline series) and strongly suggests that the samples are not cogenetic.

Volcanic rocks with similar ages are exposed in several other localities throughout Panama and Costa Rica (e.g., the Azuero and Sona Peninsulas in Panama and the Nicoya Peninsula in Costa Rica as well as other areas in the Darien of eastern Panama). We suggest that there may have been a more or less continuous arc from South America to the Chortis block of Nicaragua during the Paleocene–Early Eocene. The presence of the arc would imply that the breakup of the Farallon plate and its subduction below the new Caribbean plate in present-day southern Central America probably started some time close to the Cretaceous-Tertiary boundary.

INTRODUCTION

Clasts of Early Tertiary volcanic rocks have been collected from the streambed of the Rio Morti near the village of Morti (77°55.68′W, 8°51.34′N), eastern Panama (Fig. 1). These samples were collected during a project to identify possible recent volcanics in the Darien jungle of eastern Panama. The sampling area is located on the northern flank of the Turia-Chucunaque Basin (Escalante, 1990). The sampling area is located on the Late Eocene sandstones and shales referred to as the Corcona Formation (Shelton, 1952). The clasts are believed to have originated from the underlying Eocene (or older) Igneous Agglomerate (Shelton, 1952; Escalante, 1990) that outcrops approximately 5 km north of the sampling site

Maury, R. C., Defant, M. J., Bellon, H., de Boer, J. Z., Stewart, R. H., and Cotten, J., 1995, Early Tertiary arc volcanics from eastern Panama, *in* Mann, P., ed., Geologic and Tectonic Development of the Caribbean Plate Boundary in Southern Central America: Boulder, Colorado, Geological Society of America Special Paper 295.

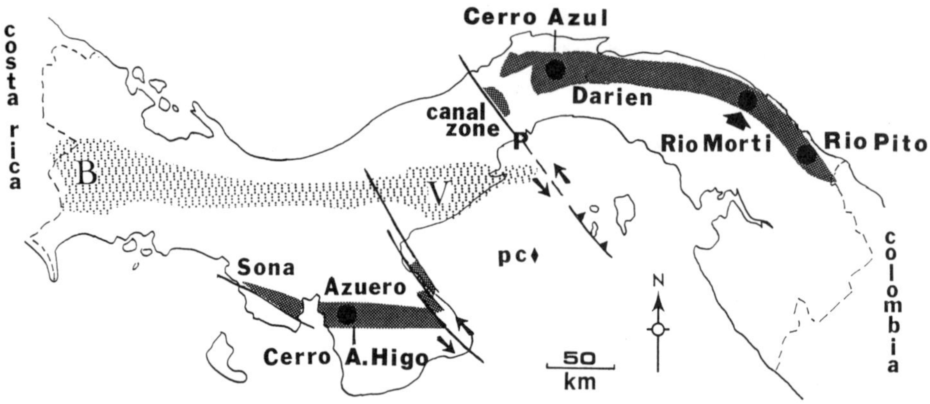

Figure 1. Map of Panama showing the Recent extension of the Central American arc in Panama (dot pattern): B-El Baru and V-El Valle stratovolcanoes. The hachured pattern represents the Darien and Azuero segments of the Paleocene–Early Eocene volcanic arc.

(United Nations Development Program, 1972). The outcrops from which the clasts came were not visited because of difficulties related to dense vegetation. The clasts, however, are probably derived from the Morti Tuffs, a local information term introduced by the Atlantic-Pacific Interoceanic Canal Study Commission (1968) and equivalent to Shelton's (1952) Igneous Agglomerate and to the Eocene Basal Complex (United Nations Development Program, 1972). The Morti Tuffs are considered to be of Early and Middle Eocene age, and their estimated thickness is 2,000 m. Although we clearly have insufficient information to draw major conclusions, we believe it useful to report petrographic and microprobe data and major- and trace-element analyses on eight selected clasts. We also report K-Ar radiometric dates on six of these samples.

PETROGRAPHY AND MINERALOGY

The samples range from basaltic andesites through andesites and dacites to rhyolites (Fig. 2). They contain up to 30% phenocrysts that are, by decreasing order of modal abundance, plagioclase, clinopyroxene, titanomagnetite, and minor orthopyroxene. Several samples have undergone greenschist-facies hydrothermal/metamorphic alteration, as shown by the occurrence of chlorite, calcite, and secondary quartz. Primary clinopyroxenes are sometimes rimmed by actinolite (sample MV10).

The composition of the primary minerals (microprobe data available upon request) does not depart from the usual range in island-arc lavas. Plagioclase phenocrysts often show complex oscillatory zoning patterns. The compositions of their cores range from An_{87} to An_{55} in andesites and from An_{60} to An_{45} in dacites and rhyolites. The compositions of the rims and groundmass plagioclases range from An_{70} (andesites) to An_{25} (rhyolites). Clinopyroxene phenocrysts exhibit a restricted range in chemical variation. They are typical augites ($Wo_{35-44}Fs_{11-15}$) that are characterized by low contents in Al_2O_3 (<2%), TiO_2 (<0.6%), and Cr_2O_3 (<0.3%). They plot in the field of clinopyroxenes from orogenic lava series in Leterrier et al.'s (1982) diagrams (not shown). With respect to the phenocryst cores, the rims and groundmass clinopyroxenes are usually iron enriched ($Wo_{33-40}Fs_{13-30}$). Pigeonite crystals ($Wo_{9.4-10.2}Fs_{32.6-39.9}$) have been analyzed from the groundmass of the andesite sample MV10.

K-Ar Radiometric Ages

Analytical data on the six dated samples are given in Table 1. The dating procedure of selected, essentially unaltered specimens is provided below the table. The corresponding ages range from 55 to 61 Ma (i.e., Late Paleocene according to recent geologic timescales). It is worth noting that four out of the seven ages cluster around 58 Ma (57.8 to 58.5 Ma). The age cluster is unlikely to have resulted from reheating. These results are consistent with the Eocene or older stratigraphic ages attributed to the Igneous Agglomerate, the likely source of the clasts. The results also fit with the previously described occurrence in Panama of tholeiitic to calc-alkaline magmatic rocks of Paleocene-Eocene ages (Kesler et al., 1977). In addition, a quartz-diorite sample from Rio Azucar, San Blas, has been dated at 63 Ma (H. Bellon and J. Tournon, unpublished results).

MAJOR- AND TRACE-ELEMENT GEOCHEMISTRY

Selected analyses of Rio Morti volcanics are presented in Table 2. The samples display a wide range of SiO_2 concentrations, and all of them are silica oversaturated. All samples have SiO_2 greater than 54% and MgO lower than 5%, indicating an evolved character with respect to basalt. The major-element analyses are typical of arc-related volcanics: low TiO_2 (<0.8 wt. %) and relatively high Al_2O_3 in the least-evolved samples. In a conventional K_2O versus SiO_2 diagram (Fig. 2), most of the samples plot near the boundary separating the low-K (island-arc tholeiitic) and medium-K (calc-alkaline) orogenic magma series fields. Major-element oxides do not vary

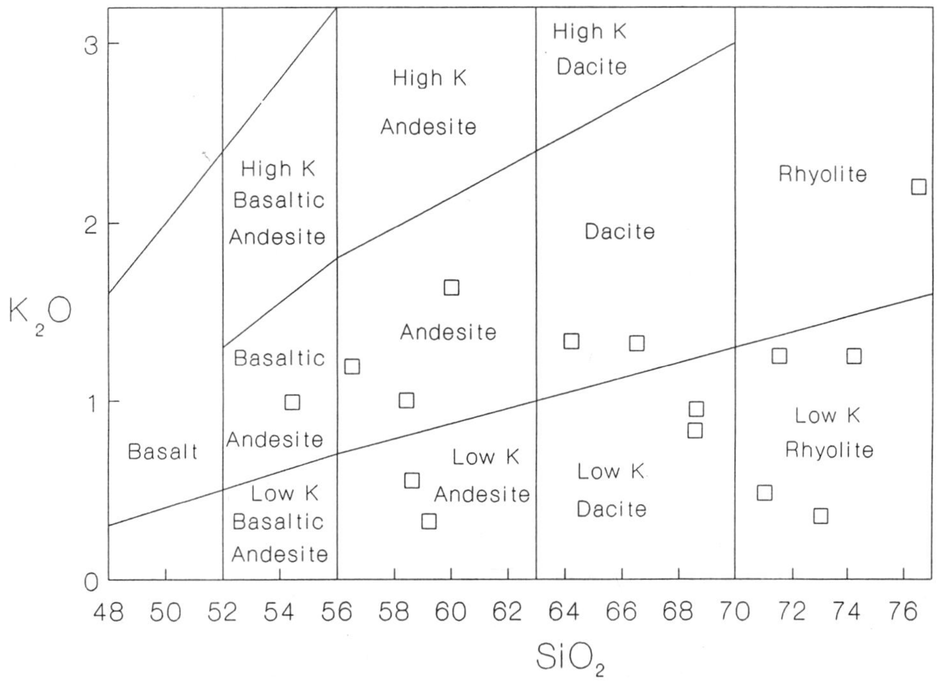

Figure 2. Graph of K_2O versus SiO_2 for all of the Rio Morti data. The fields are after Peccerillo and Taylor (1976).

TABLE 1. K-Ar RADIOMETRIC DATES†

Sample	Age ± 'error' (Ma)	$^{40}Ar^*$ (10-7 cc/g)	^{36}Ar (10-9 cc/g)	$^{40}Ar^*$ / ^{40}Ar T (%)	K_2O (%)	Analysis Number
MV4	55.22 ± 1.21	18.08	2.48	71.1	1.00	2732
MV2	57.82 ± 1.17	23.68	2.31	77.6	1.25	2721
MV13	58.00 ± 1.20	30.97	3.40	75.5	1.63	2736
MV16	58.50 ± 1.11	23.96	1.59	83.6	1.25	2731
MV14	58.35 ± 1.19	18.16	1.83	77.0	0.95	2733
MV14	60.64 ± 1.62	18.89	4.33	59.6	0.95	2735
MV10	60.81 ± 1.34	12.96	1.82	70.7	0.65	2734

†Age calculations were carried out using constants recommended by Steiger and Jager (1977) according to the following formula: t = 4154.04 log (1 +142.69 ($^{40}Ar^*$:K)); $^{40}Ar^*$ in c.c. - K in g. - t in million years; 40 K = 0.01167 K of K total in atoms. Errors were calculated using the equation proposed by Mahood and Drake (1982). Argon extraction was performed by the direct technique under vacuum (10^{-5} to 10^{-7} Torr) using induction heating on a molybdenum crucible. The argon content was measured by isotope dilution. An aluminum foil target containing a known volume of ^{38}Ar, buried as positive ions under 30 KeV energy (Bellon et al., 1981), was added to each sample at the time of weighing and was used as a spike. Consqequently, isotope dilution and homogenization were achieved during the fusion of the sample. Argon isotopes were analyzed in a 180° C stainless-steel mass spectrometer. Argon air standards, in the volume range of the samples, were run after each measurement in order to make corrections for mass discrimination. See Defant et al. (1992) for further analytical details.

TABLE 2. MAJOR AND TRACE ELEMENT DATA FROM SELECTED SAMPLES*

	MV9 BA	MV4 A	MV10 A	MV13 A	MV6 D	MV14 R	MV16 R	MV2 R
SiO_2	54.4	58.4	58.6	60.0	66.5	68.6	71.5	74.2
TiO_2	0.80	0.73	0.70	0.57	0.59	0.60	0.70	0.50
Al_2O_3	16.10	17.50	14.98	16.72	12.10	12.60	12.30	11.50
Fe_2O_3	10.70	7.75	10.12	5.30	7.30	5.65	4.65	3.40
MnO	0.14	0.11	0.24	0.14	0.10	0.15	0.14	0.14
MgO	4.80	2.35	3.63	2.00	2.25	1.25	0.83	0.53
CaO	7.02	8.20	8.28	8.54	5.35	3.46	2.94	2.93
Na_2O	2.45	2.94	2.83	3.25	2.75	4.24	3.84	3.33
K_2O	0.99	1.00	0.65	1.63	1.32	0.95	1.25	1.25
P_2O_5	0.23	0.20	0.12	0.18	0.20	0.15	0.18	0.18
LOI	2.24	1.17	0.26	1.69	1.39	1.29	0.96	0.94
Total	99.87	100.35	100.41	100.02	99.85	98.90	99.49	98.89
Rb	10.4	15.5	10.4	28.5	14.1	10.5	17.0	13.2
Sr	248	298	209	352	177	132	140	144
Ba	249	265	140	362	270	250	380	680
Sc	33	28	40	25	24	17	15.5	13
V	284	318	349	197	128	128	45	73
Cr	15	28	23	41	104	7	5	7
Co	32	25	29	16	23	13	8	7
Ni	26	25	27	15	41	8	5	7
Y	23	24	20	23	31	32	39	36
Zr	72	89	55	104	128	156	195	160
Nb	0.9	0.8	0.5	2.3	2.4	2.1	3.2	2.6
La	6.5	8.2	4.7	14.3	13.1	11.8	14.5	14.9
Ce	14.9	17.8	11.0	29.7	26.2	25.4	32.9	29.7
Pr	2.2	2.5	1.5	3.8	3.5	3.6	4.5	4.4
Nd	11.3	12.4	7.5	17.1	16.0	16.2	21.9	17.8
Sm	2.9	3.5	2.3	3.7	4.0	4.2	5.2	4.4
Eu	0.92	1.17	0.70	1.02	0.97	0.95	1.31	0.96
Gd	3.7	3.4	2.7	3.6	4.5	4.3	5.8	4.8
Dy	3.7	3.2	2.9	3.2	4.3	4.6	5.8	5.4
Ho	0.80	0.69	0.66	0.62	0.94	0.88	1.23	1.21
Er	2.6	2.2	2.1	2.0	2.9	2.9	3.9	4.1
Yb	2.2	1.9	2.2	2.0	2.8	3.1	3.9	3.5
Lu	0.30	0.29	0.29	0.31	0.41	0.39	0.56	0.52
Th	0.9	1.1	1.7	3.0	1.7	2.3	2.8	2.3
U	0.4	0.4	0.8	1.1	0.6	0.8	1.0	0.9

*Analyses by induction-coupled plasma spectrometry (ICP) for major elements and Sr, Ba, Sc, V, Cr, Co, Ni, Y, Zr, and Nb; atomic absorption spectrometry for Rb; induction-coupled plasma/mass spectrometry (ICP/MS) for REE, Th, and U. BA = basaltic andesite; A = andesite; D = dacite; R = rhyolite.

regularly with SiO_2 concentrations (Table 2), and thus the samples are unlikely to be related cogenetically through differentiation processes. They most probably belong to distinct magmatic groups.

Compatible trace-element concentrations are fairly low, in agreement with their evolved character. They tend to decrease with increasing SiO_2. Scatter among the data, however, again suggests that the samples are not cogenetic. Incompatible trace-element concentrations generally increase with differentiation. Corresponding chondrite-normalized diagrams (Fig. 3) exhibit features characteristic of rocks with volcanic arc affinities: high ratios of large-ion lithophile elements (Rb, Ba, K, etc.) to high-field strength elements (Nb, Zr, Ti, etc.), typical negative Nb and Ti anomalies, positive Ba anomalies, and relatively low Th relative to U values (Fig. 3) (e.g., Gill, 1981). The REE (Rare Earth Elements) patterns show slight enrichments in LREE (Light Rare Earth Elements), which is typical of low- to medium-K calc-alkaline rocks. The andesitic samples display positive Sr anomalies, possibly as a result of plagioclase accumulation. In contrast, the rhyolites show significant negative anomalies in Sr and P, which may indicate plagioclase and apatite fractionation, respectively.

We conclude that the rocks collected at Rio Morti were erupted in a volcanic-arc tectonic environment.

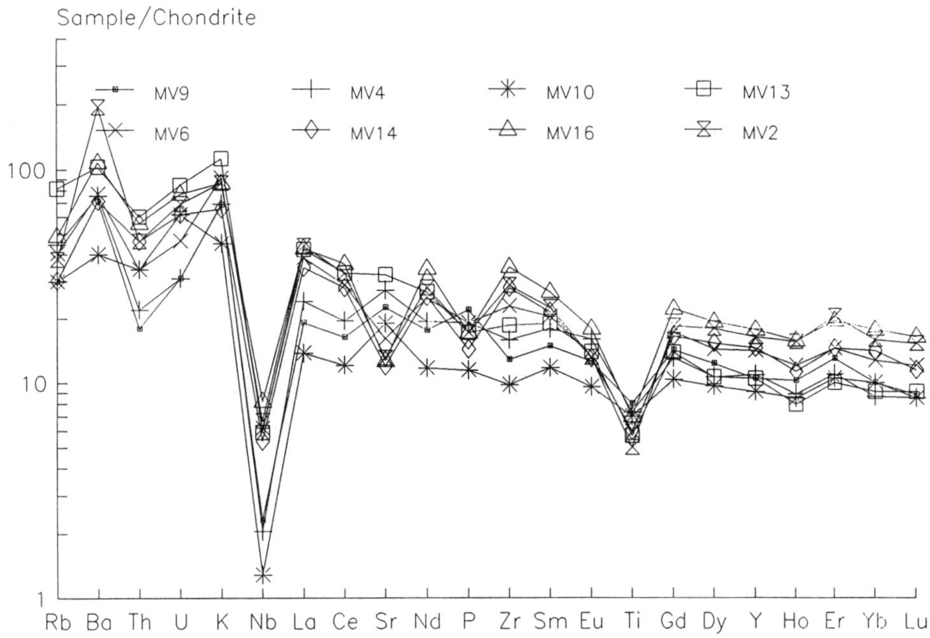

Figure 3. Chondrite-normalized diagram (normalizing values from Wakita et al., 1971; Sun and Nesbitt, 1977; Sun et al., 1979) for eight selected samples.

DISCUSSION

Recchi (1975) constructed a series of paleogeographic maps for Panama and theorized that an emerged east-west–trending volcanic arc existed between the present-day Sona Peninsula and northwest Colombia during the Late Cretaceous. This arc was intersected and offset by sinistral motion along a major north-northwest–trending fault zone(s), the remains of which should be present below the Neogene deposits in the Canal zone near Panama City. Recchi (1975) believed that virtually all strike-slip motion along the Canal fault zone was accomplished before the Late-Middle Eocene, although some motion could have occurred in the Oligocene. It is the sinistral motion along the Canal fracture zone and later deformation described by Kolarsky and Mann (this volume) that established the sigmoidal shape of the present-day isthmus. Our data appear to support Recchi's model but suggest that the earliest volcanic arc of southern Central America is probably of Paleocene and Early Eocene age.

De Boer (1979) reported the presence of a west-north-west–trending Paleocene-Eocene volcanic-arc segment, which was emplaced in/on the obducted oceanic crust of the Nicoya Peninsula (northwest Costa Rica). Recent dates on samples from this arc confirm its age (Berrangé and Thorpe, 1988; Berrangé et al., 1989).

The central segment of the Paleocene arc probably extends from southeast Costa Rica east to the southwest corner of the Azuero Peninsula (western Panama). Evidence for the presence of a Paleocene-Eocene arc below the Mio-Pliocene mountain range of southeast Costa Rica comes mainly from sediments and intercalated volcanics in the Limon Sur and Fila Costena uplifted basins in this region (de Boer et al., this volume).

In the Sona region, Tournon et al. (1989) described a sequence of Paleocene amphibolites characterized by counter-clockwise pressure-temperature paths. The metamorphism of the amphibolites is not characteristic of that related to subduction processes but is probably due to magmatic heating, the origin of which might be ascribed to the emplacement of the Paleocene volcanic arc.

The eastward extension of the western arc segment into the Azuero Peninsula is especially well shown by a series of positive west-northwest–east-southeast–trending magnetic anomalies (J. de Boer, 1973, unpublished report for United Nations Development Program). Iron-rich magmatic assemblages, including quartz-diorite plutons, are responsible for these anomalies. Good exposures occur in the Cerro Alto Higo region. The quartz diorite in this region provided mineral ages of 65 Ma (hornblende) and 53 Ma (feldspar) (Kesler et al., 1977).

Similar Paleocene-Eocene ages were obtained from quartz diorites collected in the westernmost extension of the eastern segment of the proposed Paleocene arc in eastern Panama. Quartz diorite from Cerro Azul provided dates of 62 Ma (hornblende) and 51 Ma (feldspar) (Kesler et al., 1977). In addition, a quartz-diorite sample from Rio Azucar, San Blas, has been dated at 63 Ma (H. Bellon and J. Tournon, unpublished results). The sequence of screened samples obtained in the Rio Morti farther east provided seven new K-Ar ages ranging from 61 to 55 Ma (Table 1). Quartz diorites collected near Rio Pito on the easternmost extension of the eastern-arc segment provided 49-Ma (feldspar) and 48-Ma (hornblende) ages (Kesler et al., 1977). The occurrence of the quartz diorites implies that the magmatic phase of the Paleocene arc lasted well into the Middle Eocene.

The stratigraphic section in the El Paso Corvus drill hole in the Gulf of Panama (PC on Fig. 1) indicates the nearby presence of an Early and Middle Eocene arc below the waters of the Gulf of Panama. The core obtained from the hole contains 784 m of Pliocene interbedded sandstone, shales, basaltic volcanics, and pyroclastic units (Kolarsky and Mann, this volume). These sediments and volcanics were most probably deposited in a basin bordering the Paleocene–Early Eocene arc.

The geochemical and age similarities between the Darien, Azuero, and Nicoya segments of the Paleocene–Early Eocene arc segments argue for a once continuous string of volcanic islands between South America and the Chortis block of Nicaragua. The possible existence of such an arc would indicate that the breakup of the Farallon plate and its subduction below the new Caribbean plate probably started some time close to the Cretaceous-Tertiary boundary.

REFERENCES CITED

Atlantic-Pacific Interoceanic Canal Study Commission, 1968, Engineering feasibility studies, geology: Final Report, Route 25, v. 1, 20 p.

Bellon, H., Buu, N. Q., Chaumont, J., and Philippet, J. C., 1981, Implantation ionique d'argon dans une cible support. Application au tracage isotopique de l'argon contenu dans les minéraux et les roches: Comptes Rendus de l'Académie des Sciences, v. 292, p. 977–980.

Berrangé, J. P., and Thorpe, R. S., 1988, The geology, geochemistry and emplacement of the Cretaceous/Tertiary ophiolitic Nicoya Complex of the Osa Peninsula, southern Costa Rica: Tectonophysics, v. 47, p. 193–220.

Berrangé, J. P., Bradley, D. R., and Snelling, N. J., 1989, K/Ar age dating of the ophiolitic Nicoya Complex of the Osa Peninsula, southern Costa Rica-Journal of South American Earth Sciences, v. 2, p. 49–59.

de Boer, J., 1979, The outer arc of Costa Rican orogen (oceanic basement complexes of the Nicoya and Santa Elena Peninsulas): Tectonophysics, v. 56, p. 221–259.

Defant, M. J., and seven others, 1992, The geochemistry of young volcanism throughout western Panama and southeastern Costa Rica: An overview: Journal of the Geological Society of London, v. 149, p. 569–579.

Escalante, G.,1990, The geology of southern Central America and western Colombia, *in* Dengo, G., and Case, J. E., eds., The Caribbean Region: Geological Society of America, The Geology of North America, v. H, p. 201–230.

Gill, J. B., 1981, Orogenic andesites and plate tectonics: New York, Springer-Verlag, 390 p.

Kesler, S. E., Sutter, J. F., Issignonis, M. J., Jones, L. M., and Walker, R. L., 1977, Evolution of porphyry copper mineralization in an oceanic island arc: Panama: Economic Geology, v. 72, p. 1142–1153.

Leterrier, J., Maury, R. C., Thonon, P., Girard, D., and Marchal, M., 1982, Clinopyroxene compositions as a method of identification of the magmatic affinities of paleovolcanic series: Earth and Planetary Science Letters, v. 59, p. 139–154.

Mahood, G. A., and Drake, R. E., 1982, K-Ar dating of young rhyolitic rocks: A case study of the Sierra La Primavera, Jalisco, Mexico: Geological Society of America Bulletin, v. 93, p. 1232–1241.

Peccerillo, A., and Taylor, S. R., 1976, Geochemistry of Eocene calc-alkaline volcanic rocks from the Kastamonu area, Northern Turkey: Contributions to Mineralogy and Petrology, v. 58, p. 63–81.

Recchi, J., 1975, Paleografia atlas nacional de Panama: Panama, Instituto Geografico Nacional, "Tommy Guardia," 128 p.

Shelton, B. J., 1952, Geology and petroleum prospects of the Darien, southeastern Panama [M.S. thesis]: Corvallis, Oregon State University, 62 p.

Steiger, R. H., and Jager, E., 1977, Subcommission on geochronology: Convention on the use of decay constants in geo- and cosmochronology: Earth and Planetary Science Letters, v. 36, p. 359–362.

Sun, S. S., and Nesbitt, R. W., 1977, Chemical heterogeneity of the Archaean mantle, composition of the earth and mantle evolution: Earth and Planetary Science Letters, v. 35, p. 429–448.

Sun, S. S., Nesbitt, R. W., and Sharaskin, A. Y., 1979, Geochemical characteristics of mid-ocean ridge basalts: Earth and Planetary Science Letters, v. 44, p. 119–138.

Tournon, J., Triboulet, C., and Azema, J., 1989, Amphibolites from Panama: Anti-clockwise P-T paths of a pre-Upper-Cretaceous metamorphic basement in isthmian Central America: Journal of Metamorphic Geology, v. 7, p. 539–546.

United Nations Development Program, 1972, Mineral survey of eastern Panama: New York, United Nations, DP/SF/UN/85-Technical reports 1–4, 424 p.

Wakita, H., Rey, P., and Schmitt, R. A., 1971, Abundances of the 14 rare earth elements and 12 other trace elements in Apollo 12 samples: Five igneous and one breccia rocks and four soils: Proceedings of the 2nd Lunar Science Conference, p. 1319–1329.

MANUSCRIPT ACCEPTED BY THE SOCIETY FEBRUARY 7, 1994

Printed in U.S.A.

Geological Society of America
Special Paper 295
1995

Cenozoic magmatic phases of the Costa Rican island arc (Cordillera de Talamanca)

Jelle Zeilinga de Boer
Department of Earth and Environmental Sciences, Wesleyan University, Middletown, Connecticut 06459-0139
Mark S. Drummond
Department of Geology, University of Alabama at Birmingham, Birmingham, Alabama 35294
Marc J. Bordelon and Marc J. Defant
Department of Geology, University of South Florida, Tampa, Florida 33620
Hervé Bellon
Laboratoire de Geochimie et Geochronologie, Université de Bretagne Occidentale, 29287 Brest, France
René C. Maury
Laboratoire de Petrologie, Université de Bretagne Occidentale, 29287 Brest, France

ABSTRACT

Shallow subduction of the Cocos Ridge below southeastern Costa Rica (a process that was initiated ca. 5 Ma) has elevated the central segment of the Chorotega arc (the Cordillera de Talamanca) to heights over 3,500 m. Rapid uplift and associated faulting exposed the remains of volcanic arcs predating the Quaternary volcanics that dominate in the northwestern and eastern segments of the Chorotega arc. Because of extensive cover by cloud forests and generally steep topography, collection of fresh samples remains problematic. Access has become easier, however, in the last few years because of development of a network of logging roads that enabled collection along several transects. The geochemical and isotope data indicate that at least three, possibly four, phases of igneous activity occurred in the Cordillera de Talamanca. Evidence for the earliest Paleocene–early Eocene arc is provided indirectly by the volcanic contents of both the forearc (Fila Costena) and back-arc (Limon Sur) basins. Steep subduction appears to have been responsible for formation of this presently severely segmented volcanic arc. Intermediate-angle subduction gave rise to volcanics arcs in the mid-Oligocene and mid-Miocene. Igneous rocks emplaced during these periods are similar, but their contents of the element Zr differ. On chondrite-normalized spider diagram, the Oligocene intrusives display a negative, the Miocene intrusive rocks (Talamanca suite sensu stricto) a positive Zr anomaly.

Low-angle subduction characterized the Plio-Pleistocene. Associated volcanism is quite distinct from that in the Miocene and consists predominantly of intermediate to felsic magmas with geochemical signatures suggesting partial melting of a MORB source. The four volcanic phases are separated by periods of relative quiescence that correlate with periods during which the plate tectonic configuration in the east-central Pacific changed.

de Boer, J. Z., Drummond, M. S., Bordelon, M. J., Defant, M. J., Bellon, H., Maury, R. C., 1995, Cenozoic magmatic phases of the Costa Rican island arc (Cordillera de Talamanca), *in* Mann, P., ed., Geologic and Tectonic Development of the Caribbean Plate Boundary in Southern Central America: Boulder, Colorado, Geological Society of America Special Paper 295.

INTRODUCTION

The Chorotega arc of south Central America is a slightly curved, southwest convex volcanic-plutonic belt that developed on oceanic crust commonly referred to as the Chorotega block (Escalante, 1990) (Figs. 1 and 2). The arc extends from northernmost Costa Rica (St. Elena–Hess fracture zone) to central Panama (Canal fracture zone).

The topography of the arc is characterized by an axial bilateral symmetry. Its central segment, the Cordillera de Talamanca (Fig. 3), consists predominantly of Miocene plutonic-hypabyssal volcanic complexes elevated to heights of over 3,500 m. The Cordillera is bound to the northwest and east by the massive Quaternary volcanic complexes of the Irazu-Turrialba (Costa Rica) and Fabrega-Tisingal-Baru (Panama), respectively, which reach heights of over 3,400 m. Northwest from Irazu and east from Baru, volcano height decreases systematically. The end members of the Chorotega arc, volcan Orosi (northeast Costa Rica) and volcan El Valle (central Panama), are less than 1,500 m high.

The topography of the Cordillera de Talamanca is asymmetric in cross section. Its southwestern flank is steep, but its northeastern slope is moderately inclined. Both are covered with dense cloud forests at higher elevations. With the exception of the northwestern third of this mountain range, which is crossed by the Pan-American Highway, virtually no roads are available to reach elevations above 2,000 m. Because of the dense vegetation and general absence of roads, most outcrops occur in mountain streams, through which passage is difficult and time consuming. Most geologic studies of the Cordillera have therefore been of a reconnaissance nature. The most informative studies were concentrated on outcrops along the Panamanian highway (Weyl, 1957, 1969, 1980; Weyl and Pichler, 1976). Tournon (1984) provided an excellent overview of the geochemical data base for the Talamanca Range up to 1980.

In the 1970s the Cordillera de Talamanca became the focus of a search for porphyry copper deposits. Companies such as City Services, Alusuisse, and Sancosa made studies of specific regions. The results are proprietary and generally unavailable. Some work has been referred to by Escalante (1990).

During the same period F. Maranzana and J. de Boer studied the southeasternmost segment of the Cordillera as part of a Costa Rican development project financed by the United Nations. The study consisted of an air photo reconnaissance and field checks. Several attempts to transect the range were unsuccessful because of bad terrain and weather conditions. It was, however, possible to cross the Cordillera northeast of Buenos Aires de Osa to Amubri. A preliminary report of field results in this region was published by Ballman (1976). Because of increased deforestation in the last decade it has become somewhat easier to penetrate the mountains to heights of about 2,500 m along developing logging roads. Using such roads de Boer and Drummond collected igneous and volcanic rocks from more or less equally spaced transects across the central and southeastern segments of the Cordillera. They also

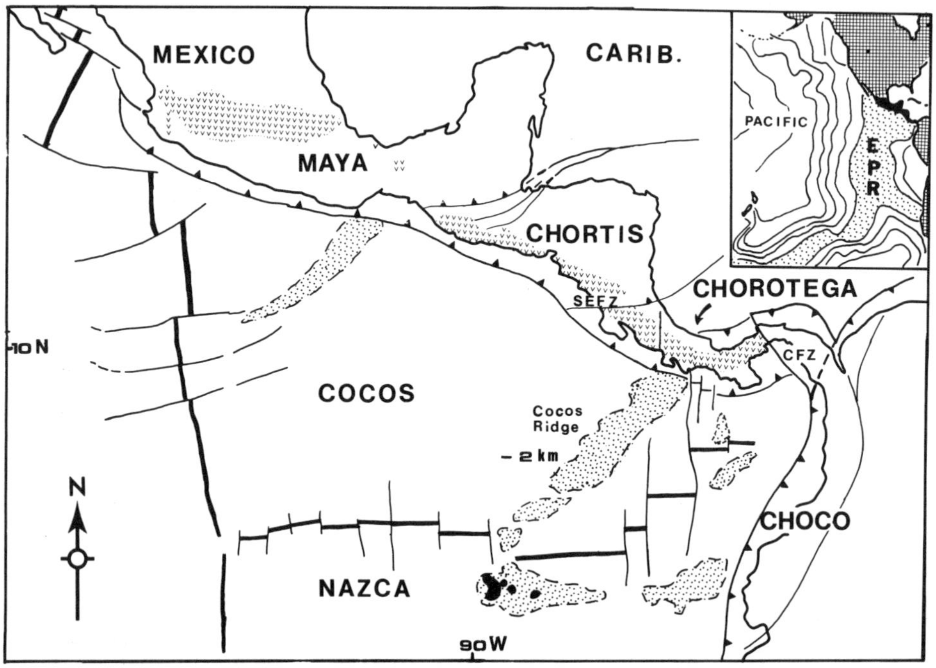

Figure 1. Regional setting of Chorotega crustal block. Inset shows regional setting of East Pacific Rise. SEFZ: Santa Elena fracture zone; CFZ: Canal fracture zone. Black: Galapagos Islands; Dots: Oceanic Ridges; v: volcanic zones.

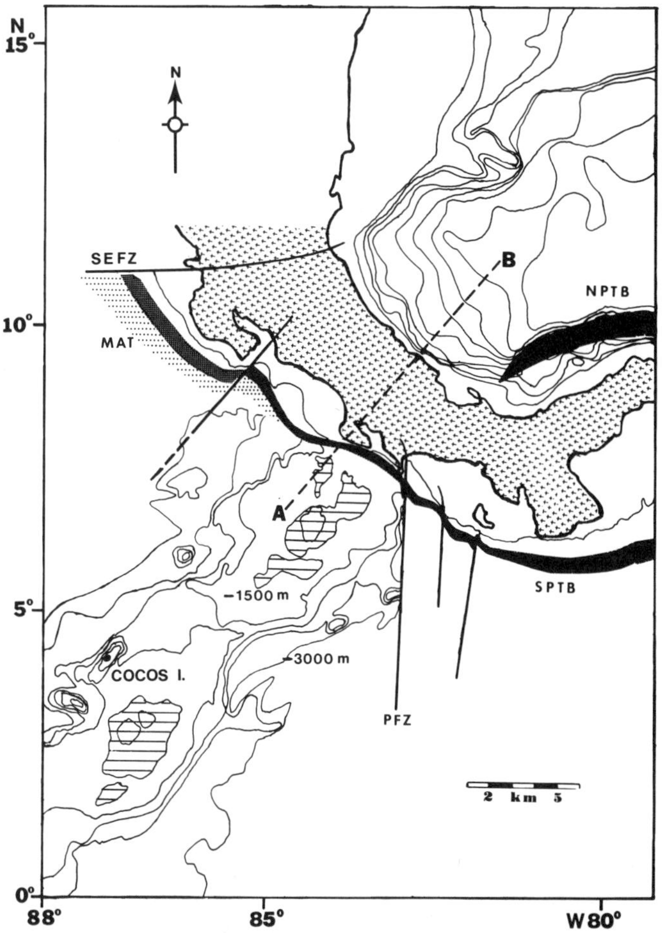

Figure 2. Regional setting of Costa Rica, western Panama, and northern segment of Cocos Ridge. A-B: Transect discussed in this chapter. MAT: Middle America Trench; NPTB: North Panama Thrust–Fold Belt; SPTB: South Panama Thrust–Fold Belt (Silver et al., 1990); PFZ: Panama fracture zone; SEFZ: Santa Elena fracture zone. Solid line = fault zone; continuation is hypothetical.

collected samples from new road cuts along the Pan American Highway and side roads that were recently developed. Bordelon finished a thesis on the Chirrippo volcanic complex that dominates the central segment of the Cordillera de Talamanca.

The Cordillera de Talamanca represents the most-elevated segment of the Chorotega volcanic arc and coincides in width with that of the Cocos Ridge sensu lato, a major northeast-trending topographic high on the Cocos plate (Fig. 2). Elevation is believed to be related to subduction of the Cocos Ridge (see Kolarsky et al., this volume).

Weyl (1980) pointed out that because of its rapid uplift the Cordillera de Talamanca represents one of the youngest volcanic belts in the world in which plutonic levels are presently exposed. Weyl (1980) referred to the Talamancas as a "block mountain," an uplifted crustal mass bounded by major fault zones. According to this author the structure of the Cordillera "resembles that of a pile of northeast tilted imbricated crustal sheets separated by thrust faults and intersected by relatively young normal faults" (Weyl, 1980). We have found little evidence for low-angle thrust faults on the Pacific side of this mountain range. Along the southwestern flank relatively steep reverse and normal faults predominate. Low-angle thrusts related to northeast décollement are common on the Caribbean side of the Talamancas. In both regions the structure is complicated significantly by landslides that originated during Pleistocene uplift and glacial events. Weyl (1980) distinguished a middle to late Miocene (15 to 6 Ma) orogenic phase characterized by weak folding and extensive magmatism, a latest Miocene to early Pliocene (6 to 3.4 Ma) period of tectonic quiescence, and a late Pliocene–Pleistocene (3.4 to present) phase of widespread normal faulting and renewed volcanism. Our data generally support his results and expand on Weyl's work.

The Talamanca segment of the Chorotega arc has been referred to as a volcanic "gap" because of the absence of Quaternary stratovolcanoes (Cross and Pilger, 1982; Nur and Ben-Avraham, 1983; McGeary et al., 1985) and as a seismic "gap" because of the relative scarcity of its seismic activity (Ponce and Case, 1987). It will be shown that although young volcanoes and earthquakes are not as common in this region as in northwest Costa Rica and western Panama, their occurrences are sufficiently common to invalidate usage of the word "gap." Although large strato-volcanic complexes of predominantly Pleistocene to Recent age are indeed absent, major volcanic complexes that contain volcanic rocks of Miocene, Pliocene, and Pleistocene age occur in the Cordillera de Talamanca. Exposures of Miocene-age rocks are restricted mainly to the axial zone and southwest flank of the Talamancas. Plio- and Pleistocene volcanic units occur predominantly along faults separating the Fila Costena, Terraba trough, and Cordillera de Talamanca and in the axial zone of the latter, along both grain-parallel and cross faults. The absence of "young" Plio-Pleistocene strato-volcanoes in the Talamanca Mountains may be a consequence of the significant elevation of the range. After all, magma surges in the Quaternary volcanoes northwest and southeast of the Cordillera de Talamanca barely reached levels to which this mountain range was elevated during the late Pliocene and Pleistocene.

TECTONIC SETTING OF THE CORDILLERA DE TALAMANCA

In order to discuss the evolution of the Talamanca Range, each of the tectonic entities adjacent to this central segment of the Chorotega arc will be treated briefly, along a northeast-southwest transect extending from the Cocos to the Caribbean plate (Figs. 2 and 3).

Cocos plate

Since subduction of the earlier Farallon and later Cocos plate controlled the geologic evolution of the Cordillera de Talamanca, a brief review of the seafloor spreading history is

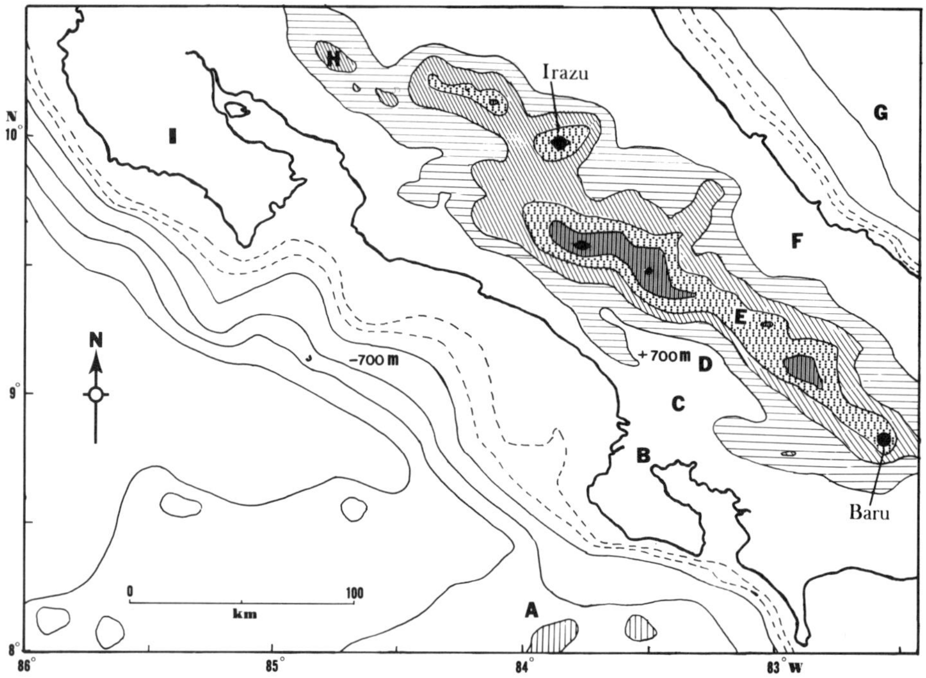

Figure 3. Digital elevation map of the Talamancas and neighboring stratovolcanic complexes of the Irazu-Turrialba and Baru-Tisingal, after Ponce and Case (1987). Patterns on land and solid lines on ocean floor delineate intervals; dashed lines indicate 170-m intervals. A, Cocos Ridge; B, Osa Peninsula; C, Fila Costena; D, Terraba trough; E, Cordillera de Talamanca; F, Limon Sur basin; G, Caribbean; H, Cordillera de Guanacaste; I, Nicoya Peninsula.

in order. Structure and age of the seafloor on the Pacific side of the isthmus are complicated by the presence of two spreading ridges and the Galapagos hot spot. Oceanic crust ranging in age from Jurassic through late middle Eocene occurs as imbricated nappes on the Pacific peninsulas in the outer arc of the Chorotega orogen (Dengo, 1962a; Weyl, 1969; de Boer, 1979; Bourgois et al., 1984; Berrange et al., 1988). Magnetic anomalies associated with the obducted seafloor fragments generally trend west-northwest–east-southeast (de Boer, 1979, and unpublished studies of the Azuero Peninsula) and appear indicative of north-south spreading, assuming that no significant rotations occurred after their development. The history of terrane accretion in the outer arc is discussed by DiMarco et al. (this volume).

Spreading patterns in the late Eocene and early Oligocene are insufficiently known. Several northwest-trending anomalies with ages ranging from 36 to 29 Ma occur in the south-central segment of the Cocos plate. The responsible spreading center has so far eluded detection. It is possible that spreading occurred along the Mathematician-Clipperton Ridge and that this segment of the Cocos plate subsequently rotated counterclockwise. Sparse but clear evidence, however, suggests that subduction occurred below southern Central America during this period and that the oldest exposed volcanic units in the Cordillera de Talamanca sensu stricto were emplaced during the Oligocene (Fisher and Pessagno, 1965, and this study).

North-south spreading inside the northeast-trending Grijalva fault zone of the Farallon plate started around 26 Ma (Lonsdale and Klitgord, 1978). Around 20 Ma east-west spreading ceased along the Mathematician-Clipperton Ridge, and a new spreading center, the East Pacific Rise, developed farther east. The period from 26 to 20 Ma thus appears to have been a time in which a general reorganization of plate motions occurred in the central eastern Pacific.

During the last 20 m.y. east-west spreading characterized tectonic processes along the East Pacific Rise, and north-south spreading occurred along a series of relatively short-lived en echelon rift zones extending east from the East Pacific Rise to the Colombian trench (Fig. 4). The spreading histories of the latter are especially important because they appear to indicate a westward plugging of the Panama trench (van Andel et al., 1971). Hardy (1991) discovered an east-west–trending rift zone of early Miocene age offshore of Colombia (unnamed rift zone, Fig. 4) that spread north-south between 22 and 16 Ma. The Buenaventura rift zone developed by ridge jump south of this early Miocene ridge and spread north-south from 16 to 12 Ma. The Malpelo rift zone represented the en echelon western continuation of the latter and spread north-south between 16 and 8 Ma. Its spreading rate slowed significantly around 12.5 Ma. North-south spreading along the Costa Rica rift zone, west of the Malpelo, has been more or less steady in the last 10 m.y. Renewed rifting appears to have been initiated along 5° north-

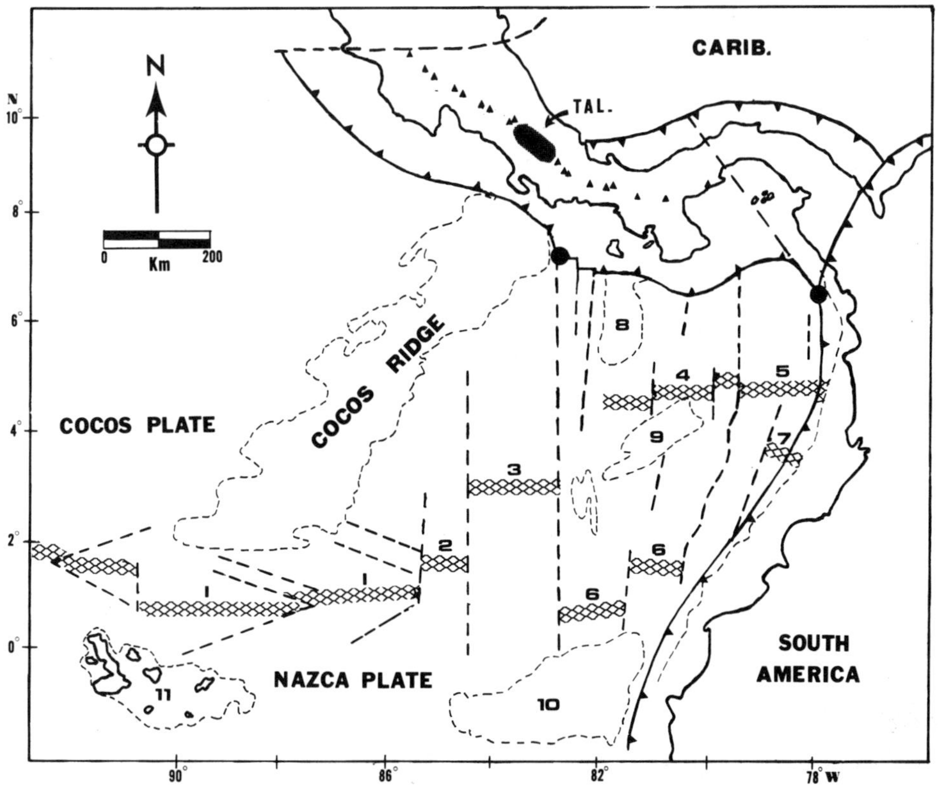

Figure 4. Plate tectonic setting of southern Central America and segments of the Nazca and Cocos plates. 1, Galapagos rift system; 2, Ecuador rift zone; 3, Costa Rican rift zone; 4, Sandra rift zone; 5, unnamed rift zone (Hardy, 1991); 6, Malpelo rift zones; 7, Buenaventura rift zone; 8, Coiba Ridge; 9, Malpelo Ridge; 10, Carnegie Ridge; 11, Galapagos Islands. TAL = Cordillera de Talamanca. Heavy dashed lines: faults; lightly dashed lines outline trench scarp and uplifted ocean-floor plateaus.

ern latitude in the upper segment of the Nazca plate during the Pliocene (de Boer et al., 1988; Hardy et al., 1990) (Fig. 4).

Major changes in seafloor spreading configurations that are of importance for understanding the magmatic evolution of the Chorotega volcanic arc thus occurred about 40 Ma and between 26 and 22 Ma; minor, more localized changes occurred 16, 12, and 8 m.y. ago.

The topography of the southern Cocos plate is dominated by the broad (200- to 300-km-wide), northeast-trending Cocos Ridge (Fig. 2), which rises to an average height of 2 km above the average Cocos plate seafloor. The ridge is believed to have originated at the Galapagos hot spot (Duncan and Hargraves, 1984; Larson and Olsen, 1991). Its shallow bathymetry has been attributed to the presence of a low-velocity, low-density volcanic layer, about 2 km thick, overlying a crust made up of mid-ocean ridge basalts (MORB) (Bentley, 1974). Major and trace elements show that the bulk of the Cocos ridge consists of tholeiitic lavas similar to those found on the Galapagos Islands (Castillo, 1984; Castillo et al., 1988). Cocos Island, which developed on the northwest flank of the Cocos ridge during the late Pliocene, is composed of an alkali basalt-trachyte sequence (Castillo et al., 1988). Trace elements in these rocks differ from those obtained from the Galapagos, suggesting a different mantle source.

The northern segment of the Nazca plate and southeasternmost corner of the Cocos plate are characterized by unusually high heat flow. Heat flow versus age and distance from the crest of the Galapagos rift zone (spreading at about 3 cm/yr) are shown in Fig. 5. The solid line depicts the theoretical heat flow predicted by the spreading rate in accordance with the model developed by McKenzie and Sclater (1969). The line formed by diamond symbols shows the arithmetic means of heat-flow values with standard errorbars and indicates that actual heat flow significantly surpasses the theoretical values north of the spreading axis at distances of more than 250 km. The pattern suggests subduction of a warm slab.

Lonsdale and Klitgord (1978) believed that the Cocos Ridge entered the Costa Rican segment of the Middle America trench during the Pleistocene, implying that the ridge should have penetrated less than 50 km below southeastern Costa Rica. In this case, uplift associated with subduction of the thickened oceanic crust would only have effected the Osa and

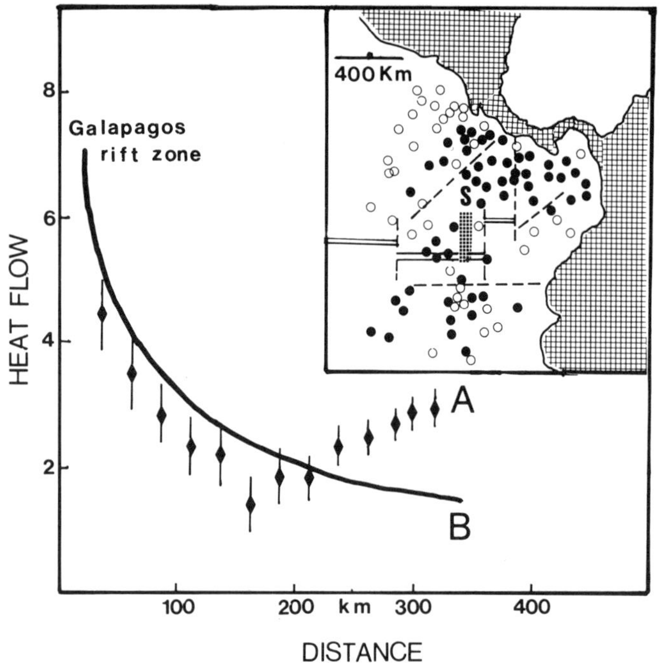

Figure 5 Heat flow in the northernmost segment of the Nazca plate and southeasternmost segment of the Cocos plate. Black dots indicate sites with heat flow greater than the theoretical value computed by using the model developed by McKenzie and Sclater (1969). The full line (B) on the section (S) shows these theoretical values; the line (A) formed by the diamonds indicates arithmetic means and standard error bars. Heat flow at site 250 km or more north of the Galapagos rift zone clearly surpasses that expected using theoretical models. Inset shows site of section S.

Burica peninsulas of the outer arc system, and the high elevation and bilateral symmetry of the Cordillera de Talamanca would remain unexplained.

Middle America trench and Cocos slab

The well-defined morphologic expression of the Middle America trench can be traced virtually uninterrupted offshore of El Salvador and Nicaragua (Fig. 1). It changes abruptly offshore of central Costa Rica. Crowe and Buffler (1983) described a major northeast-trending fault zone (East Nicoya fault zone of Corrigan et al., 1990) southeast of the Nicoya peninsula, which intersects the trench (Fig. 2). This fault zone separates oceanic crust that originated above the Galapagos hot spot (and associated spreading centers) from oceanic crust that originated along the Mathematician-Clipperton Ridge (rough-smooth ocean crust boundary). Protti et al. (this volume) believe that the northeastern continuation of this fault zone, which intersects the hanging block, overlies a major warp in the subducting plate referred to as the Quesada Sharp Contortion zone. East of this cross fault, the trench shallows (from more than 4 to less than 2 km), axes of depocenters shift landward, and its lower-slope morphology becomes very irregular (Crowe and Buffler, 1983, 1985; von Huene et al., this volume). The arc-trench gap is reduced in width from about 160 to 120 km.

Offshore of southeasternmost Costa Rica and western Panama, no clear trench can be distinguished. Its place is taken by a series of relatively narrow thrust belts separated by en echelon north-south–trending strike-slip faults (Heil, 1988; MacKay and Moore, 1990; Silver et al., 1990). The changes reflect major differences in the subduction modes of the Nazca and Cocos plates. Understanding the reason(s) for these differences, however, is complicated by the presence of the Cocos Ridge near their boundary.

The Cocos Ridge is currently moving at a rate estimated to vary from 8 to 9.6 cm/yr (Minster and Jordan, 1978; Kellogg, 1988; Kellogg et al., 1989). P-axes determined from shallow seismic activity in the hanging block above the subduction zone of southeastern Costa Rica suggest convergence at N23°E±15° (Fig. 6), which direction is supported by more-regional slip vector azimuths along the Middle America trench (DeMetz et al., 1990). The angle between the Middle America trench segment offshore of southern Costa Rica and the direction of crustal convergence is almost 90°, implying virtual orthogonal subduction of the ridge below southeastern Costa Rica.

Burbach et al. (1984) and Adamek (1986) found that the Benioff zone below southeastern Costa Rica is poorly defined

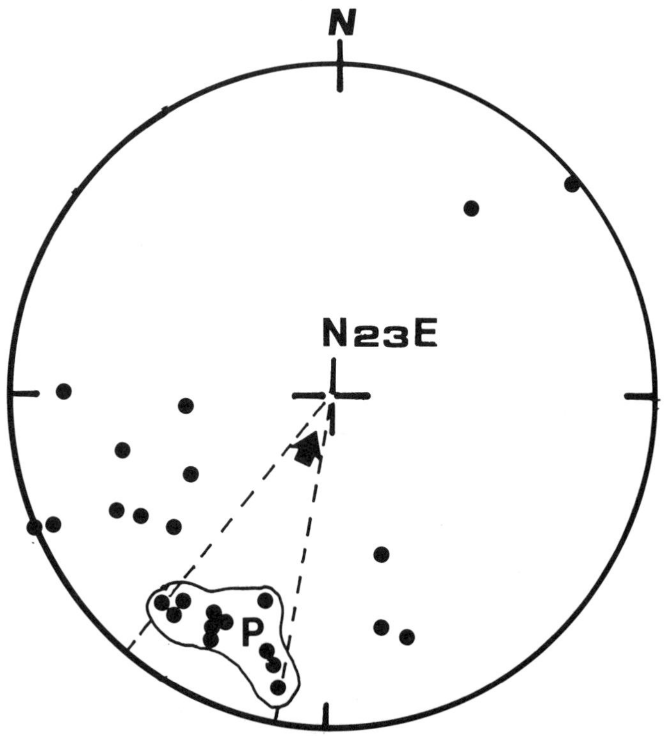

Figure 6. Distribution of P-axes determined from shallow earthquakes below southeast Costa Rica and northwest Panama (Guendel, 1986), indicating north-northeast–south-southwest (N23°E±15°) compression.

because of a general absence of earthquakes that originate at depths greater than 75 km (Fig. 7). Seismic data suggest that the shallow part of the subducted Cocos slab in this region dips from 20° to 25° northeast (Woodward-Clyde, 1980); Cross and Pilger, 1982; Adamek et al., 1987). The distribution of seismic foci below western Panama suggests a slab that dips about 10° north until it reaches a position about 25 km north of El Baru. At this point the slab may steepen to about 33° (Vergara, 1988).

Based on duration of the aftershock period and the size of the areas with aftershock activity that followed two recent thrust events, Adamek et al. (1987) inferred relatively weak, inhomogeneous coupling between the subducted slab and overlying plate. The absence of seismicity at depths greater than 75 km and weak coupling suggest subduction of relatively buoyant (warm) oceanic crust.

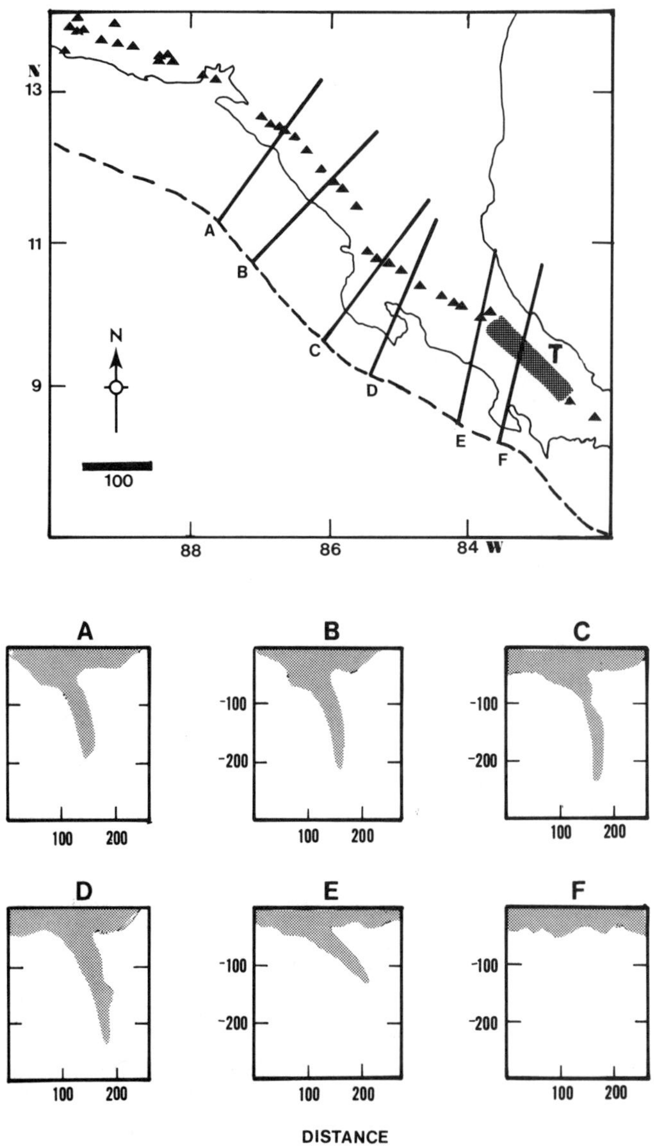

Figure 7. Outline of Benioff zone below El Salvador, Nicaragua, and Costa Rica, indicating lateral southeast shallowing of the subduction angle (Guendel, 1986). T = Cordillera de Talamanca.

Magnetic anomaly maps derived from Magsat data and reduced to sea level show that the magnetic segment of the subducted slab below Central America is generally characterized by a major positive anomaly (Counil and Achache, 1987). Secondary negative anomalies (lows) that overprint the regional positive have been interpreted to represent partially demagnetized segments of the slab near or at major tectonic discontinuities. Source models and corresponding synthetic anomaly patterns show two major discontinuities in Central America. The northern coincides with a seismic discontinuity in the Benioff plane where the Tehuantepec Ridge is subducted. The southern discontinuity coincides with the region below which the Cocos Ridge is subducted and suggests that the slab becomes demagnetized at relatively shallow depth. Thus both the seismic and magnetic signatures of the slab indicate that the shallow segment of the subducted Cocos lithosphere below southeastern Costa Rica (which includes the Cocos Ridge) is much warmer than its apparent age (>20 Ma?) would suggest.

The western boundary between the normal MORB (mid-ocean ridge basalt) lithospheric slab and buoyant Cocos Ridge lithosphere appears to be a subducted north-south–trending fault-flexure that extends from an area below the west coast of the Osa peninsula to below Turrialba, the easternmost volcano of the Cordillera Central (Costa Rica) (Guendel, 1986) (Fig. 8). We refer to this flexure-tear as the Guendel lineament. The Cocos slab below the Chorotega arc is thus divided into two segments along the Guendel lineament. The subducted oceanic crust to the northwest of this lineament is relatively old and "cold," that to the southeast is relatively young and buoyant. Northwest of the Guendel lineament the subduction mode is "normal," and volcanism is due predominantly to partial melting of the mantle wedge (Carr et al., 1990); southeast of this feature, on the other hand, subduction is accomplished by underplating, and volcanism is dominated by magmas derived predominantly from partial melting of the slab itself (de Boer et al., 1988, 1991; Defant and Drummond, 1990; Defant et al., 1991a, b).

Osa Peninsula

The Osa Peninsula (Fig. 3[B] is made up of a mosaic of fault blocks containing three major stratigraphic units: the Late Cretaceous to early Tertiary Nicoya complex, the middle to late Pliocene Osa group, and the late Pleistocene to Holocene Puerto Jimenez group (Lew, 1983; Berrange and Thorpe, 1988).

The Nicoya complex is composed of an imbricated sequence of basaltic lavas and interstratified pelagic limestones, cherts, and argillites intersected by diabase and gabbro intrusions. The sedimentary component of the complex increases from about 10% in the north to about 40% in the southern part of the region (Berrange and Thorpe, 1988).

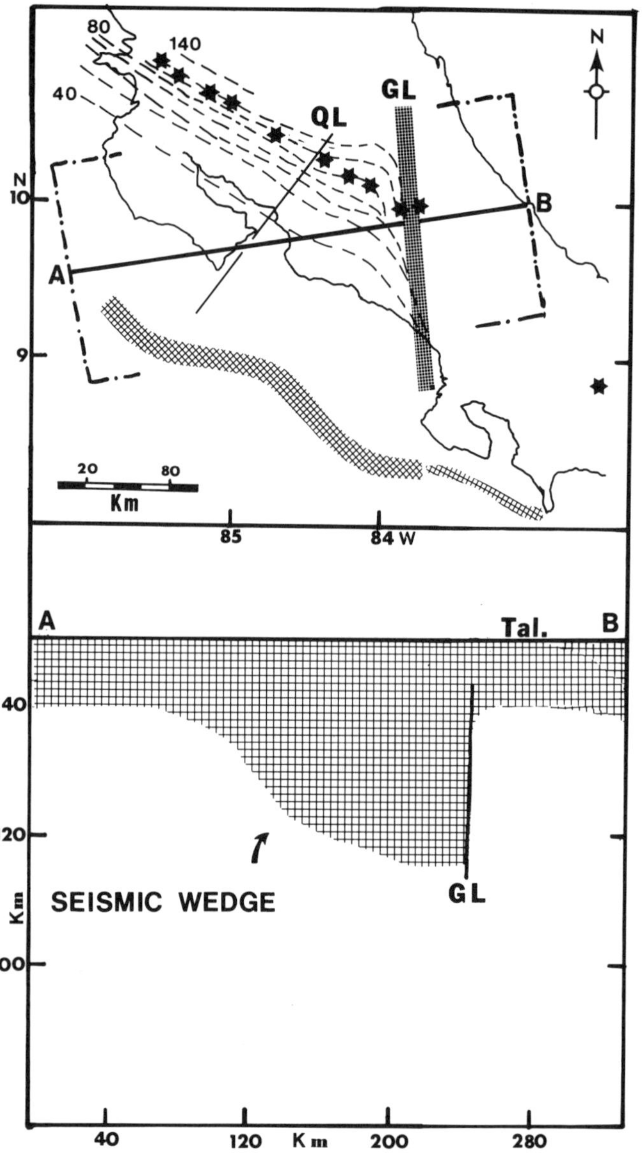

Figure 8. Benioff zone below northwestern Costa Rica and location of a major flexure-fault zone in the slab (Guendel lineament; GL) separating a northwest segment characterized by intermediate subduction angle from a southeast segment characterized by shallow subduction of relatively young buoyant oceanic crust. Dashed lines represent structured contours in the subducting slab (in km). Stars depict individual volcanic complexes. Lower diagram provides the cross section A-B indicating the Guendel discontinuity in the slab. TAL = Cordillera de Talamanca; QL = Quesada lineament.

Geochemical data indicate that "Nicoya-type" flows and intrusions are cogenetic and have affinities with transitional (T) MORB and enriched (E) MORB. Berrange and Thorpe (1988) believe that the overall element enrichment (namely, N-MORB) and enrichment of large ion lithophile (LIL) relative to high field strength (HFS) elements are consistent with their formation in a back-arc environment. It appears equally possible that geochemical enrichment was related to the nearby presence of a mantle source, specifically the Galapagos plume.

Berrange et al. (1988) obtained 20 K-Ar dates from the Osa Peninsula's Nicoya complex. The geochronologic dates concentrate in three groups: 78 to 75 Ma (n=2), 67 to 51 Ma (n=14), and 46 to 41 Ma (n=4). These groups appear to represent not different volcanic pulses but discrete fragments of oceanic crust that originated continuously over a period of more than 35 m.y. and were segmented during subsequent emplacement and imbrication in the outer arc. Obduction may have been a slow continuous process or may have been accomplished in a relatively short period during the late Eocene.

A small fault block on the southeast side of the Osa Peninsula (Punta Carbonera) contains a late Oligocene to early Pliocene sequence of biomicrite limestones and argillites interbedded with lava flows and cut by diabase dikes and sills (Hein et al., 1983; Berrange et al., 1988). The presence of volcanic rocks in the Punta Carbonera fault block implies volcanism during the late Oligocene–early Miocene. Terrigenous clastic components within the sediments containing the volcanic units suggest a shallow depositional environment in close proximity to a volcanic island. It is possible that the Punta Carbonera volcanics represent the obducted remains of a single volcanic complex comparable to Isla de Coco, which developed on the northwest flank of the Cocos Ridge (Castillo, 1984). As such it would not provide much information on the tectonic history of the Osa Peninsula and its emergence.

Middle to upper Pliocene benthic foraminiferal asssemblages in the Punta La Chancha formation (Lew, 1983), imply major eustatic changes of sea level in the late Miocene and/or early Pliocene. Sedimentary data from the Burica Peninsula suggest that it and the Osa block were rapidly uplifted in the late Pliocene and Pleistocene (Corrigan et al., 1990), a process that is continuing (Gardner et al., 1987).

Fila Costena fold/thrust belt

The Fila Costena (Fig. 3[C] is separated from the Osa tectonic unit by major northwest-trending fault zones along which the tectonic units of the Osa ophiolitic sequence have been thrust below the contents of a former forearc basin. The range (maximum height, 1,000 m) consists of a broadly folded imbricated sequence of thrust sheets composed of Eocene to early Pliocene sediments. Exposures in the northwestern and southeastern segments consist respectively of late Oligocene and Miocene clastic sediments and middle to late Eocene calcarenites, suggesting differential uplift. Henningsen (1966a, b) distinguished four formations: the middle to late Eocene Brito, the Oligocene to lower Miocene Terraba, the middle to late Miocene Curre (former Gatun), and the Pliocene Paso Real that unconformably overlies the latter. The Brito sediments were deposited in shallow water on a Nicoya-type basement and include volcanoclastic as well as carbonate debris (Yuan and Lowe, 1987). The Terraba formation is composed primarily of turbidites that contain a significant volume of volcaniclastic debris, probably derived from a volcanic arc in the Talamancas region. The Curre formation is composed mainly of tuffaceous arenites, siltstones, and conglomerates. Its lower section could include deeper-water deposits, but the upper units are fluvial and/or lacustrine (Lowery, 1982). Estimates of cumulative thickness vary from about 2 km (Weyl, 1980) to 3.5 km (Phillips, 1983). If the original basin was comparable in size and depth to the present-day Terraba trough, the latter estimate is more likely. Sedimentary evidence shows that the original forearc basin formed sometime in the middle Eocene and probably reached its maximal depth during the early Miocene, after which time it gradually shallowed. The basin deposits reached erosional levels in the late Miocene.

The Fila Costena is intersected by north-south– and northeast-southwest–trending basalt dikes and massive sills. Their age varies from 15 to 11 Ma (Woodward-Clyde, 1980; Heywood, 1984; de Boer, unpublished research). Because of hydrothermal alteration, it is difficult to determine how reliable these ages are and to establish the original geochemical affinity of the rocks, but we continue to study these occurrences. The Paso Real formation consists of about 500 m of subareal pyroclastic deposits including a few flows (Henningsen, 1966a, b; Tournon, 1984). New K-Ar ages for (fresh) flows in this unit range from 4.3 to 1.2 Ma. Plio-Pleistocene volcanic centers have been located near San Vito, in the Paso Real region (Mano de Tigre complex), and near San Isidro. These complexes are located predominantly along major fault zones separating the Fila Costena from the Terraba trough. All are relatively young, and their geochemistry differs significantly from that of most volcanic rocks in the Cordillera de Talamanca (Defant et al., 1992).

Terraba trough

The Terraba trough (Fig. 3[D]) is separated from the Fila Costena sediment wedge and from the Talamanca Range by major northwest-trending normal and reverse faults. The trough represents a forearc basin, the locus of which has migrated northeastward with regard to its original location in the Fila Costena. The migration of the axis of maximal subduction is probably related to subduction of the Cocos Ridge.

The Terraba inner deep is about 120 km long and 10 km wide and is characterized by a negative Bouguer anomaly. The gravity configuration suggests a northwest deepening of the basin (and northeast dip of basin floor). The maximal depth of the basin is about 6 km. At least two-thirds of its fill may be composed of mid-Eocene to Miocene sediments comparable to those exposed in the Fila Costena. The remainder is of Plio-Pleistocene and Quaternary age.

Limon Sur basin

Fisher and Pessagno (1965) described sections exposed along the Rio Changuinola and Rio Lari, which flow down the northeast flank of the Cordillera to the Caribbean (Fig. 3[F]). About half of the exposed 1,300-m section contains foraminiferal limestones that were deposited in moderate to bathyal depths in close proximity to a volcanic complex (Fisher and Pessagno, 1965). The limestones grade upward into volcanogenic sandstones, which are in turn succeeded by terrestrial andesitic lava flows. The limestones were dated as Late Cretaceous using foraminiferal assemblages. It appears possible therefore that the volcanic section is of early Paleocene age.

A granodioritic body crops out in the headwaters of the Rio Lari. This mass is in fault contact with a thick sequence of relatively thin (30 to 90 cm) andesitic flows and may have been the original feeder. Below these andesites a unit occurs containing at least 200 m of dacitic(?) flows alternating with tuffaceous horizons. Farther downstream Fisher and Pessagno located calcareous tuffs covered by as much as 500 m of coarse pyroclastic breccia. Microfaunas suggest that the upstream limestones are Paleocene, whereas the calcareous strata farther downstream range in age from Paleocene to middle Eocene. Reworked Cretaceous microfossils point to erosion of nearby Cretaceous highlands. Both sections provide evidence for extensive (proximal) volcanic activity during the Paleocene and Eocene. Remains of east-west–trending volcanic arcs of Paleocene age occur on the Azuero Peninsula of Panama and the Nicoya Peninsula of Costa Rica (de Boer, 1979). It appears possible that the Changuinola and Lari sections were deposited in close proximity to a segment of this arc subsequently covered by the Neogene volcanic products of the Cordillera de Talamanca.

The Rio Lari formation is unconformably(?) overlain by the late middle to upper Eocene Tuis formation, which consists of polymictite, dolomitized limestones, shales, and siltstones with an ashy matrix, and by the Oligocene Senosri formation, which consists predominantly of shaly limestones, calcareous shales, and algal limestones. The Senosri is conformably overlain by the early Miocene Dacli formation, a calcareous siltstone and micritic limestone that contains thick volcanoclastic interbeds deposited at bathyal depth (Winslow et al., 1988). Mafic dikes trending north-northeast intruded the Tuis-Senosri sequence up to and presumably including the lower Miocene Dacli, suggesting a mid-Miocene age. The Dacli formation in the Rio Banano section contains distinct ash layers, indicating the emergence of a volcanic arc in the Talamanca region some time in the early Miocene.

The Dacli formation is in turn overlain by the mid-Miocene Uscari formation, whose type locality crops out along the Quebrada Uscari (a tributary of the Rio Sixaola). The Uscari formation is composed predominantly of shales and includes a few reefal limestones and calcareous sandstone units. Its microfauna indicates deposition in intermediate- to shallow-water depth. To date no volcanic units have been located in this formation.

Basaltic units, however, do occur in the top section of the Senosri and/or basal section of the Uscari in the Quebrada Terciopelo, where the Dacli formation is absent. Basalt clasts in the basal Uscari indicate exposure and erosion, possibly related to a major late Oligocene (ca. 30 Ma) global sea-level drop. Cassell and Sen Gupta (1989) believe the basal sediments of the Uscari formation were deposited on a shallow carbonate bank near an (emerging) volcanic island chain in the late Oligocene. In the early Miocene, parts of the adjacent basin rapidly subsided to deep bathyal or even abyssal depths in which thick sections of distal turbidities accumulated (Cassell and Sen Gupta, 1989; Winslow et al., 1988). Shifts in the planktonic component of foraminiferal assemblages suggest a progressive shoaling of the basin from early Miocene to early middle Miocene (Cassell and Sen Gupta, 1989).

The late Miocene Rio Banano formation varies considerably in thickness and sediment type. Winslow et al. (1988) distinguished at least five facies, most indicative of an open marine, current-swept neritic depositional environment (Cassell and Sen Gupta, 1989). The Rio Banano formation is unconformably overlain by the Pliocene Suretka formation. It is composed mostly of coral reefs that interfinger with siltstones, sandstones, and conglomerates. Debris flows in the basal section of this formation contain basaltic, andesitic, and dioritic clasts as well as pyroclastics and tuffaceous sandstones presumably of Miocene age.

For a recent study of the stratigraphy and paleobathymetry of Neogene sediments in the Limon Sur basin, we refer to Collins et al. (this volume).

Colombian basin

Bowland (1984) and Bowland and Rosencrantz (1988) believe that the oceanic crust of the Colombian basin (Fig. 3[G]) formed in tectonic, igneous, and sedimentary settings similar to those that controlled development of aseismic oceanic rises and seamounts in the western Pacific. The Colombian oceanic crust, which is laterally heterogeneous and dominated by an oceanic plateau, developed during two principal phases of regional igneous activity (Bowland, 1984). The first phase created Early Cretaceous (or late Jurassic) oceanic crust; the second was dominated by middle to late Cretaceous intraplate volcanism.

Bowland's interpretation differs from that by Christofferson (1973), who recognized a series of east-west–trending magnetic anomalies that he believed were representative of north-south spreading in the late Cretaceous (80 to 62 Ma). Bowland (1984) concluded that the plateau crust of the western Colombian basin is probably continuous with the crust below the Cenozoic magmatic arc of south Central America. The oldest radiolarian assemblages on the Nicoya Peninsula occur in the northernmost regions and are Callovian (165 to 163 Ma). Older radiolarian deposits may be present on the Santa Elena Peninsula (Lew, 1983). The ages of exposed radiolarian deposits decrease from the Santa Elena southwestward to the Osa and Azuero Peninsulas. This decrease suggests that despite significant tectonic complexity the obducted seafloor in the outer arc of the Chorotega orogen becomes younger to the southeast. Such southward younging supports Christofferson's (1973) model, which required the presence of an east-west–trending spreading ridge somewhere south of the present-day Colombian basin. Bowland's second magmatic phase in the Late Cretaceous may have resulted from volcanic activity (and oceanic plateau building) related to the Late Cretaceous Galapagos plume (Duncan and Hargraves, 1984).

The most important sedimentation episode in the western Colombian basin began in the latest Eocene and continued into the Miocene (Bowland, 1984). The Oligocene sediment sources were presumably the volcanic terranes in Nicaragua, where tectonic uplift and intense igneous activity dominated. Erosion of the Cordillera de Talamanca supplied large volumes of coarse, volcanogenic sediment to the western Colombian basin during the Miocene (Bowland, 1984).

Bowland recognized two episodes of intraplate deformation. The earliest, probably Oligocene in age, was characterized by compression (wrench faulting and folding) along the northwest margins of the Colombian basin. This episode may coincide with the late Oligocene–early Miocene period of plate tectonic reorganization in the east-central Pacific. The second tectonic episode is post–middle Miocene and best expressed by development of the North Panama deformed belt. This belt is composed of deformed autochthonous Colombian basin sediments occupying the back-arc region of the southern Chorotega and Darien orogenic belts. Buoyant oceanic crust of the Nazca plate (Galapagos) has been and is currently being subducted below Panama (de Boer et al., 1991). A relationship may exist between the shallow subduction of buoyant oceanic crust and the development of a foldbelt in the back-arc region, especially since such a foldbelt is absent east of the northern segment of the Chorotega orogen, below which the subduction mode is normal.

AGE AND GEOCHEMISTRY OF VOLCANIC ROCKS IN THE CORDILLERA DE TALAMANCA

To obtain a better idea of the geologic evolution in the Cordillera de Talamanca, new K-Ar ages (n=14) were combined with those published earlier by Pichler and Weyl (1976), Sutter (1977), Bellon and Tournon (1978), see also de Boer (1981, unpublished research) and Tournon (1984). The new dates (Table 1) plot in three groups. A tholeiitic gabbro from the Cordillera de Talamanca provided a date of 31.6 Ma. A similar date (31.1 Ma) was obtained from a diorite in the El Baru plutonic group. Intrusive complexes of similar age occur farther east in western Panama (Kesler et al., 1977). These rock units clearly indicate the presence of a volcanic arc in the mid-Oligocene. The majority of the dates (n=25) plot between 17 and 8 Ma and represent a major magmatic event in the middle and late Miocene time. The most voluminous Miocene plutons were emplaced in a relatively short period extending

TABLE 1. K-Ar WHOLE ROCK ISOTOPIC DATES

Sample	K_2O*	$^{36}Ar_S$†	$^{40}Ar_{RS}$§	$^{40}Ar_{RS}$ (%)	Isotopic Age** (Ma)
CR17A	2.68	3.94	9.21	44.1	1.07 ± 0.04
C4-62-88‡	2.90	9.1	10.3	27.8	1.10 ± 0.06
C4-37-88	n.a.	n.a.	n.a.	n.a.	1.25 ± 0.03
C4-4-88‡	3.10	18.6	17.1	23.7	1.71 ± 0.09
CR1	2.95	5.66	32.11	65.7	3.37 ± 0.08
C4-39-88	n.a.	n.a.	n.a.	n.a.	3.56 ± 0.08
C3-16-89	n.a.	n.a.	n.a.	n.a.	4.31 ± 0.07
CR34	3.01	25.96	61.27	44.4	6.30 ± 0.24
18-3-90	n.a.	n.a.	n.a.	n.a.	7.81 ± 0.31
14-3-90	n.a.	n.a.	n.a.	n.a.	9.3 ± 0.18
C4-1-88‡	3.90	68.2	118.5	37.0	9.40 ± 0.47
CR8B	1.05	5.72	33.36	66.3	9.83 ± 0.23
CR10	4.48	10.41	147.90	82.8	10.22 ± 0.20
CR9	3.89	12.16	130.40	78.4	10.37 ± 0.21
CR13M	5.05	46.09	180.40	57.0	11.05 ± 0.31
C4-6-88‡	2.11	126.3	80.5	17.7	11.79 ± 0.59
C4-47-88‡	0.27	24.8	18.9	20.5	21.61 ± 1.08
P40	0.19	17.63	19.25	27.0	31.15 ± 2.25
CR27	0.19	15.38	19.56	30.1	31.65 ± 2.00

*Values of K_2O used in isotopic age calculations are weight percent measurements and are considered with a 1.1% precision.
†10^{-10} cm^3/g unit. Subscript S means sample (corrected from blank and wrapping aluminum foil contaminations).
§10^{-8} cm^3/g unit. Subscript R means radiogenic.
**Errors or precisions listed are for 1 sigma standard deviation and are calculated following the equations proposed by Mahood and Drake, 1982.
‡These dates are from Jackson, 1991.
n.a. = not available.

from 12 to 8 Ma. The third concentration of radiometric ages (n=8) ranges from 4 to 1 Ma. The samples that provided these dates are derived from volcanic rocks in the axial zone of the Cordillera and are exposed along the fracture zones that separate the Terraba basin from the Fila Costena. The Escazu pluton on the northwest end of our study area produced ages ranging from 6.3 to 2.2 Ma (n=3) (Kussmaul, 1987, and this study), indicating that this body is a long-lived plutonic system with multiple intrusions.

Ballman (1976) studied a northeast-southwest transect across the central Cordillera from Buenos Aires de Osa to Amubri and located a linear zone of hypabyssal/volcanic dacitic rocks emplaced in/on Miocene monzonites and diorites (Fig. 9). Their tectonic position and degree of weathering imply a Plio-Pleistocene age, which has been confirmed by a 2.8-Ma and 0.9-Ma age obtained for a dacite plug and dacitic flow, respectively (Table 2).

The period between 8 and 4 Ma has provided relatively few dates. This may be an artifact of sample collection or may indicate an actual reduction in volcanic activity.

To study the geologic evolution of the Cordillera de Talamanca in greater detail, we collected 70 samples over a 220-km along-strike expanse, ranging from the Escazu pluton in the northwest to El Baru in the southeast (Fig. 10A and B). Previous petrologic research of igneous rocks in the Cordillera de Talamanca by Weyl (1957, 1980), Dengo (1962b), Ballman (1976), Pichler and Weyl (1976), and Kussmaul (1987) was mainly of a reconnaissance nature and provided limited geochemical data. Like the others, our study is far from complete because geologic mapping is difficult in this mountain range and field relations between different igneous and/or volcanic complexes can only be ascertained occasionally. We believe, however, that our results provide an additional step forward in the knowledge of this important segment of the Chorotega arc.

The following discussion presents new geochemical and petrographic information on the plutonic and volcanic rocks of the Cordillera de Talamanca. A more detailed petrogenetic study complete with whole-rock geochemical, mineral chemistry, and Nd and Sr isotopic data will be presented by Drummond et al. and in a Master's thesis by Boredelon.

On the basis of petrographic, geochemical, and geochronologic information, four distinct igneous groups can be distinguished: tholeiitic gabbros, the El Baru plutonic rocks, calc-alkaline plutonic rocks, and an undifferentiated volcanic group (Table 3).

The mid-Oligocene plutonic group found in the Cordillera is composed primarily of tholeiitic gabbros. The tholeiitic character of these five samples is displayed on the AFM ternary diagram (Fig. 11) and also on Al_2O_3-plagioclase composition diagrams of Irvine and Baragar (1971). Petrographically, these samples are equigranular, olivine-bearing gabbros. Their geochemical signature exhibits arc affinities with LIL element enrichment (Rb, Ba, Th, U, K, La, Sr) and HFS element depletions (Nb and Zr) (refer to Fig. 14A).

Five plutonic samples were collected in the El Baru region of Western Panama. A hornblende granite, two augite-hornblende-biotite diorites, and two hornblende gabbro samples show the diversity of this mid-Oligocene group. All samples are medium grained and equigranular. Compositionally, this group is very similar to the Talamanca Intrusive Suite (Figs. 11, 12, and 13), except that the El Baru samples have negative Zr anomalies.

The greatest number of samples was collected from the Miocene calc-alkaline plutonic group that makes up the core of the Cordillera de Talamanca. Most calc-alkaline plutons are individual epizonal stocks, some of which have been given informal names, e.g., La Division and Chirripo stocks (Dengo, 1962b). These intrusions represent subvolcanic bodies emplaced into a sequence of volcanic rocks and volcanoclastic sediments. It is currently unknown if the stocks merge at depth into a major batholith, but given their geochemical continuity (e.g., Figs. 9, 11, and 12), one may suspect a common source and plumbing system. Dengo (1962b) referred to this group of intrusives as the Talamanca Comagmatic Series because he interpreted the rocks to have been derived from a common parent magma. We concur with Dengo that there is a strong

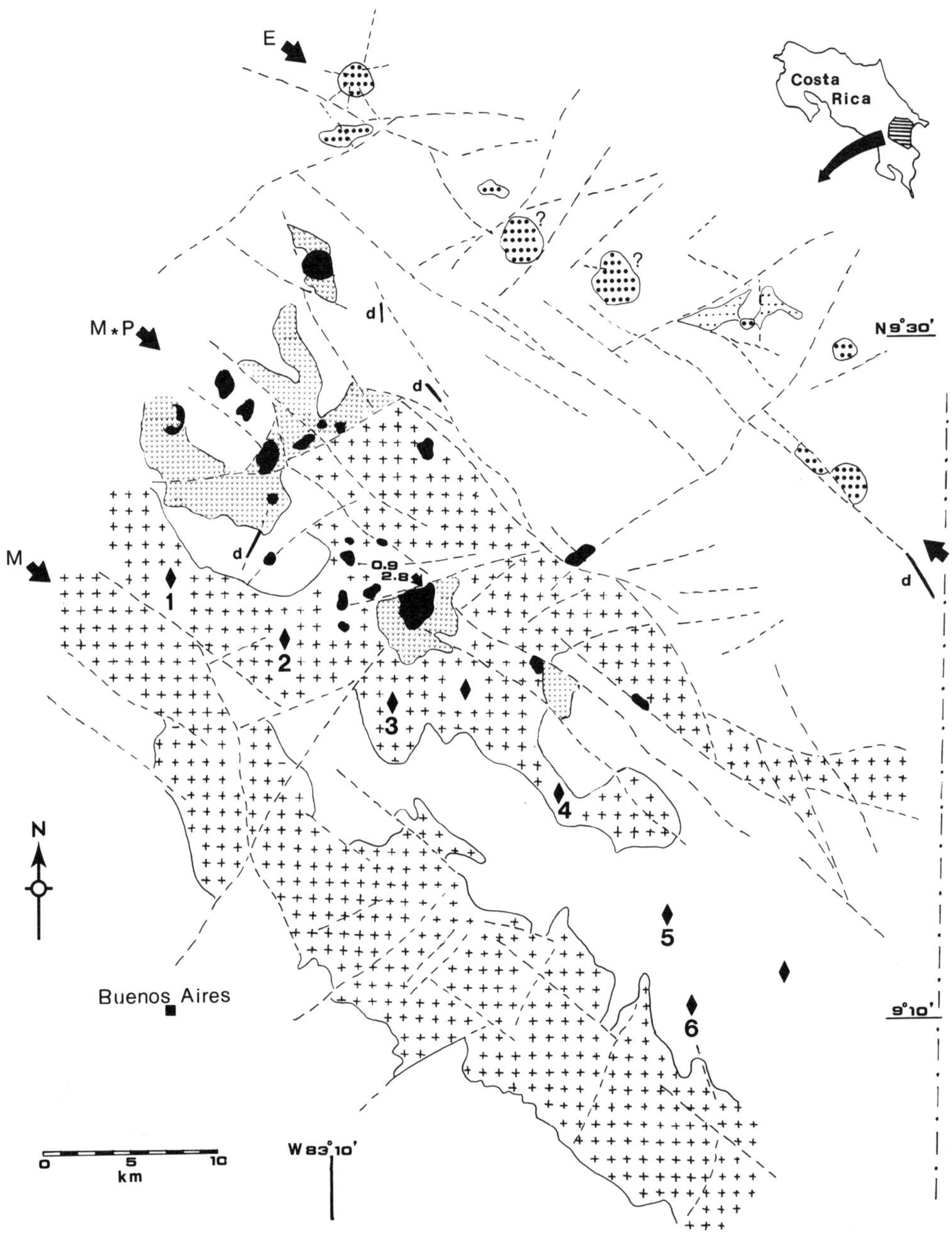

Figure 9. Geologic reconnaissance map of the southeastern segment of the Cordillera de Talamanca. Photogeologic analyses by C. Piazza and fieldwork by P. Ballman and J. de Boer for United Nations Development Project Costa Rica, 1974. E, possible Paleocene-Eocene volcanic complexes; M&P, Miocene plutonic-hypabyssal complexes intruded and covered by Plio-Pleistocene volcanics; M, predominant Miocene plutonic rock units. 1, Cerro Eli; 2, Cerru Betsu; 3, Cerro Arbolado; 4, Cerro Dika; 5, Cerro Nai; 6, Cerro Bakom. White areas: older volcanics and sediments; dashed line: hypothetical faults; d: dikes; numbers: radiometric ages.

TABLE 2. K-Ar WHOLE ROCK ISOTOPIC DATES

Sample Locality	Source	Isotopic Age (Ma)
Durika Area	de Boer, 1985 (unpublished)	0.95 ± 0.01 2.80 ± 0.05
San Vito Area	Tournon, 1984	2.6
Pan American Highway (near Division) Gabbro to granite sequence	de Boer, 1981 (unpublished)	9.3 ± 0.34 9.4 ± 0.28 10.3 ± 0.21 10.7 ± 0.35 11.2 ± 0.27 12.4 ± 0.40
Northwest Talamancas Predominantly plutonic rocks	Tournon, 1984	8.5 8.5 9.3 10.1 11.4 11.5 16.9
Baruca Dam Area Fila Costena	Heywood, 1984 Woodward-Clyde, 1980	11.1 11.6 13.1 14.8

geochemical affinity among the individual stocks; however, it is not known if these plutons came from a common magma chamber. They could represent discrete bodies derived from a common source, the cupola of a composite batholith, or nested plutonic bodies. We will refer to the group of calc-alkaline plutons as the Talamanca Intrusive Suite, following Bayly (1968) and White and Chappell (1983), meaning that these bodies appear geochemically to have been derived from a common source.

The more felsic end members of this group may be classified using the normative Ab-An-Or classification scheme of Barker (1979) (Fig. 13). The granitic rocks are composed primarily of granodiorite and quartz monzonite with subordinate tonalite and granite. The granitic rocks exhibit a variety of textures ranging from hypidiomorphic equigranular and seriate to porphyritic. The more felsic and potassic of the granitoids (quartz monzonite and granite) typically are finer grained and display abundant micrographic intergrowths of quartz and alkali feldspar, indicating low-pressure quenching of near-solidus eutectic material during ascent. Mafic silicate mineralogy includes predominant hornblende and biotite, subordinate augite, and rare orthopyroxenes (ferroan enstatite). Many of the granitic outcrops contain mafic enclaves of two common types: (1) subround to ovoid microgranular basaltic enclaves representing mingling of synplutonic basaltic dikes and (2) angular gabbroic cognate xenoliths of cogenetic gabbro, indicating the order of crystallization between gabbros and differentiated granites. Hornblende-biotite diorites and hornblende gabbros are spatially and genetically associated with the granitoids as part of the Talamanca Intrusive Suite. The texture of the diorite-gabbros is primarily equigranular with minor seriate and porphyritic varieties.

The Talamanca Intrusive Suite defines a medium- to high-K calc-alkaline magma-series trend on Figure 12. The alkali-lime index is 58.5. Most of the samples are metaluminous with the exception of the most highly differentiated granites, which are slightly peraluminous. The samples produce a tight trend on the AFM diagram (Fig. 11), corresponding to trends typical of arcs with intermediate maturity. This implies that by mid-Miocene time, the Chorotega arc was neither an immature oceanic island arc like the Tonga–Marianas–South Sandwich Island arcs nor a mature continental arc like the Cascades–northern Chile–New Guinea group but may have been similar to mature island arcs, such as the Aleutians–Lesser Antilles, or transitional (continental-island) arcs, such as those in New Zealand, Mexico, and Japan.

A chondrite normalized spider diagram (Fig. 14B) of a Talamanca Intrusive Suite composite (n=10) indicates an arc pattern with enriched LIL elements (Rb, Ba, Th, U, K, La, and Sr) and depleted HFS elements (Nb and Ti) typical for melts derived from the mantle wedge. Interestingly, the composite displays a positive Zr anomaly separating this element from the other HFS element behavior. We suggest that the positive Zr anomaly may be due to the metaluminous and relatively high-K character of these rocks, in which the solubility of Zr increases in granitic melts with decreasing aluminum saturation and increasing alkalinity (Watson and Harrison, 1983). Middle and heavy REE (rare earth elements) on the composite spider diagram are relatively flat, indicating no garnet involvement in the petrogenesis. The Talamanca Intrusive Suite also shows nonprimitive geochemical characters with Mg# ranging between 25 and 50 and Ni-Cr contents less than 75 ppm.

The volcanic group is made up of 27 samples that, as expected, show a wide chemical variation because geochronological age constraints for rocks in this group are sparse. Some mafic volcanics are found as synplutonic dikes that cross-cut the Talamanca Intrusive Suite and were dismembered into subround microgranular enclaves, indicating that some of the volcanics may be roughly coeval with the calc-alkaline plutonics. Other mafic volcanic materials have been dated and are much younger than the Talamanca Intrusive Suite. The volcanic rocks exhibit a wide range in texture from aphanitic, glassy and porphyritic to seriate. Some of the seriate and porphyritic samples (n=10) have a fine-grained groundmass and are interpreted to be hypabyssal. To simplify the classification scheme, the hypabyssal samples are given volcanic names.

Classification of the volcanic-hypabyssal group was made using the Peccerillo and Taylor (1976) diagram (Fig. 12). Mafic samples include one absarokite and six basalts (five high-Al basalts and one tholeiitic basalt according to the Alkali index versus Al_2O_3 subdivision of Middlemost [1975]). All of the mafic volcanics are plagioclase phyric. Lithologic diversity among the intermediate volcanics includes two low-K basaltic andesites, three basaltic andesites, one low-K andesite, three andesites, five high-K andesites, and one banakite. A pheno-

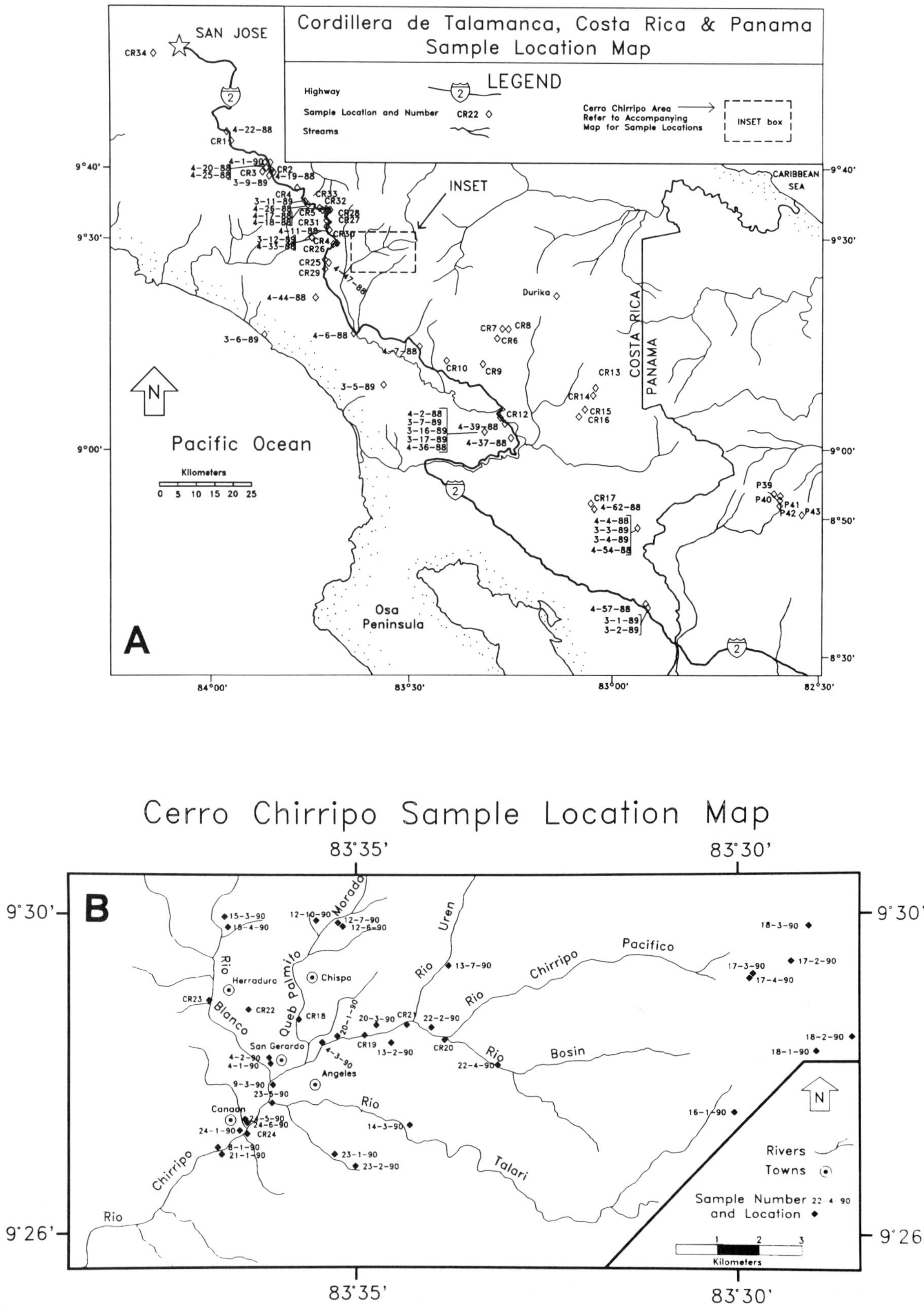

Figure 10. A, Distribution of volcanic and intrusive complexes from which samples were collected for geochemical analyses. B, Inset showing sample locations on Cerro Chirripo complex.

TABLE 3. REPRESENTATIVE WHOLE-ROCK CHEMICAL ANALYSES OF THE TALAMANCAS

Sample	CR27	P40	CR1	CR17A	TISa†
SiO_2	44.6	52.4	56.0	58.1	58.73 (5.50/33)
TiO_2	0.67	0.86	0.67	0.74	0.66 (0.14/33)
Al_2O_3	17.8	17.8	16.7	16.3	16.43 (1.05/33)
Fe_2O_3*	12.3	10.3	7.43	5.96	7.36 (2.19/33)
MnO	0.17	0.19	0.15	0.10	0.16 (0.05/33)
MgO	7.59	3.83	3.98	3.33	2.81 (1.20/33)
CaO	14.9	9.34	7.89	6.95	6.05 (2.29/33)
Na_2O	1.02	2.84	2.70	3.37	3.40 (0.38/33)
K_2O	0.19	1.88	2.95	2.68	2.84 (1.54/33)
P_2O_5	0.08	0.39	0.32	0.31	0.22 (0.06/33)
LOI	1.08	0.08	1.31	1.62	0.97 (0.41/33)
Total	100.40	99.91	100.10	99.46	99.63 (0.56/33)
Rb	20	50	80	55	78 (47/33)
Sr	620	760	1,200	1,905	616 (194/33)
Ba	160	700	1,300	1,300	1,126 (437/33)
Sc	38.1	24.0	20.1	13.3	15.7 (6.6/20)
Zr	10	90	150	120	145 (85/33)
Y	6.9	20	17	12	22 (6/33)
V	320	240	200	175	130 (71/20)
Nb	1.5	6.3	8.9	8.1	8.5 (4.1/29)
Ta	<1.0	1.0	1.0	1.0	1.2 (0.4/5)
Cr	40	20	50	60	19 (15/31)
Co	49	78	71	51	57 (21/20)
Ni	31	34	59	79	40 (20/25)
Cu	993	170	126	150	86 (84/20)
Zn	800	93	72	79	74 (29/20)
Li	10	21	19	23	25 (8/20)
Be	1.8	2.3	3.0	2.5	2.2 (0.5/20)
B	36	16	28	21	21 (6/20)
U	<1	2	5	4	2.3 (1.3/10)
Th	<1	6	15	15	6.4 (4.0/10)
La	2.9	25.3	46.6	68.9	23.6 (8.1/10)
Ce	5.9	45.1	83.6	124.5	43.0 (13.3/10)
Pr	0.9	5.5	9.3	13.5	5.15 (1.35/10)
Nd	4.9	24.7	36.9	54.9	22.2 (4.7/10)
Sm	1.4	4.7	6.0	8.2	4.3 (0.8/10)
Eu	0.50	1.39	1.73	2.17	1.07 (0.10/10)
Gd	1.5	4.3	4.5	5.1	3.9 (0.5/10)
Tb	0.3	0.6	0.6	0.6	0.65 (0.10/6)
Dy	1.4	3.7	3.1	2.7	3.5 (0.6/10)
Ho	0.32	0.72	0.56	0.45	0.70 (0.12/10)
Er	0.9	2.2	1.6	1.2	2.1 (0.4/10)
Tm	0.1	0.3	0.2	0.1	0.33 (0.08/6)
Yb	0.7	1.8	1.5	0.9	1.9 (0.4/10)
Lu	0.12	0.29	0.23	0.14	0.29 (0.06/10)

†TISa = Talamanca Intrusive Suite average (standard deviation of the items, 1 sigma/number of analyses).

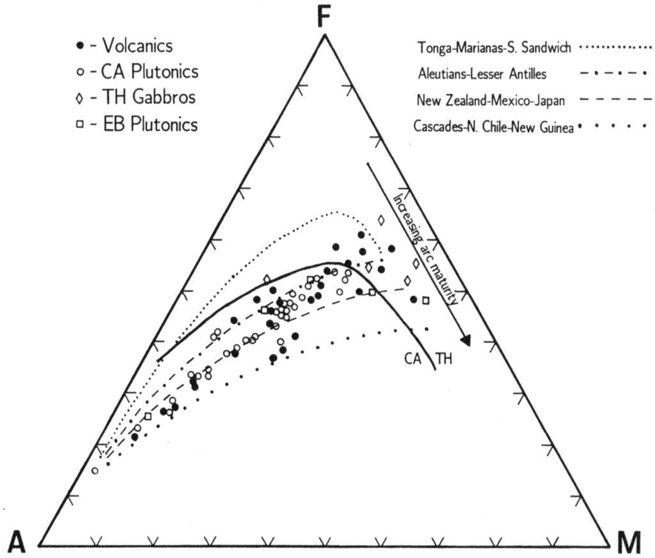

Figure 11. A (Na_2O+K_2O)-F(FeO)-M(MgO) diagram indicating tholeiitic (TH) and calc-alkaline (CA) division and relationship of Talamanca samples to generalized trends from other arc terranes (Brown, 1982). EB = El Baru.

cryst assemblage of plagioclase + hornblende ± clinopyroxene is present in the intermediate volcanics. Three dacites and two high-K dacites represent the felsic group. These felsic volcanics are all porphyritic, typically glassy, with hornblende and plagioclase phenocrysts.

A wide compositional scatter is evident from the volcanic rocks (Fig. 12). Compositional scatter is also shown by the volcanics on the AFM diagram (Fig. 11), where the majority of the samples exhibit a calc-alkaline character. The compositional scatter suggests that multiple generations of volcanic rocks from varied sources are present in the Talamancas. Most samples exhibit a chemical signature of typical mantle-derived arc volcanics. The one-million-year-old sample shown on Figure 14, however, satisfies the compositional criteria for being derived from a MORB-source under amphibolitic to eclogitic conditions, i.e., the subducted plate, as defined by Drummond and Defant (1990) and Defant and Drummond (1990); e.g., Y (12 ppm), Yb (0.9 ppm), and Nb (8.1 ppm); high Sr/Y (159) and La/Yb (77); low Rb/Sr (0.03) and K/Rb (404).

DISCUSSION

At least four major volcanic phases can be distinguished in the Cordillera de Talamanca (Fig. 15B). The volcanic content of uplifted basins that flank the Cordillera suggests that a Paleocene-Eocene volcanic arc occupied the site of the present-day Talamanca Range. So far no intrusive rocks belonging to this volcanic period have been located (and dated) in the Talamanca region sensu stricto. Exposed complexes of this arc, however, occur northwest and southeast of the Cordillera on the Nicoya and Azuero Peninsulas (de Boer, 1979; Sauermann and Lew, 1984). The relatively narrow width of the latter volcanic belts suggests steep subduction of the type currently occurring along the Tonga trench. A period of volcanic quiescence in the late Eocene–early Oligocene probably coincided with a period of general plate tectonic reorganization in the Pacific.

Subduction resumed in the mid-Oligocene. Mantle-wedge–derived magmas rose into the remains of the "Paleocene" arc and into its back-arc basins, suggesting a shallowing of the subduction zone. One of the better examples of such an Oligocene intrusive complex emplaced north of the Paleocene-Eocene arc is in the Cerro Petaquilla of western Panama (Kesler et al., 1977).

A second period of tectonic quiescence appears to have occurred in the late Oligocene–early Miocene, which was also a period of major plate tectonic reconstruction in the east-central Pacific (Fig. 15C). By mid-Miocene time, subduction resumed with full vigor and massive volumes of mantle-wedge–derived magma rose to the surface (Fig. 15B). The greatest volume was emplaced in the central segment of the Chorotega volcanic belt. Why magma generation was more intensive in this segment than in the northwest and eastern segments of this belt remains unknown. The chemistry of the calc-alkaline Oligocene and Miocene igneous rocks is quite similar. Although we have insufficient data at this stage, it nevertheless appears possible to distinguish these igneous suites geochemically by their Zr anomalies. Oligocene calc-alkaline plutonic samples are characterized by negative Zr anomalies, the Miocene plutonic samples by positive anomalies.

"Normal" calc-alkaline volcanism dominated the Miocene. About 5 m.y. ago the subduction angle of the slab started to decrease and a few million years later adakites (and perhaps Nb basalts) derived from metamorphosed MORB crust were emplaced.

The earliest volcanic rocks related to shallow subduction rose along faults on either side of the Fila Costena, indicating the "warm" slab was present below the Osa Peninsula and Fila Costena during the early Pliocene. Igneous complexes related to shallow subduction that intruded the axial zone of the Cordillera de Talamanca appear to be predominantly of late Pliocene and Pleistocene age. The Pliocene-Pleistocene volcanic products can be distinguished from the Miocene and Oligocene plutonic volcanic rocks by their relatively high Sr/Y and La/Yb ratios.

Shallowing of the subduction angle may have forced mantle-wedge material northwest across the Guendel lineament, leading to more voluminous magmatism in the Cordillera Central (Costa Rica). The cumulative volume of Quaternary volcanics in the Cordillera Central by far surpasses

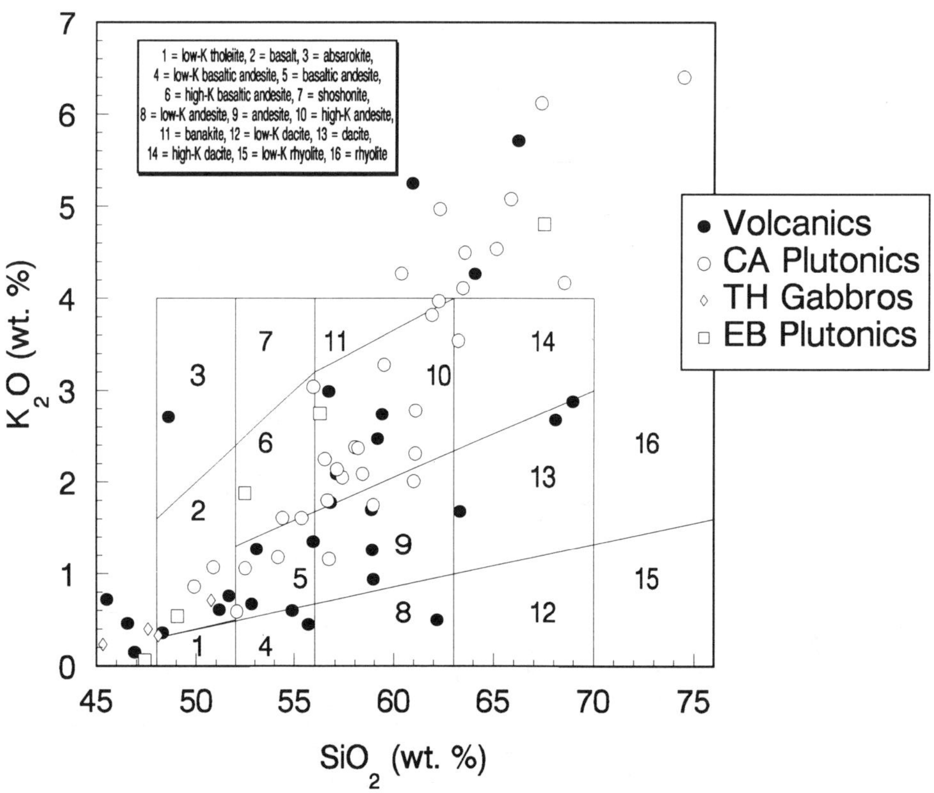

Figure 12. SiO_2 versus K_2O plot of the Talamanca samples. Volcanic rock subdivision of Peccerillo and Taylor (1976).

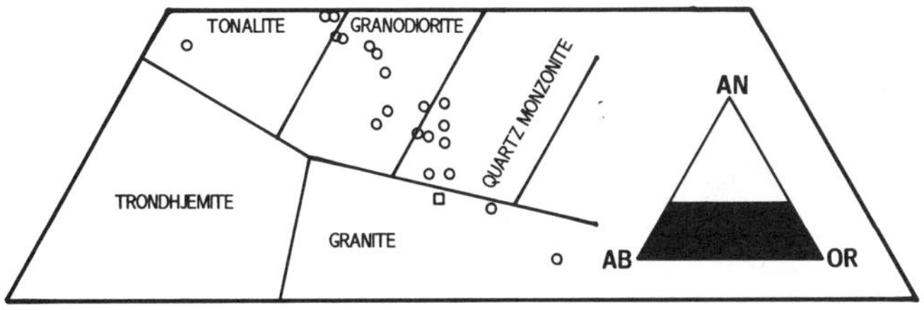

Figure 13. CIPW normative Ab-An-Or granitic rock classification scheme of Barker (1979). Symbols used coincide with those on previous diagrams.

that found in the Cordillera de Guanacaste in northwest Costa Rica, and high Nb basalts have been reported from the Turrialba and Barva volcanic complexes (Reagan and Gill, 1989: Carr et al., 1990), suggesting a mixture of MORB–and mantle-wedge–derived magmas. In a similar fashion mantle-wedge material may have been forced eastward, feeding the massive Baru complex.

The geochemical evolution of the Cordillera de Talamanca rocks thus suggests that the subduction mode of Farallon and Cocos oceanic crust below southern Central America changed over time. During the Paleocene–early Eocene, the slab is interpreted to have dipped relatively steeply north to north-northeast, and a narrow volcanic arc developed (Fig. 16c). In the mid-Oligocene the subduction angle appears to have decreased, and subduction was directed northeast. The Oligocene arc developed in northwest-southeast direction oblique to the west-northwest–east-southeast Paleocene arc, and the principal volcanic centers of the Oligocene arc are offset to the north with respect to the Paleocene arc. During the Miocene, the subduction mode was comparable to that which occurs currently below the Chortis block (Fig. 7). The slab probably dipped at an intermediate angle, and a relatively broad volcanic arc developed on top of the Paleocene-Eocene and Oligocene volcanic complexes in the Cordillera de Talamanca (Fig. 16b). By very late Miocene to early Pliocene time the slab had decreased its dip and had begun underplating the thickened crust of the Chorotega block. Underplating most probably resulted from the introduction of relatively warm (buoyant) ocean crust that originated over the Galapagos hot spot. Magmas originated primarily by partial melting of slab material, and volcanism was significantly reduced in volume (Fig. 16a). The relative sparsity of the latter volcanic units can be attributed to low magma production in the source region, the inability of those magmas to rise through the thickened and elevated crust, and/or lateral escape of magmas to adjacent arc segments, specifically the Irazu-Turrialba and Baru-Tisingal complexes.

The exact timing of the entry of the Cocos Ridge into the Middle America trench remains an enigma. Lonsdale and Klitgord (1978) believe that the Cocos Ridge entered the Costa Rican segment of the trench about 1 m.y. ago. R. H. Kesel and D. R. Lowe (*in* Gardner et al., 1987) stated that "uplift along

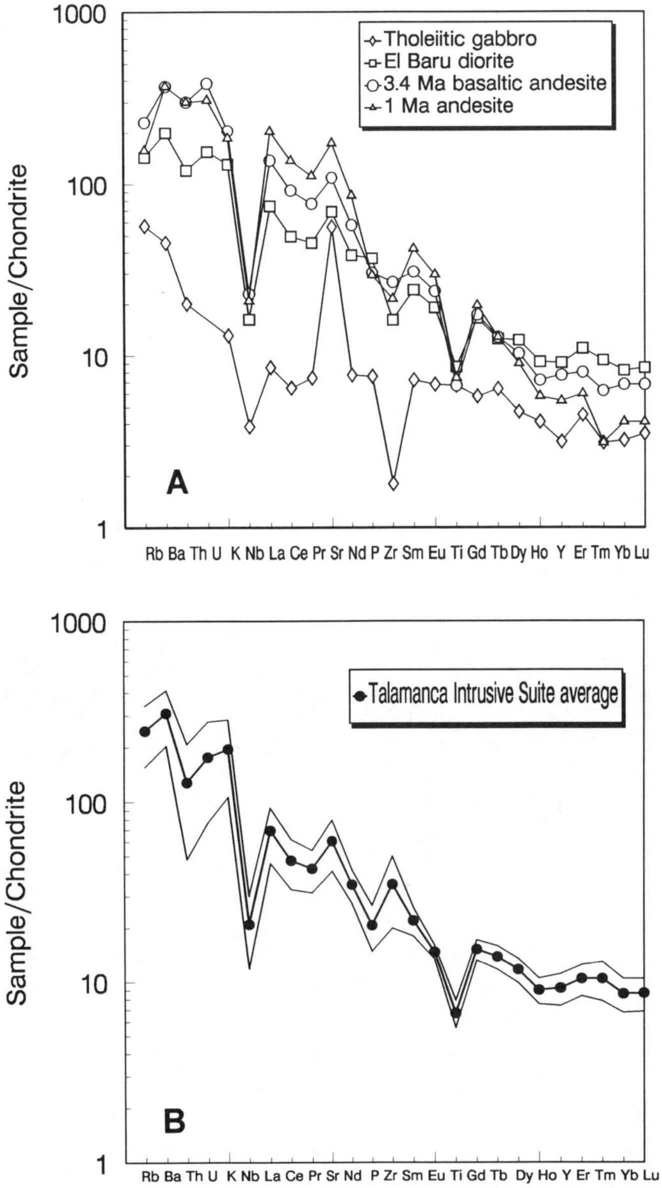

Figure 14. A, Chondrite-normalized spider diagrams of representatives from the tholeiitic gabbro (CR-27), El Baru (P-40), and volcanic groups (CR-1 and CR-17A) of the Cordillera de Talamanca (Table 3). B, Chondrite-normalized (Wakita et al., 1971; Sun and Nesbitt, 1977; Sun et al., 1979) spider diagram of average (Table 2) Talamanca Intrusive Suite plutonics. Standard deviation envelope provides an estimate of the range of each element about the mean.

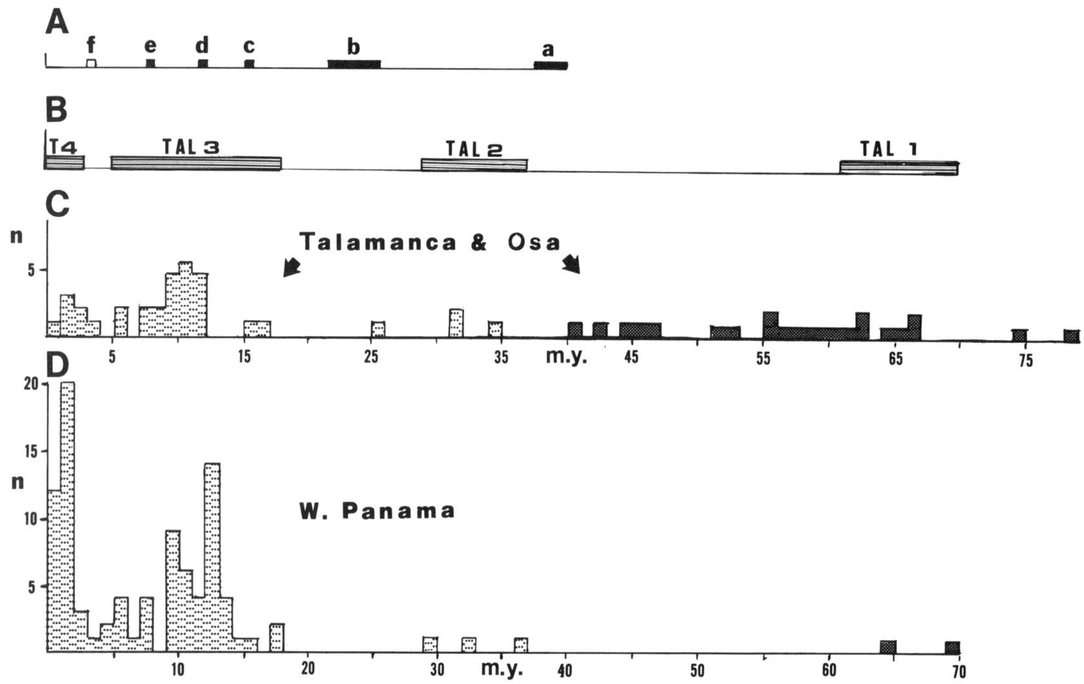

Figure 15. Chronology of tectonic (A) and volcanic (B, C, and D) events. A, Periods with major plate tectonic changes in the east-central Pacific. B, Volcanic periods (1 to 4) in the Cordillera de Talamanca using available geochronologic data for the Cordillera and previously published data for western Panama. C, Isotopic age distribution of samples collected in the Cordillera de Talamanca (light pattern) and on the Osa Peninsula (dark pattern). D, Isotopic age distribution of samples collected in western Panama. K/Ar data from Kesler et al., 1977; Bellon and Tournon, 1978; Weyl, 1980; Woodward-Clyde, 1980; de Boer (1981, unpublished research); Bourgois et al., 1984; Tournon, 1984; Zeta Analytica, 1985; Berrange et al., 1988; de Boer et al., 1988, 1991; Tables 1 and 2.

the southwestern scarps [of the Cordillera de Talamanca] raised upper Miocene rocks from elevations between 400 and 900 m, to elevations in excess of 3,000 m along the present Range crest." They accepted Henningsen's hypothesis that uplift occurred predominantly in the Quaternary, supporting the timing proposed by Lonsdale and Klitgord.

The crustal segment that includes the Burica Peninsula (Pacific coast, Panama–Costa Rica border) subsided rapidly in the period between 5 and 3.5 Ma, at which time it appears to have reached a depth of about 2,500 m. Since the early Pliocene (ca. 3.5 Ma) this region has been rapidly uplifted to exposure level (Corrigan et al., 1990). Kolarsky et al. (this volume) suggest that the Burica shallowing event may be related to either shallow subduction of the frontal segment of the Cocos Ridge or to the nearby presence of an unstable triple junction between the Caribbean, Cocos, and Nazca plates.

Using the paleobathymetry of consecutively dated stratigraphic sections in the Limon and Bocas del Toro basins, Collins et al. (this volume) calculated rates of emergence for the Caribbean coastal region and concluded that subduction of the Cocos Ridge started in the late Pliocene. Using sections on the Escudo de Verguas and Moin and a constant subduction rate, Collins et al. estimated that subduction of the ridge at the Middle America trench began around 3.6 Ma. The data they provided for the Rio Bonano section suggest, however, that subduction of the ridge started about 4.5 Ma.

Wet partial melting of the MORB slab is believed to occur at depths of about 75 to 85 km (23 to 26 kbar) and at temperatures of 700 to 775°C (Drummond and Defant, 1990). Seismic evidence supports this model, as the slab below the Cordillera de Talamanca ceases to be brittle at depth greater than 60 km. Assuming a dip angle of about 30°, the slab has to descend over a lateral distance of about 150 km to reach this depth. Assuming a subduction rate of 9 cm/yr, the descent would take 1.6 m.y. The oldest volcanic rock with characteristics indicating MORB derivation is dated 3.6 Ma (Defant et al., 1991a, b; 1992). This suggests that subduction of relatively young, thickened and buoyant ridge crust (and resulting uplift) started approximately 5 Ma. An early date for subduction of the Cocos Ridge is supported by the presence of a major angular unconformity that has been localized near the Miocene-Pliocene boundary in stratigraphic sections of the uplifted Fila Costena forearc basin (Phillips, 1983; Lowery, 1982; Yuan, 1984). This unconformity appears to reflect a ter-

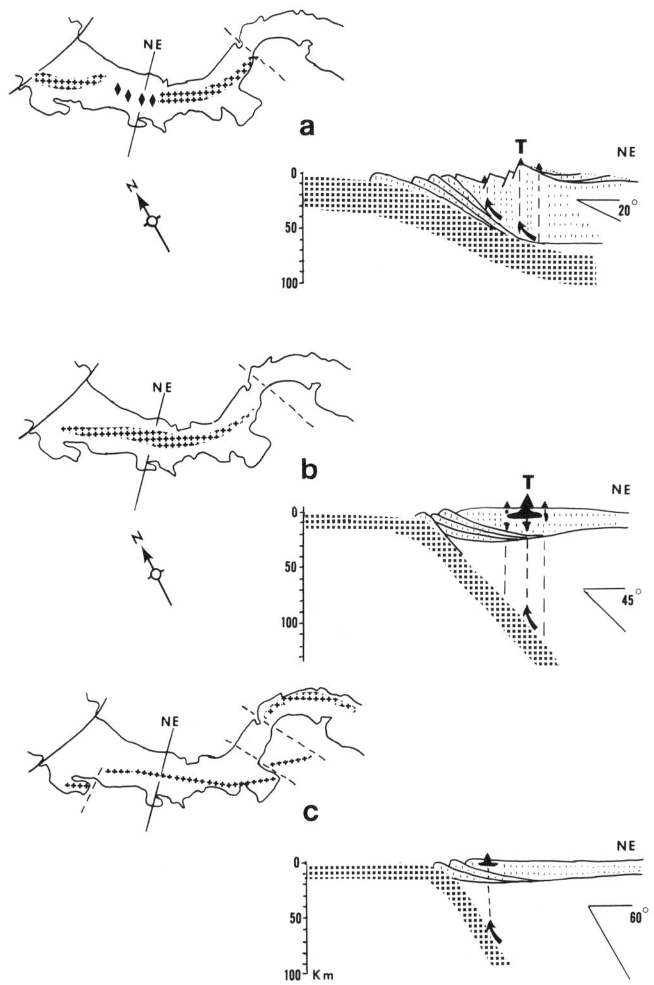

Figure 16. Evolution of the central segment (Cordillera de Talamanca; T) of the Chorotega orogen during the Cenozoic. a, Plio-Pleistocene; b, mid-Miocene and mid-Oligocene; c, Paleocene and early Eocene. Note shallowing of subduction angle over time and thickening of the crust in the hanging block.

mination of subsidence in the fore basin followed by uplift and widespread normal faulting. The earlier date is also supported by the O^{18} content of foraminifera in sediments on the Pacific and Caribbean side of the isthmus, which indicate significant shallowing and formation of the Panamanian land bridge around 4 Ma (Keigwin, 1982).

CONCLUSIONS

(1) The number of geochemical analyses and isotopic age dates for Talamancan rocks has more than doubled with this study. The new data show that this central segment of the Chorotega volcanic arc has been built up by at least three, possibly four, Cenozoic volcanic events. With exception of the earliest (Paleocene-Eocene) event, for which insufficient data are available at this time, each successive volcanic pulse can be distinguished geochemically.

(2) The geochemical evidence suggests major changes in subduction regime. The subduction angle of the Cocos plate lithosphere below southeastern Costa Rica shallowed in the Neogene, resulting in underplating during the Plio-Pleistocene. The Miocene sequence is representative of normal (i.e., mantle-wedge–derived) arc volcanism. Subduction of relatively buoyant warm crust during the Plio-Pleistocene brought the slab into a sub-arc domain where dehydration reactions released LIL element–enriched fluids, causing wet partial melting of the slab itself.

(3) Rapid (differential) uplift of the Talamanca Ridge occurred primarily in the Plio- and Pleistocene and was probably associated with rather localized subduction of a major elevated segment of the Cocos plate: the Cocos Ridge. Bathymetric data and isotopic dates of MORB-derived volcanics suggest that the frontal segment of the ridge entered the Middle America trench offshore of southeastern Costa Rica late in the Miocene or in the earliest Pliocene (ca. 5 Ma).

(4) Deformation of the Cordillera de Talamanca has been dominated by uplift and northeast tilting. The majority of exposed structures along its southwestern flank are normal faults; those on the northeastern flank are low-angle thrust faults and broad folds controlled by gravity sliding (décollement). Most structural elements are significantly influenced (and altered) by development of large gravity slides, many of them relatively young.

REFERENCES CITED

Adamek, S. H., 1986, Earthquake studies in the Panama–Costa Rica region [M.A. thesis]: Austin, University of Texas at Austin, 86 p.

Adamek, S., Tajima, F., and Wiens, D. A., 1987, Seismic rupture associated with subduction of the Cocos Ridge: Tectonics, v. 6, p. 757–774.

Ballman, P., 1976, Eine geologische traverse des ostteils der Cordillera de Talamanca, Costa Rica: Neues Jahrbuch für Geologie und Paleaontologie, v 8, p. 502–512.

Barker, F., 1979, Trondhjemite: Definition, environment and hypotheses of origin, *in* Barker, F., ed., Trondhjemites, dacites, and related rocks: New York, Elsevier, p. 1–12.

Bayly, B., 1968, Introduction to petrology: New York, Prentice-Hall, 371 p.

Bellon, H., and Tournon, J., 1978, Contribution de la geochronometrie K-Ar a l'etude du magmatisme de Costa Rica, Amerique Centrale: Bulletin de la Societe Geologique de France, v. 20, p. 955–959.

Bentley, L. R., 1974, Crustal structure of the Carnegie Ridge, Panama basin, and Cocos Ridge [M.S. thesis]: Honolulu, University of Hawaii, 49 p.

Berrange, J. P., and Thorpe, R. S., 1988, The geology, geochemistry, and emplacement of the Cretaceous-Tertiary ophiolitic Nicoya complex of the Osa peninsula, southern Costa Rica: Tectonophysics, v. 147, p. 193–220.

Berrange, J. P., Bradley, D. R., and Snelling, N. J., 1988, K-Ar dating of the ophiolithic Nicoya complex of the Osa peninsula, southern Costa Rica: Journal of South American Earth Science, v. 2, p. 49–59.

Bourgois, J., Azema, J., Baumgartner, P. O., Bournon, J., Desmet, A., and Auboin, J., 1984, The geologic history of the Caribbean-Cocos plate boundary with special reference to the Nicoya ophiolite complex (Costa Rica) and D.S.D.P. results: A synthesis: Tectonophysics, v. 108, p. 1–32.

Bowland, C. L., 1984, Seismic stratigraphy and structure of the western Colombian basin [M.A. thesis]: Austin, University of Texas at Austin, 195 p.

Bowland, C. L., and Rosencrantz, E., 1988, Upper crustal structure of the western Colombian basin, Caribbean Sea: Geological Society of America

Bulletin, v. 100, p. 534–546.

Brown, G. C., 1982, Calc-alkaline intrusive rocks: Their diversity, evolution, and relation to volcanic arcs, *in* Thorpe, R. S., ed., Andesites—Orogenic andesites and related rocks: New York, John Wiley and Sons, p. 437–464.

Burbach, G. V., Frohlich, C., Pennington, W. D., and Matumoto, T., 1984, Seismicity and tectonics of the subducted Cocos plate: Journal Geophysical Research, v. 89, p. 7719–7735.

Carr, M. J., Feigenson, M. D., and Bennett, E. A., 1990, Incompatible element and isotopic evidence for tectonic control of source mixing and melt extraction along the Central American arc: Contributions to Mineralogy and Petrology, v. 105, p. 369–380.

Cassell, D. T., and Sen Gupta, B.K.S., 1989, Foraminiferal stratigraphy and paleoenvironments of the Tertiary Uscari formation, Limon basin, Costa Rica: Journal of Foraminiferal Research, v. 19, p. 52–71.

Castillo, P. R., 1984, Geology and geochemistry of Cocos Island, Costa Rica: Implications for the evolution of the aseismic Cocos Ridge [Ph.D. thesis]: St. Louis, Washington University, 310 p.

Castillo, P. R., Batiza, R., Vanko, D., Malavasi, E., Barquero, J., and Fernandez, E., 1988, Anomalously young volcanoes on old hot-spot traces. I: Geology and petrology of Cocos Island: Geological Society of America Bulletin, v. 100, p. 1400–1414.

Christofferson, B., 1973, Linear magnetic anomalies in the Colombia basin, central Caribbean Sea: Geological Society of America Bulletin, v. 84, p. 3217–3230.

Corrigan, J., Mann, P., and Ingle, J. C., 1990, Forearc response to subduction of the Cocos Ridge, Panama–Costa Rica: Geological Society of America Bulletin, v. 102, p. 628–653.

Counil, J. L., and Achache, J., 1987, Magnetization gaps associated with tearing in the Central America subduction zone: Geophysical Research Letters, v. 14, p. 1115–1118.

Cross, T. A., and Pilger, R. H., 1982, Controls of subduction geometry, location of magmatic arc, and tectonics of arcs and back-arc regions: Geological Society of America Bulletin, v. 93, p. 545–562.

Crowe, J. C., and Buffler, R. T., 1983, Regional seismic reflection profiles across the Middle American trench and convergent margin of Costa Rica, *in* Bally, A. W., ed., Seismic expression of structural styles: American Association of Petroleum Geologists Studies in Geology, 15, v. 3, section 3.4.2., p. 147–162.

Crowe, J. C., and Buffler, R. T., 1985, Top of acoustic basement and thickness of sediments, Costa Rican margin, *in* Ladd, J. W., and Buffler, R. T., eds., Middle America Trench off western Central America, Woods Hole, Massachusetts, Atlas 7 Ocean Margin Drilling Program, Regional Atlas Series, Sheet 18.

de Boer, J. Z., 1979, The outer arc of the Costa Rican orogen: Oceanic basement complexes of the Nicoya and Santa Elena peninsulas: Tectonophysics, v. 56, p. 221–259.

de Boer, J. Z., Defant, M. J., Stewart, R. H., Restrepo, J. F., and Clark, L. F., 1988, Quaternary calc-alkaline volcanism in western Panama: Regional variations and implications for plate tectonic framework: Journal of South American Earth Science, v. 1, p. 275–293.

de Boer, J. Z., Defant, M. J., Stewart, R. H., and Bellon, H., 1991, Evidence for active subduction below western Panama: Geology, v. 19, p. 649–652.

Defant, M. J., and Drummond, M. S., 1990, Derivation of some modern arc magmas by melting of young subducted lithosphere: Nature, v. 347, p. 662–665.

Defant, M. J., and eight others, 1991a, Andesite and dacite genesis via contrasting processes: The geology and geochemistry of El Valle volcano, Panama: Contributions to Mineralogy and Petrology, v. 106, p. 309–324.

Defant, M. J., and eight others, 1991b, Dacite genesis via both slab melting and differentiation: Petrogenesis of La Yeguada volcanic complex, Panama: Journal of Petrology, v. 32, p. 1101–1142.

Defant, M. J., and seven others, 1992, The geochemistry of young volcanism throughout western Panama and southeastern Costa Rica, an overview: Journal of Geological Society of London, v. 149, p. 569–579.

DeMetz, C., Gordon, R. G., Argus, D. F., and Stein, S., 1990, Current plate motions: Geophysical Journal International, v. 101, p. 425–478.

Dengo, G., 1962a, Estudio geologico de la region de Guanacaste: San Jose, Instituto Geografico de Costa Rica, 112 p.

Dengo, G., 1962b, Tectonic-igneous sequence in Costa Rica, *in* Engel, A.E.J., and Leonard, B. F., eds., Petrologic studies: A volume to honor A. F. Buddington: Boulder, Colorado, Geological Society of America, p. 133–161.

Drummond, M. S., and Defant, M. J., 1990, A model for trondhjemite-tonalite-dacite genesis and crustal growth via slab melting: Archean to modern comparisons: Journal of Geophysical Research, v. B95, p. 21503–21521.

Duncan, R. A., and Hargraves, R. B., 1984, Plate tectonic evolution of the Caribbean region in the mantle reference frame: Geological Society of America Memoir 162, p. 81–93.

Escalante, G., 1990, The geology of southern Central America and western Colombia, *in* Dengo, G., and Case, J., eds., The Caribbean region: Geological Society of America, The geology of North America, v. H., p. 201–230.

Fisher, S. P., and Pessagno, E. A., 1965, Upper Cretaceous strata of northwestern Panama: American Association of Petroleum Geologists Bulletin, v. 49, p. 433-444.

Gardner, T. W., and 11 others, 1987, Central America and Caribbean, *in* Graf, W. L., ed., Geomorphic system of North America: Geological Society of America Centennial Special Volume 2, p. 343–401.

Guendel, F. D., 1986, Seismotectonics of Costa Rica: An analytical view of the southern terminus of the Middle American trench [Ph.D. thesis]: Santa Cruz, University of California, 156 p.

Hardy, N. C., 1991, Tectonic evolution of the easternmost Panama Basin: Some new data and inferences: Journal of South American Earth Sciences, v. 4, p. 261–269.

Hardy, N. C., Heath, R. P., and Westbrook, G. K., 1990, A complex plate boundary south of the Gulf of Panama: EOS Transactions of the American Geophysical Union, v. 71, p. 1593.

Heil, D. J., 1988, Response of an accretionary prism to transform ridge collision, south of Panama [M.S. thesis]: Santa Cruz, University of California, 88 p.

Hein, J. R., Kuypers, E. P., Denyer, J., and Sliney, E., 1983, Petrology and chemistry of Cretaceous and Paleocene cherts from western Costa Rica, *in* Iyima, A., Hein, J. R., and Siever, R., eds., Silicous deposits in the Pacific region: Amsterdam, Elsevier, p. 143–174.

Henningsen, D., 1966a, Die Pazifische Kustenkordillere (Cordillera Costena) Costa Rica's und ihre stellung innerhalb des sud-zentralamerikanischen gebirges: Geotektonische Forschungen, v. 23, p. 3–66.

Henningsen, D., 1966b, Notes on the stratigraphy and paleontology of the upper Cretaceous and Tertiary sediments in southwestern Costa Rica: American Association of Petroleum Geologists Bulletin, v. 50, p. 572–580.

Heywood, C. E., 1984, Forearc deformation in southern Costa Rica: A consequence of the collision of the aseismic Cocos Ridge [M.S. thesis]: Santa Cruz, University of California, 104 p.

Irvine, T. N., and Baragar, W.R.A., 1971, A guide to the chemical classification of common volcanic rocks: Canadian Journal of Earth Sciences, v. 8, p. 523–548.

Jackson, T. E., 1991, Neogene geochemistry of the Central American arc: Western Panama and southeastern Costa Rica [M.S. thesis]: Tampa, University of South Florida, 109 p.

Keigwin, L., 1982, Isotopic paleoceanography of the Caribbean and East Pacific: Role of Panama uplift in late Neogene time: Science, v. 217, p. 350–353.

Kellogg, J. N., 1988, Plate motion measurements with satellite geodesy in the Northern Andes: Journal of South American Earth Science, v. 1, p. 419–421.

Kellogg, J., Dixon, T., and Neilan, R., 1989, Central and South American G.P.S. geodesy: E.O.S., Transactions of the American Geophysical Union, v. 70, p. 649, 651, 655.

Kesler, S. E., Sutter, J. F., Issigonis, J., Jones, L. M., and Walker, R. L., 1977, Evolution of porphyry copper mineralization in an oceanic island arc: Panama: Economic Geology, v. 72, p. 1142–1153.

Kussmaul, S., 1987, Petrologia de las rocas intrusivas Neogenas de Costa Rica: Revue Geologica de America Central, v. 7, p. 83–111.

Larson, R. L., and Olsen, P., 1991, Mantle plumes control magnetic reversal frequency: Earth and Planetary Science Letters, v. 107, p. 437–447.

Lew, L., 1983, The geology of the Osa peninsula, Costa Rica: Observations and speculations about the evolution of part of the outer arc of the southern Central American orogen [M.S. thesis]: State College, Pennsylvania State University, 128 p.

Lonsdale, P., and Klitgord, K. D., 1978, Structure and tectonic history of the eastern Panama basin: Geological Society of America Bulletin, v. 89, p. 981–999.

Lowery, B. J., 1982, Sedimentology and tectonic implications of the middle to upper Miocene Curre formation, southwestern Costa Rica [M.S. thesis]: Baton Rouge, Louisiana State University, 100 p.

MacKay, M. E., and Moore, G. F., 1990, Variation in deformation of the south Panama accretionary prism: Response to oblique subduction and trench sediment variation: Tectonics, v. 9, p. 683–698.

Mahood, G. A., and Drake, R. E., 1982, K-Ar dating of young rhyolitic rocks: A case study of the Sierra La Primavera. Jalisco, Mexico: Geological Society of America Bulletin, v. 93, p. 1232–1241.

McGeary, S., Nur, A., and Ben-Avraham, Z., 1985, Spatial gaps in arc volcanism: The effect of collision or subduction of oceanic plateaus: Tectonophysics, v. 119, p. 195–221.

McKenzie, D. P., and Sclater, J. G., 1969, Heat flow in the eastern Pacific and seafloor spreading: Bulletin Volcanologique, v. 33, p. 101–117.

Middlemost, E.A.K., 1975, The basalt clan: Earth Science Reviews, v. 11, p. 337–364.

Minster, J., and Jordan, T., 1978, Present-day plate motions: Journal of Geophysical Research, v. 83, p. 5331–5354.

Nur, A., and Ben-Avraham, Z., 1983, Volcanic gaps due to oblique consumption of aseismic ridges: Tectonophysics, v. 99, p. 355–362.

Peccerillo, A., and Taylor, S. R., 1976, Geochemistry of Eocene calc-alkaline volcanic rocks from the Kastamonu area, Northern Turkey: Contributions to Mineralogy and Petrology, v. 58, p. 63–81.

Phillips, J. S., 1983, Stratigraphy, sedimentology and petrologic evolution of Tertiary sediments in southern Costa Rica [M.S. thesis]: Baton Rouge, Louisiana State University, 153 p.

Pichler, H., and Weyl, R., 1976, Magmatism and crustal evolution in Costa Rica (Central America): Publicaciones Geologicas del ICAITT (Guatamala), v. 5, p. 56–81.

Ponce, D. A., and Case, J. E., 1987, Geophysical interpretation of Costa Rica in mineral resource assessment of the Republic of Costa Rica: U.S. Geological Survey, Miscellaneous Investigations Series Map I-1865, scale 1:200,000, p. 8–17.

Reagan, M. K., and Gill, J. B., 1989, Coexisting calc-alkaline and high niobium basalts from Turrialba volcano, Costa Rica: Implications for residual titanates in arc magma sources: Journal of Geophysical Research, v. 94, p. 4619–4633.

Sauermann, R. P., and Lew, L. R., 1984, A redefinition of the Costa Rican outer arc stratigraphy and its tectonic implications: Geological Society of America Abstracts with Programs, v. 16, p. 644.

Silver, E. A., Reed, D. L., Tagudin, J. E., and Heil, D. J., 1990, Implications of the North and South Panama thrust belts for the origin of the Panama orocline: Tectonics, v. 9, p. 261–281.

Sun, S. S., and Nesbitt, R. W., 1977, Chemical heterogeneity of the Archaean mantle, composition of the earth and mantle evolution: Earth and Planetary Science Letters, v. 35, p. 429–448.

Sun, S. S., Nesbitt, R. W., and Sharaskin, A. Y., 1979, Geochemical characteristics of mid-ocean ridge basalts: Earth and Planetary Science Letters, v. 44, p. 119–138.

Sutter, J. F., 1977, K-Ar ages of Cenozoic volcanic rocks in Central America: Caribbean Geologic Conference, 8th, Abstracts, p. 202–206.

Tournon, J., 1984, Magmatismes du Mesozoique a l'actuel en Amerique Centrale:l'example de Costa Rica; des ophiolites aux andesites [Ph.D. thesis]: Paris, Universite Pierre et Marie Curie, 333 p.

van Andel, Tj. H, Heath, G. R., Malfait, B. T., Heinrichs, D. F., and Ewing, J. L., 1971, Tectonics of the Panama basin, eastern equatorial Pacific: Geological Society of America Bulletin, v. 82, p. 1489–1508.

Vergara, A., 1988, Tectonic patterns of the Panama Block deduced from seismicity, gravitational data and earthquake mechanisms: Implication to the seismic hazard: Tectonophysics, v. 154, p. 253–267.

Wakita, H., Rey, P., and Schmitt, R. A., 1971, Abundances of the 14 rare-earth elements and 12 other trace elements in Apollo 12 samples: Five igneous and one breccia rocks and four soils, *in* Proceedings, v. 2, Lunar Science Conference, 2nd: Geochimica et Cosmochimica Acta, Supplement no. 2, p. 1319–1329.

Watson, E. B., and Harrison, T. M., 1983, Zircon saturation revisited: Temperature and composition effects in a variety of crustal magma types: Earth and Planetary Science Letters, v. 64, p. 295–304.

Weyl, R., 1957, Contribucion a la geologia de la Cordillera de Talamanca de Costa Rica: San Jose, Instituto Geografico, 75 p.

Weyl, R., 1969, Magmatische Forderphasen und Gesteinschemismus in Costa Rica: Neues Jahrbuch für Geologie und Paläontologie no. 7, p. 423–446.

Weyl, R., 1980, Geology of Central America: Berlin, Borntraeger, 371 p.

White, A.J.R., and Chappell, B. W., 1983, Granitoid types and their distribution in the Lachlan Fold Belt, southeastern Australia, *in* Roddick, J. A., ed., Circum-Pacific plutonic terranes: Geological Society of America Memoir 159, p. 21–34.

Winslow, M. A., Vega, L. A., and Hall, R. C., 1988, Tectonic evolution of an actively deforming back-arc basin: The Limon basin of Costa Rica: Transactions, Carribean Geological Conference, 11th, Barbados: v. 19, p. 1–17.

Woodward-Clyde, 1980, Geologic and seismologic investigations for the Boruca dam site, Costa Rica: San Francisco, California (unpublished report, phase 2), 48 p.

Yuan, P. B., 1984, Stratigraphy, sedimentary, and geologic evolution of eastern Terraba trough, Costa Rica [Ph.D. thesis]: Baton Rouge, Louisiana State University, 110 p.

Yuan, P. B., and Lowe, D. R., 1987, Forearc sedimentation in Terraba trough, Costa Rica, Central America [abs.]: American Association of Petroleum Geologists Bulletin, v. 71, p. 630.

Zeta Analytica, 1985, Report on Project I: K/Ar dating of rock samples undertaken by Zeta Analytica at the instruction of I.R.H.E. (Panama): Panama City (unpublished report), 10 p.

Manuscript Accepted by the Society February 7, 1994

Geological Society of America
Special Paper 295
1995

Geochemical constraints on subduction processes in the Central American Volcanic Arc: Implications of boron geochemistry

William P. Leeman
Keith-Wiess Geological Laboratories, Rice University, Houston, Texas 77251
Michael J. Carr
Department of Geological Sciences, Rutgers University, New Brunswick, New Jersey 08903

ABSTRACT

Boron contents were measured in predominantly mafic Quaternary lavas from the Central American Volcanic Arc to evaluate along-strike variations in subduction processes. Despite the significant range in B concentrations (~2 to 46 ppm), B/La ratios vary in a systematic fashion along the arc: Higher values (>1) are typical between Guatemala and northern Costa Rica, whereas low values (most <0.5) typify central Costa Rica and western Panama. Because B/La is highly correlated with $^{10}Be/Be$ ($r^2 = 0.94$, excluding one sample), B/La may be a useful indicator of subduction contributions to the magma sources. If enrichments of both B and ^{10}Be are proportional to the flux of subducted sediment, along-strike variations in B/La suggest at least a twofold variation in this flux with maximum values below western Nicaragua and minimum values below Costa Rica and western Panama where the Cocos Ridge is being subducted. These data may also reflect significant differences in thermal state of the downgoing slab, which in turn differentially affect release patterns of fluids and fluid-mobile trace elements and possibly melting processes beneath different parts of the arc. The following scenario is suggested to explain the geochemical results.

Steep subduction of cold slab beneath the northwestern arc favors more efficient subduction of fluid components to depths beneath the volcanic front. Their release provides fluid-mobile elements to the overlying mantle, which upon melting produces calcalkalic magmas. Shallow subduction of warmer slab beneath the southeastern arc favors shallow release of fluids and limits fluid-related metasomatism of sub-arc mantle beneath the volcanic front. Under such conditions, B/La and Ba/La ratios in the sub-arc mantle vary little from values seen in oceanic island basalts. Magmas in this part of the arc nevertheless display the highest La/Yb and lowest Ba/La and B/La ratios, which are consistent with prior light rare-earth-element enrichment in the source, significantly lower degrees of melting, or a combination thereof. Because some of the largest volcanoes occur in Costa Rica, and magma flux there is nearly an order of magnitude higher than elsewhere in the arc, source enrichment is considered to be the more plausible explanation. It is proposed that Quaternary magma production below Costa Rica involved lithospheric sources containing trapped or "stored" melt components, but this enrichment process is unlikely to have involved arc magmas or subduction fluids because we see no B-enrichment.

Leeman, W. P., and Carr, M. J., 1995, Geochemical constraints on subduction processes in the Central American Volcanic Arc: Implications of boron geochemistry, *in* Mann, P., ed., Geologic and Tectonic Development of the Caribbean Plate Boundary in Southern Central America: Boulder, Colorado, Geological Society of America Special Paper 295.

INTRODUCTION

This chapter reviews salient features of the Quaternary Central American Volcanic Arc (CAVA), summarizes geochemical variations in regionally representative samples, and presents new boron concentration data for these samples. Collectively, these data provide important insights concerning the relative importance of subduction contributions, mantle source heterogeneities, and shallower processes in determining compositions of the arc magmas. These topics are discussed in terms of possible subduction and tectonic controls on the composition of mantle sources for CAVA lavas. Our principal conclusion is that variations in the style of subduction, and to some extent the flux of subducted sediment, appear to exert a major influence on the magmatic products of the arc. We begin with a review of what is known about this arc, then present our geochemical data and discuss the implications.

Physical and tectonic characteristics of the Central American Volcanic Arc

The CAVA extends 1,500 km from western Guatemala, near the Caribbean–North America boundary zone, to central Panama (Fig. 1). The region extending from Guatemala to Costa Rica has been described in numerous review papers (e.g., Carr et al., 1979; Carr, 1984; Carr and Stoiber, 1990), whereas volcanism in Panama is not as thoroughly documented (cf. De Boer et al., 1988; Defant et al., 1992); the following description is summarized from these authors. The modern arc was preceded by pulses of earlier (Pliocene to Oligocene) arc volcanism, the loci of which generally shifted with time. The present volcanic front is defined by some 50 major volcanic cones or eruptive centers, including composite or clustered volcanoes. The distribution of these volcanoes is nearly continuous with exception of a conspicuous 175-km "amagmatic gap" in southeastern Costa Rica and western Panama. Numerous smaller monogenetic vents are distributed along and locally behind the main volcanic front. Quaternary volcanism in the frontal arc region is rare but is observed in Panama. A few large rhyolitic caldera complexes occur at wide spacing near or behind the main volcanic front in Guatemala.

Extensive pre-Cretaceous sedimentary strata and pre–late Paleozoic metamorphic rocks of the Chortis block extend beneath the arc in Guatemala and possibly parts of El Salvador and Nicaragua (Donnelly and others, 1990); elsewhere along the arc the basement consists of variably deformed Cretaceous and younger marine sedimentary rocks and tectonically juxtaposed fragments of oceanic and arc crust ranging in age from Cretaceous to Neogene. Locally, the modern arc is constructed on a base of volcanic strata related to earlier pulses of arc magmatism. Volcanoes of the main front display many notable characteristics related to basement geology. Maximum elevations decrease from Guatemala to Nicaragua, corresponding to a decrease in crustal thickness (from 50 to <35 km) and variation in basement geology from continental to oceanic character. Elevations then increase across Costa Rica and, across the amagmatic gap, display a mirror image of decreasing elevation between western and central Panama. The latter pattern is symmetrical about the projected onshore extension of the Cocos Ridge, a thick, hot spot–related oceanic plateau that rises 3 km above the surrounding seafloor. Crustal thickness also reaches 45 km (estimated from gravity data) beneath this region of Central America; in comparison, thickness decreases to <25 km beneath central Panama (cf. Case and MacDonald, 1990).

Northwest of the Cocos Ridge, alignments of CAVA volcanoes define eight discrete segments, each of which is characterized by relatively uniform regional topography and structure. Stoiber and Carr (1973) defined segment boundaries by abrupt discontinuities or changes in alignment strike, some of which coincide with transverse faulting and pronounced topographic changes. They suggested that segment boundaries might correspond to discontinuities in the subducted Cocos plate and that differences in subduction geometry could account for segmentation of the arc. However, subsequent studies (Burbach et al., 1984; Protti et al., this volume) suggest that the slab is laterally continuous with a ~60° dip between Mexico and northern Costa Rica, that shallow subduction (~35 to 40° dip) occurs beneath central Costa Rica and Panama where the Cocos Ridge is being subducted, and that there is a tear or contortion in the downgoing slab beneath Costa Rica, at about 83°55′W. This transition in seismic character is dubbed the Quisada Sharp Contortion (QSC; cf. Fig. 1) by Protti et al. (this volume). These authors also note that northwest of this boundary earthquakes are concentrated within the Benioff zone and the stress regime is extensional, whereas to the southeast most seismic energy is released within the overriding plate and the stress regime is compressive.

A conspicuous volcano-tectonic depression is coincident with the volcanic front in Nicaragua and extends into El Salvador and western Costa Rica, where it is partly filled with young volcanic deposits (Mann et al., 1990). The present volcanic front in Nicaragua appears to have moved trenchward, whereas in other parts of Central America it coincides with the Pliocene volcanic axis. Such vent migration and development of extensional structures could result from local trench rollback and would be consistent with a steeper subduction dip off Nicaragua. Alternatively, segmentation, development of extensional structures, and distribution of CAVA volcanoes could be related to some combination of crustal extension and strike-slip deformation caused by plate interactions (Mann et al., 1990).

Quaternary magmatism is related to near-orthogonal convergence between the Caribbean plate and the subducting Cocos plate (along the Middle America Trench west of the Cocos Ridge) and oblique subduction of a remnant of the Nazca plate (along the Panama Trench east of the Cocos Ridge). Calculated rates for Cocos-Caribbean plate convergence vary from about 60 mm/yr off southern Mexico to nearly 90 mm/yr off Costa Rica; Nazca-Caribbean convergence is more oblique to the margin and occurs at a rate of ca. 54 mm/yr (Minster and Jordan, 1978; DeMets et al., 1990). Depending on the assumed location of the Nazca-Cocos-Caribbean triple junction, the calculated orthogonal component of subduction off Panama is estimated to be between 15 and 40 mm/yr (De Boer et al., 1988, 1991). Another significant difference between the two trench segments is that oceanic crust being subducted below Panama is hotter (average 3.4 HFU); this observation may account for the nearly aseismic subduction in that sector (De Boer et al., 1988). Protti et al. (this volume) point out that the QSC seismic boundary lies on an extension of the bathymetric rough-smooth boundary (RSB; Fig. 1) of Hey (1977), which separates older seafloor of the Cocos Plate from younger, warmer lithosphere related to spreading between the Cocos and Nazca plates. Thus, the QSC apparently represents a sharp transformlike juxtaposition of older (>30 Ma beneath Nicaragua) and younger (ca. 20 Ma beneath Costa Rica) oceanic crust in the subducting lithospheric plate (Fig. 1). The relative buoyancy of the warmer, younger lithosphere may contribute to unusual tectonic and magmatic responses in the vicinity of Costa Rica and western Panama (cf. Protti et al., this volume).

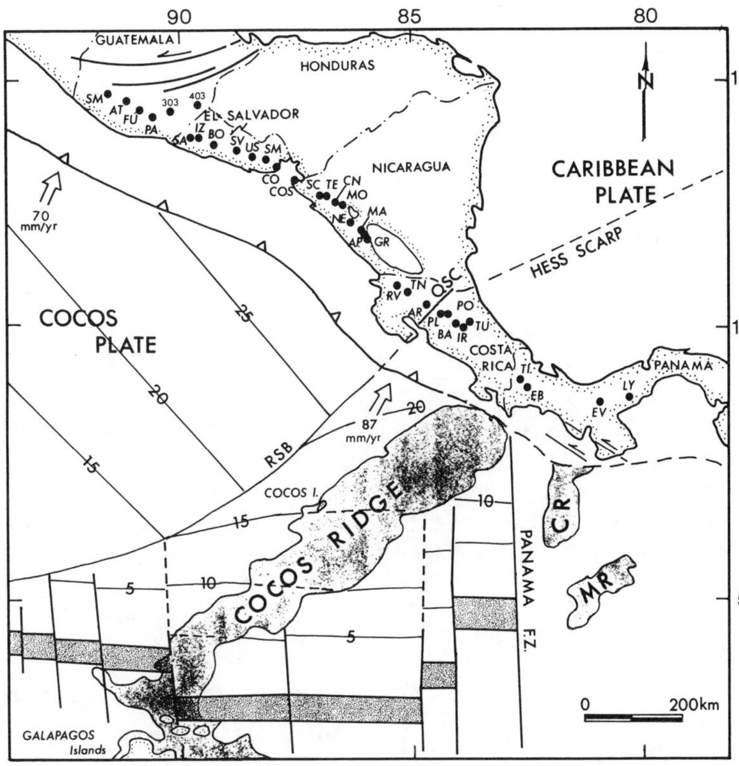

Figure 1. Tectonic framework of the Middle American region and location of volcanoes studied (solid circles). The Middle America Trench is indicated by barbed line extended to offshore Panama as proposed by De Boer et al. (1991). Prominent features include the Cocos, Coiba (CR), and Malpelo (MR) Ridges; the rough-smooth boundary (RSB) of Hey (1977) that separates mostly older crust of the Cocos plate from younger crust formed at Cocos-Nazca spreading centers (stippled); the projection of the Quisada Sharp Contortion (QSC; Protti et al., this volume); various fracture zones (e.g., Panama F.Z.); and isochrons of seafloor age (in Ma). Open arrows indicate convergence rates (Nouvel-1 model, DeMets et al., 1990) near locations where sediment fluxes were estimated. Volcanoes are keyed as follows from west to east: Santa Maria (SM), Atitlan (AT), Fuego (FU), Pacaya (PA), Santa Ana (SA), Izalco (IZ), Boqueron (BO), San Vicente (SV), Usulutan (US), San Miguel (SM), Conchagua (CO), Coseguina (COS), San Cristobal (SC), Telica (TE), Cerro Negro (CN), Momotombo (MO), Nejapa cones (NE), Masaya (MA), Apoyo (AP), Granada cones (GR), Rincon (RV), Tenorio (TN), Arenal (AR), Platanar (PL), Poas (PO), Barba (BA), Irazu (IR), Turrialba (TU), Tisingal (TI), El Baru (EB), El Valle (EV), La Yeguada (LY). Samples from Cordillera Talamancas (between Irazu and Tisingal) are not shown as they are of Pliocene age. Samples numbered 303 and 403 are from behind the volcanic front in Guatemala.

Petrogenetic characteristics

CAVA magmatism is fundamentally basaltic. Despite relatively high densities, which limit eruptive efficiency, basaltic magmas are at least as common as andesites. At most well-studied volcanoes, andesitic and more evolved lavas are readily explained as derivatives from basaltic magmas via crystal fractionation, albeit at higher pressures where crustal thickness is greatest and lower pressure where crust is thin (Carr, 1984). High-silica magmas are largely restricted to caldera complexes in Guatemala where sialic crust is thickest; they likely represent crustal anatectic melts formed in response to injections of mafic magmas. Notable exceptions are Neogene rhyodacitic magmas from the El Valle and La Yeguada complexes in Panama, which Defant et al. (1991a, b) have interpreted as direct melts of unusually hot subducted slab.

Minimum SiO_2 and $(Na_2O)_{52}$ (interpolated at 52% SiO_2) values for individual volcanoes generally correlate with changes in elevation and crustal thickness; both reach minima in Nicaragua and increase in Guatemala and Costa Rica. Magma compositions may be controlled in part by depth of storage and fractionation, most likely near the crust-mantle interface (Carr, 1984; Carr and Stoiber, 1990). Plank and Langmuir (1988) suggest that such compositional variations may result from differences in degree of source melting—i.e., low incompatible element contents reflect high degrees of melting, which in turn correlate inversely with crustal thickness. According to their model, maximum degree of melting would occur beneath Nicaragua. However, incompatible trace element contents (normalized to constant SiO_2 to minimize variation due to fractionation) generally correlate with edifice volume and reach maximum levels in central Costa Rica (Carr, 1984; Carr et al., 1990). Ratios between such elements (e.g., Ba/La, La/Yb) also correlate positively with volume. Yet another important observation is that eruptive volumes for individual volcanic front volcanoes and cumulative magma flux for different segments of the arc both are highest in central Costa Rica where subduction zone dip is the shallowest (Carr et al., 1990); there must exist significant along-strike variations in source composition for this fact to jibe with the model of Plank and Langmuir (1988).

Subduction-zone characteristics

Available seismic profiles and Ocean Drilling Program transects across the Middle America Trench suggest that subduction style varies along strike. Offshore of southern Mexico the convergent margin is characterized by progressive accretion and underplating of trench deposits via imbricate thrusting; only about a third of the incoming sediment is subducted (von Huene, 1984; Watkins, 1989). In contrast, the Guatemala margin is characterized by late Cenozoic tectonic erosion and/or sediment subduction. There are no accreted deposits younger than Eocene and much of the younger trench-fill and underlying plate sediment has apparently been subducted (Moore et al., 1986; von Huene, 1989). During the late Tertiary, sediment accretion and underplating has occurred off Costa Rica, where about three-quarters of the incoming plate sediment is subducted (Shipley and Moore, 1986; Shipley et al., 1990). Using these estimates with current convergence rates (DeMets et al., 1990) and sediment thickness, it appears that sediment was subducted at rates of ~30 km^3/Ma per km of arc off Guatemala and ~22 km^3/Ma per km of arc off western Costa Rica. Approximate sites of these two studies are indicated in Figure 1. Off Panama the accretionary prism thins toward the west (Silver and others, 1990), suggesting diminished availability of sediment approaching the Cocos Ridge. Since at least 1 m.y. ago (possibly 5 m.y.; DeBoer et al., this volume), attempted subduction of the Cocos Ridge may have resulted in a significant decrease in subduction dip and shallowing of Benioff zone seismicity compared to nearby regions in western Costa Rica, development of a 175-km-wide magmatic gap, rapid isostatic uplift, and structural deformation of the forearc region between central Costa Rica and western Panama (Lonsdale and Klitgord, 1978; Corrigan et al., 1990; DeBoer et al., 1991; Gardner et al., 1992; Kolarsky et al., this volume). However, Protti and others (this volume) note that if Cocos subduction began as recently as 1 Ma, at present convergence rates only about 100 km would have been subducted by now. They infer that this process would not in itself account for the tectonic effects observed as far inland as 200 to 300 km and suggest that subduction of younger lithosphere beneath Costa Rica and western Panama is a more plausible impetus. Also, as pointed out by von Huene et al. (this volume), subduction of the Cocos Ridge and numerous seamounts (cf. Castillo and others, 1988) along its flanks likely resulted in a high degree of subduction erosion offshore of south-central Costa Rica, compared to other sectors of the Middle America Trench.

GEOCHEMICAL EVALUATION OF SUBDUCTION PROCESSES

In this study we attempt to further constrain processes of magma generation using the element boron. Boron is useful for the following reasons. It is concentrated in marine sediments and altered oceanic crust and is a potential tracer for subduction of these materials. B is also quite soluble in aqueous fluids compared to the rare earth elements (REE; e.g., La, Nd. Yb) or high-field-strength elements (HFSE; e.g., Zr, Nb, Ti) and therefore may be used in conjunction with other geochemical data to discriminate between effects of magmatic as opposed to hydrothermal processes. In reviewing previous work, we will make the case that ratios such as B/La are controlled mainly by the amount of sediment and/or altered oceanic crust that is subducted and by the temperature-dependent release of fluids and B during progressive subduction. Thus, a geochemical approach may be useful in evaluating relative variations in subduction fluxes and thermal conditions in subduction zones (Moran and others, 1992).

The samples analyzed for boron include 65 basalt to basaltic andesite (<55% SiO_2) lavas, 20 andesites (mainly from Costa Rica), and two dacitic lavas from Panama; all are fresh. Locations of the volcanoes sampled are shown in Figure 1. The primary emphasis of this study is on mafic rocks, which most likely reflect mantle source compositions. Nevertheless, relatively low contents of Ni, Cr, and Sc indicate that most of the mafic lavas studied have evolved by fractional crystallization during their ascent. Phenocryst assemblages in the mafic rocks typically consist of plagioclase + olivine ± augite ± magnetite, and compositional variations in lavas from most volcanoes are consistent with fractionation of these moderate- to low-pressure minerals (Carr, 1984).

All samples were previously analyzed for major and selected trace elements (cf. Reagan and Gill, 1989; Carr et al., 1990; Walker et al., 1990; Defant et al., 1991a, b, 1992; De Boer et al., this volume; M. J. Carr, unpublished). Strontium and neodymium isotopic data are available for many of these samples (Feigenson and Carr, 1986; Carr et al., 1990; Walker et al., 1990; Defant and others, 1991a, b). ^{10}Be and Be data are available for 14 samples (Tera and others, 1986; Morris and Tera, 1989; and J. Morris, oral communication, 1992). Boron contents were redetermined in some samples studied by Morris et al. (1990); both sets of data are in close agreement.

ANALYTICAL RESULTS

Boron concentrations were determined via nondestructive prompt-gamma neutron activation using the reactor facility at McMaster University. Leeman (1988) reports data for geological reference samples analyzed in the course of this study. These data and replicate analyses of selected samples indicate overall precision and accuracy on the order of 5% at concentrations exceeding 10 ppm and ca. 20% at concentrations approaching 2 ppm. Data for boron and other selected elements are presented in Table 1.

Boron concentrations

Most of the observed variation in B content (2 to 37 ppm) occurs in rocks with less than 55% SiO_2, as seen in Figure 2a. Only two andesites have more B; otherwise concentrations do not exceed 20 ppm, even in the dacites. B is anomalously enriched in many CAVA lavas, as seen by comparison with the trend for Hawaiian lavas (basalt to trachyte) from Haleakala volcano (arrow in Fig. 2a). There is a clear geographic variation along the arc front, with distinct B-SiO_2 covariation for Guatemala–El Salvador–Nicaragua versus Costa Rica and Panama. Samples from two Guatemalan volcanoes located behind the volcanic front have lower B contents than corresponding volcanic front volcanoes. This pattern is seen in all cross-arc traverses documented so far (Morris et al., 1990; Ryan et al., 1990; Patino et al., 1993; W. P. Leeman, unpublished). High-Ti basalts erupted from the Nejapa-Grenada cinder cones in Nicaragua (cf. Walker, 1984; Walker et al., 1990) also have lower B contents (<5 ppm) than calcalkalic basalts from nearby larger volcanic front volcanoes. Most basalts from Costa Rica and Panama contain less than 8 ppm B.

Variations between boron and other incompatible trace elements

Because absolute trace element concentrations reflect differing degrees of magma evolution, it is useful to monitor boron enrichments relative to contents of other highly incompatible and relatively immobile trace elements (e.g., rare earths, Zr, Hf, Be). La was used for this purpose because it is well determined in most of the samples studied, whereas data for other incompatible trace elements are not as extensive or have greater analytical uncertainties. Correlations between B and La (Fig. 2b) mimic those seen for B-SiO_2, with high B/La (>1.0) in lavas from Guatemala–El Salvador–Nicaragua and low values (<0.5) for most lavas from Costa Rica and Panama. This geographic control is more evident in Figure 2c. Exceptions to the along-strike trend are the two lavas behind the Guatemalan volcanic front that have lower B/La (0.3 to 0.8) compared to the volcanic front samples.

Despite wide variations in the estimated thickness of the underlying crust (Carr, 1984; Case et al., 1990), there is no obvious systematic relation between this parameter and B-enrichment (e.g., B/La; Fig. 2d). For example, both high (Guatemala) and low (Costa Rica) B/La values occur in regions of relatively thick crust (45 to 50 km), and low B/La and thin crust (<27 km) occur in western Panama. Nearly the entire range of B/La values, and the highest ones observed, are found in western Nicaragua, which has a median crustal thickness (ca. 35 km).

Boron and ^{10}Be systematics

Morris et al. (1990) observed that enrichment of ^{10}Be (as the $^{10}Be/^{9}Be$ ratio) is highly correlated with B/Be in four different volcanic arcs, including Central America ($r^2 = 0.87$, excluding one sample from Masaya volcano; Fig. 3a). These arrays were interpreted in terms of addition of a high $^{10}Be/^{9}Be$–high B/Be subduction component to sub-arc mantle domains in which these ratios originally are relatively small (e.g., B/Be <5). Similar correlations are seen for $^{10}Be/^{9}Be$ versus B/La (r^2 = 0.94; Fig. 3b) or Ba/La ($r^2 = 0.86$; Fig. 3c). Again, the only exception to these trends is the high ^{10}Be sample from Masaya; this volcano is distinctive in other aspects (Carr and Stoiber, 1990), and further work is needed to understand this discrepancy. Finally, $^{10}Be/^{9}Be$ ratios display a geographic dependence similar to that seen for B/La (cf. Fig. 3d and Fig. 2d).

Pursuing the hypothesis of binary source mixing, the low $^{10}Be/^{9}Be$ mantle reservoir seemingly has low B/La (<<1) and moderately high Ba/La (≥20). It has been suggested (e.g., Carr et al., 1990; Lin et al., 1990; Stern et al., 1991) that high Ba/La ratios reflect an increased "slab component" in arc magma

TABLE 1. SELECTED ANALYTICAL DATA FOR CENTRAL AMERICA VOLCANIC ROCKS

Sample	Volcano	Comment*	SiO_2†	MgO†	K_2O†	B	Ba	La
GU SM126	S Maria	VF	52.2	5.28	0.77	12.8	439	9.0
GU AT50	Atitlan	VF	52.2	4.60	0.90	15.2	448	10.5
GU VF74-2	Fuego	VF	49.3	6.35	0.59	12.9	296	5.9
GU E1	Pacaya	VF	51.7	3.52	0.84	8.0	442	10.1
GU P701	Pacaya	VF	51.7	3.52	0.84	8.2	478	11.1
GU C303	Cuilapa	38 km BVF	50.7	7.80	0.58	2.4	179	8.5
GU C403	....	56 km BVF	49.2	6.82	0.74	9.7	327	10.8
SAL SA204	Santa Ana	VF	52.3	4.29	1.41	15.0	549	12.0
SAL IZ108	Izalco	VF	52.7	4.32	0.94	18.8	444	6.6
SAL IZ112	Izalco	VF	52.5	4.65	0.93	17.1	449	7.4
SAL B21	Boqueron	VF	59.6	2.07	2.25	37.0	804	16.5
SAL B6	Boqueron	VF	52.9	4.51	1.26	12.7	513	8.8
SAL AP3	S Vicente	VF	49.6	6.01	0.35	5.7	280	5.3
SAL US2	Usulutan	VF	49.5	5.14	0.71	9.8	378	5.5
SAL SM7	San Miguel	VF	51.9	3.43	0.88	20.0	531	8.6
SAL SM8	San Miguel	VF	50.5	4.78	0.64	11.8	414	6.1
SAL CO3	Conchagua	VF	51.3	5.54	0.53	7.7	514	6.4
NI COS9A	Coseguina	VF	58.6	2.19	1.26	46.0	762	8.6
NI SC2	S Cristobal	VF	49.0	5.93	0.58	7.4	449	5.0
NI SC3	S Cristobal	VF	51.3	3.26	0.80	9.6	605	6.7
NI TE1	Telica	VF	52.7	4.48	1.34	36.0	818	7.0
NI TE3	Telica	VF	49.6	5.91	0.94	30.0	570	5.9
NI TE6	Telica	VF	50.6	6.50	0.89	16.0	396	6.2
NI CN1	C. Negro	VF	50.8	4.73	0.46	13.3	407	3.5
NI CN10	C. Negro	VF	49.5	6.01	0.37	11.0	336	3.2
NI MT101	Momotombo	VF	55.0	4.52	0.92	25.0	625	6.0
NI MT8	Momotombo	VF	53.0	5.06	0.73	19.5	566	5.6
NI NE202	Nejapa	SV	48.2	9.64	0.24	2.9	136	3.4
NI NE203	Nejapa	SV	47.7	11.09	0.14	3.2	89	3.2
NI NE3	Nejapa	SV	50.9	5.60	0.63	6.8	455	6.2
NI NE5	Nejapa	SV	48.9	8.17	0.42	3.6	365	6.6
NI NE6	Nejapa	SV	48.5	8.93	0.41	4.3	363	5.0
NI MS4	Masaya	VF	51.7	5.35	1.18	24.5	781	10.2
NI AP3	Apoyo	VF	48.3	8.65	0.30	5.0	199	4.6
NI GR3	Granada	SV	48.9	7.00	0.14	3.4	117	2.8
NI GR5	Granada	SV	48.7	8.08	0.10	2.1	117	3.4
NI GR101	Granada	SV	47.6	8.85	0.30	4.8	212	5.2
CR RV1	Rincon	VF	57.3	4.52	1.39	12.0	712	18.3
CR TE1	Tenorio	VF	57.8	4.09	1.37	6.8	780	21.1
CR TE9	Tenorio	VF	53.6	4.76	0.76	5.3	484	11.2
CR AR82	Arenal	VF	54.6	4.76	0.64	11.5	504	13.0
AR 102069	Arenal	VF	54.5	4.49	0.59	11.5	512	....
AR 111168	Arenal	VF	56.0	3.23	0.70	14.0	558	....
AR 91968	Arenal	VF	56.0	3.06	0.70	12.0	593	....
AR JT	Arenal	VF	56.4	2.89	0.72	13.0	549	....
AR LE	Arenal	VF	54.1	4.93	0.56	11.0	497	....
AR 185	Arenal	VF	63.3	4.79	0.65	15.0	558	....
AR3	Arenal	VF	54.5	5.10	0.61	10.0	510	....
AR30	Arenal	VF	54.9	4.77	0.65	12.0	534	....
AR57	Arenal	VF	55.2	4.87	0.64	13.0	553	....
AR8	Arenal	VF	54.6	5.08	0.62	12.0	517	....
AR983	Arenal	VF	55.2	4.88	0.65	14.5	565	....
CR PP11	Platanar	VF	50.3	4.63	2.84	6.3	1988	103.0
CR PP3	Platanar	VF	52.9	4.41	1.71	7.9	959	37.4
CR PP7	Platanar	VF	46.1	14.47	1.21	3.6	2863	60.3
CR PP8	Platanar	VF	47.4	10.27	1.33	2.4	646	59.9
CR PP9	Platanar	VF	49.8	8.83	0.81	2.7	552	33.4
CR PO2	Poas	VF	51.3	5.57	0.75	4.7	377	10.7
CR PO10	Poas	VF	53.1	3.94	1.26	6.1	802	29.5
CR B2	Barba	VF	51.0	7.85	1.69	5.3	744	52.9

TABLE 1. SELECTED ANALYTICAL DATA FOR CENTRAL AMERICA VOLCANIC ROCKS (continued)

Sample	Volcano	Comment*	SiO_2†	MgO†	K_2O†	B	Ba	La
CR 12	Irazu	VF	56.1	4.77	1.90	14.0	715	34.9
CR IZ63	Irazu	VF	55.2	5.69	1.98	14.5	747	38.1
CR D2	Irazu	VF	51.3	9.16	1.59	6.0	676	41.4
CR D7	Irazu	VF	51.5	8.55	1.68	7.1	773	44.4
CR T24-2	Turrialba	VF	53.0	6.29	1.48	10.0	670	33.0
CR T35-2	Turrialba	VF	58.5	4.12	2.36	17.0	949	47.0
CR T47	Turrialba	VF	58.5	2.47	3.43	11.0	1218	55.0
CR T59	Turrialba	VF	55.1	6.24	2.39	7.7	858	68.0
CR T7	Turrialba	VF	57.3	4.58	1.42	6.8	794	26.0
CR T85	Turrialba	VF	52.2	6.54	1.10	3.3	524	42.0
CR T91	Turrialba	VF	51.6	6.37	1.35	8.2	643	37.7
CR T37-5	Turrialba	VF	52.8	6.01	1.47	5.8	680	46.9
CR T64	Turrialba	VF	57.9	3.99	2.16	10.0	859	35.4
CR T38-5A	Turrialba	VF	62.0	2.85	2.96	14.1	1059	46.7
CR T70-12	Turrialba	VF	58.3	3.83	1.58	11.9	722	18.0
CR T129	Turrialba	VF	64.4	2.23	3.10	18.8	1117	43.8
CR C8	Turrialba	VF	63.6	3.49	3.10	17.7	1027	38.9
CR CA22	Talamancas	OA	50.2	4.56	0.57	3.4	465	8.3
CR CA37	Talamancas	OA	52.5	5.07	1.37	4.4	715	24.1
CR CA55	Talamancas	OA	50.5	7.33	1.08	4.3	524	40.0
PA T8-6	Tisingal	VF	53.1	6.17	1.46	2.1	913	45.9
PA B10-11	El Baru	VF	51.8	9.46	0.94	4.7	388	16.6
PA Y19-5	Yeguada	VF	53.9	4.49	0.91	10.2	483	10.9
PA Y21-4	Yeguada	VF	51.3	4.60	0.45	2.4	366	9.4
PA Y26-3	Yeguada	VF	50.0	6.36	2.25	6.0	1170	91.8
PA Y9-1	Yeguada	VF	68.0	1.01	2.65	9.7	1630	9.5
PA Y11-10	Yeguada	VF	75.3	0.45	3.21	9.3	1350	20.1
PA V27-5	El Valle	VF	58.1	3.83	1.06	4.4	511	18.6

*VF = volcanic front volcano; BVF = behind volcanic front; SV = high titanium lava from satellite vents adjacent to volcanic front; OA = extinct Piocene arc.
†Major element oxide data are based on analyses recalculated to 100% on a dry basis with all iron as FeO.

sources. However, Costa Rican lavas with Ba/La values of 20 to 40 also have low $^{10}Be/^{9}Be$. This condition could be explained in several ways. First, they could represent melts of a relatively primitive (i.e., "unmodified") mantle source. If so, this source would appear to be two- to fourfold enriched in Ba/La compared to estimates for a primitive oceanic island basalt (OIB) mantle source (Ba/La = 9.9; Hofmann, 1989). Alternatively, low $^{10}Be/^{9}Be$ in Costa Rican lavas could reflect selective offscraping of the uppermost ^{10}Be-rich part of the subducted sediment column, such that the sub-arc mantle was never enriched in ^{10}Be. Finally, the enrichment of Ba/La, if subduction related, could have happened sufficiently long ago (>~10 Ma) to allow decay of any ^{10}Be originally present. It should be noted that many basalts from Costa Rica and Panama have Ba/La ratios (between 10 and 20) that overlap with OIB values. High-Ti lavas from Nicaragua also have relatively small $^{10}Be/^{9}Be$, B/La, and Ba/La ratios, in which respects they could be samples of only slightly modified mantle; yet nearby volcanic front lavas (e.g., Masaya) are enriched in these ratios, thus raising the possibility that subduction processes may have a heterogeneous impact on the mantle beneath volcanic arcs.

Because the only likely source of ^{10}Be is subducted young oceanic sediment, Morris et al. (1990) concluded that both ^{10}Be and B are derived largely from this source in the form of a "subduction component" that has been added to the arc magma sources. The similarities in behavior between $^{10}Be/^{9}Be$ and B/La suggest that these ratios are controlled by similar sub-arc mantle processes, and both may be used to monitor subduction contributions. Based on our much larger B/La data base, arc magma sources beneath Costa Rica and western Panama apparently received smaller subduction contributions than more westerly regions of Central America. Likewise, low-B/La lavas behind the Guatemalan volcanic front contain less slab component than nearby volcanic front lavas. Although other source components or contaminants (e.g., crust) cannot be precluded in explaining other geochemical characteristics of CAVA lavas, their impact on the B systematics may be limited. We will attempt to evaluate relative variations in such processes for the CAVA.

B/La variations with strontium and neodymium isotopic compositions

Costa Rica and Panama lavas with the smallest subduction component (B/La <0.5) have $^{87}Sr/^{86}Sr$ (~0.7035 to 0.7039) and $^{143}Nd/^{144}Nd$ (0.51285 to 0.51305) ratios typical of many OIB (Fig. 4). Slightly higher $^{87}Sr/^{86}Sr$ and lower $^{143}Nd/^{144}Nd$ ratios found in Guatemala volcanic front lavas have been attributed to assimilation of crustal material (Carr et al., 1990). However, it would be difficult to distinguish effects of this process from those associated with subduction of old continent-derived sediment or possible source heterogeneity. The lack of significant isotopic variations with magmatic differentiation at Pacaya volcano (Guatemala) led Bardintzeff and Deniel (1992) to conclude that high-level contamination was not significant there. Slightly elevated B/La ratios in the Guatemala lavas suggest the involvement of some subducted material and thereby support subduction of terrigenous sediment as a possible explanation of the Sr and Nd isotopic data (Carr et al., 1990).

Nicaragua and El Salvador samples with varied B/La ratios have relatively high $^{87}Sr/^{86}Sr$ (most between 0.7038 and 0.7042) compared to their mid-ocean ridge basalt (MORB)–like $^{143}Nd/^{144}Nd$ ratios (~0.5130 to 0.5131). As a group, these lavas define an unusual positive correlation between Sr and Nd isotopic ratios that has been attributed to addition of subducted

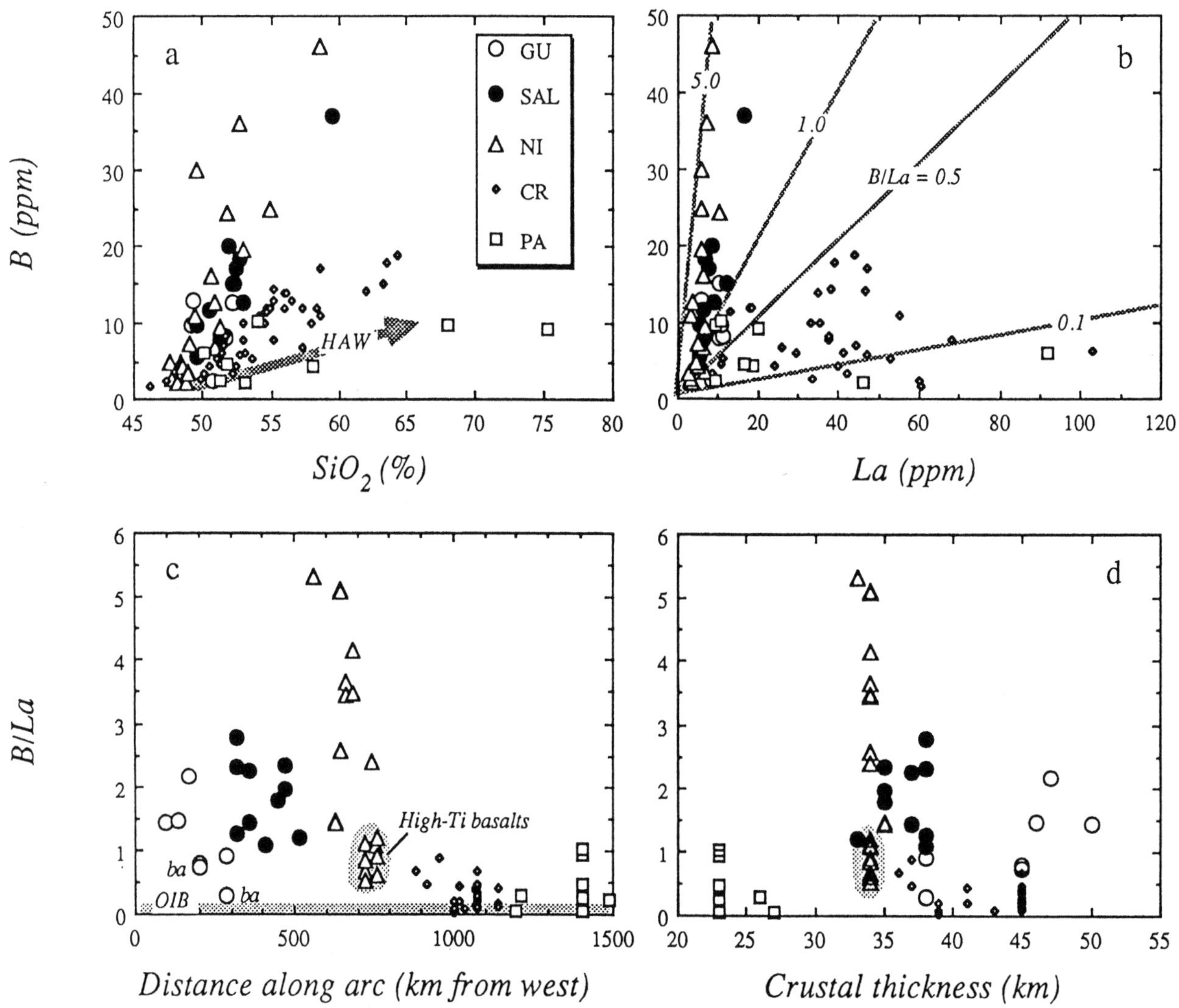

Figure 2. Variation in B content of Central American arc lavas with (a) SiO_2 and (b) La contents and variations in B/La ratios with (c) projected geographic position along the volcanic front (measured from the westernmost volcano in Guatemala) and (d) with crustal thickness. Samples in this and subsequent diagrams are subdivided by country: Guatemala (GU), El Salvador (SAL), Nicaragua (NI), Costa Rica (CR), and Panama (PA). Most samples are from volcanic front localities. SiO_2 values are based on major element analyses recalculated to total 100% on a dry basis with all iron as FeO. Crustal thickness estimates are from Carr (1984) and Case and MacDonald (1990). In a, the arrow labeled HAW shows B-SiO_2 variations for a typical ocean island basalt-trachyte suite (Haleakala). In c and d, high-Ti basalts from Nicaragua are distinguished. B/La ratios in fresh oceanic island and mid-ocean ridge basalts fall within the stippled band in c. OIB is oceanic island basalt.

sedimentary material to fundamentally MORB-like magma sources (Feigenson and Carr, 1986; Carr et al., 1990). This interpretation is supported by a weak positive correlation between B/La and Nd isotopic ratios, which suggests that the most MORB-like source (highest $^{143}Nd/^{144}Nd$) has the greatest subduction contribution (highest B/La). However, Sr isotopic ratios for this group show no significant correlation with B/La. The lowest B/La lavas from this sector (high-Ti basalts) display the greatest range in $^{87}Sr/^{86}Sr$. There is no evidence for significant involvement of a high-$^{87}Sr/^{86}Sr$ seawater or sedimentary component, but we cannot rule out contributions from subducted oceanic crust.

Alternatively, Nicaragua and El Salvador lavas may sample a spectrum of sources approximated by varied "mixing" proportions between (1) an initially MORB-like source that was variably overprinted by additions of B, ^{10}B, and related components brought in via a fluid or melt phase and (2) an OIB-like source with little or no subduction overprint. By "mixing" it is implied that such sources may either be physically mixed on the length-scale of melt formation and extraction or that melts produced from such sources may subsequently mix during ascent and storage. Similar scenarios have been proposed for arcs elsewhere (e.g., Ellam and Hawkesworth, 1988; Stern et al., 1991).

B/La variations with Ba/La, light rare earth element (LREE)–enrichment, and K/B

Carr et al. (1990) found that much of the geochemical variation seen in CAVA magmas can be effectively represented by two parameters: (1) $(La/Yb)_n$ (subscript indicates normalization to chondritic values) and (2) Ba/La. The first reflects the enrichment-level of incompatible elements in the mantle and ranges from values less than one for MORB-source (or depleted) mantle to values greater than two for

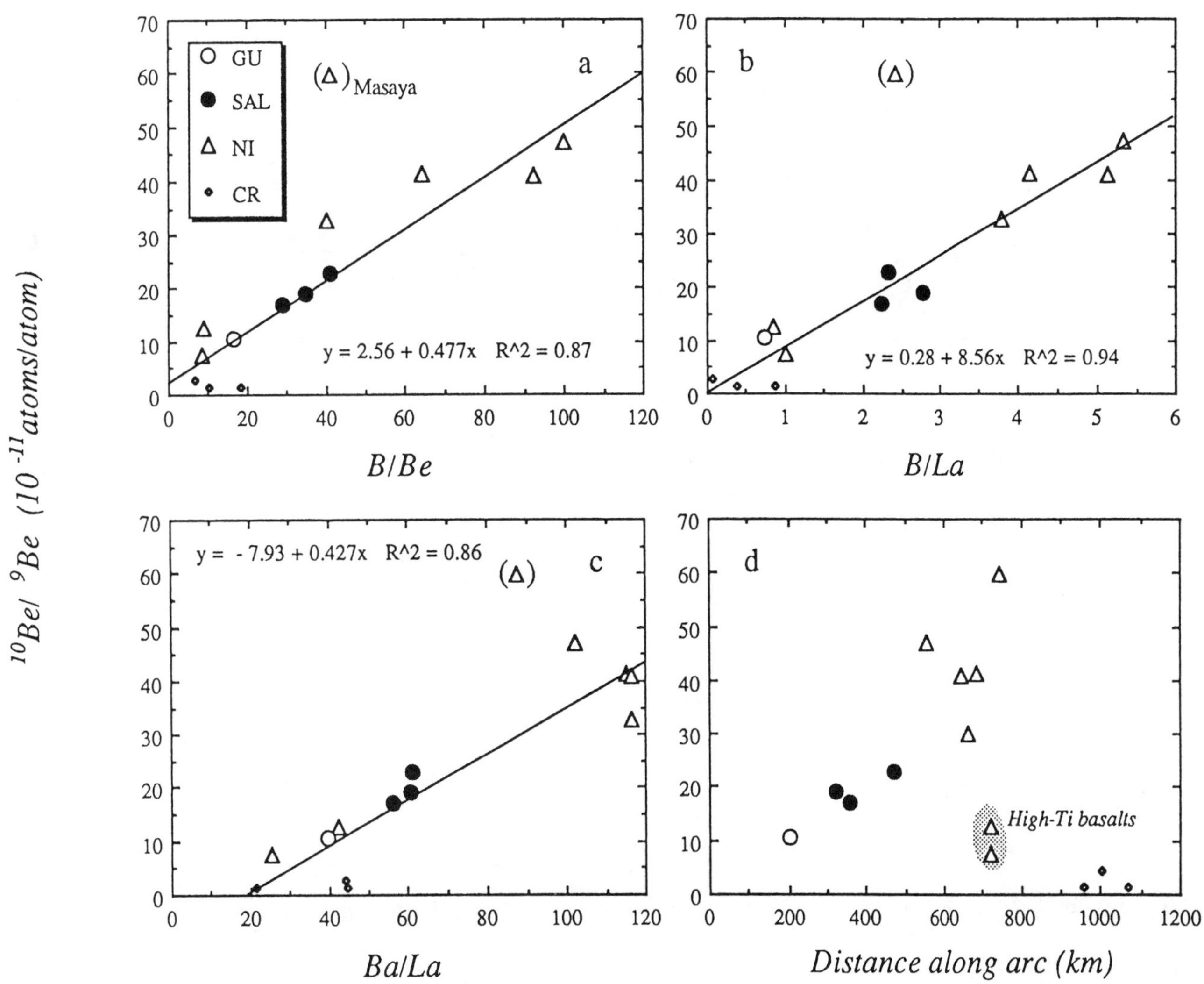

Figure 3. Variation in $^{10}Be/^{9}Be$ ratios of Central American arc lavas with (a) B/Be, (b) B/La, (c) Ba/La, and (d) geographic position along the volcanic front. Regression lines and equations shown include all samples except the one from Masaya volcano.

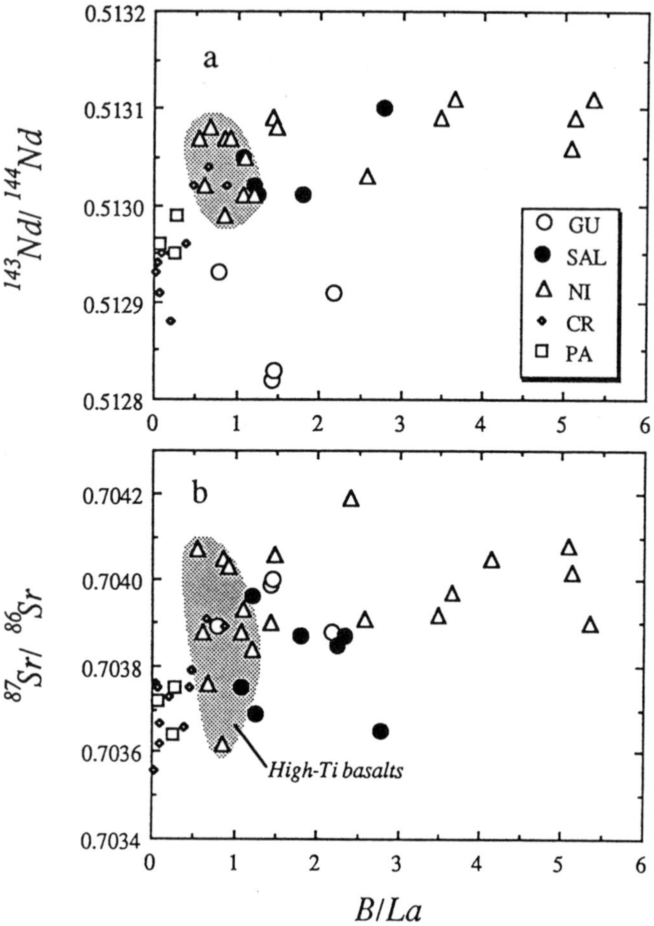

Figure 4. Variation in B/La ratios of Central American arc lavas with (a) neodymium isotopic ratios and (b) strontium isotopic ratios.

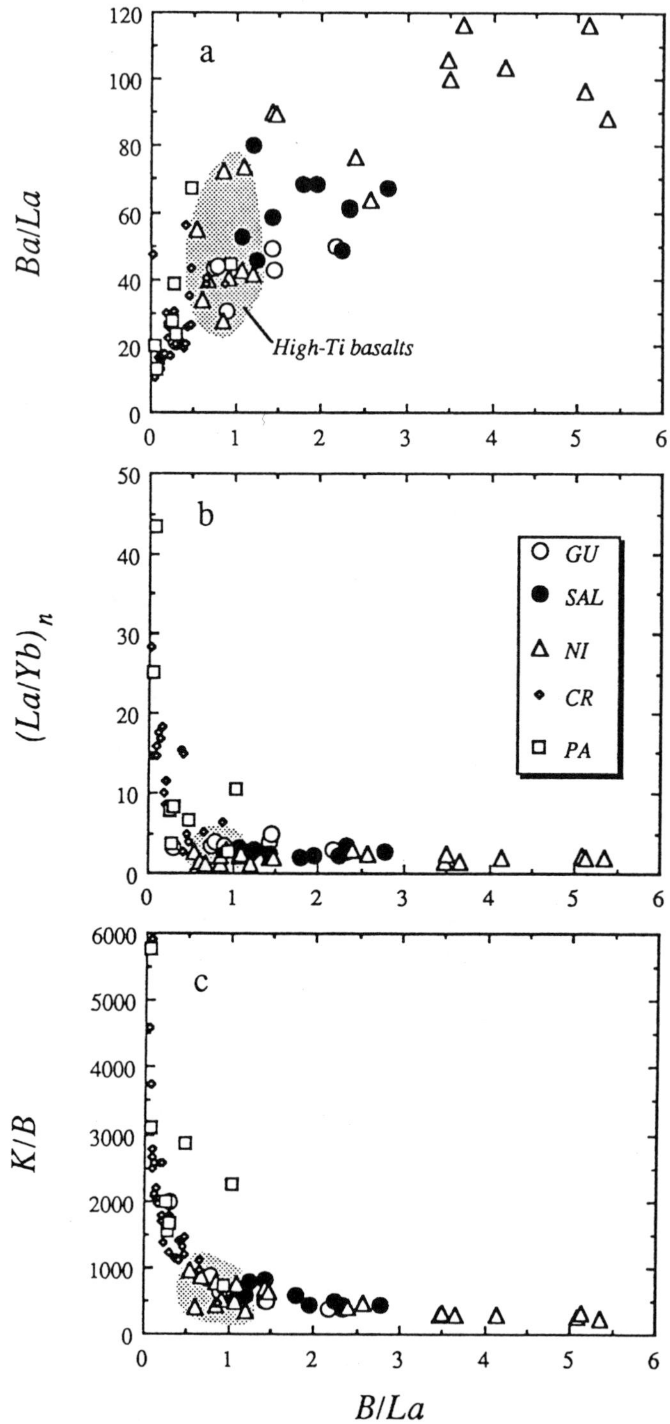

Figure 5. Variation in B/La ratios of Central American arc lavas with (a) Ba/La, (b) $(La/Yb)_n$, and (c) K/B.

OIB-source (or enriched) mantle. These extremes can also be viewed as reflecting greater versus lesser degrees of previous melt extraction. The second ratio monitors degree of enrichment of arc magma sources by incorporation of subducted Ba-rich pelagic sediments (cf. Kay, 1980). Relations between these parameters and B/La are illustrated in Figure 5.

B/La and Ba/La ratios are moderately correlated (r^2 = 0.75; Fig. 5a). In detail, the lavas from Nicaragua define two distinct groups. High-Ti basalts from the Nejapa-Grenada cinder cones define a low B/La (<1.2)–low Ba/La (<70) group that is distinct from the one defined by volcanic front stratocones. Among the latter group, all samples with B/La > 3 are from the western Nicaragua segment of Stoiber and Carr (1973); these lavas show no significant correlation between B/La and Ba/La.

B/La (like Ba/La) is inversely correlated with $(La/Yb)_n$ (Fig. 5b). Most lavas west of Costa Rica have low $(La/Yb)_n$ (<2–3) and relatively low La contents (most <10 ppm), whereas most lavas from Costa Rica and Panama have $(La/Yb)_n$ >3 to 4 and La contents above 20 ppm. K/B ratios also correlate inversely with B/La (Fig. 5c). Potassium is also systematically enriched in lavas from the eastern part of the arc. Nejapa-Grenada high-Ti basalts have slightly higher K/B (ca. 300 to 1000) than basaltic lavas from the nearby volcanic front stratocones (most <500).

Regional variations in lava chemistry

It is also useful to examine the variation in magma chemistry with respect to the difference in subduction geometry and specifically the QSC boundary, which seems to mark a significant change in many physical characteristics of the arc. Figure 6 shows that many geochemical characteristics of CAVA lavas also change at the QSC. To the east, nearly all lavas have Ba/B > 50 and Ba/La< 30. The only conspicuous exceptions are two rhyolitic lavas from La Yeguada (Panama) and some basaltic lavas from Platanar (Costa Rica), all of which have very high Ba (see Table 1). In contrast, nearly all lavas west of the QSC have Ba/B < 50 and Ba/La > 30; exceptions include lavas behind the Guatemala volcanic front and the Nicaragua high-Ti basalts, all of which have low B and high Ba/B ratios compared to their neighbors. Furthermore, relative enrichment of these elements differs between the two geographic areas: La > Ba > B to the east versus B > Ba> La to the west. Because these elements are similarly incompatible in magmatic systems, processes other than melting and crystallization are required to explain these observations.

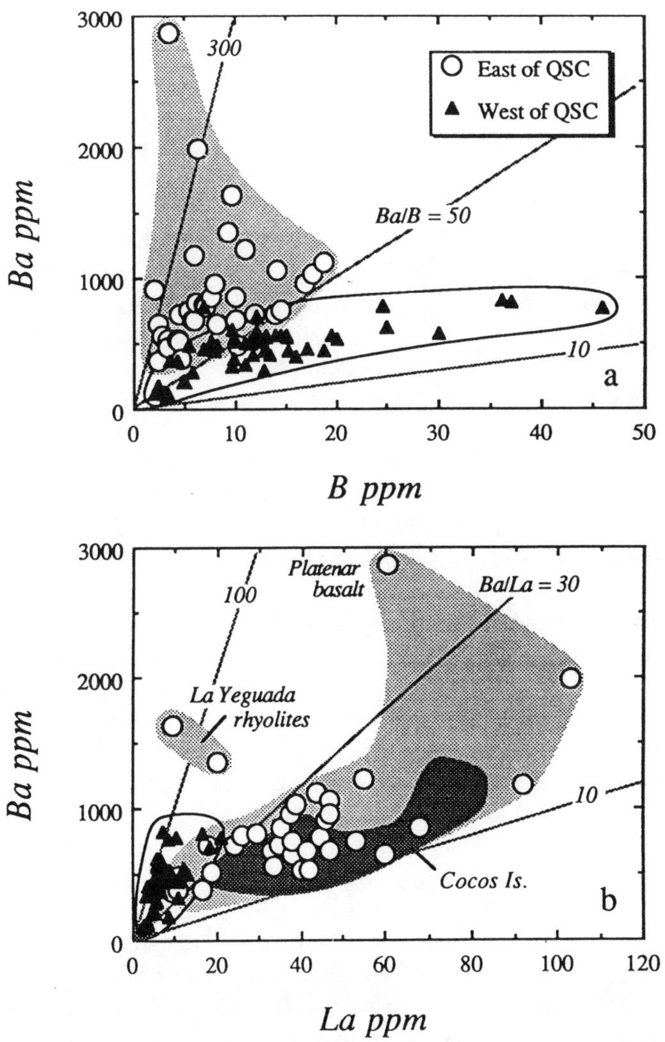

Figure 6. Covariation of Ba with (a) B and (b) La, showing the dramatic differences with respect to the Quisada Sharp Contortion (QSC) geophysical transition. In b the dark stippled field for lavas from Cocos Island is based on data of Castillo and others (1988).

The marked regional differences could be attributed to systematic source heterogeneities beneath the arc, in which case we need to evaluate how they formed and what they mean. The existence of regional-scale source heterogeneities is likely and, judging from the data for Nicaragua basalts, probably occur on a local scale as well. Differential enrichments of incompatible trace elements could also result from differences in degree of partial melting at very low melt fractions. However, we immediately face the conundrum of why magmatic fluxes and incompatible element enrichments both are highest in Costa Rica; this seemingly requires that low-degree melts be derived from a relatively large source domain below Costa Rica compared with that for volcanoes further west (Carr et al., 1990; Feigenson and Carr, 1993). Moreover, the behavior of B is decoupled from that of Ba and La in a manner that is inconsistent with magmatic processes. Other processes are probably important, such as differential mobility of B (relative to Ba and especially La) in metamorphic or hydrothermal fluids (Moran et al., 1992).

DISCUSSION

Regional geochemical variations in CAVA lavas have been discussed most recently by Carr et al. (1990). They suggested that Sr and Nd isotopic ratios of these lavas can be described by mixing between four reservoirs or source components, including marine sediment, relatively depleted MORB-source mantle (DM), relatively enriched mantle (EM), and continental crust. The sub-arc mantle wedge may exist as variably enriched (e.g., E-MORB- or OIB-source) domains (EM) within an otherwise depleted DM matrix. Addition of a small amount of sediment (or fluid derived therefrom) to such a heterogeneous mantle could produce a variably modified mantle (MM) that upon melting would produce lavas having a distinct subduction signature (e.g., high B/La). Lavas lacking a subduction signature could be derived from relatively unmodified parts of the sub-arc mantle. A similar model was proposed for the Cascades arc (Leeman and others, 1990), except that development of MM domains was extremely limited owing to subduction of a relatively hot, fluid-depleted slab and significant offscraping of marine sediment in the Cascadia accretionary prism. Heterogeneous mantle sources were also proposed by Reagan and Gill (1989) for the mantle beneath Turrialba volcano (Costa Rica).

Using their estimated dips for the subducted slab, Carr et al. (1990) suggested that regional variations in incompatible element and isotopic ratios in CAVA lavas correlate with both slab dip and flux of erupted magma. In essence, steep dip and low flux correspond to the volcanoes that tap MM sources, whereas shallow dip and high flux correspond to the central Costa Rica volcanoes that tap relatively unmodified enriched mantle sources (i.e., high EM:DM proportions). It was proposed that EM-dominated magmas formed by decompression melting of fertile domains where convection caused mantle material to flow toward the wedge corner (cf. Spiegelman and McKenzie, 1987). In contrast, MM-dominated magmas were attributed to solidus reduction (or "fluxing") of metasomatized domains by ascending fluids released from the subducted slab. Based on Sr-Nd isotopic systematics, basaltic lavas with intermediate compositions were attributed to mixing of magmas produced from EM and MM sources. Regional variations in the proportions of these magma types were attributed to local convective flow patterns in the sub-arc mantle. It was proposed that where inferred dip of the subducted slab is steep, as in western Nicaragua, restricted lateral flow limits movement of EM-rich mantle (or its melts) toward the volcanic front, so that MM-derived magmas can ascend more or less unmodified. Where shallow subduction occurs, lateral movement of EM-rich mantle (or melt) was considered to be more pronounced. The lack of an MM signature in lavas from Costa Rica thus seemed to reflect the overwhelming influence of EM-derived magma moving into the mantle below the volcanic front.

This model does not completely resolve the nature of subduction contributions to sub-arc mantle. Nor is it clear why low-degree melts are associated with the largest volcanoes or why they are dominant in high magma flux sectors of the arc. Finally, along-strike variations in subduction style were not addressed, and it is desirable to consider alternative models to explain the regional geochemical and petrologic variations. The following discussion addresses these points in the light of our boron data and their relation to other geochemical information.

Evidence for source-control of boron systematics

An important question is, to what extent do the B systematics in CAVA lavas reflect compositional features of the magma sources? First, B is a highly incompatible trace element (Leeman et al., 1992), and its concentration in magmas varies in response to differing degrees of partial melting and fractional crystallization as well as source heterogeneities. Boron may also be affected by interaction between magmas and crustal or mantle-wall rocks. However, the strong correlations of B/La and Ba/La with $^{10}Be/^{9}Be$ suggest that these ratios are probably representative of sub-arc magma source regions.

Magma-wall rock interaction at deep crustal levels is unlikely to produce the elevated B concentrations seen in CAVA lavas because moderate- to high-grade metamorphic rocks have low B contents (< 3 ppm) and low B/La (<0.1) ratios (Leeman et al., 1992); partial melts of such rocks would have comparable B/La. Relatively primitive mantle rocks have even lower B (<~0.1 ppm; cf. Leeman et al., 1992). Assimilation of such materials would likely result in decreases in B and B/La in a contaminated magma. MM domains may be characterized by significant B enrichment and high B/La, but the fact that $^{10}Be/^{9}Be$–B/Be or –B/La arrays for CAVA lavas project through the origin (Fig. 3) suggests that these lavas did not interact with B-enriched domains produced by old (> 5 to 10 Ma) subduction-related modification (cf. Morris et al., 1990). We conclude that regional and local variations in incompatible element ratios (B/La, Ba/La, $(La/Yb)_n$, K/B, etc.) and Sr, Nd, and Be isotopic ratios in CAVA lavas are not readily explained by any plausible contamination or fractionation models starting with uniform sources or parental magmas. On the assumption that these variations indeed reflect source heterogeneities, we proceed to examine how they might have developed.

Relation of geochemical variations to subduction and mantle heterogeneities

In a parallel study, we reexamined possible interactions between subducted slab components and two extreme mantle variants (Leeman et al., 1994); relevant conclusions of that study are summarized here. Covariations between B/La and selected geochemical parameters in CAVA lavas are consistent with additions of a subduction component (e.g., ~10% bulk sediment or ~3% fluid) to either DM or EM. However, addition of more than 10% of bulk sediment similar to that on the Cocos plate would result in significantly higher $^{87}Sr/^{86}Sr$ than seen in the lavas. Furthermore, metamorphism accompanying subduction of marine sediments (or altered oceanic crust) will likely result in dehydration, significant loss of boron (Bebout et al., 1991; Moran et al., 1992), and low B/La ratios (e.g., <<1, as seen in high-grade metapelites: Leeman et al., 1992). Thus, bulk addition of metamorphosed sediment cannot produce B/La ratios as high as those seen in most Nicaragua and El Salvador lavas. Addition of as little as 2 to 3% of sediment-derived fluid could increase source B/La to the maximum levels observed in the arc lavas. Melting of variably metasomatized mantle sources could produce the entire spectrum of CAVA lavas. Similarly, fluid metasomatism is also effective in producing the observed ranges in $^{10}Be/^{9}Be$ and Sr and Nd isotopic ratios. Ba/La and $(La/Yb)_n$ are also reasonably well matched by fluid additions to the mantle wedge, except below central Costa Rica and western Panama, where they are at least a factor of two to four higher in the lavas than calculated for metasomatized DM or EM sources.

Excess enrichment of La and Ba and production of high Ba/La in arc magma sources can be achieved by repeated influx of Ba-rich fluids or magmas into the sub-arc mantle over a prolonged period of subduction. Ba and REE are significantly less mobile than B during progressive metamorphism (cf. Leeman et al., 1992) and could be concentrated efficiently in the sub-arc mantle as "stored components." Thus, pronounced incompatible element enrichments in eastern versus western CAVA lavas could reflect increased involvement of "stored component"–rich mantle domains toward the east. Presumably this would occur if melting is induced in shallower parts of the mantle (e.g., lithosphere), where "stored components" likely reside. The close correspondence of incompatible element enrichment and high magma flux with subduction of the Cocos ridge suggests a causal relation.

Figure 7 shows schematic sections across the arc in Costa Rica and Nicaragua and provides an informative comparison between the two distinctive arc segments. Most important is the fact that, unless the Costa Rica lithosphere is thinner than most oceanic plates, the shallow subduction dip requires that melting take place within the mantle of the overriding plate or within the slab itself. Inasmuch as seismicity in that area is limited to depths shallower than ca. 50 km (Protti et al., this volume), it may be that temperatures are very high in the lithosphere mantle. Unless this mantle domain can convect, it is difficult to see how there can be efficient migration of low-degree melts as suggested by Feigenson and Carr (1993) to account for the high magma flux in Costa Rica. The lack of significant B- and ^{10}Be-enrichment in Costa Rican arc magmas can be explained by dehydration of the downgoing slab before it reaches a point below the volcanic front or by off-scraping of part of the sediment pile. Either effect would limit subduction-related modification of lithospheric mantle, thus helping to preserve its original OIB-like characteristics. Extensive melting of this reservoir may be necessary to match magma fluxes, in which case preservation of the high La/Yb and other geochemical characteristics of Costa Rican lavas requires that the source was pre-enriched in meltlike constituents rather than subduction-related fluids.

Steep subduction below Nicaragua also carries interesting implications. First, the cool slab may not dehydrate completely until it reaches great depths. If arc magmatism is indeed triggered by fluid release in this area, the subduction geometry indicates this must occur at depths in excess of 200 km. Whether melting occurs by decompression of rising diapirs or some other process, asthenospheric mantle (plus subduction contributions) will be the prevalent magma source, and lithospheric mantle may produce little or no magma. In this view, the main difference in magma production along the arc depends upon which mantle reservoir is melted.

Slab melting is yet another consideration, particularly in Costa Rica or Panama. Such melts are likely to be fairly silicic and relatively poor in B because the necessarily steep thermal gradient should promote extensive dehydration and B mobility (Moran et al., 1992). Low B/La (<1.0) in silicic lavas from El Valle and La Yeguada seems consistent with this view (cf. Defant et al., 1991a, b). However, their high Ba/La ratios (cf. Fig. 6) make these lavas poor candidates as metasomatizing agents for basalt sources in this sector of the arc.

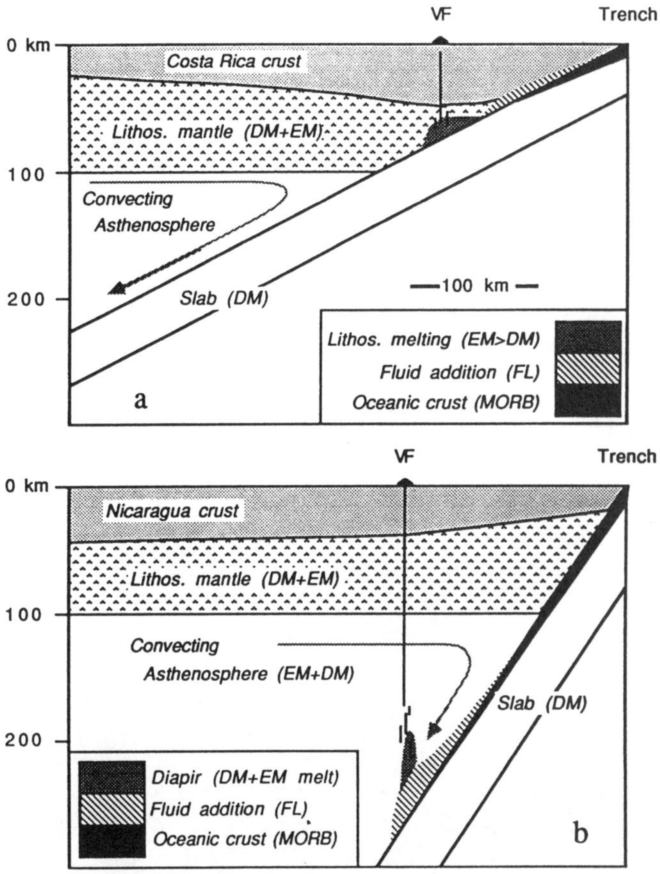

Figure 7. Cross sections perpendicular to the volcanic front (VF) in (a) Costa Rica and (b) Nicaragua (no vertical exaggeration), illustrating dramatically different boundary conditions for melt formation. Width of arc-trench gap from Figure 1, dip of slab from Protti et al. (this volume), crustal thickness from Carr (1984); slab thickness is arbitrary but is thinner below Costa Rica where it is younger and warmer; lithosphere thickness is arbitrarily set at 100 km; other details are schematic. Distribution of geochemical reservoirs assumes that lithosphere contains some fraction of EM domains (e.g., stored magmatic components) in a DM matrix. Asthenosphere may also consist of a heterogeneous mixture of DM+EM. Thinning of oceanic crust is intended to portray progressive dehydration of the uppermost part of the slab; dehydration of slab occurs at shallower depth where it is warmer (Costa Rica). Zone of fluid addition portrays where dehydration fluids may be added to the overlying mantle or crust. Possible melting zones are shaded. Subduction geometry below Costa Rica confines melting to the lithosphere unless an improbably thin lithosphere is assumed; presumably EM-rich domains will melt preferentially, as there would appear to be limited fluid available to lower the solidus of refractory DM-rich material. If melting is triggered by fluid release from the slab below Nicaragua, this could occur very deep (>200 km as shown); subduction geometry allows melting over a broad depth range within convecting asthenosphere if a diapiric model is assumed.

Subduction and tectonic controls on magma source compositions

Geochemical indicators for CAVA lavas define two distinct domains in the arc. From western Costa Rica to Guatemala, all volcanic front samples contain a subduction component (e.g., B/La > 0.5, $^{10}Be/^{9}Be > 5 \times 10$–11, Ba/La > 20), whereas basaltic samples from central Costa Rica to western Panama generally have little or no subduction signature (B/La < 0.5, $^{10}Be/^{9}Be < 5 \times 10$–11, Ba/La < 20). As discussed in the previous section (cf. Leeman et al., 1994), transfer of subducted material to the sub-arc mantle most likely occurs via fluid transport. The fluid flux is determined by subduction rate and fluid content of the combined sediment and altered oceanic crust section that is subducted and by the thermal state of the downgoing slab, which influences the release of its fluids. The amount of fluid required to produce appropriate magma sources is relatively small (≤3%) but varies with position along the arc and correlates with B/La. Based on Figure 2c, qualitative strength of the subduction signature in volcanic front lavas varies from moderate in Guatemala and El Salvador (B/La = 0.8 to 2.8), highest in western Nicaragua (B/La = 2.3 to 5.3), and moderate in eastern Nicaragua (B/La = 0.5 to 2.4) to small in western Costa Rica (B/La = 0.5 to 0.9), almost negligible in central Costa Rica (B/La < 0.5), and small in western Panama (<0.1 to 1.0). Variation in sediment flux could contribute to the observed along-strike geochemical variation in CAVA lavas and is an important parameter to quantify. However, as discussed earlier, constraints on the fraction of incoming sediment that is ultimately subducted are available only off central Guatemala and western Costa Rica. Combining our estimated fluxes (3 and 2.2 $km^3/10^5$ yrs per km of arc, respectively) with average B/La for these two arc segments provides a linear relationship (Fig. 8a) that can be used tentatively to infer approximate sediment fluxes elsewhere along the arc. Projections of average B/La for other arc sectors onto this line suggest the following fluxes ($km^3/10^5$ yrs per km of arc): western Panama (~2.0), central Costa Rica (~1.8), eastern Nicaragua (~2.9), western Nicaragua (~4.5), El Salvador (~3.1), and eastern Guatemala (~2.4). Of course, independent verification of these estimates would be useful, especially given the large range in B/La for some segments (e.g., western Nicaragua).

B/La correlates rather poorly with most of the subduction parameters tabulated by Carr et al. (1990), including slab dip, cumulative eruptive volume, and crustal thickness. It appears to show a moderate logarithmic correlation ($r^2 = 0.78$) with calculated eruptive flux (Figure 8b), but this plot could equally be interpreted as showing little correlation for the region between western Costa Rica and Guatemala with a discontinuous increase in flux in central Costa Rica. The main characteristics of the central Costa Rica–western Panama segment are the dramatic decrease in subduction component (B/La < ~0.5) and the dominance of enriched source characteristics. There is no clear explanation for the unique magmatic style that characterizes

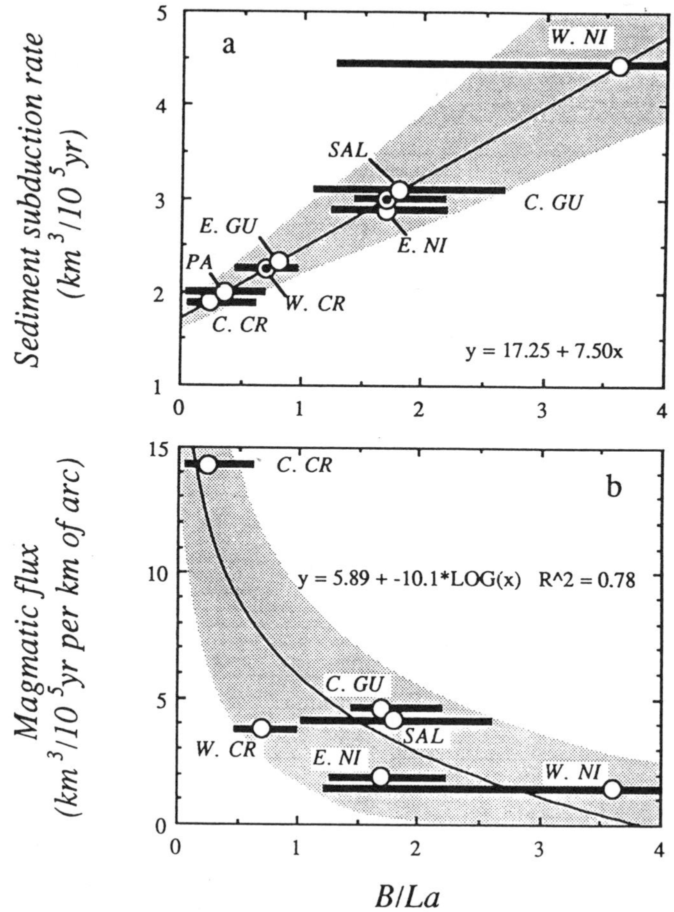

Figure 8. Variation between average B/La for different arc segments (cf. Carr et al., 1990) and (a) flux of subducted sediment, (b) eruptive flux (both in units of $km^3/10^5$ yr per km of arc). Points are identified as follows: central (C. GU) and eastern (E. GU) Guatemala; El Salvador (SAL); western (W. NI) and eastern (E. NI) Nicaragua; and western (W. CR) and central (C. CR) Costa Rica. In a, sediment fluxes were calculated for western Costa Rica and central Guatemala (bulls-eye symbol), and the equation given represents the line through these points. All other points represent projections of average B/La onto this line for the purpose of estimating sediment flux at each segment. In a, solid bars show the range in B/La for each arc segment (using basaltic samples only). In b, negative correlation of B/La and magma flux is approximated by a logarithmic fit, and the resulting equation shown.

this part of the CAVA, but it appears to be related to subduction tectonics of the Cocos Ridge and subjacent lithosphere.

As alternatives to the postulated increase in convective flow of EM-rich mantle into the sub-arc mantle wedge (Carr et al., 1990), we suggest the following. First, subduction of the Cocos Ridge and/or the relatively warm lithosphere apparently caused a shallowing of the subduction zone dip and a decrease in the depth of seismicity (Protti et al., this volume). The latter effect in particular suggests that temperatures in the slab are higher than elsewhere along the arc. If so, melting part of the subducted Cocos Ridge could tap EM-like domains that were formed in association with the hot-spot history of the ridge. Basaltic to trachytic lavas from Cocos Island have MORB-like isotopic compositions and are moderately enriched in the incompatible elements Ba and La (Castillo et al., 1988); B data are not available, but if they are similar to other oceanic island suites (ca. 1 to 2 ppm for basalts; Leeman et al., 1992), such magmas would have low B/La (~0.1) and Ba/La (~10). Intrusive equivalents of such magmas within the downgoing slab could remelt at substantially lower temperatures than more refractory peridotitic matrix of the lithosphere mantle; such melts could rise and metasomatize the mantle of the overriding plate, leaving low B/La and Ba/La and generally OIB-like trace element signatures.

Furthermore, subduction of a relatively hot slab could raise temperatures in a broad region of the overlying mantle wedge and promote melting at shallower depths than elsewhere along the CAVA, perhaps encountering lithosphere domains (Fig.7a) that were infiltrated by earlier magmas (i.e., rich in "stored components"). This process of source enrichment below central Costa Rica and western Panama could account for the slightly elevated Ba/La and $(La/Yb)_n$ and for general enrichment of incompatible elements in modern arc magmas without requiring unrealistically small degrees of partial melting. Shallow subduction of warm lithosphere could also promote melting within a broader than usual region of the sub-arc and back-arc mantle wedge, possibly contributing to the high magma flux in that area. However, models invoking large source volumes ultimately must be reconciled with realistic rates and processes of melt accumulation and transfer; for example, a given eruptive flux requires that lateral transfer rates increase on the order of the cubed radius of the melt extraction domain.

Yet another feature that requires explanation is the apparently thick crust of the overriding plate beneath central Costa Rica. Because the basement geology of the area consists of accreted oceanic and older arc terranes, it is unlikely that there is a thick continental root. If so, thickened crust may be related to tectonic processes such as imbrication; certainly there is evidence for the operation of such processes in eastern Panama and in the deformed region along the Caribbean margin of Panama (Silver et al., 1990). If lithospheric thickening resulted from the same processes, then previously melt-infiltrated mantle could be depressed to depths at which melting of "stored component"-rich domains contributed to magmatism in the modern arc.

CONCLUSIONS

Concentrations of boron and ratios of boron to other incompatible elements provide important constraints on the role of subducted material in genesis of CAVA lavas. A strong correlation exists between relative B enrichment (e.g., as B/La) and $^{10}Be/^{9}Be$, which is an unequivocal indicator for involvement of young marine sediment in arc magma sources. Thus B/La was used to monitor subduction-related enrichment of arc magma sources. Calibrated against available estimates of sediment subduction flux, B/La reflects along-strike relative variations in subduction contributions elsewhere in the arc. Despite large uncertainties in such estimates, sediment subduction flux appears to vary along the arc by at least a factor of two, with maximum rates off Nicaragua, El Salvador, and parts of Guatemala and minimum rates off Costa Rica. However, variations in B/La over short arc segments, particularly in Nicaragua, suggest that subduction processes have heterogeneously modified parts of the sub-arc mantle.

Regional geochemical variations in CAVA lavas reveal a major compositional discontinuity in the arc in western Costa Rica. This break corresponds with a sharp transition in seismic character, subduction angle, and stress regime and the projection of a transformlike boundary between older and younger lithosphere in the subducting slab (i.e., the QSC discontinuity of Protti et al., this volume). The geochemical discontinuity is believed to reflect a combination of source heterogeneities (related to lateral differences in age and evolution of the subducted lithosphere), along-strike variations in sediment subduction flux, geometry of subduction, and thermal state of the downgoing slab.

There is little geochemical evidence for direct addition of bulk subducted sediment to the mantle wedge to produce CAVA magma sources. However, progressive metamorphism of subducted sediments (±oceanic crust) can produce fluids that, if added to OIB- or MORB-source mantle domains, can produce realistic arc magma sources. It is concluded that selective enrichment of B (like ^{10}Be) in arc magmas depends principally on the availability of B-rich fluids derived from the subducted slab. Where such fluids are not available or are present only in limited amounts, arc magmas can be produced having compositions very similar to OIB or even MORB (cf. Leeman et al., 1990); some of the Costa Rican basalts fall in this category (Reagan and Gill, 1989). Relative enrichments of fluid-mobile elements in arc magma sources will be greatest where sediment subduction rates and fluid fluxes are high, as west of the QSC. Low initial slab temperatures in that sector will also favor deeper initiation of dehydration reactions and promote more efficient transfer of fluids to sub-arc depths (cf. Moran and others, 1992).

Unusual incompatible element enrichments and limited subduction contributions in lavas from central Costa Rica and western Panama suggest that magmatism in this part of the arc is related to special conditions associated with subduction of the Cocos Ridge and/or relatively warm lithosphere formed at the Nazca-Cocos spreading center. Higher temperatures in the downgoing plate may induce melting at shallower depths than elsewhere along the arc. If lithospheric mantle in the overriding plate contains "stored components" of older intruded magmas, these could locally result in moderate Ba/La, high $(La/Yb)_n$, and general incompatible element enrichment relative to typical primitive mantle (i.e., OIB-source) compositions. Remelting of such enriched domains could produce magmas with OIB affinities or add an OIB-like metasomatic flavor to shallower mantle domains. Perhaps even more important is a combination of low sediment subduction rates and warm temperatures that may have caused premature slab dehydration and depletion of fluid-mobile elements beneath the forearc region and limited transfer of ^{10}Be or B to arc magma sources east of the QSC.

ACKNOWLEDGMENTS

This research was supported by grants from the National Science Foundation awarded to W. P. Leeman (EAR90-14802 and EAR90-18996). Some samples and analytical data were provided by Marc Defant (Panama and the Cordillera Talamanca region of Costa Rica) and Mark Reagan (Arenal and Turrialba volcanoes); we are grateful for their generous assistance. This research benefited from our discussions with colleagues Julie Morris, Fouad Tera, Jeff Ryan, and Marc Defant. In addition, we thank Diane Smith, Bob Stern, and Paul Mann for their helpful reviews and Seth Stein for recalculating convergence rates.

REFERENCES CITED

Bardintzeff, J.-M., and Deniel, C., 1992, Magmatic evolution of Pacaya and Cerro Chiquito volcanological complex, Guatemala: Bulletin of Volcanology, v. 54, p. 267–283.

Bebout, G. E., Ryon, J. G., and Leeman, W. P., 1991, Boron and beryllium concentrations in subduction-related metamorphic rocks of the Catalina Schist: Implications for subduction zone recycling: Carnegie Institution of Washington Yearbook 90, p. 23–29.

Burbach, G. V., Frohlich, C., Pennington, W. D., and Matamoto, T., 1984, Seismicity and tectonics of the subducted Cocos plate: Journal of Geophysical Research, v. 89, p. 7719–7735.

Carr, M. J., 1984, Symmetrical and segmented variation of physical and geochemical characteristics of the Central American volcanic front: Journal of Volcanology and Geothermal Research, v. 20, p. 231–252.

Carr, M. J., and Stoiber, R. E., 1990, Volcanism, *in* Dengo, G., and Case, J. E., eds., The Caribbean Region: Geological Society of America, The Geology of North America, v. H, p. 375–391.

Carr, M. J., Rose, W. I., and Stoiber, R. E., 1979, Central America, *in* Thorpe, R. S., ed., Andesites: Orogenic andesites and related rocks: New York, John Wiley and Sons, p. 149–166.

Carr, M. J., Feigenson, M. D., and Bennett, E. A., 1990, Incompatible element and isotopic evidence for tectonic control of source mixing and melt extraction along the Central American arc: Contributions to Mineralogy and Petrology, v. 105, p. 369–380.

Case, J. E., and MacDonald, W. D., 1990, Crustal structure of the Caribbean region, *in* Dengo, G., and Case, J. E., eds., The Caribbean Region: Geological Society of America, The Geology of North America, v. H, Plate 3.

Case, J. E., MacDonald, W. D., and Fox, P. J., 1990, Caribbean crustal provinces; seismic and gravity evidence, *in* Dengo, G., and Case, J. E., eds., The Caribbean Region: Geological Society of America, The Geology of North America, v. H, p. 15–36.

Castillo, P., Batiza, R., Vanko, D., Malavassi, E., Barquero, J., and Fernandez, E., 1988, Anomalously young volcanoes on old hot-spot traces. I: Geology and petrology of Cocos Island: Geological Society of America Bulletin, v. 100, p. 1400–1414.

Corrigan, J., Mann, P., and Ingle, J. C., Jr., 1990, Forearc response to subduction of the Cocos Ridge: Panama–Costa Rica: Geological Society of America Bulletin, v. 102, p. 628–652.

De Boer, J. Z., Defant, M. J., Stewart, R. H., Restrepo, J. F., Clark, L. F., and Ramirez, A. H., 1988, Quaternary calc-alkaline volcanism in western Panama: Regional variation and implication for the plate tectonic framework: Journal of South American Earth Sciences, v. 1, p. 275–293.

De Boer, J. Z., Defant, M. J., Stewart, R. H., and Bellon, H., 1991, Evidence for active subduction below western Panama: Geology, v. 19, p. 649–652.

Defant, M. J., and eight others, 1991a, Andesite and dacite genesis via contrasting processes: The geology and geochemistry of El Valle volcano, Panama: Contributions to Mineralogy and Petrology, v. 106, p. 309–324.

Defant, M. J., and eight others, 1991b, Dacite genesis via both slab melting and differentiation: Petrogenesis of La Yeguada volcanic complex, Panama: Journal of Petrology, v. 32, p. 1101–1142.

Defant, M. J., and seven others, 1992, The geochemistry of young volcanism throughout western Panama and southeastern Costa Rica: An overview: Journal of the Geological Society of London, v. 149, p. 569–579.

DeMets, C., Gordon, R. G., Argus, D. F., and Stein, S., 1990, Current plate motions: Geophysics Journal International, v. 101, p. 425–478.

Donnelly, T. W., Horne, G. S., Finch, R. C., and Lopez-Ramos, E., 1990, Northern Central America; the Maya and Chortis blocks, *in* Dengo, G., and Case, J. E., eds., The Caribbean Region: Geological Society of America, The Geology of North America, v. H, p. 37–76.

Ellam, R. M., and Hawkesworth, C. J., 1988, Elemental and isotopic variations in subduction-related basalts: Evidence for a three-component model: Contributions to Mineralogy and Petrology, v. 98, p. 72–80.

Feigenson, M. D., and Carr, M. J., 1986, Positively correlated Nd and Sr isotope ratios of lavas from the Central American volcanic front: Geology, v. 14, p. 79–82.

Feigenson, M. D., and Carr, M. J., 1993, The source of Central American lavas: Inferences from geochemical inverse modelling: Contributions to Mineralogy and Petrology, v. 113, p. 226–235.

Gardner, T. W., and six others, 1992, Quaternary uplift astride the aseismic Cocos Ridge, Pacific coast, Costa Rica: Geological Society of America Bulletin, v. 104, p. 219–232.

Hey, R., 1977, Tectonic evolution of the Cocos-Nazca spreading center: Geological Society of America Bulletin, v. 88, p. 1404–1420.

Hofmann, A. W., 1989, Chemical differentiation of the Earth: The relationship between mantle, continental crust, and oceanic crust: Earth and Planetary Science Letters, v. 90, p. 297–314.

Kay, R. W., 1980, Volcanic arc magmas: Implications of a melting-mixing model for element recycling in the crust–upper mantle system: Journal of Geology, v. 88, p. 497–522.

Leeman, W. P., 1988, Boron contents in selected international geochemical reference samples: Geostandards Newsletter, v. 12, p. 61–62.

Leeman, W. P., Smith, D. R., Hildreth, W., Palacz, Z. A., and Rogers, N. W., 1990, Compositional diversity of late Cenozoic basaltic magmas of the southern Washington Cascades: Journal of Geophysical Research, v. 95, p. 19561–19582.

Leeman, W. P., Sisson, V. B., and Reid, M. R., 1992, Boron geochemistry of the lower crust: Evidence from granulite terranes and deep crustal xenoliths: Geochimica et Cosmochimica Acta, v. 56, p. 775–788.

Leeman, W. P., Carr, M. J., and Morris, J. D., 1994, Boron geochemical constraints on genesis of subduction-related magmas of the Central American volcanic arc: Geochimica et Cosmochimica Acta, v. 58, p. 149–168.

Lin, P. N., Stern, R. J., Morris, J., and Bloomer, S. H., 1990, Nd- and Sr-isotopic compositions of lavas from the northern Mariana and southern Volcano arcs: Implications for the origin of island arc melts: Contributions to

Mineralogy and Petrology, v. 105, p. 381–392.

Lonsdale, P., and Klitgord, K. D., 1978, Structure and tectonic history of the eastern Panama Basin: Geological Society of America Bulletin, v. 89, p. 981–999.

Mann, P., Schubert, C., and Burke, K., 1990, Review of Caribbean neotectonics, *in* Dengo, G., and Case, J. E., eds., The Caribbean Region: Geological Society of America, The Geology of North America, v. H, p. 307–338.

Minster, J. B., and Jordan, T. H., 1978, Present-day plate motions: Journal of Geophysical Research, v. 83, p. 5331–5354.

Moore, G. F., Shipley, T. H., and Lonsdale, P. F., 1986, Subduction erosion versus sediment offscraping at the toe of the Middle America Trench off Guatemala: Tectonics, . 5, p. 513–523.

Moran, A. E., Sisson, V. B., and Leeman, W. P., 1992, Boron in subducted oceanic crust and sediments: Effects of metamorphism and implications for arc magma compositions: Earth Planetary Science Letters, v. 111, p. 331–349.

Morris, J., and Tera, F., 1989, ^{10}Be and ^{9}Be in mineral separates and whole rocks from volcanic arcs: Implications for sediment subduction: Geochimica et Cosmochimica Acta, v. 53, p. 3197–3206.

Morris, J., Leeman, W. P., and Tera, F., 1990, The subducted component in island arc lavas: Constraints from Be isotopes and B-Be systematics: Nature, v. 344, p. 31–36.

Patino, L. C., Carr, M. J., and Feigenson, M. D., 1993, A geochemical arc-transect: Honduras, Central America: EOS (American Geophysical Union Transactions), v. 74, p. 349.

Plank, T., and Langmuir, C. H., 1988, An evaluation of the global variations in the major element chemistry of arc basalts: Earth and Planetary Science Letters, v. 90, p. 349–370.

Reagan, M. K., and Gill, J. B., 1989, Coexisting calcalkaline and high-niobium basalts from Turrialbo volcano, Costa Rica: Implications for residual titanates in arc magma sources: Journal of Geophysical Research, v. 94, p. 4619–4633.

Ryan, J. G., Morris, J. D., Tera, F., Leeman, W. P., and Tsvetkov, A., 1990, Petrologic and geochemical characteristics of recent lavas from the Kurile-Kamchatka arc [abs.]: EOS (American Geophysical Union Transactions), v. 71, p. 1714.

Shipley, T. H., and Moore, G. F., 1986, Sediment accretion, subduction, and dewatering at the base of the trench slope off Costa Rica: A seismic reflection view of the decollement: Journal of Geophysical Research, v. 91, p. 2019–2028.

Shipley, T. H., Stoffa, P. L., and Dean, D. F., 1990, Underthrust sediments, fluid migration paths, and mud volcanoes associated with the accretionary wedge off Costa Rica: Middle America trench: Journal of Geophysical Research, v. 95, p. 8743–8752.

Silver, E. A., Reed, D. L., Tagudin, J. E., and Heil, D. J., 1990, Implications of the north and south Panama thrust belts for the origin of the Panama orocline: Tectonics, v. 9,p. 261–281.

Spiegelman, M., and McKenzie, D., 1987, 2-D models for melt extraction at mid-ocean ridges and island arcs: Earth and Planetary Science Letters, v. 83, p. 137–152.

Stern, R. J., Morris, J., Bloomer, S. H., and Hawkins, J. W., Jr., 1991, The source of the subduction component in convergent margin magmas: Trace element and radiogenic isotope evidence from Eocene boninites, Mariana forearc: Geochimica et Cosmochimica Acta, v. 55, p. 1467–1481.

Stoiber, R. E., and Carr, M. J., 1973, Quaternary volcanic and tectonic segmentation of Central America: Bulletin of Volcanology, v. 37, p. 304–325.

Tera, F., Brown, L., Morris, J., and Sacks, I. S., 1986, Sediment incorporation in island-arc magmas: Inferences from ^{10}Be: Geochimica et Cosmochimica Acta, v. 50, p. 535–550.

von Huene, R., 1984, Tectonic processes along the front of modern convergent margins—Research of the past decade: Annual Reviews of Earth Science, v. 12, p. 359–381.

von Huene, R., 1989, The Middle America convergent plate boundary, *in* Winterer, E. L., Hussong, D. M., and Decker, R. W., eds., The eastern Pacific Ocean and Hawaii: Geological Society of America, The Geology of North America, v. N, p. 535–550.

Walker, J. A., 1984, Volcanic rocks from the Nejapa and Grenada cinder cone alignments, Nicaragua: Central America: Journal of Petrology, v. 25, p. 299–342.

Walker, J. A., Carr, M. J., Feigenson, M. D., and Kalamarides, R. I., 1990, The petrogenetic significance of interstratified high- and low-Ti basalts in central Nicaragua: Journal of Petrology, v. 31, p. 1141–1164.

Watkins, J. S., 1989, The Middle America Trench off southern Mexico, *in* Winterer, E. L., Hussong, D. M., and Decker, R. W., eds., The eastern Pacific Ocean and Hawaii: Geological Society of America, The Geology of North America, v. N, p. 523–533.

MANUSCRIPT ACCEPTED BY THE SOCIETY FEBRUARY 7, 1994

Printed in U.S.A.

Geological Society of America
Special Paper 295
1995

Tectonic development of Panama, Costa Rica, and the Colombian Andes: Constraints from Global Positioning System geodetic studies and gravity

James N. Kellogg and Victor Vega
Department of Geological Sciences, University of South Carolina, Columbia, South Carolina 29208

With an appendix by:
T. C. Stallings and Carlos L.V. Aiken
Center for Lithospheric Studies, University of Texas at Dallas, Plano, Texas 75074
James N. Kellogg

ABSTRACT

Global Positioning System (GPS) measurements suggest the existence of a rigid Panama–Costa Rica microplate that is moving northward relative to the stable Caribbean plate. Northward motion of Central America relative to the Caribbean plate is independently suggested by the April 1991 Costa Rica earthquake, active folding in the North Panama deformed belt, and a south-dipping Wadati-Benioff zone beneath Panama. Panama may also be continuing to collide eastward with the northern Andes. Rapid subduction is occurring at the Middle America (72 mm/yr), Ecuador (70 mm/yr), and Colombia (50 mm/yr) trenches. The northern Andes are moving northeastward relative to stable South America. Preliminary GPS results also suggest Caribbean–North Andean convergence and an independent North Nazca plate.

About 6 Ma the Panama-Choco island arc collided with the northwestern margin of South America, eventually forming a land bridge between the Americas; closed the Pacific-Caribbean seaway, changing ocean circulation patterns and perhaps the world's climate; folded the East Panama deformed belt; and uplifted the Eastern Cordillera of Colombia. An interpretation of the paleo-Romeral suture in southern Colombia as a low-angle fault dipping to the west into the lower crust under the Cordillera Occidental is compatible with seismic velocity and gravity data. During the Late Cretaceous the Western Cordillera oceanic terrane was obducted eastward on the fault system over continental crust.

INTRODUCTION

The tectonic development of southern Central America and the northern Andes has been the subject of a number of studies (e.g., Case et al., 1971; Jordan, 1975; Toussaint, 1978; Duque-Caro, 1979; Case and Holcombe, 1980; Pennington, 1981; Wadge and Burke, 1983; Dengo, 1985; Escalante, 1990; Mann and Corrigan, 1990; Silver et al., 1990). Particularly controversial questions are the timing and geometry of the accretion of the Western Cordillera of Colombia and Ecuador, the location of the Panama arc–South America suture, and present-day plate boundaries and relative motions. The rela-

Kellogg, J. N., and Vega, V., 1995, Tectonic development of Panama, Costa Rica, and the Colombian Andes: Constraints from Global Positioning System geodetic studies and gravity, *in* Mann, P., ed., Geologic and Tectonic Development of the Caribbean Plate Boundary in Southern Central America: Boulder, Colorado, Geological Society of America Special Paper 295.

tive velocity of the Caribbean plate with respect to the Cocos, Nazca, and South American plates is poorly constrained because the Caribbean plate has no spreading ridges on its western, southern, and eastern boundaries. The only direct rate information available for the Caribbean plate is from magnetic lineations produced by the small spreading center within the Cayman Trough (Rosencrantz and Sclater, 1986).

Caribbean–Nazca–South America triple junctions have been shown as far east as Colombia's Eastern Cordillera (Dewey, 1972), as far west as the Panama fracture zone (Shagam, 1975), and as far south as Ecuador's Gulf of Guayaquil (Aggarwal, 1983). Deformation of Panama by distributed slip on a number of left-lateral strike-slip faults has been suggested by Mann and Corrigan (1990), based on field mapping and on air photo and satellite image analysis. Silver et al. (1990) explained the kinematics of Panama with a flexural beam model. Their model assumed an original east trend to the isthmus and an eastward movement relative to the South American plate. Jordan (1975) and Wadge and Burke (1983) noted the diffuse seismicity in Panama and northwestern South America (Fig. 1) and interpreted the seismic zone as a diffuse plate boundary accommodating deformation over a broad area. Pennington (1981) and Adamek et al. (1988) explained the tectonics as the interaction of three large plates and one to three microplates. Kellogg et al. (1985, 1989) have interpreted the tectonics in the region by two additional blocks or microplates, Panama and the North Andes (Fig. 2). In this chapter we use geologic, gravity, seismic, and Global Positioning System (GPS) geodetic data to derive a tectonic model for the North Andes and southern Central America.

One of the most controversial problems in Caribbean tectonics is the location of the current Caribbean–South American plate boundary. Seismologists noted that most of the historical seismicity was located along the Bocono fault in the Venezuela Andes and put the Caribbean–South American plate boundary there (e.g., Dewey, 1972; Aggarwal, 1983). However, multichannel seismic reflection profiles across the South Caribbean deformed belt show that Caribbean acoustic basement has underthrust the deformed belt, and folding of the youngest sediments suggests that the deformation is active (Edgar et al., 1971; Silver et al., 1975; Lu and McMillen, 1982; Lehner et al., 1983; Ladd et al., 1984). Furthermore, Dewey (1972) and Pennington (1981) recognized a zone of earthquakes dipping about 20 degrees to the southeast under the Santa Marta massif and Sierra de Perija and terminating 200 km below the Maracaibo Basin. Kellogg and Bonini (1982) and Toto and Kellogg (1992) interpreted these earthquakes as a Benioff zone produced by slow amagmatic subduction of Caribbean lithosphere. A convergence rate of 1.9 ± 0.3 cm/yr was estimated from the 390-km length of the seismic zone, assuming a thermal equilibration time of 10 m.y. A similar rate of 1.7 ± 0.7 cm/yr was calculated from velocity vector closure between the North Andean block and the Caribbean and South American plates (Fig. 2; Kellogg et al., 1985; Kellogg and

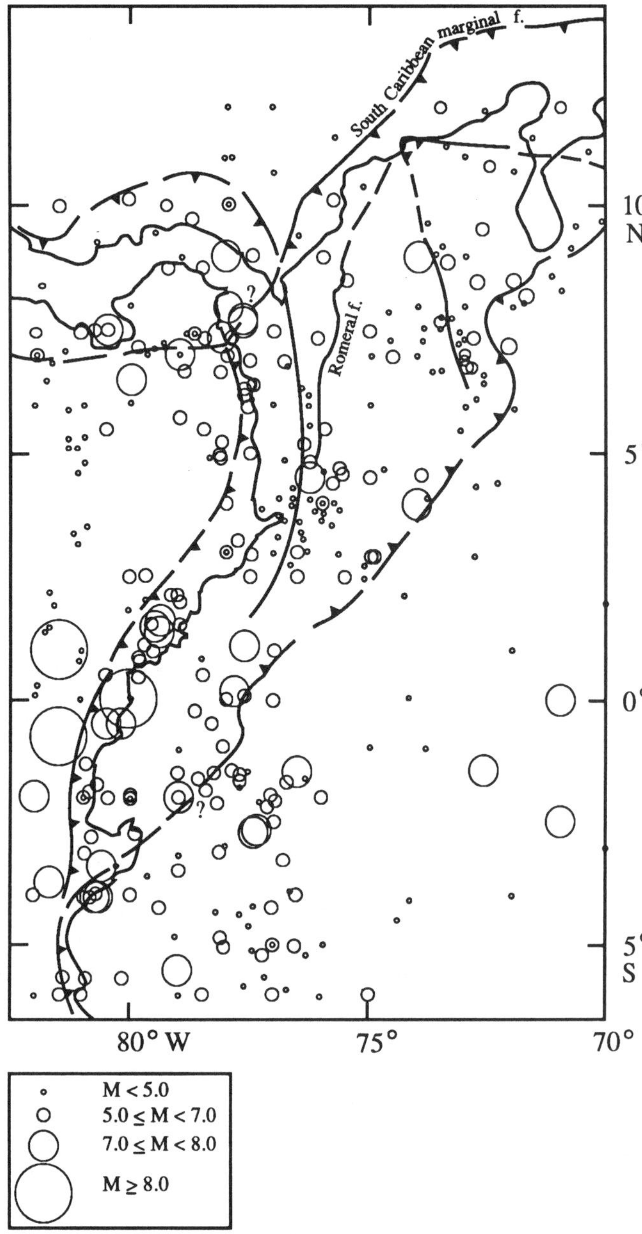

Figure 1. Earthquake epicenters for northwestern South America from the NOAA-EDIS Earthquake Data File. Most of the earthquakes are from the period 1962 to 1981.

Bonini, 1985). This predicted rate is about twice as fast as the 1.0 cm/yr right-lateral displacement on the seismically active Bocono fault.

CRUSTAL MOTIONS MEASURED WITH THE GLOBAL POSITIONING SYSTEM

The Central and South America (CASA) Global Positioning System (GPS) project was inaugurated in 1988 to study plate motions and crustal deformation in Central America and

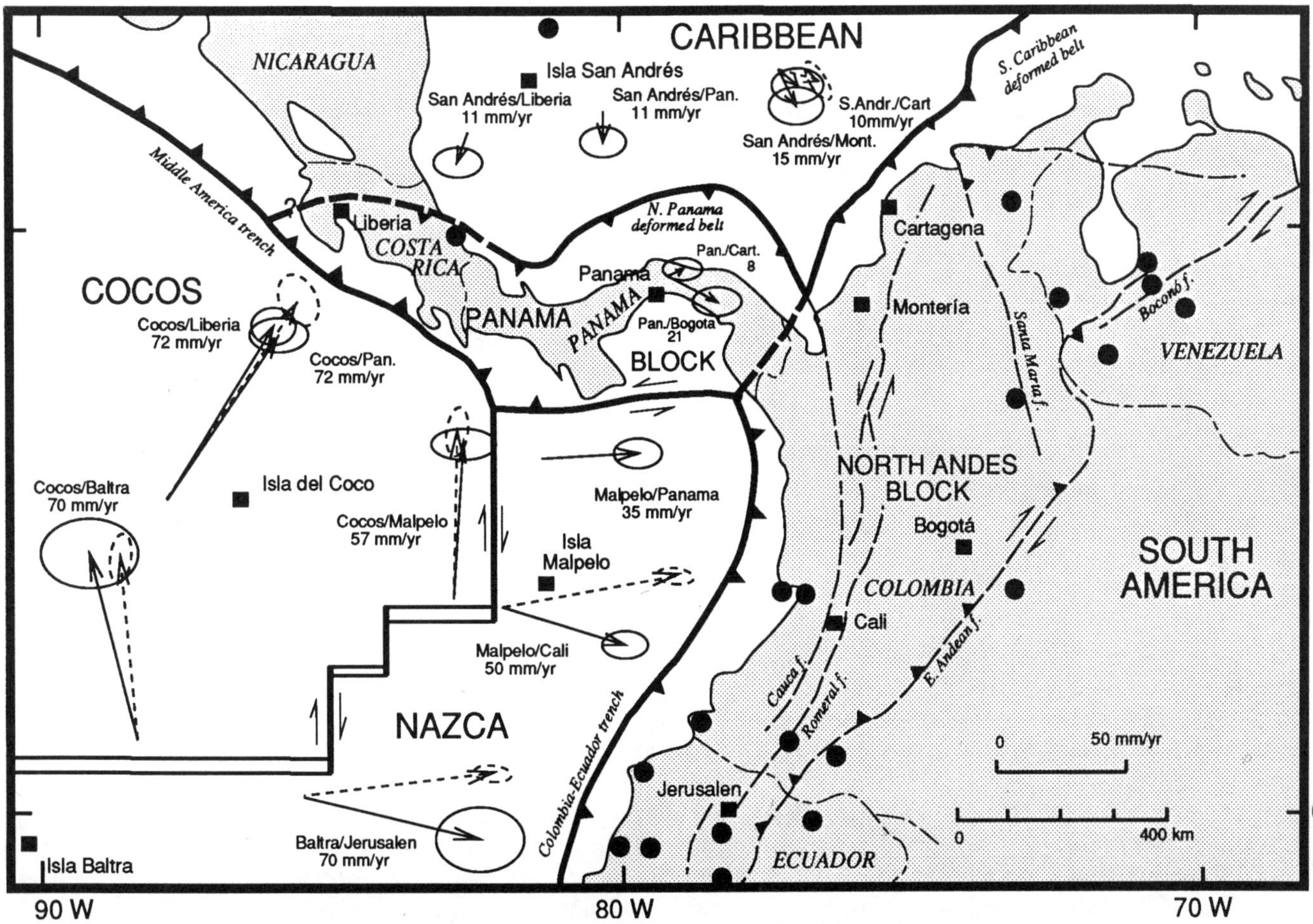

Figure 2. Observed GPS (solid lines) and model NUVEL-1 (dashed lines) baseline rates of change, with their 95% confidence ellipses, between the sites indicated by solid squares. The solid circles indicate the remaining sites in the CASA network.

northern South America. GPS is a satellite-based positioning system allowing centimeter-level geodesy to be performed with low-cost portable receivers. Centimeter-level precision has been demonstrated on baselines up to 2,000 km in length (e.g., Beutler et al., 1987; Blewitt, 1989; Tralli and Dixon, 1988; Dong and Bock, 1989). Repeated geodetic measurements with GPS provide direct measurements of displacements due to active plate motions and intraplate deformation, which can provide important constraints on regional kinematic plate motions and on the active tectonics of the deforming zones of the convergent margins. Dynamic problems such as the relation between the earthquake cycle and large-scale plate motions or intraplate deformation at convergent margins can be addressed once a sufficiently long-term span of data has been collected.

The CASA network currently consists of 30 primary sites in five countries (Fig. 2), most of which have now been occupied at least twice (Kellogg et al., 1989; Kellogg and Dixon, 1990). Preliminary computed baseline vector changes in mm/yr are shown with their 95% confidence ellipses in Figure 2 (Vega and Kellogg, 1992). The measurements cover a three-year period, 1988 to 1991. Freymueller et al. (1993) presented results from four sites that were occupied in all three campaigns (Isla Baltra and Jerusalen, Ecuador; Isla del Coco and Liberia, Costa Rica) and two sites (Isla San Andrés and Cartagena, Colombia) that were occupied in 1988 and 1991. This chapter presents results with 10 additional station days of data from Baltra and Jerusalen and six additional station days of data from Isla San Andrés and Cartagena. We also present new results based on 103 station days of data for five additional sites that were occupied in 1988 and 1991 (Panama City, Panama; Monteria, Bogotá, Cali, and Isla Malpelo, Colombia).

Details of the GPS data analysis are given in Freymueller (1991). Data from a global tracking network were used to estimate corrections to the GPS satellite orbits. The coordinates of three tracking sites were constrained to the coordinates determined from Very Long Baseline Interferometry (VLBI). The 95% confidence ellipses for baselines that were occupied three times (Fig. 2) are based on the formal uncertainties and estimates of other substantial systematic errors such as fiducial, global, and local network effects. For example, the 585-km baseline from Isla del Coco to Liberia, Costa Rica, was occupied in 1988, 1990, and 1991. The baseline solutions as a function of time (Figs. 2 and 3) give a shortening rate of 72 mm/yr, 40 ± 9.4 mm/yr in the east and 59 ± 6.5 mm/yr in the north, at the 95% confidence level. For baselines that were occupied in only two campaigns, 1988 and 1991, the displacement rate is the difference between the two solutions divided by the time elapsed between the experiments (3.05 years). The uncertainties in the rates for these baselines are based on the mean formal precision of each experiment's solution.

Isla del Coco–Liberia baseline

The 585-km Isla del Coco–Liberia baseline crosses the Middle America trench. Relatively rapid subduction occurs at the Middle America trench, with the Cocos plate being subducted beneath the Caribbean and North American plates. DeMets and others (1990) predicted a subduction rate of 91 mm/yr, nearly normal to the trench. The submarine Cocos Ridge is being subducted under southern Costa Rica (Fig. 4; Kolarsky et al., this volume). North of the ridge, the geometry of the subducted slab changes, and Burbach et al. (1984) suggest that the slab may have been torn as a result of the effects of the attempted subduction of the buoyant Cocos Ridge. Subduction may be slowing or stopping in the Costa Rica segment of the subduction zone (van Andel et al., 1971; Pennington, 1981; Burbach et al., 1984). In March and April 1990, a series of five subduction zone earthquakes, the largest being $M_s \sim 6.9$, and three others being larger than $M_s \sim 6.0$, ruptured the trench from southern Nicaragua south to Panama. The first and largest event occurred just south of the tip of Nicoya Peninsula, approximately on a line between Liberia and Isla del Coco (Fig. 4).

The 72 mm/yr computed displacement rate for the Isla del Coco–Liberia baseline is based on a linear fit of the GPS solutions as a function of time (Fig. 3). The reduced chi-squared statistic for the linear fit was 1.29, indicating that the scatter of the data about the best fit line is consistent with the given uncertainties. The observed rate is only 79% of the 91 mm/yr Cocos-Caribbean convergence predicted by the NUVEL-1 plate motion model (DeMets et al., 1990). The discrepancy may be partly explained by the 11 mm/yr convergence rate measured between Liberia, Costa Rica, and Isla San Andrés, Colombia, located on the stable Caribbean plate (Fig. 2).

Isla San Andres–Liberia and Isla San Andres–Panama City baselines

Similar convergence rates were measured for the Isla San Andres–Liberia and the Isla San Andres–Panama City baselines (11 mm/yr; Fig. 2). No significant deformation was measured between Liberia, Costa Rica, and Panama City, Panama.

Earthquake focal mechanisms and seismicity suggest a poorly defined Wadati-Benioff zone dipping about 50° to the southwest beneath the Panama fold belt (Wolters, 1986; Adamek et al., 1988). The April 1991 Costa Rica earthquake ($M_s \sim 7.4$) on the Caribbean coast of Costa Rica was a thrust event on a plane dipping 17° to 30° to the southwest (Fig. 4; Guendel et al., 1991; Goes et al., 1993; Tajima, this volume). It occurred on one of a system of thrust faults that make up the northern extension of the North Panama fold belt. The earthquake produced as much as 157 cm of uplift along the coast, measured by uplift of coral reefs and other shoreline features (Plafker and Ward, 1992) and by repeated geodetic leveling (De Obaldia et al., 1991). Vertical coseismic displacements of a similar magnitude also have been measured using GPS satellite geodesy (Lundgren et al., 1993). The center of uplift due to the earthquake was located near the Limon GPS site, and over 2 m of horizontal coseismic displacement was observed for this site (Lundgren et al., 1993).

Bowin (1976), Case et al. (1984), Kellogg et al. (1985), and Adamek et al. (1988) predicted a Panama block or microplate to help explain the regional tectonics. The GPS results are consistent with this model if the microplate includes most of Panama and Costa Rica. Northward motion of Central America relative to the stable Caribbean plate is independently suggested by the April 1991 Costa Rica earthquake (Fig. 4), a south-dipping Wadati-Benioff zone beneath Panama (Wolters,

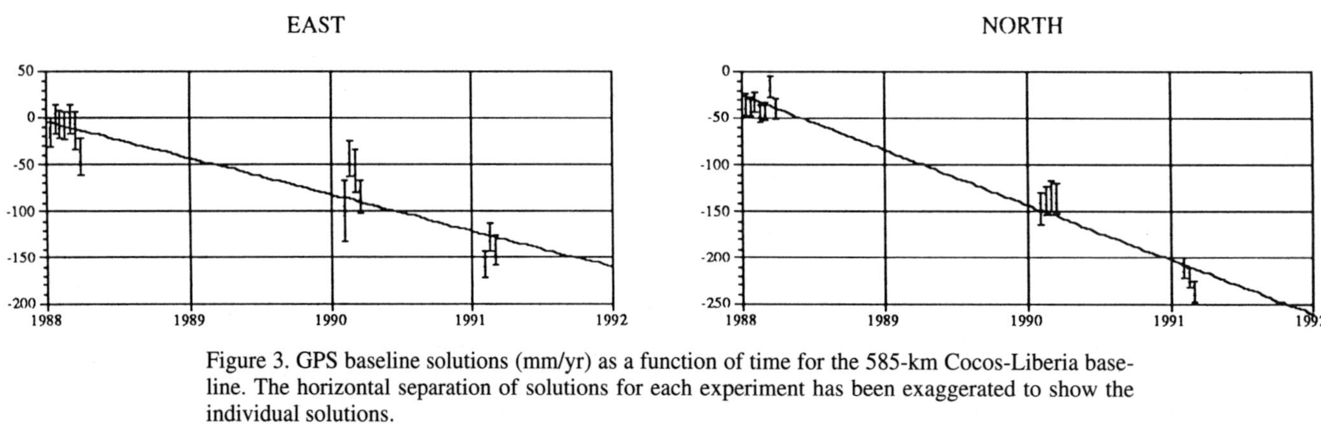

Figure 3. GPS baseline solutions (mm/yr) as a function of time for the 585-km Cocos-Liberia baseline. The horizontal separation of solutions for each experiment has been exaggerated to show the individual solutions.

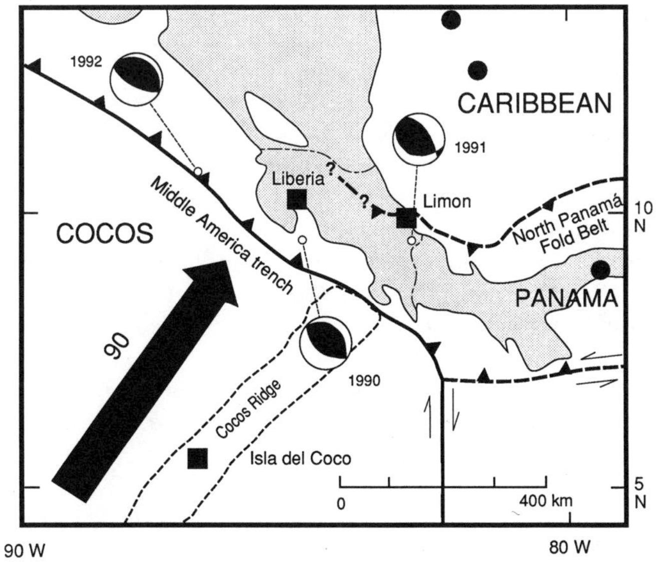

Figure 4. The Costa Rica subduction system (Freymueller and others, 1993). Focal mechanisms are shown for the 1990 (M ~ 6.9; Harvard CMT solution), 1991 (M ~ 7.4; Plafker and Ward, 1992), and 1992 (M ~ 7.0; Harvard CMT solution) earthquakes. Solid squares and solid circles as in Figure 2.

1986; Adamek et al., 1988), and active folding in the North Panama deformed belt (e.g., Silver et al., this volume). A SeaMARC II survey of the North Panama fold belt shows that it consists of four nearly straight segments (Silver et al., 1990, this volume). The Central Panama segment is the longest and maintains a nearly constant orientation of 075°. Silver et al. (1990, this volume) showed that this orientation is consistent with a model of bending of the Isthmus of Panama as a result of the collision with South America. Previous workers have proposed that the northwestern boundary of the Panama microplate passed through the middle of Costa Rica (e.g., Kellogg et al., 1989; Fan et al., 1993). The lack of motion observed between Liberia and Panama City, however, suggests that the northwestern boundary of the Panama–Costa Rica microplate must be north of Liberia and may be the Santa Elena suture along the Nicaragua–Costa Rica border (Case and Dengo, 1982; Escalante, 1990). Alternatively, the Liberia observations can be explained by elastic strain accumulation with a locked subduction zone (Dixon, 1993).

Isla Malpelo–Panama City baseline

Malpelo Island was displaced 35 mm/yr eastward relative to Panama City (Fig. 2). The vector azimuth is consistent with left-lateral strike-slip movement south of Panama predicted from plate vector closure by Jordan (1975) and from the seismicity by Adamek et al. (1988). The observed vector azimuth is also consistent with oblique convergence at the Panama fracture zone complex southwest of Panama observed in seismic reflection profiles and SeaMARC II imagery by Moore and Sender (this volume). The observed motion does not, however, support active northward spreading in the northern part of the Nazca plate as proposed by de Boer and others (1988).

Panama City–Cartagena–Bogota baselines

CASA GPS measurements at Panama City, Panama, and Cartagena and Bogota, Colombia, suggest ongoing collision between the Panama–Costa Rica microplate and the North Andes at about 8 to 21 mm/yr (Fig. 2). The Panama-Colombia border area is one of the most seismically active areas in northwestern South America (Fig. 1; Bermudez, 1985). More than 64 earthquakes of magnitude 5.0 or larger occurred in the area between 1963 and 1981. A shallow earthquake on January 20, 1904, measured 7.9 on the Richter scale. The seismic zone coincides with an active Quaternary fault zone identified from aerial photographs (W. Page, Woodward-Clyde, personal communication, 1984) and may mark a Panama–Nazca–North Andes triple junction (Fig. 2). A number of focal mechanisms from the seismically active Panama-Colombia border area indicate east-west compression (e.g., Pennington, 1981; Kolarsky and Mann, this volume). Focal mechanisms for two large October 1992 earthquakes in northwestern Colombia (M_s = 6.6 and 7.2) are consistent with northwest-southeast compression.

Isla San Andres–Cartagena and Isla San Andres–Monteria baselines

Repeat GPS measurements in 1988 and 1991 at Isla San Andrés, Cartagena, and Monteria, Colombia, show 10 and 15 mm/yr of convergence in a northwest-southeast direction. These preliminary results are in agreement with predictions of 17 to 19 mm/yr of northwest-southeast convergence at the South Caribbean deformed belt by Kellogg and Bonini (1982) and Kellogg et al. (1985). Thus, although the Caribbean–South American plate boundary may be a broad zone of deformation, most of the deformation is occurring at the South Caribbean deformed belt.

Isla Baltra–Jerusalen baseline

Nazca–North Andes convergence was measured by the 1,323-km baseline from Baltra to Jerusalen, Ecuador, and the 585-km baseline from Malpelo Island to Cali, Colombia. Active subduction of Nazca oceanic lithosphere at the Colombia trench is indicated by a Benioff zone dipping 30° to the east (Galvis, 1980; Pennington, 1981) and by andesitic volcanism in the Central Cordillera. The observed convergence rate of 70 mm/yr between the Baltra and Jerusalen sites is based on a linear fit of the 1988, 1990, and 1991 GPS solutions as a function of time, with a reduced chi-squared statistic of 0.44. The observed displacement rate is in general agreement with the NUVEL-1 model.

On the east side of the North Andes, right-lateral movement is taking place on the Bocono fault zone (Figs. 2 and 5). From the short-term seismicity Aggarwal (1983) deduced a seismic-slip rate of 1 ± 0.2 cm/yr. Using a measured 100-m right-lateral offset of glacial moraines, Schubert (1980) estimated a similar average rate of strike slip motion on the fault, 0.8 cm/yr for the last 12,000 years. If the predicted northward motion of the North Andes block relative to South America is added to the displacement rate, the NUVEL-1 model would lie within the 95% confidence ellipse of the GPS estimate (Freymueller et al., 1993).

Isla Malpelo–Cali baseline

A much slower convergence rate at the Colombia trench (50 mm/yr) was measured between Malpelo Island and Cali from 1988 to 1991. The difference between the Ecuador and Colombia trench convergence rates may be related to east-west–trending left-lateral strike-slip motion in the northern part of the Nazca plate (Wolters, 1986; Adamek et al., 1988), although the east-west uncertainty in the measurements is rather large (±14 mm/yr at the 95% confidence level).

Future measurements

The GPS results for Liberia, Isla del Coco, Isla Baltra, and Jerusalen are based on measurements in 1988, 1990, and 1991. The scatter of the data about the best-fit displacement rate is consistent with the given uncertainties. All other GPS results discussed in this chapter are based on measurements in 1988 and 1991, and there are potential systematic errors. Additional occupations of the network will significantly improve the uncertainties in the GPS-derived plate motion rates. With the first three occupations, the uncertainty in the Cocos-Caribbean vector from GPS is comparable to the uncertainty for the same vector in the NUVEL-1 model; with future occupations, GPS can be expected to determine relative plate motions with a precision comparable to or better than that of the NUVEL-1 plate motion model (Freymueller et al., 1993). Past plate motion determinations and plate margin geometries must be constrained by other methods, however. The gravity field (Fig. 5; Stallings et al., Appendix, this volume), for example, can help to constrain the geometry of paleo-plate margins.

GRAVITY

Density model of the North Panama fold (deformed) belt

The northern margins of Panama and Colombia consist of extensive fold belts (Case, 1974; Bowin, 1976) that are part of a deformed zone extending from Costa Rica to eastern Venezuela (Fig. 6). As much as 10 km of Tertiary turbidites, carbonates, and fluvial and lacustrine sediments have been preserved in the deformed belt in northern Colombia (Duque-

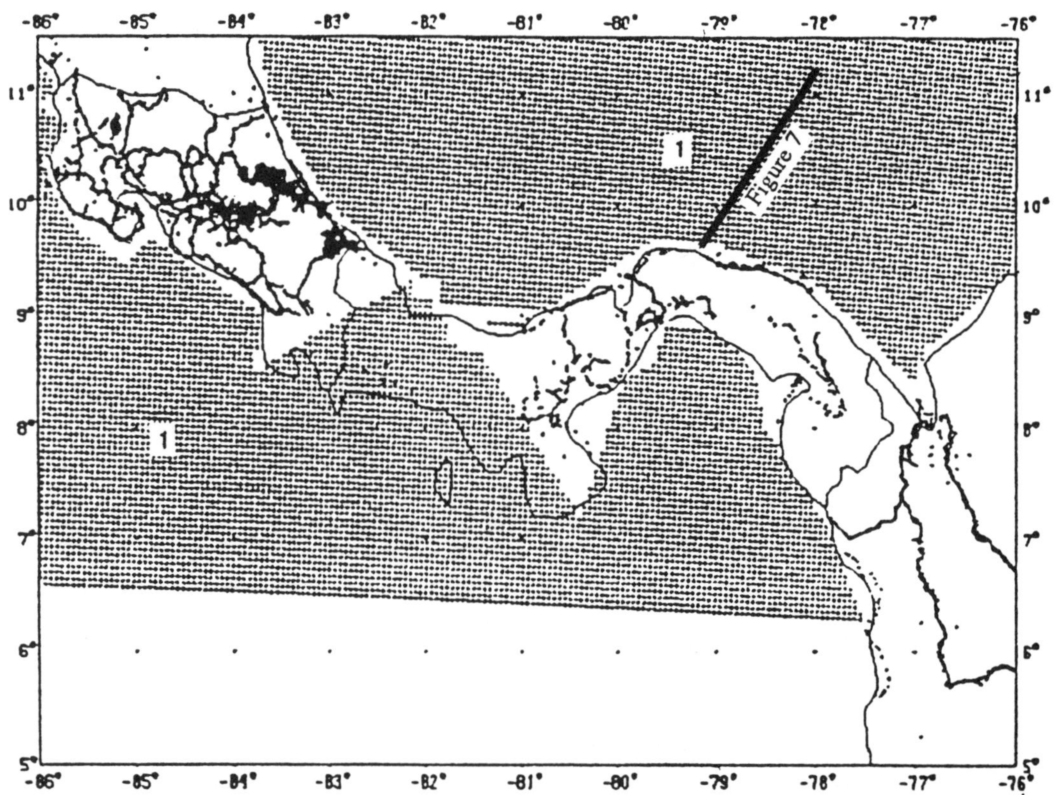

Figure 5. Data source location map for gravity anomaly map of southern Central America (Stallings et al., Appendix, this volume). 1: National Geophysical Data Center (1990). Figure 7: Profile of the North Panama fold belt.

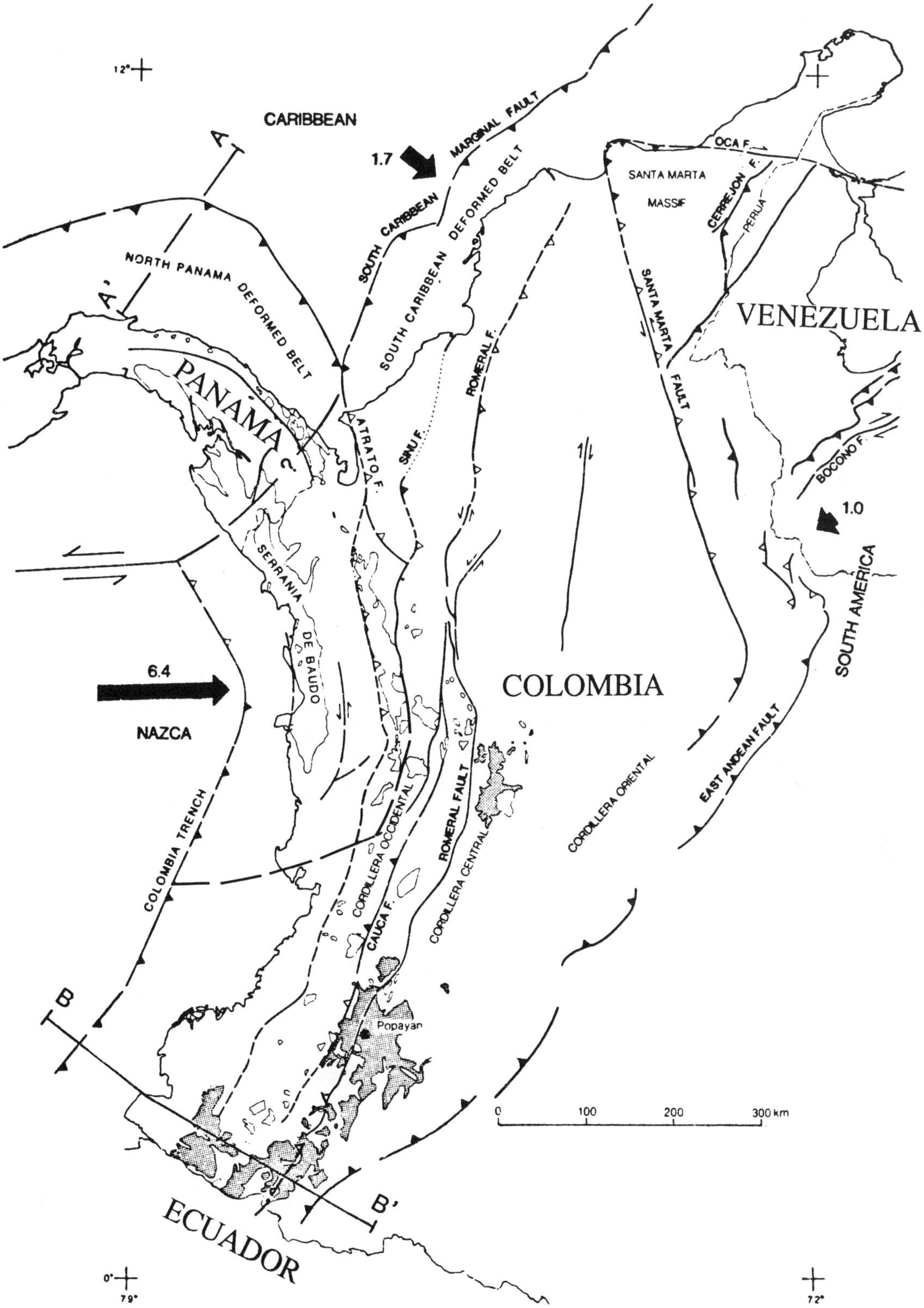

Figure 6. Tectonic map of northwestern South America. Enclosed areas are exposed Tertiary igneous rocks; stippled areas are the youngest andesites and dacites (Tertiary-Quaternary). Dark barbs: active overthrusts. Open barbs: inactive or uncertain. Left-lateral slip shown for the northern Romeral fault zone is based on the results of a 17-month microearthquake seismic study (Hutchings et al., 1981). Present-day plate motions (bold arrows) relative to the northern Andes showing average slip rates (cm/yr) during the last 5 to 10 m.y. are after Minster and Jordan (1978). A-A′: Profile of North Panama fold belt (Figure 7). B-B′: Profile of the Andean margin in southern Colombia (Figures 8 and 9).

Caro, 1979). The North Panama fold belt is bounded on the south by uplifted oceanic basalts, andesites, silicic intrusives, and scattered Tertiary and Quaternary volcanic deposits of eastern Panama (Terry, 1956; Case, 1974).

Seismic refraction measurements near the North Panama fold belt indicate that the Colombian basin crust is oceanic, although its 15-km thickness is greater than typical oceanic crust (Ewing et al., 1960; Houtz and Ludwig, 1977). A multichannel seismic survey shows undeformed Caribbean acoustic basement dipping landward beneath folded sediments of the deformed belt (Lu and McMillen, 1982). Lu and McMillen (1982) pointed out the similarities of the deformed belt with active margins elsewhere and noted that the youngest sediments were involved in the folding.

A two-dimensional density model (Fig. 7; Ogujiofor, 1985) was constructed along multichannel seismic profile CT1-21 of the North Panama fold belt (Lu and McMillen, 1982). Densities were derived from refraction interval velocities (Ewing et al., 1960; Houtz and Ludwig, 1977) using the empirical velocity-density relationships of Nafe and Drake (1957) and Gardner et al. (1974).

The observed gravity field is similar to those of other active accretionary wedges with a forearc basin and landward-dipping oceanic crust beneath the arc (Bowin, 1976; Briceño-Guarupe, 1978). Arcward (south) of the basin, crustal rocks have high densities suggestive of oceanic crustal basement. Mechanically, the arc could therefore be expected to act as a very rigid backstop relative to the weak accretionary wedge.

From the cross-sectional area and geometry of the accretionary prism as shown in Figure 7, an order of magnitude estimate can be made for the age of the prism. We assume that the thickness of sediments on the underthrusting Caribbean plate is constant (1.9 km), the Panama-Caribbean convergence vector is constant (11 mm/yr approximately north/south; Fig. 2), and the contribution of river sediment to the prism is 1,000 metric tons/yr/km^2 of area drained (Milliman and Meade, 1983). If we assume that all of the sediments on the underthrusting Caribbean plate are accreted to the prism, we calculate the river sediment input as 24% of the total and the age of the prism as approximately 18 Ma. Total Panama-Caribbean convergence would be on the order of 200 km. Seismic stratigraphic analysis of the southern Colombian basin (Bowland, 1984) independently suggests that the North Panama fold belt has been active since middle Miocene time (11 to 17 Ma).

Geological and gravity cross sections of the Andean margin in southern Colombia

The Cordillera Occidental (West Andes or Western Cordillera) of Colombia (Fig. 6) is part of the Basic Igneous Complex, one of the world's largest ophiolitic complexes, which extends from Costa Rica through Panama and Colombia to Ecuador. The Western Cordillera consists of Cretaceous marine strata, ultramafic rocks, tholeiitic basalts, calc-alkaline mafic andesites, andesites, and Tertiary quartz diorite intrusives (Case et al., 1973; Galvis, 1980; Arango and Ponce, 1982; Bourgois et al., 1982; Grosser, 1989). In the Central and Eastern Cordilleras of southern Colombia, Quaternary andesitic volcanic deposits cover Precambrian migmatites and Paleozoic metamorphic rocks of greenschist to amphibolite grade (Arango and Ponce, 1982).

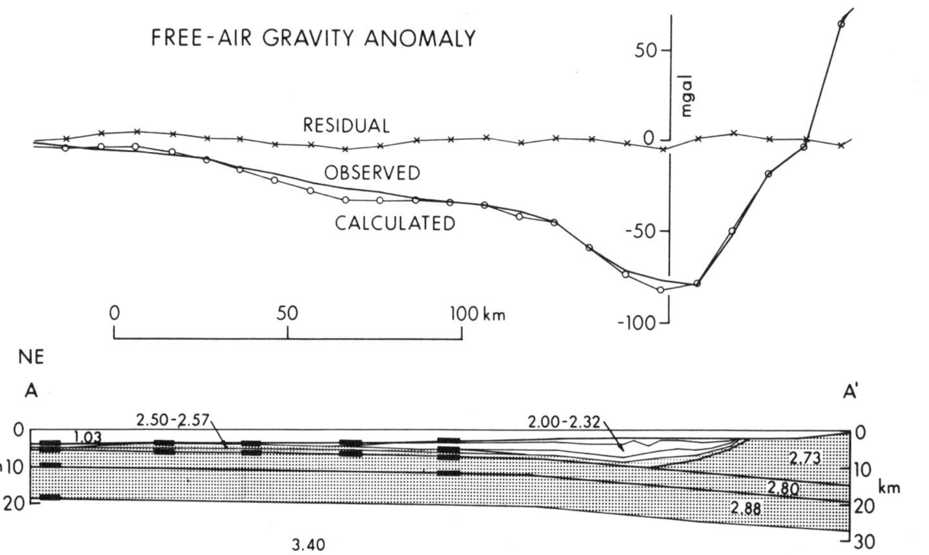

Figure 7. Section of North Panama fold belt showing observed free-air gravity anomalies (Kellogg et al., 1991), two-dimensional density model, and calculated and residual anomalies. Pattern: oceanic crustal basement, 2.73 to 2.88 g/cm^3; dense sediments, 2.50 to 2.57 g/cm^3. The bars indicate approximate locations of seismic refraction velocity discontinuities (Ewing and others, 1960; Houtz and Ludwig, 1977): 2.50 to 2.57 g/cm^3 (1.9 to 3.8 km/s), 2.80 g/cm^3 (4.6 to 6.1 km/s), 2.88 g/cm^3 (7.0 km/s), 3.4 g/cm^3 (8.2 km/s). For location, see Figs. 5 and 6 (A-A′).

To explain the pronounced gravity and seismic velocity gradients across the Romeral fault system, Case et al. (1971, 1973) interpreted the fault as separating dense, high-velocity oceanic crust to the west from lighter, lower-velocity continental crust to the east. Meissner et al. (1976) reached a similar conclusion based on deep seismic refraction results from the Nariño Project. If the paleo-Romeral zone marks a Cretaceous trench, however, why is the fault system as close as 15 km to the Cretaceous batholiths? Bourgois and others (1982) concluded from a structural field study (4°N) that the Cretaceous ophiolitic suite of the Western Cordillera was emplaced by thrusting to the southeast.

To model the paleosuture geometry, Kellogg et al. (1985) constructed a schematic geologic cross section (Fig. 8) from the surface geology (Arango and Ponce, 1982; Ingeominas, 1988). The oceanic-continental suture zone is referred to in this chapter as the paleo-Romeral or Dolores suture zone. The trace of the suture zone was drawn between outcropping oceanic and continental rocks within the Central Cordillera of Colombia (Fig. 6; Arango and Ponce, 1982; Ingeominas, 1988). This is about 30 km east of where Case et al. (1973) placed the suture. Subsurface structure was partly constrained by seismic refraction data (Meissner et al., 1976). A density model (Fig. 9) was then constructed from the geologic cross section and the seismic velocities. The nonuniqueness problem of density models was thus minimized by seismic constraints and the requirement that the model make geological and structural sense. The thickness of sediments in the trench (Colombia fault), shown as 5 or 6 km in Figure 8, is actually no more than 3 km (Westbrook et al., this volume). Oceanic crust is all shown as metaophiolite in this schematic cross section (Fig. 8). The descending oceanic crust can transform to eclogite at 30-km depth, resulting in a local density excess. Grow and Bowin (1975) estimated the regional gravity contribution due to the descending lithosphere at the Chile margin as over 160 mGals at the shore line, decreasing gradually to the east. However, the steep gravity gradients associated with the Western Cordillera gravity high, both parallel to and perpendicular to the margin (Kellogg et al., 1991), indicate that the density anomaly has a crustal origin not associated with the descending lithosphere. Because the regional gravity contribution of the descending lithosphere does not significantly affect the Andean crustal density interpretation, it has been ignored in the density model (Fig. 9). Terrain corrections were not attempted because of inadequate topographic map coverage, but Case et al. (1973) estimated them as up to 15 mGals. Terrain corrections should not significantly affect our interpretation of the 275-mGal anomaly associated with the ocean-continent suture zone.

The crustal thickness model derived from the gravity data by Case et al. (1973) was supported by the deep seismic refraction data from the Nariño Project (Meissner et al., 1976). The crust is as thick as 45 km under the altiplano and as thin as 28 km beneath the Western Andes (Figs. 8 and 9). However, our

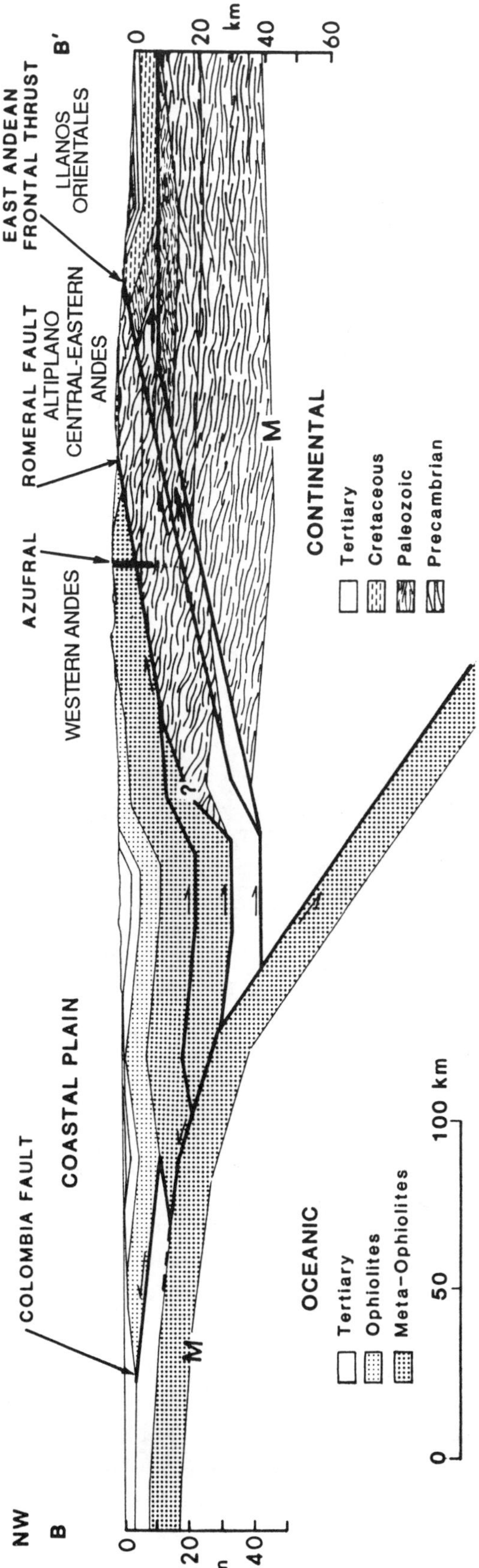

Figure 8. Geologic cross section of the Andean margin in southern Colombia. The vertical bar is a schematic feeder conduit for Azufral volcano. M: Moho boundary. No vertical exaggeration. For location see Figure 6 (B-B').

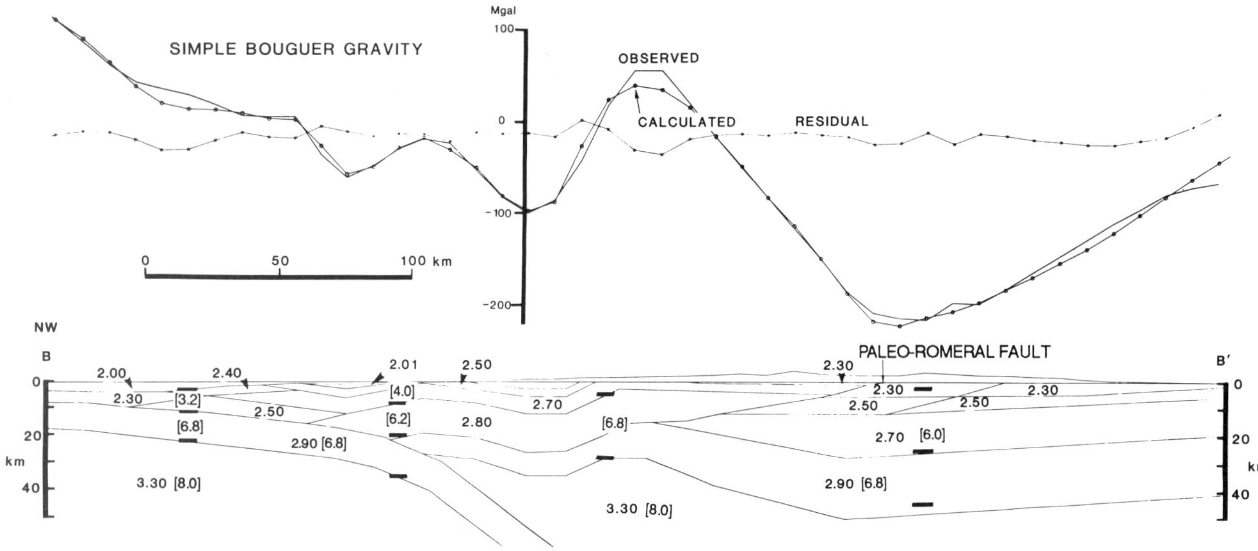

Figure 9. Section of the Andean margin in southern Colombia showing the observed Bouguer gravity anomalies (Kellogg et al., 1991), two-dimensional sea-level density model (g/cm^3), and calculated and residual anomalies. Bars indicate approximate locations of seismic velocity discontinuities from Project Nariño refraction studies (Meissner and others, 1976) with the velocities (km/s) in brackets. No vertical exaggeration. Same location as Figure 8.

model differs from that of Case et al. (1973) in the shallow structure of the suture zone. The difference is primarily caused by our relocation of the surface trace of the paleo-Romeral suture zone based on geologic mapping by Arango and Ponce (1982) and Ingeominas (1988). The observed geology, gravity, and seismic velocities are compatible with a low-angle (approximately 12°) northwest dip on the paleo-Romeral fault (Figs. 8 and 9). Our model supports the conclusion of Toussaint (1978) and of Bourgois and others (1982) that during the Cretaceous period the Western Cordillera oceanic terrane was obducted eastward over continental crust. Ophiolite slabs dipping into the lower crust and upper mantle (Coleman, 1971) are also suggested by geophysical studies of the Troodos massif of Cyprus (Gass and Masson-Smith, 1963), the Eastern Papua New Guinea belt (Milsom, 1973; Finlayson et al., 1977), and the New Caledonia ultramafic complex (Collot et al., 1987). Our geologic cross section (Fig. 8) suggests at least 70 km of horizontal shortening on the paleo-Romeral fault. Such convergence might help explain the proximity of the suture and the Cretaceous batholiths.

Paleozoic and Precambrian crystalline rocks and Cretaceous units overthrust Tertiary sediments on the East Andean frontal thrust fault (Arango and Ponce, 1982), but there is no large gravity anomaly associated with the fault zone. The densities of the metamorphic units may have been reduced by fracturing, or the East Andean décollement may be shallower than we have proposed.

DISCUSSION

Present plate motions

Observed GPS Isla San Andres–Panama City baseline rates of change (Fig. 10) support Panama-Caribbean convergence proposed by Adamek et al. (1988) and Silver et al. (1990). The measured convergence rate (11 mm/yr) is not significantly different at the 95% confidence level from the 7 mm/yr proposed by Silver et al. (1990). The lack of significant movement between Panama City, Panama, and Liberia, Costa Rica, agrees with the rigid Panama microplate model proposed by Adamek et al. (1988) if the microplate includes most of Costa Rica (Fig. 2) but does not support the active Panama bending model (Silver et al., 1990) or the left-shear model (Mann and Corrigan, 1990). However, this preliminary conclusion is only based on observations at two sites in Central America, and the geology of the isthmus and North Panama fold belt suggest that both bending and left-shear were important deformation mechanisms in the past. Alternatively, the Liberia observations can be explained with elastic strain accumulation with a locked subduction zone (Dixon, 1993).

The measured Panama City–Isla Malpelo velocity vector (Figs. 2 and 10) indicates left-lateral strike-slip movement on a transform fault south of Panama, as suggested by Jordan (1975). The small magnitude of the vector (35 mm/yr) and the slower convergence at the Colombia trench (50 mm/yr, Malpelo-Cali,

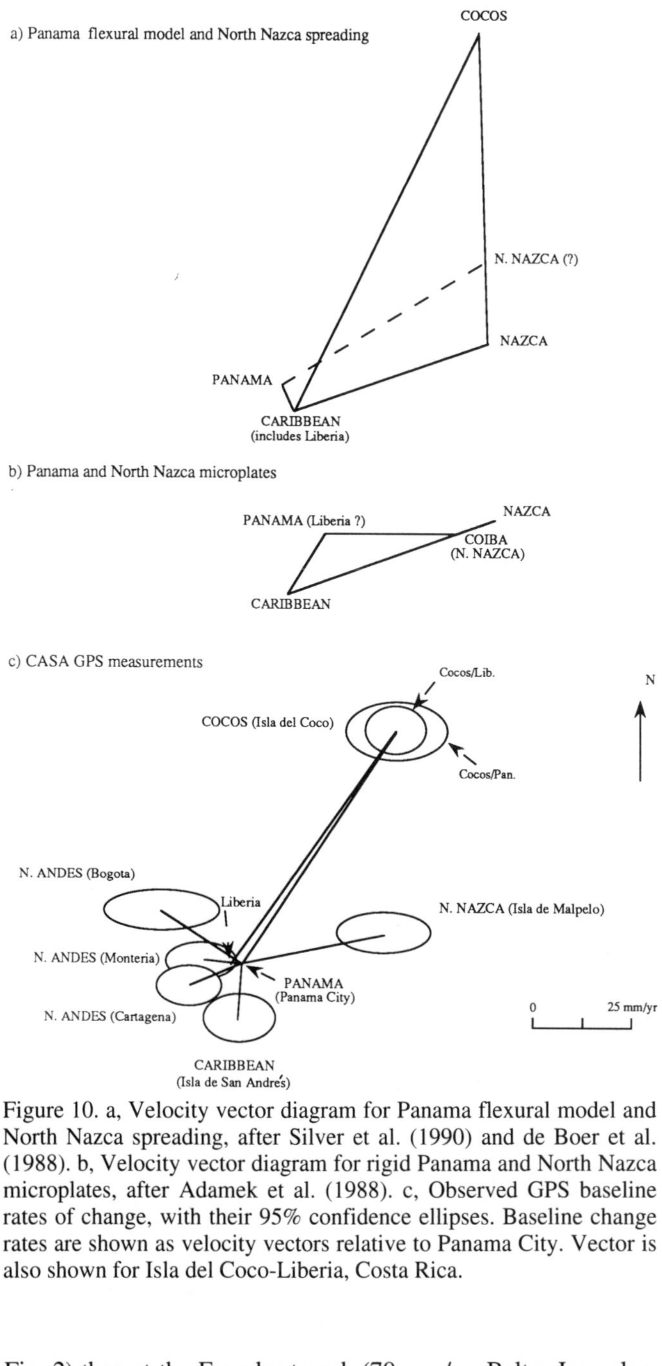

Figure 10. a, Velocity vector diagram for Panama flexural model and North Nazca spreading, after Silver et al. (1990) and de Boer et al. (1988). b, Velocity vector diagram for rigid Panama and North Nazca microplates, after Adamek et al. (1988). c, Observed GPS baseline rates of change, with their 95% confidence ellipses. Baseline change rates are shown as velocity vectors relative to Panama City. Vector is also shown for Isla del Coco-Liberia, Costa Rica.

Fig. 2) than at the Ecuador trench (70 mm/yr, Baltra-Jerusalen, Fig. 2) supports the existence of a separate North Nazca plate, similar to those proposed by Wolters (1986) and Adamek et al. (1988) but extending south of Isla Malpelo. The GPS observations do not indicate significant active northward spreading in the northern part of the Nazca plate as proposed by de Boer et al. (1988, Fig. 10; Silver et al., 1990).

GPS measurements also suggest that the Panama–Costa Rica microplate is continuing to collide eastward with the northern Andes. Rapid subduction is occurring at the Middle America and the Colombia-Ecuador trenches. The northern Andes are moving northeast relative to stable South America. Preliminary GPS results also suggest Caribbean–North Andean convergence, with most of the deformation occurring at the South Caribbean deformed belt.

Miocene collision of Panama arc and South America

The Panama-Choco arc terrane (Figs. 6 and 11) can be recognized in the Western Cordillera by positive gravity anomalies that are continuous over the Serrania de San Blas–Darien in Panama (Case et al., 1971; Case, 1974; Kellogg et al., 1991; Stallings and others, Appendix, this volume); exotic Paleocene planktonic foraminifera, which resemble the Paleocene fauna of northern Central America in Guatemala and Mexico (Duque-Caro, 1990); and Eocene-Oligocene age (34 to 47 Ma) calc-alkaline mafic andesites and andesites (Grosser, 1989). The suture zone separating the Panama-Choco arc from South America passes southeastward along the Atrato fault, through the Western Cordillera of Colombia, then turns southwestward along the Garapatas or Istmina fault zone to meet the Colombia trench (Figs. 6 and 11).

About 6 to 12 Ma the Panama-Choco island arc arrived on the Caribbean plate at the northwestern margin of South America (Fig. 11; Keigwin, 1982; Keller et al., 1989; Duque-Caro, 1990). The arc-continent collision eventually formed a land bridge between the Americas, allowing the migration of mammals; closed the Pacific-Caribbean seaway, drastically changing ocean circulation patterns and perhaps the world's climate, folded the East Panama deformed belt (Kolarsky and Mann, this volume); and uplifted the Eastern Cordillera of Colombia. The correlation between the Panama arc–South America collision and the uplift of the Eastern Cordillera is based on the geometry (the maximum shortening is east-southeast of the suture zone) and the Late Miocene–Pliocene timing of the uplift (Dengo and Covey, 1993). Estimates of shortening in the Eastern Cordillera based on balanced cross sections range from 100 to 150 km (Colletta et al., 1990; Dengo and Covey, 1993). Over a 12-m.y. period this would imply an average shortening rate of 8 to 13 mm/yr. Measurements over a three-year period with GPS satellite geodesy suggest ongoing collision at a rate of 10 to 20 mm/yr between Panama City and Bogota, Colombia.

The proposed Panama-Choco allochthonous arc also helps explain the occurrence of westward coarsening of quartzose turbidites of Late Cretaceous age in the Western Cordillera (Calle and Salinas, 1986; Duque-Caro, 1990). The observed westward coarsening is not in agreement with the eastern continental lands, the Central Cordillera, as the quartz source for these Cretaceous basins, the explanation that has previously been generally accepted.

Late Cretaceous accretion of Western Cordillera oceanic terrane

During Late Cretaceous time, the suture zone between oceanic and continental crust was 250 km east of the present trench at the paleo-Romeral fault (Fig. 12). The geologic and gravity model in Figures 8 and 9 support the conclusion of Toussaint (1978), and Bourgois et al. (1982) that the Western Cordillera oceanic terrane was obducted eastward over continental crust. The trace of the suture zone in Figure 12 was drawn between outcropping oceanic and continental rocks within the Central Cordillera of Colombia (Arango and Ponce, 1982; Ingeominas, 1988) and thus represents schematically the position of the terrane boundaries at the end of obduction. The Western Cordillera (West Andean) oceanic terrane may have been derived from the Caribbean plate and been continuous with the Aves Ridge terrane (Fig. 12; Pindell and Barrett, 1990). The Western Cordillera terrane was an island arc south in Ecuador and an immature oceanic island arc (e.g., Barrero, 1977; McCourt et al., 1984) or ocean crust (Bourgois et al., 1982) north in Colombia. The Aves Ridge terrane was subsequently transported eastward to its present location east of the Venezuelan Basin by Tertiary Caribbean plate motion. The andesitic arc rocks recognized by Barrero (1977) and Grosser (1989) in the northwest flank of the Western Cordillera are part of the Panama-Choco arc accreted during the last 6 to 12 m.y.

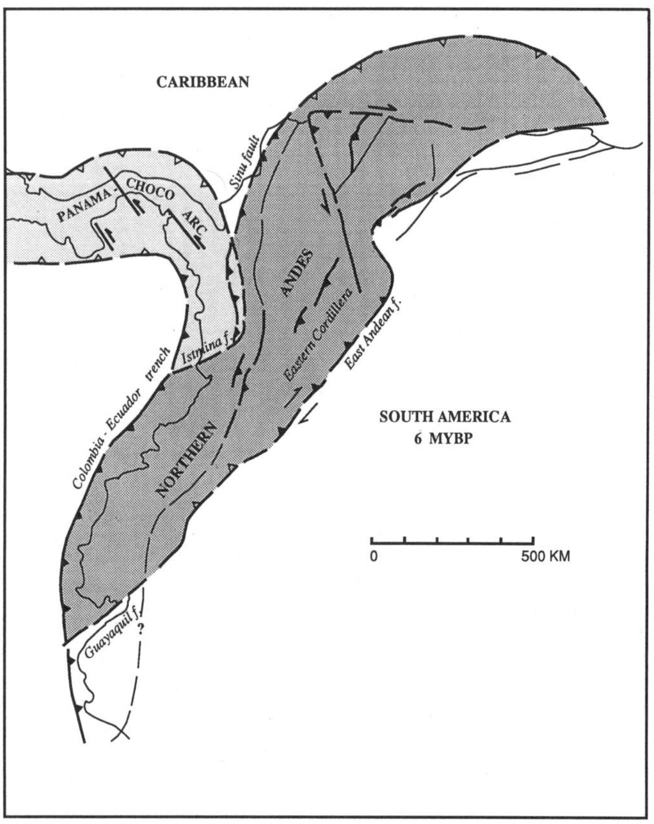

Figure 11. Schematic reconstruction of the northern Andean margin 6 Ma showing the Panama-Choco arc (light pattern) and the North Andean block or microplate (dark pattern). Plio-Pleistocene right-lateral strike-slip movement on the East Andean fault has been removed, but no attempt has been made to remove Plio-Pleistocene crustal shortening in the northern Andes. Open barbs on thrust zones: location uncertain. The present coastlines are shown for location.

SUMMARY

We propose the following simplified sequence of events for the accretionary development of southern Central America and the northern Andes:

1. During Late Cretaceous time the Western Cordillera oceanic terrane was obducted eastward on the paleo-Romeral suture zone over South American continental crust.

2. The accretion of the Sinu–San Jacinto sedimentary wedge (South Caribbean deformed belt) commenced in the Eocene epoch (Duque-Caro, 1979; Toto and Kellogg, 1992) contemporaneous with a major change in Caribbean plate motion.

3. In late Oligocene time uplift of the Central Cordillera of Colombia (unpublished apatite fission-track ages), the Santa Marta massif, and the Sierra de Perijá (Kellogg, 1984) began.

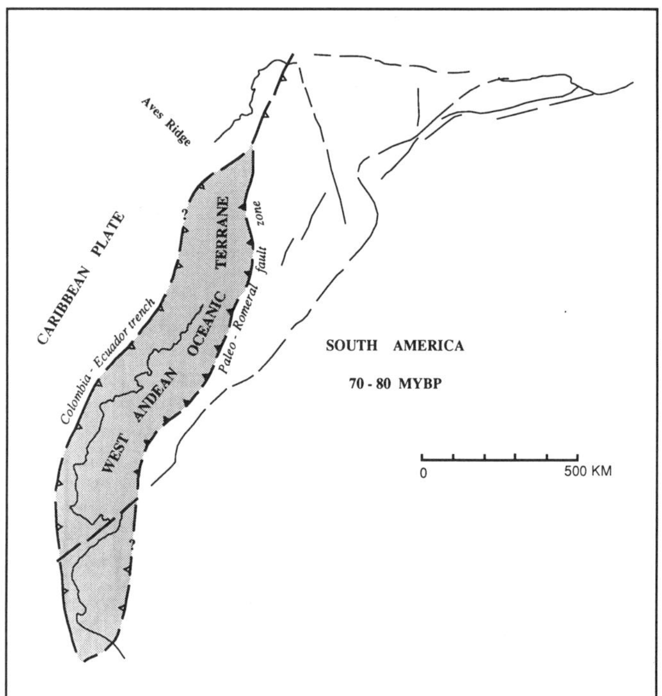

Figure 12. Schematic reconstruction of the northern Andean margin 70 to 80 Ma showing the positions of the Caribbean plate, the accreted West Andean oceanic terrane (dark pattern), the paleo-Romeral fault zone, and the Aves Ridge. Open barbs on thrust zones: location uncertain. The present coastlines and fault traces are shown for location.

4. The late Miocene-Pliocene collision of the Panama-Choco island arc with South America eventually formed a land bridge between the Americas, allowing the migration of mammals; closed the Pacific-Caribbean seaway, drastically changing ocean circulation patterns and perhaps the world's climate; folded the East Panama deformed belt; and uplifted the Eastern Cordillera of Colombia. The coastal range (Serrania de Baudo) was accreted to the continental margin.

5. GPS measurements suggest the existence of a rigid Panama–Costa Rica microplate that is moving northward relative to the stable Caribbean plate and continuing to collide eastward with the northern Andes. Measurements do not indicate active left-slip or bending of Panama. However, the geology of the isthmus and North Panama fold belt suggest that both left-slip and bending were important deformation mechanisms in the past. Rapid subduction is occurring at the Middle America and the Colombia-Ecuador trenches. The northern Andes are moving northeast relativc to stable South America. Preliminary GPS results also suggest an independent North Nazca plate and Caribbean–North Andean convergence, with most of the deformation occurring in the South Caribbean deformed belt. We expect that additional occupations of the CASA GPS network will significantly improve the uncertainties in the GPS-derived plate motion rates and will help resolve details of active slip partitioning in southern Central America and the northern Andean margin.

ACKNOWLEDGMENTS

The CASA GPS Project is a continuing collaborative effort involving scientists from over 30 institutions in Europe and the Americas. Key scientists involved in CASA include Timothy Dixon, Ruth Neilan, and Steven Fisher (Jet Propulsion Laboratory); James Stowell (University NAVSTAR Consortium); Clemente Ropaín (Instituto Nacional de Investigaciones Geológico-Mineras, Colombia); Sergio Camargo (Instituto Geográfico Agustín Codazzi, Colombia); José Sáenz (Instituto Geográfico Nacional, Panamá); Marino Protti (Universidad Nacional, Heredia, Costa Rica); Wolfgang Niemeier (Technical University, Surveying Institute, Braunschweig, Germany); Aníbal Salazar (Instituto Geográfico Militar, Ecuador); Jeffrey Freymueller (Stanford University); Hugh Caddess, Kenneth Barger, Charles Klimicek, Thomas Hughes, and Vernon Perdue (Inter-American Geodetic Survey); Brian Tallman and Robert Russman (Defense Mapping Agency); Hermann Drewes (German Geodetic Research Institute, Munich); and Melvin Hoyer (University of Zulia, Venezuela). Careful reviews by James Case, Graham Westbrook, Paul Mann, and an anonymous reviewer improved the manuscript. The work presented here was supported by NSF grants EAR-8617485 and EAR-8904657, NASA, the University of South Carolina, and the University NAVSTAR Consortium.

APPENDIX. GRAVITY ANOMALY MAP OF SOUTHERN CENTRAL AMERICA

T. C. Stallings, Carlos L. V. Aiken, and James N. Kellogg

The data used to construct the gravity anomaly map (Plate 1) were from several bases: (1) Aiken, C. L. V., University of Texas at Dallas, and Kellogg, J. N., University of South Carolina, Latin America gravity data base, 1992, (2) U.S. Geological Survey (Ponce, 1986), (3) Defense Mapping Agency/Gravity Library, St. Louis, 1990, and (4) Committee for Gravity Anomaly Map of North America (1987). All these data sets had principle facts except for (4), which consisted of a 6-km grid. In order to edit the data for the map the above order was used for prioritizing the preference of the station values if there were data value conflicts.

The offshore data consist exclusively of grid values from the Gravity Anomaly Map of North America, distributed by the National Geophysical Data Center (NGDC). Examining the station location map from the North American map, it can be seen that the offshore data are made up of some tracklines in the Caribbean, digitized contour maps over much of the area, and a grid of Seasat altimeter data for the remaining areas. This gridded data set was considered the most complete for the marine areas of the map, so part of the North American grid was extracted and used. The land areas of the new map (Plate 1) include data from Aiken and Kellogg that were used in the original North American map along with additional data from Aiken and Kellogg and new data from the Defense Mapping Agency and Ponce, 1986. The data coverage of Costa Rica, and to a lesser extent of Panama, is better than that used for previously published maps of the area. Part of the land coverage also includes the gridded North American data, because there were some areas in western Panama where the coverage from the grid was slightly superior. There are areas of very sparse gravity coverage in the map, especially along the Caribbean coasts of Nicaragua and western Panama.

The data were gridded at a 10-km interval and then contoured at 5-mGal intervals by the RADIAN COS/PC graphics software. Areas with very sparse data are left blank, and hachures are on the side of more negative gravity values. This marine free-air and land Bouguer anomaly map extends farther west and south than most published gravity maps of the area (e.g., Case et al., 1971; Bowin, 1976; Kellogg et al., 1991).

All the main tectonic elements discussed in the text (Fig. 2) have apparent gravity expressions. This is especially true for the marine free-air anomalies that reflect the bathymetry, and hence the first-order tectonics. For example, the trenches, which are great bathymetric lows, result in elongate gravity lows off the Pacific coasts of Nicaragua and Costa Rica and off the Colombian coast. The high Cocos Ridge, trending northeasterly perpendicular to the Middle America trench (Fig. 4), is associated with a gravity high. The Middle America trench is intersected by the north-trending Panama fracture zone, which is characterized by a series of parallel gravity highs and lows. To its east is a parallel gravity high over the Coiba Ridge. The gravity low north of Panama reflects the thick sedimentary wedge of the North Panama fold belt (Fig. 7).

On land, Bouguer gravity anomaly lows are associated with the volcanoes of northern Costa Rica. High Bouguer gravity anomalies reflect the uplift of oceanic crust in Panama (Case et al., 1971) and the continuation of the Panama-Choco arc (Fig. 11) into northwestern Colombia.

REFERENCES CITED

Adamek, S., Frohlich, C., and Pennington, W. D., 1988, Seismicity of the Caribbean-Nazca boundary: Constraints on microplate tectonics of the Panama region: Journal of Geophysical Research, v. 93, p. 2053–2075.

Aggarwal, Y., 1983, Seismic slip rates and earthquake rupture zones in the southern Caribbean: Implications for plate motions and earthquake hazard in this region, Abstracts Caribbean Geological Conference, 10th: Cartagena, Colombia, Instituto Nacional de Investigaciones Geológico-Mineras, p. 16.

Arango C., J. L., and Ponce M., A., 1982, Mapa geológico generalizado del Departamento de Nariño: Bogotá, Colombia, Instituto Nacional de Investigaciones Geológico-Mineras, scale: 1:400,000.

Barrero, D., 1977, Geology of the Central Western Cordillera, west of Buga and Roldanillo, Colombia [Ph.D. thesis]: Golden, Colorado School of Mines, 157 p.

Bermudez G., A., 1985, Estudio frecuencial de terremotos del noroccidente de Sur America y areas vecinas: Bogotá, Colombia, Instituto Nacional de Investigaciones Geológico-Mineras, 25 p.

Beutler, G., Bauersima, I., Gurtner, W., Rothacher, M., and Schildknecht, T., 1987, Evaluation of the 1984 Alaska Global Positioning System campaign with the Bernese GPS software: Journal of Geophysical Research, v. 92, p. 1295–1303.

Blewitt, G., 1989, Carrier phase ambiguity resolution for the Global Positioning System applied to geodetic baselines up to 2,000 km: Journal of Geophysical Research, v. 94, p. 10187–10203.

Bourgois, J., Calle, B., Tournon, J., and Toussaint, J-F., 1982, The Andean ophiolitic megastructures on the Buga-Buenaventura transverse (Western Cordillera–Valle Colombia): Tectonophysics, v. 82, p. 207–299.

Bowin, C., 1976, Caribbean gravity field and plate tectonics: Geological Society of America Special Paper 169, 79 p.

Bowland, C. L., 1984, Seismic stratigraphy and structure of the western Colombian Basin, Caribbean Sea [M.S. thesis]: Austin, University of Texas at Austin, 248 p.

Briceño-Guarupe, L. A., 1978, The crustal structure and tectonic framework of the Gulf of Panama [M.S. thesis]: Corvallis, Oregon State University, 71 p.

Burbach, G., Frolich, C., Pennington, W., and Matumoto, T., 1984, Seismicity and tectonics of the subducted Cocos plate: Journal of Geophysical Research, v. 89, p. 7719–7735.

Calle, B., and Salinas, R., 1986, Geología y geoquímica de la plancha 165, Carmen de Atrato: Bogota, Instituto Nacional de Investigaciones Geológico-Mineras, Report Bogota 1987.

Case, J. E., 1974, Oceanic crust forms basement of eastern Panama: Geological Society of America Bulletin, v. 85, p. 645–652.

Case, J. E., and Dengo, G., 1982, The Caribbean Region, *in* Palmer, A. R., ed., Perspectives in regional geological synthesis: Planning for The Geology of North America: Geological Society of America DNAG Special Publication 1, p. 163–170.

Case, J. E., and Holcombe, T. L., 1980, Geologic-tectonic map of the Caribbean region: U.S. Geological Survey Miscellaneous Investigation Series Map I-1100, 2 sheets, scale 1: 2,500,000.

Case, J. E., Duran S., L. G., Lopez R., A., and Moore, W. R., 1971, Tectonic investigations in western Colombia and eastern Panama: Geological Society of America Bulletin, v. 82, p. 2685–2712.

Case, J. E., Barnes, L., Paris, G., Gonzalez, H., and Viña, A., 1973, Trans-Andean geophysical profile, southern Colombia: Geological Society of America Bulletin, v. 84, p. 2895–2904.

Case, J. E., Holcombe, T. L., and Martin, R. G., 1984, Map of geological provinces in the Caribbean region, *in* Bonini, W. E., Hargraves, R. B., and Shagam, R., Caribbean–South American plate boundary and regional tectonics: Geological Society of America Memoir 162, p. 1–30; map, scale 1:5,000,000.

Coleman, R. G., 1971, Plate tectonic emplacement of upper mantle peridotites along continental edges: Journal of Geophysical Research, v. 76, p. 1212–1222.

Colletta, B., Gebrard, F., Letouzey, J., Werner, P., Rudkiewicz, J.-L., 1990, Tectonic style and crustal structure of the Eastern Cordillera (Colombia) from a balanced cross-section, *in* Letouzey, J., ed., Petroleum and tectonics in mobile belts: Paris, Editions Technip, p. 81–100.

Collot, J. Y., Malahoff, A., Recy, J., Latham, G., and Missegue, F., 1987, Overthrust emplacement of New Caledonia ophiolite: Geophysical evidence: Tectonics, v. 6, p. 215–232.

Committee for Gravity Anomaly Map of North America, 1987, Gravity anomaly map of North America: Geological Society of America Continent/Ocean Scale Map-002, scale 1:5,000,000.

de Boer, J., Defant, M. J., Stewart, R. H., Restrepo, J. F., Clark, L. F., and Ramirez, A., 1988, Quaternary calc-alkaline volcanism in western Panama: Regional variation and implication for plate tectonic framework: Journal of South American Earth Sciences, v. 1, p. 275–293.

DeMets, C., Gordon, R. G., Argus, D. F., and Stein, S., 1990, Current plate motions: Geophysical Journal International, v. 101, p. 425–478.

Dengo, C. A., and Covey, M. C., 1993, Structure of the Eastern Cordillera of Colombia: Implications for trap styles and regional tectonics: American Association of Petroleum Geologists Bulletin, v. 77, p. 1315–1337.

Dengo, G., 1985, Mid America: Tectonic setting for the Pacific margin from southern Mexico to northwestern Colombia, *in* Nairn, A.E.M., and Stehli, F. G., eds., The ocean basins and margins: New York, Plenum Press, v. 7, p. 123–180.

De Obaldia, F., Marino, T., van der Laat, R., Hernandez, F., Malavassi, E., and McNally, K. C., 1991, Coseismic uplift associated with the Costa Rica earthquake (Ms 7.5) of April 22, 1991: EOS (American Geophysical Union Transactions), v. 72, p. 301.

Dewey, J. F., and Pindell, J. L., 1985, Neogene block tectonics of eastern Turkey and northern South America: Continental application of the finite difference method: Tectonics, v. 4, p. 71–83.

Dewey, J. W., 1972, Seismicity and tectonics of western Venezuela: Bulletin of the Seismological Society of America, v. 62, p. 1711–1751.

Dixon, T. H., 1993, GPS measurement of relative motion of the Cocos and Caribbean plates and strain accumulation across the Middle America trench: Geophysical Research Letters, v. 20, p. 2167–2170.

Dong, D. N., and Bock, Y., 1989, Global Positioning System network analysis with phase ambiguity resolution applied to crustal deformation studies in California: Journal of Geophysical Research, v. 94, p. 3949–3966.

Duque-Caro, H., 1979, Major structural elements of northern Colombia: American Association of Petroleum Geologists Memoir 29, p. 329–351.

Duque-Caro, H., 1990, The Choco Block in the northwestern corner of South America: Structural, tectonostratigraphic, and paleogeographic implications: Journal of South American Earth Sciences, v. 3, p. 71–84.

Edgar, N. T., Ewing, J. I., and Hennion, J., 1971, Seismic refraction and reflection in Caribbean Sea: American Association of Petroleum Geologists Bulletin, v. 55, p. 833–870.

Escalante, G., 1990, The geology of southern Central America and western Colombia, *in* Dengo, G., and Case, J. E., eds., The Caribbean Region: Geological Society of America, The Geology of North America, v. H., p. 201–230.

Ewing, J., Antonie, J., and Ewing, M., 1960, Geophysical measurements in the western Caribbean Sea and in the Gulf of Mexico: Journal of Geophysical Research, v. 65, p. 4087–4126.

Fan, G.-W., Beck, S. L., and Wallace, T. C., 1993, The seismic source parameters of the 1991 Costa Rica aftershock sequence: Evidence for a transcurrent plate boundary: Journal of Geophysical Research, v. 98, p. 15759–15778.

Finlayson, D. M., Drummond, B. J., Collins, C.D.M., and Connelly, J. B., 1977, Crustal structures in the region of the Papuan ultramafic belt: Physics of Earth and Planetary Interiors, v. 14, p. 13–29.

Freymueller, J. T., 1991, CASA–Central and South America GPS Geodesy: Crustal motions determined from 1988 and 1990 epoch measurements in Colombia, Costa Rica, and Ecuador [Ph.D. thesis]: Columbia, University of South Carolina, 172 p.

Freymueller, J. T., Kellogg, J. N., and Vega, V., 1993, Plate motions in the North Andean Region: Journal of Geophysical Research, v. 98, p. 21853–21863.

Galvis Vergara, J., 1980, Un arco de islas Terciario en el occidente Colombiano: Geologia Colombiana, v. 11, p. 7–43.

Gardner, G.H.F., Gardner, L. W., and Gregory, A. R., 1974, Formation velocity and density—The diagnostic bases for stratigraphic traps: Geophysics, v. 39, p. 770–780.

Gass, I. G., and Masson-Smith, D., 1963, The geology and gravity anomalies of the Troodos massif, Cyprus: Philosophical Transactions of the Royal Society of London, Series A, v. 255, p. 417–467.

Goes, S.D.B., Velasco, A. A., Schwartz, S. Y., and Lay, T., 1993, The April 22, 1991, Valle de la Estrella, Costa Rica (M_w = 7.7) earthquake and its tectonic implications: A broadband seismic study: Journal of Geophysical Research, v. 98, p. 8127–8142.

Grosser, J. R., 1989, Geotectonic evolution of the Western Cordillera of Colombia: New aspects from geochemical data on volcanic rocks: Journal of South American Earth Sciences, v. 2, p. 359–369.

Grow, J. A., and Bowin, C. O., 1975, Evidence for high-density crust and mantle beneath the Chile trench due to the descending lithosphere: Journal of Geophysical Research, v. 80, p. 1449–1458.

Guendel, F., and 13 others, 1991, Mainshock-aftershock sequence associated with the Costa Rica earthquake of April 22, 1991: EOS (American Geophysical Union Transactions), v. 72, p. 301.

Houtz, R. E., and Ludwig, W. J., 1977, Structure of Colombian Basin, Caribbean Sea, from profiler sonobuoy measurements: Journal of Geophysical Research, v. 82, p. 4861–4867.

Hutchings, L., Turcotte, T., McBride, J., and Ochoa, H., 1981, Microseismicity along and near the Dolores shear zone in Antioquia, Colombia: San Francisco, Woodward-Clyde Consultants for Interconexion Electrica S.A., 12 p.

Ingeominas, 1988, Mapa geológico de Colombia: Bogotá, Colombia, Instituto Nacional de Investigaciones Geológico-Mineras, 2 sheets, scale 1:1,500,000.

Jordan T., 1975, The present-day motions of the Caribbean plate: Journal of Geophysical Research, v. 80, p. 4433–4439.

Keigwin, L., 1982, Isotopic paleoceanography of the Caribbean and east Pacific: Role of Panama uplift in late Neogene time: Science, v. 217, p. 350–353.

Keller, G., Zenker, C. E., and Stone, S. M., 1989, Late Neogene history of the Pacific-Caribbean gateway: Journal of South American Earth Sciences, v. 2, p. 73–108.

Kellogg, J. N., 1984, Cenozoic tectonic history of the Sierra de Perijá and adjacent basins, *in* Bonini, W. E., Hargraves, R. B., and Shagam, R., eds., Caribbean–South American plate boundary and regional tectonics: Geological Society of America Memoir 162, p. 239–261.

Kellogg, J. N., and Bonini, W. E., 1982, Subduction of the Caribbean plate and basement uplifts in the overriding South American plate: Tectonics, v. 1, p. 251–276.

Kellogg, J. N., and Bonini, W. E., 1985, Reply *to* Comment *on* 'Subduction of the Caribbean plate and basement uplifts in the overriding South American plate': Tectonics, v. 4, p. 785–790.

Kellogg, J. N., and Dixon, T., 1990, Central and South America GPS geodesy—CASA UNO: Geophysical Research Letters, CASA UNO Special Issue, v. 17, p. 195–198.

Kellogg, J. N., Ogujiofor, I. J., and Kansakar, D. R., 1985, Cenozoic tectonics of the Panama and North Andes blocks, *in* Memoirs, Latin American Congress on Geology, 6th, v. 1: Bogota, Consejo Consultivo de Directores de Servicios Geologicos de Latinoamerica y el Caribe, p. 40–59.

Kellogg, J. N., Dixon, T., and Neilan, R., 1989, Central and South America GPS Geodesy-CASA: EOS (American Geophysical Union Transactions), v. 70, p. 649–656.

Kellogg, J. N., Godley, V. M., Ropain, C., Bermudez, A., and Aiken, C.L.V., 1991, Gravity field of Colombia, eastern Panama, and adjacent marine areas: Geological Society of America Map and Chart Series MCH070, scale 1:1,500,000.

Ladd, J. W., and seven others, 1984, Seismic reflection profiles acrossthe southern margin of the Caribbean, *in* Bonini, W. E., Hargraves, R. B., and Shagam, R., eds., Caribbean–South American plate boundary and regional tectonics: Geological Society of America Memoir 162, p. 153–159.

Lehner, P., Doust, H., Bakker, G., Allenbach, P., and Gueneau, J., 1983, Active margins–Caribbean margin of South America, profiles C-1422, C-1412, and C-1413, *in* Bally, A. W., ed., Seismic expression of structural styles: Tulsa, American Association of Petroleum Geologists, Studies in Geology ser. 15, v. 3, p. 3.4.2-111–3.4.2-128.

Lu, R. S., and McMillen, K. J., 1982, Multichannel seismic survey of the Colombia Basin and adjacent margins, *in* Watkins, J. S., and Drake, C. L., eds., Studies in continental margin geology: American Association of Petroleum Geologists Memoir 34, p. 395–410.

Lundgren, P. R., Kornreich Wolf, S., Protti, M., and Hurst, K. J., 1993, GPS measurements of crustal deformation associated with 22 April 1991, Valle de la Estrella, Costa Rica earthquake: Geophysical Research Letters, v. 20, p. 407–410.

Mann, P., and Corrigan, J., 1990, Model for late Neogene deformation in Panama: Geology, v. 18, p. 558–562.

McCourt, W. J., Aspden, J. A., and Brook, M., 1984, New geological and geochronological data from the Colombian Andes: Continental growth by multiple accretion: Journal of the Geological Society, London, v. 141, p. 831–845.

Meissner, R. O., Flueh, E. R., Stibane, F., and Berg, E., 1976, Dynamics of the active plate boundary in southwest Colombia according to recent geophysical measurements: Tectonophysics, v. 35, p. 115–136.

Milliman, J. D., and Meade, R. H., 1983, World-wide delivery of river sediment to the oceans: The Journal of Geology, v. 91, p. 1–21.

Milsom, J. S., 1973, Papuan ultramafic belt: Gravity anomalies and the emplacement of ophiolites: Geological Society of America Bulletin, v. 84, p. 2243–2258.

Minster, J. B., and Jordan, T. H., 1978, Present-day plate motions: Journal of Geophysical Research, v. 83, p. 5331–5354.

Nafe, J. E., and Drake, C. L., 1957, Variation with depth in shallow and deep water marine sediments of porosity, density, and velocity of compressional and shear waves: Geophysics, v. 23, p. 523–552.

National Geophysical Data Center, 1990, Decade of North American Geology/ Bouguer (land) and free-air (marine) 6 km anomaly grid: Boulder, Colorado, Geological Society of America, Geophysics of North America CD-ROM.

Ogujiofor, I. J., 1985, Active convergent margins of northwest South America [M.S. thesis]: Honolulu, University of Hawaii, 100 p.

Pennington, W. D., 1981, Subduction of the eastern Panama basin and the seismotectonics of northwestern South America: Journal of Geophysical Research, v. 86, p. 10753–10770.

Pindell, J. L., and Barrett, S. F., 1990, Geological evolution of the Caribbean region: A plate-tectonic perspective, *in* Dengo, G., and Case, J. E., eds., The Caribbean Region: Geological Society of America, The Geology of North America, v. H, p. 405–432.

Plafker, G., and Ward, S. N., 1992, Backarc thrust faulting and tectonic uplift along the Caribbean sea coast during the April 22, 1991 Costa Rica earthquake: Tectonics, v. 11, p. 709–718.

Ponce, D. A., 1986, Gravity observations in northwest Costa Rica: U.S. Geological Survey Open-File Report 86-556, 9 p.

Rosencrantz, E., and Sclater, J. G., 1986, Depth and age in the Cayman Trough: Earth and Planetary Science Letters, v. 79, p. 133–144.

Schubert, C., 1980, Late-Cenozoic pull-apart basins, Boconó fault zone,

Venezuelan Andes: Journal of Structural Geology, v. 2, p. 463–468.

Shagam, R., 1975, The northern termination of the Andes, *in* Nairn, A.E.M., and Stehli, E. G., eds., The ocean basins and margins: New York, Plenum, v. 3, p. 325–420.

Silver, E. A., Case, J. E., and MacGillavry, H. J., 1975, Geophysical study of the Venezuelan borderland: Geological Society of America Bulletin, v. 86, p. 213–226.

Silver, E. A., and six others, 1990, Implications of the North and South Panama deformed belts for the origin of the Panama orocline: Tectonics, v. 9, p. 261–282.

Terry, R. A., 1956, A geological reconnaissance of Panama: California Academy of Science Occasional Paper 23, 91 p.

Toto, E. A., and Kellogg, J. N., 1992, Structure of the Sinu–San Jacinto fold belt—An active accretionary prism in northern Colombia: Journal of South American Earth Sciences, v. 5, p. 211–222.

Toussaint, J-F., 1978, Grandes rasgos geológicos de la parte septentrional del occidente Colombiano: Medellín, Universidad Nacional de Colombia, Boletin de Ciencias de la Tierra, v. 3, 231 p.

Tralli, D. M., and Dixon, T. H., 1988, A few parts in 10^8 geodetic baseline repeatability in the Gulf of California using the Global Positioning System: Geophysical Research Letters, v. 15, p. 353–356.

van Andel, T., Heath, G., Malfait, B., Heinrichs, D., and Ewing, J., 1971, Tectonics of the Panama Basin, eastern equatorial Pacific: Geological Society of America Bulletin, v. 82, p. 1489–1508.

Vega, V., and Kellogg, J. N., 1992, Repeat observations over northwestern South America, the Caribbean, and Panama from the CASA Global Positioning System Project: EOS (American Geophysical Union Transactions), v. 73, p. 86.

Wadge, G., and Burke, K., 1983, Neogene Caribbean plate rotation and associated Central American tectonic evolution: Tectonics, v. 2, p. 633–643.

Wolters, B., 1986, Seismicity and tectonics of southern Central America and adjacent regions with special attention to the surroundings of Panama: Tectonophysics, v. 128, p. 21–46.

Manuscript Accepted by the Society February 7, 1994

Printed in U.S.A.

Geological Society of America
Special Paper 295
1995

Structure and tectonics of the Panama-Nazca plate boundary

Graham K. Westbrook, Nicholas C. Hardy*, and Roger P. Heath*
School of Earth Sciences, University of Birmingham, Edgbaston, Birmingham, B15 2TT, United Kingdom

ABSTRACT

The structure and the tectonic development of the southern Panama plate boundary have been derived from an interpretation of marine geophysical data, including GLORIA (Geological LOng Range Inclined Asdic) long-range side-scan sonar and seismic reflection profiles, most of which were acquired in 1989 from a cruise of the RRS *Charles Darwin*. The northern boundary of the Nazca plate runs within the continental margin of southern Panama. South of the Gulf of Panama this boundary, the Southern Panama fault zone, is predominantly left-lateral strike-slip and occupies an elongate sedimentary basin. South and southwest of the Azuero Peninsula the boundary becomes one of oblique subduction, with active formation of an accretionary complex. The eastern part of this accretionary complex slips around the bend in the overriding Panama block to the purely strike-slip portion of the plate boundary, where it ceases to accrete sediment. South of the Gulf of Panama, the fault zone is flanked on its southern side by a bathymetric ridge, containing rocks of a high density, that was once a part of the Panamanian continental margin. The ridge has been displaced 140 km eastward by the motion between the Panama block and the Nazca plate. The eastern end of this ridge is being subducted beneath South America, and at the ridge crest, the deformation front of the Colombian accretionary complex meets the Southern Panama fault zone. The inactive trench, filled with sediment, that lies at the foot of the Southern Panama continental margin owes its existence, in the west, to the downward flexure of the Nazca plate beneath the overriding Panama block at the oblique subduction boundary, and in the east, where the lithosphere of the Nazca plate is "broken" along its transform boundary, to the flexural load of the displaced basement ridge. A continuation of the Southern Panama fault zone runs southeastward behind the Colombian accretionary complex, separating it from forearc basin sediments deposited on the "continental" basement of northernmost Colombia. Deformed mud diapirs indicate a component of left-lateral strike-slip motion on the fault zone. This pattern of tectonics around this northernmost corner of the Nazca plate has probably been active since the collision of Panama and South America about 3.5 Ma.

INTRODUCTION

The Panama block appears to be an independent microplate at the "junction" of the Cocos, Nazca, and Caribbean plates (Pennington, 1981; Adamek et al., 1988; Vergara Muñoz, 1988b) (Fig. 1). The plate boundary south of the Panama block was first described as a left-lateral transform by Jordan (1975), which he envisaged separated the Caribbean and Nazca plates. Evidence from teleseismic earthquakes for such a transform fault is inconclusive (Wolters, 1986; Adamek et al., 1988). There is, however, no evidence for a well-defined Benioff zone that would support the occurrence of subduction

*Present addresses: Hardy, Amerada Hess Ltd., 33 Grosvenor Place, London, SW1X 7HY, United Kingdom; Heath, Amoco U.K. Exploration Company, Amoco House, Westgate, London, W5 1XL, United Kingdom.

Westbrook, G. K., Hardy, N. C., and Heath, R. P., 1995, Structure and tectonics of the Panama-Nazca plate boundary, *in* Mann, P., ed., Geologic and Tectonic Development of the Caribbean Plate Boundary in Southern Central America: Boulder, Colorado, Geological Society of America Special Paper 295.

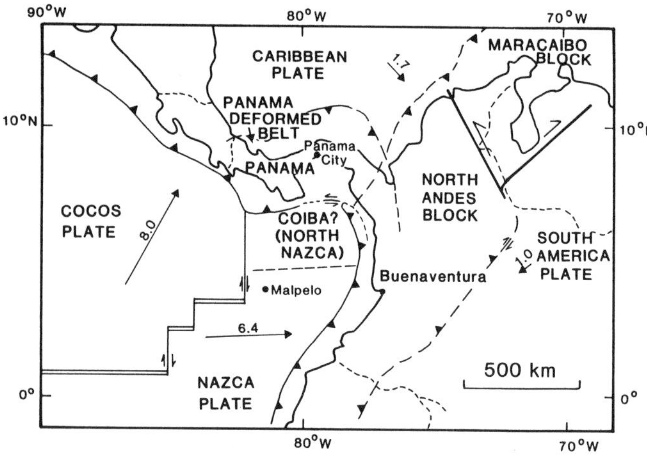

Figure 1. The location of the southern Panama plate boundary in relation to the surrounding plates (adapted from Kellogg and Dixon, 1990). Arrows indicate motions of plates with respect to the North Andes block in cm/yr, except for the Cocos plate, for which the motion is shown relative to the Caribbean plate.

at the margin for any large period of time. A series of northwest-southeast–trending left-lateral faults is known to exist, in a zone stretching from Coiba Island to the eastern side of the Azuero Peninsula (Mann and Corrigan, 1990; Vergara Muñoz, 1988a; Okaya and Ben-Avraham, 1987). It has been proposed that these form part of a diffuse boundary (Mann and Burke, 1984) rather than the simple transform of Jordan (1975). A hybrid of these two models proposed by Vergara Muñoz (1988b) comprises a left-lateral transform south of the Gulf of Panama that passes westward into a broad left-lateral shear zone between the Gulf of Panama and the Coiba Ridge. Analysis of the tectonics of this region is further complicated by the postulated existence of a Coiba microplate (Adamek et al., 1988) or North Nazca microplate (Silver et al., 1990) immediately to the south of the Panama block and forming the northernmost part of what is generally regarded as the Nazca plate.

Active subduction along the southern margin of Panama west of 80°30′W, shown by seismic reflection profiles (MacKay and Moore, 1990; Heil, 1988; Moore and Sender, this volume), proves a degree of oblique convergence between the Panama block and the Nazca plate (or Coiba microplate). Data on volcanism in the region by de Boer et al. (1988) appear to support this view, showing that, although there is currently little active volcanism in Panama, significant volcanic activity occurred in western Panama during the Quaternary. This Quaternary volcanism has led authors (e.g., Silver et al., 1990) to assume that active subduction occurs along the full length of the south Panama boundary.

Silver et al. (1990) have attributed the shape and the deformation pattern of the Isthmus of Panama to processes similar to those induced by bending a flexible beam under east-west compression. Mann and Corrigan (1990) have further described diffuse deformation within the Panama block involving the complex interaction of oroclinal bending with left-lateral strike-slip faulting. The importance of a degree of nonrigid deformation is acknowledged, especially in eastern Panama. The mechanism for deformation of the Panama block is probably a combination of "tectonic escape" (Mann and Burke, 1984) and oroclinal bending (Silver et al., 1990) as the block is squeezed between Central and South America.

In July 1989, the continental margin in the region south of the Gulf of Panama (Fig. 2) was investigated with GLORIA (Geological LOng Range Inclined Asdic) long-range side-scan sonar, seismic reflection, 3.5 kHz profiling, and measurements of the Earth's gravity and magnetic fields from Cruise 40 of the RRS *Charles Darwin* (Hardy, 1991a). These data shed new light on the structure and evolution of the continental margin south of Panama, and together with other extant data, provide the basis for the interpretation presented in this chapter.

NEW SIDE-SCAN SONAR DATA

The GLORIA data collected from *Charles Darwin* Cruise 40 were processed to correct the slant range and the variation of reflected amplitude with range, depth, and slope of the seabed and were digitally composed into the mosaic displayed as Plate 2. Coverage of the area is not complete. (The time available in the cruise was cut by 40% from that originally planned because of unforeseen logistic and political problems.) Even so, the major features of the margin can be seen in the mosaic, an interpretation of which is given in Plate 2.

Submarine canyons

In the eastern and northern margins of the mosaic a large number of canyons are developed in the continental slope of Panama and northern Colombia, including the eastern part of the southern boundary of the Gulf of Panama. The canyons have fairly straight courses, reflecting the steep slope, and tend to join together downslope. They terminate at the base of the slope, generally in some form of forearc basin. (The term "forearc basin" is used throughout this chapter in the general sense of a basin on the margin of the overriding plate at a subduction zone, situated landward of an accretionary wedge, where an accretionary wedge is present.)

Spreading fabric in the Nazca plate

Along the southwestern edge of the southern part of the mosaic, the comparatively featureless floor of the Panama basin, with its pelagic and hemipelagic sediment cover that is seen on seismic reflection profiles to be around 500 m thick, shows east-west–trending lineations that are associated with ridges and troughs in the oceanic crust. The lineations represent a structural fabric acquired from the east-west–oriented spreading center from which this portion of the Nazca plate was formed (Lonsdale and Klitgord, 1978; Hardy, 1991b).

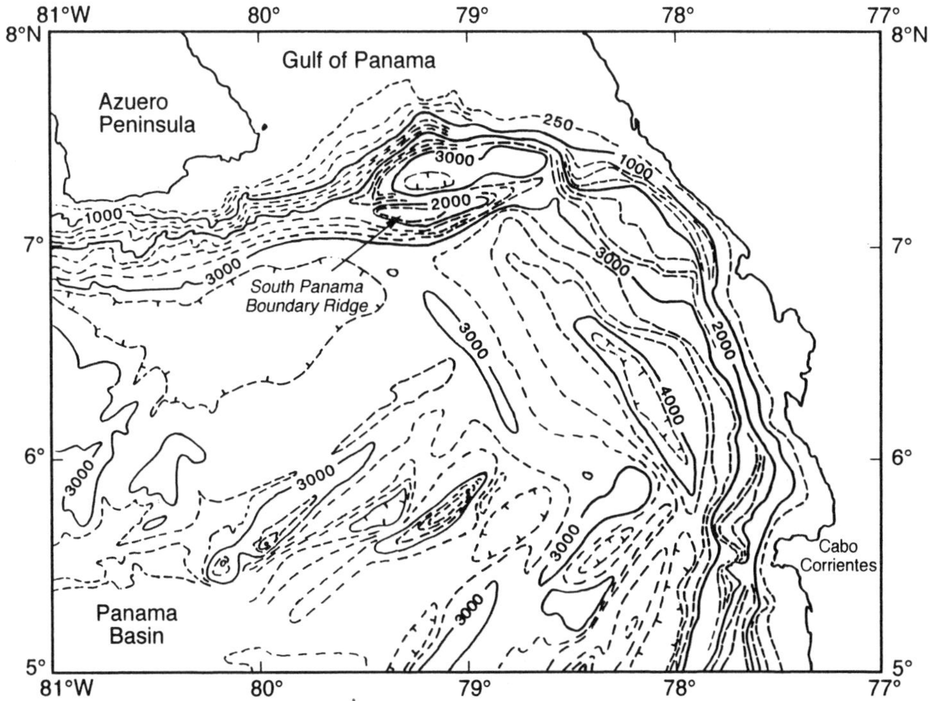

Figure 2. Bathymetric map of the region. Solid contours every 1,000 m; dashed contours every 250 m.

Colombian accretionary complex

The seaward boundary of the accretionary complex forming the landward flank of the Colombian Trench is imaged well along most of its length. At its northwestern end the front of the complex bulges westward where the thicker sedimentary section filling a trough described by Lowrie (1978) as the "buried Panama trench" is accreted. East of the front lies a zone 20 to 30 km wide with lineated fabric that is semiparallel to the front. The lineations come from westward-facing slopes that are normally gently convex to the west and range in their length from 5 to 30 km. Seismic sections show that in many (perhaps all) cases these lineations represent the outcrops of thrust faults in the accretionary complex. The faults commonly coalesce or terminate against each other in a pattern reminiscent of overlapping shingles. The typical separation of these faults is 2 km. At 6°43′N, the bright lineament that marks the deformation front (seaward boundary) of the accretionary complex is absent over about a 10-km length. In this gap, and in the trench 20 km north and south of it, is a region of generally bright seafloor reflectivity with many smaller brighter targets. The seismic reflection and 3.5 kHz profiles crossing this bright region show a layer of incoherent internal structure beneath a rough seafloor. This feature is a large slump from the landward slope of the trench, the more fluidized parts of which have flowed along the trench in both directions from the source of the slump.

Colombian forearc basin and fault zone

Between the lineated part of the accretionary complex and the continental slope lies a region of smoother seafloor that is underlain by a thick slope-drape and forearc basins. The region is widest at 7°N and narrows southward. Within it lies a major lineament comprising several individual anastomosing lineaments that lie close to the western (seaward) margin of the forearc basin. The lineament is a major oblique-slip fault system downthrowing to the west, with left-lateral horizontal displacement. The vertical movement is shown in the seismic sections. The evidence for the horizontal movement comes from the interpretation of the outcrop shape of mud diapirs that lie on or east of the fault between 6°25′N and 7°15′N, especially those between 6°55′N and 7°10′N. The mud diapirs are identified from the seismic reflection and 3.5 kHz records as well as from their bright reflective appearance as seafloor features. Ten km away from the fault they are almost round in shape. Closer to the fault, the diapirs become progressively more elongate and semiparallel to the fault trend, so that they have an arcuate form overall. Assuming that there was no preferential alignment prior to movement of the fault, then a minimum displacement on the fault of about 8 km can be inferred.

Southern Panama fault zone

The fault system can be followed to the northwest beyond the accretionary complex, where it curves around to an east-west trend and runs through a sedimentary basin that lies immediately south of the continental slope of the Gulf of Panama. Between the sedimentary basin and the Panama Basin lies a ridge, the Southern Panama Boundary Ridge, with a steep (20°) north-facing slope that is a very bright feature on the GLORIA sonographs. The high reflectivity is probably a consequence of a rough eroded surface and the exposure of consolidated rocks. The southern side of the ridge has a lower reflectivity with substantial sediment cover. Running down the southern slope are many canyons similar to, but not as bright as, those on the continental margin. These canyons do not appear to be active. The 3.5 kHz records show them to be mantled with a thin sediment cover. The canyons come to the crest of the ridge, where they have no hinterland to drain. At the eastern end of the ridge there is no feature that marks the base of the slope, other than the termination of the canyons. West of 79°15′W, the foot of the slope is occupied by wavy, westerly trending features that seismic and 3.5 kHz data show to be gently folded sediments probably displaced by small thrusts. These features are not cut by the canyons, and so postdate them. Between 79°15'W and 79°00'W a single lineament runs along the foot of the slope with another fainter feature 4 to 6 km north of it that is cut by the canyons and farther west, lies north of the northernmost of the younger folds.

The eastern end of the ridge noses down eastward beneath the accretionary wedge and the basinal sediment behind (as shown by the seismic reflection data; see below), and forms the northern boundary of the accretionary wedge. The front of the wedge trends north-northeast as it climbs the southern flank of the ridge and cuts across the canyons as the wedge overrides them, giving further evidence that they are inactive.

The basin lying north of the ridge narrows progressively to the west and shallows, as shown by seismic profiles by Lowrie (1978), by unpublished GLORIA data and bathymetry from Cruise 110 of the RRS *Discovery* in 1980, and by the data presented here. The basin eventually disappears west of 80°W, passing into the rear of the accretionary complex that lies south of the Azuero Peninsula (MacKay and Moore, 1990).

NEW SEISMIC REFLECTION DATA

Multichannel seismic reflection profiles 15 and 17 (Figs. 3 and 4, and Plates 3 and 4) show that normal subduction, with some accretion, is occurring along the Colombia trench as far north as 7°05′N. In contrast, single-channel seismic reflection profiles, extending from the oceanic crust of the east Panama Basin onto the basement high at the mouth of the Gulf of Panama, image only a filled trough in which a largely undisturbed sedimentary sequence is seen to overlie more chaotic reflectors at depth, with no indications of active subduction.

Northern Colombian accretionary complex

Line 15. This 24-fold stacked and migrated seismic section shows a structure that is typical of the northern part of the Colombian continental margin (Fig. 4a). A sequence of pelagic and hemipelagic sediments, about 500 m thick, on the ocean crust is covered by a turbidite fill up to 250 m thick and 3 km wide that occupies the bottom of the trench. At the leading edge of the accretionary complex, the turbidites and uppermost hemipelagic sediments are deformed and stripped away from the remainder of the hemipelagic and pelagic sequence below, along a décollement which carries them attached to the subducting oceanic lithosphere beneath the accretionary complex for a distance of at least 40 km from the trench. At the toe of the accretionary complex its structure is a fairly simple one of westward-verging folded thrust slices, but farther east the structure is more difficult to resolve. There are many landward- (eastward-) dipping reflectors within the complex. Near the surface there are many short reflectors that are close to parallel to the seabed and are likely to be part of a slope-drape sequence. With the exception of the bottom-simulating reflector (BSR) believed to be produced by gas hydrate, none of these reflectors is longer than about 1 or 2 km in the region extending to 25 km landward of the trench. What slope-drape sediments exist in this portion of the complex have been sufficiently disturbed by the tectonics of the deforming complex for it to be virtually impossible to distinguish any clear boundary between them and accreted sediment. From 25 km to the landward end of the seismic section a substantial thickness of sediment overlies the accretionary complex. East of 40 km from the trench (CDP 1966), the overlying sediments occupy a well-defined forearc basin, reaching a maximum thickness of about 3 km in the vicinity of CDP 2550. The sediments in the basin lap onto a high-amplitude seaward-dipping reflector that is the surface of the basement of the continental margin. Two mud diapirs at CDPS 2225 and 2280 have come to the surface through the forearc basin and are clearly visible on the GLORIA sonographs as part of a group of diapirs arranged en echelon along a major fault zone. The major fault zone with left-lateral strike-slip motion is intersected by the section between CDPS 1940 and 2060. The structure displayed by the seismic section in this region is obscure, and no clear dip or sense of vertical displacement on the fault can be determined.

Line 17. This seismic section crosses the accretionary complex in a direction that is oblique to the direction of subduction, although it is locally orthogonal to structure at the front of the accretionary wedge (Fig. 4b). There is little, or no, trench fill where the section crosses the trench. The décollement beneath the accretionary wedge appears to have formed close to what may be the boundary between pelagic and hemipelagic sediments on the subducting oceanic lithosphere. Without many clear marker horizons, the structure within the toe of the wedge is not well defined. Farther landward, northeast of CDP 1750, the upper part of the section contains substantial thicknesses (up to 1 sec two-way time (twt) of well-layered sequences with significant lateral continuity (5 km). These sediments could represent a slope basin that has subsequently been disturbed, in which case it is not clear how the sediments had a long enough time to accumulate on such an apparently young portion of the

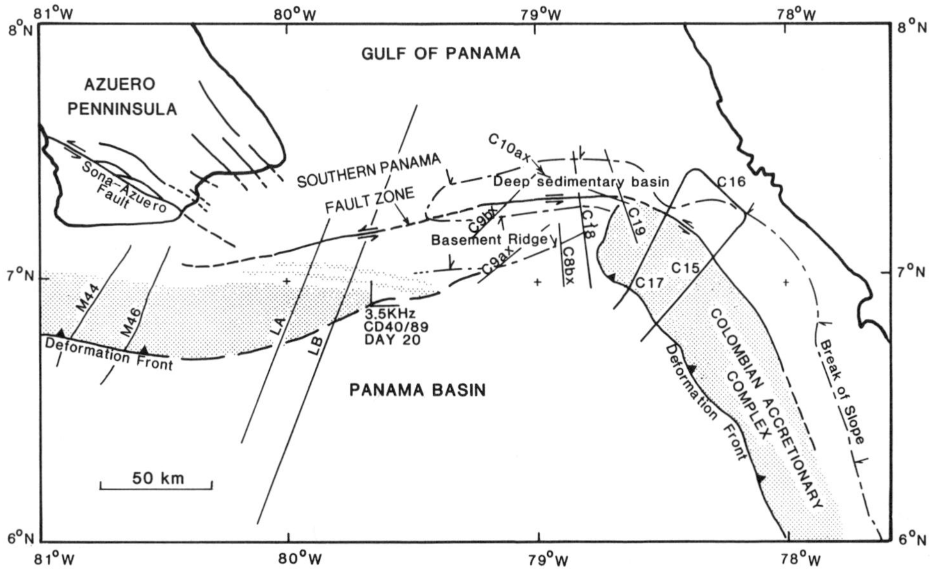

Figure 3. Map of seismic reflection lines described in the text, in relation to the major structural and bathymetric features. The extent of the seismic reflection lines shown on the map is what is given in Figures 4, 5, and 7 through 11 (with exception of lines C16, M44, M46, LA, and LB). Accretionary complexes are shown by stipple.

wedge. Alternatively the sediments could be a trench sequence accreted during a period when there was substantial trench fill. In the southern part of the Colombian trench there is a thick turbidite fill, which is currently prevented from filling the northern part of the trench by a subducting basement ridge blocking the trench at 5°50′N (Hardy, 1991a). The section also crosses into a region of the accretionary complex that is currently accreting the thick sediment sequence filling the basement trough along the southern edge of the southern Panamanian margin, described by Lowrie (1978) as a buried trench. Northeast of CDP 1300 it appears that slope-drape–basin deposits occupy the upper part of the section.

Beneath the accretionary complex, the subducting igneous ocean crust with a layer of sediment overlying it can be traced to 12 km from the trench, and the oceanic crust alone can be traced to 30 km (CDP 1000). Farther northeast there is a major change in the structure. The structure in the upper part of the section becomes very broken and incoherent between CDPS 910 and 820, beneath part of the seabed that is locally rough. This ma;y be the fault zone identified on the GLORIA sonographs, although locally the main fault trace appears from the sonographs to run through CDP 730, where there is a small bump in the seabed. Northeast of the fault zone, a slope sequence of sediments onlaps onto a shallowing basement reflector. This reflector has an irregular surface and may have been faulted, although such faults do not pass clearly into the overlying sedimentary section. The absence of the basement reflector southwest of the fault zone is a measure of the significance of the fault zone as a major tectonic feature. The landward part of Line 17 climbs a bathymetric spur that lies at the foot of the main continental slope. A positive gravity anomaly and short-wavelength magnetic anomalies are associated with this spur, which appears to be underlain by a locally uplifted segment of the basement of the hinterland. The spur lies along the strike of a subducting basement ridge on the northern edge of the Nazca plate (see below), and it is probable that the spur is produced by the local uplift caused by the relief of this ridge.

Southern Panama Boundary Ridge and fault zone

Line 18. Both the deep and shallow structure of this part of the margin are well imaged by Line 18 (Fig. 5). Normally faulted oceanic basement at the southern end of the line (CDPS 300–750) lies beneath 0.4 to 0.6 sec twt of sediments on the oceanward side of the buried trench. The thickness of the trench-fill sediment sequence increases northward toward the margin as the depth of basement increases. Initially the bedding is subhorizontal and largely undisturbed, but broken chaotic reflectors are characteristic of the deepest part of the trench, which extends to >2 sec twt beneath the seabed. This disturbed section is overlain by bedded reflectors, apparently from a northern source, because reflectors with continuity upslope lap onto the disturbed zone and the units between the reflectors thin onto and over the zone. A slope apron deposit of up to 1 sec twt (approximately 1 km) in thickness blankets the southern flank of the Southern Panama Boundary Ridge that lies directly to the north of the buried trench. This sequence thickens down slope and onlaps onto an acoustic basement within the ridge; it appears, at least in part, to have a northern source. The junction between oceanic basement and the edge

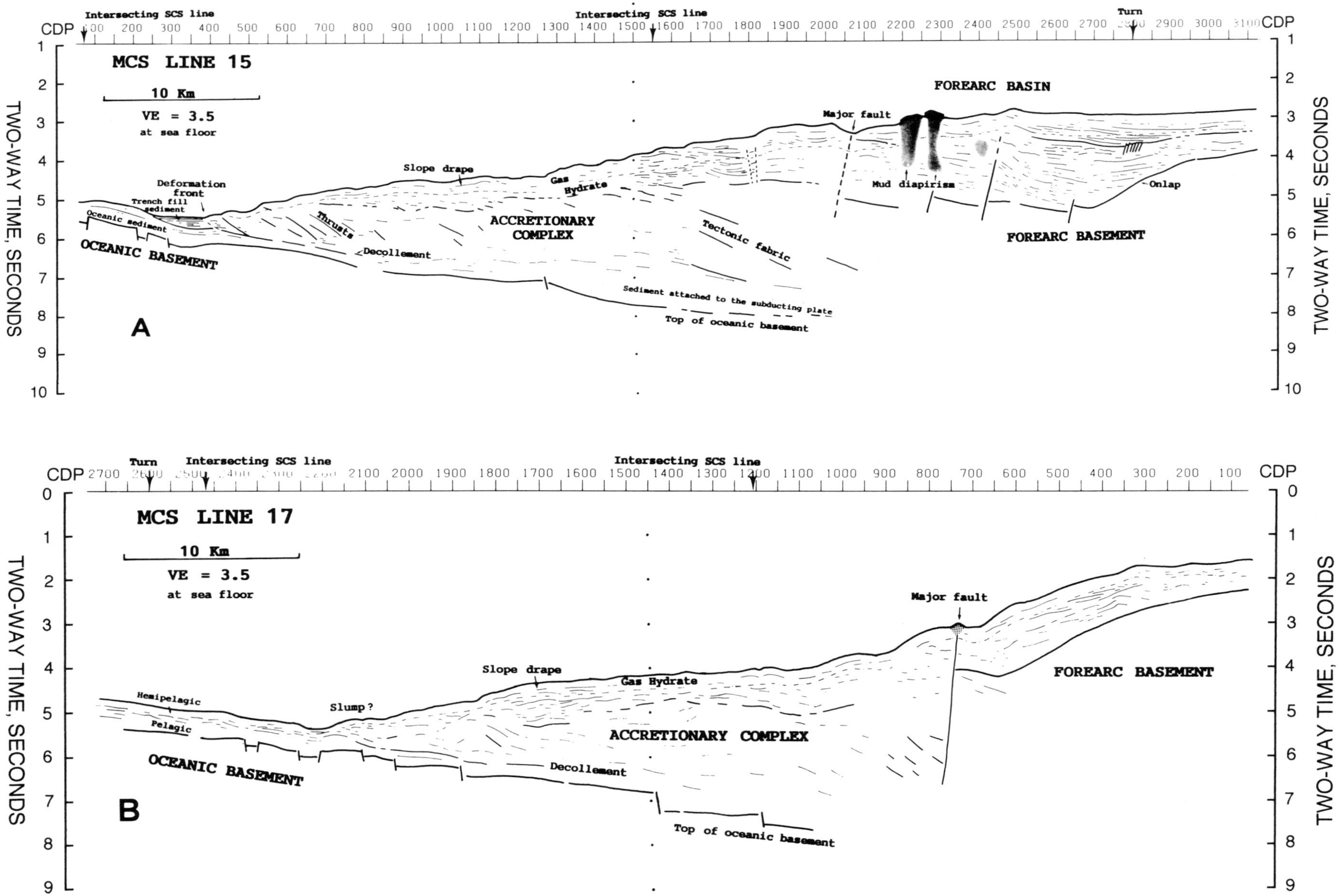

Figure 4. A, line drawing of multichannel seismic section 15 (24-fold stacked and migrated), showing a cross section through the northernmost part of the Colombian accretionary complex. The structure is typical of the complex north of 5°50′N. Location of line is shown in Figure 3. B, line drawing of multichannel seismic section 17. This section runs obliquely across the ridge on the northern edge of the subducting Nazca plate, which is believed to be responsible for the higher than normal elevation of the forearc (continental) basement. (Seismic sections for lines 15 and 17 are displayed as Plates 3 and 4.)

of the acoustic basement reflector of the ridge is not clear, but oceanic basement appears to extend beneath it a little way.

The basement of the ridge outcrops along its northern flank. Separating the ridge from the continental shelf, at this longitude, is an elongate east-west–oriented sedimentary basin. A narrow (approximately 2 km) strip of undeformed, well-bedded sediments directly abuts the northern ridge flank. A folded and deformed zone, approximately 7 km wide, rises 300 m above this strip, where the profile crosses the fault zone identified from GLORIA data. This style of deformation indicates an element of convergence across the fault at this location. In the most active part of the fault zone, few coherent reflectors are visible, and the data are affected by numerous migration smiles that are in part the result of spatial aliasing. It is proposed that this highly deformed zone is part of an anastomosing fault system that forms the active plate boundary at the southern edge of the Panama block. It is likely that faulting within the basement itself is less complex and that it fans out into a flower structure as it extends upward through relatively soft basin sediments (e.g., Naylor et al., 1986). Although coherent reflectors are generally absent in the vicinity of the fault zone, a BSR is prominent. This reflector is attributed to the base of a gas hydrate layer within the sediment.

Between the fault zone and the continental slope lies a wedge-shaped sequence, 12 km wide at the surface, of sediments that onlap the continental slope. The sequence is at least 3.0 sec thick in reflection time (nearly 4 km). The sediments become increasingly, although gently, deformed with increasing depth and proximity to the fault zone. It is clear that they have become progressively deformed during deposition. During the most recent phase of deposition, the raised relief of the fault zone has acted as the seaward margin of the basin.

A satisfactory model of crustal structure that could produce the gravity anomaly may be obtained if the basement of the Southern Panama Boundary Ridge south of the basin is assigned a density of 2,860 kg/m^3 and the basement of the continental slope 2,800 kg/m^3, with the sediments of the basin only 2,030 kg/m^3. The model (Fig. 6) confirms the V-shaped cross-sectional shape of the basin containing the fault zone indicated by the seismic sections. The basement densities would allow the basement to be composed of rocks similar to the Cretaceous mafic volcanic rocks of the Azuero Peninsula and eastern Panama. A magnetic anomaly running along the ridge, with a peak-to-peak amplitude of about 200 nT, is also indicative of the presence of mafic igneous rocks within it.

Line 19. This profile crosses the region of transition between the faulted margin south of the Gulf of Panama and the accretionary complex along the Pacific margin of Colombia (Fig. 7). From north to south, the profile starts on the continental slope, crosses an area of likely slumping at the slope foot, and passes out onto the deep sedimentary basin described above. Bedded basin-sediments are imaged down to 2 sec twt beneath the seabed. The first 0.5 to 1.0 sec twt of sediment is largely undeformed, but the broken nature of deeper reflectors is indicative of relatively intense deformation. Reflector separation increases toward a depocenter close to the middle of the basin, suggesting synsedimentary subsidence. The resolution of features at depth is not really adequate to determine whether sediments in the deeper basin are faulted, as is the case on Line 18.

Continuing southward, the profile passes up onto the northern edge of the Colombian accretionary complex. Reflectors within the complex are disturbed and often chaotic. A BSR is again prominent, implying the presence of a gas hydrate layer. Beneath the northern part of the deformed complex, basement is imaged at approximately 2 sec twt beneath the seabed. This is interpreted as the top of the basement ridge that outcrops on Line 18, which, together with the buried trench, is drawn down beneath the accretionary complex as it enters the Colombian subduction zone, as the northernmost part of the Nazca plate. The sediments forming the upper part of the fault zone are apparently pushed relatively southward over the northern flank of the basement ridge so that they directly abut the accretionary complex. Sediments that lay to the south of the fault zone have probably been deformed and accreted to the accretionary complex. A décollement must lie between the accretionary complex and the basement ridge, although we have not been able to identify its position on the seismic section from Line 19. Sediment beneath this décollement will be subducted along with the ridge.

Lines 9bx and 10ax. Further coverage of the basin situated at the mouth of th Gulf of Panama is provided by single-channel profiles 9bx and 10ax. Line 9bx (Fig. 8) crosses obliquely from the basement ridge to the south and strikes northeastward across the basin. The ridge basement is essentially seismically opaque, showing no indication of the internal structure. The basin sediments are well bedded, with reflectors clearly imaged down to a maximum of 1.5 sec twt beneath the seabed. Active faulting and associated folding are seen to be present in the vicinity of the South Panama fault zone, although the sea-bottom relief in the fault zone is less pronounced than that observed on Line 18. Line 10ax (Fig. 9) lies in the eastern half of the basin and intersects Lines 18 and 19. Much of the profile is over the fault zone with its rough seafloor, so penetration is generally poor. The submerged basement ridge is imaged at the eastern end of the line, coming to within 1.5 sec twt of the sea bottom, where it is overlain by deformed sediments in the flank of the fault zone.

Line 8bx. In addition to Line 18, the buried trench south of the Gulf of Panama is also crossed by single-channel reflection profiles 8bx and 9ax. Line 8bx, which is nearly perpendicular to the trench, shows the characteristic normal faulting in the oceanic crust at the southern flank of the trench (Fig. 10). The trench-fill deposits are again well bedded, except in the deepest part of the trench, where coherent reflectors dome up over a zone of more chaotic reflectors similar to those imaged on Line 18. The southern flank of the basement ridge is covered by a thick apron of sediments, apparently sourced from

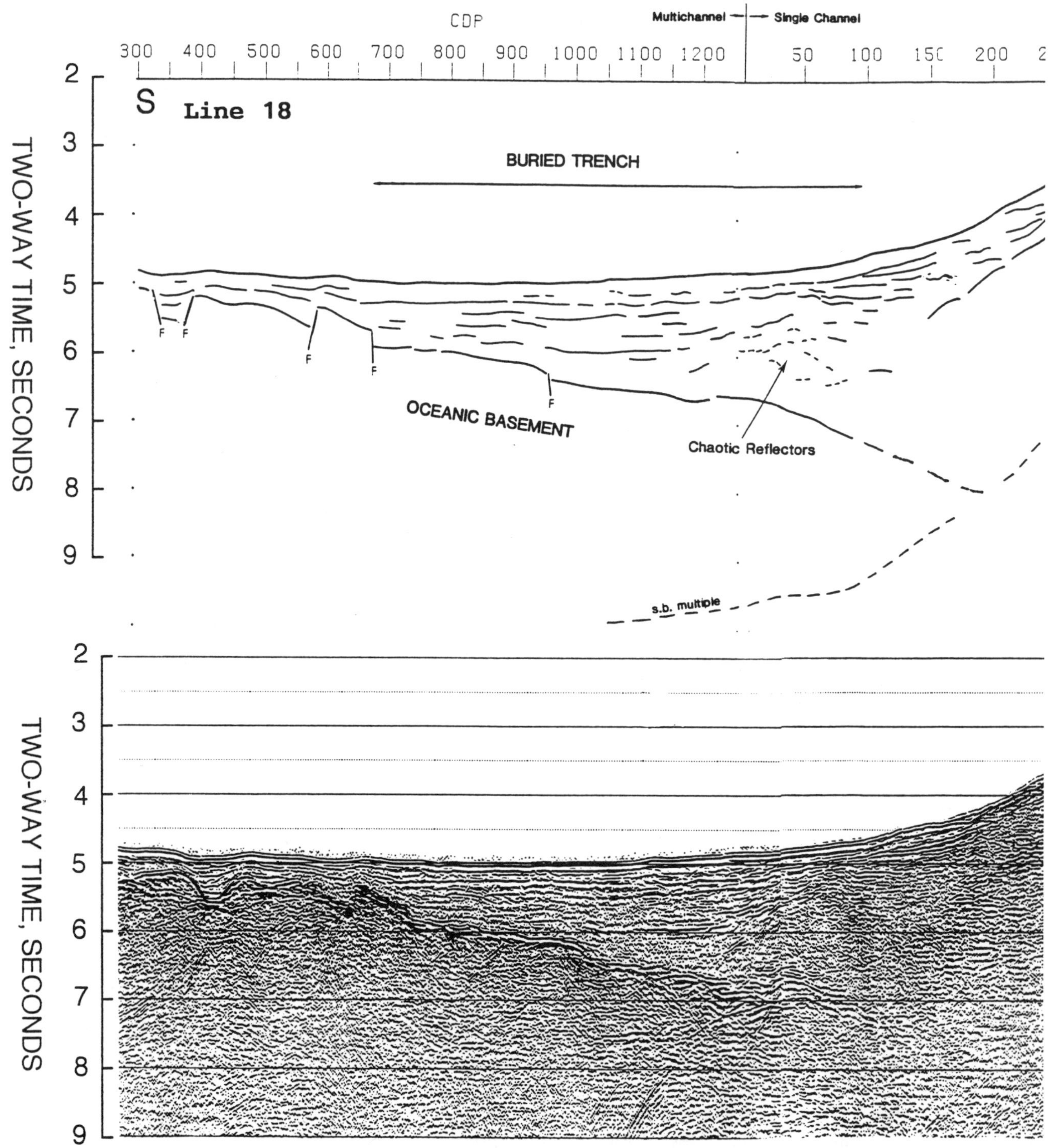

Figure 5. Seismic section 18 (lower) with its interpretation displayed above. Section is migrated with a constant velocity of 1,500 m/sec. This line runs from the Panama Basin across a filled inactive trench and a sediment-covered basement ridge on the northern edge of the Nazca plate to cross the plate boundary, a transpressive strike-slip fault running through a sedimentary basin at the edge of the Gulf of Panama.

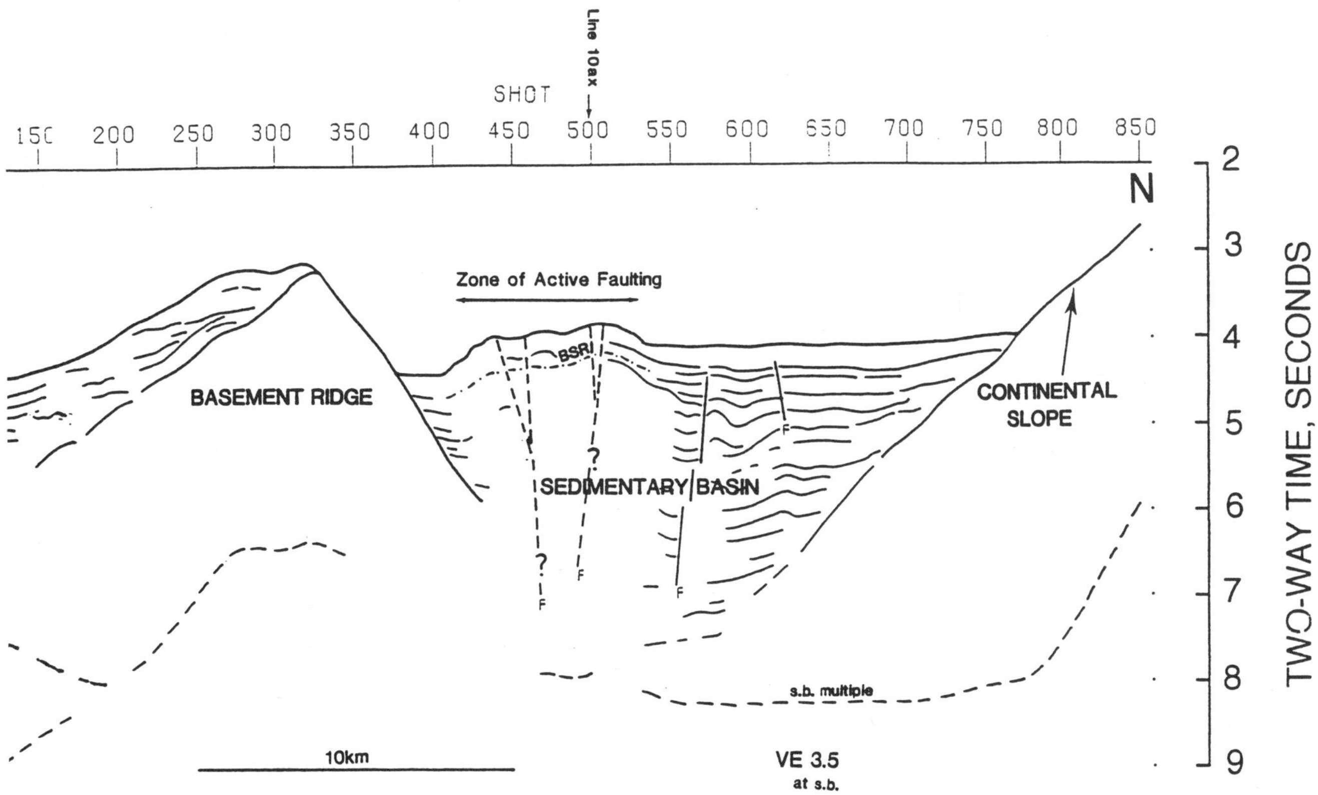
Line 10ax
SHOT
150 200 250 300 350 400 450 500 550 600 650 700 750 800 850
N
Zone of Active Faulting
BSR
BASEMENT RIDGE
SEDIMENTARY BASIN
CONTINENTAL SLOPE
s.b. multiple
10km
VE 3.5 at s.b.
2 3 4 5 6 7 8 9
TWO-WAY TIME, SECONDS

2 3 4 5 6 7 8 9
TWO-WAY TIME, SECONDS

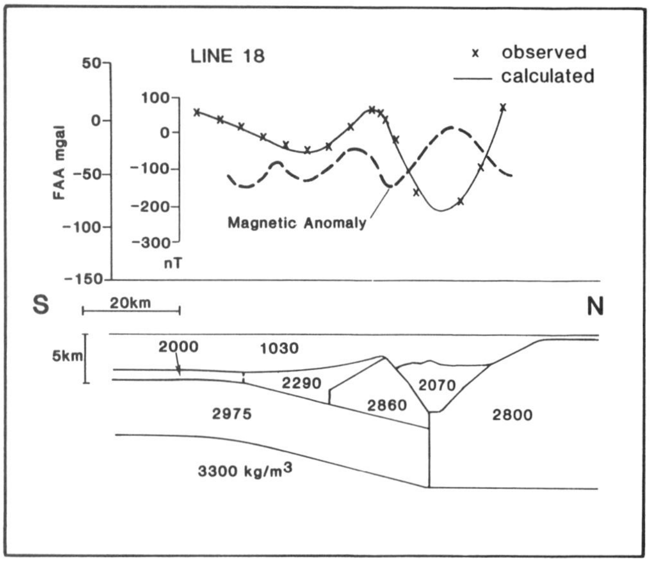

Figure 6. Two-dimensional model of the crustal structure across the southern Panamanian margin. The model has a free air gravity anomaly (solid line) that matches the anomaly measured (crosses) across the margin. Densities of the different components of the model are shown in kg/m^3. The model is along the section of seismic line 18, and the boundaries between sediment and basement are taken from the seismic section (Fig. 5), where they are visible. Two important features of the model are that the ridge south of the Panamanian continental edge is composed principally of high density rock and that the sedimentary basin between the ridge and the continental edge is composed principally of high density rock and that the sedimentary basin between the ridge and the continental edge has a maximum thickness of fill of at least 4 km. (The density used for the sediment fill is probably lower than its true average density.) The magnetic anomaly (shown as dashed line) indicates that the core of the ridge is magnetized and therefore likely to contain mafic igneous rock.

the north. A striking feature of Line 8bx is a marked irregularity of the ridge basement adjacent to the trench. This appears to be faulting, although it is not obvious from the section whether this faulting is of a dominantly normal or reverse nature.

DISCUSSION

A reevaluation of earlier data

Single-channel seismic reflection lines crossing the Panamanian margin, at approximately 80°W, were used by Lowrie (1978) to infer the presence of a buried, inactive trench. These profiles apparently show no deformation of the sediments in the trench or at its boundary with the continental slope (accretionary wedge). A 3.5 kHz profile collected during RRS *Charles Darwin* cruise 40 in 1989 crosses the margin just to the east of these seismic lines, at 79°40′W. It shows the uppermost 50 to 100 m of bedded trench sediments, gently folded at the foot of the slope (Fig. 11). Weak deformation of this type suggests that there is a small amount of active convergence, although interpretation of the data is hampered by a right-angle turn in the ships track.

West of the Gulf of Panama and south of the Azuero Peninsula, SeaMARC II side-scan sonar and single-channel seismic reflection data were collected by the Hawaii Institute of Geophysics, from R/V *Moana Wave* in 1985. These data show active sediment accretion occurring along an apparently convergent margin between 80°30′W and 81°10′W (Moore et al., 1985; MacKay and Moore, 1990; Moore and Sender, this volume). In addition to the observations made by MacKay and Moore (1990) regarding the differences in style of deformation as a response to changes in trench-sediment thickness and lithology, these data show evidence of a possible gradual eastward reduction in the rate of convergence. Seismic lines crossing the margin west of 81°W show clear evidence of convergence with the folding and thrusting of even the most recent sediments as they are accreted. Farther east, Line M44 (Fig. 3), crossing the deformation front at 80°53′W (MacKay and Moore, 1990), shows a turbidite trench fill about 500 m thick, becoming gently folded and displaced by thrusts, with the most recent and most southerly fold uplifting the seabed. East of Line M44, at 80°39′W, Line M46 shows that turbidite fill, 1 km thick, is also accreted in a series of thrust-faulted anticlines but that the seabed above the outermost fold is nearly flat. The uppermost 300 m of fill (at its thickest) thins over the crest of the outermost fold, showing that the rate of growth of the fold has been sufficiently slow for sedimentation to keep pace with it.

Tectonic boundaries and kinematics

The interpretation of the tectonic major elements of the region is illustrated in Figure 12. The fault zone south of the Panamanian continental edge (Southern Panama fault zone) is the plate boundary between the Nazca plate and the Panama block (microplate). There is no zone of concentrated active deformation and displacement to the south or north of it. The trend of the fault zone is very close to the direction of relative motion between the Panama block and the Nazca plate measured from relocation of GPS (Global Positioning System) stations (Kellogg et al., this volume) and that inferred from seismicity by Adamek et al. (1988). The fault zone has predominantly left-lateral strike-slip motion, but there is some convergence across it that is shown by the positive flower structure of the fault zone and by deformation of the sediments in the basin at the foot of the continental slope of the Gulf of Panama (Fig. 5). Shortening across the fault zone itself is at least 600 m (from measuring the cross-sectional area of the flower structure above the seabed to the south of it). Over the history of the basin, the shortening in the sediments north of the fault is at least 300 m (from measuring the length of the most folded horizon), but without accurate knowledge of the strike of the folds in the sediments, the direction of shortening is not well defined. South of the Azuero Peninsula, the South Panama fault zone passes into the rear of an accretionary complex, where it must change its trend from 256° to 280° as it is no longer clearly distinguishable as a separate feature from the

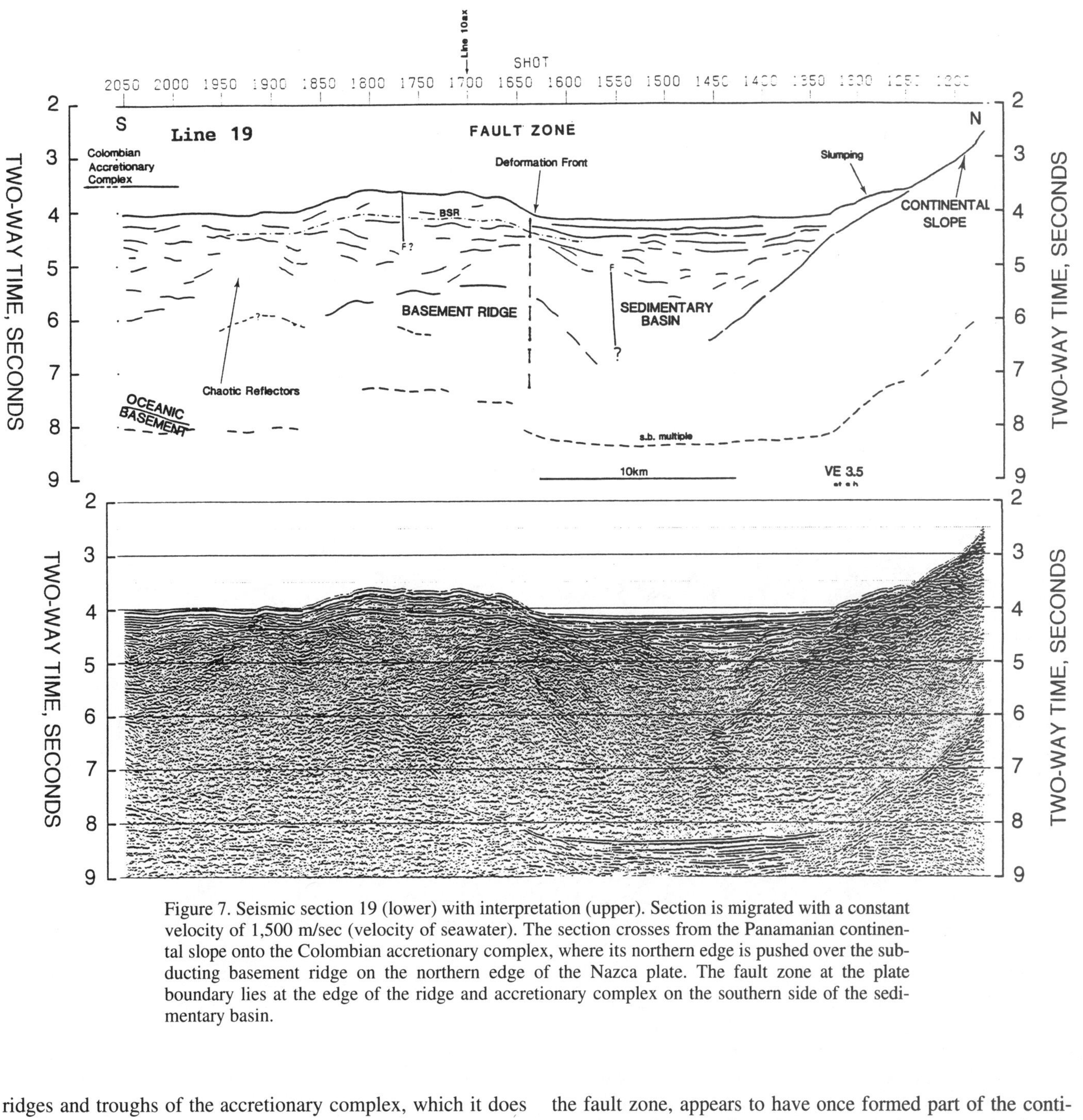

Figure 7. Seismic section 19 (lower) with interpretation (upper). Section is migrated with a constant velocity of 1,500 m/sec (velocity of seawater). The section crosses from the Panamanian continental slope onto the Colombian accretionary complex, where its northern edge is pushed over the subducting basement ridge on the northern edge of the Nazca plate. The fault zone at the plate boundary lies at the edge of the ridge and accretionary complex on the southern side of the sedimentary basin.

ridges and troughs of the accretionary complex, which it does not crosscut in the seismic and sonar data presented by MacKay and Moore (1990). We believe that the oblique convergence across the accretionary complex southwest of the Azuero Peninsula is partitioned into strike-slip motion parallel to the complex and convergence orthogonal to it, in a manner similar to that first proposed by Fitch (1972) for the whole of the Sumatran margin. This minimizes the work done between the two plates, as the surface area of the vertical strike-slip fault is much less than that of the gently dipping décollement at the base of the accretionary complex.

The Southern Panama Boundary Ridge, to the south of the fault zone, appears to have once formed part of the continental margin of southern Panama near the Azuero Peninsula and to have become detached from it by the left-lateral strike-slip faulting. The high density core of the ridge and the only minor degree of deformation in the foot of its southern slope indicate that the ridge is not primarily an accretionary wedge. The presence of canyons that are no longer active, on the southern slope of the ridge, is strongly suggestive of this slope's having once been part of the continental slope, which currently exhibits many similar active canyons.

The extension to the west-southwest of this ridge is an inactive accretionary complex that passes into the active accre-

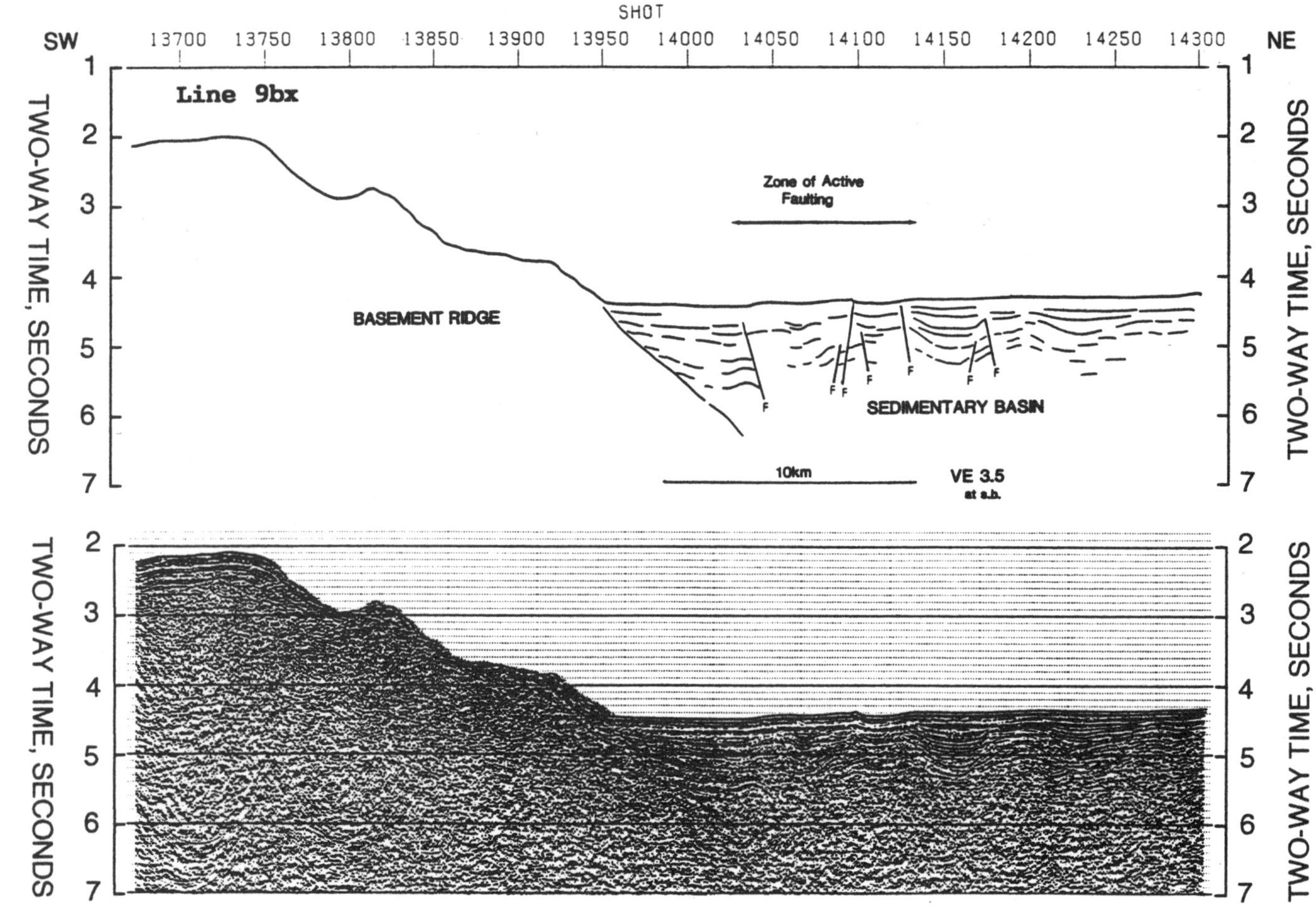

Figure 8. Seismic line 9bx (lower) with line drawing interpretation (upper). Section has been migrated with a constant velocity of 1,500 m/sec.

tionary complex southwest of the Azuero Peninsula. The transition, from active accretion to a relict accretionary complex carried passively on one plate past the other along a transform, is produced by the migration of the corner (gentle bend) between the oblique subduction and strike-slip margins of the Panama block relative to the Nazca plate. The convergent motion at the subduction boundary accretes sediment that is progressively thickened into the accretionary complex, which is translated along the margin by the strike-slip component of motion until it reaches the corner with the transform boundary, beyond which there is no further accretion, because of the absence of any convergent motion across the transform boundary. The principles of this process are illustrated in Figure 13. In southern Panama, this simple scheme is modified by the effect of strike-slip faults within the Panama block, such as the Sona-Azuero fault, which displace segments of the block across the fault line of Panama-Nazca transform and produce further compressional deformation in what would otherwise be an inactive accretionary complex. The mild deformation of the accretionary complex east of 80°20′W is presumably of this origin.

A reconstruction of the margin prior to the strike-slip motion on the southern Panama Fault Zone has been made by moving the Southern Panama Boundary Ridge westward along the fault zone so that the ridge, including that part which is buried in the Colombian subduction zone, no longer overlaps the continental slope now forming the northern side of the sedimentary basin. This requires a displacement of at least 140 km, which, using the rate of motion between the Panama block and Nazca plate given by Kellogg et al. (this volume) of 37 mm/yr, places the onset of displacement at 3.8 Ma. This is about the time commonly quoted for the collision of the Panamanian arc with South America (Coates et al., 1992). This arc included the Panama block and the Choco block of northwestern Colombia (Duque-Caro, 1990) which forms the hinterland to the accretionary complex off northern Colombia.

The reconstruction is illustrated in Figure 14. An implication of the time of collision of the arc with South America is that if the arc was attached to the Nazca plate at the time of collision, forming its eastern boundary, then the present continental margin west of it has developed in the last 3.5 m.y., and

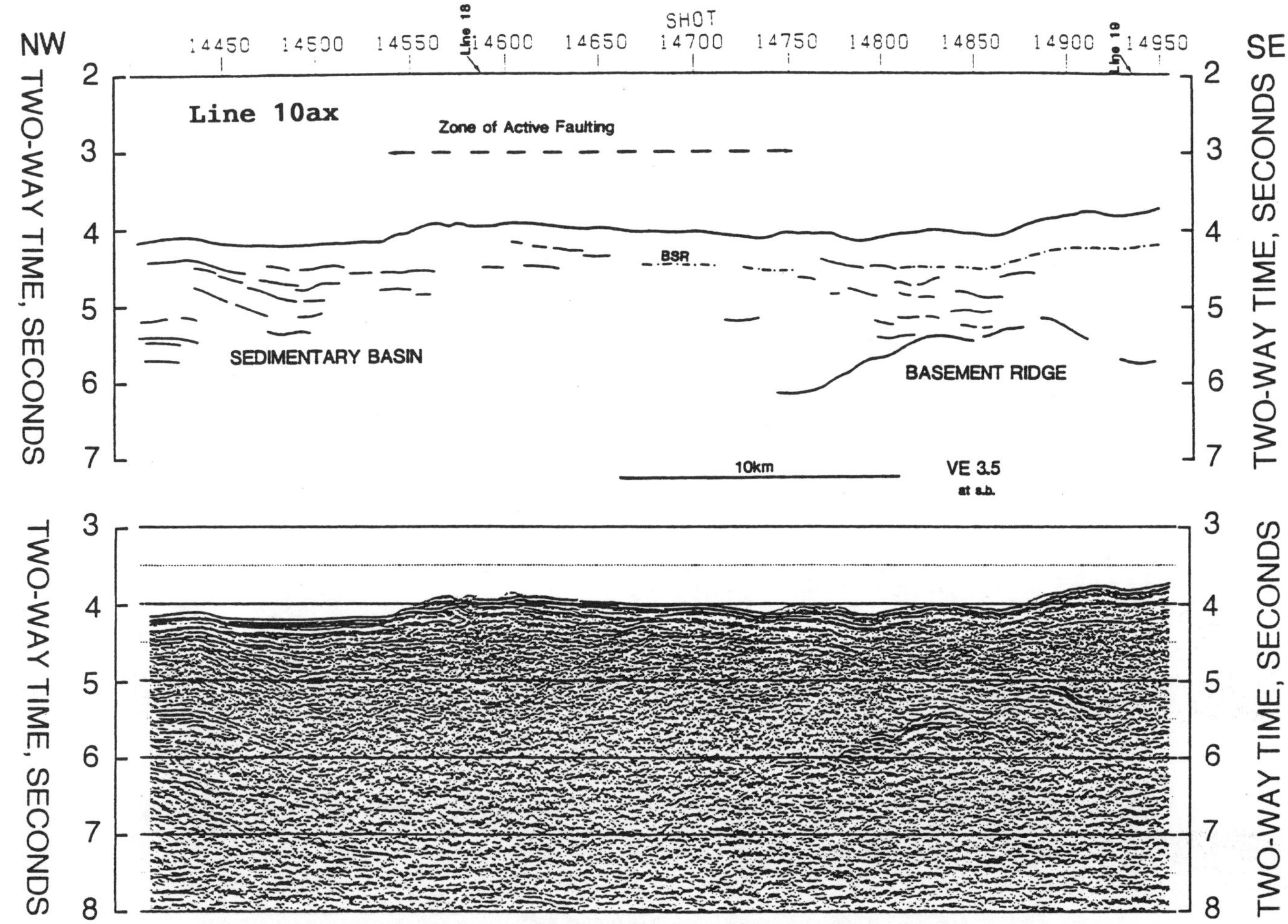

Figure 9. Seismic line 10ax (lower) with line drawing interpretation (upper). Section has been migrated with a constant velocity of 1,500 m/sec.

the entire accretionary complex of the northern Colombia subduction zone has grown in that period. The cross-sectional area of the accretionary complex is about 140 km^2, and if the average thickness of the accreted layer of sediment was about 1.5 km (allowing for compaction), then it is quite possible for this to have occurred. The current thickness of sediment accreted at the toe of the wedge is only a few hundred meters, but as mentioned above, at present the trench is starved of sediment by the damming of the trench by a basement ridge farther south (Hardy, 1991a).

The Southern Panama fault zone continues southeastward through the convergent margin of Colombia, where it controls the transition from the accretionary complex to the forearc basin at the foot of the steep continental slope. It acts to partition strain and displacement between the component normal to the margin and the component parallel. Left-lateral displacement along the fault zone is shown by the shearing of mud diapirs in the fault zone just north of 7°N. The rate of strike-slip displacement between the Nazca plate and the margin trending 320° at latitude 7°N that is predicted from the motion between the Nazca plate and the North Andes block (Kellogg et al., this volume) is about 44 mm/yr. The plate motion model of Minster and Jordan (1978) for Nazca and South America gives a rate of strike-slip displacement of about 38 mm/yr. The components of convergence normal to the margin are 32 mm/yr and 51 mm/yr, respectively. Even if northern Colombia were rigidly attached to Panama, the Nazca-Panama vector of Kellogg et al. (this volume) gives 18 mm/yr of strike-slip and 32 mm/yr of convergent motion. Most of the convergent displacement will be taken up by slip along the décollement between the accretionary complex and the subducting Nazca plate, but a small amount may be expressed in the forearc basin overriding the accretionary wedge, or possibly vice versa. Evidence from the seismic reflection data of the amount and sense of vertical displacement in the fault zone is not always clear and is complicated by the horizontal movement, but where it does seem to show displacement, it is the landward side that is uplifted. The presence of a basement on the landward side of the fault zone and its absence on the seaward side indicate that the fault zone is a fundamental boundary that is

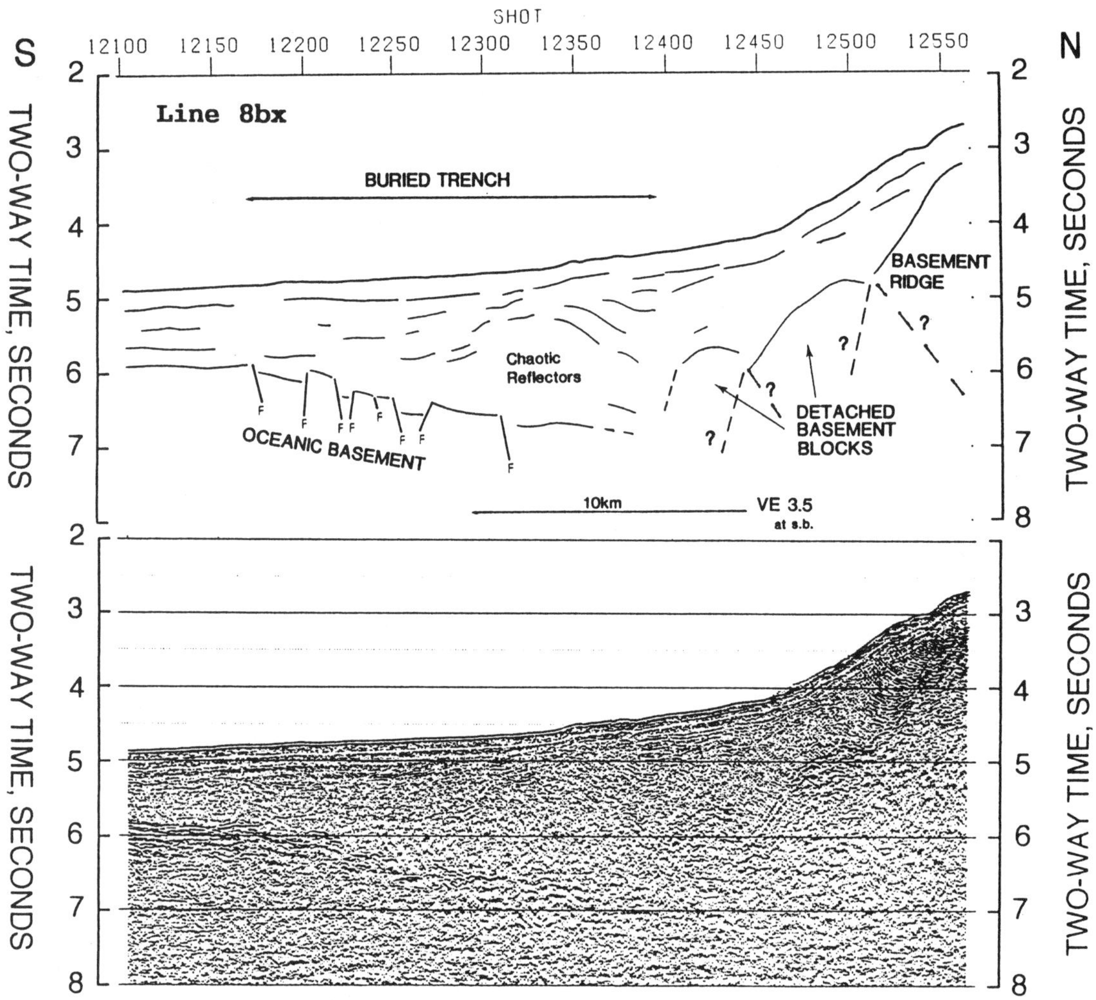

Figure 10. Seismic line 8bx (lower) with line drawing interpretation (upper). Section has been migrated with a constant velocity of 1,500 m/sec.

the seaward limit of the crystalline crust of Colombia and eastern Panama.

If the plate motion vectors of Kellogg et al. (this volume) are the correct ones to use, then the strike-slip motion is a consequence of the northwestward convergence of the North Andes block with Panama. If, however, there is no local convergence in this area (see below) and the Colombian hinterland to this part of the margin is not part of the North Andes block, then the accretionary complex is moving southeastward away from the Gulf of Panama. The plough action of the southern flank of the Southern Panama Boundary Ridge and the additional input of sediment from the trough on the ridge's southern flank, added to the northern end of the accretionary complex, drive it southeastward, complementing the drive that comes from shear stresses along the décollement beneath the complex, which are oblique to the trend of the fault zone.

A plate boundary between the Panama block and northern South America (North Andes block) was drawn in a position close to the international boundary between Panama and Colombia, joining the edge of the subduction zone offshore, by Kellogg and Dixon (1990), primarily on the basis of seismicity. There is, however, no evidence from the GLORIA or seismic reflection data to suggest that a major tectonic discontinuity lies in this position. Seismic Line 16 (not illustrated), which connects the northeastern ends of multichannel Lines 15 and 17, shows a well-bedded undisturbed sedimentary sequence, overlying basement. The predicted plate boundary should be crossed by Line 16. There is just some room to place a boundary north of Line 17, without violating the data (Fig. 12), but southward it is necessary to go at least as far south as 7°N before any feature that might be interpreted as an offshore continuation of this tectonic boundary might be observed. More realistically, only south of 6°N are many candidate features for this to be observed. The offshore continuation of the southern end of the Serrania de Baudo (the coastal mountain range extending into eastern Panama) at 5°30′N appears a plausible position, from the major change in the structure and character of the continental margin (Hardy, 1991a). Kellogg et al. (this vol-

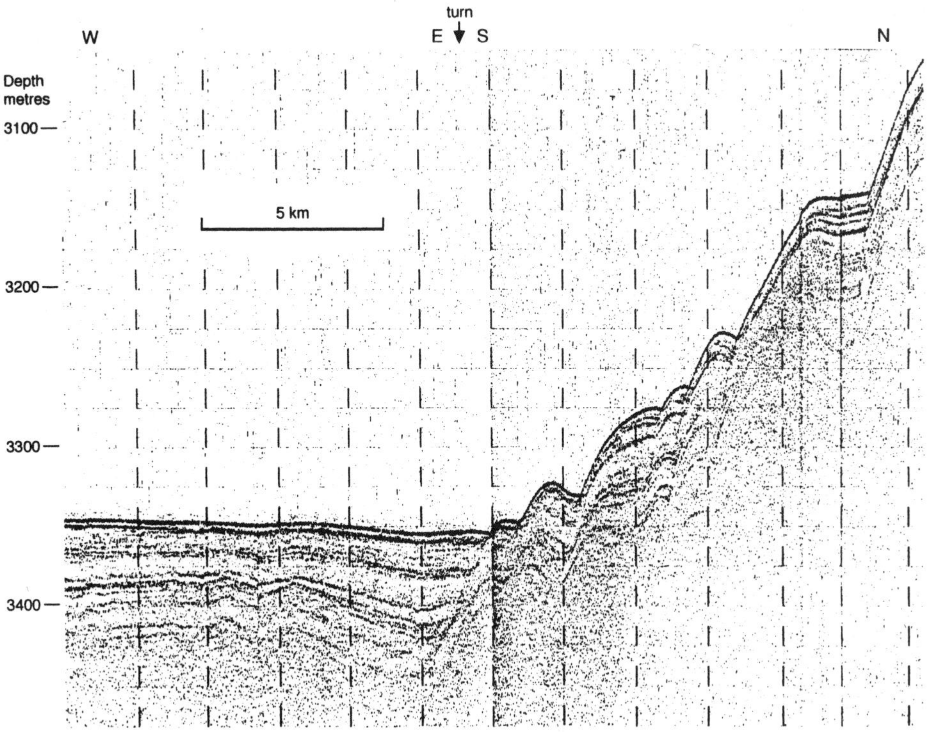

Figure 11. 3.5 kHz profiler record, showing the foot of the continental slope of southern Panama at 79°40′W. Vertical exaggeration is 43:1.

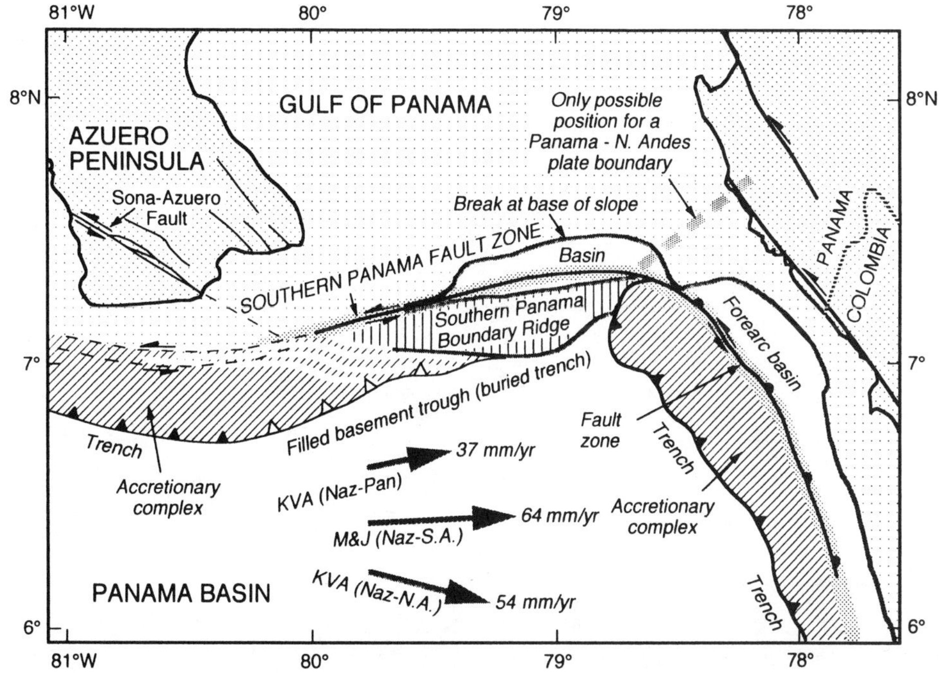

Figure 12. Map of the major tectonic elements around the northern boundary of the Nazca plate. The arrows show the relative motions between four plates or microplates in the region. (Naz = Nazca plate; Pan = Panama block; N.A. = North Andes block; SA = South American plate), taken from Minster and Jordan (1978) (M&J) and Kellogg, Vega, and Aiken (this volume) (KVA).

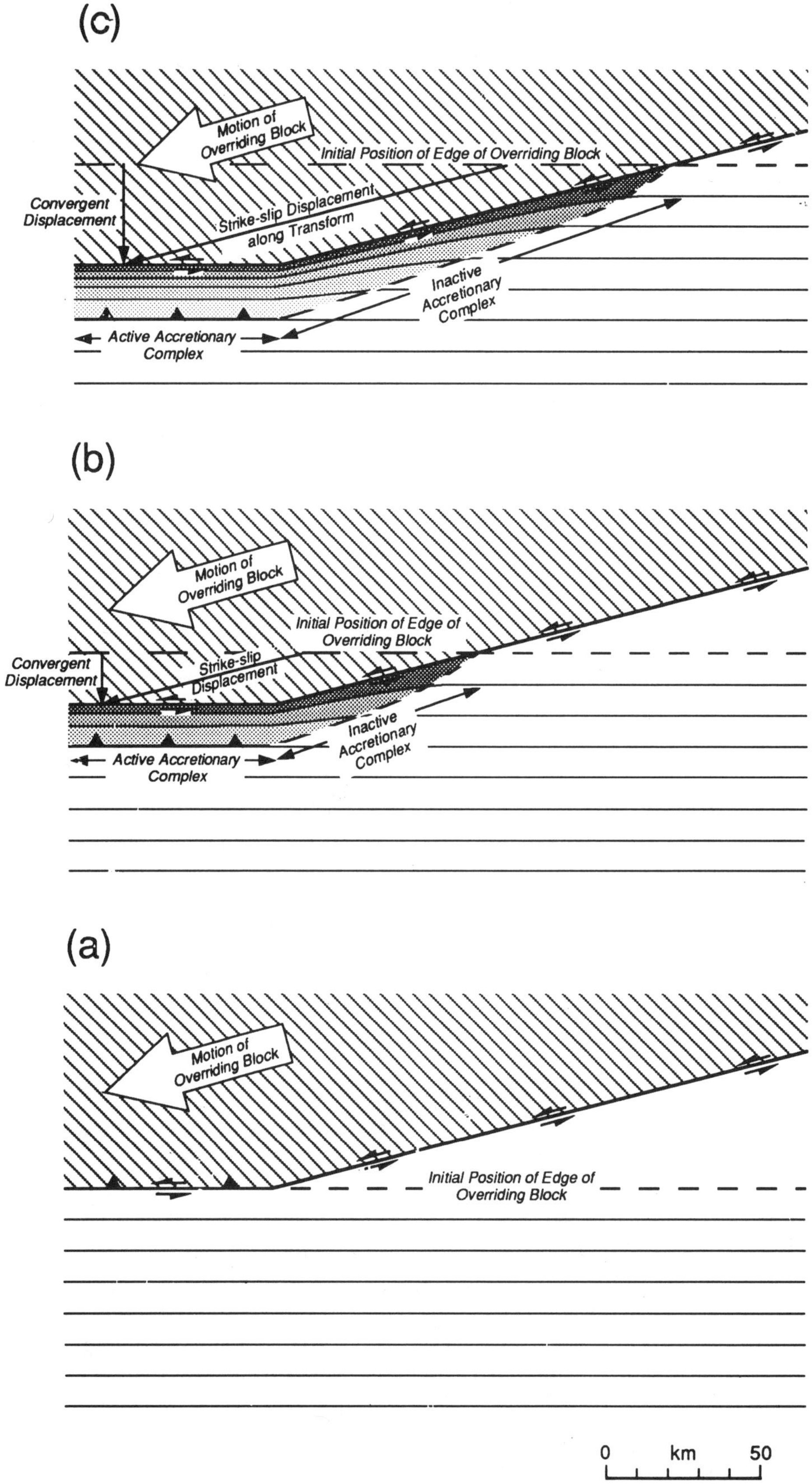
(c)
Motion of Overriding Block
Initial Position of Edge of Overriding Block
Convergent Displacement
Strike-slip Displacement along Transform
Inactive Accretionary Complex
Active Accretionary Complex
(b)
Motion of Overriding Block
Initial Position of Edge of Overriding Block
Convergent Displacement
Strike-slip Displacement
Inactive Accretionary Complex
Active Accretionary Complex
(a)
Motion of Overriding Block
Initial Position of Edge of Overriding Block
0 km 50

Figure 13. Illustration of the effect of the oblique subduction and transform margins between the Panama Block and Nazca plate upon the development of an accretionary complex. a, initial case. Overriding Panamanian block is shown in diagonal pattern. Lines of equal spacing (10 km) are marked on the sediment of the Nazca plate to illustrate the effect of accretion and displacement by the overriding block. b, situation after 65 km of motion between the two plates. Accreted sediment has been formed into an accretionary complex against the obliquely convergent margin of the overriding block. (The sediment accreted from each 10-km-wide strip is indicated by a different density toned pattern, darkest being the first to be accreted.) As the sediment thickens into the wedge its surface area decreases. (The model shown uses an angle of taper of wedge of 6° and an initial thickness of sediment accreted from the subducting plate of 700 m.) The accretionary complex is displaced relative to the overriding block along a strike-slip fault until, after passing the corner between the obliquely convergent and strike-slip boundaries of the overriding block, the accretionary complex ceases to widen and is carried passively past the overriding block. c, situation after 130 km of motion between the two plates. The edge of the active accretionary complex has advanced 50 km across the subducting plate, and the edge of the overriding block has similarly moved 32 km. The trend of structures within the inactive accretionary complex is oblique to its margin and asymptotic toward the strike-slip fault.

ume) propose that at the time of collision, the Panamanian arc extended as far south as 5°30′. The mafic rocks of the Serrania de Baudo are generally thought to have formed part of that arc (Duque-Caro, 1990). It may be the case that movement on the plate boundary proposed near the border between Panama and Colombia is so recent that is does not yet show as a discontinuity in seismic reflection records or a feature on sonographs. No GPS stations were located on the Choco block (Kellogg et al., this volume), however, so it is possible that there is motion between the Choco block and the North Andes block to the west.

Flexure and rupture of the lithosphere

The reason for the existence of the buried trench south of Panama is perhaps not immediately apparent. As shown by seismic Line 18 (Fig. 5), the oceanic crust descends beneath the southern flank of the Southern Panama Boundary Ridge. The initial cause of the active trench in the western part of the margin by downward flexure of the lithosphere is evident, but why should the trench persist once it has been moved along the transform margin? The continued presence of the trench is

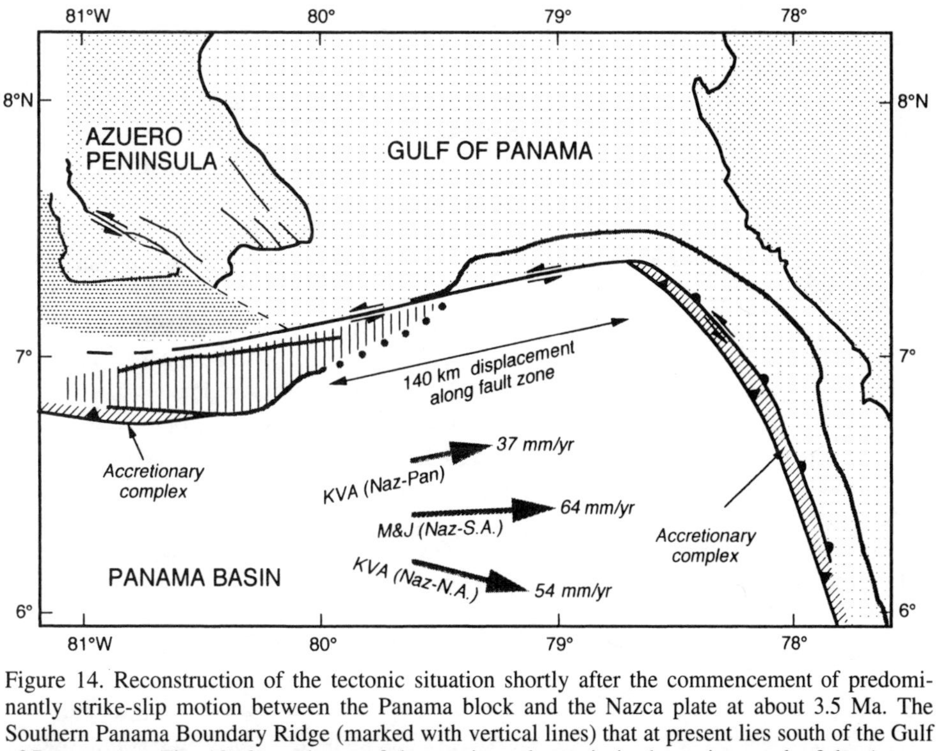

Figure 14. Reconstruction of the tectonic situation shortly after the commencement of predominantly strike-slip motion between the Panama block and the Nazca plate at about 3.5 Ma. The Southern Panama Boundary Ridge (marked with vertical lines) that at present lies south of the Gulf of Panama (see Fig. 12) formed part of the continental margin in the region south of the Azuero Peninsula. Behind it (marked with denser toning) lies the region of the margin that was subsequently displaced southeastward along the Sona-Azuero fault, into the rear of the accretionary complex. Abbreviations as in Figure 12.

partly explained by the loading of the accreted sediments, but the negative free-air gravity anomaly over the accretionary complex south of the Azuero Peninsula demonstrates that the load causing flexure there still comes principally from the Panama block, and so the subducting Nazca plate is not broken to any significant degree in this part of the boundary. There is, however, a large positive anomaly over the Southern Panama Boundary Ridge, which is underlain by the Nazca plate, so perhaps the plate has broken into a transform in this part of the boundary. If one assumes that the plate is broken at the transform, one can use the distance to the flexural bulge to derive the flexural rigidity of the plate and hence assess whether the load of the ridge is sufficient to deflect it. The distance from the transform to the flexural bulge is about 100 km, which (following Turcotte and Schubert, 1982) yields a flexural parameter of 42 km and a flexural rigidity of 1.8×10^{22} Nm (assuming a density contrast between the mantle and seawater of 2,300 kg/m^3). The load-per-unit cross section imposed by the ridge is about 1.5×10^{12} N (taken from the gravity model of Fig. 6). This should produce a deflection beneath the load (assuming that it is all applied at the edge of the plate) of 3.0 km. The deflection in the gravity model is 4.0 km. Consequently it appears that a substantial part of the loading comes from the Southern Panama Boundary Ridge and that the Nazca plate is broken in this segment of the boundary. The analysis applied here is very simple; it assumes a two-dimensional case. In reality, a component of the loading will still come from the Panama block along strike to the west, and the situation is further complicated by the flexure of the lithosphere into the Colombian subduction zone close by. Farther west, across the zone of active oblique subduction, the distance of the fore bulge from the gravity high near the edge of the overriding Panama block is 140 km. If it is assumed that the plate is unbroken here, then it has a flexural parameter of 44 km (a broken plate assumption would give a flexural parameter of 59 km). If values for Young's modulus of 70 GPa and for Poisson's ratio of 0.25 are assumed for the subducting lithosphere, then its elastic thickness is 14 km. This value is consistent with its age of about 25 Ma (Watts, 1978). The rupture of the subducting Nazca plate provides a reason for the relative subsidence of the Gulf of Panama region, which unlike the region to the west, does not experience the uplift provided by overriding the subducting lithosphere, although Mann and Kolarsky (this volume) attribute the subsidence to the development of the Gulf of Panama as a foreland basin produced by eastward underthrusting beneath eastern Panama.

History of the southern Panama boundary

It is interesting to speculate on the nature of the southern Panama boundary prior to the most recent Plio-Pleistocene phase of tectonics. The only segment in which evidence of its older structure is preserved is the Southern Panama Boundary Ridge. The landward dip of the oceanic basement is suggestive of a subduction margin, although about 2 km of the landward deepening could be attributed to subsidence of the ridge once it became detached from the Panama block. Lonsdale and Klitgord (1978) have suggested that prior to 8 to 12 Ma there was northward subduction beneath Panama, which was halted by the arrival on the subducting oceanic plate of a basement scarp between older lithosphere generated at the East Pacific Rise and younger, more-elevated, and rugged lithosphere formed at the Galapagos (early Cocos) spreading center. Hardy (1991b), in a development of Lonsdale and Klitgord's model, suggests that the initiation of strike-slip motion south of Panama dates from about that time. There is a small area of deformed sediments beneath the foot of the southern flank of the ridge, but it is hardly suggestive of a long history of subduction-accretion, although tectonic erosion at the boundary could have removed an older accretionary complex, as off Peru (von Huene and Lallemand, 1990). The palaeogeography and tectonics of the boundary before the late Miocene are still very poorly known.

SUMMARY AND CONCLUSIONS

The existence of a major fault zone, the Southern Panama fault zone, that runs along the southern Panamanian margin and forms the northern boundary of the Nazca plate has been demonstrated from GLORIA sonographs and seismic reflection profiles. For an 80-km segment of its length, the fault zone occupies a sedimentary basin, filled with up to 5 km of sediment, that lies south of the Gulf of Panama. South of this basin lies a ridge with a high density core and with inactive canyons running down its southern flank that appears to have been translated about 140 km east of its former position on the continental margin of southern Panama by left-lateral strike-slip motion along the fault. This Southern Panama Boundary Ridge, with a sediment-filled trench south of it, is subducted eastward beneath the continental margin of northernmost Colombia (southernmost Panama) at the northern limit of the Colombian subduction zone. The deformation front of the northern Colombian accretionary complex terminates against the ridge, climbing its flank northeastward until it joins the fault zone that is the northern boundary of the Nazca plate. The fault zone continues around to the southeast behind the bulk of the Colombian accretionary complex, separating it from a forearc basin overlying "continental" basement. The deformation of mud diapirs indicates a component of left-lateral strike-slip motion on this part of the fault zone.

If there is a discrete tectonic boundary between the Panama block and the North Andes microplate, then its position is constrained to be north of the Southern Panama fault zone, trending southwest and meeting it at the eastern end of the sedimentary basin. Otherwise, it seems probable that the boundary lies east of the Serrania de Baudo, running south until it cuts the Colombian continental margin in the region of the Cabo Corrientes, around 5°30′N.

To the south and southwest of the Azuero Peninsula, the southern Panamanian margin is one of oblique subduction, with the active accretion of sediment from the Nazca plate. Passing eastward along the margin, the accretionary complex becomes predominantly inactive where it lies parallel to the strike-slip (transform) portion of the margin. This has developed with continued motion between the Panama block and the Nazca plate, as the accretionary complex slips around the bend in the plate boundary between the oblique subduction and transform portions of the margin. Left-lateral movement on the Sona-Azuero fault within the Panama block leads to displacement across the line of strike-slip motion, producing some continued deformation in the otherwise inactive portion of the accretionary complex.

The Nazca plate remains unbroken beneath the western part of the strike-slip margin, south of the Azuero Peninsula, as indicated by the negative gravity anomaly over the margin and the greater flexural wavelength of the depressed plate, but south of the Gulf of Panama a fault is developed in the plate, resulting in a smaller flexural wavelength. There, the principal load depressing the plate's edge comes from the Southern Panama Boundary Ridge with its associated positive gravity anomaly, rather than from the Panama block.

The present pattern of tectonics appears to have begun at about 3.5 ma, judging from the displacement of 140 km along the South Panama fault zone and published estimates of the rates of motion between the Panama block and the Nazca plate.

ACKNOWLEDGMENTS

This chapter was improved by informative reviews from James Case, James Kellogg, and Paul Mann. The research was supported by the Natural Environment Research Council, U.K., through research grant GR3/6663, the provision of shiptime on the RRS *Charles Darwin,* and the award of a research studentship to Roger Heath.

REFERENCES CITED

Adamek, A., Frolich, C., and Pennington, W., 1988, Seismicity of the Caribbean-Nazca boundary: Constraints on microplate tectonics of the Panama region: Journal of Geophysical Research, v. 93, p. 2053–2075.

Coates, A. G., and seven others, 1992, Closure of the Isthmus of Panama: The nearshore marine record of Costa Rica and western Panama: Geological Society of America Bulletin, v. 104, p. 814–828.

de Boer, J., Defant, M. J., and Stewart, R. H., 1988, Quaternary calc-alkaline volcanism in western Panama: Regional variation and implication for plate tectonic framework: Journal of South American Earth Sciences, v. 1, p. 275–293.

Duque-Caro, H., 1990, The Choco block in the northwestern corner of South America: Structural, tectonostratigraphic and paleogeographic implications: Journal of South American Earth Sciences, v. 3, p. 71–84.

Fitch, T. J., 1972, Plate convergence, transcurrent faults, and internal deformation adjacent to southeast Asia and the western Pacific: Journal of Geophysical Research, v. 77, p. 4432–4460.

Hardy, N., 1991a, A marine geophysical study of the Pacific margins of Colombia and southeast Panama [Ph.D. thesis]: University of Birmingham, United Kingdom, 253 p.

Hardy, N., 1991b, Tectonic evolution of the easternmost Panama Basin: Some new data and inferences: Journal of South American Earth Sciences, v. 4, p. 261–270.

Heil, D. J., 1988, Response of an accretionary prism to transform ridge collision south of Panama [M.Sc. thesis]: Santa Cruz, University of California.

Kellogg, J. N., and Dixon, T. H., 1990, Central and South American GPS geodesy—CASA UNO: Geophysical Research Letters, v. 17, p. 195–198.

Lonsdale, P., and Klitgord, J. D., 1978, Structure and tectonic history of the eastern Panama Basin: Geological Society of America Bulletin, v. 89, p. 981–999.

Lowrie, A., 1978, A buried trench south of the Gulf of Panama: Geology, v. 6, p. 434–436.

MacKay, M. E., and Moore, G. F., 1990, Variation in deformation of the south Panama accretionary prism: Response to oblique subduction and trench sediment variation: Tectonics, v. 9, p. 683–698.

Mann, P., and Burke, K., 1984, Neotectonics of the Caribbean: Review of Geophysics and Space Physics, v. 22, p. 309–362.

Mann, P., and Corrigan, J., 1990, Model for late Neogene deformation in Panama: Geology, v. 18, p. 558–562.

Minster, J. B., and Jordan, T. H., 1978, Present-day plate motions: Journal of Geophysical Research, v. 83, p. 5331–5354.

Moore, G. F., and six others, 1985, Structure of the south Panama continental margin: A zone of oblique convergence [abs.]: EOS Transactions of American Geophysical Union, v. 66, p. 1087.

Naylor, M. A., Mandl, G., and Sijpestey, C.H.K., 1986, Fault geometries in basement-induced wrench faulting under different initial stress states: Journal of Structural Geology, v. 8, p. 737–752.

Okaya, D. A., and Ben-Avraham, Z., 1987, Structure of the continental margin of southern Panama: Geological Society of America Bulletin, v. 99, p. 792–802.

Pennington, W. D., 1981, Subduction of the eastern Panama Basin and seismotectonics of northwestern South America: Journal of Geophysical Research, v. 86, p. 10753–10770.

Silver, E. A., Reed, D. L., Tagudin, J. E., and Heil, D., 1990, Implications of the north and south Panama thrust belts for the origin of the Panama Orocline: Tectonics, v. 9, p. 261–281.

Turcotte, D. L., and Schubert, G., 1982, Geodynamics: Application of continuum physics to geological problems: New York, John Wiley and Sons, 450 p.

Vergara Muñoz, A., 1988a, Seismicity of the Panama Block. I: Magnitudes and spatial distribution of epicentres: Tectonophysics, v. 145, p. 213–224.

Vergara Muñoz, A., 1988b, Tectonics patterns of the Panama Block deduced from seismicity, gravitational data and earthquake mechanisms: Implications to seismic hazard: Tectonophysics, v. 154, p. 253–269.

von Huene, R., and Lallemand, S., 1990, Tectonic erosion along the Japan and Peru convergent margins: Geological Society of America Bulletin, v. 102, p. 704–720.

Watts, A. B., 1978, An analysis of isostasy in the world's oceans. 1: Hawaiian-Emperor seamount chain: Journal of Geophysical Research, v. 83, p. 5989–6004.

Wolters, B., 1986, Seismicity and tectonics of southern Central America and adjacent regions with special attention to the surroundings of Panama: Tectonophysics, v. 128, p. 21–46.

Manuscript Submitted August 1992
Manuscript Accepted by the Society February 7, 1994

Printed in U.S.A.

Geological Society of America
Special Paper 295
1995

East Panama deformed belt: Structure, age, and neotectonic significance

Paul Mann
Institute for Geophysics, University of Texas at Austin, 8701 Mopac Boulevard, Austin, Texas 78759
Radim A. Kolarsky*
Department of Geological Sciences, University of Texas at Austin, Austin, Texas 78712

ABSTRACT

The East Panama deformed belt is defined here as a seismically active, on- and offshore belt of deformed Late Cretaceous–Neogene rocks that forms a diffuse tectonic boundary between the eastern Isthmus of Panama and South America. Satellite imagery and geologic maps of eastern Panama compiled from previous studies indicate that northwest-striking, left-lateral strike-slip faults form the major Neogene tectonic structures in the onshore part of the deformed belt. The best studied onland strike-slip fault is characterized by left-stepping, *en echelon* folds that deform rocks ranging in age from Cretaceous to late Miocene and terminates at a high angle of intersection on north-striking reverse faults in the Pirre Hills on the Panama-Colombia border. North- to north-east-striking normal faults appear to terminate left-lateral strike-slip motion in the area of the Panama Canal at the northwestern edge of the deformed belt.

Interpretation of marine multichannel seismic profiles and two well logs from the Gulf of Panama suggests that east-dipping reverse faults of middle Miocene through Plio-Pleistocene age form a largely buried, 90-km-wide thrust belt beneath the broad shelf area of the eastern half of the Gulf of Panama. We cannot prove or disprove a left-slip component on offshore reverse faults mapped on widely spaced seismic profiles. Exploration wells document the presence of an eastward-thickening Plio-Pleistocene foredeep basin overlying a middle Eocene section in the east-central Gulf of Panama. Post–middle Eocene sedimentary units were deposited in marine environments, range in age from middle Miocene to Plio-Pleistocene, and are locally deformed by east-dipping thrust faults. The previously mapped geology of the Pearl Islands at the eastern edge of the Panama deformed belt suggests that the islands are the exposed axis of a hanging-wall anticline associated with an east-dipping reverse fault zone. We correlate Plio-Pleistocene strike-slip faulting of the onshore Panama deformed belt and reverse faulting in the offshore Panama deformed belt and Pearl Islands with increasingly restricted surface water exchange and eventual Pliocene closure of the Caribbean-Pacific seaway in Panama, determined from micropaleontological studies by previous workers of Pacific and Caribbean sedimentary sections.

---

*Present address: Kolarsky, Eastern E&P Region, Texaco USA, P.O. Box 60252, New Orleans, Louisiana 70160.

Mann, P., and Kolarsky, R. A., 1995, East Panama deformed belt: Structure, age, and neotectonic significance, *in* Mann, P., ed., Geologic and Tectonic Development of the Caribbean Plate Boundary in Southern Central America: Boulder, Colorado, Geological Society of America Special Paper 295.

INTRODUCTION

The collision of the Panama island arc with the South American continent in Colombia and terminal closure of the Caribbean-Pacific seaway is a critical event in the late Neogene tectonic, biogeographic, and paleoceanographic history of the Caribbean and eastern Pacific region (see Duque-Caro, 1990a, and Coates et al., 1992, for recent reviews) (Fig. 1). The precise age of closure and the structural record produced by closure of the seaway remain poorly understood because of a lack of high-quality geologic and geophysical data from the Panama arc–South America collision zone in eastern Panama. The purpose of this study is improve our understanding of the Caribbean-Pacific closure event by a comprehensive examination of the style and age of on- and offshore structures in eastern Panama and offshore areas.

We here define the East Panama deformed belt as the seismically active, on- and offshore belt of deformed Late Cretaceous–Neogene rocks that now occupies the former site of the Pacific-Caribbean seaway and forms a diffuse tectonic boundary between the eastern Isthmus of Panama and South America. The East Panama deformed belt forms the eastern edge of the Panama block or microplate that is detached from the Caribbean plate to the north by the North Panama deformed belt (Adamek et al., 1988; Silver et al., 1990; Reed and Silver, this volume) and is detached from the Nazca plate to the south by the South Panama deformed belt (Kolarsky and Mann, this volume; Moore and Sender, this volume; Westbrook, this volume) (Fig. 2).

In this chapter, we describe structures in the onshore parts of the deformed belt in eastern and east-central Panama using the results of previous mapping by other workers combined with interpretations of satellite radar imagery by Mann and Corrigan (1990). We also describe structures from the offshore part of the deformed belt beneath the shallow (average water depth 20 to 200 m) of the Gulf of Panama using multichannel seismic data and two well logs provided by industry (Fig. 3).

TECTONIC SETTING OF THE EAST PANAMA DEFORMED BELT

Regional seismicity

Maps of regional seismicity in eastern Panama show a diffuse zone of seismicity in the area of the Panama–South America collision zone in eastern Panama and western Colombia (Pennington, 1981) (Fig. 4). Most of these earthquake events in eastern Panama are shallow, have magnitudes less than 5, and are clustered aftershocks of the July 13, 1974 (M = 5.3) and July 11, 1976 (M = 6.2) earthquakes (Adamek et al., 1988). The depths of these two larger events was 23 and 25 km, respectively. Adamek et al. (1988) carefully relocated 36 events in eastern Panama and western Colombia but were unable to determine any correlation of epicenters to mapped surface faults. The only zone of deep focus earthquakes with hypocenters >70 km is found in a 1,250-km^2 area near the Panama-Colombia border and may mark a zone of active or recently active underthrusting in the collision zone (Fig. 4).

Four focal mechanism solutions by Pennington (1981) have nodal planes that strike approximately northward; principal axes suggest that the greatest stress is east-west compression. Additional focal mechanisms studied by Vergara-Muñoz (1988) indicate compression along northwest-southeast–striking fault planes. Focal mechanisms from the North Panama deformed belt beneath the Caribbean are associated with underthrusting of the Caribbean plate beneath Panama and are

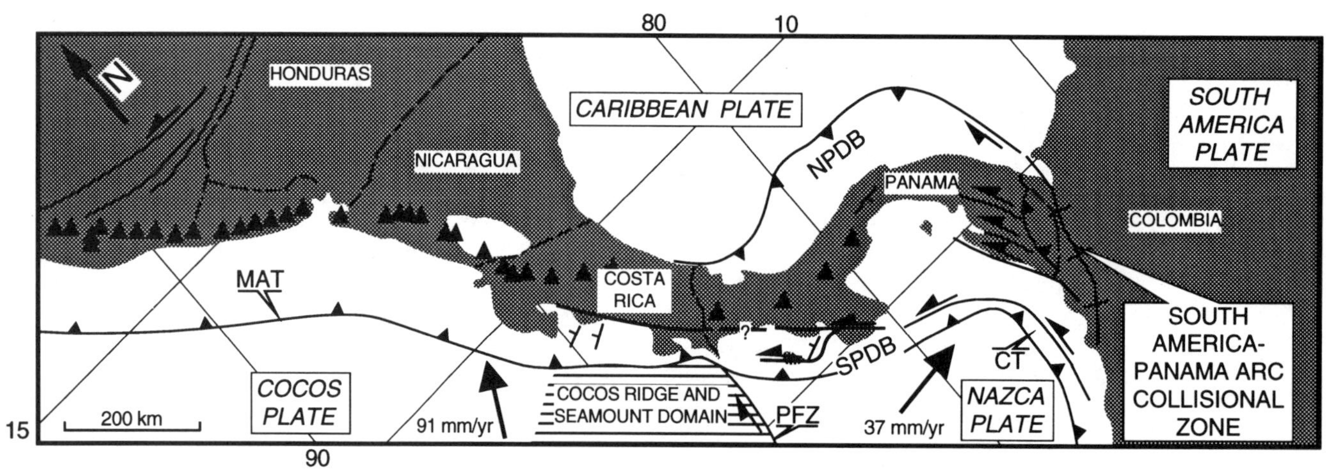

Figure 1. Tectonic setting of Central America, Middle America trench (MAT), Colombia trench (CT), North Panama deformed belt (NPDB), South Panama deformed belt (SPDB), and Panama fracture zone (PFZ). Black triangles represent active calc-alkaline volcanoes; arrows represent plate convergence directions with numbers indicating rates of relative plate motion in mm/yr from DeMets et al. (1990). Note oblique convergence direction predicted for the Nazca-Caribbean plate boundary along the southern margin of Panama.

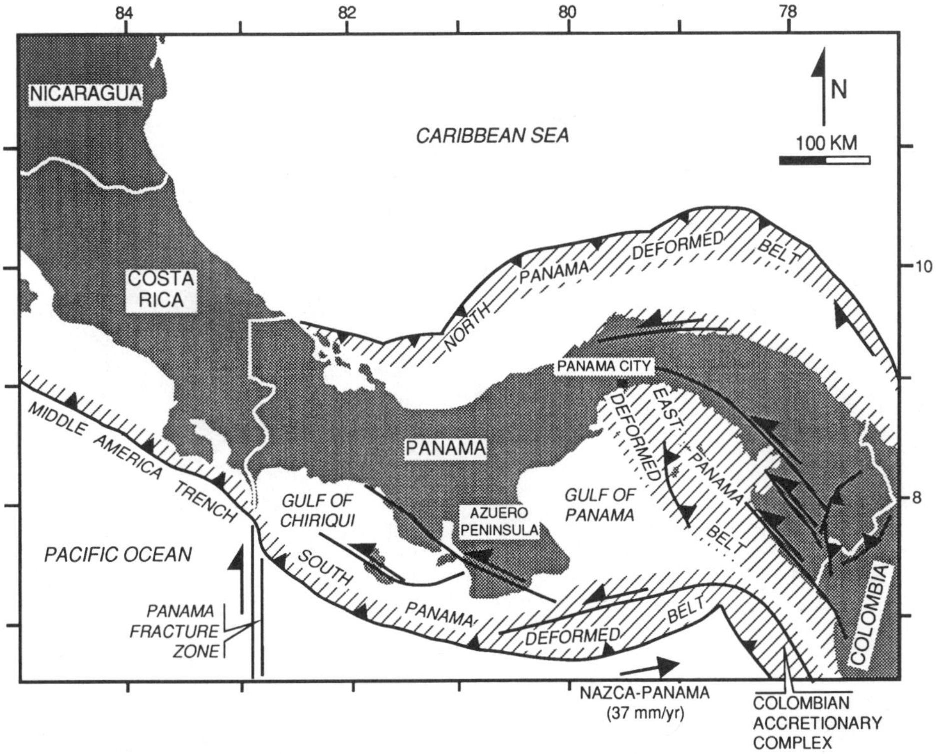

Figure 2. Location map and terminology for belts of Neogene deformation (hatched pattern) that border the Isthmus of Panama. In this chapter we define the age and style of deformation in the onshore and offshore (Gulf of Panama) parts of the East Panama deformed belt.

mainly shallow thrust faults and normal faults related to internal plate deformation (Wolters, 1986; Adamek et al., 1988).

Plate tectonic development

Most estimates of the age and nature of the deformation in the East Panama deformed belt are indirectly based on one or more of the following methods: (1) qualitative plate reconstructions based on regional geology, (2) semiquantitative plate reconstructions based on both analysis of magnetic anomalies in the eastern Pacific Ocean and regional geology, (3) determinations of the age of biologic divergence between Caribbean and Pacific marine microfauna or determination of the age of biologic interchange of fauna between North and South America, and (4) determinations of the age of uplift and deformation in Panama and northwestern South America that is inferred to be related to the collisional event. All of these methods indicate deformation and uplift in eastern Panama and Colombia sometime during the Miocene-Pliocene interval with final emergence of the present-day land area of the Panama Isthmus in Plio-Pleistocene time.

Lonsdale and Klitgord (1978) used magnetic anomalies in the eastern Pacific Ocean (East Panama basin) to make semiquantitative reconstructions of the Panama area from the Oligocene to the present. They interpreted the Panama island arc as a west-northwest–trending linear arc that was juxtaposed with the Colombian margin about 4 Ma along a left-lateral transform fault. Pindell and Dewey (1982) published the first synthesis of onshore Caribbean geology that emphasized the post-Miocene collision of a northwest-trending Panama island arc with the South American continental plate in Colombia. This collision was envisioned as a local complication in the long-term eastward displacement of the Caribbean plate relative to northwestern South America.

Wadge and Burke (1983) presented qualitative reconstructions of Panama that were based mainly on a rigid body rotation of Central America along the strike-slip faults of the Cayman trough and the Motagua-Polochic fault system of northern Central America (Fig. 1) (Mann, Preface, this volume, Fig. 14). They inferred that the Panama arc may originally have been east-trending and have been internally deformed and bent into its present-day reclined S-shape during collision with South America about 10 Ma. Burke et al. 1984), Mann and Burke (1984), Burke and Sengor (1986), and Stéphan et al. (1986) all emphasized that the late Neogene Panama arc collision may have been a tectonic mechanism for regional deformation and northward and northwestward "escape" of strike-slip fault-bounded blocks in northwestern South America and Panama itself. Mann and Corrigan (1990) emphasized that strike-slip faults may play an important role in segmenting the bending orocline.

Silver et al. (1990) based a model for the bending of the

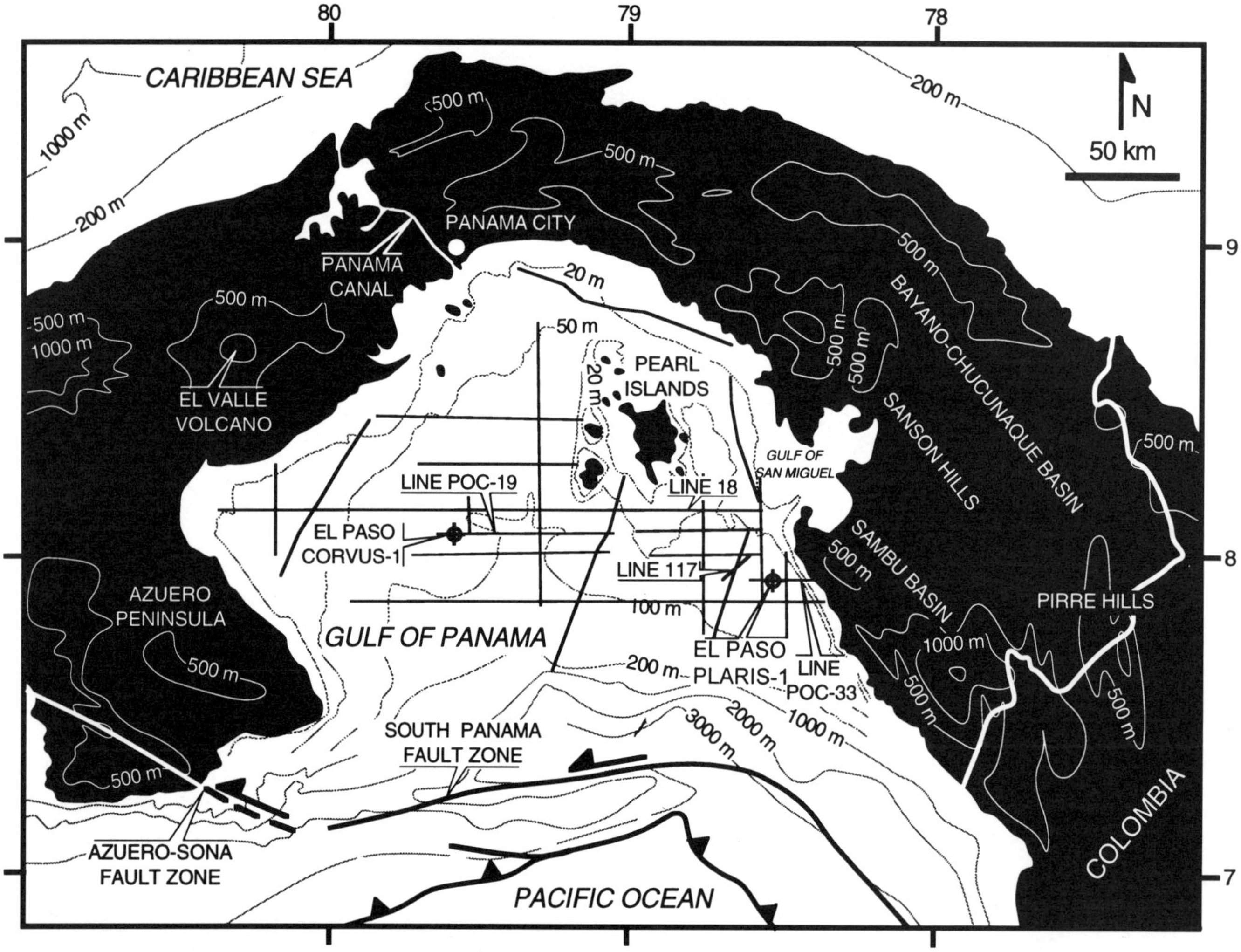

Figure 3. Location map for places mentioned in this chapter. Offshore bathymetric contours are taken from U.S. Navy bathymetric charts. Lines give locations of Mobil multichannel seismic lines used in this study, and Mobil wells El Paso Plaris-1 and El Paso Corvus-1 are indicated. "Buried trench" is from Lowrie (1978) and South Panama fault zone is from Westbrook et al. (this volume).

Isthmus of Panama on the observed pattern of deformation in the North and South Panama deformed belts. They related observed convergence across the North Panama deformed belt to oroclinal bending of the isthmus and convergence along the South Panama deformed belt to relative convergence of the Nazca plate on Panama. The structure of the western part of the North Panama deformed belt was attributed to either left-lateral strike-slip faults or to the proximity of the slope of the Panama arc, whereas the structure of the central part of the belt was attributed to bending of Panama following its Middle to Late collision with South America.

Previous estimates of the age of Panama-South America collision

Previous workers have suggested that the age of the suturing of Panama with South America is indicated by: (1) Early Pliocene uplift of the three Cordillera of Colombia, or the so-called Andean orogeny (Van der Hammen et al., 1973; Irving, 1975; Pindell and Dewey, 1982); (2) regional latest Miocene to Early Pliocene uplift and increased clastic sedimentation in sedimentary basins in Panama and northwestern South America (Mann and Burke, 1984; Duque-Caro, 1990b); (3) Pliocene contrasts in deep-marine marine foraminifera species and isotopic data (Keigwin, 1978, 1982; Keller et al., 1989); (4) the Pliocene (~3 Ma) "Great American Interchange" of terrestrial faunas between North and South America (for example, migration of raccoons and their allies to South America and migration of ground sloths to North America) (Marshall, 1988); and (5) early Pliocene (~3.5 Ma) contrasts in nearshore marine calcareous nannofossils and planktonic foraminifera collected from onshore sections around Panama and Costa Rica (Coates et al., 1992).

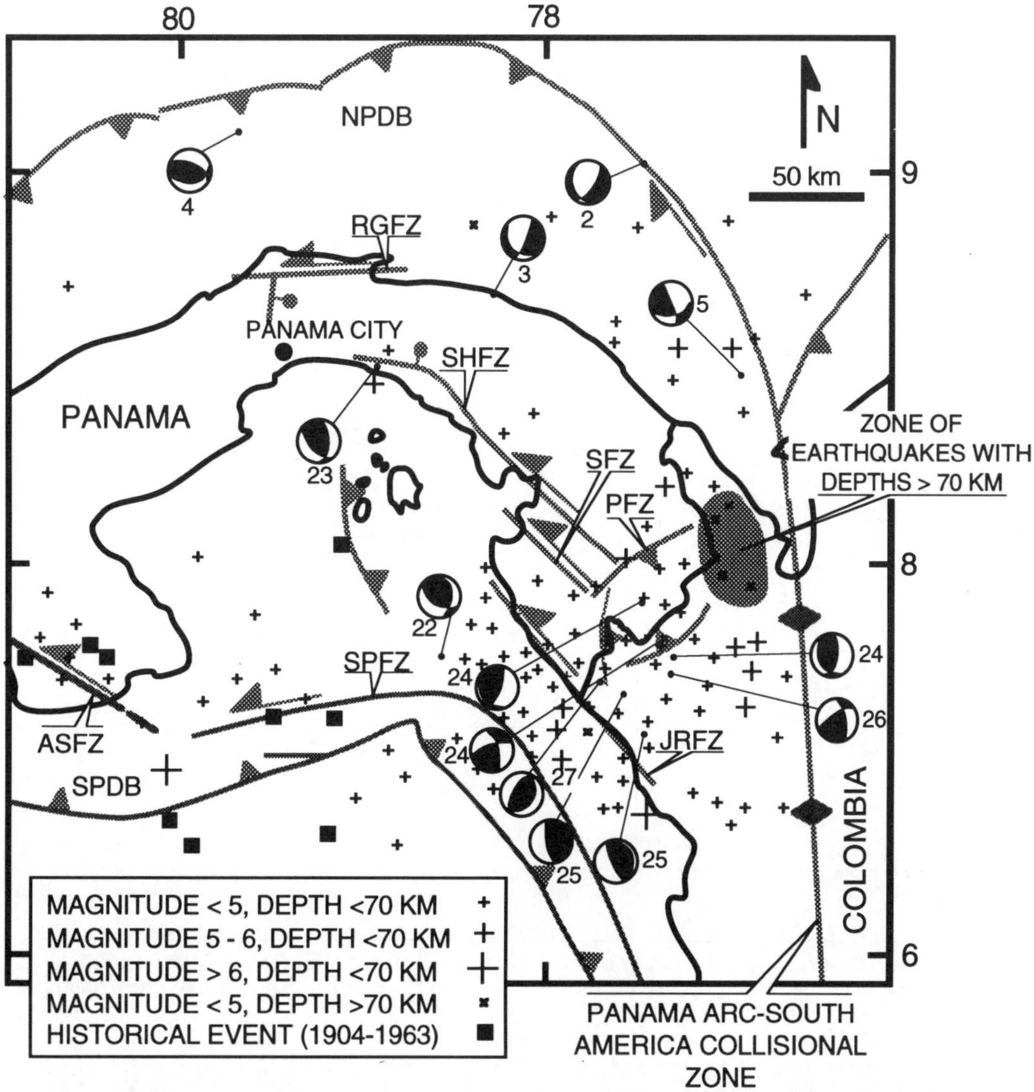

Figure 4. Plot of all published earthquake focal mechanisms and epicenters in the eastern Panama area for the years 1964 to 1988 (small numbers indicate the depth in km to earthquakes with focal mechanisms as compiled by Adamek et al., 1988). The source of epicenters is a file from the International Seismological Center (ISC). Historical events from 1904 to 1963 were also taken from Adamek et al. (1988). Note the zone of earthquakes with depths greater than 70 km in eastern Panama. Key to abbreviations of faults: SPFZ = South Panama fault zone; NPDB = North Panama deformed belt; SPDB = South Panama deformed belt; SHFZ = Sansón Hills fault zone; RGFZ = Río Gatún fault zone; ASFZ = Azuero-Soná fault zone; JRFZ = Jaqué River fault zone; SFZ = Sambú fault zone; PFZ = Pirre fault zone.

GEOLOGY OF THE ONSHORE EAST PANAMA DEFORMED BELT

Faults of the East Panama deformed belt in onshore eastern Panama

Major faults in the onshore part of the East Panama deformed belt include mapped (Sansón Hills) and inferred (Sambú, Jaqué River) left-lateral strike-slip faults bounding northwest-trending highlands near the Pacific Coast and mapped (Pirre) and inferred (Unguía) reverse faults bounding northeast-trending highlands (Pirre Hills) along the Panama-Colombia border (Fig. 5). Because much of this area is shrouded by clouds and covered by dense tropical forest, studies of airborne and satellite radar imagery has proven valuable for regional geologic mapping (e.g., MacDonald, 1969) (Fig. 6B).

Pirre and Unguía faults. MacDonald (1969) reported that major faulting is evident on airborne radar imagery along the Pirre fault, and Terry (1956) reported that the Pirre fault is a high-angle reverse fault cutting vertical strata (Fig. 6A). On the eastern flank of the Pirre Hills in Colombia, the sinuous trace of the Unguía fault suggested reverse movement to W. D. Page

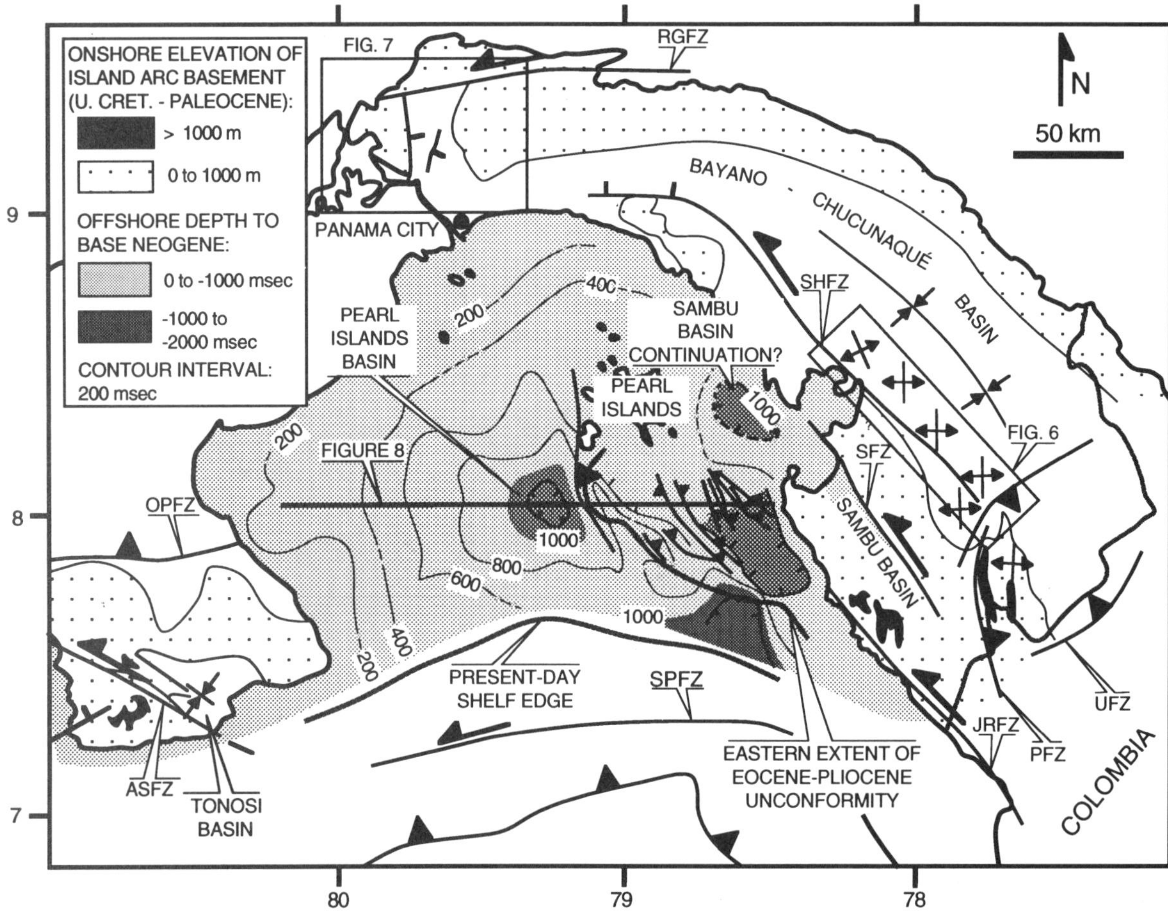

Figure 5. Location map for onshore structure maps of eastern Panama shown in Figures 6 and 7 and structural time-map to the base of the Neogene section in the Gulf of Panama. Contours are based on Mobil seismic and well data shown in Figure 3. Depth contour interval is 200 milliseconds. Offshore contour intervals are shaded gray and represent the top surface of middle Eocene (west of the gray line representing the eastern extent of Eocene-Pliocene unconformity) and Oligocene? (east of the line representing the eastern extent of Eocene-Pliocene unconformity) sedimentary rocks. Onshore dotted and black areas represent topographic elevations of upper Cretaceous–early Tertiary crystalline basement rocks (mostly basalt). Heavy black line shows line of regional cross section shown in Figure 8. Key to abbreviations: ASFZ = Azuero-Soná fault zone; OPFZ = Ocú-Parita fault zone; RGFZ = Río Gatún fault zone; SHFZ = Sansón Hills fault zone; SFZ = Sambú fault zone; JRFZ = Jaqué river fault zone; PFZ = Pirre fault zone; UFZ = Unguía fault zone; SPFZ = South Panama fault zone.

(1986, unpublished). Within the uplifted highland block of the Pirre Hills defined by the two faults, two anticlines show dips ranging from 10° to 40° and affect rocks as young as late Miocene (Shelton, 1952). The north-south–trending Pirre Hills abruptly truncate northwest-striking faults and stratigraphic units of the Bayano-Chucunaqué basin (Fig. 6A).

Jaqué River and Sambú faults. The Jacqué River and Sambú faults define a rectangular, steep-sided mountain block with elevations up to 1,700 m (Serranía del Sapo) (Fig. 5). Both faults define linear coastlines or occupy linear river valleys and strike 135° to 155°. Because of difficult access to this area, detailed mapping along the Sambú and Jaqué River faults has not been done. Using airborne radar imagery, MacDonald (1969) extended the linear trace of the Sambú fault 25 km farther southeast than had been previously reported and identified the Jaqué River fault as a continuous lineament that extended across the Colombian border. Pending further field studies, Mann and Corrigan (1990) proposed that both the Sambú and Jaqué River faults are left-lateral strike-slip because of (1) their straight and prominent physiographic expression in varied topography, (2) the presence of the apparent offshore continuation of the Sambú basin at a left step in the Sambú fault, and (3) the parallelism of both faults to the known left-lateral Sansón Hills fault to the northeast (Fig. 5).

The onshore Sambú basin and its inferred offshore continuation contain up to 5 km of Tertiary clastic sedimentary

A.

B.

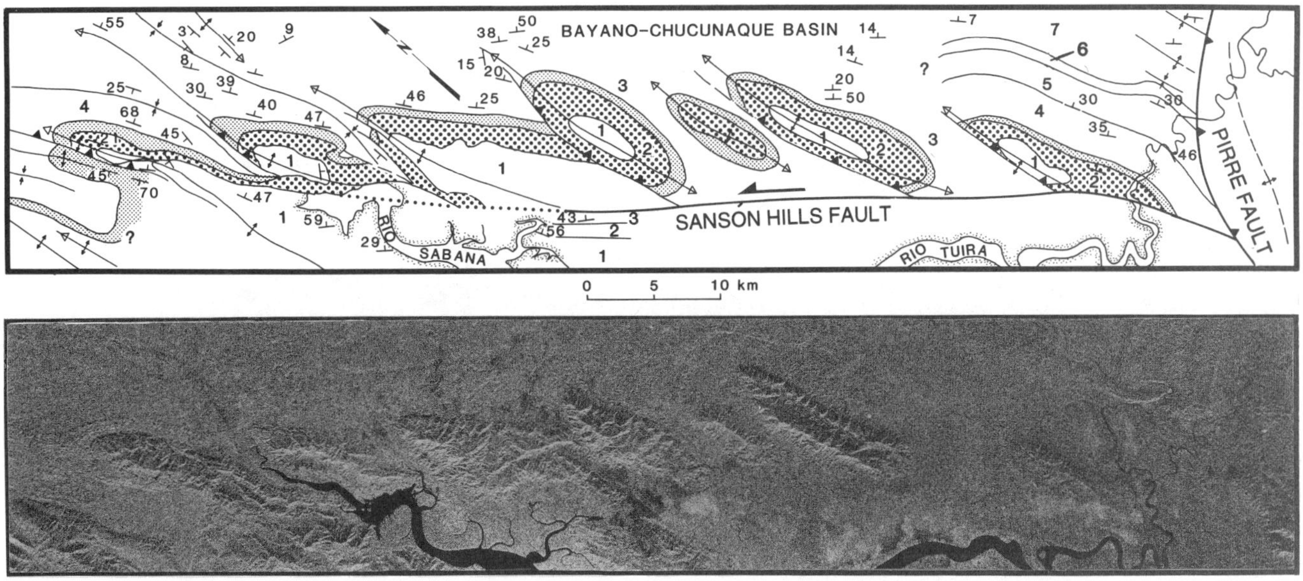

Figure 6. A, Geologic map of the Sansón Hills and Pirre fault zones in eastern Panama from Mann and Corrigan (1990) using unpublished data from Shelton (1952) and Panama Dirección General de Hidrocarburos (1972, unpublished). Key to stratigraphic units (boldface numbers): 1 = Cretaceous volcaniclastic sedimentary rock, tuffs, and limestone; 2 = middle to upper Eocene volcaniclastic sedimentary rock; 3 = lower Miocene limestone; 4 = middle to upper Miocene mudstone; 5 = middle to upper Miocene volcaniclastic sedimentary rock; 6 = middle to upper Miocene limestone; 7 = middle to upper Miocene mudstone. Other numbers indicate dips of strata. B, Photo is Shuttle Imaging Radar-A (SIR-A) (data taken 24C, original scale 1:500,000).

rocks (Lowrie et al., 1982) that are marked by a strongly negative Bouguer gravity anomaly (Case, 1974; Kellogg et al., 1991, and this volume). The onland Sambú basin is the site of natural onshore oil seeps and several exploration wells (Johnson and Headington, 1971). We interpret the Sambú basin and its inferred offshore continuation as a pull-apart or fault termination basin along the left-lateral Sambú fault because the basinal depocenters appear to occupy left-steps in the fault trace.

Sansón Hills fault zone. A series of seven doubly plunging or truncated *en echelon* folds is present to the northeast of the 135° striking Sansón Hills fault zone and records an unknown amount of left-lateral displacement between an oceanic assemblage to the southwest and thick sedimentary rocks of the synclinal Bayano-Chucunaqué basin to the northeast (Shelton, 1952) (Fig. 6). Southwest of the fault, Bandy and Casey (1973) and Case (1974) have described an upper Cretaceous basalt-chert-abyssal sedimentary sequence that crops out in intertidal coastal exposures around the Gulf of San Miguel (Fig. 3). Much of the stratigraphy of the Bayano-Chucunaqué basin is exposed in breached, *en echelon* anticlines along the Sansón fault zone. The oldest rocks in the cores of the anticlines are upper Cretaceous sedimentary and igneous rocks, and the youngest rocks are upper Miocene mudstones (Shelton, 1952). The structural pattern of the Sansón Hills was interpreted by Wilcox et al. (1973) in their global examination of strike-slip tectonics to illustrate an intermediate stage of strike-slip displacement during which a throughgoing strike-slip fault transects earlier-formed, *en echelon* folds.

To the northwest, the Sansón Hills fault curves to westward along the southern edge of the Bayano-Chucunaqué basin (Fig. 5). In this area the basin becomes an asymmetric half-graben, the south side of the basin being downfaulted (R. Stewart, 1966, unpublished). We suggest that this pattern may reflect extensional termination of left slip along the Sansón Hills fault as the fault curves into a more westward extensional orientation. To the southwest, the Sansón Hills fault terminates abruptly at the Pirre reverse fault (Fig. 6A).

Faults of the East Panama deformed belt in the area of the Panama Canal

In the area of the Panama Canal, the Rio Gatún fault strikes 070° and forms a prominent topographic lineament that extends offshore where it remains unmapped (Fig. 5). The fault forms the steep southern margin of a coastal highland that extends up to 600 m in elevation and exposes pre-Tertiary basement rocks. To the south of the Rio Gatún fault zone, a large number of north-striking and generally east-facing scarps are easily visible on radar imagery and can be identified with faults mapped by W. P. Woodring (unpublished) and Stewart and Stewart (1980) during canal widening projects and in artificial and natural outcrops in the former Panama Canal Zone (Fig. 7).

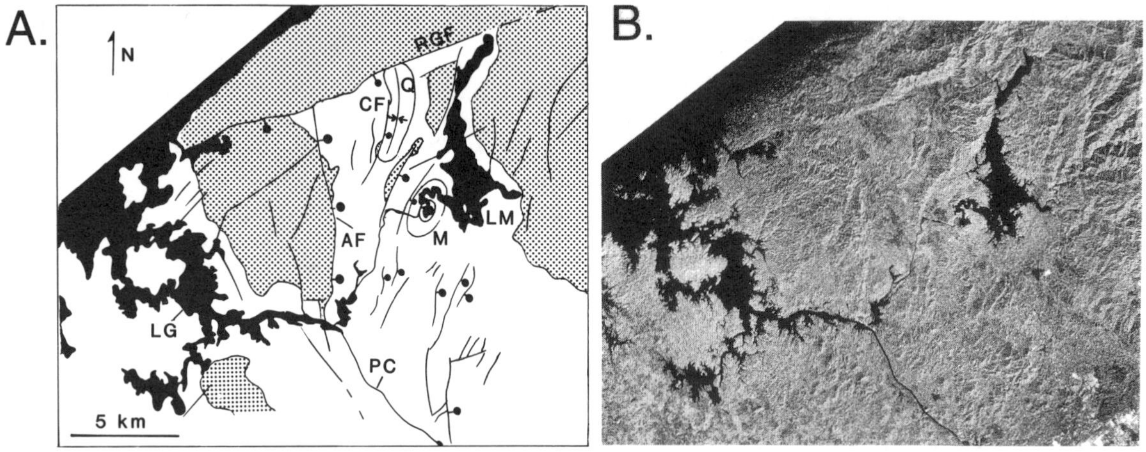

Figure 7. A, Structural-geologic sketch map of the Panama Canal–Lake Gatún–Lake Madden area of east-central Panama from Mann and Corrigan (1990). Map interpretation combines field observations and structural names from mapping by Stewart and Stewart (1980) with interpretation of satellite image by Mann and Corrigan (1990). Key to abbreviations: LM = Lake Madden; LG = Lake Gatún; PC = Panama Canal; RGF - Rio Gatún fault; AF = Azota fault; CF = Caballo fault; Q = Quebrancha syncline; M = Madden syncline; patterned areas represent pre-Tertiary basement rock; black areas are lakes or Caribbean Sea; ball and ticks on downthrown sides of fault scarps. B, Photo is SIR-A image, data take 37, original scale 1:500,000.

This family of fault scarps defines and occupies a topographically low region (<300 m elevation) of mainly Cenozoic volcaniclastic deposits and intrusive bodies that are preserved between topographically high blocks of pre-Tertiary basement rocks. Lake Madden forms a reservoir impounded in the lowest part of the valley, and Lake Gatún (actually an arm of the Caribbean Sea) naturally infills the seaward part of the valley.

On the basis of the 1:100,000 scale geologic map by Stewart and Stewart (1980), we interpret north- and northeast-striking faults to be normal faults, which generally place Cenozoic volcaniclastic sedimentary rocks against pre-Tertiary basement rocks that are commonly exposed along the fault. The Cenozoic section contains rocks well dated to be as young as late Miocene (Gatún Formation of Coates et al., 1992). "Dog-leg" fault patterns are present along some of the scarps in map view, and large, doubly plunging synclines or structural sags (Quebrancha and Madden) are within the volcaniclastic hanging-wall sequence adjacent to two of the fault scarps (Fig. 7). "Dog-leg" fault patterns and structural sags above hanging-wall blocks are common in better-studied extensional environments (Harding, 1984).

On the basis of these data, Mann and Corrigan (1990) suggested that the Panama Canal area is undergoing active east-west extension across a series of normal faults that generally dip east. We suggest that these normal faults may represent accommodation structures that transform and terminate postulated left-lateral shear along the Rio Gatún fault zone to the north (Fig. 7). The Rio Gatún fault may bend to a more northeast strike offshore and extend into the previously described left-lateral strike-slip province of eastern Panama (Fig. 5). The pattern of proposed normal fault termination structures in the Panama Canal area is similar to (1) the eastern Bayano-Chucunaqué basin, where the Sansón Hills left-lateral strike-slip fault zone appears to terminate on the half-graben structure of the basin, and (2) the Sambú basin, where the left-lateral Sambú fault appears to terminate on the left-stepping Sambú basin (Fig. 7A).

GEOLOGY OF THE OFFSHORE EAST PANAMA DEFORMED BELT IN THE GULF OF PANAMA

Previous geophysical studies in the Gulf of Panama

The Gulf of Panama extends between the Azuero Peninsula in southwest Panama and the border of Colombia and covers an area of approximately 35,000 km^2 (Fig. 3). Lowrie (1978) published a seismic reflection profile acquired using a 30 kJ sparker system from the slope and trench area of the southeastern end of the South Panama deformed belt and identified a buried, inactive trench (Fig. 5A). Case (1974) and Briceño-Guarupe (1978) presented gravity maps of the Gulf of Panama and surrounding onshore region. To our knowledge, no seismic reflection data have been published from the shelf area of the Gulf of Panama. Lowrie et al. (1982) published a map showing offshore sedimentary depocenters in the Gulf of Panama but did not publish supporting geophysical data. Case and Holcombe (1980) showed northwest-striking left-lateral strike-slip faults crossing the Gulf of Panama on their regional geologic-tectonic map of the Caribbean region but were unable to reveal the proprietary sources of oil company reports on which this interpretation was based (J. E. Case, personal communication, 1990). Kellogg et al. (this volume) have compiled a Bouguer gravity anomaly map at scales of 1:1,500,000 for the Gulf of Panama and adjacent land areas. A. J. Kudrna (1989, unpublished) presented a poster session at the 1989 American Association of Petroleum Geologists Circum-Pacific conference in San José, Costa Rica, in which he presented two seismic sections and maps of two horizons made from the Mobil data set. This poster was the first documentation of thrust faulting beneath the Gulf of Panama.

Data used in this study of the offshore East Panama deformed belt (Gulf of Panama)

This study represents the first comprehensive study of the Gulf of Panama using a grid of 1,600 km of multichannel seismic data tied to two wells (Fig. 3). Distance between adjacent seismic lines ranges from 10 to 30 km. Data used in this study were provided by two principal sources. (1) Mobil Exploration and Production Services, Inc., Dallas, Texas, provided 1,600 km of multichannel seismic data from the Gulf of Panama tied to two well logs of exploration wells drilled in the Gulf of Panama in 1969 and 1971 (Fig. 3). (2) Department of Natural Resources, Panama Ministry of Public Affairs, Panama City, Panama, provided lithologic and paleontologic field data collected by industry geologists in the Darien Province of eastern Panama and on the Pearl Islands of the Gulf of Panama between 1970 and 1971. Uninterpreted paper copies of the seismic data were made available to us by Mobil along with the two well logs and well completion reports. The original 24-fold Mobil data was collected in 1969–1972 and processed by Western Geophysical in 1988, using the parameters listed on Table 1. Because we judge that data quality is adequate, no effort was made by us to obtain the tapes and reprocess the data.

TABLE 1. FIELD AND PROCESSING PARAMETERS FOR OFFSHORE MULTICHANNEL SEISMIC DATA USED IN THIS CHAPTER

FIELD PARAMETERS	MOBIL
Recorded	1969, 1971, 1972
System	DFS III
Sampling interval	4 msec
Filter	8-62 Hz
Air gun	n.a.
Pressure	n.a.
Shotpoint interval	70 m
Number of channels	24
Channel length	58 m
Near trace	208 m
PROCESSING PARAMETERS	**MOBIL**
Processed	1988
	Trace editing and correction
	Demux
	CDP-Gathers sort
	Velocity analysis
	NMO
	24 fold CDP-stack
	Finite difference migration
	Time-variant scaling
	Display (normal polarity)*

*A compression at the geophone produces a negative number represented by a black peak on plot.
n.a. = not available.

TECTONIC ELEMENTS OF THE EAST PANAMA DEFORMED BELT IN THE GULF OF PANAMA

We interpreted approximately 1,600 km of multichannel seismic profiles provided by Mobil Exploration and Production Services from the Gulf of Panama (Fig. 5). Two offshore lithologic well logs (El Paso Corvus-1, El Paso Plaris-1) provided age control for the seismic reflectors seen on the profiles. These data were integrated into a map showing depth in time to the base of the Neogene section (Fig. 5). Only a time version of this map is presented because sonic logs and velocity information were not available from the area.

The Neogene surface was selected for a structural contour map of the Gulf of Panama for the following reasons: (1) The base Neogene surface is the deepest horizon that can be identified and correlated throughout the Gulf of Panama using the available Mobil seismic profiles. Middle Eocene and older sedimentary rocks, which underlie this surface in the east-central Gulf of Panama, are visible on some of our seismic lines, and have been drilled at the El Paso Corvus-1 well. (2) Igneous basement, known to crop out over large areas of eastern Panama (Bandy and Casey, 1973; Case, 1974) (Fig. 5), could not be identified on the seismic lines because of thick sedimentary cover and the poor resolution of deeper horizons on the available profiles. (3) Igneous basement offshore was not penetrated in either of the two offshore wells available for this study or other industry wells reported from the Gulf of Panama and eastern Panama (Johnson and Headington, 1971).

These data allow recognition and description of most of the major tectonic elements of the East Panama deformed belt in the Gulf of Panama. Many of these features are described and named for the first time in this chapter. These tectonic elements include:

1. Active South Panama deformed belt and its inactive eastern extension. The active South Panama deformed belt south of the Azuero Peninsula has been described by MacKay (1989), MacKay and Moore (1990), and Moore and Sender (this volume) using multichannel seismic profiles and SeaMARC II sidescan images. The zone of active deformation extends eastward to at least longitude 80°30′W. Using single-channel profiles, Lowrie (1978) showed that the eastern extension of this feature from 80°W to 79°45′W is a buried (inactive) trench segment (Fig. 3). The buried trench segment appears to be continuous with the zone of active accretion along the Colombian trench (Hardy et al., 1991; Westbrook et al., this volume). Because we lack new data from this area, we do not discuss the South Panama deformed belt further in this chapter.

*2. **South Panama fault zone*** is named by Westbrook et al. (this volume) for the fault occupying a major east-trending canyon north of the buried trench segment of Lowrie (1978) (Fig. 3). Using GLORIA sidescan imagery and seismic profiles, Hardy et al. (1991) and Westbrook et al. (this volume) provide evidence for the fault to be an active left-lateral fault zone that appears to intersect the left-lateral Azuero-Soná fault zone to the west (this approximate geometry is shown in Fig. 3). Because we lack new data from this area, we do not discuss the South Panama deformed belt further in this chapter.

3. The inferred offshore extension of the onland Sambú basin is a 2,000-km^2, rectangular Neogene depocenter located east of the Pearl Islands and bounded on its northeastern edge by the Sambú left-lateral strike-slip fault zone (Fig. 5). Because of the sparsity of Mobil lines over this basin (Fig. 3), we do not discuss this basin further in this chapter.

4. Pearl Islands basin is named here for the 2,500-km^2, subcircular Neogene depocenter located to the east of the Pearl Islands. The basin is defined here using the Mobil seismic data as a depocenter with a two-way travel to basement of 1,000 to 2,000 milliseconds (Fig. 5). The location of this basin does not correspond to any of the depocenters with sediment thicknesses exceeding 2 km previously defined in the Gulf of Panama by Lowrie et al. (1982; their Figure 5). It is possible that these depocenters were defined by Lowrie et al. (1982) in the middle Eocene and older stratigraphic section considered "basement" in this study and are therefore beyond the resolution of our seismic reflection data.

5. Offshore East Panama deformed belt is a 30,000-km^2 area of northwest-striking fold and thrust deformation in the East Panama deformed belt that is defined here in the area between the Pearl and Sambú basins. The deformed belt is not well defined north of the Pearl Islands or in the area near the Colombian border because of the lack of seismic coverage (Fig. 3).

Faults in the offshore East Panama deformed belt strike to the northwest and are comparable to the strike of faults mapped in the onshore part of the deformed belt in eastern Panama and on the Azuero Peninsula (Fig. 5). Mobil seismic lines allow definition of two unnamed depocenters within the East Panama deformed belt with a two-way travel time to basement of 1,000 to 2,000 milliseconds. We use the name "deformed belt" rather than "fold-thrust belt" to describe the area east of the Pearl Islands basin because some of these apparent thrust structures seen on seismic profiles may accommodate significant left-lateral strike-slip displacement as observed for well-mapped, subparallel onland faults like the Samsón Hills fault zone (Fig. 6).

REGIONAL TRANSECT ACROSS THE EAST PANAMA DEFORMED BELT AND PEARL ISLANDS BASIN

We constructed a regional transect through the East Panama deformed belt and the Pearl Islands basin (Fig. 8) using Mobil seismic lines 1 and 18 that extend for 230 km in an east-west direction across the Gulf of Panama (Fig. 3). The age of reflectors was determined using ties to El Paso Corvus-1 well in the central Gulf of Panama and the El Paso Plaris-1 well in the eastern Gulf of Panama.

Western Gulf of Panama and Pearl Islands basin

The Pearl Islands basin and western Gulf of Panama make up a shelf area that has an average water depth between 20 and 200 m and is underlain by an unfaulted Plio-Pleistocene section that thins westward in the direction of the Azuero Peninsula (Fig. 8). The Plio-Pleistocene section is characterized by laterally continuous reflectors that onlap the underlying middle Eocene section along the prominent unconformity taken as the base of the structure contour map (i.e., "basement") shown in Figure 5 (see seismic sections 1 and 2 in Fig. 8B). The thickness of the Plio-Pleistocene section generally does not exceed 1,000 milliseconds except in the Pearl Islands basin, where thicknesses up to 1,200 milliseconds are observed. This area of thicker sedimentary rocks beneath the Gulf of Panama is reflected in an area of slightly deeper water, as shown by northward reentrant of the 100-m bathymetric contour (Fig. 3).

The well log of the El Paso Corvus-1 well that is tied to Mobil seismic line POC-19 in the western Gulf of Panama reveals that the faunally-dated middle Eocene section consists of interbedded basaltic volcanic rocks, associated pyroclastic material, sandstone, and shale and that the overlying Plio-Pleistocene section consists of interbedded sandstone and shale (Fig. 9). Eastward thickening of the Plio-Pleistocene sandstone-shale section of the Pearl Islands basin suggests that the basin formed as a foredeep when Eocene acoustic basement tilted eastward in front of east-dipping folds and thrusts of the East Panama deformed belt.

Plio-Pleistocene basinal rocks in the western Gulf of Panama are undeformed by faulting. We see no evidence in this area for the northwest-striking left-lateral strike-slip faults postulated by Case (1974) and Case and Holcombe (1980) and more recently shown by Silver et al. (1990) in their bending model of the Panama orocline. It is possible that such faults exist below the Eocene-Pliocene unconformity and below the resolution of our seismic data. These deeper structures would be pre-Pliocene in age and therefore would not be responsible for the active deformation of the Isthmus of Panama.

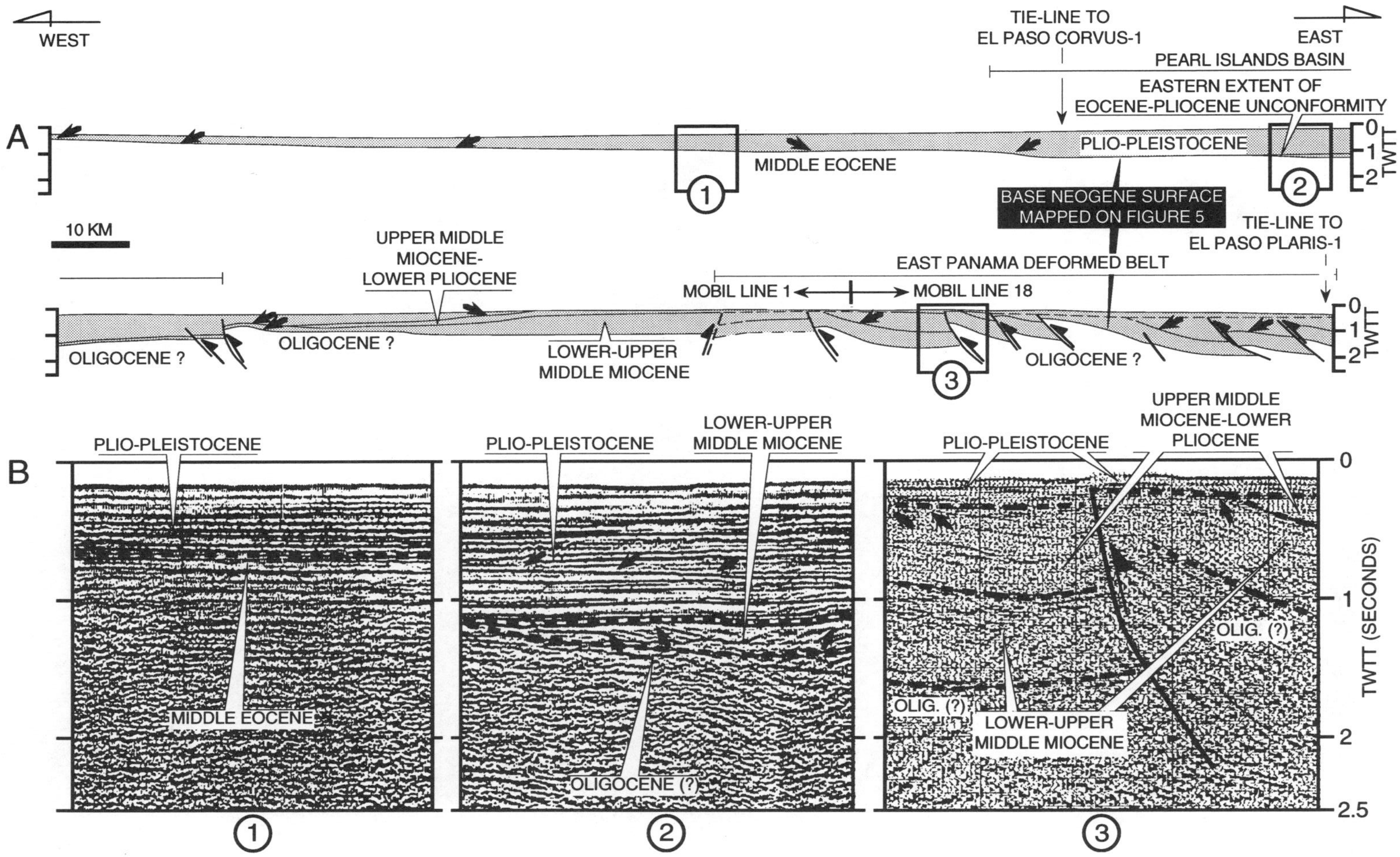

Figure 8. A, East-west regional cross section through the Gulf of Panama based on Mobil lines 1 and 18 (see Fig. 5 for line of section). Vertical scale is in two-way travel time; horizontal scale in km is shown in upper left corner. Seismic horizon indicating base of the Neogene section mapped in Figure 5 is indicated. Small arrows represent reflection terminations. B, Insets 1, 2, 3 show representative seismic examples taken from Mobil line 1 (insets 1, 2) and line 18 (inset 3). Note eastward thickening of Miocene to Pleistocene foredeep basin section in front of reverse faults, eastward increase in reverse fault deformation, and eastward tilting of Eocene and older basement.

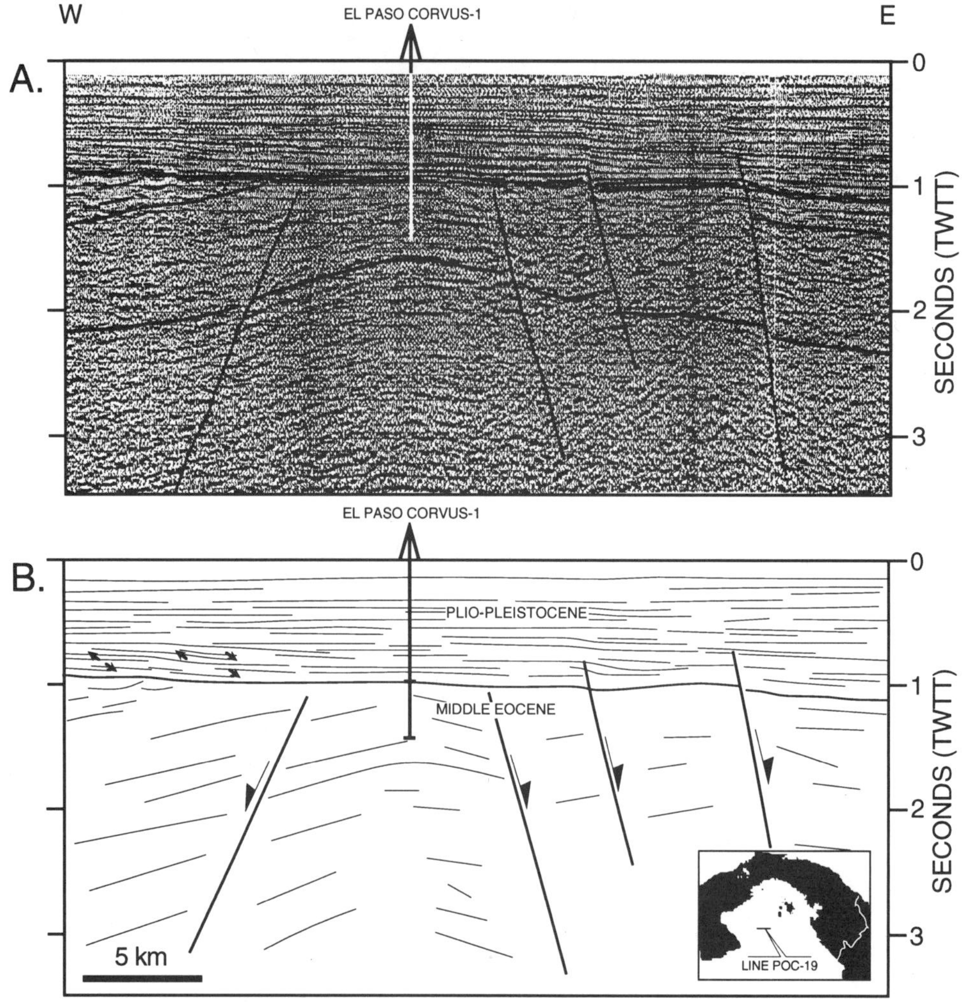

Figure 9. A, Segment of unmigrated Mobil 24-fold seismic line (line POC-19 on Fig. 3) from the East Panama deformed belt in the west-central Gulf of Panama. Note that the line is tied to the El Paso Corvus-1 well. B, Interpretation showing block-faulted and truncated Middle Eocene sedimentary sequence overlain unconformably by a flat-lying Plio-Pleistocene sedimentary section.

Relation of the Pearl Islands anticline to offshore structure

The western edge of the Pearl Islands basin is unfaulted and is arbitrarily taken at the point where two-way travel time to acoustic basement exceeds 1 second (Fig. 5). The eastern edge of the basin is defined by two east-dipping reverse faults that elevate acoustic basement by about 0.5 seconds two-way travel time. A hanging-wall anticline onlapped by upper middle Miocene–lower Pliocene strata is observed east of the two faults (Fig. 8A).

The Pearl Islands, an archipelago in the central Gulf of Panama are located about 25 km to the north of the Line 1 and Line 18 transect shown in Figure 8A and appear to be an along-strike topographic expression of the hanging-wall anticline at the eastern limit of the Pearl Island basin. We compiled maps and descriptions of the structure and stratigraphy of the basement and overlying sedimentary cover of the Pearl Islands from unpublished reports provided by the Panama Dirección General de Hidrocarburos for use in this study (Fig. 10). Basement rocks crop out over 60% of the larger islands of Isla del Rey and all of Isla del San José and are composed of basalt, associated pyroclastic material, and chert. To our knowledge, there are no published fossil or radiometric ages from the basement rocks of the Pearl Islands. Igneous basement rocks exposed in the Pearl Islands consist of Cretaceous basaltic rocks that may correlate with (1) Late Cretaceous basalt-chert-abyssal sedimentary sequence described by Bandy and Casey (1973) and Case (1974) from intertidal coastal exposures around the Gulf of San Miguel (Fig. 3); (2) acoustic basement of the western Colombian basin that has been interpreted by Bowland and Rosencrantz (1988) as Late Cretaceous basalt (Fig. 2); and (3) basalt and chert sequences known from the Azuero Peninsula (Kolarsky et al., this volume); and (4) probably middle Eocene basalts and interbedded sediments encountered in the El Paso Corvus-1 well to the west of the islands (Fig. 5).

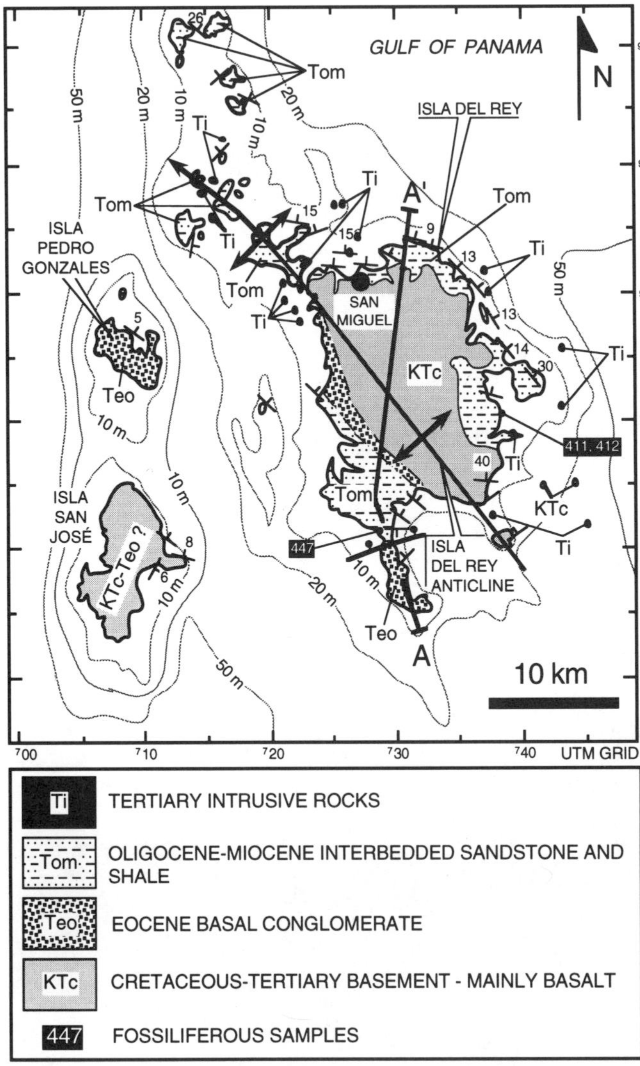

Figure 10. Geologic compilation map of the Pearl Islands in the Gulf of Panama (from Dirección General de Hidrocarburos, 1972, unpublished). Map coordinates refer to Universal Transverse Mercator (UTM) grid given on the 1:50,000 topographic sheet of the islands. Key to abbreviations: KTc = Cretaceous-Tertiary basement rocks; Teo = Eocene basal conglomerate; Tom = Oligocene-Miocene interbedded sandstone and shale; Ti = Tertiary intrusive rocks. Numbers indicate fossiliferous samples reported by the Panama Dirección General de Hidrocarburos (1972, unpublished) (Table 2). Heavy line A-A′ indicates line of cross section shown in Figure 11.

Unpublished reports provided by the Dirección General de Hidrocarburos describe the sedimentary section on the Pearl Islands as less than 500 m of an Eocene basal conglomerate overlying basaltic basement and up to 1,000 m of interbedded marine sandstone and shale of Oligocene-Miocene age that have been dated with microfauna. On Isla del Rey, both basement and the overlying sedimentary rocks are intruded by small plugs of Tertiary basalt (Fig. 11).

The map pattern of lithologic units and the orientations of bedding indicate that the Pearl Islands form a large anticline with a northwest-striking axial trace that is probably associated with east-dipping thrust faults at the eastern offshore margin of the Pearl Islands basin (Fig. 10). The dominance of northeast dips of bedding suggests that this structure is highly asymmetric, an interpretation consistent with its association with underlying, east-dipping thrust faults (seismic section 3 in Figure 8B). Rocks exposed on the Pearl Islands are correlated with pre-Oligocene rocks known from the El Paso Çorvus-1 well in the east-central Gulf of Panama and correlated to acoustic basement over the area (Fig. 8). The existence of marine sedimentary rocks of Miocene age exposed on the Pearl Islands suggests that the age of apparent acoustic basement in the western and central Gulf of Panama may be locally younger than Eocene in age.

East Panama deformed belt

Nine apparent thrust faults deforming Oligocene(?) to Plio-Pleistocene rocks were mapped offshore using the Mobil seismic profiles. The largest thrust offsets lower- to upper-middle Miocene reflections by about 1 second of two-way travel time (TWTT) (Fig. 8, section 3). The thrust faults segment the East Panama deformed belt into a series of east-tilted fault blocks that range in width from 5 to 20 km. Displacements on the faults elevate acoustic basement thought here to be of Oligocene? age to as close as 0.5 to 0.25 seconds TWTT from the seafloor.

Age of deformation in the East Panama deformed belt

Line 117. Mobil seismic line 117 shows imbricated thrust faults in the East Panama fold and thrust belt south of the Pearl Islands (Fig. 12). Thrust faults have up to 0.5 seconds of apparent displacement on fault planes. A prominent high-amplitude reflection defines the top of a lower to middle Miocene sequence that is disrupted by thrust faults.

The middle Miocene to lower Pliocene section is faulted and onlaps the underlying anticlines produced by thrusting (Fig. 12). Plio-Pleistocene reflections are undisturbed by deformation and completely bury thrust-related structures. On line 18, 25 km to the north of line 117 (Fig. 8, section 3), lower Pliocene reflections in the hanging wall of a reverse fault are truncated by erosion and are unconformably overlain by the Plio-Pleistocene section. Thickening of the Plio-Pleistocene unit in the footwall and arching of the unit and seafloor suggests that thrusting remains active to the present.

Line POC-33 and El Paso Plaris-1. Line POC-33 is useful for evaluating the age of deformation in the eastern Gulf of Panama because it is tied directly to El Paso Plaris-1 (Fig. 13). The sedimentary sequence can be divided into three sedimentary sequences. The lower seismic unit consists of an eastward-tilted Oligocene? to middle Miocene aggradational shelf margin. The base of the aggradational shelf margin is a prominent unconformity on the flank of a large anticline interpreted to be in the hanging wall of a thrust fault. Reflections of the overlying westward-prograding middle Miocene–lower Plio-

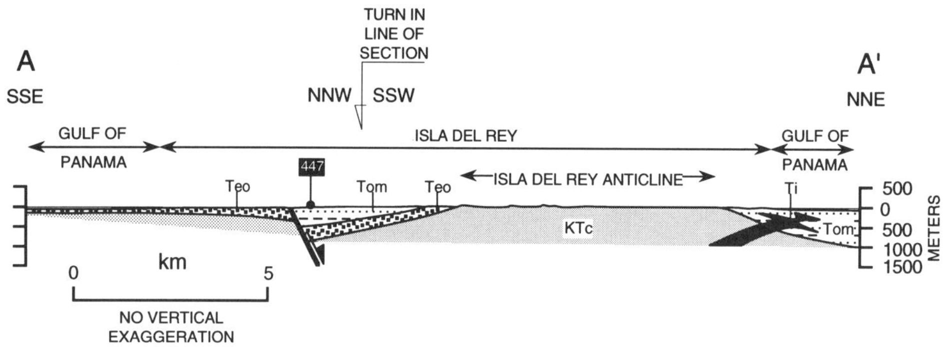

Figure 11. Geologic cross section of the Pearl Islands in the Gulf of Panama (from Dirección General de Hidrocarburos, 1972, unpublished). See Figure 10 for line of section. Abbreviations for rock units are given in caption of Figure 10. Note axis of anticline and block faulting. Black number 447 shows location of dated, fossiliferous rock sample used to constrain age (Table 2).

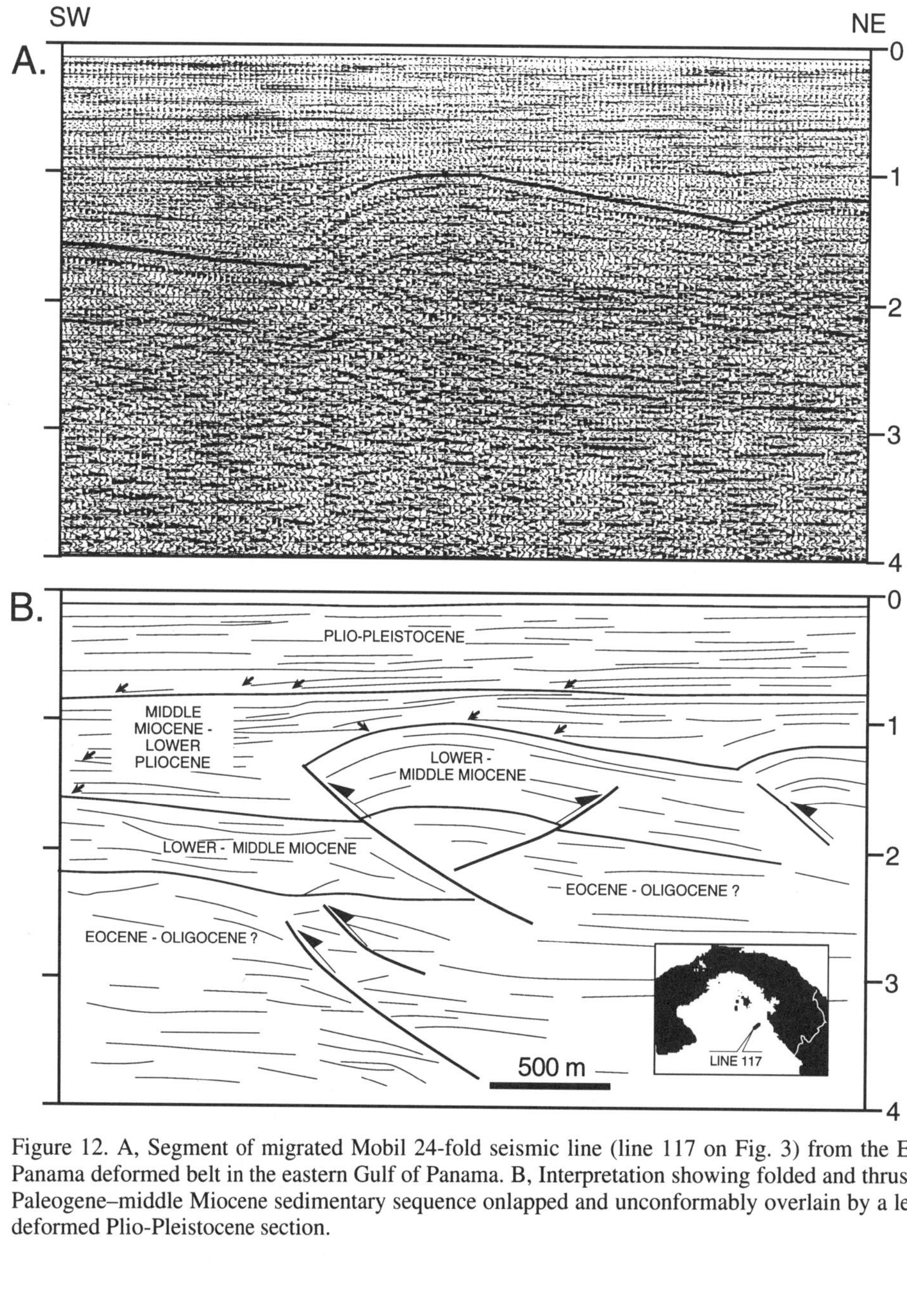

Figure 12. A, Segment of migrated Mobil 24-fold seismic line (line 117 on Fig. 3) from the East Panama deformed belt in the eastern Gulf of Panama. B, Interpretation showing folded and thrusted Paleogene–middle Miocene sedimentary sequence onlapped and unconformably overlain by a less-deformed Plio-Pleistocene section.

cene shelf margin downlap the flank of the anticline that continued to be uplifted until the Pliocene, as shown by downlap and onlap relationships of overlying reflectors known to be of Pliocene age in the El Paso Plaris-1 well.

Line 18. Line 18 shown on the structural transect (section 3 of Figure 8) shows seafloor arching above the thrust fault. This arching suggests that this fault is an active structure that is presently deforming the seafloor and may be responsible for some of the observed shallow earthquakes in the Gulf of Panama (Fig. 4).

Pearl Islands anticline. The emergence of basement rocks to elevations of 20 m above sea level in the Pearl Islands in an otherwise submerged marine embayment suggests the possibility of recent movement along the north-striking thrust fault thought to underlie these islands. Studies of uplifted late Quaternary shoreline features are needed to confirm this hypothesis.

Uplift history of the East Panama deformed belt based on microfauna on the Pearl Islands and lithologic logs and microfossil data from the El Paso Corvus-1 and Plaris-1 wells

Pearl Islands. According to unpublished reports of the Dirección General de Hidrocarburos, Eocene conglomerate above basaltic basement on the Pearl Islands does not contain microfossils with age or paleoenvironmental information. Micro-

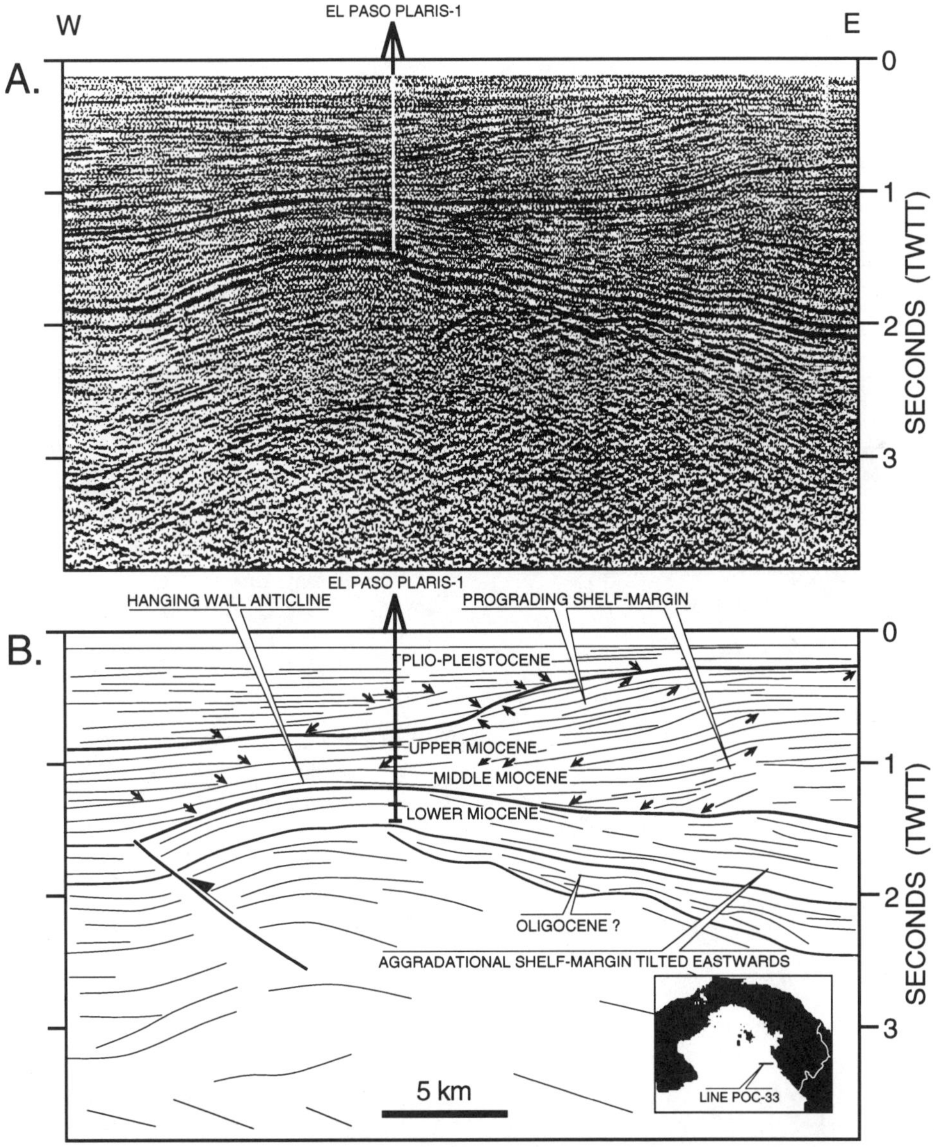

Figure 13. A, Segment of migrated Mobil 24-fold seismic line (line POC-33 on Fig. 3) from the East Panama deformed belt in the eastern Gulf of Panama. B, Interpretation showing an Oligocene?–lower Miocene aggradational shelf margin that is tilted eastward in the hanging wall of an east-dipping thrust fault and middle Miocene–Pliocene progradational shelf margin.

fauna in samples 411 and 412 from the sedimentary section along the east coast of Isla del Rey (Fig. 10) indicate a late Miocene–Pliocene? age and a middle neritic or deeper environment of deposition (Table 2). Larger foraminifera in sample 477 yielded a late Oligocene age and a very shallow, clastic-dominated inner neritic environment of deposition.

El Paso Corvus-1. This well was drilled to a total depth of 2,654 m below sea level and was continuously cored from a depth of 246 m (Fig. 14A). The well penetrated a sequence of Plio-Pleistocene sandstone and shale with a sandstone-to-shale ratio much less than 1 down to depth of 784 m. Fossil-based paleobathymetry indicates two deepening-upward cycles within restricted inner neritic to outer neritic environments in the Plio-Pleistocene section. The base of this Plio-Pleistocene section rests unconformably on a sequence of interbedded sandstone and shale with a sandstone-to-shale ratio of approximately 1 and minor limestone, volcanic rocks, and associated pyroclastic material of middle Eocene age (foraminiferal biozone *Globigerapsis kuglieri*). Sandstone is fine to coarse-grained feldspathic arenite and quartzarenite characterized by high carbonate content and rare pyrite concretions.

El Paso Plaris-1. El Paso Plaris-1 was drilled to a depth of 2,894 m below sea level and was continuously cored from 172 m (Fig. 14B). An interbedded sequence of Plio-Pleistocene sandstone and shale with a sandstone-to-shale ratio much less than 1 was drilled down to a depth of 677 m. Microfossils indicate that this section shallows upward and is characterized by upper bathyal conditions at its base and middle neritic conditions in its upper part. Sandstone is mainly quartzarenite. The

TABLE 2. MICROPALEONTOLOGICAL DATA FROM THE PEARL ISLANDS, GULF OF PANAMA*

Samples 4ll and 412 (Eastern Isla del Rey, Pearl Islands)

Pulleniatina aff. obilquiloculata obliquiloculata
Globigerina venezuelena
Globigerina triloba immatura
Globorotalia cf. acostanensis acostanensis
Globigerionoides sacculifera
Uvigerina sp.
Lenticulina (Robulus) sp.
Sphaeroidina cf. bulloides
Nonion pompilioides
Nonion sp.
Sphogenerina transversa

Paleoenvironment: Middle neritic or deeper
Age Upper Miocene or Pliocene? (Zones N18-N21)

Sample 447 (Southern Isla del Rey, Pearl Islands)

Lepidocyclina (Nephrolepidina) tournieri
Lepidocyclina canellei

Paleoenvironment: Very shallow inner neritic
Age: Upper Oligocene

*From an unpublished 1971 report by Direccíon General de Hidrocarburos, Panama City. See Figure 10 for sample locations.

inner-neritic sandstone tends to be coarse grained with abundant microfossil debris and the outer-neritic sandstone tends to be very fine grained.

Middle and upper Miocene rocks are represented by outer neritic and upper bathyal interbedded sandstone and shale with a sandstone-to-shale ratio much less than 1 and minor limestone. Sandstone is mainly very fine grained glauconitic quartzarenite. Corresponding Miocene foraminiferal biozones are indicated on the well log in Figure 14.

Lower Miocene rocks are represented by a deepening-upward (middle to outer neritic) sequence of interbedded sandstone and shale with sandstone-to-shale ratio greater than 1 and interbedded basaltic agglomerate and minor conglomerate. The sandstone is mainly fine to medium-grained volcaniclastic litharenite and feldspathic arenite. Corresponding Miocene biozones are indicated on th well log.

Correlation between wells. The Plio-Pleistocene section of interbedded sandstone and shale can be correlated between both wells (Fig. 14). The base of the Plio-Pleistocene section is a major unconformity in the El Paso Corvus-1 well but a conformable transition with underlying upper Miocene rocks in the El Paso Plaris-1.

Middle Eocene rocks found over an interval of 1,869 m in the bottom of the El Paso Corvus-1 were not penetrated in the El Paso Plaris-1, which bottomed in Miocene. Based on the pronounced regional dip to the east in the Gulf of Panama, the Eocene section of the El Paso Corvus-1 well would project far below the base of the El Paso Plaris-1 well in the eastern part of the Gulf of Panama (Fig. 14).

Lower to upper Miocene rocks were found over an interval of 2,217 m at the bottom of the El Paso Plaris-1 but are missing from the El Paso Corvus-1. Regional seismic stratigraphic mapping summarized in Figure 5 shows that Miocene rocks pinch out approximately 8 km to the southwest of El Paso Plaris-1 and are absent from a large area of the western part of the Gulf of Panama, including the site of El Paso Corvus-1. Late Miocene and older deposition in the eastern Gulf of Panama corresponds chronostratigraphically to a hiatus at a middle Eocene to early Pliocene unconformity in the western part of the gulf.

CONCLUSIONS

Structure of the East Panama deformed belt in the Gulf of Panama

Seismic data from the Gulf of Panama show the presence of east-dipping thrust faults and an associated foredeep basin (Pearl Islands basin) that had been previously recognized using these data (A. J. Kudrna, unpublished) but not been recognized from previous onshore mapping of the margins of the Gulf of Panama. A regional transect based on two long seismic lines suggests that at least nine thrust faults segment the East Panama deformed belt into a series of east-tilted fault blocks that range in width from 5 to 20 km. Thrust motion

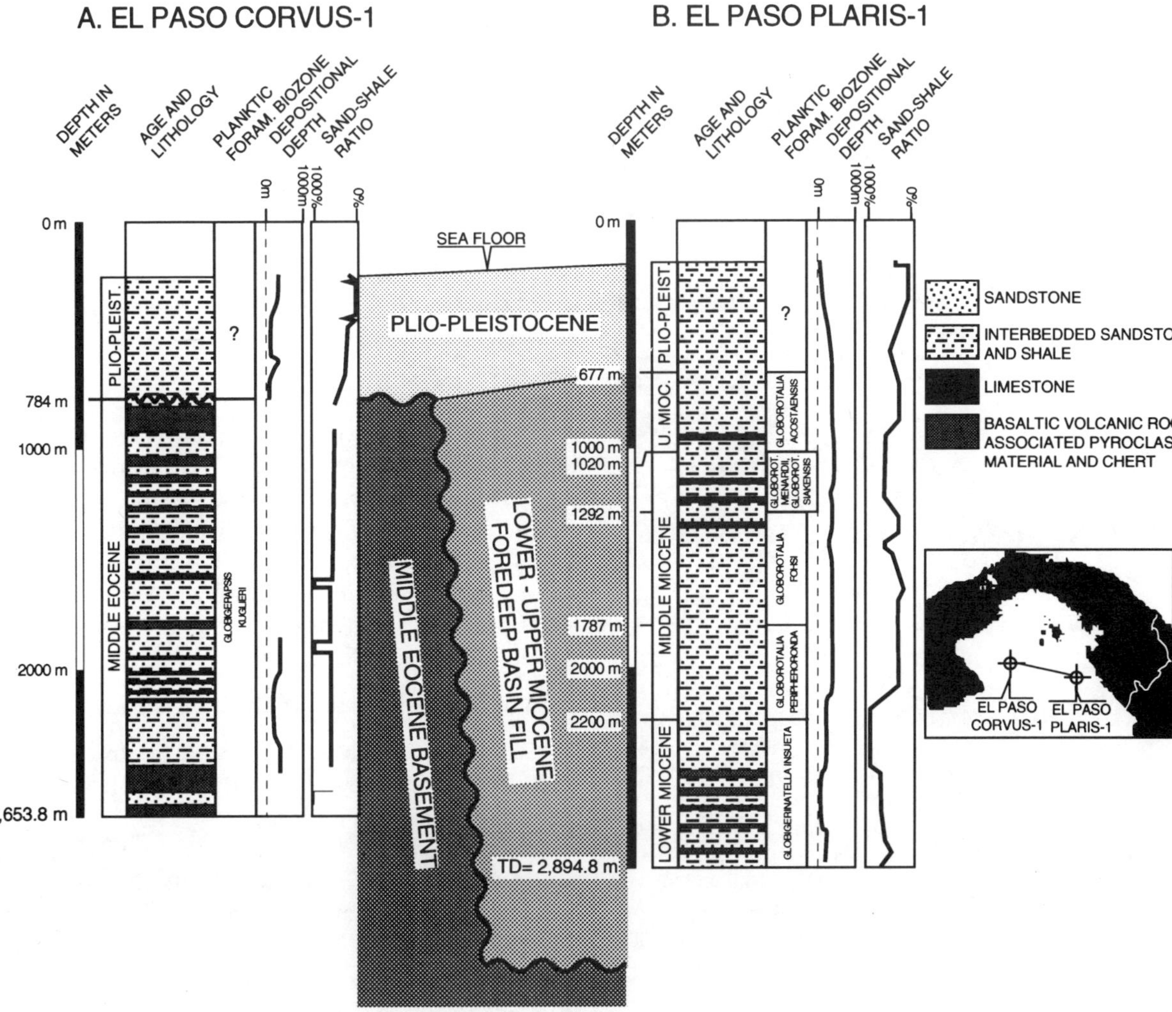

Figure 14. A, Well log of Mobil exploratory well El Paso Corvus-1 in the west-central Gulf of Panama. Inset map shows well location. Note the prominent unconformity separating middle Eocene igneous and sedimentary section from a Pliocene section containing planktic foraminifera characteristic of shallow water. B, Well log of Mobil exploratory well El Paso Plaris-1 in the eastern Gulf of Panama. Inset map shows well location. Well correlation shows the Miocene section pinching out to the west but thickening to the east in the foredeep basin.

tends to elevate basement to higher structural levels than observed in the central and western Gulf of Panama and may explain the presence of exposures of Late Cretaceous basement rocks along the eastern margin of the Gulf of Panama and on the Pearl Islands. The Pearl Islands appear to be the surface expression of a hanging-wall anticline formed at the boundary of the Pearl Islands foreland basin and the East Panama deformed belt.

Age of the East Panama deformed belt in the Gulf of Panama

Interpretation of line 117 shows onlap of folds in middle Miocene to lower Pliocene rocks deformed by underlying west-verging thrust faults (Fig. 12). Folds are overlain by unfolded and unfaulted rocks of Plio-Pleistocene age. This suggests that the age of deformation in this part of the deformed belt occurred prior to Pliocene time (~5.3 Ma). This age of deformation is consistent with pre-Pliocene folding interpreted from line POC-33 (Fig. 13). Arching of the seafloor and thrust deformation of Plio-Pleistocene rocks that are seen on line 18 (Fig. 5B) and the anomalously elevated Pearl Islands in the central Gulf of Panama suggest that active displacement may continue on individual segments of faults in the deformation zone and may contribute to scattered shallow earthquakes recorded in the area (Fig. 4).

Uplift history and neotectonic significance of the East Panama deformed belt

The El Paso Plaris-1 well drilled in the eastern Gulf of Panama documents the persistence of water depths of several

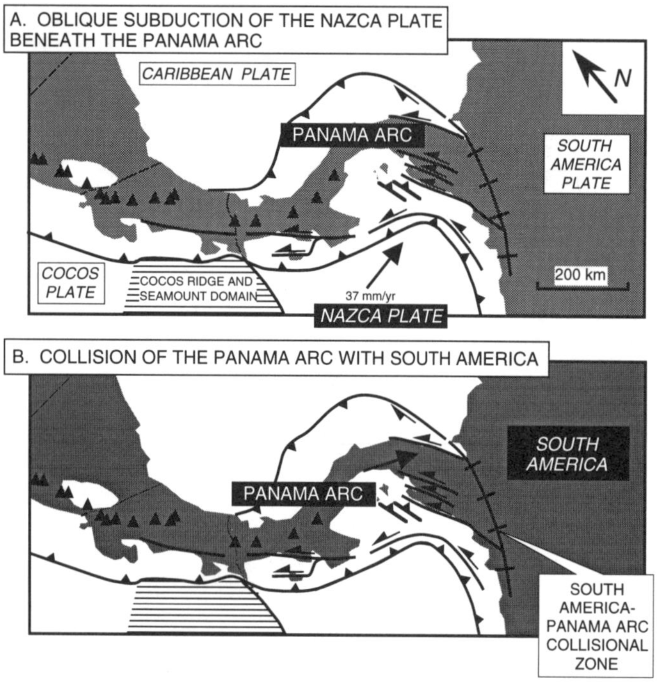

Figure 15. Two possible models for Neogene deformation of the on- and offshore parts of the East Panama deformed belt. A, Oblique subduction of Nazca plate beneath the Panama arc produces strike-slip and transpressional faulting subparallel to the margin. B, Oroclinal bending of the Panama Isthmus results from collision of an east-west–trending Panama arc with the South America plate. Bending is accommodated both by ductile, oroclinal bending and segmentation along northwest-striking strike-slip faults. The largely submerged and inactive thrust belt in the Gulf of Panama suggests that north-south thrust faults may have formed as an early response to east-west shortening and bending of the Isthmus of Panama during the Miocene and early Pliocene.

hundred meters from early Miocene to Pliocene time (Fig. 14B). It is interesting to compare this sedimentary record with other areas of the Caribbean-Pacific seaway in the Gulf of Panama. The marine sedimentary record of El Paso Plaris-1 is characterized by deeper marine rocks that contrast to shallower-water facies or hiatuses of the same age in the (1) Pearl Islands (middle neritic or deeper conditions in late Miocene or Pliocene? based on limited sampling results given in Table 2), and (2) El Paso Corvus-1 well and western Gulf of Panama (hiatus separating rocks of middle Eocene age from rocks of Plio-Pleistocene age; Fig. 14A).

Uplift and shallowing of the East Panama deformed belt certainly must have accompanied late Neogene folding and thrusting but is only suggested in the uppermost part of the paleobathymetric curve from the El Paso Plaris-1 well located within the eastern part of the deformed belt (Fig. 14B). One explanation for the persistence of deeper facies in the eastern part of the deformed belt is a flexural effect caused by eastward underthrusting of much of the Gulf of Panama beneath the East Panama deformed belt. Rocks in the eastern area remain deep in an underthrust zone, whereas rocks in the west are elevated by flexure that accompanies underthrusting. Localized structural highs that occur in the eastern and central zone deformed belt, such as the Pearl Islands, are associated with the hanging walls of reverse faults (Fig. 8).

Keller et al. (1989) have concluded that the first significant divergence of planktic foraminiferal provinces of the eastern equatorial Pacific (DSDP hole 503A) and the Colombian basin in the western Caribbean (DSDP hole 502A) occurred at 6.2 Ma (latest Miocene). Increasingly restricted surface water exchange across the Panama seaway occurred between 4.2 Ma (early Pliocene) and 2.4 Ma (late Pliocene), and littoral-neritic leakage occurred through the seaway between 2.4 and final closure at 1.8 Ma (latest Pliocene). Coates et al. (1992) examined the nannofossils and planktic foraminifera from 31 sections of near-shore marine rocks exposed by uplift in Panama and Costa Rica and concluded that final closure of the seaway occurred about 3.5 Ma (early Pliocene). We propose that Pliocene restriction at the Caribbean-Pacific seaway may be directly related to the late Neogene formation and elevation of folds by thrust faults observed in the East Panama deformed belt.

Models for formation of the East Panama deformed belt

We propose two possible models to explain the observed pattern of deformation observed in the Gulf of Chiriquí and the Gulf of Panama (Fig. 15).

Oblique subduction of the Nazca plate beneath the southern margin of Panama. Oblique convergence between Panama and the Nazca plate in the Gulf of Panama area may be accommodated by left-lateral strike-slip faults and folds and thrust faults of the East Panama deformed belt (Fig. 15A). In western Panama, the oblique direction of Nazca plate convergence (DeMets et al., 1990; Kellogg et al., this volume) is consistent with the uniform direction of northwest maximum extensional strain on the land areas around the Gulf of Chiriquí and the presence of large strike-slip faults like the Azuero-Soná fault zone (Kolarsky and Mann, this volume). An argument against application of this model to eastern Panama is the fact that some of the fold and thrust structures appear to have become inactive in early Pliocene time (for example, Fig. 13). If the oblique subduction model is relevant, one might expect all related structures to remain active and accompany present-day oblique-subduction along this margin.

Bending of the Panama island arc following collision with the South American continent. Wadge and Burke (1983) and Silver et al. (1990) suggested that the collision of the Panama island arc with the South American continent resulted in S-shaped bending of the central and eastern parts of the isthmus (Fig. 15B). Mann and Corrigan (1990) pointed out that the bending process may not have been entirely ductile because of the presence of northwest-striking left-lateral strike-slip faults like the Sansón Hills fault zone that segment the isthmus. The presence of north-striking thrust faults in the Gulf of Panama is consistent with east-west shortening of the Isthmus of Panama and earthquake studies by Pennington (1981) that suggest that the present-day maximum compressive stress is oriented east-west.

We cannot distinguish the amount of strike-slip faulting occurring on apparent thrust faults in the Gulf of Panama with the present data. We can show that strike-slip and thrust structures must be limited to the eastern edge of the Gulf of Panama and do not affect the western gulf as shown on the compilation map of Case and Holcombe (1980) and in the tectonic interpretation of Silver et al. (1990). If the onshore strike-slip faults are active structures and the main episode of thrusting in the Gulf of Panama ended in Early Pliocene time, we can speculate that approximately east-west thrusting was an early response to Panama–South America collision and that early thrusting was succeeded by strike-slip faulting in more interior parts of the orocline (Fig. 15B). These and other problems associated with the East Panama deformed belt await more detailed, future studies.

ACKNOWLEDGMENTS

This study would have not been possible without geophysical and well data from the Gulf of Panama that was kindly provided by Mobil for use in this study. We thank C. Petter, T. Christiansen, and S. Pierce at Mobil for locating and allowing us to use these data. We also thank R. Miranda, A. Burgos, and L. Campos of the Dirección General de Recursos Minerales and Ing. N. Villareal of the Dirección General de Hidrocarburos for providing us with unpublished reports related to the geology of eastern Panama and giving us permission to use these data in this publication. Special thanks to A. Kudrna of Western Geophysical for alerting us to the existence and tectonic significance of the Mobil offshore data in the Gulf of Panama, to W. Page for providing a copy of his unpublished report on the seismic geology and seismicity of northwestern Colombia, and to R. Stewart for providing a copy of his unpublished report on the geology of the Bayano-Chucunaqué basin. We appreciate reviews of different versions of this chapter by R. Buffler, J. Case, A. Coates, J. Kellogg, A. Kudrna, M. Lagoe, and G. Westbrook. Funding for this project was provided by the Petroleum Research Fund of the American Chemical Society (grant number 2076-AC2-2). UTIG contribution no. 1063.

REFERENCES CITED

Adamek, S., Frohlich, C., and Pennington, W., 1988, Seismicity of the Caribbean-Nazca boundary: Constraints on microplate tectonics of the Panama region: Journal of Geophysical Research, v. 93, p. 2053–2075.

Bandy, O. L., and Casey, R. E., 1973, Reflector horizons and paleobathymetric history, eastern Panama: Geological Society of America Bulletin, v. 84, p. 3081–3086.

Bowland, C. L., and Rosencrantz, E., 1988, Upper crustal structure of the western Colombian basin, Caribbean Sea: Geological Society of America Bulletin, v. 100, p. 534–536.

Briceño-Guarupe, L. A., 1978, The crustal structure and tectonic framework of the Gulf of Panama [M.S. thesis]: Corvallis, Oregon State University, 71 p.

Burke, K., and Sengor, A.M.C., 1986, Tectonic escape in the evolution of the continental crust, *in* Reflection seismology: The continental crust: Washington, D.C., American Geophysical Union Geodynamics Series, v. 14, p. 41–53.

Burke, K., Cooper, C., Dewey, J. F., and Pindell, J. L., 1984, Caribbean tectonics and relative plate motions, *in* Bonini, W. E., Hargraves, R. B., and Shagam, R., eds., The Caribbean–South American plate boundary and regional tectonics: Geological Society of America Memoir 162, p. 31–64.

Case, J. E., 1974, Oceanic crust forms the basement of eastern Panama: Geological Society of America Bulletin, v. 85, p. 645–652.

Case, J. E., and Holcombe, T. L., 1980, Geologic-tectonic map of the Caribbean region: U.S. Geological Survey Miscellaneous Investigation Map I-1100, scale 1:2,500,000.

Coates, A. G., and seven others, 1992, Closure of the Isthmus of Panama: The near-shore marine record of Costa Rica and western Panama: Geological Society of America Bulletin, v. 104, p. 814–828.

DeMets, C., Gordon, R. G., Argus, D. F., and Stein, S., 1990, Current plate motions: Geophysical Journal International, v. 101, p. 425–478.

Duque-Caro, H., 1990a, Neogene stratigraphy, paleoceanography, and paleobiogeography in northwest South America and the evolution of the Panama Seaway: Palaeogeography, Palaeoclimatology, and Palaeoecology, v. 77, p. 203–234.

Duque-Caro, H., 1990b, The Choco block in the northwestern corner of South America: Structural, tectonostratigraphic, and paleogeographic implications: Journal of South American Earth Sciences, v. 3, p. 71–84.

Harding, T. P., 1984, Graben hydrocarbon occurrences and structural style: American Association of Petroleum Geologists Bulletin, v. 68, p. 333–362.

Hardy, N. C., 1991, Tectonic evolution of the easternmost Panama basin: Some new data and inferences: Journal of South American Earth Sciences, v. 4, p. 261–269.

Irving, E. M., 1975, Structural evolution of the northernmost Andes, Colombia: U.S. Geological Survey Professional Paper 846, 47 p.

Johnson, M. S., and Headington, E., 1971, Panama: Exploration history and petroleum potential: Oil and Gas Journal, v. 69, p. 96–100.

Keigwin, L. D., Jr., 1978, Pliocene closing of the Isthmus of Panama, based on biostratigraphic evidence from nearby Pacific Ocean and Caribbean Sea cores: Geology, v. 6, p. 630–634.

Keigwin, L. D., Jr., 1982, Isotopic paleoceanography of the Caribbean and East Pacific: Role of Panama uplift in Late Neogene time: Science, v. 217, p. 350–353.

Keller, G., Zenker, C., and Stone, S., 1989, Late Neogene history of the Pacific-Caribbean gateway: Journal of South American Earth Sciences, v. 2, p. 73–108.

Kellogg, J. N., Godley, V. M., Ropain U. C., Bermudez, A., and Aiken, C.L.V., 1991, Gravity field map of Colombia, eastern Panama, and adjacent marine areas: Geological Society of America Map and Chart Series MCH070, scale 1:1,500,000.

Lonsdale, P., and Klitgord, K. D., 1978, Structure and tectonic history of the eastern Panama basin: Geological Society of America Bulletin, v. 89, p. 981–999.

Lowrie, A., 1978, Buried trench south of the Gulf of Panama: Geology, v. 6, p. 434–436.

Lowrie, A., Stewart, J., Stewart, R. H., van Andel, R. J., and McRaney, L., 1982, Location of the eastern boundary of the Cocos plate during the Miocene: Marine Geology, v. 2, p. 73–108.

MacDonald, H. C., 1969, Geologic evaluation of radar imagery from Darien Province, Panama: Modern Geology, v. 1, p. 1–63.

MacKay, M. E., 1989, Accretionary response to trench sediment variation along the south Panama margin [M.S. thesis]: Tulsa, Oklahoma, University of Tulsa, 44 p.

MacKay, M. E., and Moore, G. F., 1990, Variation in sediment of the south Panama accretionary prism: Response to oblique subduction and trench sediment variation: Tectonics, v. 9, p. 683–698.

Mann, P., and Burke, K., 1984, Neotectonics of the Caribbean: Reviews of Geophysics and Space Physics, v. 22, p. 309–362.

Mann, P., and Corrigan, J., 1990, Model for late Neogene deformation in Panama: Geology, v. 18, p. 558–562.

Marshall, L. G., 1988, Land mammals and the Great American Interchange: American Scientist, v. 76, p. 380–388.

Pennington, W. D., 1981, Subduction of the eastern Panama basin and the seismotectonics of northwestern South America: Journal of Geophysical Research, v. 86, p. 10753–10770.

Pindell, J. L., and Dewey, J. F., 1982, Permo-Triassic reconstruction of western Pangea and the evolution of the Gulf of Mexico/Caribbean region: Tectonics, v. 1, p. 179–211.

Shelton, B. J., 1952, Geology and petroleum prospects of Darien, southeastern Panama [M.S. thesis]: Corvallis, Oregon State College, 59 p.

Silver, E. A., Reed, D. L., Tagudin, J. E., and Heil, D. J., 1990, Implications of the North and South Panama deformed belts for the origin of the Panama orocline: Tectonics, v. 9, p. 261–282.

Stéphan, J., Blanchet, R., and Mercier de Lépinay, B., 1986, Northern and southern Caribbean festoons (Panama, Colombia-Venezuela, and Hispaniola–Puerto Rico) interpreted as pseudosubductions induced by east-west shortening of the Peri-Caribbean frame, *in* Wezel, F., ed., The origin of arcs (Developments in Geotectonics, volume 21): Amsterdam, Elsevier, p. 401–422.

Stewart, R. H., and Stewart, R. L., 1980, Geologic map of the Panama Canal and vicinity, Republic of Panama: U.S. Geological Survey Miscellaneous Investigation Series Map I-1232, scale 1:100,000.

Terry, R., 1956, A geological reconnaissance of Panama: California Academy of Sciences Occasional Papers, v. 23, 91 p.

Van der Hammen, T., Werner, J. H., and Van Dommeln, H., 1973, Palynological record of the upheaval of the northern Andes: A study of the Pliocene and Lower Quaternary of the Colombian Eastern Cordillera and the early evolution of its high-Andean biota: Review of Palaeobotany and Palynology, v. 16, p. 1–122.

Vergara-Muñoz, A., 1988, Tectonic patterns of the Panama block deduced from seismicity, gravitational data, and earthquake mechanisms: Implications to the seismic hazard: Tectonophysics, v. 154, p. 253–267.

Wadge, G., and Burke, K., 1983, Neogene Caribbean plate rotation and associated Central American tectonic evolution: Tectonics, v. 2, p. 633–643.

Wilcox, R. E., Harding, T. P., and Seely, D. R., 1973, Basic wrench tectonics: American Association of Petroleum Geologists Bulletin, v. 57, p. 74–96.

Wolters, B., 1986, Seismicity and tectonics of southern Central America and adjacent regions with special attention to the surroundings of Panama: Tectonophysics, v. 128, p. 21–46.

Manuscript Accepted by the Society February 7, 1994

Printed in U.S.A.

Geological Society of America
Special Paper 295
1995

Structure and neotectonics of an oblique-subduction margin, southwestern Panama

Radim A. Kolarsky*
Department of Geological Sciences, University of Texas at Austin, Austin, Texas 78712
Paul Mann
Institute for Geophysics, University of Texas at Austin, 8701 Mopac Boulevard, Austin, Texas 78759

ABSTRACT

We present on- and offshore structural data from the Nazca-Panama plate boundary zone in the Gulf of Chiriquí and surrounding onshore areas of southwest Panama. Major offshore structures interpreted on multichannel seismic profiles from the Gulf of Chiriquí include Cébaco basin complex, a series of northeast-striking, Plio-Pleistocene half-grabens, and Montuosa basin, an asymmetric Plio-Pleistocene sag basin associated with a major strike-slip fault. We interpret Cébaco basin complex as a pull-apart basin between two major, left-lateral strike-slip faults that accommodate oblique motion between the Nazca plate and the mainland of southwestern Panama. Interpretation of regional seismic stratigraphic data indicates that the Plio-Pleistocene extensional phase that produced the Cébaco basin complex extended the area by about 7%.

We studied outcrop-scale, conjugate strike-slip fault systems exposed on landmasses surrounding the Gulf of Chiriquí in order to place kinematic and age constraints on large-scale faults mapped on seismic profiles. Fault systems deforming Eocene to Lower Miocene sedimentary rocks on Coiba Island and the Azuero and Soná Peninsulas suggest an approximately northwest-southeast orientation of maximum extensional strain in an area that encompasses the offshore Cébaco basin complex.

We propose three possible models to explain the observed pattern of strike-slip deformation observed in the Gulf of Chiriquí: (1) Neogene oblique subduction of the Nazca plate beneath Panama produces left-lateral strike-slip faulting and related northwest-oriented extension within the forearc (Gulf of Chiriquí) (2) Plio-Pleistocene shallow subduction/collision between the Cocos ridge and Costa Rica produces southwestward motion or "escape" of a Gulf of Chiriquí block that is detached from the rest of Panama by left-lateral strike-slip faults, and (3) Neogene bending of the Panama island arc following collision with the South American continent is accommodated in part by strike-slip motion and underthrusting along the southwest margin of Panama. Observed deformation may be a composite effect of more than one of these tectonic mechanisms.

*Present address: Eastern E&P Region, Texaco USA, P.O. Box 60252, New Orleans, Louisiana 70160.

Kolarsky, R. A., and Mann, P., 1995, Structure and neotectonics of an oblique-subduction margin, southwestern Panama, *in* Mann, P., ed., Geologic and Tectonic Development of the Caribbean Plate Boundary in Southern Central America: Boulder, Colorado, Geological Society of America Special Paper 295.

INTRODUCTION

The southwestern margin of Panama marks a diffuse zone of active motion between the Caribbean plate in Panama and the Nazca plate of the eastern Pacific Ocean (Fig. 1). The relative amount of Neogene to Recent strike-slip motion and underthrusting along this boundary has remained controversial since the pioneer study by Jordan (1975) of plate motions in the area.

Earthquake studies

Earthquake studies of western Panama and eastern Costa Rica have shown that most of the seismicity occurs at depths of less than 70 km and is strike-slip or thrust in nature (Fig. 2). The Nazca plate south of the South Panama deformed belt contains three seismically active north-trending fracture zones: the Panama, Balboa, and Coiba. Adamek et al. (1988) proposed that the three fracture zones form a 70-km right-lateral transform boundary between the Nazca and Cocos plates, with transfer of active plate motion from east (Coiba fracture zone) to west (Panama fracture zone) in accord with plate models based on oceanic magnetic anomalies (Lonsdale and Klitgord, 1978). Right-lateral and oblique-slip shallow earthquakes at depths of 8 to 16 km also occur along the northward projection of the Balboa fracture zone in the South Panama deformed belt (Fig. 2). Focal mechanisms of earthquakes near the Burica Peninsula at the Panama–Costa Rica border indicate convergent first motions interpreted by Adamek and Tajima (1987) as the result of shallow underthrusting of the Cocos plate beneath the Caribbean plate.

In western Panama, most shallow seismicity is concentrated in the Gulf of Chiriquí area or in the area near the Panama–Costa Rica border (Fig. 2). Large historical events compiled by Adamek et al. (1988) are possibly associated with movement on strike-slip faults of the Azuero Peninsula and Coiba Island that are described by Mann and Corrigan (1990) and in this chapter. One oblique-slip focal mechanism presented by Muñoz (1988) from the Gulf of Chiriquí is consistent with the interpretation presented in this chapter for oblique slip faults in the offshore area of the Gulf of Chiriquí.

Plate motion models

On the basis of theoretical plate modeling, Jordan (1975) first proposed that the southern Panama margin at the contact between the Caribbean plate in Panama and the Nazca plate in the Pacific Ocean may be a dominantly left-lateral strike-slip boundary. He proposed that the Nazca plate converges obliquely in an east-northeasterly (071) direction on the southern margin of Panama at a predicted velocity of 5.4 cm/yr (Fig. 1). Studies of magnetic anomalies on the oceanic Nazca plate in the Pacific Ocean south of Panama by Lonsdale and Klitgord (1978) constrained a northwestward jump in the Cocos-Nazca plate boundary at 11 to 8 Ma from about 80.5W to 83°W (distance of 275 km). Triple-junction migration terminated subduction and island arc magmatism and initiated an active regime of left-lateral strike-slip faulting in a manner similar to that described along the San Andreas fault in California. Lonsdale and Klitgord (1978) and Lowrie et al. (1982) discussed later migration of the plate boundary from the Coiba fracture zone to the Panama fracture zone; this is a distance of about 75 km at the trench. Models of present-day plate motion by Jordan (1975) and DeMets et al. (1990) suggest that the Cocos-Nazca plate boundary must move east relative to the Caribbean plate and Panama.

Okaya and Ben-Avraham (1987) presented interpretations

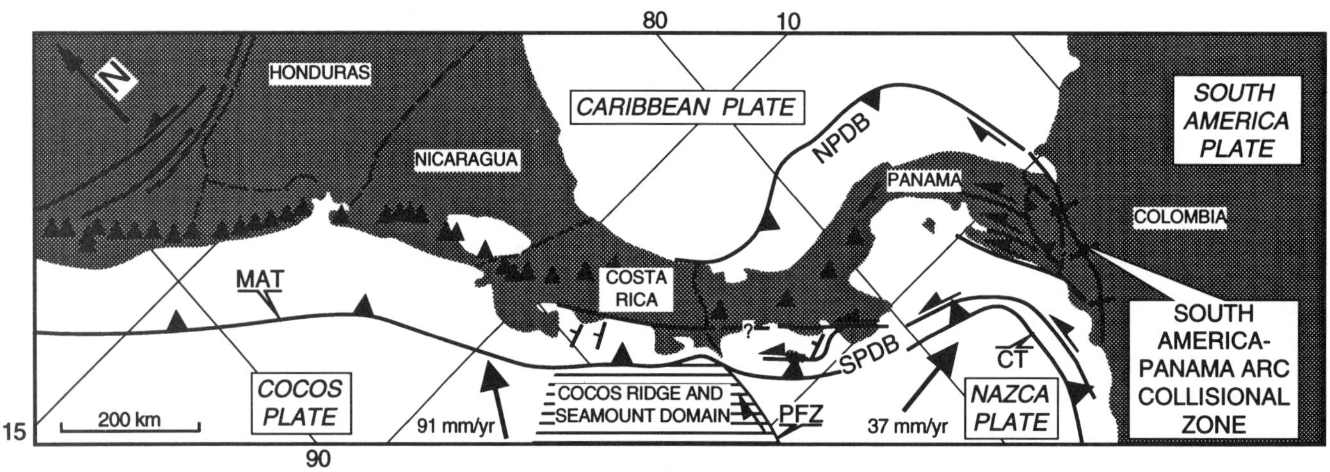

Figure 1. Tectonic setting of Central America, Middle America trench (MAT), Colombia trench (CT), North Panama deformed belt (NPDB), South Panama deformed belt (SPDB), and Panama fracture zone (PFZ). Black triangles represent active calc-alkaline volcanoes; arrows represent plate convergence directions with numbers indicating rates of relative plate motion in cm/yr from DeMets et al. (1990). Note oblique convergence direction predicted for the Nazca-Caribbean plate boundary along the southern margin of Panama.

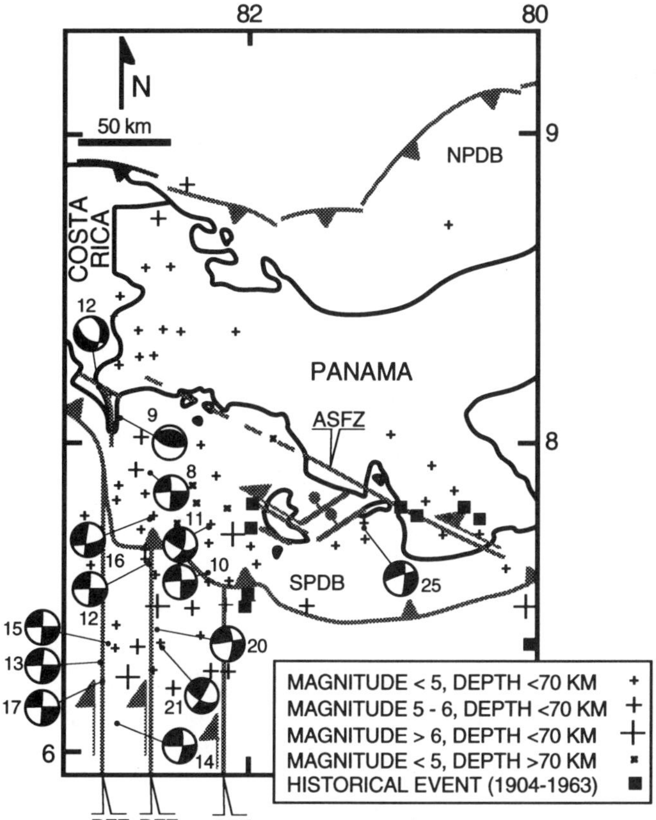

Figure 2. Plot of all published earthquake focal mechanisms and earthquake epicenters in the Panama–Costa Rica area (small numbers indicate depths of focal mechanisms as compiled by Adamek et al., 1988). Historical events are also compiled from Adamek et al. (1988). Major faults are shown in gray. Earthquake epicenters are for the years 1964 to 1988 and are from the files of the International Seismological Center (ISC). Abbreviations to faults and fracture zones: ASFZ = Azuero-Soná fault zone; NPDB = North Panama deformed belt; SPDB = South Panama deformed belt; PFZ = Panama fracture zone; BFZ = Balboa fracture zone; CFZ = Coiba fracture zone.

of Texaco seismic lines from the Gulf of Chiriquí that supported the idea that an active phase of strike-slip motion had followed an early phase of island-arc tectonics. Mann and Corrigan (1990) presented a compilation of onland geologic data and proposed that active and widespread strike-slip tectonics accommodated collision-related bending of the isthmus of Panama. MacKay and Moore (1990) presented marine data from the southwest margin of Panama and suggested an oblique-slip model in which plate motion is partitioned into a normal component of mainly offshore convergence (as observed in their study area) and a mainly onshore component of strike-slip faulting. De Boer et al. (1988) and de Boer et al. (1991) presented evidence for widespread Late Miocene to Quaternary volcanism in western Panama. They suggested that subduction, not strike-slip, dominated the interaction of the Nazca plate and Panama because of the presence of a subducted Nazca plate well beneath the mainland of Panama. However, there is no evidence of deep seismicity associated with a subducted slab beneath the Miocene to Quaternary arc rocks of western Panama (Fig. 2).

PREVIOUS WORK AND DATA USED IN THIS STUDY

Previous marine geophysical work in the Gulf of Chiriquí

The Gulf of Chiriquí extends between Burica and Azuero Peninsulas in southwest Panama and covers an area of approximately 25,000 km^2 (Fig. 3). There have been several previous marine geophysical studies within the shallow (~100 m water depth) Gulf of Chiriquí shelf and along the active, deeper-water (~1,000 to 2,000 m) accretionary margin (South Panama deformed belt) separating the Nazca plate and Panama landmass.

Ross and Shor (1965) published one reconnaissance single-channel seismic profile from the outer shelf and slope of the Gulf of Chiriquí. Okaya and Ben-Avraham (1987) presented a structural interpretation of a Texaco multichannel seismic profile 3 recorded in 1971 and donated by Texaco for use in their study (TEX-3 in Fig. 3). They identified the presence of active strike-slip faults and related these faults to the conversion of the margin from a subduction to a strike-slip zone following the passage of the Caribbean-Nazca-Cocos triple junction (Lonsdale and Klitgord, 1978). Heil (1988) and Silver et al. (1990) produced regional structural maps of the South Panama deformed belt between Burica Peninsula and Coiba Island, using SeaMARC II side-scan data and single-channel data collected in 1985. MacKay (1988), MacKay and Moore (1990), and Moore and Sender (this volume) described the structure of the accretionary margin southwest of the Azuero Peninsula using single-channel seismic data and SeaMARC II side-scan data collected during the same cruise. They discussed the possibility of partitioning of oblique interplate motion into a margin normal and a margin parallel component because the accretionary prism of the South Panama deformed belt resembles prisms formed along normal subduction margins.

Well control on offshore seismic data

There are three poorly documented onshore wells and no offshore wells in the Gulf of Chiriquí region. Corrigan et al. (1990) integrated the well log of the Corotú-1A well (drilled 1958 by Ball Associates to T. D. 2,623 m) of the northern Burica Peninsula into their cross section of the area based on surface mapping. The Dirección General de Hidrocarburos in Panama City provided us the well logs from the Corotú-1A (drilled 1949 by Sinclair Panama Oil Company, T. D. 2,841 m and Corotú-1B (drilled 1958 by Petrolea Chiricana, T. D. 506 m) wells located along the northwestern edge of the Gulf of Chiriquí (Fig. 3) (see Appendix 5 of Kolarsky, 1992). To our

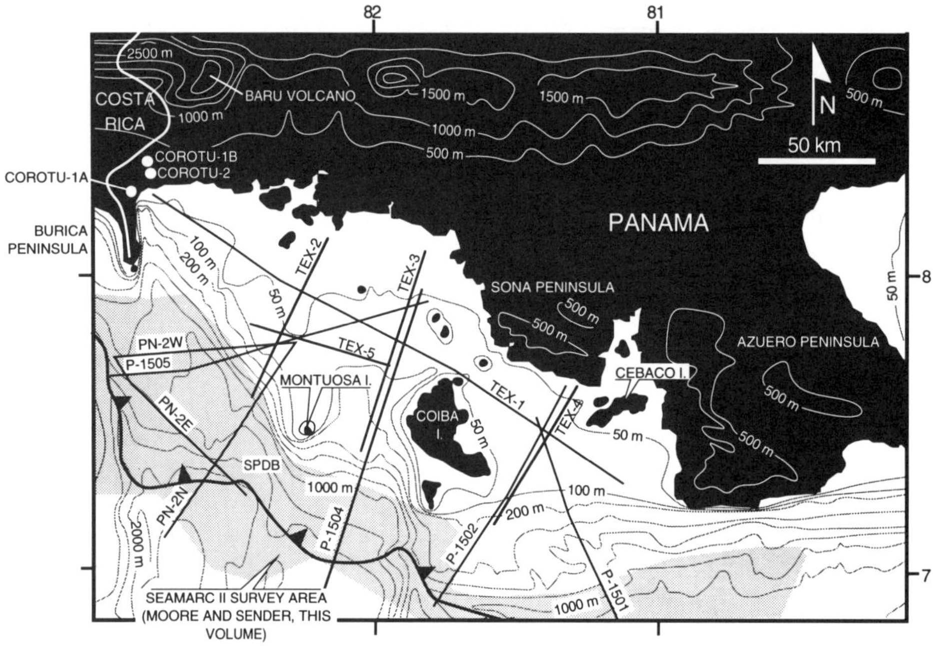

Figure 3. Bathymetric and topographic map of the Gulf of Chiriquí area, southwestern Panama. Bathymetric contours are taken from U.S. Navy bathymetric charts. Lines and numbers give locations and line designations of Texaco, Shell, and Institute for Geophysics, University of Texas at Austin (UTIG) multichannel seismic lines used in this study. Shaded area along the South Panama deformed belt (SPDB) shows area of UC/Santa Cruz and University of Hawaii single-fold seismic and SeaMARC II study (Heil, 1988; MacKay and Moore, 1990; Silver et al., 1990). Multichannel seismic lines provided by Shell Internationale Petroleum Mij. B.V. are P-1501, P-1502, P-1504, and P-1505; lines provided by Texaco are lines TEX-1, 2, 3, 4, and 5 in the Gulf of Chiriquí; lines provided by UTIG are PN-2W, PN-2E, and PN-2N in the Gulf of Chiriquí and the South Panama deformed belt. Well logs from Corotú-1A, Corotú-1B, and Corotú-2 include wireline and lithologic logs provided by the Department of Natural Resources, Ministry of Public Affairs in Panama City.

knowledge, no gravity or aeromagnetic data have been collected or published from the Gulf of Chiriquí.

Data used in this study

This chapter represents the first comprehensive offshore study of the Gulf of Chiriquí using a data set of 1,500 km of multichannel seismic data (Fig. 3). Distance between seismic lines ranges from 5 to 60 km. Data used in this study also include results from previous on- and offshore studies that were compiled for this study, satellite imagery, and field data collected by us in 1990. These data were provided by five principal sources: (1) Amoco Production Company, Houston, Texas, provided single- and multiband LANDSAT multi-spectral scanner (MSS) imagery of land areas around the Gulf of Chiriquí, Panama (Fig. 4). (2) Department of Natural Resources (Dirección General de Recursos Minerales), Panama Ministry of Commerce and Industry (Ministerio de Comericio y Industría), Panama City, Panama, provided lithologic and paleontologic field data collected on the Azuero Peninsula during the 1970s and 1980s (Fig. 3). (3) Shell Internationale Petroleum Mij. B.V., The Hague, The Netherlands, provided 700 km of multichannel seismic data collected from the Gulf of Chiriquí in 1973 (Fig. 3). (4) Texaco Production Company, Bellaire, Texas, provided 600 km of multichannel seismic data collected from the Gulf of Chiriquí in 1971 (Fig. 3). (5) Institute for Geophysics, The University of Texas at Austin, provided 200 km of multichannel seismic data collected from the Gulf of Chiriquí in 1977 (Fig. 3).

This data set includes all of the multichannel seismic data ever collected from the Gulf of Chiriquí (1,500 km). We reprocessed Texaco lines 1 and 4 using the University of Texas at Austin Department of Geological Sciences VAX 11/780 computer featuring DISCO processing software version 6.5. Field tapes were provided by Texaco for reprocessing in this study. Field and processing parameters for our offshore data set are summarized in Table 1.

TECTONIC ELEMENTS OF THE GULF OF CHIRIQUÍ

We converted two-way travel time to the top of acoustic basement seen on Texaco and Shell International multichannel seismic profiles in the Gulf of Chiriquí and South Panama deformed belt to depth in meters (Fig. 5). Depths were derived using seismic interval velocities obtained from the Texaco and

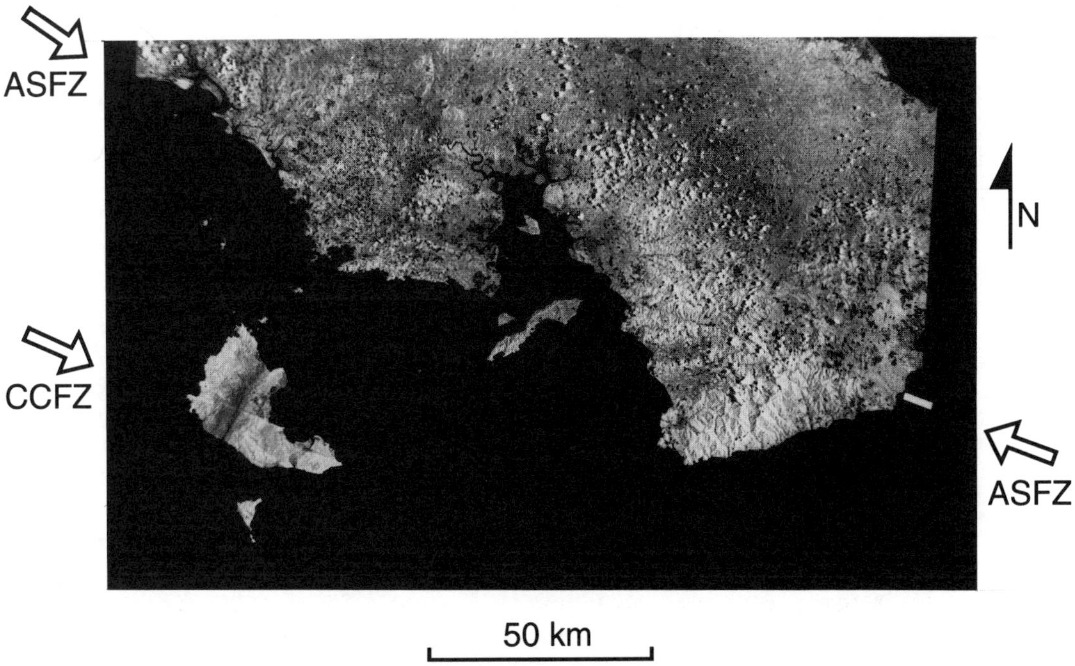

Figure 4. Band 7 (near-infrared) LANDSAT multi-spectral scanner image of the Gulf of Chiriquí area. Azuero-Soná (ASFZ) and Central Coiba fault zones (CCFZ) appear as prominent northwest-striking lineaments (see Figs. 9 and 18 for detailed interpretations). Scattered white patches above the Azuero Peninsula are low-lying clouds. Light tones on the ground are heavily forested areas; darker tones are mainly cleared grasslands and populated areas.

Shell seismic profiles. The offshore depth to basement map was integrated with onshore geologic field data from the Auzero Peninsula, Soná Peninsula, Burica Peninsula, and Panama landmass and the results of seismic profiling and SeaMARC II side-scan studies of deeper-water areas of the South Panama deformed belt by Heil (1988), MacKay (1988), MacKay and Moore (1990), Silver et al. (1990), and Moore and Sender (this volume) (this combined study area is shaded in gray in Fig. 3). The integrated data set reveals five major tectonic elements beneath and surrounding the Gulf of Chiriquí. These tectonic elements include the following:

1. South Panama deformed belt, an active accretionary prism characterized by folds, reverse faults, and thrust faults, extends from at least 80°W to 83°W and covers an area of 5,000 km^2 south of the Gulf of Chiriquí. Silver et al. (1990) refer to this feature as the "South Panama thrust belt," but we prefer the name "South Panama deformed belt" because this name has fewer genetic connotations and because this name is consistent with previous usage of the term "North Panama deformed belt" to describe the broad belt of deformation north of the Isthmus of Panama (Bowland, 1984; Bowland and Rosencrantz, 1988; Reed et al., this volume).

2. Parida and Ranchería arches are expressed as relief on acoustic basement and the overlying sedimentary cover of the central Gulf of Chiriquí. The arches trend north to north-northeast, are 100 to 120 km long and 15 to 40 km wide, and cover a total area of 5,000 km^2. There is up to 3,000 m of structural relief across the Parida arch on the top of the basement surface. Although discussed by Heil (1988), both arches are named and described in detail in this chapter. The name of the Parida arch is taken from Parida Island located at 82°30′W and 8°20′N near the crust of the arch. The name of the Ranchería arch is taken from Ranchería Island located at 81°42′W and 7°44′N northeast of Coiba Island (Fig. 5).

3. Northwest-striking left-lateral strike-slip and oblique-slip faults. From north to south, these faults include the Azuero-Soná fault zone, Central Coiba fault zone, and South Coiba fault zone. The Azuero-Soná fault zone can be traced as a continuous feature for 150 km from the southeastern corner of the Azuero Peninsula to the northwestern corner of the Soná Peninsula (Mann and Corrigan, 1990). The Azuero-Soná fault zone may extend offshore into the southern Gulf of Panama and South Panama deformed belt as the South Panama fault zone (Hardy et al., 1990; Hardy, 1991; Mann and Kolarsky, this volume; Westbrook et al., this volume). The Central Coiba fault zone can be traced for 100 km from the southwestern end of the Cébaco basin complex to the southeastern end of the Montuosa basin. The onland extensions of all three fault zones are visible on LANDSAT imagery where they cross the Azuero and Soná Peninsulas and Coiba Island (Fig. 4).

4. Montuosa basin is an asymmetric sag basin that covers an area of 2,100 km^2 and is formed by subsidence along the

TABLE 1. FIELD AND PROCESSING PARAMETERS FOR OFFSHORE MULTICHANNEL SEISMIC DATA*

Field Parameters	Texaco (Lines TEX-2, TEX-3, TEX-5)	Field Parameters	Texaco (Lines TEX-1, TEX-4)
Recorded	1971	Reccorded	1971
System	DFS III	System	DFS III
Sampling interval	4 msec.	Sampling interval	4 msec
Filter	8-62 Hz	Filter	8-62 Hz
Air gun	930 cu in array	Air gun	930 cu in aray
Pressure	n.a.†	Pressure	n.a.
Shotpoint interval	100 m	Shotpoint interval	100 m
Number of channels	48	Number of channels	48
Channel distance	50 m	Channel distance	50 m
Near trace	150 m	Near trace	150 m
Far trace	2,350 m	Far trace	2,350 m
Processing Parameters	**Mobil**	**Processing Parameters**	**UTIG§ (Kolarsky, 1992)**
Year processed at Texaco	1971	Year reprocessed at UTIG	1990
	Demultiplex		Data recovery
	Gain modification		Demultiplex
	CDP sorting		CDP sorting
	Velocity analysis		Velocity analysis
	NMO correction		Spherical divergence correction
	Stack		Water column muting
	Display (normal polarity)**		Deconvolution test
			Gap deconvolution
			Bandpass filtering (33-55 Hz)
		See Figures 6 and 8 of this chapter for lines TEX-1 and TEX-4, respectively.	NMO correction
			Stack
			Display (normal polarity)**

northern edge of the Central Coiba fault zone and between the Parida and Ranchería arches. The basin is 35 km wide in its northeast dimension by 60 km wide in its northwest dimension. Although first recognized by Okaya and Ben-Avraham (1987) using Texaco Line 3 (Line Tex-3 in Fig. 3), this basin is described in detail in this chapter and is named for Montuosa Island located at 82°15′W and 7°25′N, near the Central Coiba fault zone bounding the southern edge of the basin.

5. Cébaco basin complex is an *en echelon* series of five northeast– to east-northeast–striking half-grabens. The basin complex is named here after Cébaco Island at the northeastern edge of the complex in the eastern Gulf of Chiriquí (Fig. 5). The Cébaco basin complex remained undocumented prior to this study for two main reasons. First, the previous study by Okaya and Ben-Avraham (1987) of the Gulf of Chiriquí did not recognize the Cébaco basin complex because Texaco lines 1 and 4 and Shell lines P-1501 and P-1502 were not available for use in their study (Fig. 3). Second, the half-grabens and adjacent horsts extend onshore as prominent fault-bounded topographic depressions and ridges in only two cases. Cébaco Island in the eastern Gulf of Chiriquí forms a narrow horst block adjacent to one of the submarine half-grabens. The Rio Flores fault zone and the adjacent flat-bottomed alluvial valley on the southwest Azuero Peninsula (Fig. 4) appear to be the onshore extension of one of the easternmost down-to-the-northwest normal faults and associated half-grabens of the Gulf of Chiriquí (Fig. 5).

South Panama deformed belt

Ridges on the subducting Nazca plate. The bathymetric ridges associated with the Panama, Balboa, and Coiba fractures zones are fault-bounded horsts rising more than 2,000 m above the surrounding seafloor (Lonsdale and Klitgord, 1978; Moore and Sender, this volume). Heil (1988), MacKay and Moore (1990), and Silver et al. (1990) noted that the Panama, Balboa, and Coiba Ridges are subducted with the Nazca plate at the South Panama deformed belt and deform both the belt and the Gulf of Chiriquí area by margin indentation and localized reverse faulting and uplift. Our interpretation of Shell line 1504 (refer to Fig. 10A, B) shows that the subducting Nazca plate dips northeast at approximately 6° beneath the South Panama deformed belt and that the slopes of the outer forearc high dip steeply down to a broad, 30-km-wide accretionary wedge below which sedimentary rocks of the subducting Nazca plate appear to be underplated beneath the overriding plate.

Structural reentrants in the South Panama deformed belt. The most prominent structural reentrants in the South Panama deformed belts are associated with the Balboa and

TABLE 1. FIELD AND PROCESSING PARAMETERS FOR OFFSHORE MULTICHANNEL SEISMIC DATA* (continued)

Field Parameters	Shell (Lines P-1501, 1504, 1506, 1600)	Field Parameters	UTIG (Lines PN-2W, PN-2N, PN-2E)
Recorded	1973	Recorded	1977
System	DFS III	System	DFS III
Sampling interval	4 msec	Sampling interval	2 msec
Filter	8-62 Hz	Filter	
Airgun	1,200 cu in	Air gun	1,200 cu in
Pressure	140 atm	Pressure	140 atm
Shotpoint interval	50 m	Shotpoint interval	50 m
Number of channels	24	Number of channels	24
Channel distance	n.a.	Channel distance	100 m
Near trace	413 m	Near trace	413 m
Far trace	2,713 m	Far trace	2,713 m
Processing Parameters	**Shell**	**Processing Parameters**	**Mobil**
Year processed at Shell	1973	Year processed at UTIG	1977
	Demultiplex		Demultiplex
	Trace editing and correction		Trace editing and correction
	CDP sorting		CDP sorting
	Velocity analysis		Velocity analysis
	NMO		NMO
	24 fold CDP-stack		24 fold CDP-stack
See Figure 10 of this chapter	Deconvolution filtering		Time-variant scaling
for line 1504 (parts 1 and 2);	Time-variant scaling		Display (normal polarity)**
see Figure 5 of chapter 13	Water column muting		
for line 1600 (parts 1 and 2).	Display (normal polarity)**		

*Data used in this chapter and in chapter 13 (Fig. 5, line P-1600).
†n.a. = not available.
§UTIG = Univeristy of Texas Institute for Geophysics.
**A compression at the geophone produces a negative number represented by a black peak on plot.

Coiba ridges. Both reentrants are 17 to 25 km wide, correspond well with the width of the subducting Balboa and Coiba ridges, and deflect the thrust front of the South Panama deformed belt northward by 8 to 15 km (Silver et al., 1990; Moore and Sender, this volume) (Fig. 5). The Coiba Ridge, the ridge with the greatest topographic relief on the surface of the subducting Nazca plate (2,000 m), produces the greatest (15 km) marginward deflection of the thrust front of the accretionary prism indentation.

Parida arch, Gulf of Chiriquí

The top of the acoustic basement in the Gulf of Chiriquí defines two northerly oriented regional arches which we name here the Parida and Coiba arches (Fig. 5). The Parida arch is roughly defined by the southward deflection of the 1,000-m and 2,000-m depth contours to the top of the basement surface. The Parida arch exhibits vertical basement relief of 3,000 m, forms the western edge of the Montuosa basin, and extends 110 km across the northwestern half of the Gulf of Chiriquí (Fig. 6).

Age correlation across the Parida arch. At the crest of the Parida arch, acoustic basement interpreted as basalt and hemipelagic deposits interbedded with basalt (Kolarsky et al., this volume) rises to 200 m beneath the seafloor of the Gulf of Chiriquí and is overlain by 200 m of Plio-Pleistocene sedimentary rocks (Fig. 6). In order to infer the age of sedimentary rocks onlapping the Parida arch and, therefore, the age of the arching, we projected the dip of outcrops of Plio-Pleistocene sedimentary rocks and a basal upper Paleocene–lower Eocene gravel and limestone unit above basaltic basement on the Burica Peninsula (Corrigan et al., 1990) to the western part of Texaco line 1 (Fig. 7). Although an 8.4-km gap separates the onshore data from the end of Line 1, the interpretation appears spatially and seismically reasonable. In this interpretation, the arch is onlapped by sedimentary rocks of late Pliocene–early Pleistocene age and is, therefore, inferred to have formed in late Pliocene time.

It is interesting that the western Gulf of Chiriquí appears to be largely unaffected by the type of large-scale normal faulting seen in the Cébaco basin complex to the east. It is the absence of high-angle faults that allows us to make long-distance correlations between rock units exposed on the Burica Peninsula at the westernmost margin of the Gulf of Chiriquí and rocks overlying the Parida arch.

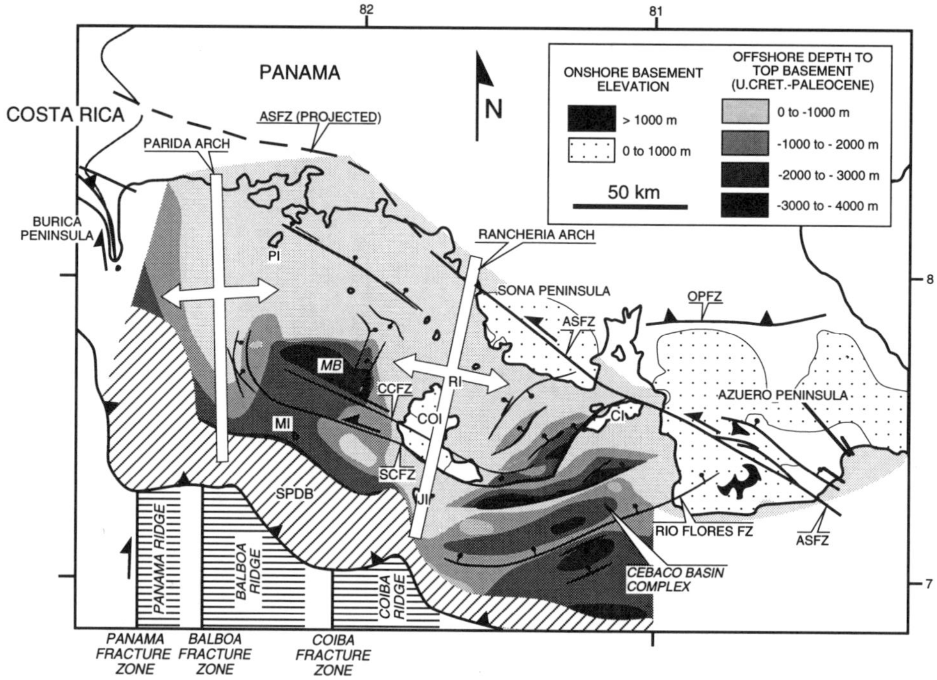

Figure 5. Structural map of the top of the acoustic basement surface in the Gulf of Chiriquí based on seismic lines shown in Figure 3. Depth conversion is based on seismic interval velocities of seismic lines. Offshore contour intervals of depth to basement are shown in tones of gray. Cross-hatched offshore area is the South Panama deformed belt (SPDB), an accretionary wedge. Onshore topographic elevation of onshore basement rocks is shown in black or stipple. Abbreviations: ASFZ = Azuero-Soná fault zone; OPFZ = Ocú-Parita fault zone; CI = Cebaco Island; COI = Coiba Island; MI = Montuosa Island; JI = Jicarón Island; PI = Parida Island; MB = Montuosa basin; SCFZ = South Coiba fault zone; CCFZ = North Coiba fault zone.

Ranchería arch, Gulf of Chiriquí

The Ranchería arch is roughly defined by the southward deflection of the 1,000-m depth contour to the top of the basement surface and the land areas of Coiba, Jicarón, and Ranchería Islands (Fig. 5). Basaltic basement rocks and interbedded pelagic limestone are the primary rock types found cropping out on all of the islands overlying the Ranchería arch (del Giudice and Recchi, 1969; Kolarsky et al., this volume). The Ranchería arch exhibits a vertical basement relief of 3,000 m, forms the western edge of the Cébaco basin complex, and extends approximately 100 km across the central part of the Gulf of Chiriquí (Fig. 6). At the crest of the Ranchería arch, acoustic basement forming the seafloor is interpreted as the offshore extension of basalt and hemipelagic deposits interbedded with basalt found on the islands (Kolarsky et al., this volume). Upper Pliocene–Pleistocene sedimentary units onlap both an older middle Eocene–lower Miocene unit and basaltic basement.

Strike-slip fault zones in the Gulf of Chiriquí

Previous mapping of faults in the Gulf of Chiriquí. Okaya and Ben-Avraham (1987) interpreted four, west-northwest–striking, high-angle faults in the Gulf of Chiriquí on Texaco Line 3. On the basis of the structural configuration of the fault zones interpreted from seismic data and their geologic and tectonic setting, Okaya and Ben-Avraham (1987) interpreted the four faults as left-lateral strike-slip faults and named them faults F1, F2, F3, and F4. We recognize their Fault F2, which we rename the South Coiba fault zone (Fig. 8). We also recognize their Fault F3 ~10 km to the north of the South Coiba fault zone, which we renamed the Central Coiba fault zone (Fig. 8). We do not recognize faults in the locations where Okaya and Ben-Avraham (1987) identify their faults F1 and F4.

Central and South Coiba fault zones. Central and South Coiba fault zones can be traced for a distance of approximately 100 km and form prominent lineaments seen on satellite images of Coiba Island (Fig. 9). To the east of Coiba Island, both faults curve to the northeast over a distance of 15 km to become faults of the Cébaco basin complex (Fig. 5). On Texaco line 4, the South Coiba fault is interpreted as a vertical flower structure composed of one strand 2.3 to 1.6 seconds two-way time and seven upward branching strands at 1.2 to 0.2 seconds two-way time (Fig. 8). The southwestern side of the block appears to be upthrown by approximately 1 second, and sequence of reflectors interpreted as Plio-Pleistocene sedimentary rocks thickens by approximately 0.2 seconds two-way time to the north of the fault. Similarly, the Central Coiba fault zone is interpreted as a flower structure with three upward fault branches. The Central Coiba fault zone does not have significant vertical offset at the location of Texaco line 4.

To the west of Coiba Island, both faults can be traced to the south and southwest of Montuosa basin, where they appear to curve to the north and merge with the eastern boundary of the Parida arch (Fig. 5). Neither fault appears to have any significant offset on the north-northeast–trending Ranchería arch. On Shell line 1504, both faults are interpreted as upward-splaying flower structures that downthrow basement rocks to the northwest (Fig. 10).

Azuero-Soná fault zone. This fault zone extends from the southwestern end of Azuero Peninsula, across the Gulf of Montijo separating the Azuero and Soná Peninsulas and continues onshore across the Soná Peninsula (Fig. 5). No seismic lines extend far enough to the north to intersect this fault in the Gulf of Montijo separating the Azuero and Sona Peninsulas. The offshore extension of the fault to the west of the Soná Peninsula is unclear because this area of coastal estuary or shallow shelf is not crossed by seismic lines (Fig. 5). Mann and Corrigan (1990) proposed that the western prolongation of the Azuero-Soná fault may correlate to a topographic fault scarp with up to 100 m of relief that extends across the central lowland area of westernmost Panama. This scarp appears continuous with the Ballena-Celmira fault zone of Costa Rica, a strike-slip or oblique-slip fault separating two distinct basement blocks of the Costa Rican forearc (Corrigan et al., 1990; Kolarsky et al., this volume).

In the offshore area southeast of the Azuero Peninsula, Hardy et al. (1990) and Hardy (1991) proposed that the Soná-Azuero fault may extend to 78°10′W on the basis of single-channel seismic profiles and GLORIA side-scan sonar data from this area (Fig. 5). Using Landsat MSS and SIR-A (Shuttle Imaging Radar-A) images, Mann and Corrigan (1990) proposed the following evidence for left-lateral strike-slip displacement along the Soná-Azuero fault zone: (1) Regional linearity and topographic prominence of the fault trace, (2) topographic and structural dome along a gentle, right-stepping curvature in the center of the Soná Peninsula, and (3) 150-m apparent left-lateral offset of the Rio Quebro on the Azuero Peninsula.

The geometry of normal faults of the Cébaco basin complex in the Gulf of Chiriquí discussed below is also consistent with our interpretation of Neogene to Recent left-lateral strike-slip displacement along the Soná-Azuero fault zone.

Cébaco basin complex, Gulf of Chiriquí

Structure of the Cébaco basin complex. The Cébaco basin complex consists of five half-grabens that underlie one-third of the area of the eastern Gulf of Chiriquí (Fig. 5). The basin complex occurs in a 7,000-km^2 overlap area between the Central Coiba fault zone and the Azuero-Soná fault zones. The largest half-graben is 70 km long by 16 km wide and has a sedimentary fill of more than 2,600 m. The smallest half-graben is 30 km long by 5 km wide and has a sedimentary fill of more than 1,300 m. Maximum depth to basement on Texaco Line 1 that crosses the five half-grabens is 2,600 m (Fig. 6). Throw on the four basin-bounding faults decreases from southwest to northeast. The maximum amount of throw (2,600 m) is seen on the fault bounding the southeasternmost basin, and the least amount of vertical throw (1,200 m) is seen on the northwesternmost basin.

On Texaco line 1, the character of the half-graben–bounding faults can be seen to differ from the "flower structure" character of the Central and South Coiba fault zones (Fig. 6). The basin-bounding faults of the Cébaco basin complex are not associated with upward splaying faults typical of strike-slip flower structures but instead form uniformly northwest-dipping fault planes more typical of normal faults. The true dip of the basin-bounding faults ranges from 45° to 60°. Given a more dense grid of seismic data in the curving area of the Central and South Coiba fault zones, it might be possible to establish the changing oblique-slip nature of the fault surfaces as they transform from dominantly normal faults in a northeast-striking orientation to dominantly strike-slip faults in a west– to west-northwest–striking orientation.

Basement stratigraphy of the Cébaco basin complex. There are two distinct basement types in the landmasses and islands around the Gulf of Chiriquí (Fig. 11). To the north of the Azuero-Soná fault zone, a poorly studied heterogeneous basement on the northern parts of the Azuero and Soná Peninsulas is characterized by: (1) Upper Cretaceous basalt and chert on the southeastern coast of the Azuero Peninsula near Playa Venado (Bourgois et al., 1982) (Fig. 5), (2) small outcrops of low-grade metamorphic rocks that occur as slivers in the Azuero-Soná fault zone near the west coast of the Azuero Peninsula and in the interior of the Soná Peninsula (Tournon et al., 1989), (3) Upper Cretaceous pelagic limestone and terrigenous turbidites of the Ocú Formation (Recchi and Miranda, 1977), (4) granodiorite plutons with late Cretaceous to Eocene K-Ar cooling ages (del Giudice and Recchi, 1969), and (5) undifferentiated Eocene to upper Tertiary volcanic, volcaniclastic, and related sedimentary rocks (del Giudice and Recchi, 1969).

To the south of the Azuero-Soná Peninsula, the basement is composed mainly of upper Cretaceous basalts with interbedded pelagic rocks and terrigenous turbidites of the Ocú Formation (Kolarsky et al., this volume) (Fig. 11). This type of basement corresponds to descriptions of the Nicoya Complex on the Burica Peninsula (Corrigan et al., 1990; Obando, 1986), the Osa Peninsula of Costa Rica (Obando, 1986; DiMarco et al., this volume), and the Nicoya Peninsula of Costa Rica (Gursky et al., 1982). Basement rocks to the north and south of the Azuero-Soná fault zone are unconformably overlain by shallow-marine Eocene carbonate and clastic rocks (Recchi and Miranda, 1977; Mann and Corrigan, 1990; Kolarsky et al., this volume). Because of the location of the Cébaco basin complex south of the Azuero-Soná fault zone

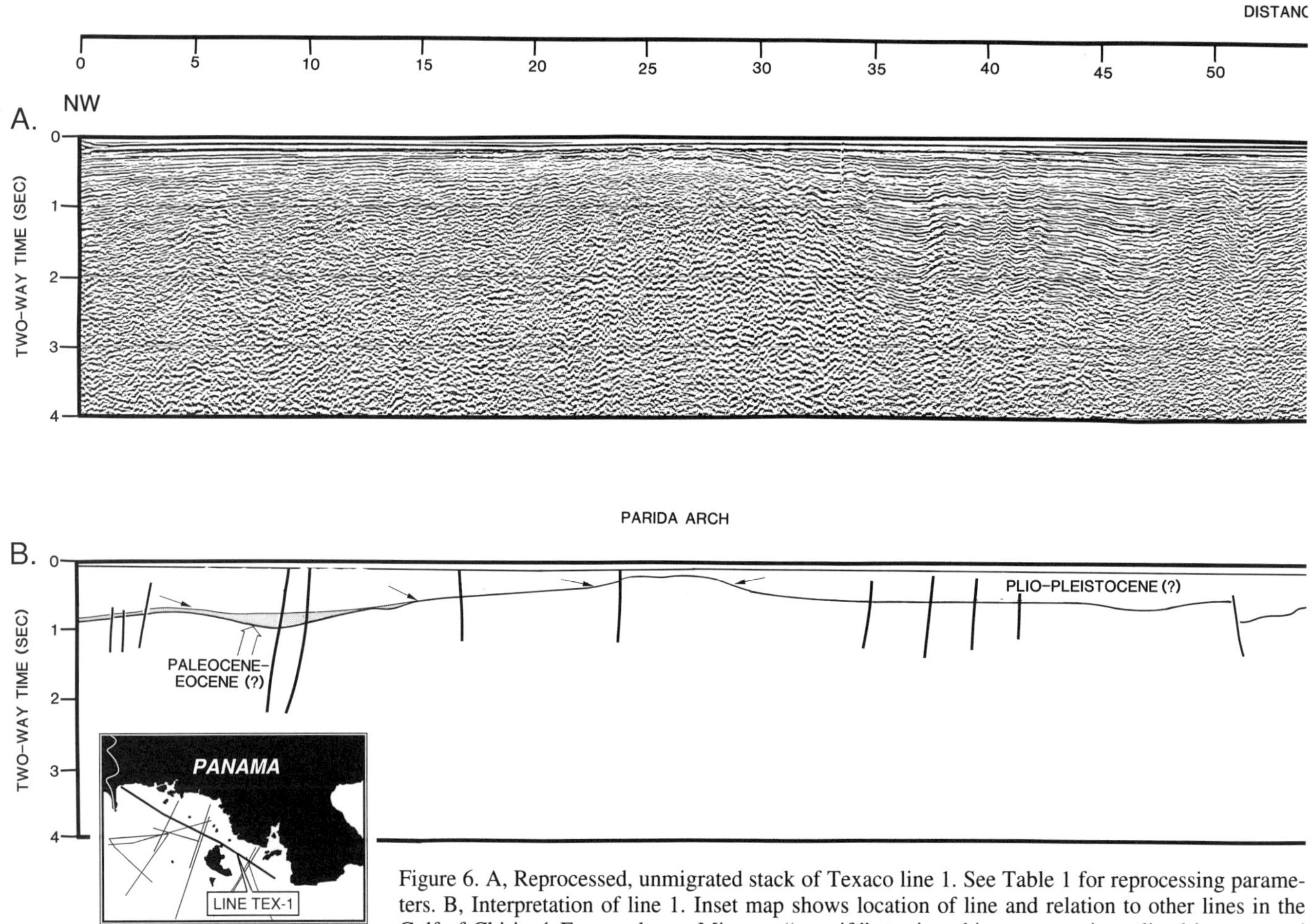

Figure 6. A, Reprocessed, unmigrated stack of Texaco line 1. See Table 1 for reprocessing parameters. B, Interpretation of line 1. Inset map shows location of line and relation to other lines in the Gulf of Chiriquí. Eocene–lower Miocene "pre-rift" stratigraphic sequence is outlined in gray, and arrows indicate directions of stratigraphic onlap.

and its seismic character, we interpret acoustic basement beneath the Cébaco basin complex to be basalts with interbedded pelagic rocks and terrigenous turbidites of the Ocú Formation.

Basin stratigraphy of the Cébaco basin complex. Sedimentary thicknesses in the half-grabens of the Cébaco basin complex range from less than 500 m to 2,600 m (Fig. 5). Sedimentary thicknesses are greater in the central two half-grabens of the complex and less in the more exterior of the half-grabens in the eastern and western part of the complex. Onshore stratigraphic studies on Coiba Island and along the eastern coast of the Azuero Peninsula suggest that the age of the pre-rift sedimentary forearc apron sequence preserved offshore in the half-graben lows of the Cébaco basin complex is middle Eocene to lower Miocene (Kolarsky et al., this volume). Although this sequence thickens slightly toward the half-graben controlling faults (Fig. 6), it does not onlap the underlying rocks and is therefore interpreted as a pre-rift sequence. See Kolarsky et al. (this volume) for a detailed seismic stratigraphic description of these offshore units.

Correlation of reflectors seen on Texaco line 1 with the stratigraphy of the Burica Peninsula suggests that the age of the syn-rift fill of the half-grabens is upper Pliocene to Pleistocene (Corrigan et al., 1990) (Fig. 7). The half-grabens are therefore interpreted to have formed at the same time (late Pliocene-Pleistocene) as the Parida and Ranchería arches.

Montuosa basin, Gulf of Chiriquí

Montuosa basin is an elongate, 1,250-km^2 northeast-trending basin formed between the Parida and Ranchería arches and the North Coiba fault zone (Fig. 5). The northeastern and northwestern slopes of the Montuosa basin are very steep and exhibit more than 3,000 m of relief. The top of acoustic basement along the northeastern margin dips steeply upward to the east and northeast toward exposures of basaltic basement rocks on Coiba Island. The top of acoustic basement is more than 3,000 m deep in the center of the basin and is unfaulted. North-northeast–striking faults at the northeastern margin of the basin show apparent vertical offsets of 100 to 1,000 m. The Montuosa basin contains more than 2,700 m of sedimentary rocks interpreted to be of Plio-Pleistocene age based on correlations on Texaco line 1 over a distance of 120

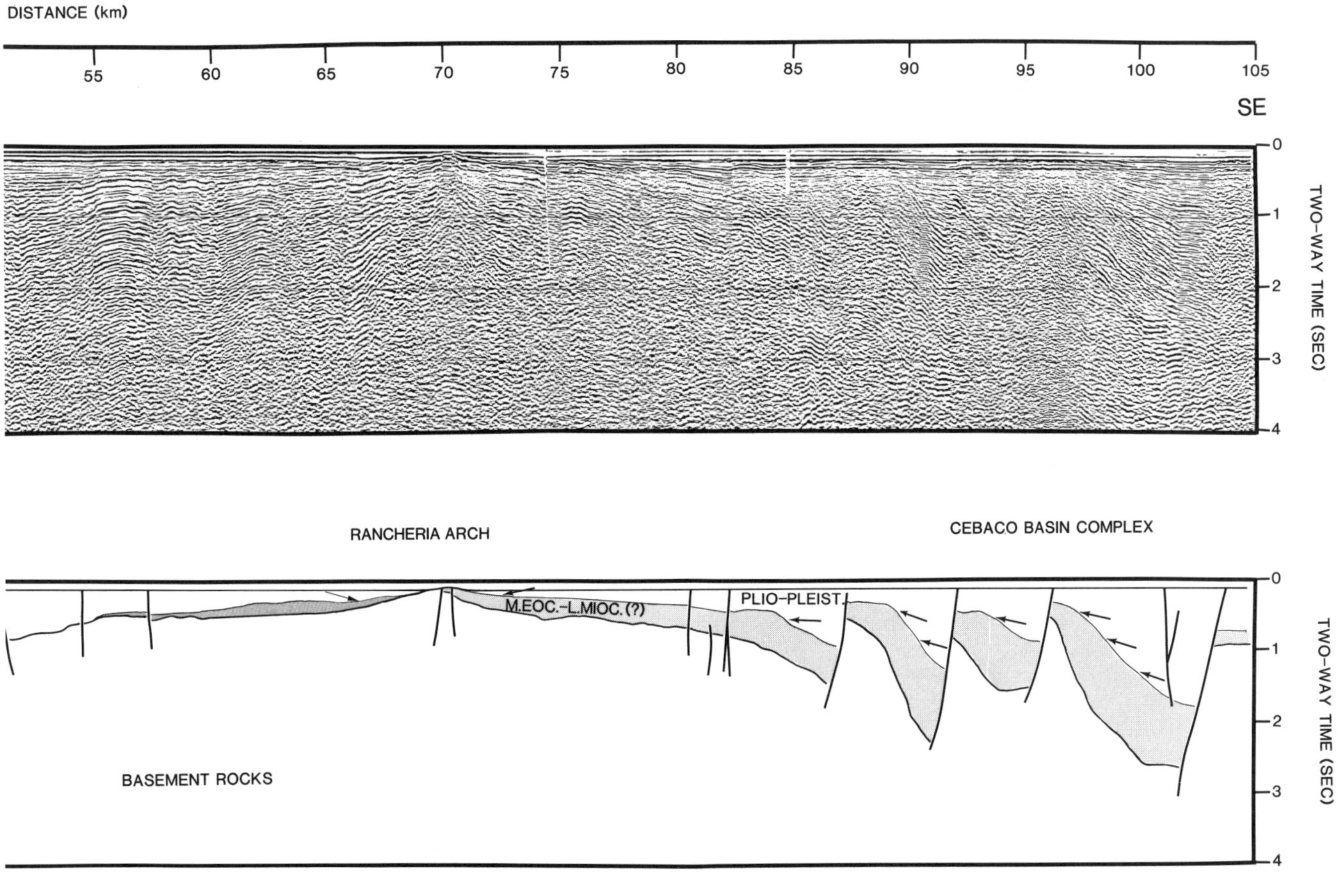

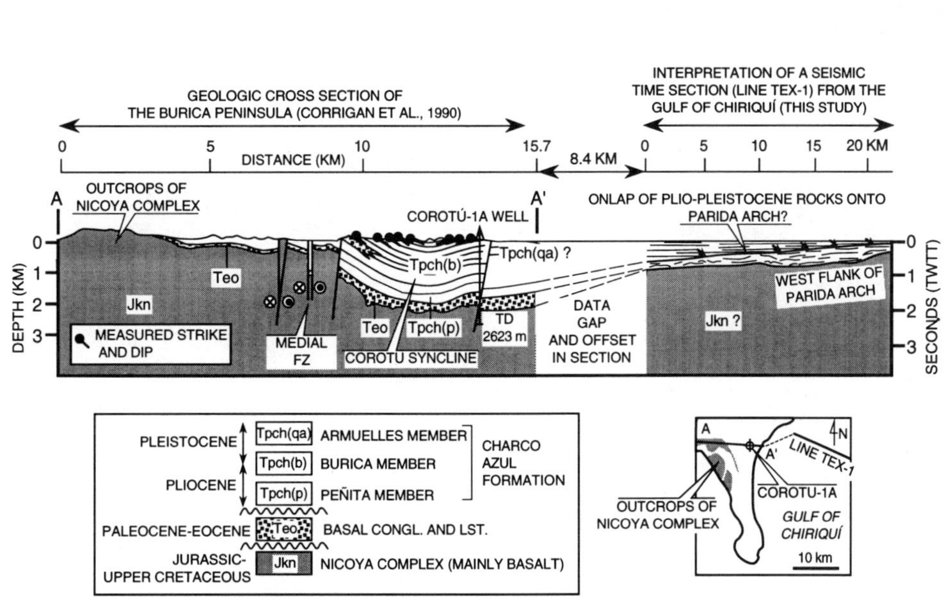

Figure 7. Interpreted stratigraphic correlation between onshore exposures on the Burica Peninsula of Panama and Costa Rica and the western end of Texaco line 1 in the Gulf of Chiriquí. Onshore cross section and ages of rocks on the Burica Peninsula are taken from Corrigan et al. (1990). Note 8.4-km gap in the line of section separating Texaco line 1 and the onshore cross section and the difference in vertical time scale for Texaco line 1 and the depth scale for the cross section. Inset in lower left shows stratigraphy of the Burica Peninsula from Corrigan et al. (1990). Inset in lower right shows location of onshore section, Texaco line 1, and Corotú-1A well used in cross section. See Figure 6B for an interpretation of the entire Texaco line 1.

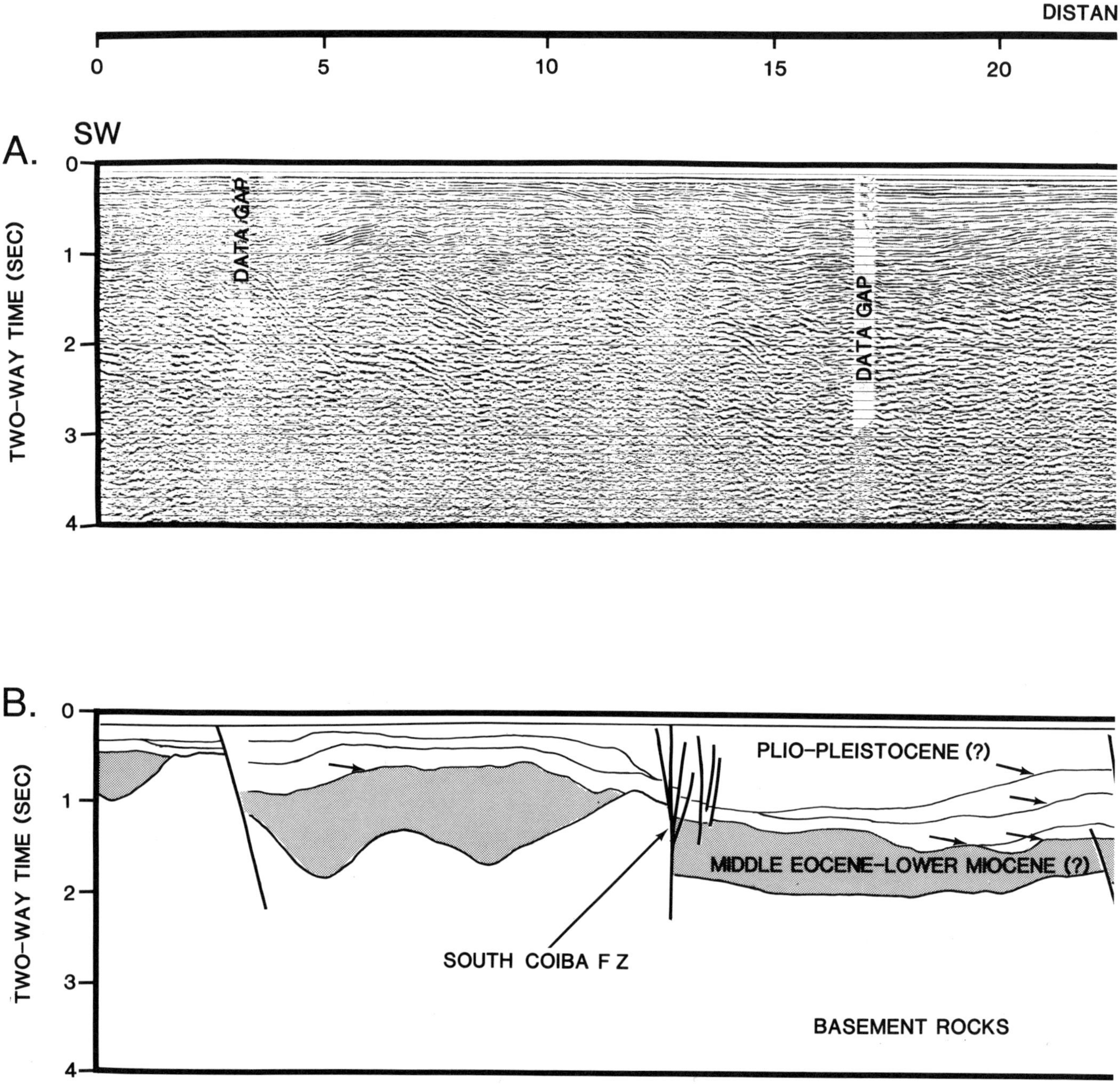

Figure 8. A, Unmigrated stack of Texaco line 4. See Table 1 for reprocessing parameters. B, Interpretation of line 4. Inset map shows location of line and relation to other lines in the Gulf of Chiriquí. Eocene–early Miocene "pre-rift" stratigraphic sequence is outlined in gray, and arrows indicate directions of stratigraphic onlap. Note flower structure at the South and Central Coiba fault zones.

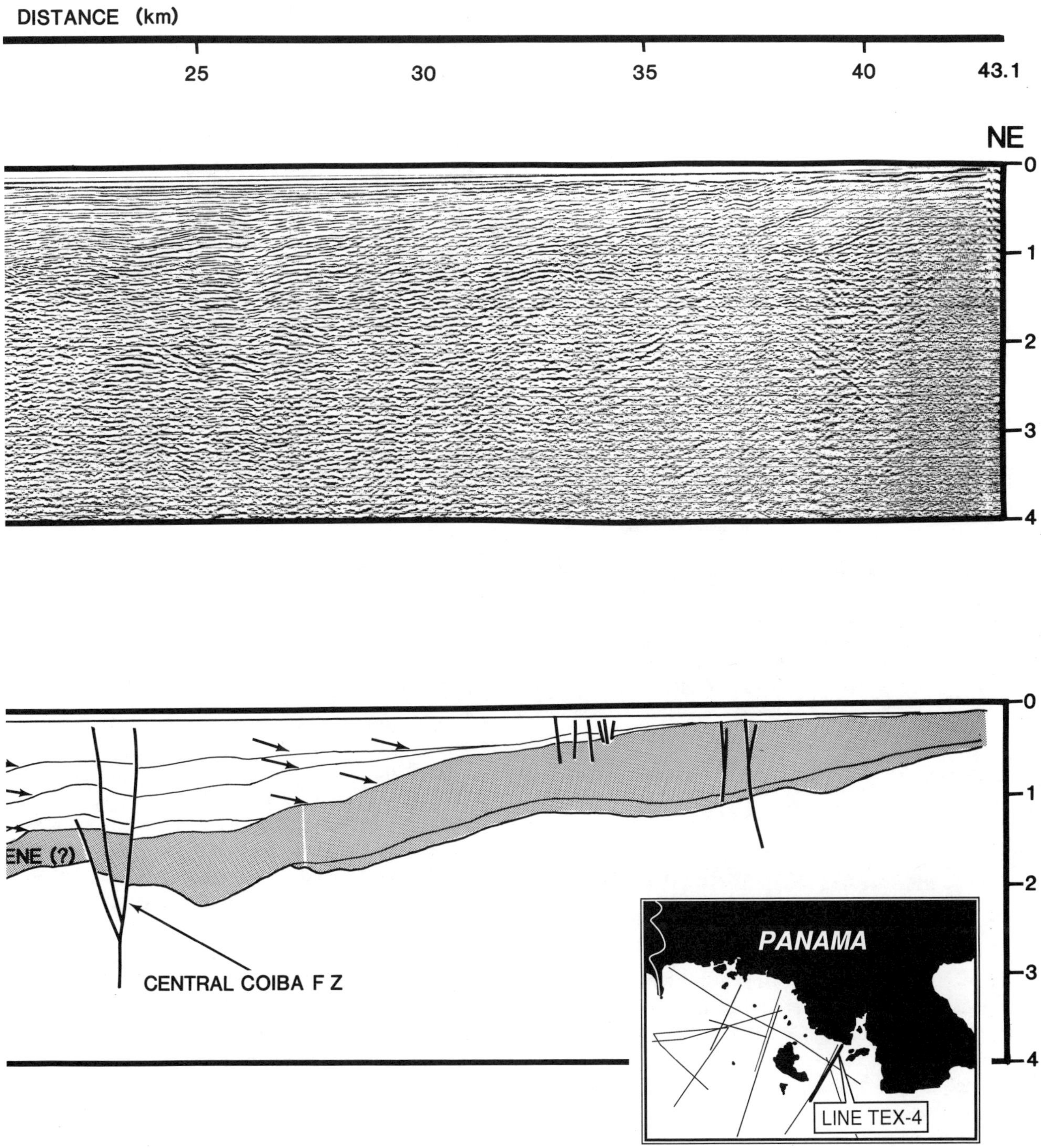

DISTANCE (km)
25
30
35
40
43.1
NE
0
1
2
3
4
TWO-WAY TIME (SEC)
ENE (?)
CENTRAL COIBA F Z
0
1
2
3
4
TWO-WAY TIME (SEC)
PANAMA
LINE TEX-4

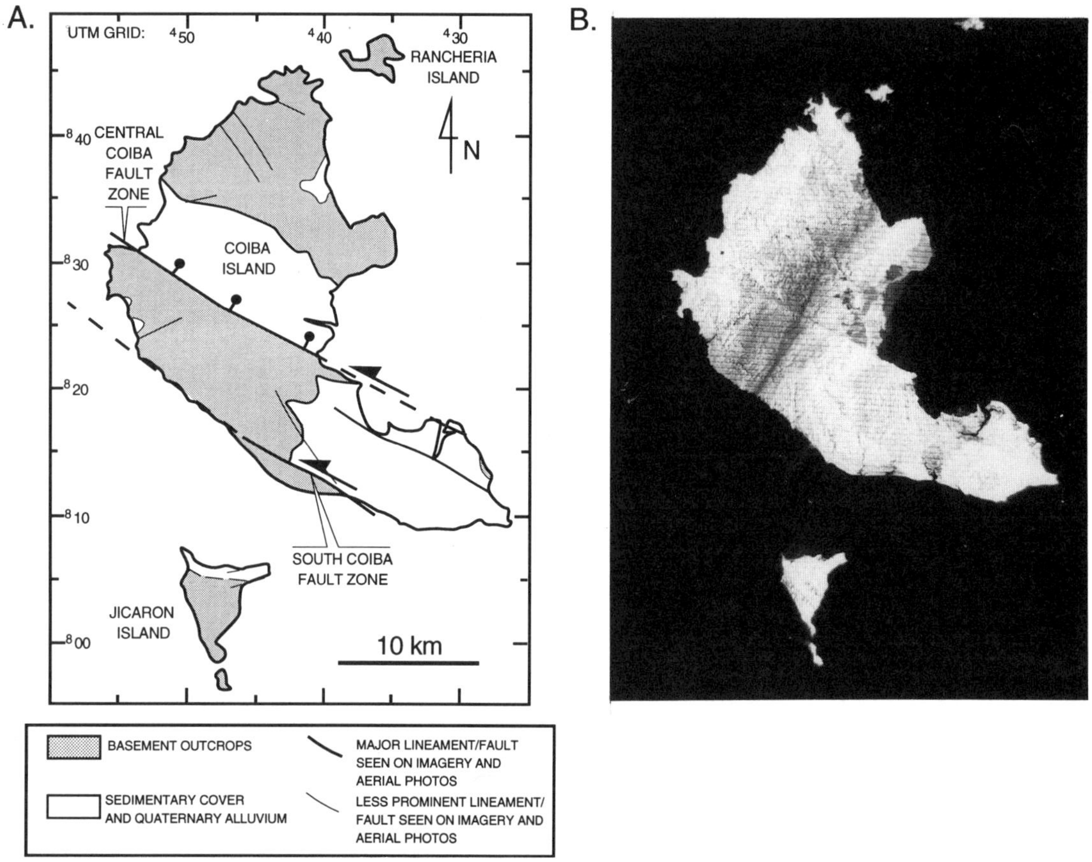

Figure 9. A, Geologic sketch map of Coiba Island showing distribution of basement rocks and sedimentary cover modified from del Giudice and Recchi (1969) and this chapter. B, Band 7 (near-infrared) LANDSAT MSS image of Coiba Island. Note prominent lineaments corresponding to mapped strike-slip faults on Coiba Island and in the Gulf of Chiriquí (see Fig. 5). Less-prominent lineaments visible on image are shown as thinner lines.

km to the upper Pliocene-Pleistocene sedimentary rocks of the Burica Peninsula (Fig. 7).

STRUCTURES IN ROCK OUTCROPS SURROUNDING THE GULF OF CHIRIQUÍ

Purpose of fault study

We obtained regional fault kinematic data from outcrop-scale and regional-scale faults in areas surrounding the Gulf of Chiriquí to better interpret the age and character of large-scale faults seen on offshore seismic profiles (Fig. 5). Outcrop-scale faults are defined here as faults that range in size from less than a centimeter to a few meters and are observable at ground level in single, continuous exposures. Regional-scale faults are defined here as faults seen at scales of hundreds of meters to kilometers that are visible from low-altitude aircraft in continuous tidal flat exposures.

We measured the attitudes of 56 outcrop-scale faults with known offsets at outcrops in wave-cut exposures and tidal flats on the east coast of Coiba Island, the south coast of the Soná Peninsula, and the east and south coasts of the Azuero Peninsula (Fig. 12). Areas of both scales of faulting lie to the south of the Azuero-Soná fault zone and are underlain by a basement composed mainly of Upper Cretaceous basalt and interbedded pelagic limestone and turbidites. Sedimentary cover rocks include Middle Eocene to Lower Miocene sedimentary rocks of the Tonosí Formation. New micropaleontological ages of the basement and overlying Tonosí Formation are presented by Monechi in Kolarsky et al. (this volume).

Method of determination of principal strain orientations from fault data

Determination of principal stress orientations σ1, σ2, and σ3 just prior to faulting can be reasonably estimated using conjugate fault sets in isotropic-appearing rocks (Hobbs et al., 1976). The faults can be assumed to be conjugate if there is evidence of movement in opposite senses directed normal to their line of intersection and if neither set of faults consistently offsets the other. In such a case, σ1 is parallel to the bisector of the "compressional angle" (usually the acute angle) between the two fault orientations, and σ2 is parallel to the line of intersection of the two faults. In most real rock bodies there is sufficient anisotropy of one type or another so that the above method yields erroneous results. However, in the case of minor faults around the Gulf of Chiriquí, the method yielded reasonably consistent results over an area of 5,000 km^2 (Fig. 12). This consistency suggests that, first, principal strain orientations derived from faults around the Gulf of Chiriquí may be close to the orientations of the principal stresses and, second, that there is insufficient anisotropy in the rocks of the Gulf of Chiriquí area to produce errors in the interpretation of the fault data.

STRUCTURES ON COIBA ISLAND

Regional significance

Because of its central location within the Texaco and Shell offshore seismic survey area, Coiba Island is a critical area for collecting onshore structural data to compare with and provide age constraints on large-scale structures seen on offshore seismic profiles (Fig. 5). Coiba Island is an emergent part of the Ranchería arch and provides subaerial exposures of rocks affected by strike-slip faulting along the Central and South Coiba fault zones (Fig. 10).

General structure of Coiba Island

Despite its complex tectonic setting in the Gulf of Chiriquí, large-scale structure of Coiba Island is relatively simple (Fig. 13). The northwest-trending Central Coiba fault zone (CCFZ) divides the island into high-standing northern and southern basement blocks that are separated by a lower-standing valley. The central valley is underlain by less resistant clastic sedimentary rocks of the Tonosí Formation, whereas more resistant basalts underlie the higher-standing northern and southern blocks (Kolarsky et al., this volume). Basement rocks of the northern and southern blocks are similar to those on the Azuero and Soná Peninsulas south of the Azuero-Soná fault zone and consist of massive basalt flows interbedded with layers of pillow-lavas or beds of the upper Cretaceous Ocú Formation (Kolarsky et al., this volume). The southern basement block between the Central Coiba and South Coiba fault zones is tilted 10 to 30° to the southeast and thereby exposes basement basaltic rocks along the south and southwest coasts of the island. Clastic rocks of the Tonosí Formation are also exposed along the southeastern coast (Fig. 13).

Access

The attitudes of minor faults and layering were mapped in coastal exposures of basement and overlying sedimentary rocks along the eastern coast of Coiba Island. Measurements were made only along the east coast of Coiba Island because of the difficulty in access to other areas of the island and surrounding smaller islands like Jicarón and Ranchería (Fig. 13). The road system that radiates from the central penal colony extends along the coast from Punta Damas to the steep escarpment of the Central Coiba fault zone and does not penetrate the interior of the island (Fig. 13). Access to other areas along the east coast of the island is possible only by foot or by outboard motor boats operated by the central prison colony to supply provisions and to ferry guards and prisoners to semiautonomous prison camps scattered along the northeast and southeast coasts of the island. The interior and west coast of the island are uninhabited and can only be reached by boat or by hiking along jungle trails crossing the interior of the island. The island of Jicarón is uninhabited, with unprotected coastlines that are difficult to approach in small boats.

Exposures

The large tidal range in the Gulf of Chiriquí reveals semicontinuous, wave-cut exposures of both basement and sedimentary rocks in almost all areas except the central valley separating the two high-standing basement blocks (Fig. 13). Layering in basement rocks was defined using horizons of interbedded upper Cretaceous pelagic limestone and terrigenous turbidites (Ocú Formation) and sheetlike, pillowed, and nonpillowed basalt flows (Kolarsky et al., this volume).

Bedding in basement rocks

Bedding north of the Central Coiba fault zone generally dips to the southwest (Fig. 14) and is consistent with the map pattern of the island (Fig. 13). The attitude of basement rocks to the south of the Central Coiba fault zone, based on three measurements, is to west or north.

Bedding in sedimentary cover

Scattered bedding measurements in the Tonosí Formation north of the Central Coiba fault zone may reflect diverse bedding attitudes, probably indicative of small-scale folding related to the incompetence and thickness of these beds relative to the more massive basement rocks (Fig. 14). South of the Central Coiba fault zone bedding shows a predominance of shallow dips to the north and northeast. Scattering of bedding orientations in discontinuous outcrops of this area may reflect localized deformation along the Central Coiba fault zone (Fig. 13).

Regional-scale faulting

A regional-scale (~3-km^2) outcrop of faults affecting well-bedded clastic rocks of the Tonosí Formation was photographed on the intertidal zone along the southeast coast of Coiba Island from an aircraft flying at an altitude of 150 m (Fig. 15). The type of exposure and style of conjugate faulting is very similar

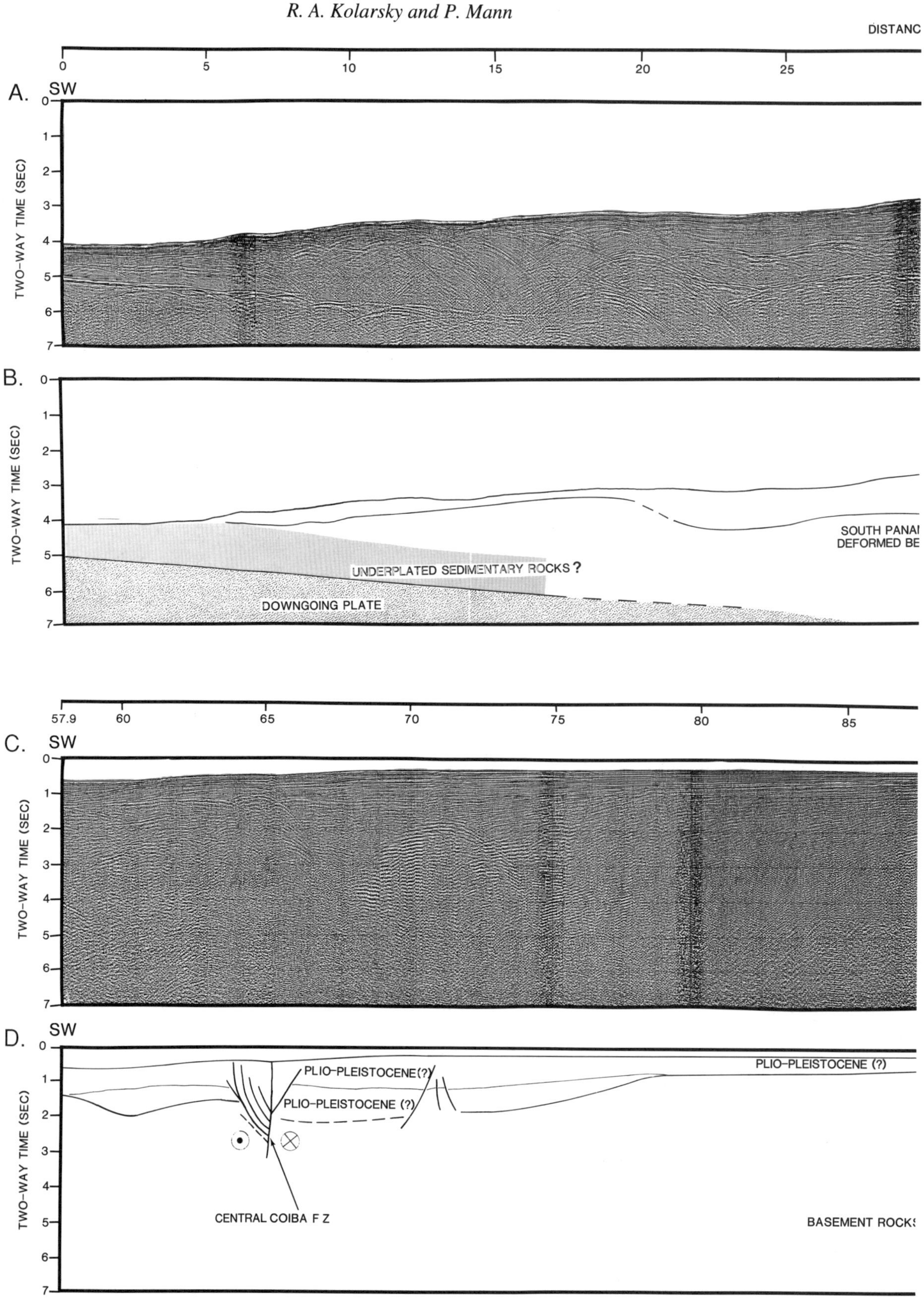

Figure 10. A, Unmigrated stack of southern part (Part 1) and C, northern part (Part 2) of Shell line P-1504. See Table 1 for processing parameters. B, Interpretation of southern part (Part 1) and, D, northern part (Part 2) of line P-1504. Area of possible underplating of oceanic sediments is indicated by light gray. Subducting oceanic crust of the Nazca plate is shown with dark gray.

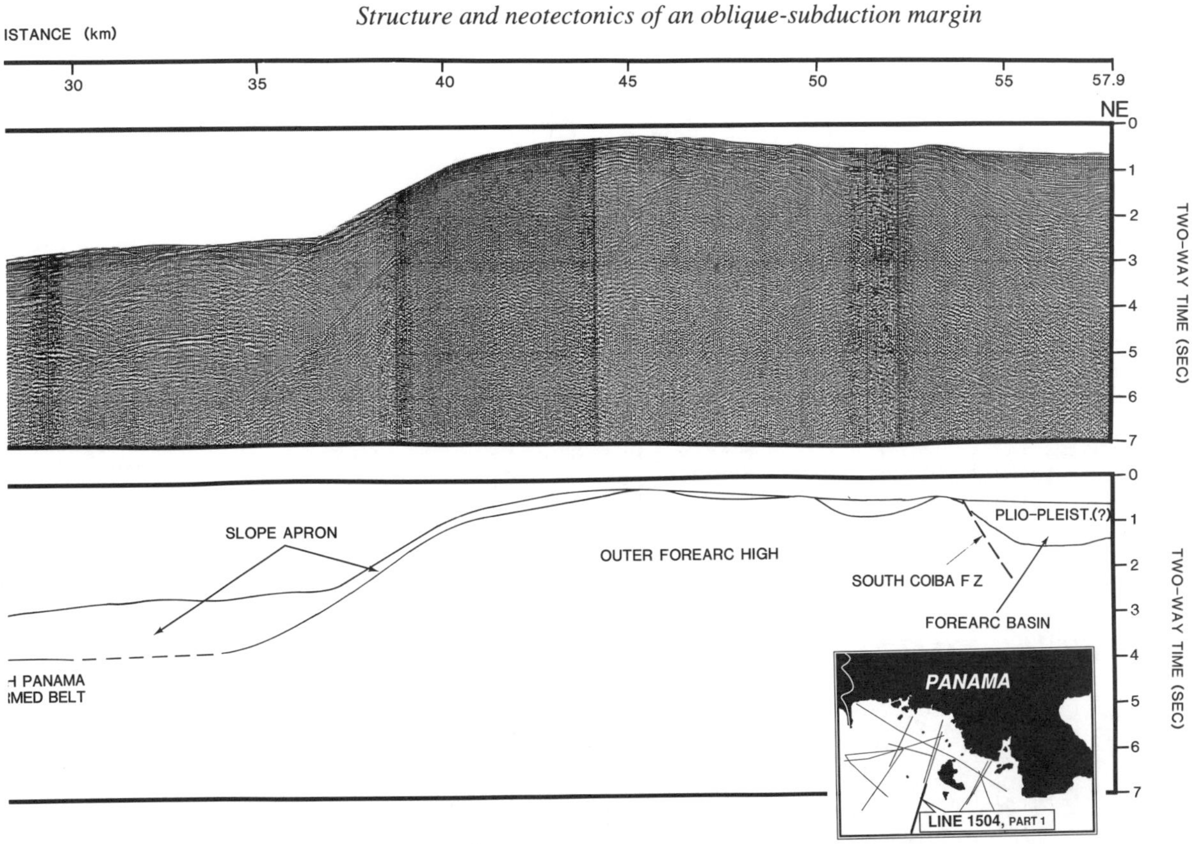
ISTANCE (km)
30
35
40
45
50
55
57.9
NE
TWO-WAY TIME (SEC)
SLOPE APRON
OUTER FOREARC HIGH
PLIO-PLEIST.(?)
SOUTH COIBA F Z
FOREARC BASIN
H PANAMA
MED BELT
PANAMA
LINE 1504, PART 1

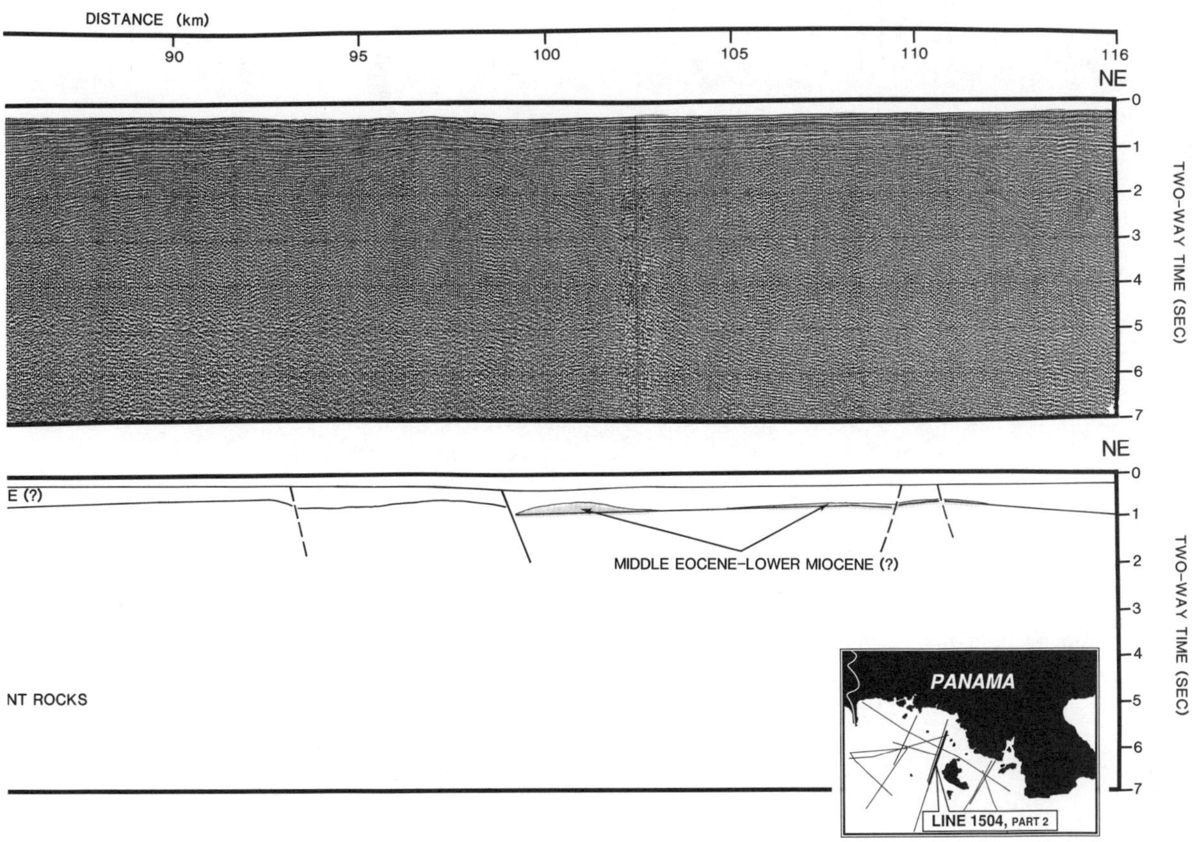
DISTANCE (km)
90
95
100
105
110
116
NE
TWO-WAY TIME (SEC)
E (?)
MIDDLE EOCENE-LOWER MIOCENE (?)
NT ROCKS
PANAMA
LINE 1504, PART 2

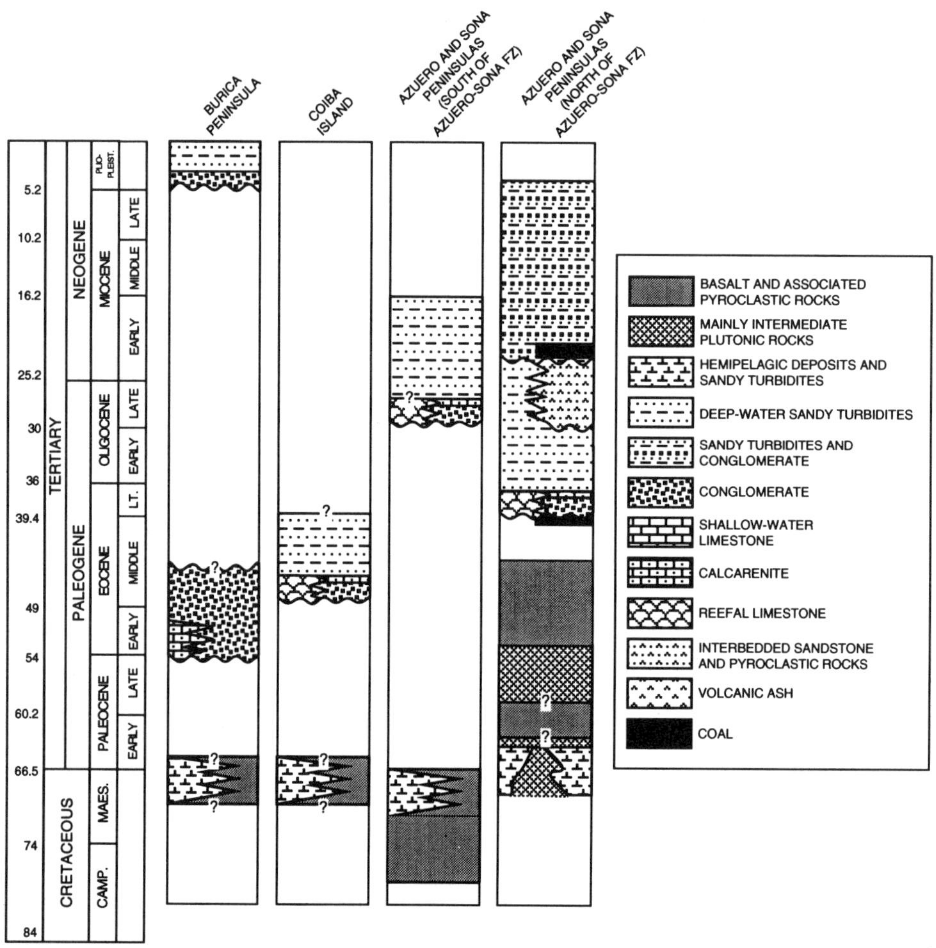

Figure 11. Summary of the stratigraphy of the Pacific margin of southwestern Panama from Corrigan et al. (1990) and Kolarsky et al. (this volume). Note lithologic differences north and south of the Azuero-Soná fault zone that may reflect large-scale displacement and juxtaposition of two different terranes along the fault. Rocks on Coiba Island and the Burica Peninsula lie south of the fault and are similar to rocks found south of the fault on the Azuero Peninsula.

to that mapped on the Burica Peninsula by Corrigan et al. (1990; their Fig. 11).

A line drawing of the photograph reveals a conjugate pattern of faults affecting the unfolded section with a dip later measured to be 43° to the south (Fig. 15). Based on ground studies of bedding offsets along nearby outcrop-scale faults, the conjugate fault pattern is interpreted as northeasterly striking (030) right-lateral and westerly striking (085) left-lateral strike-slip faults (Fig. 15B, inset). Interpretation of principal strain orientations indicates that the direction of maximum extension in these rocks has an approximate northwest-southeast orientation.

Outcrop-scale faulting

Orientations of 64 outcrop-scale faults were measured in basement and sedimentary cover rocks along the eastern coast of Coiba Island. Faults were usually exposed on the intertidal flat or in the cutbank at the landward edge of the intertidal flat (Fig. 16). Fifty-seven of these faults lack fault-plane striations or other indicators of fault sense and are subdivided according to basement or cover occurrence in Figure 17.

Locality 3. At two localities on the east coast of Coiba Island, we measured the attitudes of sets of conjugate faults with well-developed fault-plane striations that constrain the sense of slip. Conjugate faults affecting rocks of the upper Cretaceous Ocú Formation from locality 3 on the northeastern coast of the island near Punta Damas include steeply dipping faults that strike north and are right-lateral and steeply dipping faults that strike west and are left-lateral. Northeast-striking faults are normal faults. This fault geometry indicates a northwest-southeast orientation of maximum post-Cretaceous extension.

Locality 4. Conjugate faults affecting rocks of the middle Eocene to lower Miocene Tonosí Formation from locality 4 on the southeastern coast of the island include steeply dipping faults that strike north-northeast and are right-lateral and

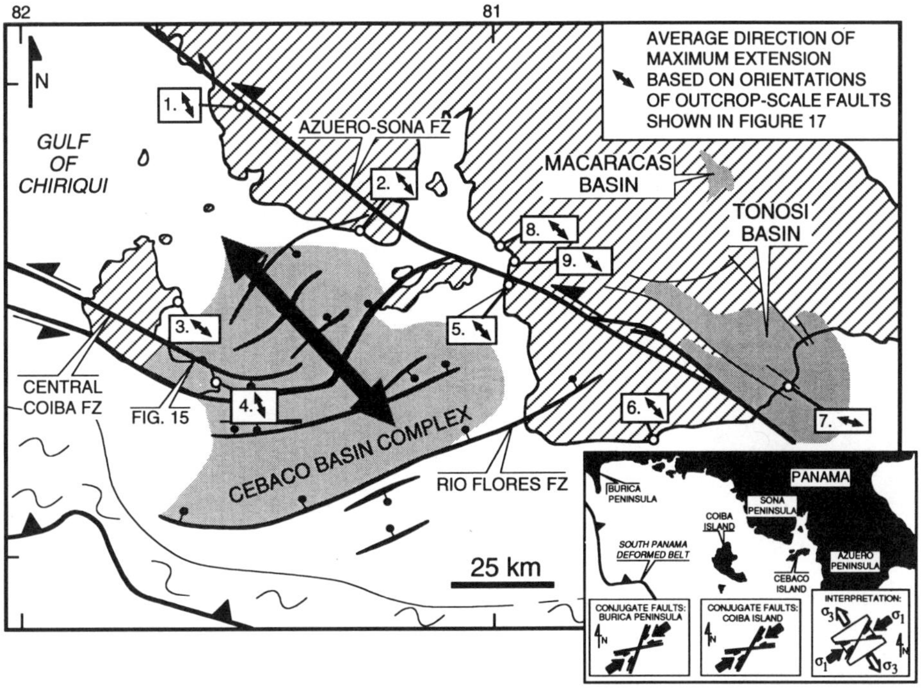

Figure 12. Summary of onshore outcrop-scale structural data in tectonic context of larger-scale structures seen on offshore seismic reflection profiles (Fig. 3). Small arrows give approximate directions of maximum extension based on outcrop measurements of minor conjugate strike-slip and normal faults plotted on Figure 17. Large arrow gives the average direction of opening of the Cébaco basin complex based on the orientation of half-grabens interpreted from the seismic data set. Inset diagram summarizes orientations of minor conjugate strike-slip fault sets from the Burica Peninsula (Corrigan et al., 1990) and Coiba Island and provides an interpretation of a northeast-directed maximum compressional strain and a northwest-directed maximum extensional strain.

steeply dipping faults that strike west-southwest and are left-lateral. One northeast-striking fault is a normal fault. This fault geometry indicates an approximate northwest-southeast orientation of maximum post–early Miocene extension.

STRUCTURES ON THE AZUERO PENINSULA

Regional significance

Although the Azuero Peninsula lies to the east of the Texaco and Shell offshore seismic grid, it is a critical area for collecting onshore structural data to compare with and provide age constraints on large-scale structures seen on offshore seismic profiles (Fig. 5). The Rio Flores fault zone and the easternmost half-graben appear to strike onshore along the southwestern margin of the peninsula and the Azuero-Soná fault zone crops out along the northwest part of the coast (Fig. 12).

General structure of Azuero Peninsula

The northwest-trending Azuero-Soná fault zone divides the peninsula into two basement terranes of differing composition (Fig. 18). Basement rocks southwest of the Azuero-Soná fault zone are similar to those on Coiba Island, whereas those northeast of the fault are more characteristic of an island arc terrane (Fig. 11). The major valleys are generally underlain by less-resistant clastic rocks of the Tonosí Formation or Quaternary alluvium, whereas more-resistant crystalline rocks underlie mountainous areas. Rugged coastlines expose mainly basaltic basement rocks except for small areas of Tonosí Formation outcrop on the northwest, south, and southeast coasts.

Access

The attitudes of minor faults and layering were mapped in coastal exposures of basement and overlying sedimentary rocks along the southeastern, southern, and western coasts of the peninsula. Access is good along paved roads to the northeast, southeast, and south coasts but more difficult along a poorly maintained dirt road that follows the west coast (Fig. 18).

Exposures

As on Coiba Island, the large semidiurnal tidal range in the Pacific Ocean reveals semicontinuous, wave-cut exposures of both basement and sedimentary rocks in almost all areas ex-

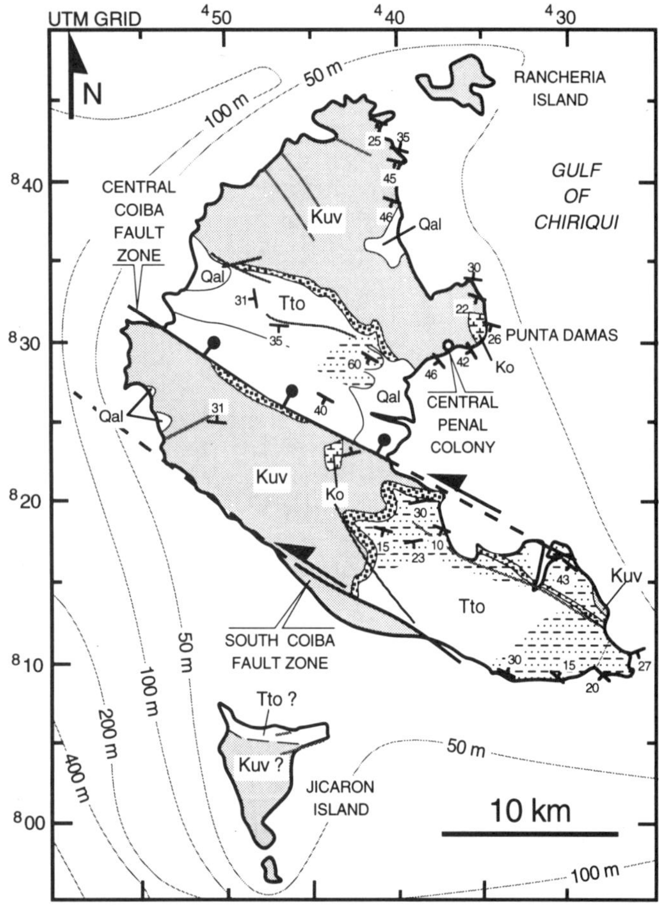

Figure 13. Structural-geologic map of Coiba Island showing basaltic basement (gray unit labeled Kuv) and interbedded Ocú Formation (dashed unit labeled Ko) overlain by Eocene conglomerate and shallow-marine limestone (stippled facies of unit Tto), Eocene to lower Miocene turbidites of the Tonosí Formation (dashed facies of unit Tto), and Quaternary alluvium (uncolored unit labeled Qal). Map is based on del Giudice and Recchi (1969) and this chapter. Queried occurrence of Tonosí Formation on the north coast of Jicarón Island is based on interpretation of aerial photographs.

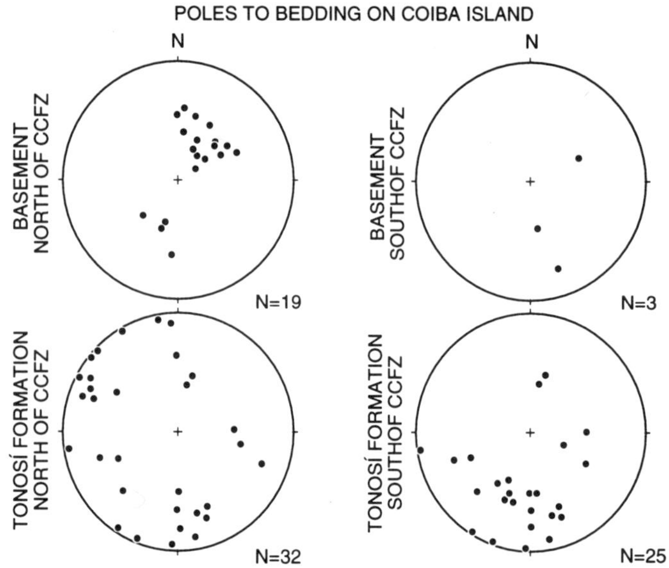

Figure 14. Lower hemisphere stereoplots of poles to minor bedding from Coiba Island. Data are grouped according to geologic units (basaltic basement or sedimentary cover) and according to geographic location north of the Central Coiba fault zone (CCFZ) or south of the fault zone (see Fig. 13 for fault locations and distribution of basement and cover rocks). Layering in basaltic basement is defined by interbedded sedimentary horizons (Ocú Formation), sheetlike basalt flows, or horizons of flattened pillow basalts.

cept near alluvial valleys (Fig. 18). Layering in basement rocks was defined using sheetlike, pillowed, and nonpillowed basalt flows (Kolarsky et al., this volume).

Bedding in sedimentary and basement rocks

Bedding of sedimentary rocks north of the Azuero-Soná fault zone reflects a northeast-trending girdle consistent with small-scale post-Eocene northwest-trending fold axes observed in outcrop (Fig. 19). Most of these data were collected along road sections and from coastal exposures in the area of the Tonosí basin. Bedding south of the Azuero-Soná fault zone that was collected mainly along the south coast suggests a subhorizontal sedimentary cover basement. Attitudes of layered basement are scattered and appear to have no regional significance.

Outcrop-scale faulting

No regional-scale faults were photographed in coastal exposures around the Azuero Peninsula. Orientations of 90 outcrop-scale faults were measured in basement and sedimentary cover rocks along coastal sections. Faults were usually exposed on the intertidal flat or in the cutbank at the landward edge of the intertidal flat of the Azuero Peninsula and were found to be similar in orientation and slip to conjugate fault sets on Coiba Island (Fig. 17). Seventy of these faults lack fault-plane striations or other indicators of fault sense and are subdivided according to basement or cover occurrence in Figure 17.

Locality 6. At four localities on Azuero Peninsula, we measured the attitudes of sets of conjugate faults with well-developed fault-plane striations that constrain the sense of slip. Conjugate faults affecting rocks of the middle Eocene to lower Miocene Tonosí Formation from locality 6 at Punta Blanca on the southern coast of the peninsula include steeply dipping faults that strike north and are right-lateral and steeply dipping faults that strike west and are left-lateral. This fault geometry indicates an approximate northwest-southeast orientation of post–early Miocene maximum extension.

Locality 7. Conjugate faults affecting rocks of the middle Eocene to lower Miocene Tonosí Formation from locality 7 on the southeastern coast of the peninsula at Punta Bucaró include steeply dipping faults that strike north-northeast and are

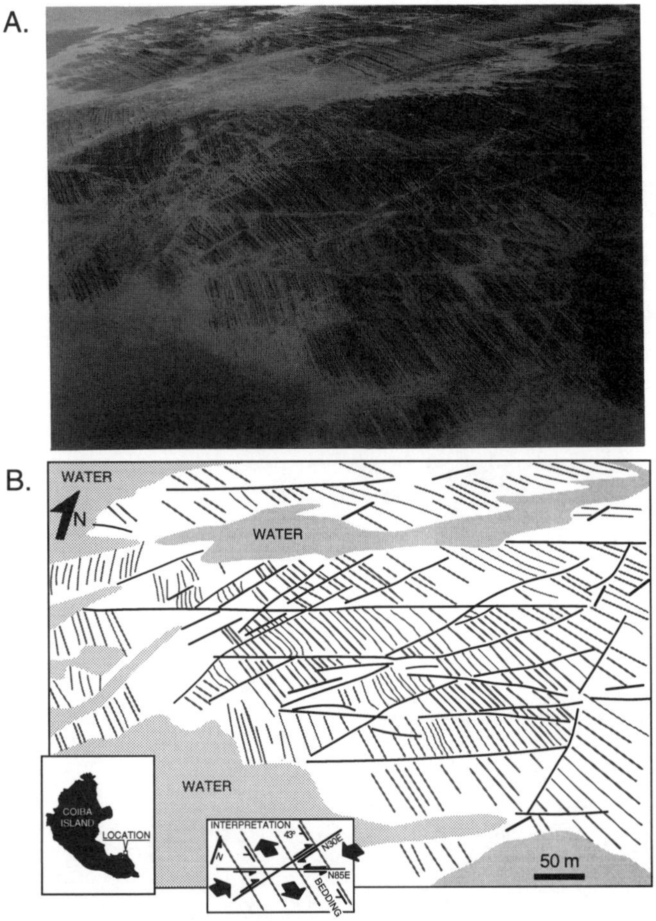

Figure 15. A, Oblique aerial photograph (taken at an altitude of 150 m from a private plane) of a coastal exposure of thin-bedded south-dipping Eocene–lower Miocene sandstone and shale of the Tonosí Formation on the southeast coast of Coiba Island. B, Location, line drawing, and interpretation of the outcrop shown in 15A. The conjugate fault system is interpreted as post–Tonosí Formation (i.e., post–early Miocene) conjugate strike-slip faults accommodating northwest-southeast extension of the Cébaco basin complex. Interpretation of slip-senses along photographed strike-slip faults is based on ground studies of bedding offsets along nearby outcrop-scale faults. Note that inset "Interpretation" is drawn in 3-D perspective, and therefore angle between faults appears smaller than 55°.

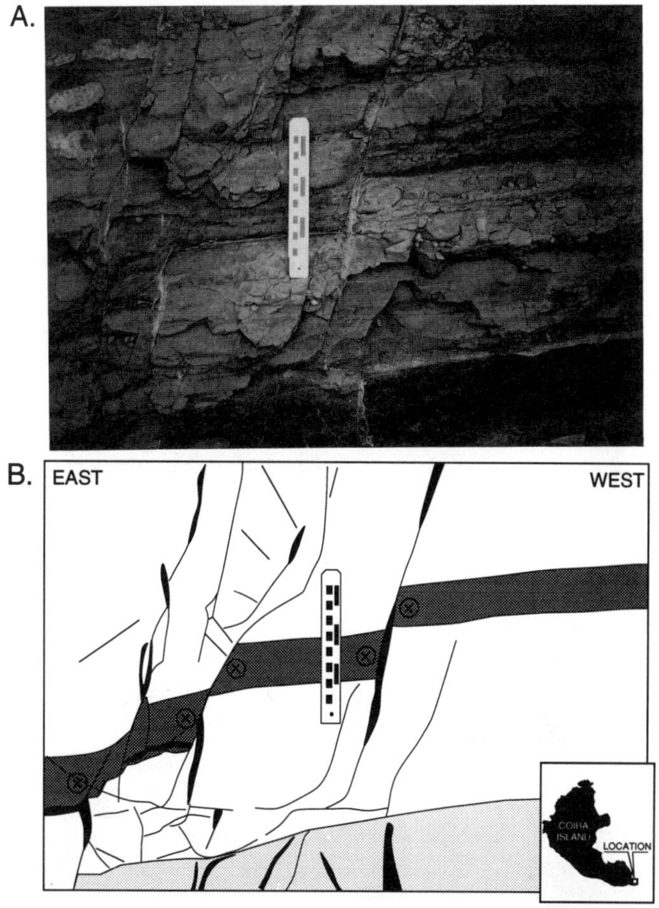

Figure 16. A, Photograph of east-dipping normal faults developed in Eocene–lower Miocene turbidite sandstone and siltstone of the Tonosí Formation at the southeast coast of Coiba Island (see inset for location). B, Line drawing of outcrop. Scale is 10 cm; white represents siltstone beds; light and dark gray represents sandstone beds. X's mark originally continuous points along the upper sandstone bed. Note calcite veining along fault planes (shown in black) indicating extensional nature of fault displacement and listric nature of fault in slightly siltier horizon.

right-lateral and steeply dipping faults that strike northeast and are left-lateral. This fault geometry indicates a northwest-southeast orientation post–early Miocene maximum extension.

Localities 5, 8, and 9. Conjugate faults affecting rocks of the middle Eocene to lower Miocene Tonosí Formation from locality 8 on the northwestern coast of the peninsula at Punta Malena include moderately eastward-dipping faults that strike north and are right-lateral and moderately northward-dipping faults that strike west and are left-lateral. This fault geometry indicates an approximate northwest-southeast orientation of post–early Miocene maximum extension. Conjugate faults affecting rocks of the middle Eocene to lower Miocene Tonosí Formation from locality 9 on the northwestern coast of the peninsula at Punta Malena include moderately dipping faults that strike north and are right-lateral and moderately dipping faults that strike west and are left-lateral. This fault geometry indicates an approximate northwest-southeast orientation of post–early Miocene maximum extension. Conjugate faults affecting rocks of the Tonosí Formation south of the Azuero-Soná fault zone also display a geometry indicative of approximately northwest-southeast extension.

STRUCTURES ON THE SONÁ PENINSULA

Regional significance

The Soná Peninsula lies to the north of the Texaco and Shell offshore seismic grid and exposes a highly embayed coastline suggestive of recent coastal drowning (Fig. 5). Recent

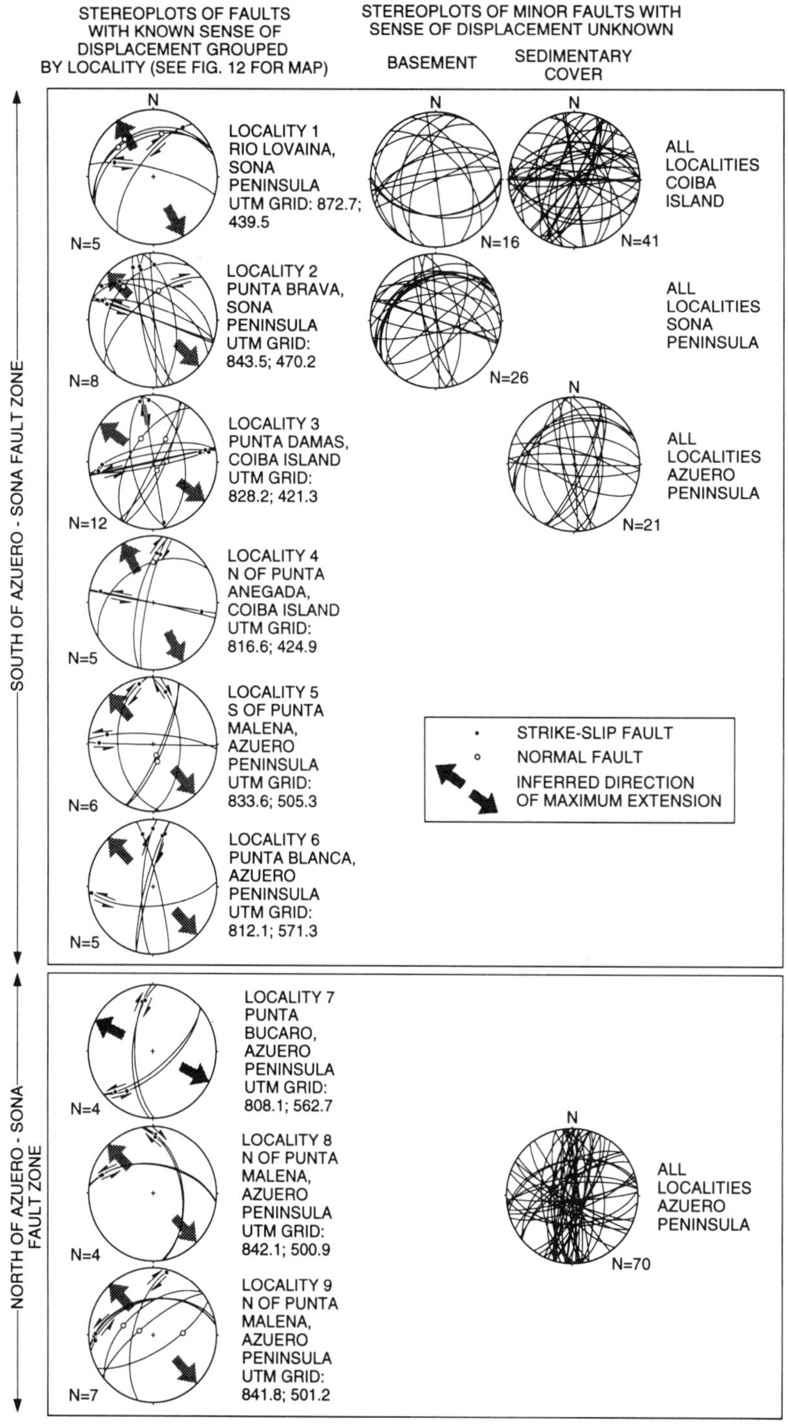

Figure 17. Summary of outcrop-scale fault data from the Soná Peninsula (localities 1 and 2), Coiba Island (localities 3 and 4), and the Azuero Peninsula (localities 5, 6, 7, 8, and 9) (see Fig. 12 for all locations of all sites). UTM grid coordinates provided next to each locality number allow precise relocation by future workers of each outcrop using 1:50,000 topographic maps. Faults from the nine localities are collected from mainly sedimentary rocks overlying basaltic basement and constrain directions of maximum extension that are indicated by the gray arrows and directions in upper left corner. Nine localities and faults with unknown sense of slip are grouped according to their location north or south of the Azuero-Soná fault zone.

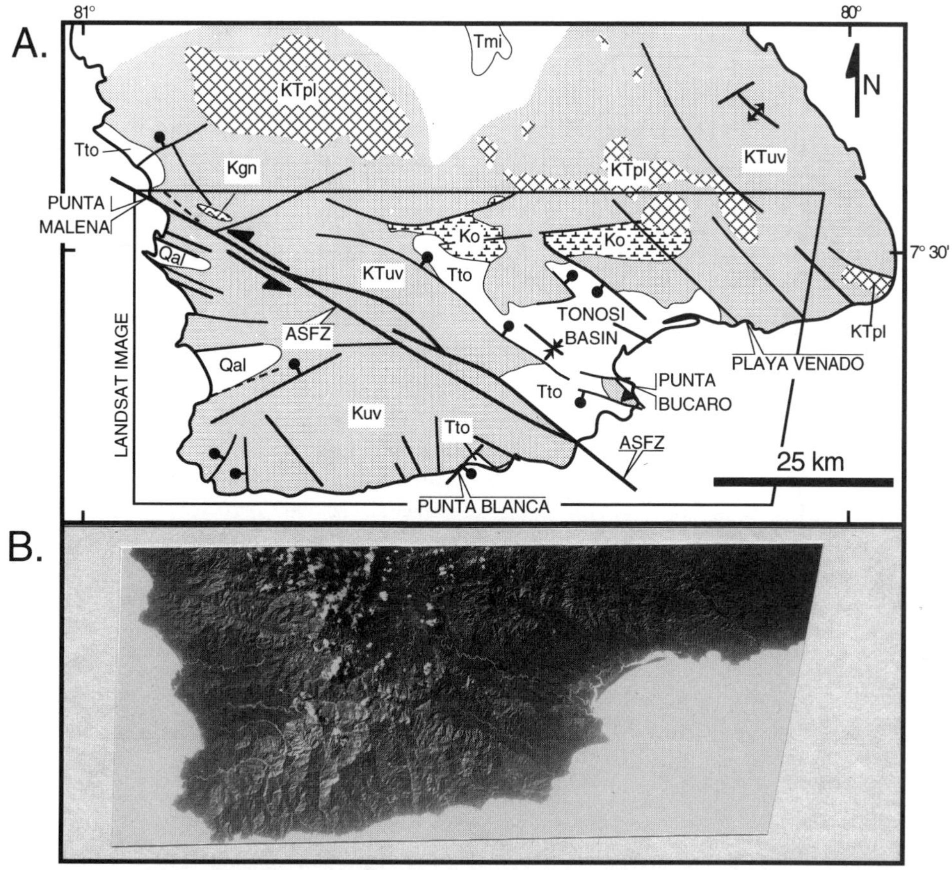

Figure 18. A, Simplified tectonic-geologic map of the Azuero peninsula showing mainly basaltic basement south of the Azuero-Soná fault zone (Kuv); mainly volcanic basement (KTuv), plutonic intrusions (KTpl), metamorphic rocks (Kgn) and Ocú Formation (Ko) north of the Azuero-Soná fault zone (ASFZ). Eocene–lower Miocene sedimentary rocks of the Tonosí Formation (Tto) overlie contrasting basement types on both sides of the Azuero-Soná fault zone. Areas of Miocene clastic rocks (Tmi) and Quaternary alluvium (Qal) are also shown. B, Band 7 (near-infrared) LANDSAT MSS image of the boxed area in Figure 18A.

drowning is consistent with the position of this coastline adjacent to the northern edge of the Cébaco basin complex and the eastern flank of the Ranchería arch. It appears that the westernmost normal fault of the Cébaco basin complex extends onshore the Soná Peninsula near our fault locality 2 (Fig. 12).

General structure of Soná Peninsula

Although poorly mapped, the northwest-trending Azuero-Soná fault zone appears to divide the Soná Peninsula into two basement terranes of differing composition in the same manner as discussed for the Azuero Peninsula (Fig. 11). Basement rocks southwest of the Azuero-Soná fault zone are similar to those on Coiba Island, whereas those northeast of the fault are more characteristic of an island-arc terrane. Rugged coastlines expose mainly basaltic basement rocks except for small areas of Tonosí Formation reported by del Giudice and Recchi (1969) along the southwestern coast.

Access

The attitudes of minor faults and layering in basement were mapped in coastal exposures along the southeastern coast and valley of the Azuero-Soná fault zone. We did not visit exposures of sedimentary rocks reported along the southwestern coast because of difficult access by boat or car (Fig. 18).

Exposures

As on Coiba Island and the Azuero Peninsula, the large tidal range in the Pacific Ocean reveals semicontinous, wave-cut exposures of both basement rocks in almost all areas except near alluvial valleys. Layering in basement rocks was defined using sheetlike, pillowed and nonpillowed basalt flows (Kolarsky et al., this volume). No sedimentary rocks were found along the southeast coast.

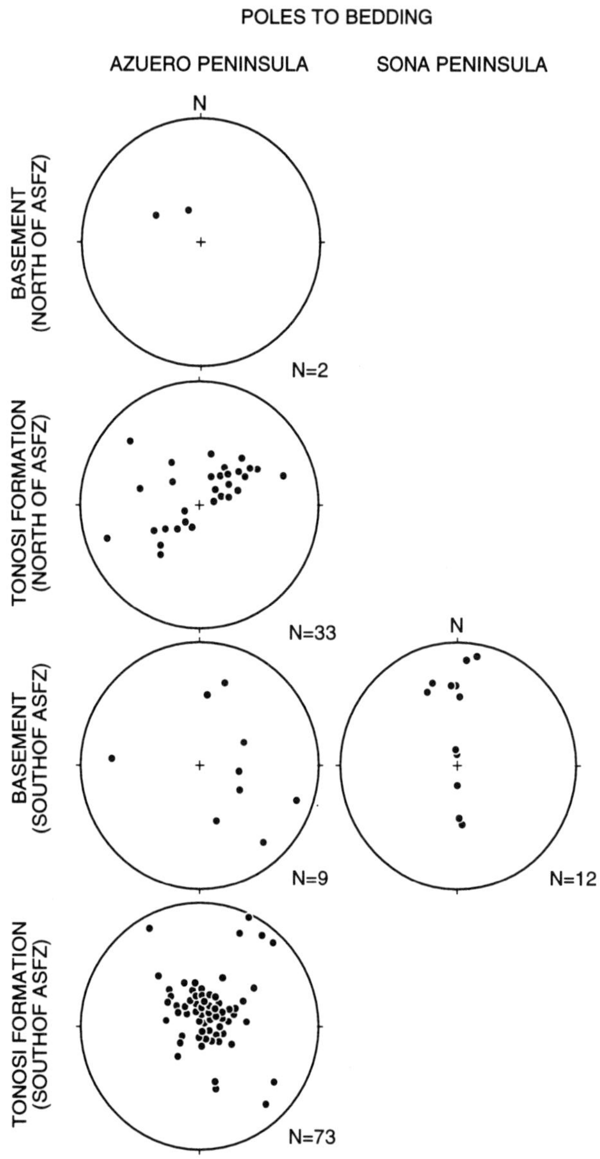

Figure 19. Lower hemisphere stereoplots of poles to bedding from Azuero and Soná Peninsulas. Data are grouped according to geologic units (basaltic basement or sedimentary cover) and according to geographic location north or south of the Azuero-Soná fault zone (ASFZ) (see Fig. 12 for fault location and distribution of basement and cover rocks).

Bedding in basement rocks

Bedding of basement rocks south of the Azuero-Soná fault one suggests folding about an east-trending axis (Fig. 19).

Outcrop-scale faulting

No regional-scale faults were photographed in surf exposures around the Soná Peninsula. Orientations of 39 outcrop-scale faults were measured in basement and sedimentary cover rocks along the eastern coast of Soná Peninsula (Fig. 17). Faults were usually exposed on the intertidal flat or in the cutbank at the landward edge of the intertidal flat of the Soná Peninsula and were found to be similar in orientation and slip to conjugate fault sets on Coiba Island and the Azuero Peninsula. Of these faults, 26 lack fault-plane striations or other indicators of fault sense and are subdivided according to basement or cover occurrence in Figure 17.

Locality 1. At two localities on Soná Peninsula, we measured the attitudes of sets of conjugate faults with well-developed fault-plane striations that constrain the sense of slip. Conjugate faults affecting Upper Cretaceous(?) basement rocks in the western valley of the Azuero-Soná fault zone (Rio Lovaina valley) include steeply dipping faults that strike northeast and are right-lateral, steeply dipping faults that strike west and are left-lateral, and low-dipping normal faults that strike northeast. This fault geometry indicates an approximately northwest-southeast orientation of post-Cretaceous (?) maximum extension.

Locality 2. Conjugate faults affecting Upper Cretaceous basaltic basement from locality 2 on the southeastern coast of the peninsula at Punta Brava include scattered steeply dipping faults that range in strike from north to northwest. Northeast faults are right-slip, and northwest faults are left-slip. This fault geometry indicates an approximately northwest-southeast orientation of post–Late Cretaceous maximum extension.

STRUCTURAL AND TECTONIC INTERPRETATION OF SEISMIC AND OUTCROP DATA FROM THE GULF OF CHIRIQUÍ

Comparison of Coiba Island, Azuero Peninsula, and Soná Peninsula fault data with fault data from the Burica Peninsula

Regional and outcrop-scale conjugate fault patterns on tidal flats along the east coast of the Burica Peninsula (Corrigan et al., 1990) are very similar to those seen at these scales in the study area and are interpreted as reflecting a northwest-southeast orientation of maximum extension across the Gulf of Chiriquí (Fig. 12, inset). On the Burica Peninsula, Corrigan et al. (1990) mapped the orientations of approximately 70 faults using low-altitude aerial photographs and outcrop observations. The fault pattern in well-dated upper Pliocene to lower Pleistocene rocks is composed of two sets: (1) a fault set consisting of subparallel, steeply dipping (65° to 90°) faults, striking N50° ± 20°E and spaced tens of meters apart, and (2) a fault set consisting of subparallel, steeply dipping (60° to 90°) faults, striking either N80° ± 10°E or N20° ± 10°E and spaced hundreds of meters apart. The oppositely dipping faults of the first set were interpreted as normal faults, whereas the X-shaped pattern of the second fault set was interpreted as conjugate strike-slip faults. Both sets of faults were interpreted as reflecting post–early Pleistocene shortening orientation of N50°E–55°E and extension orientation of N40°W–S40°E.

Age and direction of extension in the eastern Gulf of Chiriquí

The northwest-southeast orientation of post–early Pleistocene extension from the Burica Peninsula is consistent with the orientation of maximum extension in the land areas studied around the eastern Gulf of Chiriquí. Based on this data, the calculated average orientation of maximum extension (e_1) for the onshore areas is N44°W. This orientation matches well with inferred orientation of maximum extension interpreted from the geometry of the faults bounding half-grabens in the Cébaco basin complex (Fig. 12, inset). The similarity of the Burica direction of maximum extension (N40°W–S40°E) to the average direction of extension of all nine outcrops in the eastern Gulf of Chiriquí suggests that the regional extension affecting the Gulf of Chiriquí is post–early Pleistocene. We suggest that this younger post–early Pleistocene age of inferred faulting may be a late brittle faulting event that culminates the formation and filling of the offshore half-grabens of the Cébaco basin complex thought to be of Plio-Pleistocene age.

Amount of extension in the Cébaco basin complex, eastern Gulf of Chiriquí

To estimate the amount of extension in the eastern Gulf of Chiriquí, we palinspastically restored the extensional deformation responsible for the four half-grabens seen on the eastern end of Texaco Line 1 (Fig. 20). A simplified version of the time section shown in Figure 5 is given in Fig. 20A. A depth section with vertical exaggeration was produced using seismic interval velocities calculated during reprocessing of Line 1; it is given in Fig. 20B. This depth section is shown without vertical exaggeration in Figure 20C and is palinspastically restored in Figure 20D. Restoration was carried out by removing the inferred Plio-Pleistocene sedimentary fill of the half-grabens, removing the throw on the four normal faults, and restoring the inferred horizon of Eocene–lower Miocene rocks of the Tonosí Formation to a semicontinous horizon. This restoration assumes that the throw on all the faults is entirely dip-slip.

The restoration suggests that ~7% extension in a northwest direction formed the Cébaco basin complex (Fig. 20D). Large areas up to 900 m thick were removed by erosion in the area of the Ranchería arch and the easternmost part of the basin. The large uplifted area around the Ranchería arch suggests that some of the arching may have been related to flexural effects that accompanied normal faulting in the Cébaco basin complex.

DISCUSSION

Summary of deformation in the Gulf of Chiriquí

On- and offshore data from the Gulf of Chiriquí area constrain a Plio-Pleistocene extensional phase that extended the area by about 7% and produced the Cébaco basin complex and

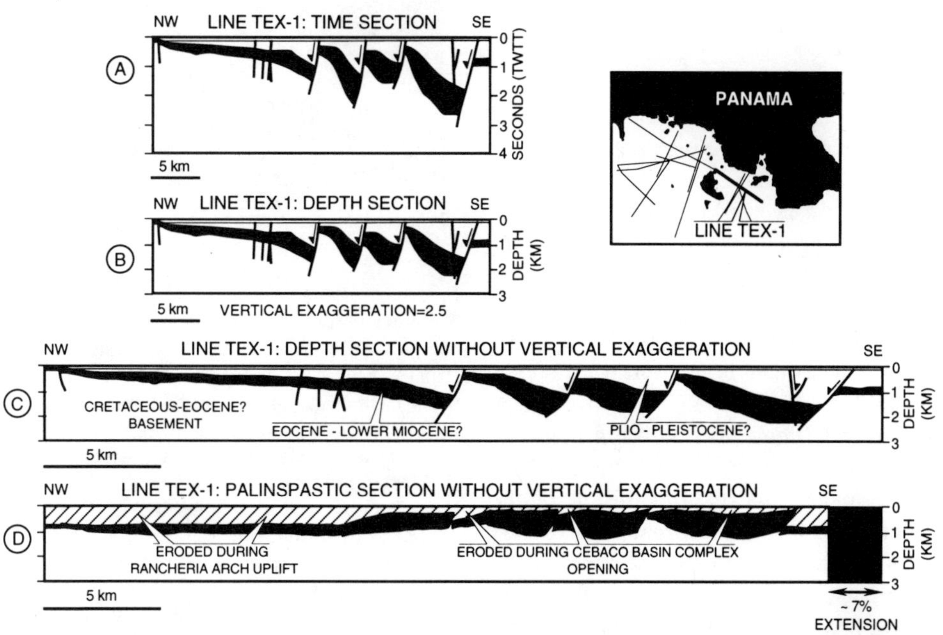

Figure 20. A, Time section of southeastern part of Texaco line 1 crossing the Cébaco basin complex (see Fig. 6 for original seismic data and interpretation). B, Depth section of the same line. C, Depth section plotted without vertical exaggeration. D, Palinspastic section without vertical exaggeration produced by removing normal fault offsets. Restoration suggests 7% extension. Note striped areas that are assumed to have been eroded off fault blocks during opening of the Cébaco basin complex and formation of the Ranchería Arch.

Montuosa basin (Fig. 5). Fault data from onshore areas suggest an approximately northwest-southeast orientation of maximum extensional strain extending from the Burica Peninsula to the eastern coast of the Azuero Peninsula (Fig. 20). Correlation of the style and orientation of brittle faulting with faults mapped previously by Corrigan et al. (1990) on the Burica Peninsula suggests that the age of faulting is post–early Pleistocene. Brittle faulting may represent the culmination of the extensional event responsible for the formation of the Plio-Pleistocene Cébaco basin complex.

Models for neotectonics of southwestern Panama

We propose three possible models to explain the observed pattern of deformation observed in the Gulf of Chiriquí (Fig. 21).

Oblique subduction of the Nazca plate beneath the southern margin of Panama. Oblique convergence between Panama and the Nazca plate in the Gulf of Chiriquí area causes left-lateral strike-slip faults with northwest-southeast–oriented extension within the forearc (MacKay and Moore, 1990) (Fig. 21A). The oblique direction of Nazca plate convergence is consistent with the approximately northwest-southeast orientation of maximum extensional strain on the land areas around the Gulf of Chiriquí (N40°W–S40°E) and the presence of large strike-slip faults like the Azuero-Soná fault zone. Oblique slip of the Nazca plate must be accompanied by sufficient underthrusting of the Nazca slab to create an overlying Miocene to Quaternary volcanic arc in western Panama (De Boer et al., 1988, 1991) and to produce arching of the forearc area landward of the subducted extensions of the Coiba Ridge (Ranchería Arch) and the Balboa Ridge (Parida Arch) (Fig. 5).

Lateral escape of the Gulf of Chiriquí block from the collision zone between the Cocos Ridge and Panama and Costa Rica. The collision of the Cocos Ridge with the Costa Rica forearc had profound effects on basement structure of the entire area of eastern Costa Rica (Kolarsky et al., this volume). One effect may have been the decoupling of the entire Gulf of Chiriquí block from the rest of the Panama landmass along the Azuero-Soná fault zone (Fig. 21B). During southeastward extrusion of the Gulf of Chiriquí block, it is possible that the block experienced internal deformation and northwest-southeast–oriented extensional strain.

Bending of the Panama island arc following collision with the South American continent. Wadge and Burke (1983) and Silver et al. (1990) suggested that the collision of the Panama island arc with the South American continent resulted in S-shaped bending of the central and eastern parts of the isthmus. Mann and Corrigan (1990) pointed out that the bending process may not have been entirely ductile because of the presence of northwest-striking left-lateral strike-slip faults that segment the isthmus. Left-lateral faulting and accompanying pull-apart basin formation in the Cébaco basin complex may accommodate the bending of the Panama orocline. Geologic data from Panama and northwestern South America suggest

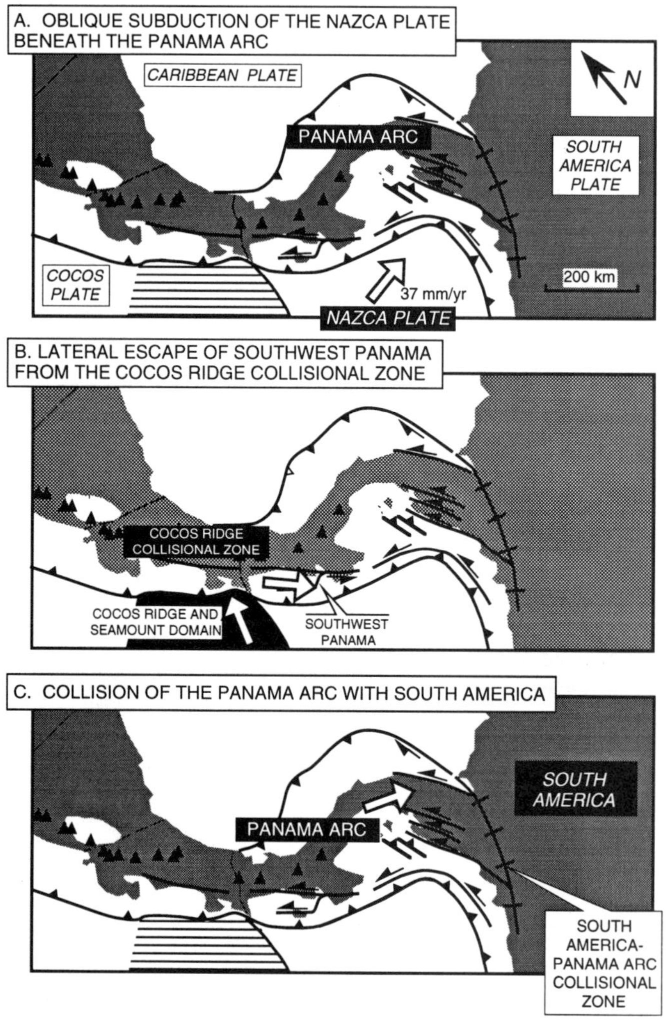

Figure 21. Three mutually compatible tectonic driving forces for Neogene deformation of the Isthmus of Panama and its offshore basins. A, Oblique subduction of Nazca plate beneath southwestern Panama causes decoupling of the Gulf of Chiriquí forearc area from the volcanic arc and its displacement along the Azuero-Soná fault zone. Part of the oblique subduction is accommodated by trench-parallel strike-slip faults like the Azuero-Soná fault zone (ASFZ). B, Margin indentation and convergence of the Cocos Ridge and seamount domain induce southeastward displacement or "extrusion" of the Gulf of Chiriquí forearc area of southwest Panama relative to the Cocos Ridge "indentor." C, Oroclinal bending of the Panama Isthmus is induced by the collision of an east-west–trending Panama arc with the South America plate. Bending is accommodated in part by left-lateral strike-slip fault formation throughout Panama.

that this bending event initiated in the Late Miocene and may continue to the present day (Mann and Corrigan, 1990).

Data from this study cannot distinguish between these three models. Because these models are not mutually exclusive, we propose that the observed deformation may be a composite effect of more than one of these tectonic mechanisms.

ACKNOWLEDGMENTS

This study would not have been possible without geophysical data kindly provided by Q. Nary of Texaco and A. J. Bol of Shell International. We also thank R. Miranda, A. Burgos, and L. Campos of the Dirección General de Recursos Minerales for their logistic support and field assistance support in Panama and N. Villareal of the Dirección General de Hidrocarburos for providing unpublished well data. We also thank the Panama Ministerio de Gobierno y Justicia and the Panama Air Service for permission to visit Coiba Island. Special thanks to the staff and prisoners on Coiba for their hospitality and to M. Mohageri for providing a low-altitude overflight of the island. Reviews by J. de Boer, M. Gordon, K. McIntosh, G. Moore, and R. Stewart significantly improved this chapter. Funding for this project was provided by the Petroleum Research Fund of the American Chemical Society (grant number 20676-AC2-2). UTIG contribution no. 1064.

REFERENCES CITED

Adamek, S., and Tajima, F., 1987, Seismic rupture associated with subduction of the Cocos Ridge: Tectonics, v. 6, p. 757–774.

Adamek, S., Frohlich, C., and Pennington, W. D., 1988, Seismicity of the Caribbean-Nazca microplate boundary: Constraints on microplate tectonics of the Panama region: Journal of Geophysical Research, v. 93, p. 2053–2075.

Bourgois, Jacques, and 10 others, 1982, Ages et structures des complexes basiques et ultrabasiques de la façade pacifique entre 3°N et 12°N (Colombie, Panama et Costa Rica): Bulletin Societe Géologique France, v. 24, p. 545–554.

Bowland, C. L., 1984, Seismic stratigraphy and structure of the western Colombian basin, Caribbean Sea [M.A. thesis]: Austin, University of Texas, 238 p.

Bowland, C. L., and Rosencrantz, E., 1988, Upper crustal structure of the western Colombian basin, Caribbean sea: Geological Society of America Bulletin, v. 100, p. 534–546.

Corrigan, J. D., Mann, P., and Ingle, J. C., Jr., 1990, Forearc response to subduction of the Cocos ridge, Panama–Costa Rica: Geological Society of America Bulletin, v. 102, p. 628–652.

de Boer, J. Z., Defant, M. J., Stewart, R. H., Restrepo, J. F., Clark, L. F., and Ramirez, A. H., 1988, Quaternary calc-alkaline volcanism in western Panama: Regional variation and implications for the plate tectonic framework: Journal of South American Earth Sciences, v. 1, p. 275–293.

de Boer, J. Z., Defant, M. J., Stewart, R. H., and Bellon, H., 1991, Evidence for active subduction below western Panama: Geology, v. 19, p. 649–652.

del Giudice, D., and Recchi, G., 1969, Geologia del area del proyecto minero de Azuero: Administracion de Recursos Minerales, Republica de Panama, Map 5, scale 1:250,000.

De Mets, C., Gordon, R. G., Argus, D. F., and Stein, S., 1990, Current plate motions: Geophysical Journal International, v. 101, p. 425–478.

Gursky, Hans-Jürgen, Schmidt-Effing, R., Strebin, M., and Wildberg, H., 1982, The ophiolite sequence in northwestern Costa Rica (Nicoya complex): Outline of stratigraphical, geochemical, sedimentological and tectonic data: Quinto Congreso Latinoamericano de Geología, v. 3, p. 607–619.

Hardy, N. C., 1991, Tectonic evolution of the easternmost Panama basin: Some new data and inferences: Journal of South American Earth Sciences, v. 4, p. 261–269.

Hardy, N. C., Heath, R. P., and Westbrook, G. K., 1990, A complex plate boundary south of the Gulf of Panama: Eos (Transactions of the American Geophysical Union), v. 71, p.1593.

Heil, Darla J., 1988, Response of an accretionary prism to transform ridge collision south of Panama [M.S. thesis]: Santa Cruz, University of California, 88 p.

Hobbs, B. E., Means, W. D., and Williams, P. F., 1976, An outline of structural geology: New York, John Wiley and Sons, 571 p.

Jordan, T. H., 1975, The present-day motions of the Caribbean plate: Journal of Geophysical Research, v. 80, p. 4433–4449.

Kolarsky, R. A., 1992, Late Cretaceous–Cenozoic tectonics and sedimentation in southern Central America (Costa Rica and Panama) [M.A. thesis]: Austin, University of Texas, 351 p.

Lonsdale, P., and Klitgord, K. D., 1978, Structure and tectonic history of the eastern Panama basin: Geological Society of America Bulletin, v. 89, p. 981–999

Lowrie, A., Stewart, J., Stewart, R. H., Van Andel, R. J., and McRaney, L., 1982, Location of the eastern boundary of the Cocos plate during the Miocene: Marine Geology, v. 45, p. 261–279.

MacKay, M. E., 1988, Accretionary response to trench sediment variation along the south Panama margin [M.S. thesis]: Tulsa, Oklahoma, The University of Tulsa, 44 p.

MacKay, M. E., and Moore, G. F., 1990, Variation in deformation of the South Panama accretionary prism: Response to oblique subduction and trench sediment response: Tectonics, v. 9, p. 683–698.

Mann, P., and Corrigan, J. D., 1990, Model for late Neogene deformation in Panama: Geology, v. 18, p. 558–562.

Muñoz, A., 1988, Tectonic patterns of the Panama block deduced from seismicity, gravitational data, and earthquake mechanisms: Implications to the seismic hazard: Tectonophysics, v. 154, p. 253–267.

Obando, J. A., 1986, Sedimentología y tectónica del Cretácico y Paleógeno de la region de Golfito, Península de Burica y Península de Osa, Provincia de Puntarenas, Costa Rica [B.A. thesis]: San José, Costa Rica, University of Costa Rica, 211 p.

Okaya, D. A., and Ben-Avraham, Z., 1987, Structure of the continental margin of southwestern Panama: Geological Society of America Bulletin, v. 99, p. 792–802.

Recchi, G., and Miranda, R., 1977, Calizas de los planes—Guaniquito (Tonosí): Panama City, Panama, Dirección General de Recursos Minerales (unpublished report) 27 p.

Ross, D. A., and Shor, G. G., Jr., 1965, Reflection profiles across the Middle America Trench: Journal of Geophysical Research, v. 70, p. 5551–5572.

Silver, E., Reed, D. L., Tagudin, J. E., and Heil, D. J., 1990, Implications of the North and South Panama thrust belts for the origin of the Panama orocline: Tectonics, v. 9, p. 261–281.

Tournon, J., Triboulet, C., and Azema, J., 1989, Amphibolites from Panama: Anticlockwise P-T paths from a pre–Upper Cretaceous basement in the isthmian Central America: Journal of Metamorphic Petrology, v. 7, p. 539–546.

Wadge, G., and Burke, K., 1983, Neogene Caribbean plate rotation and associated Central American tectonic evolution: Tectonics, v. 2, p. 633–643.

MANUSCRIPT ACCEPTED BY THE SOCIETY FEBRUARY 7, 1994

Printed in U.S.A.

Geological Society of America
Special Paper 295
1995

Stratigraphic development of southwestern Panama as determined from integration of marine seismic data and onshore geology

Radim A. Kolarsky*
Department of Geological Sciences, University of Texas at Austin, Austin, Texas 78712
Paul Mann
Institute for Geophysics, University of Texas at Austin, 8701 Mopac Boulevard, Austin, Texas 78759
Simonetta Monechi
Dipartimento de Scienze della Terra, Università degli Studi di Firenze, Via La Pira 4, 50121 Firenze, Italy

Appendix 1 by:
Radim A. Kolarsky and Paul Mann

Appendix 2 by:
Donna Meyerhoff-Hull and Emile A. Pessagno, Jr.
Programs in Geosciences, University of Texas at Dallas, P.O. Box 830688, Richardson, Texas 75083-0688

Appendix 3 by:
Simonetta Monechi

ABSTRACT

We correlate seismic map units identified on industry seismic lines in the Gulf of Chiriquí, southwestern Panama, with onland igneous rocks and sedimentary formations described in this chapter. We propose six principal stages in the stratigraphic development of southwestern Panama based on our results and the results of previous workers in Costa Rica, westernmost Panama, and the western Colombian basin. The first stage in southwestern Panama is represented by basaltic basement rocks of Jurassic?–Late Cretaceous age interbedded with Upper Cretaceous pelagic sedimentary rocks. Following previous workers and data presented here, we suggest that these rocks formed in an intraoceanic, oceanic plateau setting. A second stage is represented by a major stratigraphic hiatus inferred to represent an erosional event that affected the basaltic basement of Panama in Paleocene time. A third stage is represented by a widespread basal transgressive section of coarse clastic rocks and reefal carbonate rocks of early to middle Eocene age. This section records initiation of clastic sedimentation over much of southern Central America. A fourth stage is represented by a thick section of mainly marine turbidites that appears to represent continued erosion of the land areas in southern Central America and upward deepening of adjacent marine environments. A fifth stage is repre-

*Present address: Eastern E&P Region, Texaco USA, P.O. Box 60252, New Orleans, Louisiana 70160.

Kolarsky, R. A., Mann, P., and Monechi, S., 1995, Stratigraphic development of southwestern Panama as determined from integration of marine seismic data and onshore geology, *in* Mann, P., ed., Geologic and Tectonic Development of the Caribbean Plate Boundary in Southern Central America: Boulder, Colorado, Geological Society of America Special Paper 295.

sented by a stratigraphic hiatus in middle Miocene to late Miocene time that may represent the "breakup" unconformity associated with initiation of strike-slip faulting and rifting in the Gulf of Chiriquí. A sixth stage is represented by early Pliocene to Pleistocene rifting and syn-rift sedimentation in the Gulf of Chiriquí. Thick sedimentary fill of rift basins may reflect accelerated uplift of southern Central America and increased activity of the Middle America arc. The regional extent of the stratigraphic record of several of these stages across large areas of southern Central America and the western Colombian basin supports the previously proposed hypothesis that the crust of southern Central America represents the western upturned edge of a Late Cretaceous Caribbean oceanic plateau known from deep-sea drilling and seismic stratigraphic studies in the Colombian and Venezuelan basins of the Caribbean Sea.

INTRODUCTION

The purpose of this chapter is to clarify the stratigraphic and tectonic development of the southwestern margin of Panama by integrating onshore stratigraphic information with offshore geophysical data supplied by industry sources (Fig. 1). The neotectonic setting and significance of deformation in the Gulf of Chiriquí, surrounding land areas, and the South Panama deformed belt is discussed by Kolarsky and Mann (this volume) and Moore and Sender (this volume).

Stratigraphic and tectonic questions we address in this study are:

(1) How well can the Late Cretaceous–Cenozoic onshore stratigraphy of the landmasses surrounding the Gulf of Chiriquí be correlated with seismic map units seen on offshore seismic profiles?

(2) What are the major Late Cretaceous–Cenozoic eustatic-tectonic episodes affecting this margin of Panama, and are these episodes observed in other parts of southern Central America and the Colombian basin?

(3) Can stratigraphic and tectonic events on the Pacific margin of southwestern Panama be correlated with events described by Bowland and Rosencrantz (1988) and Bowland (1993) from the northeastern (Caribbean) margin of southern Central America in the western Colombian basin?

(4) Can the stratigraphy of southwestern Panama help distinguish between contrasting models for the origin of the crust beneath southern Central America: (a) The crust represents the western upturned edge of Jurassic–Early Cretaceous? oceanic crust and overlying Late Cretaceous oceanic plateau crust known from deep sea drilling and seismic stratigraphic studies in the Colombian and Venezuelan basins (Bowland and Rosencrantz, 1988; Bowland, 1993). (b) The crust represents a collage of fragments of seamounts and other features accreted to the west- and southwest-facing Panama island arc during Late Cretaceous and early Tertiary time (DiMarco et al., this volume).

Because the onshore and marine geology of southwestern Panama has been studied at only a reconnaissance scale, to answer these fundamental questions has required us to collect, interpret, and summarize a large amount of geophysical and geologic information. We present three appendices of biostratigraphic information to better constrain the age of several key stratigraphic units.

PREVIOUS WORK

Onshore mapping

Terry (1956) and a team of geologists from the United Nations Development Programme and the Dirección General de Recursos Minerales (del Giuduce and Recchi, 1969; Metti et al., 1972; Metti and Recchi, 1976; Recchi and Miranda, 1977) carried out early geologic field mapping in the onshore areas of southwestern Panama. These studies produced reasonably detailed, colored geologic maps of the area at scales of 1:250,000, established stratigraphic nomenclature for individual areas, placed micropaleontologic age constraints on the major rock units, and attempted general stratigraphic correlations with nearby areas in Panama and Costa Rica. More recent geologic field studies include those by Corrigan et al. (1990) on the geology of the Burica Peninsula of Panama and Costa Rica, de Boer et al. (1988) and de Boer et al. (1991) on the Quaternary island arc of central and western Panama, and Coates et al. (1992) on Neogene stratigraphy in western Panama and eastern Costa Rica.

Between the 1950s and early 1970s southwestern Panama became the site of sporadic exploration by several oil companies, including Sinclair, Texaco, Mobil, Shell International, and Ball Associates. Extensive areas of the Azuero Peninsula were mapped on a reconnaissance level to site several wildcat wells drilled onland near the northwestern margin of the Gulf of Chiriquí and offshore in the Gulf of Panama to the east (see Kolarsky and Mann, this volume) (Fig. 1B). All wells were dry holes, and exploration was abandoned by the mid-1970s.

Previous marine studies

Terry (1941) made early bathymetric observations along the southwestern margin of Panama and mapped several large submarine canyons east of the Burica Peninsula (Fig. 1). Between the mid-1960s and late 1980s, several academic and

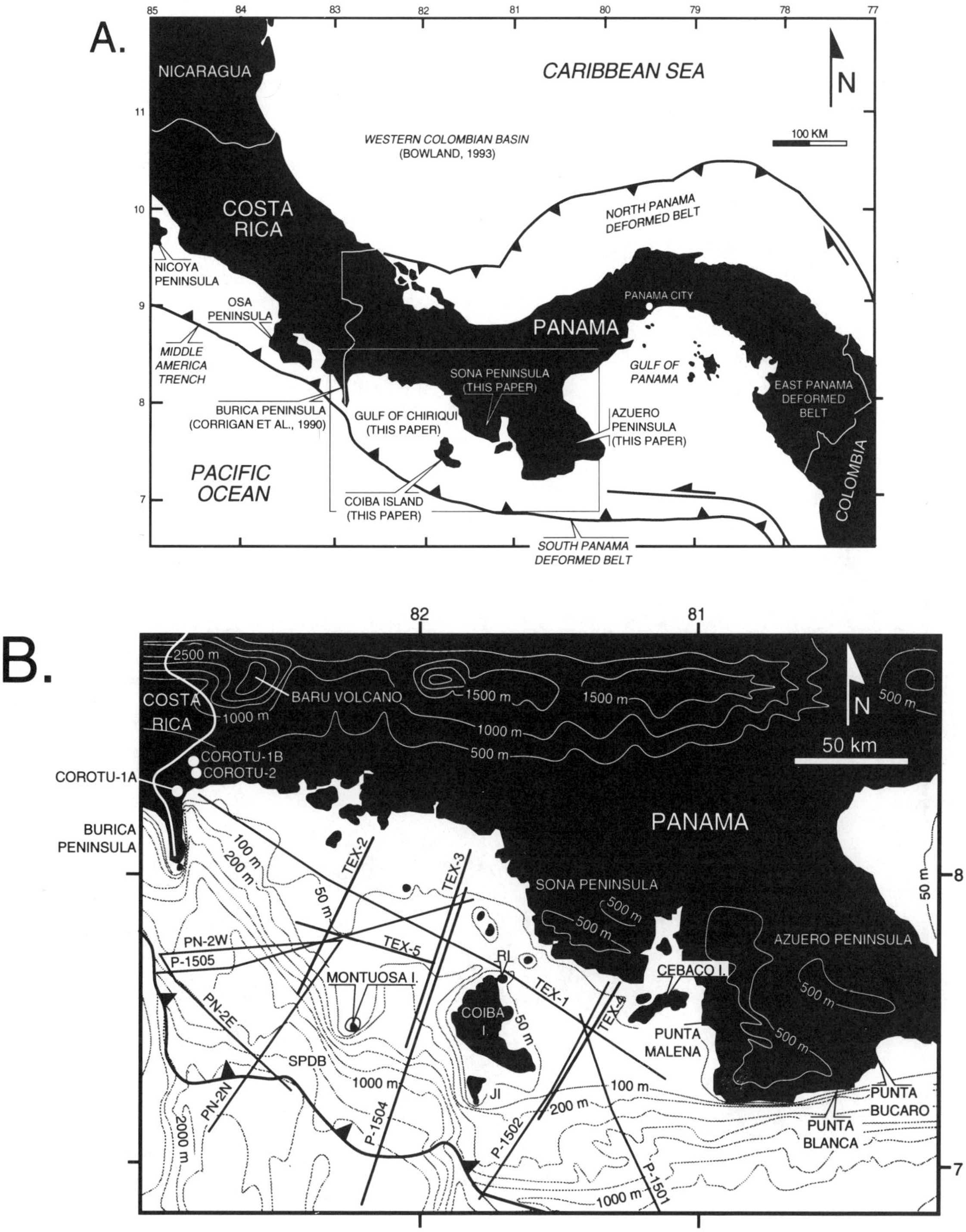

Figure 1. A, Map of Panama–Costa Rica region showing location of Gulf of Chiriquí study area and adjacent study areas of Corrigan et al. (1990) and Bowland (1993). B, Bathymetric and topographic map of the Gulf of Chiriquí area. Bathymetric contours are taken from U.S. Navy bathymetric charts. Lines and numbers give locations and line designations of Texaco (TEX), Shell (P), and University of Texas Institute for Geophysics (PN) seismic lines used in this study. Well logs from Corotú 1-A, Corotú 1-B, and Corotú-2 include wirelines and lithologic logs provided by the Department of Natural Resources, Ministry of Public Affairs, in Panama City.

private company surveys were conducted in the offshore area of southern Panama to provide constraints on the tectonic history of the area (Fig. 1B). These studies were mainly focused on mapping offshore structures but made little progress in understanding the relation between marine geology and the rocks exposed in land areas of southwestern Panama. Ross and Shor (1965) published a single-channel seismic profile normal to the outer shelf and slope of southwestern Panama. Heil (1988) and Silver et al. (1990) presented results of a SeaMARC II/single-channel seismic study of the deeper-water trench and accretionary prism of the South Panama deformed belt south of the Burica Peninsula and Coiba Island, and MacKay (1988) and MacKay and Moore (1990) described the structure of offshore south Azuero Peninsula using multichannel seismic lines (Fig. 1B). The results of Heil and MacKay from the South Panama deformed belt are reviewed by Silver et al. (1990) and Moore and Sender (this volume).

In addition to these studies by academic groups, approximately 3,100 km of multichannel seismic reflection data were collected in the early 1970s along the southern margin of Panama by Texas, Shell International, and the University of Texas Marine Science Institute (now Institute for Geophysics) under contract to a consortium of industry sponsors. Okaya and Ben-Avraham (1987) interpreted one of the Texas multichannel seismic profiles from the shelf of the Gulf of Chiriquí (line Tex-3) (Fig. 1B). Results from this same Texaco data set (lines Tex-1, 2, 3, and 4) are included in this chapter and in the chapter by Kolarsky and Mann (this volume). With the exception of the paper by Okaya and Ben-Avraham (1987), the results of marine surveys of the southwestern Panama margin have not been integrated with the results of onshore stratigraphic studies. Part of this shortcoming reflects the lack of progress made in onshore mapping since the early studies of Terry (1956) and the United Nations mapping team in the 1960s and 1970s.

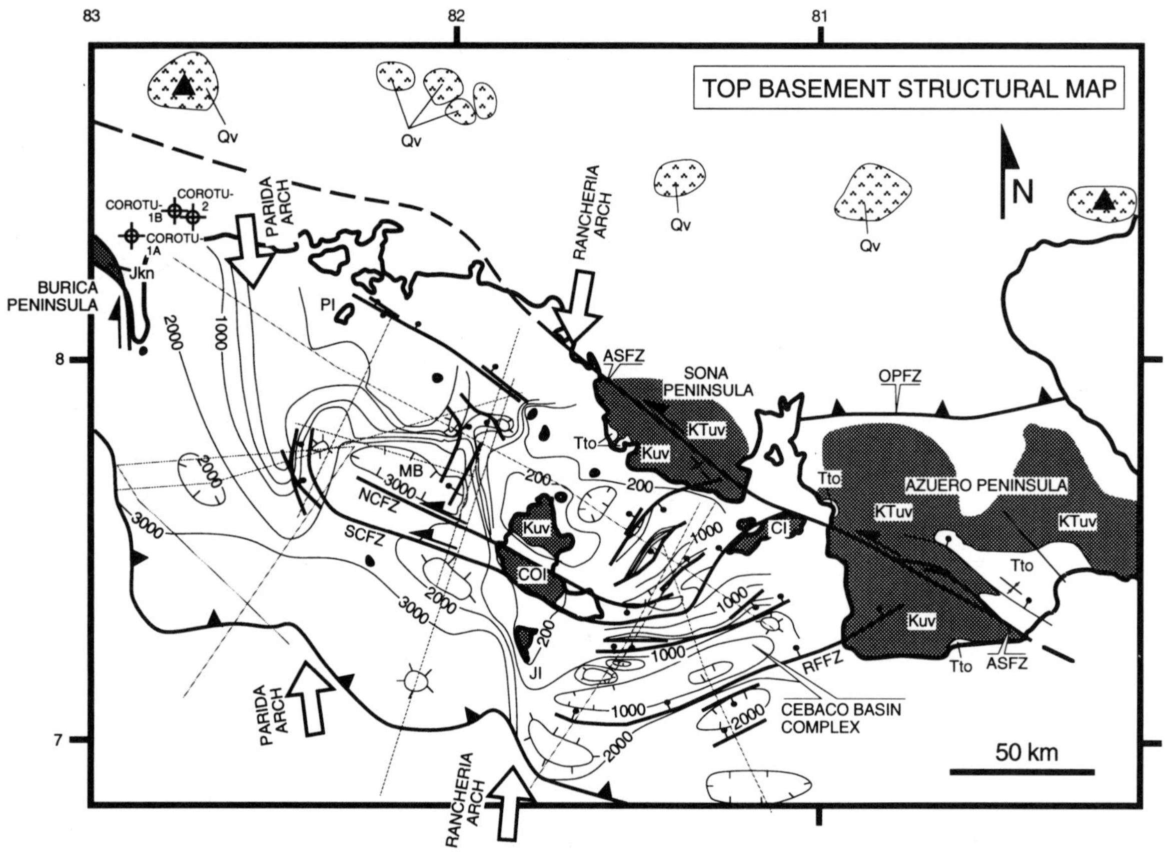

Figure 2. Structural map showing depth in meters to top surface of acoustic basement in the Gulf of Chiriquí based on seismic grid shown in dotted lines. Contour interval is 200 m for values less than 1,000 m. Depths are based on seismic interval velocities from seismic data. Onshore basement exposures correlated with acoustic basement are shaded: Jkn = Jurassic-Cretaceous Nicoya Complex; Kuv = Cretaceous undifferentiated volcanic rocks; KTuv = Cretaceous-Tertiary undifferentiated volcanic rocks; Tto = Eocene-Miocene Tonosí Formation; Qv = Quaternary volcanic rocks (black triangles indicate late Quaternary stratovolcanoes); NCFZ = North Coiba fault zone; SCFZ = South Coiba fault zone; RFFZ = Rio Flores fault zone; OPFZ = Ocú-Parita fault zone; arrows indicate Parida and Rancheria arches; PI = Parida Island; JI = Jicarón Island; MB = Montuosa basin.

DATA USED IN THIS CHAPTER

Onshore data

Onshore data sources for this chapter include compilation of existing published data mentioned in the previous section and original onland field data collected during field seasons in 1987 (Mann and Corrigan), 1988 (Corrigan), and 1991 (Kolarsky and Mann) (Fig. 1).

Marine data

The locations of marine seismic lines interpreted in this study are plotted on Figure 2. These lines include a total of approximately 3,100 km of multichannel seismic lines shot by Mobil in the Gulf of Panama and by Shell International Petroleum in the Gulf of Chiriquí. To our knowledge, this data set represents all marine seismic data that has been collected in the Gulf of Chiriquí. In addition, the Dirección General de Hidrocarburos provided us with reports from three onshore coastal wells (Corotú 1A, 1B, and 2). Corotú 1A is reported and interpreted by Corrigan et al. (1990). The other wells are described by Kolarsky (1992) but are shallow (<1,000 m) and poorly documented holes that are not used in this study.

METHODS

The offshore seismic data are described first and provide a regional framework for interpreting the onshore geologic map data from the Azuero and Soná Peninsulas and Coiba and Cébaco Islands. Using these seismic lines we recognize acoustic basement and two major seismic stratigraphic map units beneath the Gulf of Chiriquí that can be correlated with onshore sedimentary formations cropping out in the surrounding land areas. We collected numerous samples from onland outcrops for micropaleontological studies to improve our onshore stratigraphic control of the age of offshore units and to better constrain the timing of regional eustatic-tectonic events (Appendixes 1, 2, 3). Well-dated onland outcrops compensate in some respects for the total lack of offshore well control in the Gulf of Chiriquí.

A structural depth-to-basement map was prepared by a depth conversion (using seismic interval velocities) of an original time-map constructed from the available seismic data (Fig. 2). Isopach maps were made in a similar manner for both seismic map units GC1 and GC2. Although unit GC1 was mapped as a single unit, we distinguish two subunits on some individual profiles. Maps of the thickness of units GC1 and GC2 allow improved stratigraphic correlations to be made between onshore rock units and offshore seismic map units.

ACOUSTIC BASEMENT BENEATH THE GULF OF CHIRIQUÍ

The seismic expression of acoustic basement beneath the Gulf of Chiriquí consists of chaotic, discontinuous, and variable-amplitude reflections overprinted by numerous multiple reflections (Fig. 3). This seismic expression is uniform on lines throughout the Gulf of Chiriquí and is typical of the seismic expression of crystalline basement rocks on unmigrated, multichannel seismic lines.

The top of the acoustic basement is disrupted by high-angle faults that mainly strike northeast and west-northwest (Fig. 2) (Kolarsky and Mann, this volume). The top of the basement defines deep sedimentary basins that lack bathymetric expression because they are almost completely filled with sedimentary rocks: the Cébaco basin complex in the eastern Gulf of Chiriquí (Fig. 3) and the Montuosa basin to the west. The top of acoustic basement shallows gradually west-northwestward from the Montuosa basin toward exposures of Jurassic?–Late Cretaceous? basaltic basement rocks (Nicoya Complex) on the Burica Peninsula of Panama and Costa Rica (Kolarsky and Mann, this volume) (Fig. 2).

The Cébaco Basin complex consists of several northeast-striking, *en echelon* half-grabens extending between the Azuero-Soná and Coiba left-lateral strike-slip fault zones (Mann and Corrigan, 1990; Kolarsky and Mann, this volume) (Fig. 3). The basin complex is interpreted as a pull-apart basin produced by slip transfer between the two parallel, strike-slip fault zones (Kolarsky and Mann, this volume). The basin complex contains a maximum of 2,600 m of sedimentary fill and is subdivided by basement horst blocks such as Cébaco Island, which reaches a topographic elevation of 349 m above sea level (Fig. 2).

The Cébaco basin complex and the Montuosa basin are separated by the Ranchería arch, a northward-trending arch in acoustic basement that exposes Coiba and Ranchería Islands as landmasses (Kolarsky and Mann, this volume) (Fig. 2). The Montuosa basin is a deeply subsided sag basin and separates basement highs to the north, east, and west and an outer forearc high bordering the accretionary wedge of the South Panama deformed belt to the south (Moore and Sender, this volume). To the east of the Montuosa basin is the Parida arch, another north-trending arch of lesser magnitude than the Ranchería arch (Kolarsky and Mann, this volume). Both the Ranchería and Parida arches are aligned with bathymetric ridges on the subducting Nazca plate and appear to have formed as a response to localized shallow subduction (Heil, 1988; Moore and Sender, this volume). The lack of faults in the western part of the Gulf of Chiriquí allows correlations to be made between the basement and Neogene sedimentary cover exposed on the Burica Peninsula and seismic map units in the Gulf of Chiriquí.

SEISMIC MAP UNIT GC1 BENEATH THE GULF OF CHIRIQUÍ

Map unit GC1 is a sequence of variable thickness that is found underlying the area of half-grabens in the Cébaco basin complex (Fig. 4). Unit GC1 is as much as 1 km thick beneath the larger half-grabens. We informally subdivide unit GC1 into upper and lower subunits that are distinguishable by reflection character (Fig. 3). The lower subunit exhibits high-amplitude,

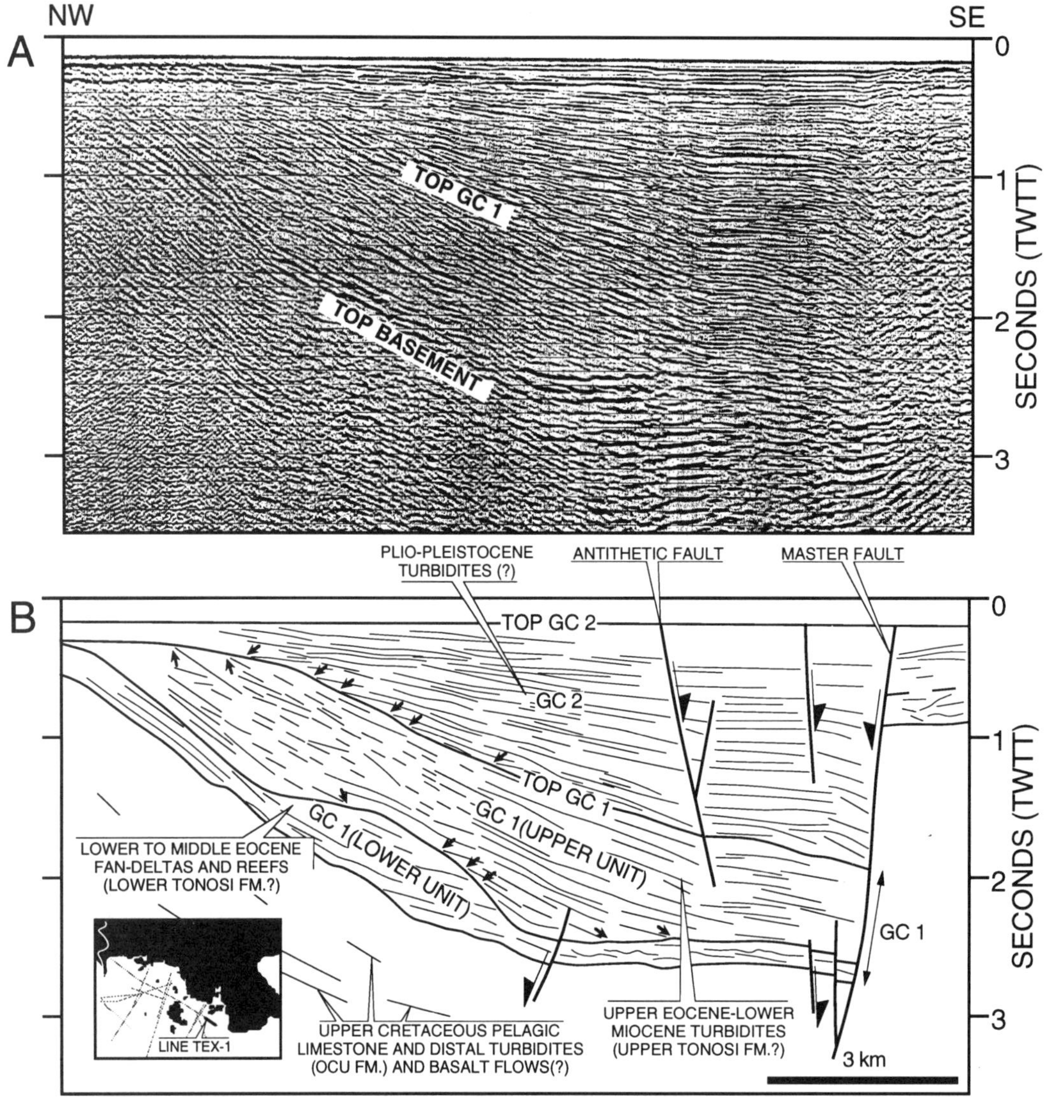

Figure 3. A, Southeastern end of the reprocessed 24-fold migrated stack of Texaco line 1, eastern Gulf of Chiriquí (location shown in inset box). See Kolarsky and Mann (this volume) for structural interpretation of entire line. B, Interpretation of line showing half-graben structure, pre-rift Upper Cretaceous to Lower Miocene unit (seismic map unit GC1), and syn-rift Pliocene-Recent? sedimentary fill (seismic map unit GC2). Black arrows indicate reflection onlaps. Note that unit GC1 is divided into two subunits: a basal unit above acoustic basement with high-amplitude, continuous reflections and minor hummocky reflections and an upper unit of variable amplitude semicontinuous reflections that downlap and onlap onto the top of the basal subunit.

continuous reflections and minor hummocky reflections. The basal contact forms a rough, uneven surface on acoustic basement. Reflections at the base of the upper subunit of GC1 downlap and onlap onto the top of the lower subunit (Fig. 3). The upper subunit consists of mainly variable amplitude, semicontinuous reflections.

Although both the upper and lower subunits become thin in the direction of elevated horst blocks in the Cébaco basin complex (Fig. 4), the reflection terminations across these horst blocks appear to be truncations rather than toplap or offlap and therefore do not appear to record syndepositional growth faulting during deposition of seismic map unit GC1. Based on these observations, we interpret unit GC1 as a pre-rift sequence that was not affected by the younger normal faults bounding the half-grabens.

Unit GC1 is thin (<200 m) to the west and missing in the central gulf in the Montuosa basin and across the Parida and Ranchería arches (Fig. 4). To the west, where GC1 is thin, the unit contains higher-amplitude reflections than in the area of the Cébaco basin complex to the east.

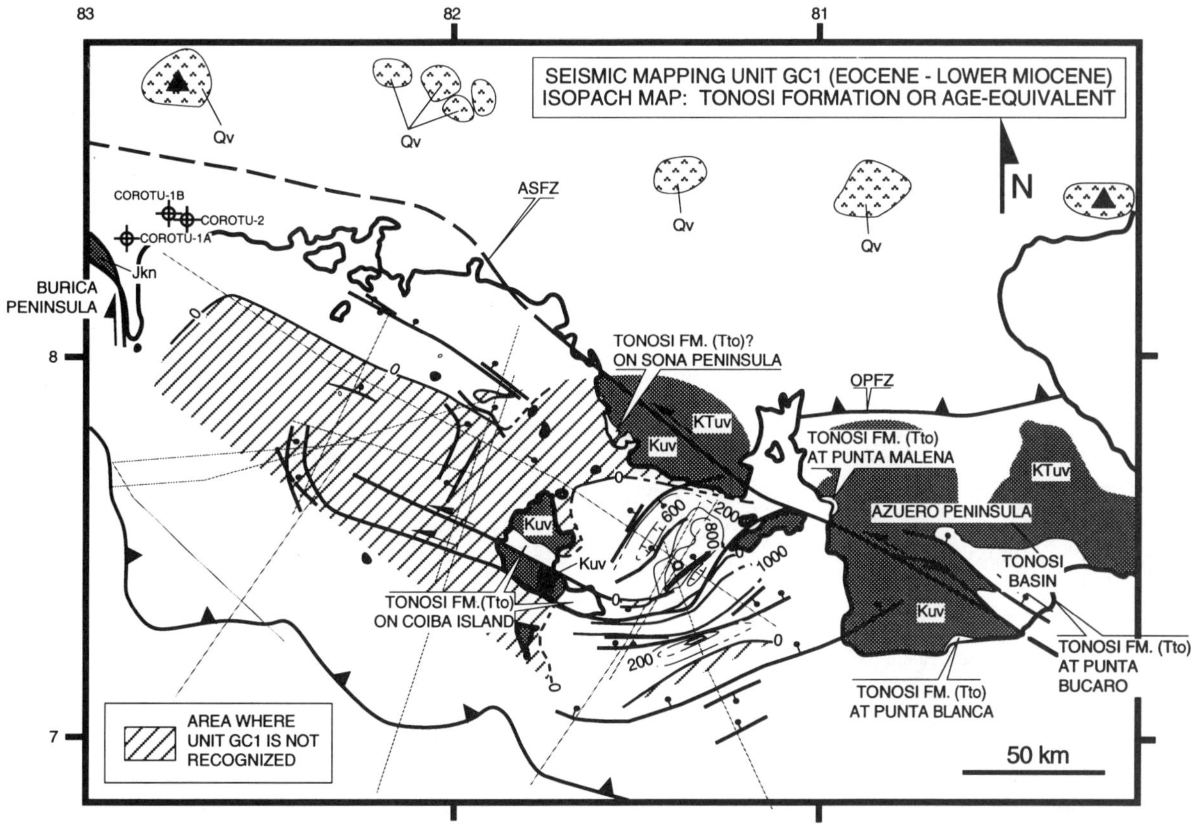

Figure 4. Isopach map of seismic map unit GC 1 (Middle Eocene? to Lower Miocene?). Depths are based on seismic interval velocities from the dotted lines. Contour interval is 200 m. Onshore basement outcrops correlated with acoustic basement are shaded (see caption of Fig. 2 for key to abbreviations). Onshore outcrops of Tonosí Formation (Tto) correlated with seismic unit GC1 are indicated. Queried outcrops of the Tonosí Formation on the Soná Peninsula were not visited during this study but were mapped by del Giudice and Recchi (1969).

SEISMIC MAP UNIT GC2 BENEATH THE GULF OF CHIRIQUÍ

Map unit GC2 comprises syn-rift fill of highly variable thickness that is found in half-grabens of the Cébaco basin complex (Fig. 5). Unit GC2 ranges from 2 to 3 km in thickness in the larger half-grabens (Fig. 3). West of the Cébaco basin complex, unit GC2 onlaps the Parida arch, pinches out across the Ranchería arch, and thickens toward the Burica Peninsula. The wedge-shaped geometries of unit GC2 contrast with the more sheetlike geometry of the underlying, pre-rift GC1 seismic unit.

Divergent reflections and wedge-shaped geometries of unit GC2 indicate fault control during deposition of the unit by normal and oblique-slip faults bounding the half-grabens of the Cébaco basin complex (Kolarsky and Mann, this volume). In the Cébaco basin complex, syn-rift fault control is also indicated by stratigraphic relationships at the base of unit GC2 where reflections within GC1 onlap the underlying seismic unit GC1 along a prominent unconformity surface on Texaco line 1 (Fig. 3).

CORRELATION OF SEISMIC MAP UNIT GC2 IN THE GULF OF CHIRIQUÍ TO THE BURICA PENINSULA

Field studies by ourselves and previous workers, including Terry (1956) and del Giudice and Recchi (1969), in the land areas surrounding the eastern Gulf of Chiriquí have not mapped any exposed half-graben structures containing syn-rift sediments that can be correlated to seismic map unit GC2 in the Cébaco basin complex (Fig. 5). The youngest sedimentary rocks mapped to date on the Azuero and Soná Peninsulas and Coiba Island are early Miocene in age and are placed in the upper part of the Tonosí Formation, which we correlate to the upper part of the pre-rift seismic unit GC1. It is likely that the large vertical fault movements that created the half-grabens of the Cébaco basin complex and exposed basement rocks in land areas around the eastern Gulf of Chiriquí acted to strip post–early Miocene sedimentary off of basaltic basement and underlying older sedimentary rocks.

To attempt a direct stratigraphic correlation between unit GC2 and the Burica Peninsula, the only area of exposed late

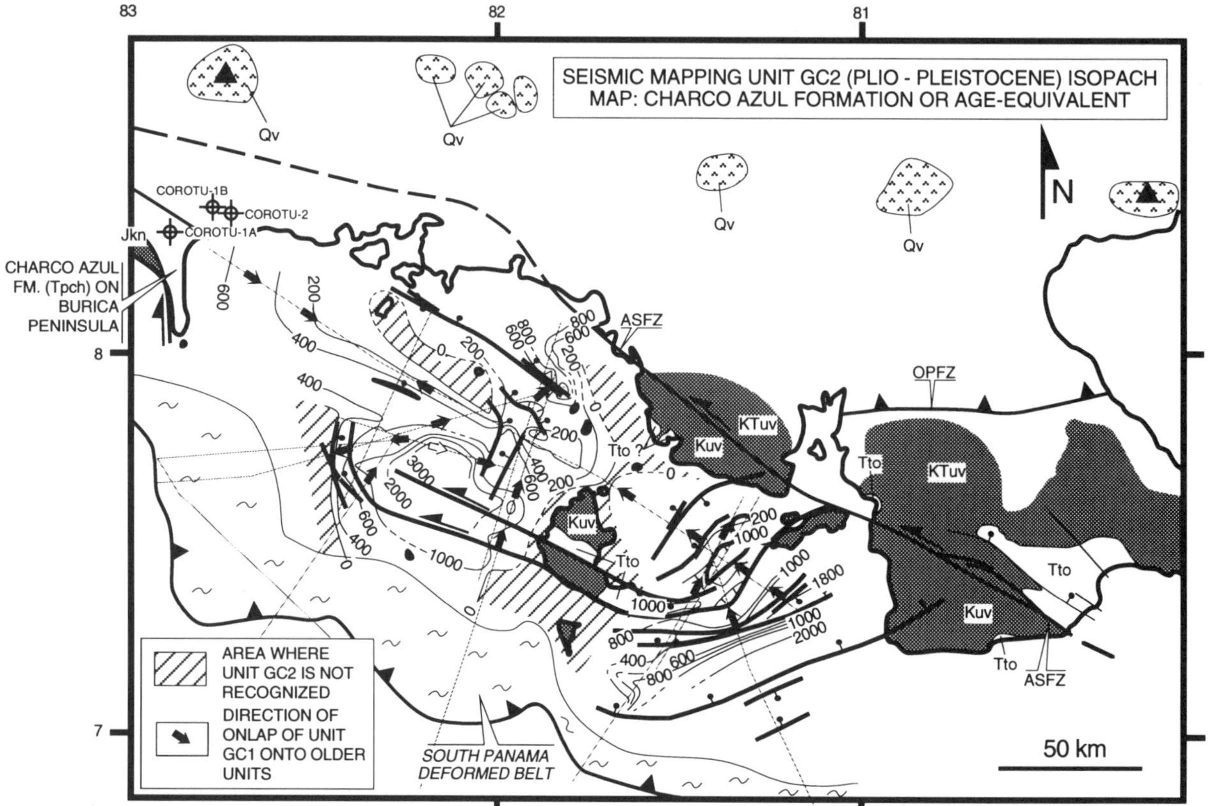

Figure 5. Isopach map of seismic map unit GC2 (Plio-Pleistocene?). Depths are based on seismic interval velocities of the dotted lines. Contour interval is 200 m for values less than 1,000 m. Onshore basement outcrops correlated with acoustic basement are shaded (see caption of Fig. 2 for key to abbreviations). Onshore outcrop of Charco Azul Formation (Tpch) mapped by Corrigan et al. (1990) on the Burica Peninsula and correlated with seismic map unit GC2 is indicated.

Neogene rocks in the circum–Gulf of Chiriquí area, we constructed a regional cross section using an onshore cross section of the Burica Peninsula of Panama and Costa Rica published by Corrigan et al. (1990) and the western end of Texaco line 1 (Fig. 6). The lack of high-angle faulting on seismic lines in the western Gulf of Chiriquí makes the Burica Peninsula the best area to correlate on- and offshore sedimentary sequences despite a 8.4-km gap between the western end of Line 1 and the eastern end of the Corrigan et al. (1990) cross section (Fig. 6).

Stratigraphy of the Burica Peninsula

The stratigraphy of the Burica Peninsula of Panama and Costa Rica has been described by Corrigan et al. (1990) and Coates et al. (1992). It consists of three distinctive lithologic units:

(1) Highly fractured and veined basalt exposed in the northwestern part of the peninsula (Fig. 6, inset) that is generally considered to be part of the Jurassic?–Late Cretaceous? Nicoya Complex described in more detail on the Nicoya Peninsula of Costa Rica (Schmidt-Effing et al., 1982), the Osa Peninsula of Costa Rica (Berrangé and Thorpe, 1988), and the mainland of eastern Costa Rica (DiMarco et al., this volume) (Fig. 1).

(2) A discontinuous and unnamed bioclastic limestone, overlying basaltic basement, that contains a foraminiferal assemblage of late Paleocene to early Eocene age (Corrigan et al., 1990).

(3) An overlying 3,000-m-thick section of mainly early Pliocene to early Pleistocene marine sandstone and shale that Corrigan et al. (1990) have called the Charco Azul Formation after Terry (1956).

Corrigan et al. (1990) subdivide the Charco Azul Formation into three members (Fig. 6):

(1) Peñita Member, a 300- to 330-m-thick basal section of glauconite-bearing sandstone, siltstone, and conglomerate containing shallow-marine macrofossils. Foraminifera indicate deposition in an early Pliocene inner or outer neritic environment. Mapping by Corrigan et al. (1990) suggests that the member unconformably overlies basaltic basement and was probably deposited during the initial stages of transgression. More recent study of foraminifera by Coates et al. (1992) confirms that the unit is older than 3.5 Ma.

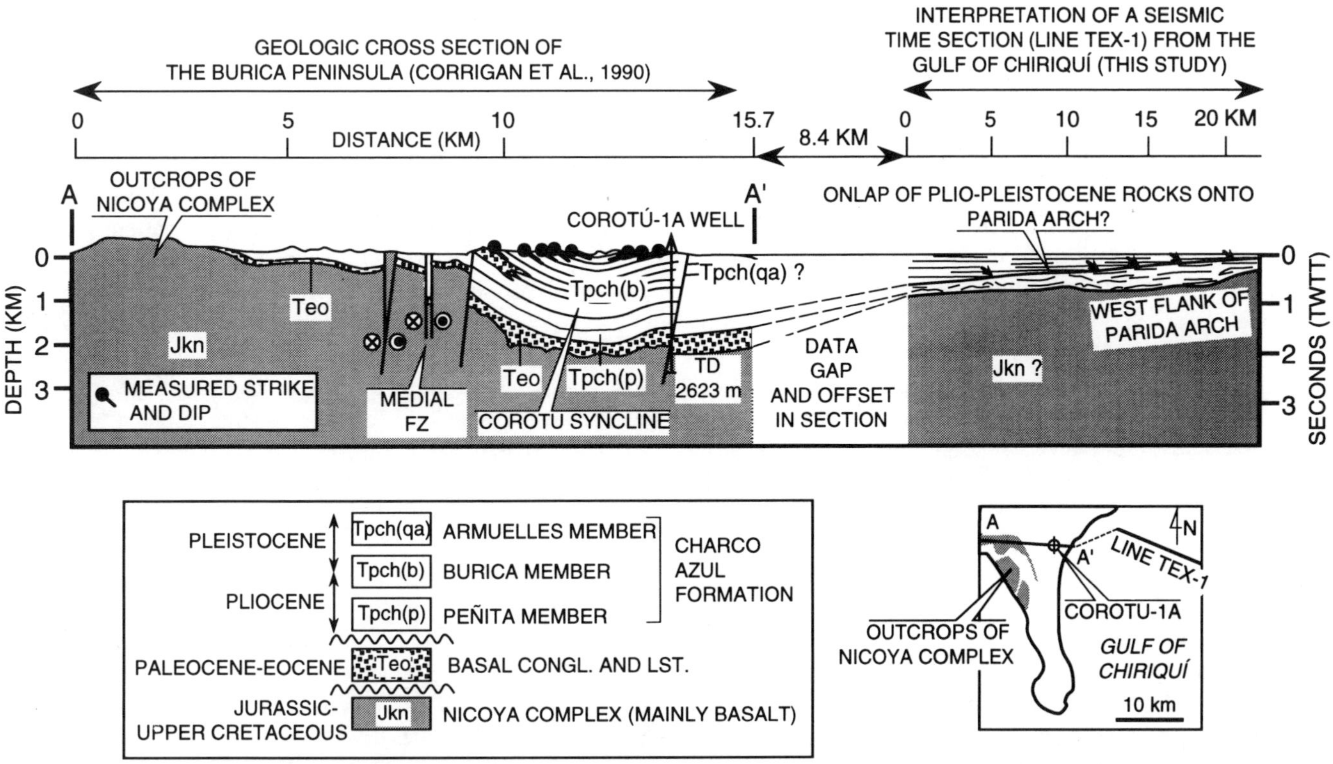

Figure 6. Stratigraphic and age correlation between rocks exposed on the Burica Peninsula (Corrigan et al., 1990) and reflectors seen on the western end of Texaco line 1. Inset shows location of cross section and basement exposures on the Burica Peninsula correlated to acoustic basement in the Gulf of Chiriquí.

(2) Burica Member, a 3,000-m-thick section of fine-grained turbidites and less-common coarser-grained rocks including channel-fill, slump, and debris-flow conglomerate deposited on a south-dipping paleoslope. Detailed study of foraminifera of the Burica Member by Corrigan et al. (1990) and Coates et al. (1992) indicates that the age of the basal Burica Member is early Pliocene (3.6 to 3.5 Ma) and the top of the member is earliest Pleistocene (1.7 to 1.5 Ma). Foraminifera indicate rapid deepening from the early Pliocene neritic environments (Peñita Member) to the Pliocene lower bathyal environments (Burica Member).

(3) Armuelles Member, a 200-m-thick section of shallow-water sandstone, siltstone, and pebbly sandstone that contains abundant fragments of shells and molluscs. Strata in the lower part of the early Pleistocene Armuelles Member contain abundant middle-bathyal foraminifera that record rapid shoaling from slope to shelf depositional environments (Corrigan et al., 1990; Coates et al., 1992; Collins et al., this volume).

Correlation between onland rocks and offshore area

Comparison of the eastern end of Texaco line 1 with the cross section by Corrigan et al. (1990) suggests that the youngest seismic map unit in the Gulf of Chiriquí (GC2) may correlate with the early Pliocene to early Pleistocene Charco Azul Formation and that basaltic basement of the Burica Peninsula may correlate with acoustic basement (Fig. 6). The lower part of unit GC2, which exhibits hummocky, discontinuous reflections, is correlated with the Peñita Member, the basal, shallow-marine transgressive section of the Charco Azul Formation. The upper part of unit GC2, exhibiting continuous parallel reflections, is correlated with the well-bedded Burica Member (marine turbidites) and Armuelles Member (shallow marine sands and silts of the Charco Azul Formation). The lower subunit of seismic map unit GC1 is correlated with limestone and conglomerate that separates basaltic basement from the base of the Charco Azul Formation (Corrigan et al., 1990) (Fig. 6). We interpret the age of the lower subunit of GC1 as late Paleocene to early Eocene based on microfossil ages obtained from samples of the unnamed limestone unit on the Burica Peninsula (Corrigan et al., 1990).

CORRELATION OF ACOUSTIC BASEMENT AND SEISMIC MAP UNITS IN THE GULF OF CHIRIQUÍ TO COIBA ISLAND

To correlate acoustic basement and seismic units in the Gulf of Chiriquí with onshore rocks, we carried out mapping, and fossil sampling on Coiba Island. The island is critical for regional correlation because it occupies a central location in the Gulf of Chiriquí (Fig. 1).

Stratigraphy of Coiba Island

The stratigraphy of Coiba Island has been previously described by Terry (1956), del Giudice and Recchi (1969), Metti et al. (1972), and Metti and Recchi (1976) and has been correlated to offshore units interpreted on a single Texaco seismic line (Line Tex-3) in the Gulf of Chiriquí by Okaya and Ben-Avraham (1987) and by Kolarsky and Mann (this volume) (Fig. 1B). We limited our field studies to the more accessible eastern coast of the island. The stratigraphy consists of two lithologically distinct units: (1) massive and pillowed basalt of Late Cretaceous(?) age underlying approximately 60% of the area of the island and well exposed in sea cliffs and tidal flats (Fig. 7); (2) Tonosí Formation, an ~800-m-thick section of mainly early to middle Eocene reefal limestone, conglomerate, sandstone, and shale overlying basaltic basement. A 10-m-thick conglomerate and limestone layer was found at one locality, separating basalt from the overlying sandstone and shale section. The conglomerate and limestone was assumed to be the base of the Tonosí Formation.

Basaltic basement and Ocú Formation

Our mapping along the east coast verified the basic outcrop pattern of basement rocks shown in maps by Terry (1956), Metti et al. (1972), and Metti and Recchi (1976) and also confirmed the lithologic similarity of the basement of Coiba Island to that found along the west coast of the Azuero Peninsula and the southern coast of Soná Peninsula and on Cébaco Island.

Basalt on Coiba Island consists of massive flows, massive but intensively fractured flows, horizons of pillow basalts (Fig. 8), and horizons exhibiting columnar jointing. Layering in the basaltic flows and flattening of individual pillows shows a general south-to-southwest dip and was interpreted as primary layering (Kolarsky and Mann, this volume).

At several localities on Coiba Island, basalt flows are interbedded with limestone and calcareous siltstone of the Ocú Formation. This relationship had not been previously recognized on the Azuero Peninsula where the Ocú Formation is poorly exposed (del Giudice and Recchi, 1969).

Measured section of interbedded basalt and Ocú Formation at Punta Damas

In order to document the interbedded relation between the Ocú Formation and basaltic basement, a 450-m section of the Ocú Formation was measured along the coast east near the penal colony at Punta Damas (Fig. 9). The lithologic assemblage, fracturing, and slight degree of metamorphism in the section are identical to those in outcrops of Ocú Formation seen on the Azuero Peninsula.

Lithology. The 350-m section consists of fractured and veined, recrystallized dark-gray laminated micritic limestone, marl, and shale (Fig. 9). Fine to coarse-grained mainly thin-bedded interbedded sandstone and shale are exposed between 200 and 245 m of the section. Beds of fine- to medium-grained sandstone and hemipelagic/pelagic limestone make up less than 5% of the section. The hemipelagic/pelagic limestone exhibits fine laminations defined by color variations (Fig. 10). Grading is not visible in this rock type because of the fine grain sizes. A single 130-m-thick massive sandstone unit occurs in the middle part of the section. All lithologies are highly indurated, presumably because of low-grade metamorphism accompanying eruptions of interbedded basalt flows. The upper contact of the Ocú Formation is stratigraphically overlain by basalt; the lower contact is covered, but strike and dip measurements suggest that it is conformable on nearby outcrops of basalt.

Facies interpretation. The fine grain sizes and laminations indicate a quiet deep-marine environment punctuated by infrequent turbidity flows of sandstone. The 130-m-thick sandstone unit appears to represent a large slump into the otherwise quiet and fine-grained deep-marine environment. The conformable stratigraphic relationship of these rocks with the underlying and overlying basalt indicates deposition during recurring submarine volcanic activity. Interbedding of the Ocú Formation with basalt flows on Coiba Island indicates that deep-marine sedimentation was coeval with submarine volcanism.

Age. Two shale samples collected from the Ocú Formation were barren of microfossils (Appendix 1). We tentatively assume that the Ocú Formation on Coiba Island and the interbedded basalt are the same age as outcrops of the Ocú Formation on the Azuero Peninsula that have been dated using planktic foraminifera as Late Campanian to Maastrichtian (del Giudice and Recchi, 1969; Recchi and Miranda, 1977).

Tonosí Formation (lower to middle Eocene)

Subdivision of the formation. Excellent coastal exposures of clastic rocks of the Tonosí Formation were mapped along the southeast coast of Coiba Island (Fig. 7). The sections are accessible only by foot or launches operated by guards from the central penal colony to outlying prison camps in the area. Outcrops of the Tonosí Formation on Coiba Island are mainly shallow-water transgressive units composed of conglomerate, coarse sandstone, and reefal limestone; this lower unit forms a ~150-m-thick unit unconformably resting on Cretaceous basaltic basement. The coarser-grained basal section grades upward into a finer-grained unit composed of finer-grained massive, trough cross-bedded sandstone and interbedded sandstone and shale. Three samples from the coarser-grained and finer-grained units contain nannofossils of early middle Eocene age and are, therefore, placed in lower part of the Tonosí Formation (Appendices 1, 3).

Lithologic description of basal, coarser-grained unit. Metti et al. (1972) and Metti and Recchi (1976) mapped conglomerate and limestone in thin, discontinuous layers overlying basement on Coiba Island. We observed the limestone-conglomerate association at three localities above basaltic basement along the southwestern coast of the island at UTM (Universal Transverse Mercator) coordinates $^{8}21$, $^{4}37$; $^{8}18$, $^{4}35$; and $^{8}17$, $^{4}30$ (Fig. 7). Conglomerate consists of massive, commonly unbedded breccias composed almost entirely of clasts of angular basalt identical to the underlying basaltic basement. Because the matrix and clasts are all basalt and the clasts are angular, it is often difficult to distinguish the basaltic breccia from underlying and fractured *in situ* basalt, particularly in outcrops within the North Coiba fault zone (Fig. 7).

At one locality (UTM coordinates $^{8}21$, $^{4}37$), the basaltic conglomerate is conformably overlain by approximately 10 m of a cream-colored, massive to rubbly-bedded reefal limestone composed of shell debris, *Lepidocyclina* foraminifera, and algal debris (Fig. 11A). The basement-limestone contact is unfaulted and forms an uneven surface in which limestone fills

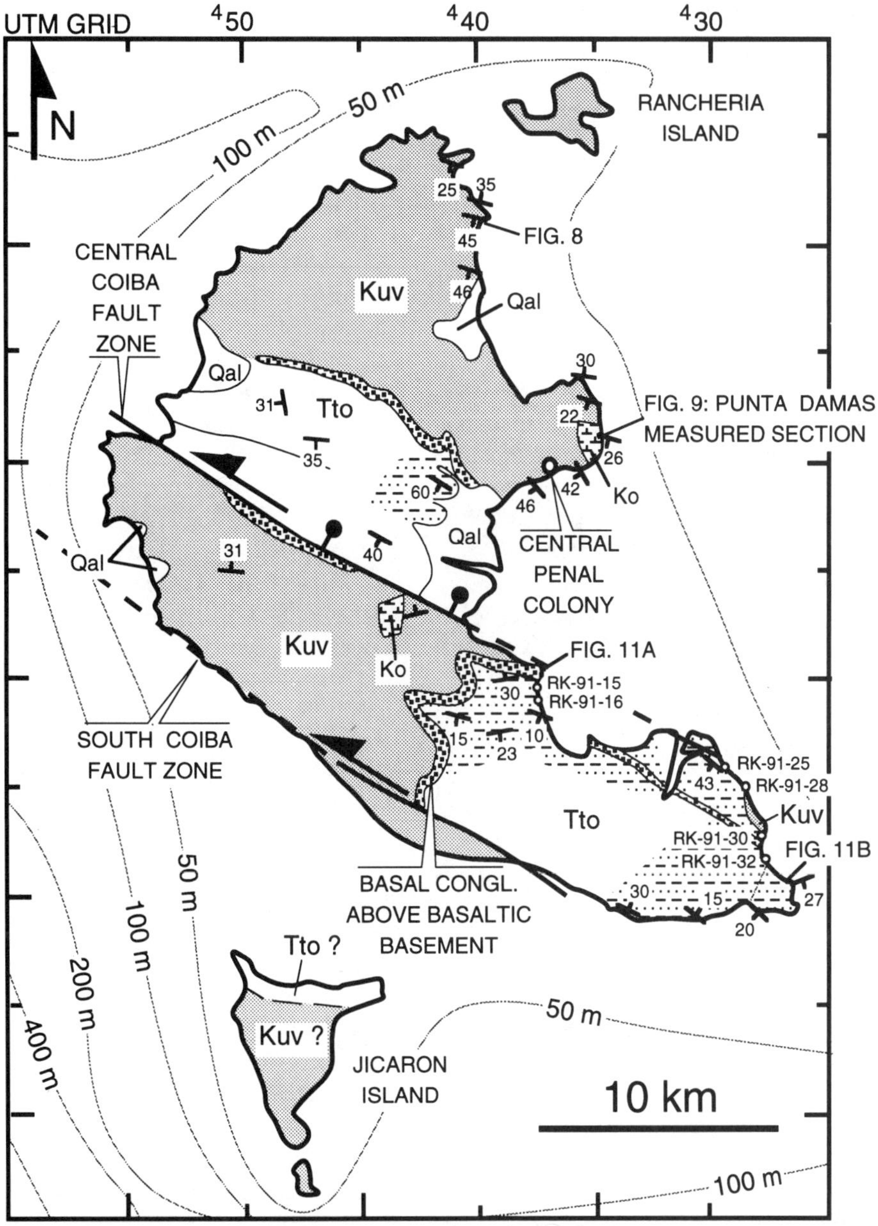

Figure 7. Structural geologic map of Coiba Island showing basaltic basement in shading (Kuv), overlying sedimentary rocks of the Tonosí Formation (Tto) shown where seen in dash-dot pattern, and Quaternary alluvium (Qal). Only outcrops along the eastern coast of the islands were visited during this study. Geology of other parts of the island, Rancheria Island, and Jicarón Island is from del Giudice and Recchi (1969) and our interpretation of aerial photographs. Geographic coordinates given by the Universal Transverse Mercator grid (UTM).

Figure 8. Photograph of Late Cretaceous? pillow basalts along the southern coast of Coiba Island. We correlate these rocks to acoustic basement mapped beneath the Gulf of Chiriquí (see Fig. 2). Plane of flattening in pillows corresponds to stratigraphic bedding in interbedded rocks of the Ocú Formation and is represented by strike and dip measurements in Figure 7. Mechanical pencil is 14 cm long. See Figure 7 for location of outcrop.

voids in the upper surface of basaltic basement. The conglomerate is composed entirely of angular basalt clasts that range in size from 1 to 4 cm. The limestone crops out in isolated areas and is not distinguished as a separate unit of the Tonosí Formation on the map of Figure 7.

Facies interpretation. The massive conglomerate overlying basement exhibits sedimentary features indicative of debris-flow deposits (e.g., poor sorting, angular clasts, lack of internal fabric, dominance of clasts of one composition). The massive to rubbly bedding and presence of shell and algal material in the overlying reefal limestone are consistent with a shallow-marine environment.

Age. Although the age of the conglomerate is not known, the limestone is of Middle Eocene age based on the presence of *Lepidocyclina.* Three samples collected in shales cropping out less than 100 m above the limestone outcrop were submitted to S. Monechi for nannofossil dating. One sample was barren and two contained early to middle Eocene nannofossils (Appendices 1, 3).

Lithologic description of finer-grained upper unit. A finer-grained unit containing early to middle Eocene nannofossils was examined in small exposures along the Rio San Juan in the central part of the island at UTM coordinates $^{8}30$, $^{4}42$ and in excellent coastal exposures along southeastern end of the island (Fig. 7).

In the Rio San Juan exposures, the finer-grained unit of the Tonosí Formation consists of thin- to medium-bedded, graded and planar laminated turbiditic sandstone (Facies D of Mutti and Ricci Lucchi, 1978). Along the southern coast of the island (UTM coordinates $^{8}17$, $^{4}30$), a massive 20- to 30-m-thick, fine- to coarse-grained sandstone of facies B is present along with facies D turbiditic sandstone (Fig. 11B). The volcanogenic sandstone is well sorted, and dominant sedimentary structures include trough cross-bedding and planar stratification.

Facies interpretation. Turbiditic sandstone and shale are interpreted as deep-marine sedimentation that followed shallow-marine transgressive sedimentation represented by the conglomerate and limestone of the basal Tonosí Formation.

Age of upper unit. The single sample from the Rio San Juan area submitted to S. Monechi from the upper unit of the Tonosí Formation was barren of nannofossils (Appendices 1, 3). Of the four samples submitted from the southeastern coast of the island, one yielded early to middle Eocene nannofossils (Appendixes 1, 3). This age date correlates with the early to middle Eocene ages of samples collected close to the basalt-sediment contact and suggests that the coarser-grained and reefal units may represent lateral facies equivalents of the finer-grained massive volcanogenic sandstone and sandstone and shale horizons 5 to 10 km to the southwest (Fig. 7).

Sandstone petrography of upper and lower units of the Tonosí Formation. Two hundred point counts were made on three sandstone samples from the sandstone on Coiba Island to determine the relative content/abundance of rock fragments, cement, matrix, macroporosity, organic matter, and fossil debris (Table 1). The rocks are volcaniclastic litharenites composed of abundant feldspar, over 50% altered volcanic rock fragments, scarce angular volcanic quartz, and scarce organic matter.

CORRELATION OF ACOUSTIC BASEMENT AND SEISMIC MAP UNITS IN THE GULF OF CHIRIQUÍ AND ROCKS ON COIBA ISLAND

Acoustic basement

Acoustic basement in the Gulf of Chiriquí is correlated with the basalt-sedimentary complex exposed on Coiba Island (Fig. 7). Discontinuous reflections in acoustic basement are attributed to energy scattering by fractured, pillowed, and massive basalt layers. We interpret the age of this basement as Late Cretaceous, based on a single Coniacian age obtained from radiolarians in chert interbedded with basalt along the west coast of the Azuero Peninsula (Appendix 2) and foraminifera-based ages of Late Campanian to Maastrichtian obtained by del Giudice and Recchi (1969) and Recchi and Miranda (1977) from inland outcrops of the Ocú Formation on the Azuero Peninsula. Moderately continuous intrabasement reflectors, such as those seen on the northwestern end of Texaco line 1 in the Cébaco basin complex (Fig. 3), are correlated with sedimentary units of the Ocú Formation interbedded in basalt, including those measuring up to 200 m in thickness at Punta Damas (Fig. 9).

Lower subunit of GC1

Fairly continuous, high-amplitude reflectors of lower subunit of seismic map unit GC1 are correlated with conglomer-

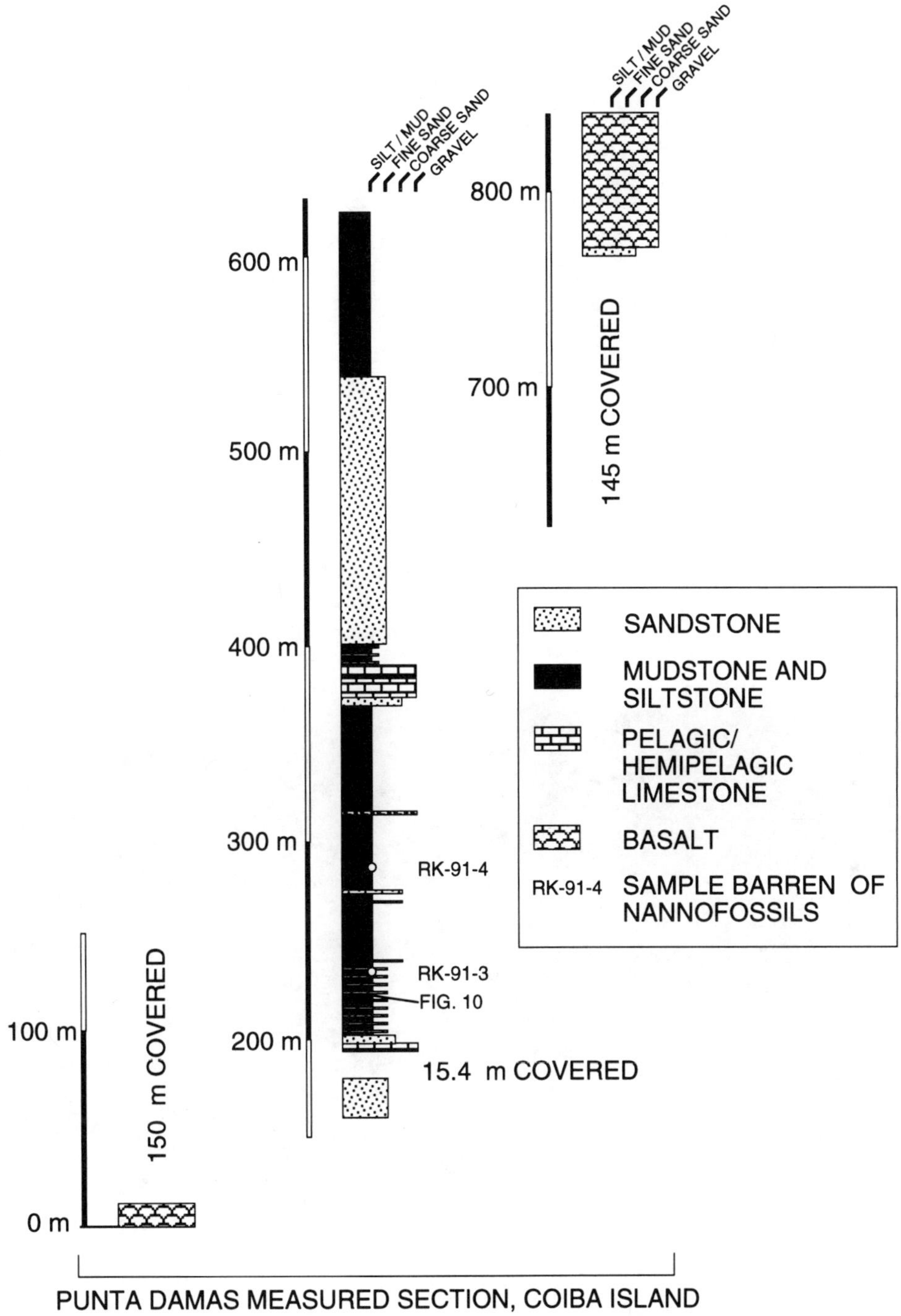

Figure 9. Measured stratigraphic section of the Upper Cretaceous Ocú Formation at Punta Damas on the eastern coast of Coiba Island. We correlate these rocks with acoustic basement mapped beneath the Gulf of Chiriquí (see Fig. 2). Note interbedding of sedimentary rocks of the Ocú Formation and basalt that may account for the induration and slight metamorphism of the Ocú Formation. See Figure 7 for location of section.

Figure 10. Photograph of fractured and slightly metamorphosed hemipelagic limestone and siltstone of the Ocú Formation on the northeastern coast of Coiba Island near Punta Damas. The induration and lack of preferential weathering between layers of differing composition suggests that the rock is slightly metamorphosed. Note mechanical pencil for scale. See Figure 9 for location of this outcrop in Punta Damas measured section.

ate, reefal limestone, and overlying sandstone and shale of early to middle Eocene age on Coiba Island (Fig. 7). This correlation is based on: (1) the stratigraphic position of both unit GC1 and the Tonosí Formation directly above basement, and (2) the discontinuous character of reflections within unit GC1 that may represent conglomeratic and reefal facies of limited lateral continuity.

CORRELATION OF ACOUSTIC BASEMENT AND SEISMIC MAP UNITS IN THE GULF OF CHIRIQUÍ WITH THE GEOLOGY OF THE AZUERO AND SONA PENINSULAS AND CEBACO ISLAND

To correlate acoustic basement and seismic units seen on seismic lines in the Gulf of Chiriquí with onshore rocks, we carried out mapping and fossil sampling on the Azuero and Soná Peninsulas and Cébaco Island (Fig. 1). The mainland areas of the Azuero and Soná Peninsulas are critical for correlation because they form the eastern and northeastern edge of the Gulf of Chiriquí and are adjacent to the grid of offshore seismic data. Cébaco Island is a footwall uplift along one of the normal faults of the Cébaco basin complex (Kolarsky and Mann, this volume). The southern extension of the Cébaco horst block can be seen on the depth-to-basement map of the Gulf of Chiriquí (Fig. 2).

Figure 11. Photographs of Eocene clastic and carbonate rocks stratigraphically overlying basaltic basement on the southeastern coast of Coiba Island. See Figure 7 for outcrop locations. A, Rapid upward transition between basal basaltic conglomerate (B) above basalt and algal limestone (L). We correlate this horizon to the contact between acoustic basement and lower subunit of seismic map unit GC1 seen on Figure 3. Nannofossil age dates of sandstone and shale stratigraphically above this horizon are Early to Middle Eocene (see Appendixes 1, 3). B, Laminated, fine-grained, massive sandstone of the basal Tonosí Formation. We correlate this horizon to lower subunit of seismic map unit GC1 seen on Figure 3. Nannofossil age dates of sandstone and shale stratigraphically above this horizon are Early to Middle Eocene (see Appendices 1, 3). Note conjugate normal fault sets described by Kolarsky and Mann (this volume).

Stratigraphy of the Azuero and Soná Peninsulas and Cébaco Island

The stratigraphy of these areas has been previously described by Recchi and Miranda (1977) and correlated to offshore units in the Gulf of Chiriquí interpreted on Texaco line 3 by Okaya and Ben-Avraham (1987) (Fig. 1B). We focused our field studies on the more accessible western and southeastern coasts of the Azuero Peninsula, the southwestern coast of the Soná Peninsula, and Cébaco Island. The stratigraphy consists of two main units: (1) Poorly dated, massive and pillowed basalt of Late Cretaceous age is well exposed along sea cliffs and tidal flats, and (2) Tonosí Formation, a ~900-m-thick section of mainly middle Eocene to lower Miocene sandstone and shale conformably overlies acoustic basement. A 10- to 20-m-thick basal conglomerate and limestone layer is present locally.

Basaltic basement of the Azuero and Soná Peninsulas and Cébaco Island

Poorly studied basement rocks are exposed over approximately 75% of the Azuero Peninsula and even more of the Soná Peninsula (Fig. 2). Basement rocks can be divided into two distinct basement assemblages that are separated by the Azuero-Soná fault zone (Fig. 12). On both peninsulas, the fault forms a sharp boundary between a homogeneous, poorly dated basaltic basement to the southwest and a more heterogeneous, Early Cretaceous through Eocene volcanic arc? basement intruded by Early Cretaceous–Eocene plutons to the north (del Giudice and Recchi, 1969) (Fig. 12). Structural trends in the volcanic arc basement and plutons northeast of the fault appear to have a more westerly strike than the fault itself on geologic maps (del Giudice and Recchi, 1969) and on unpublished aeromagnetic maps (J. de Boer, personal communication, 1991). This relationship suggests that the fault may have offset large-scale lithologic belts. Much of the exposures of basement rocks on both peninsulas may be related to right-stepping gentle restraining bends along the active, left-lateral Azuero-Soná fault zone (Mann and Corrigan, 1990).

Basement along the west coast of the Azuero Peninsula, Cébaco Island, and the southeast coast of the Azuero Peninsula

Lithology. Reconnaissance mapping, mainly along the coast and Rio Torio valley of the western Azuero Peninsula, revealed six distinctive basement lithologies: (1) schistose amphibolites and minor metasedimentary rocks exposed as slivers along the Azuero-Soná fault zone (Tournon et al., 1989), (2) metabasalt with a pronounced shear foliation, (3) tectonic mélange composed of sheared basalt and chert blocks, (4) basalt flows exhibiting columnar jointing, (5) pillow basalts; (6) massive basalt lacking evidence for columnar jointing or pillows (Fig. 13). South of the town of Malena and north of Punta Duartes on the west coast of the Azuero Peninsula, basement

TABLE 1. SANDSTONE MODAL COMPOSITIONS, IN PERCENT, UPPER TONOSÍ FORMATION*

Location	Age	Sample	Quartz (%)	Carbonate (%)	Plagioclase (%)	Volcanic Rock Fragments (%)	Opaque Grains (%)	Cement (%)	Matrix (%)	Macroporosity (%)	Organic Matter (%)	Fossil Debris (%)	Total (%)
SE Coiba Island	Late Eocene	RK-91-25	4	3	21	58	5	0	9	0	0	0	100
SE Coiba Island	Late Eocene	RK-91-28	10	0	10	57	4	13	0	0	1	5	100
SE Coiba Island	Late Eocene	RK-91-32	3	0	25	51	4	10	0	<5	2	0	100
Punta Malena	Late Oligocene-Early Middle	RK-91-104	10	0	11	56	0	0	15	0	0	8	100
Punta Malena	Late Oligocene-Early Middle	RK-91-105	6	0	23	47	0	2	7	<5	0	10	100
Punta Malena	Late Oligocene-Early Middle	RK-91-108	10	0	17	59	4	0	7	3	0	0	100
Punta Malena	Late Oligocene-Early Middle	RK-91-110	>5	5	13	56	0	0	8	8	5	0	100

*Azuero Peninsula and Coiba Island, Southwest Panama.

types 2 through 6 form an apparent 5,000-m-thick section that dips to the northeast about 40° (Fig. 13, section A-A′).

Age. Fourteen samples of sheared and interbedded chert from basaltic rocks in these areas south of the Azuero-Soná fault zone on the Azuero and Soná Peninsulas and Cébaco Island were collected by P. Mann and J. Corrigan in 1987 and were sent for radiolarian analysis at the laboratory of E. Pessagno, Jr., and D. Meyerhoff Hull of the University of Texas at Dallas. Only one of these samples (JC-86-A7), from a tectonic mélange of sheared basalt and chert (refer to unit 3 on Fig. 13) along the west coast of the Azuero Peninsula, contained recognizable radiolaria of Coniacian age (Appendixes 1, 2). To our knowledge, this is the first faunal age determination on basement rocks south of the Azuero-Soná fault zone.

North of the Azuero-Soná fault zone, Bourgois et al. (1982) reported a 98-Ma (Albian) K-Ar radiometric age from an exposure of pillow basalts at Playa Venado on the southeastern coast of the Azuero Peninsula (Bourgois et al., 1982) (Fig. 12). K-Ar dates from plutonic intrusions into basement rocks north of the Azuero-Soná fault zone have yielded cooling ages of 53 ± 5, 69 ± 10, and 65 ± 1.5 Ma for these intrusions (del Giudice and Recchi, 1969).

Ocú Formation (Upper Campanian–Maastrichtian) of the Azuero Peninsula

Lithology. The Ocú Formation on the Azuero Peninsula (Fig. 12) has an apparent thickness of up to 1,500 m and consists of two facies: (1) thin to medium-bedded grayish-white limestone and calcareous siltstone; and (2) light brown, fine-grained calcareous siltstone and sandstone. Rare graded beds and ripple drift cross-lamination in the clastic facies suggest deposition by turbidity currents. The Ocú Formation appears to be slightly metamorphosed in fresh outcrops and is commonly intensively fractured, faulted, and folded, with bedding dips generally greater than 45°. The lateral relations between the carbonate and clastic facies of the Ocú Formation will remain unknown until the regional structure of the formation can be established.

The relation of the Ocú Formation to basaltic basement is not clear from inland exposures of the Ocú Formation on the Azuero Peninsula. Del Giudice and Recchi (1969) and Bourgois et al. (1982) suggested that the Ocú Formation unconformably overlies basaltic basement rocks. However, our work on Coiba Island, discussed above, suggests that the Ocú Formation is mainly interbedded with basaltic rocks.

Facies interpretation. The presence of planktonic foraminifera, graded bedding, and the small average grain size suggest distal, low-density turbidite deposition in a deep-marine environment.

Age. Del Giudice and Recchi (1969) report planktonic foraminifera (*Globotruncana*) from samples of the Ocú Formation from the central Azuero Peninsula indicative of a Late Campanian–Maastrichtian age.

Tonosí Formation (Eocene to lower Miocene) of the Azuero Peninsula

Introduction. Recchi and Miranda (1977) introduced the name Tonosí Formation for middle Eocene to early Oligocene

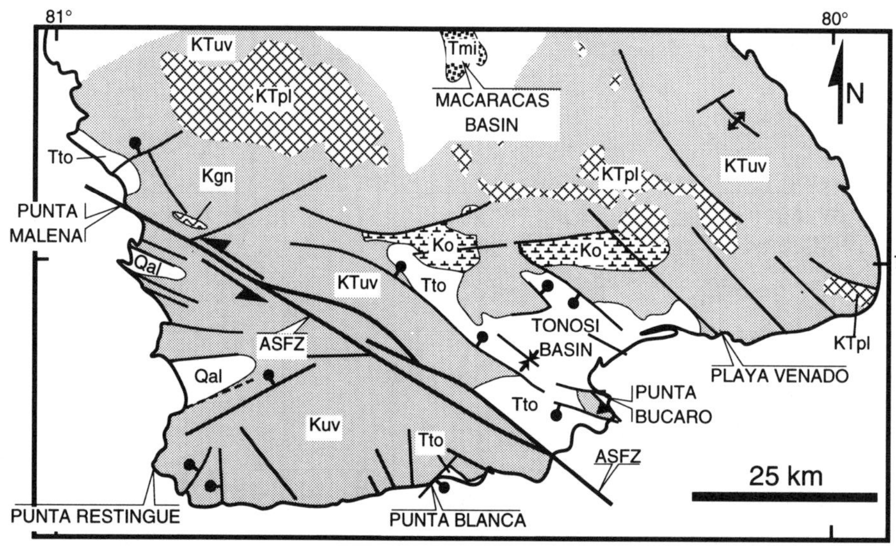

Figure 12. Geologic map of the Azuero Peninsula showing basaltic basement (shaded) overlain by Eocene to Quaternary sedimentary cover (uncolored). Key to abbreviations: Kgn = Three-dimensional schistose amphibolite and minor metasedimentary rocks exposed as slivers along the Azuero-Soná fault zone; KTpl = Cretaceous-Eocene plutons; KTuv = Cretaceous-Tertiary undifferentiated volcanic and sedimentary basement northeast of the Soná-Azuero fault zone; Kuv = Cretaceous-Tertiary undifferentiated volcanic and sedimentary basement southwest of the Azuero-Soná fault zone (ASFZ); Ko = Upper Cretaceous Ocú Formation; Tto = Middle Eocene–Lower Miocene Tonosí Formation; Tmi = Miocene clastic rocks; Qal = Quaternary alluvium.

conglomerate, reefal limestone, and associated calcarenite overlying basaltic basement northeast of the Azuero-Soná fault zone on the Azuero Peninsula (Fig. 12). The Tonosí Formation on the Azuero Peninsula had been originally proposed as a distinct lithologic unit by del Giudice and Recchi (1969).

Subdivision of the formation. As on Coiba Island, we subdivide the Tonosí Formation on the Azuero Peninsula into two major lithologic units. (1) The first is a lower, shallow-water transgressive unit composed of minor coal seams, conglomerate, coarse sandstone, and reefal limestone. This lower unit lies unconformably on Cretaceous basaltic basement, and all lithologies are considered lateral facies equivalents. (2) The second is an upper unit composed of deeper-marine interbedded sandstone, siltstone, shale, and calcarenite.

Type locality. The type locality of an upper Eocene–Oligocene section of the Tonosí Formation is the Tonosí basin, a 500-km^2, fault-bounded synclinal depression in the southeastern part of the Azuero Peninsula (del Giudice and Recchi, 1969; Recchi and Miranda, 1977) (Fig. 14). Road cuts and bare ridge tops expose numerous examples of the forma-

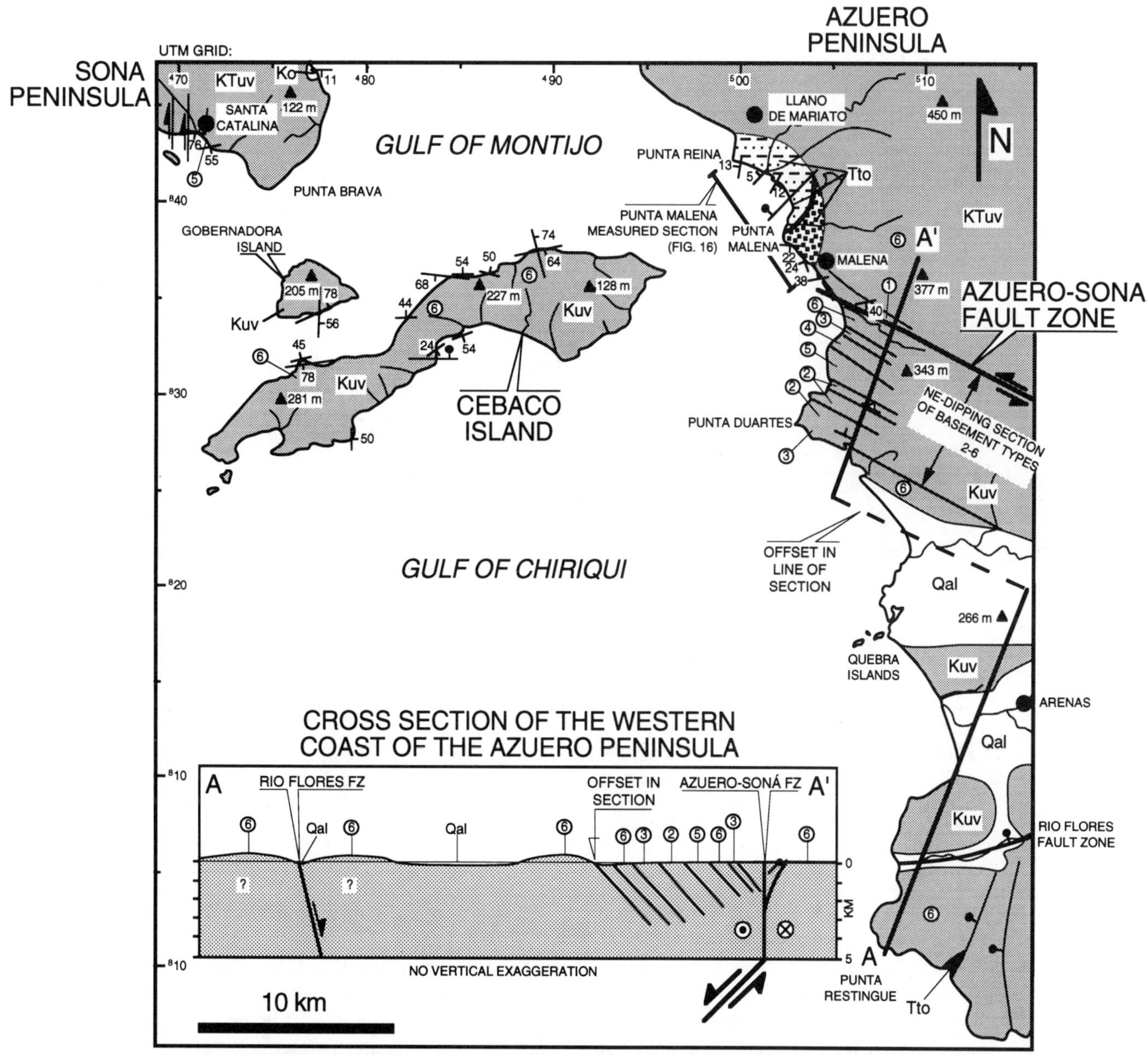

Figure 13. Geologic map of the western coast of the Azuero Peninsula, the southeast coast of the Soná Peninsula, and Cébaco Island (see Fig. 12 for location map). Black triangles give highest elevations in meters. Key to basement lithologies: 1 = high-grade metamorphic rocks; 2 = sheared metabasalt; 3 = tectonic mélange of basalt and chert; 4 = basalt flows with columnar jointing; 5 = pillow basalt; 6 = massive basalt. A sample of chert from unit 3 yielded a Coniacian age (see Appendixes 1, 2).

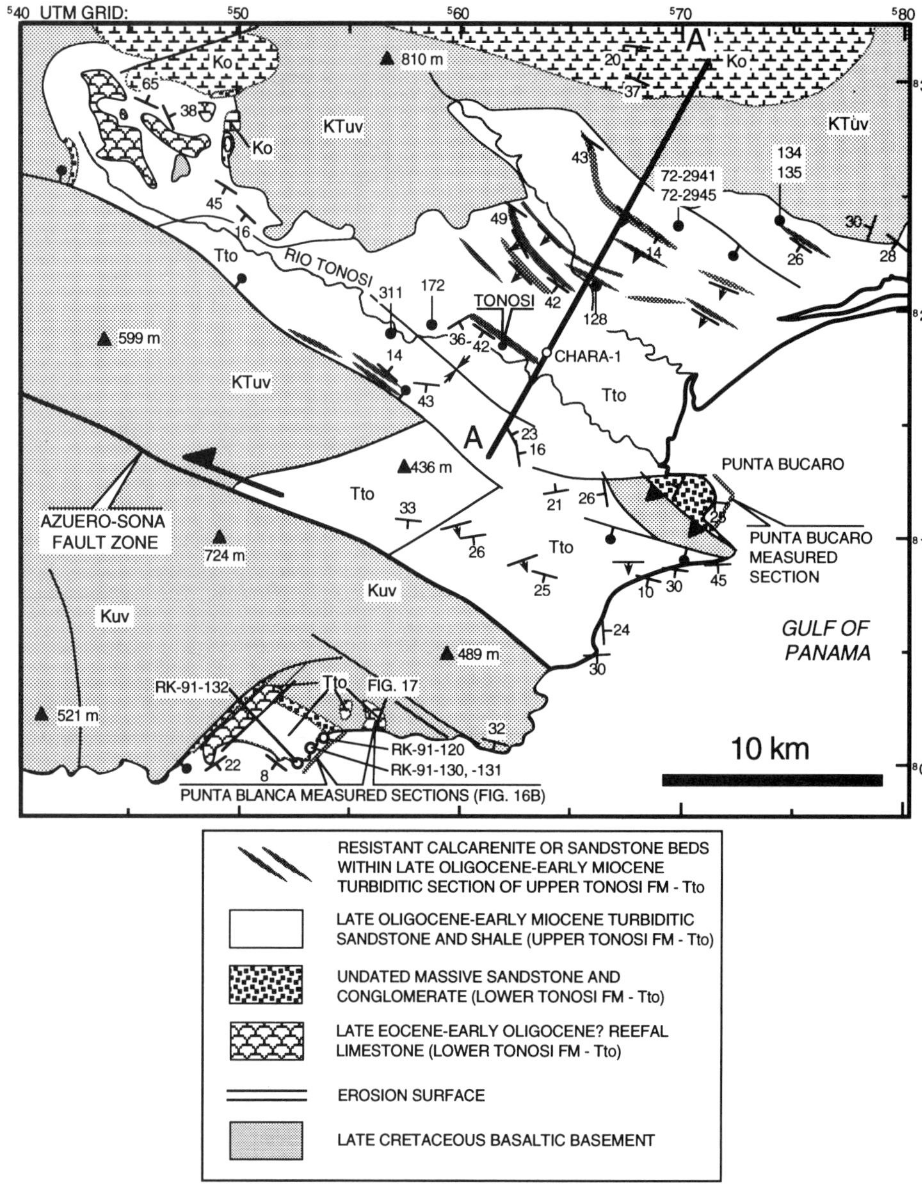

Figure 14. Geologic map of the Tonosí basin on the southeastern Azuero Peninsula (see Fig. 12 for regional map). Heavy black line labelled A-A′ indicates cross section shown in Figure 15. Gray lines indicate measured sections shown in Figure 16. Ages and paleoenvironments of numbered samples are described in Appendix 1, Tables 1, 2.

tion from the town of Tonosí in the center of the basin. Sedimentary rocks of the Tonosí Formation are faulted against erosionally resistant, high-standing mountains of Upper Cretaceous–lower Tertiary? basalts along the Azuero-Soná, Tonosí, and unnamed fault zones along the northern edge of the basin (Fig. 15; section A-A′). Based on the projections of surface dips along the line of section, the rocks of the Tonosí Formation are interpreted to reach up to 2,000 m in thickness in the southern part of the basin. One shallow exploration well, Chara-1, penetrated approximately 400 m of the Tonosí Formation in the central part of the basin (Fig. 15).

Lithology. The most common rock type found in the central Tonosí basin is brown-weathering, medium- to thin-bedded sandstone and siltstone of the upper and lower parts of the Tonosí Formation. About a dozen northwest-striking erosionally resistant ridges composed mainly of massive sandstone, conglomerate, and calcarenite are found to the northeast and southwest of the town of Tonosí and are interbedded in the medium- to thin-bedded sandstone and siltstone (Fig. 14). The town of Tonosí is built on a low elevation in alluvium produced by a 10-m-thick, southwest-dipping calcarenite bed containing abundant coral and echinoid fragments (Fig. 15;

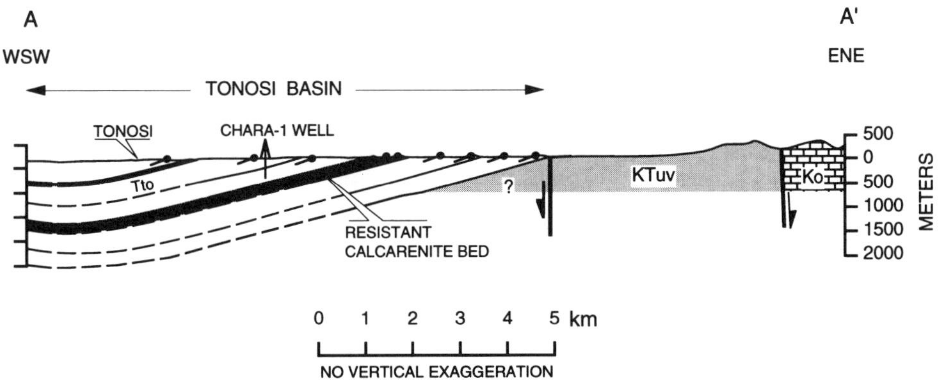

Figure 15. Cross section A-A′ across the Tonosí basin (see Fig 14 for location and lithologic abbreviations). Foraminifera listed in Appendix 1, Table 1, suggest that the base of the section is late Eocene and the upper part is Oligocene in age.

section A-A′). This and other outcrops of calcarenite interbedded in deep-water turbidite deposits are informally called "Tonosí limestone" by Panamanian geologists.

Coal. Thin (<1 m) coal seams were reported by R. K. Jain of the Dirección General de Hidrocarburos (Appendix 1) from the northeastern part of the Tonosí Basin but were not seen by us in the field. Coal samples have yielded pollen dated as late middle to early late Eocene (samples 72-29241 and 72-29245; Appendix 1). The pollen assemblage studied by R. K. Jain (unpublished report to Dirección General de Hidrocarburos, 1974) suggests that the coal formed in a fresh-water, forested swamp close to the paleocoastline. The presence of coal suggests that the northeastern margin of the basin may be stratigraphically low in the Tonosí Formation section and represent a completely different environment of deposition than overlying marine turbiditic rocks.

Igneous rocks. Recchi and Miranda (1977) reported intrusive rocks of andesitic composition that intrude both basement rocks and sedimentary rocks of the Tonosí Formation in the northeastern area of the Tonosí basin. They observed contact metamorphic effects in adjacent rocks of the Tonosí Formation.

Areas of measured sections of the Tonosí Formation. Because the Tonosí Formation exhibits facies changes in different areas across its extensive outcrop area on the Azuero Peninsula, we describe the lower and upper units of the formation separately, using sections measured in three different coastal areas of the Azuero Peninsula: (1) a section at Punta Bucaró along the southeastern coast (Fig. 14), (2) two sections at Punta Blanca along the southern coast (Fig. 14), and (3) a section measured at Punta Malena along the western coast (Fig. 13).

Lower unit of Tonosí Formation (middle Eocene?) at Punta Bucaró

Geologic setting. Excellent coastal exposures of clastic rocks of the Tonosí Formation occur adjacent to a fault-bounded basalt block at Punta Bucaró in the southeastern part of the Tonosí basin (Fig. 14). The section is easily accessible by car on a road from the town of Tonosí to a fishing village at Punta Bucaró. Based on his early observations of coastal outcrops at Punta Bucaró, Terry (1956) proposed the name Bucaró Formation for the clastic section that Recchi and Miranda (1977) later renamed the Tonosí Formation.

Lithologic description. A 560-m-thick section of basal conglomerate and minor sandstone was measured in the coastal exposure at Punta Bucaró (Fig. 16A). The clastic section is assumed to directly overlie basement, although at this locality the base of the section is not exposed and the lowermost exposed beds are assumed to be faulted against basaltic basement.

Conglomerate beds range from massively bedded to thinly bedded and exhibit no thinning- or fining-upward trends. Thinly bedded conglomerate is more common than massive conglomerate, with beds of sand and gravel commonly 4 to 5 cm thick. Beds define both planar parallel stratification and tabular foresets. In decreasing abundance, conglomerate clasts include basalt, sandstone, and shale fragments similar in lithology to the Ocú Formation, and shell fragments.

Minor sandstone beds make up less than 5% of the section. They are commonly 5 to 200 cm thick, poorly sorted, fine- to coarse-grained, and rarely pebbly. Sedimentary structures in sandstone beds include normal grading and planar cross stratification and trough cross-stratification.

Facies interpretation of lower unit. Dominance of stratified conglomerate over massive conglomerate and the complete absence of mudstone interbeds suggests deposition of the lower Tonosí Formation by traction currents or sheet flow in a high-energy shallow-water environment. The presence of rare fossil-rich beds and lenses in conglomerate and sandstone is interpreted as the product of storm reworking and supports a shallow-water environment of deposition.

Age of lower unit. Two samples of finer-grained material collected from the Punta Bucaró section were barren of microfossils (Appendix 3).

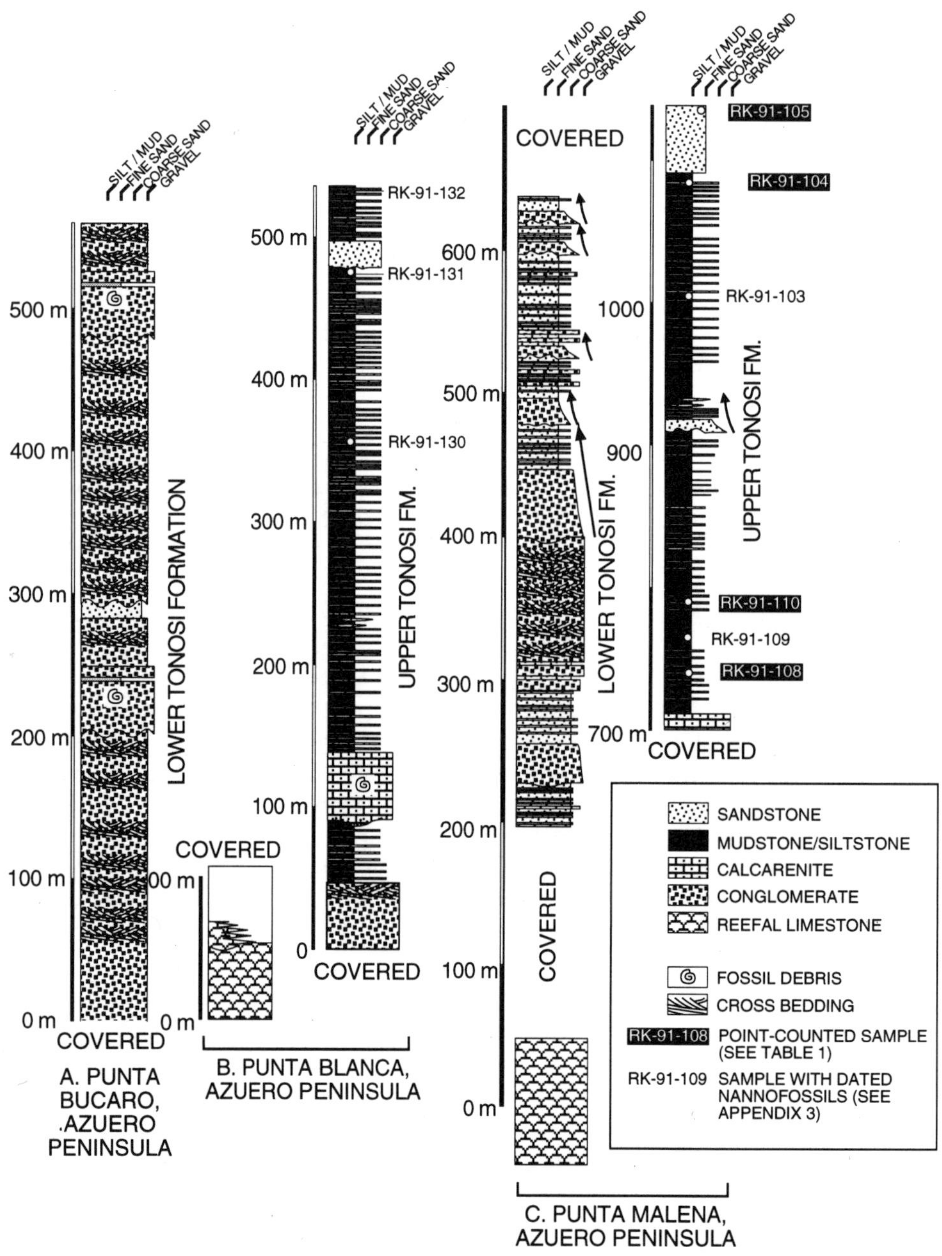

Figure 16. Measured stratigraphic sections of the lower part of the Tonosí Formation on the Azuero Peninsula. We correlate these units to seismic map unit GC1 in the Gulf of Chiriquí (see Fig. 3). A, Measured section at Punta Bucaró on the southeast coast of the Azuero Peninsula (location shown in Fig. 14). B, Two measured sections at Punta Blanca on the southern coast of the Azuero Peninsula: The section to the left shows Middle Eocene reefal limestone unconformably overlying basaltic basement (photograph of this contact is shown in Fig. 17; location shown in Fig. 14); the section to the right shows basal conglomerate and turbiditic section overlying a covered interval assumed to be basaltic basement similar to that seen at the section on the left. C, Measured section at Punta Malena on the west coast of the Azuero Peninsula (location shown in Fig. 13).

Lower unit of Tonosí Formation (upper Eocene–lower? Oligocene) in the northeastern Tonosí basin

Lithologic description. A carbonate facies of the lower Tonosí Formation is exposed in the valley of the Rio Tonosí in the northeastern part of the Tonosí basin near the head waters of the Rio Tonosí (Fig. 14; UTM coordinates $^{5}45$, $^{8}25$). This carbonate facies is considered to be laterally equivalent to the lower clastic facies described at Punta Bucaró and possibly the coal seams reported from the northeastern part of the basin by R. K. Jain (Appendix 1). Carbonate rocks consist of approximately 20 m of reef limestone and calcarenite. Recchi and Miranda (1977) suggested that some of the calcarenite beds consist of reworked clasts from nearby reefal units. Upper Oligocene tuffaceous sandstone overlies the carbonate rocks in the northwestern part of the Tonosí Basin (Recchi and Miranda, 1977).

Facies interpretation. The faunal characteristics of the limestone described from the northwestern part of the Tonosí Basin suggest a reefal environment (Recchi and Miranda, 1977). Based on petrographic study of calcarenite beds within deep-marine turbiditic rocks of the Tonosí basin, Recchi and Miranda (1977) proposed that these reefs along the northwestern edge of the basin shed carbonate debris in the form of fore-reef talus to deeper parts of the basin. This reef-derived talus mixed with a volumetrically larger amount of clastic debris derived from erosion of basaltic basement and the Ocú Formation to the north, west, and south of the Tonosí basin.

Age. Foraminifera from limestone samples have yielded ages of late Eocene to early (?) Oligocene (Recchi and Miranda, 1977; Appendix 3).

Tonosí Formation (middle Eocene–lower Miocene) at Punta Blanca

Geologic setting. Punta Blanca is on the southern coast of Azuero Peninsula to the south of the Tonosí basin (Fig. 14). Two sections were measured on the coast east of Punta Blanca in the intertidal zone or in coastal cliffs (Fig. 16B). The first section is an approximately 20-m-thick section of reef limestone located at UTM coordinates $^{5}48$, $^{8}02$. The second measured section is a 523-m-thick section of interbedded conglomerate, calcarenite, siltstone, and shale located approximately 5 km east of the first section. Both sections are easily accessible by car on a road from the town of Tonosí to a small fishing village near the coast at Punta Blanca.

Lithologic description of Punta Blanca reef section. The section exposes approximately 20 m of reef limestone unconformably overlying basement basalt (Fig. 16B). The limestone is massive and rubbly at its base and grades upward into a flaggy white limestone (Fig. 17).

Facies interpretation of Punta Blanca reef section. Lithologic characteristics and fauna suggest reef facies transgression across basaltic basement. A similar fossiliferous, reefal limestone was observed as a fault sliver in basalt approximately 30 km west of this outcrop at Punta Rastingue (UTM GR $^{5}09$, $^{8}09$) on the southwest coast of the Azuero Peninsula (Fig. 12). Based on these widespread exposures, it is possible that an Eocene, transgressive shallow-marine limestone may have extended over much of the southern Azuero Peninsula and have been eroded off of basaltic basement in most areas following Neogene faulting.

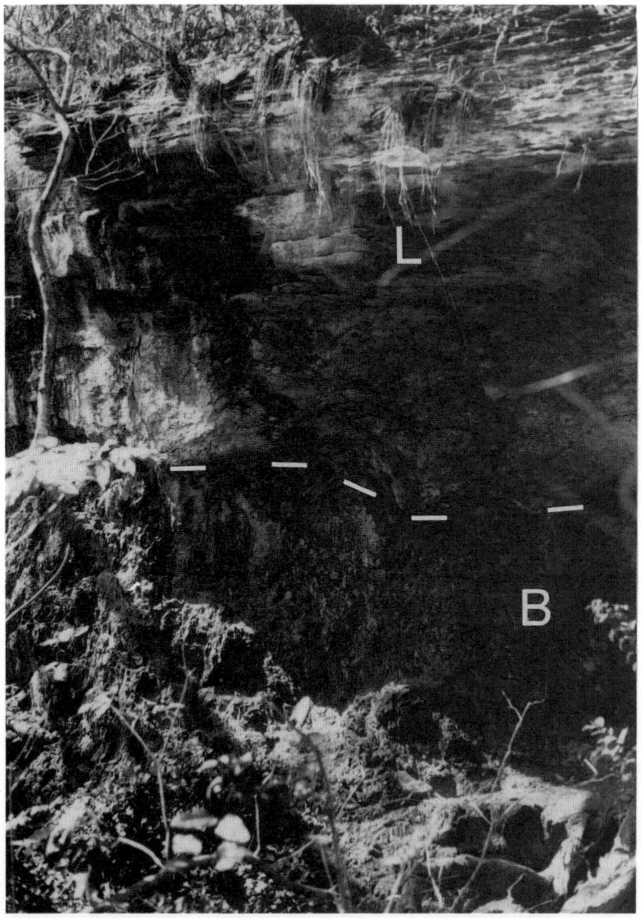

Figure 17. Photo of Middle Eocene reefal limestone (L) unconformably overlying Late Cretaceous basalt (B) at Punta Blanca on the southern coast of the Azuero Peninsula (see Fig. 14 for outcrop location; dashed line shows approximate position of contact). We correlate this horizon to the contact between acoustic basement and the overlying seismic unit GC-1 in the Gulf of Chiriquí (see Fig. 3). Overlying sandstone and shale of the Punta Blanca area are dated by nannofossils to be in the age range from the late Oligocene to near the Oligocene-Miocene boundary (see Appendices 1, 2).

Age of Punta Blanca reef section. The reef limestone at Punta Blanca contains *Lepidocyclina* foraminifera of Eocene age.

Lithologic description of lower unit of Punta Blanca clastic section. The lower unit consists of a 40-m-thick, thick-bedded to massive conglomerate, rich in fossil debris, that is abruptly overlain by a 500-m-thick section of thinly bedded sandstone and siltstone (Fig. 16B). The top of this coarse lower unit is taken as the top of the cross-bedded sandstone unit that immediately overlies the conglomerate.

Facies interpretation of the lower Tonosí Formation. Abundant fossil fragments and cross-bedding in the basal conglomerate suggest a high-energy marine environment that may have coexisted laterally to the reef observed 5 km to the east at Punta Blanca. The overlying section is a much finer grained turbiditic sandstone and shale that suggests a deepening-upward, transgressive event.

Lithologic description of the Upper Tonosí Formation. The lower unit is conformably overlain by a 450-m-thick sequence of interbedded sandstone, siltstone, and shale that show no well-defined coarsening- or fining-upward trends (upper Tonosí Formation in Fig. 16B). It is composed mainly of thin to medium beds of coarse sandstone interbedded with thinner beds of shale and siltstone (Fig. 18). The thicker beds of sandstone have scoured bases and are usually normally graded and contain planar laminations and rare cross-bedding. Sandstone beds show good lateral continuity along strike and are rarely lenticular. Clasts in coarser sandstones are mainly basalt fragments with lesser amounts of vein quartz, feldspar, and wood fragments.

Facies interpretation of the upper Tonosí Formation. Grading and scour features of sandstone beds suggest deposition by turbidity currents (facies D and E of Mutti and Ricci Lucchi, 1978). Gravity-driven deposition along a sloping surface is indicated by a single 15-m-thick slump unit (Facies F) near the top of the section (Fig. 16B).

Age of the upper Tonosí Formation. Nannofossils from four shale samples indicate a late Oligocene to early Miocene age for the upper unit of the Tonosí Formation at Punta Blanca (Appendixes 1, 3).

Figure 18. Photograph of steeply dipping Eocene-Oligocene siltstone and turbidite sandstone of the Tonosí Formation cut by several minor faults on Punta Blanca, Azuero Peninsula (location shown in Fig. 14). We correlate this unit to seismic unit GC-1 in the Gulf of Chiriquí. These rocks have been dated by nannofossils to be in the age range from the late Oligocene to near the Oligocene-Miocene boundary (see Appendices 1, 2). Beds young to left; 1 m of tape measure for scale.

Tonosí Formation (middle Eocene) at Punta Malena

Geologic setting. Punta Malena is located north of the village of Malena and south of the town of Llano de Mariato on the western coast of the Azuero Peninsula (Fig. 13). Semi-continuous exposures of the Tonosí Formation are present for several kilometers along the coast at Punta Malena at low tide. Conglomerate forms the lower portion of the 1,100-m-thick section that was measured at UTM coordinates $^{8}35$-$^{8}40$, $^{4}97$-$^{5}03$.

Lithologic description of the lower Tonosí Formation. The basal pebbly conglomerate and coarse-grained sandstone of the Punta Malena section is approximately 680 m thick and appears to directly overlie basement, although the contact is covered at this locality (Fig. 16C). Conglomerate clasts include basalt and intraclasts of sandstone and siltstone and range in size from 1 to 3 cm. Conglomerate beds are massive and range from 5 to 200 cm thick. In the lower part of the section, from 200 to 400 m, conglomerate beds commonly exhibit scoured and channeled bases (e.g., bed at 230 m). Between 310 and 400 m is a massive conglomerate bed with planar stratification. Planar stratification is common in both sandstone and conglomerate beds, but graded bedding and cross-stratification are rare. This interval (200 to 400 m on Fig. 16C) demonstrates a gradual coarsening-upward trend defined by an increase in grain size from fine sand to gravel. The upper part of the section (400 to 640 m) shows five fining-upward cycles over vertical distances of 20 to 80 m, each defined by a decrease in grain size (gravel to fine sand) and a decrease in the frequency of beds of coarse sandstone.

Facies interpretation of the lower Tonosí Formation. The coarsening-upward trend between 200 and 400 m, common planar stratification in conglomerate and sandstone beds, and lack of mudstone interbeds suggest clastic deposition in a marginal marine, high-energy environment similar to the one interpreted for the Punta Bucaró section (Fig. 16A).

Age of the lower Tonosí Formation. Samples collected from the basal Punta Malena section were barren of nannofossils (Appendix 2).

Lithologic description of the upper Tonosí Formation. The upper finer-grained section at Punta Malena is 432 m thick (Fig. 16C). A coarsening-upward trend over this interval is defined by: (1) an upward increase in the sand to shale ratio, and (2) an upward increase in sandstone grain size from fine to coarse (Fig. 16C). Sandstone in the uppermost part of the section is fine grained, well sorted, and thick bedded, with siltstone interbeds. The sandstones show good lateral continuity and fairly constant bedding thickness. The sandstones are commonly normally graded and contain rare laminations and planar cross-beds. Complete Bouma sequences are rarely developed, but $T_{a\text{-}b}$ intervals are common.

The much finer grained upper part of the section is in fault contact with sandstone and conglomerate of the lower, coarser section (Fig. 16C). The base of this finer-grained section is a 12-m-thick cream-colored calcarenite bed that is in direct fault contact with underlying conglomerate and sandstone.

Petrography of sandstones from the upper Tonosí Formation. Two hundred point counts were made on four sandstone samples from the upper unit to determine the relative abundance of minerals, rock fragments, cement, matrix, macroporosity organic matter, and fossil debris (Table 1). Sandstone of the upper unit at Punta Malena is volcaniclastic litharenite composed of abundant feldspar, altered glass fragments, altered volcanic rock fragments, scarce angular volcanic quartz, and scarce organic matter. Macroporosity is low, owing primarily to the abundance of clay matrix and minor pore-filling carbonate cement.

Facies interpretation of the upper Tonosí Formation. Grading and sole marks indicate that the upper unit comprises turbidite facies D of Mutti and Ricci Lucchi (1978). The overall coarsening-upward trend of this section reflects a sand-rich submarine fan system prograding over coarser-grained shallow-marine rocks of the lower unit.

Age of the Upper Tonosí Formation. Nannofossils in two shale samples from this section yielded a late upper Oligocene to early Miocene age for this section (Appendix 3).

Santiago Formation (upper Oligocene–Miocene?) of the Macaracas basin, north-central Azuero Peninsula

Geologic setting. The Macaracas basin is an approximately 30-km^2, subcircular basin in the mountainous area of the north-central Azuero Peninsula and prior to this study contained the youngest clastic sedimentary rocks reported by Recchi and Miranda (1977) from the Azuero Peninsula (Fig. 12, unit labeled Tmi). The area was visited and sampled during this study because of possible correlation of these rocks to offshore Neogene clastic rocks in half-grabens beneath the Gulf of Chiriquí. Outcrops are small and mainly restricted to recent road cuts. No sections have been measured here, either by us or by Recchi and Miranda (1977).

Lithologic description. Rocks of the Macaracas basin comprise an unknown thickness of interbedded sandstone, siltstone, shale, and coal and were considered by Recchi and Miranda (1977) to be part of the Santiago Formation described from the area north of the Azuero Peninsula.

Facies interpretation. Palynological analysis of one coal sample by R. K. Jain for the Dirección General de Hidrocarburos (Appendix 1) suggests deposition in a freshwater swamp.

Age. Palynological analysis of one coal sample by R. K. Jain indicates a possible late Oligocene to early Miocene age for the coal (Appendix 1, Table 2). Foraminiferal dating by the Dirección General de Hidrocarburos of one sandstone sample in 1974 from the northern part of the basin yielded a Miocene age (Appendix 1, Table 1). Nannofossil analysis of one sample collected by R. Kolarsky and analyzed by S. Monechi indicates a late Oligocene age (Appendix 3). Based on these ages, the age of sedimentation in the Macaracas basin could range from late Oligocene to Miocene and is therefore coeval with sedimentation of the upper Tonosí Formation in coastal areas to the southeast, south, and southwest (Fig. 12).

Equivalent units to the Tonosí Formation on the Soná Peninsula and Cébaco Island

Soná Peninsula. Outcrops of marine clastic rocks similar to the Tonosí Formation of the Azuero Peninsula are present on the southwestern coast of the Soná Peninsula (Metti et al., 1972; Metti and Recchi, 1976) (Fig. 4). We did not visit this area because of a lack of roads from populated areas to the north and unprotected coastlines that prohibit the landing of small boats.

Cébaco Island. Mapping by J. Corrigan showed that almost all of the island consisted of layered basaltic basement, with the exception of one small outcrop of undated coarse sandstone and conglomerate (UTM grid reference $^{8}32.5$, $^{4}84$) found on the north side of a west-striking, north-dipping normal fault (Fig. 13).

CORRELATION BETWEEN ACOUSTIC BASEMENT AND SEISMIC MAP UNITS IN THE GULF OF CHIRIQUÍ AND ROCKS ON THE AZUERO PENINSULA

Acoustic basement

Acoustic basement in the Gulf of Chiriquí is correlated with the layered basalt complex that is well exposed on the western and southern coast of the Azuero Peninsula (Fig. 13). We interpret the age of this basement as Late Cretaceous, based on radiolarian age data from the western coast (Appendix 2) and the Late Campanian–Maastrichtian age dates of the inland Ocú Formation by del Giudice and Recchi (1969). Intrabasement reflectors seen on the eastern end of Texaco line 1 may correlate to either sedimentary units of the Ocú Formation or to layering in igneous units as mapped along the western coast and on Cébaco Island (Fig. 13).

Lower subunit of GC1

The discontinuous character of seismic reflections within the lower subunit of GC1 and its position directly above what is proposed to be basaltic basement suggest that GC1 correlates to conglomerate, reefal limestone, and massive sandstone of the lower Tonosí Formation. This lower unit reaches a thickness of about 500 to 700 m at Punta Bucaro, Punta Blanca, and Punta Malena (Fig. 16).

Upper subunit of GC1

The more continuous reflection of this seismic unit correlates with the finer-grained and thinner-bedded rocks of the upper Tonosí Formation as seen at Punta Malena and Punta

Blanca. The upper unit reaches thicknesses of 400 to 500 m. The facies of the upper unit suggests deposition in a submarine fan setting.

Unit GC2

Rocks younger than early Miocene were not found on the Azuero Peninsula. Erosion of the topographically elevated, older rocks of the Azuero and Sona Peninsulas was probably the source area for rapid deposition of unit GC2 during Pliocene-Pleistocene time in the offshore half-grabens (Fig. 3). Erosion of arc volcanoes in western Panama (e.g., Barú, Fig. 1B) may also have contributed to the formation of unit GC2.

DISCUSSION

Introduction

Proposed stratigraphic correlations between seismic units in the Gulf of Chiriquí and onshore rocks on the Azuero and Soná Peninsulas and on Coiba and Cébaco Islands are summarized in Figure 19. Stratigraphic data used to make these correlations and regional interpretations are compiled in Table 2A–D, which also includes the results of Bowland and Rosencrantz (1988) and Bowland (1993) for the western Colombian basin (Table 2E). On the basis of our stratigraphic interpretations, we subdivide the geologic history of the southwest margin of Panama into seven stages.

Jurassic?–Late Cretaceous period of basaltic volcanism and pelagic sedimentation

Panama. We have documented the interbedded nature of massive and pillowed basalt with pelagic and hemipelagic limestone, sandstone, and shale at Punta Damas on the east coast of Coiba Island (Fig. 9). Sedimentary samples from this area were barren of microfauna (Appendixes 1, 3), but samples of lithologically identical limestone (Ocú Formation) in the central Azuero Peninsula have yielded Late Campanian–Maastrichtian foraminifera (del Giudice and Recchi, 1969), and radiolarians from chert in a basalt-chert tectonic mélange on the western coast of the Azuero Peninsula yielded a Coniacian age (Appendix 2). We conclude that the crust of southwestern Panama consists of basaltic crust formed in a submarine setting, partly in Late Cretaceous time and isolated from major terrigenous sediment sources. Bandy and Casey (1973) and Case (1974) reached a similar conclusion about the basaltic crust forming the basement of eastern Panama.

Costa Rica. In the well-studied Nicoya Complex of Costa Rica, Lundberg (1982) and Schmidt-Effing et al. (1982) recognized a basal sequence of massive basalt flows and pillow basalt overlain by abyssal radiolarite of Jurassic to late Early Cretaceous age. They interpreted this section as upper Jurassic oceanic crust and associated deep-water sedimentary rocks.

The upper Nicoya Complex comprises a basal sequence of gabbroic and dioritic stocks overlain by a series of basaltic sheet flows of Aptian to Early Campanian age. The flows are interbedded with mafic agglomerate, volcanogenic turbidites, and pelagic sedimentary rocks that become increasingly calcareous upward. The sedimentary rocks of the lower and upper Nicoya Complex are mainly pelagic deposits and coarse sedimentary breccias that contain no significant continent-derived quartzo-feldspathic detritus (Lundberg, 1991). It is not clear whether the contact between the upper and lower Nicoya complex is tectonic or stratigraphic. Gursky et al. (1982) and Schmidt-Effing et al. (1982) proposed that the upper Nicoya Complex is a Late Cretaceous oceanic plateau constructed upon oceanic crust of the Late Jurassic–Early Cretaceous Nicoya Complex. Shallowing of sedimentation surfaces in the upper Nicoya Complex represents a buildup of successive volcanic flows that formed the plateau.

Western Colombian basin. Undrilled, acoustic basement in the western Colombian basin is interpreted by Bowland and Rosencrantz (1988) and Bowland (1993) as laterally heterogeneous crust dominated by the Mono Rise, an oceanic plateau similar to aseismic oceanic rises in the western Pacific. They interpreted the crust of the Colombian basin as the product of two phases of regional igneous activity. The first phase created oceanic crust inferred to be Late Jurassic–Early Cretaceous. Later intraplate volcanism in the Early to Late Cretaceous created the Mono Rise. They proposed that the modified oceanic crust of the Colombian basin is continuous with the crust of Panama and Costa Rica because seismic lines suggest that basement rises continuously toward onshore outcrops.

Bowland (1993) has mapped a sheet-draped seismic unit (map unit CB5) that lies directly on acoustic basement of the Colombian basin and is restricted to topographically high areas. The maximum thickness of the unit is about 0.5 km near the crest of the Mono Rise. He interprets the unit as mainly biogenic deposits. Following the interpretation of Bowland (1993), we correlate his seismic map unit CB5 with Upper Cretaceous pelagic sedimentary rocks described from the Nicoya Peninsula (Schmidt-Effing, 1982), the southwestern margin of Panama (Ocú Formation of this chapter) and from eastern Panama (Bandy and Casey, 1973; Case, 1974).

Paleogene erosional event

Southwestern Panama. We have identified a stratigraphic hiatus inferred to represent a major erosional event that affected the southwestern margin of Panama from Paleocene to early Eocene time (Fig. 19). The age of this event is constrained by Late Campanian–Maastrichtian foraminifera from the underlying Ocú Formation of the Azuero Peninsula (del Giudice and Recchi, 1969), Coniacian radiolarians from the basaltic basement of the Azuero Peninsula (Appendix 2), and early to middle Eocene nannofossils from the basal Tonosí Formation on Coiba Island (Appendix 3). The erosional sur-

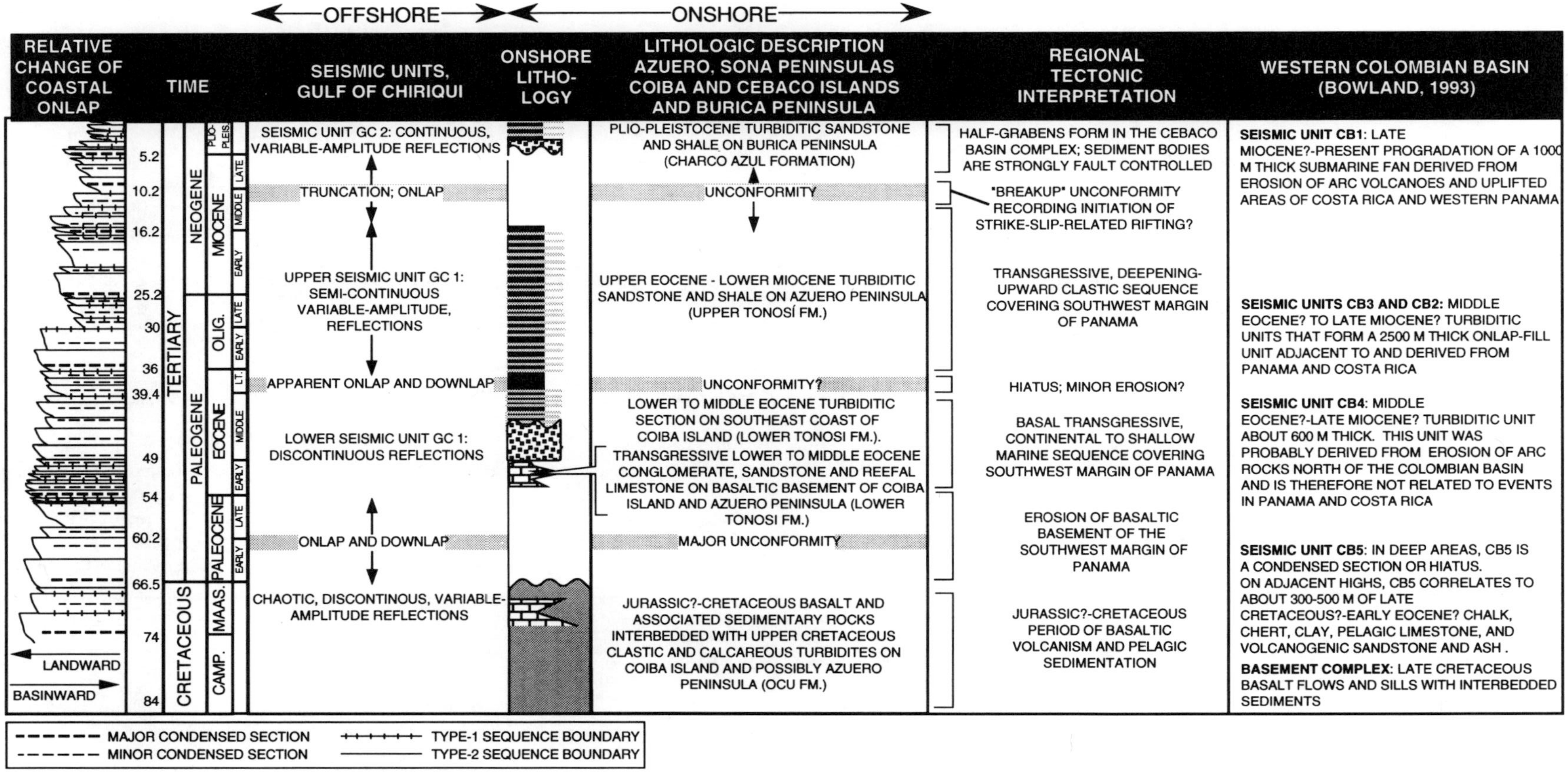

Figure 19. Stratigraphic, age, and tectonic correlation between rocks exposed on the Azuero Peninsula and reflectors seen on seismic lines across the Cébaco Basin complex of the eastern Gulf of Chiriquí. Note Paleogene and Miocene erosional events are interpreted to have affected a large area that now includes both offshore and onshore areas. Relatiave change of coastal onlap curve, condensed sections, and sequence boundaries is from Haq et al. (1987). Note that the geologic history of the western Colombian basin from Bowland (1993) is broadly similar to the history of the Gulf of Chiriquí area.

TABLE 2. COMPARISON OF SEQUENCE STRATIGRAPHY OF THE GULF OF CHIRIQUI AND WESTERN COLOMBIAN BASIN WITH THE ONLAND STRATIGRAPHY OF COIBA ISLAND AND THE AZUERO PENINSULA, SOUTHWESTERN PANAMA

Seismic Unit	Seismic Facies	Stratigraphic Thickness	Relative Sea Level Trend	Interpreted Age and Correlation with Onshore Units	
A. Gulf of Chiriqui, Southwestern Panama (this chapter)					
GC2	Continuous, variable amplitude reflections	0 - 3,000 m	Deepening upward	Plio-Pleistocene; Charco Azul Formation Time-equivalent	
GC1 (Upper subunit)	Variable-aplitude, semi-continuous reflections	0 - 1,000 m	Deepening upward	Time-equivalent of upper unit of Tonosi Formation	Eocene Lower Miocene
GC1 (Lower subunit)	High amplitude, continuous reflections		Deepening upward	Time-equivalent of lower unit of Tonosi Formation	

Formation	Age	Environment	Relative Sea Level Trend	Depositional System	Stratigraphic Thickness
B. Burica Peninsula (from Corrigan et al., 1990; Coates et al, 1992)					
Charco Azul Formation -Armuelles Member	Upper Pleistocene	Lower Bathyal (1,200 - 1,300 m)	Shallowing upward	Slope turbidites	200 m
Charco Azul Formation -Burica Member	Pliocene - Lower Pleistocene	Lower Bathyal (1,400 - 2,000 m)	Deepening upward	Slope basin turbidite fill	3,000 m
Charco Azul Formation - Peñita Member	Lower Pliocene	Inner-outer Neritic	Deepening upward	Fan delta (?)	300 - 330 m
Unnamed	Upper Paleocene-Lower Eocene	Shallow marine?	?	Carbonate platofrm	<50 m
Nicoya Complex	Upper Cretaceous-Upper Paleocene(?)	Deep marine?	?	Pelagic-hemipelagic	Variable
C. Coiba Island (this chapter)					
Tonosi (Upper unit)	Middle Eocene	Neritic Bathyal?	Deepening upward	Slope turbidites	-800 m
Tonosi (Lower unit)		Neritic?	Deepening upward	Marginal marine, reefal	-150 m
Ocú	Upper Cretaceous	Bathyal?	High	Pelagic/hemipelagic	>450 m
D. Azuero Peninsula (thic chapter)					
Santiago	Miocene	Inner-outer Neritic	Deepening upward?	Continental marginal marine	?
Tonosi (Upper unit)	Eocene-Lower Miocene	Middle-outer Neritic	Deepening upward	Slope turbidites	>500 m
Tonosi (Lower unit)		Mainly inner Neritic	Deepening upward	Continental, reefalmarginal marine	>400 m
Ocú	Upper Cretaceous	Bathyal?	Deepening upward?	Pleagic/hemipelatic	?

TABLE 2. COMPARISON OF SEQUENCE STRATIGRAPHY OF THE GULF OF CHIRIQUI AND WESTERN COLOMBIAN BASIN WITH THE ONLAND STRATIGRAPHY OF COIBA ISLAND AND THE AZUERO PENINSULA, SOUTHWESTERN PANAMA (continued)

Seismic Unit	Age	Seismic Facies	Calculated Stratigraphic Thickness	Depositional Mechanism	Relative Sea Level Trend
E. Western Colombian Basin (From Bowland, 1984, and Bowland, in press)					
CB 1	Pliocene Recent	Completely mounded, variable amplitude and continuous reflections.	1,000 m	Proximal turbidites	Shallowing upward
CB 2	Middle Eocene-Upper Miocene	Sub-parallel-parallel, variable amplitude and continuous reflections	2,500 m	Distal turbidites	Gradual shallowing upward
CB 3	Middle Eocene-Upper Miocene	Sub-parallel-parallel, variable amplitude and continuous reflections		Distal turbidites	Gradual shallowing upward
CB 5	Upper Cretaceous	Hummocky mounded, highly disrupted, variable amplitude, discontinuous reflections	<100 m	Pelagic fallout	Deepening?

face is overlain by transgressive conglomerate and reefal limestone on Coiba Island and the Azuero Peninsula. On seismic lines in the Gulf of Chiriquí, the unconformity is marked by a prominent irregular surface separating acoustic basement from seismic map unit GC1 (Fig. 3). We are unsure of the tectonic significance of this event beyond the obvious conclusion that it marks the end of submarine basaltic volcanism and pelagic sedimentation and the beginning of terrigenous clastic sedimentation in this area of Panama.

Comparison to Costa Rica and westernmost Panama. Paleocene sedimentation on the Nicoya Peninsula is characterized by a 1,000-m-thick, coarsening-upward turbidite wedge (Las Palmas Formation) that prograded northeastward (Lundberg, 1982). Thick Paleocene turbidites are not present on the seaward part of the Nicoya Peninsula, where a thin (200 to 300 m) sequence of thin-bedded turbidites and calcareous mudstone is found instead. Lundberg (1982) interpreted this relationship as a Paleocene outer arc high along the Pacific margin of Costa Rica. Berrangé and Thorpe (1988) have dated a major volcanic event of Paleocene age on the Osa Peninsula of Costa Rica. Corrigan et al. (1990) reported a bioclastic limestone from above basaltic basement on the Burica Peninsula with an upper Paleocene to lower Eocene foraminiferal assemblage.

Colombian basin. Bowland (1993) infers that the Late Campanian to early middle Eocene is represented by a hiatus in sedimentation, a condensed section, or erosion in deeper-water areas. He infers a change from biogenic pelagic sedimentation to clastic sedimentation in early Tertiary time, probably during early to middle Eocene. Therefore, it is possible that the Paleocene–early Eocene erosional interval of southwestern Panama correlates in part with the gap in sedimentation proposed by Bowland (1993) for the western Colombian basin.

Lower to Middle Eocene basal stratigraphic section

Southwestern Panama. We have identified a ~1,000-m-thick basal transgressive sequence, which we dated using nannofossils, as early to middle Eocene age on Coiba Island and assume it to be the same age on the Azuero Peninsula (Appendixes 1, 3). The Eocene basal section appears to be thicker on Coiba Island than on the Azuero Peninsula, where marine rocks of Oligocene age occur almost directly above the basal limestone and conglomerate (Fig. 16). It is possible that this basal section is highly variable in thickness and lateral facies as suggested from its proposed seismic equivalent, the lower subunit of unit GC1, in the Gulf of Chiriquí (Fig. 3). For example, unit GC1 is much thinner (<200 m) in the western gulf and missing in the central gulf in the Montuosa basin and across the Parida and Ranchería arches (Kolarsky and Mann, this volume) (Fig. 4). In the western gulf, unit GC1 exhibits higher-amplitude reflections than does the thicker GC1 unit in the Cébaco basin complex, which may reflect lateral changes in lithology and sedimentary facies (facies variations between carbonate rocks on highs and clastic rocks in lows?).

It is possible that a second erosional event occurred at the interval between the deposition of the lower and upper units of seismic unit GC1 in the Gulf of Chiriquí, based on a prominent reflection surface that is downlapped and onlapped by the upper subunit of unit GC1 (Fig. 3). We suggest that this surface is an unconformity representing erosion of part or all of the lower to middle Eocene section. This interpretation could explain the localized nature of the early to middle Eocene basal section on the southeastern end of Coiba Island and the presence of younger Oligocene and Miocene rocks immediately above the basement surface on the Azuero Peninsula.

Costa Rica and westernmost Panama. Lundberg (1982) has described a distinctive white, shallow-water Eocene limestone that is several tens of meters thick and that unconformably overlies deep-marine deposits of the Nicoya basement complex and overlying sedimentary formations of Cretaceous to Paleocene age. The limestone forms the base of a shallow-water section composed of limestone and sandstone (Mal Pais unit of Lundberg, 1982). In the Terraba belt of Costa Rica, Eocene deposits are represented by a maximum of 1,400 m of the Brito Formation, composed of bioclastic limestone, calcareous sandstone, marl, tuff, and volcanic breccias formed in a shallow-water environment (Phillips, 1983). The basal contact with underlying basaltic basement rocks is faulted, and the great thickness of the unit may be caused by repetition along thrust faults (Kolarsky et al., this volume). The Mal Pais unit of the Nicoya Peninsula and the Brito Formation of the Terraba belt are lithologically similar to the limestone found above basement on the Azuero Peninsula and Coiba Island, which suggests that shallow-marine deposition covered a large area of southwestern Panama and Costa Rica during Eocene time.

Western Colombian basin. Bowland (1993) recognized two seismic units (CB2, CB3) that he interpreted to be a 2,500-m-thick section of middle Eocene to middle Miocene clastic rocks, probably deposited in the form of unconfined turbidity-flow deposits. Thick units of ponded turbidites of units CB2 and CB3 indicate that important sedimentary-source terranes developed along the northern and western margins of the western Colombian basin. We correlate the lower section of the principal onlap-fill unit (CB3) in the western Colombian basin with the deposition of the lower unit of the Tonosí Formation on the Azuero and Soná Peninsulas and with deposition of lower subunit GC1 in the Gulf of Chiriquí (Fig. 4). These clastic units appear to reflect the initial uplift and erosion of the volcanic arc of southern Central America.

Upper Eocene to lower Miocene stratigraphic section

Southwestern Panama. We have identified a ~500-m-thick late Eocene to lower Miocene section in the Tonosí area of the Azuero Peninsula and along the western coast of the Azuero Peninsula. Rocks in this section include mainly marine turbiditic sandstone and shale that exhibit thinning- and fining-upward trends. The oldest faunas recovered include on Eocene nannofossil assemblage from the western coast of the Azuero Peninsula and late Eocene, Oligocene, and early Miocene nannofossil and foraminifera assemblage from the Tonosí area and the southern coast of the Azuero Peninsula (Appendixes 1, 3). We propose that this well-bedded, mainly turbiditic section correlates with the upper subunit of seismic unit GC1 in the Gulf of Chiriquí, where the upper subunit is characterized by more continuous reflections than the lower subunit (Fig. 4).

Costa Rica and westernmost Panama. The upper Eocene to lower Miocene interval of the Nicoya Peninsula is represented by poorly dated marine rocks (Lundberg, 1982). This interval in the Terraba belt is represented by early Oligocene–middle Miocene Terraba Formation deposited mainly as turbidity or debris flows in submarine fan or basin plain environments (Phillips, 1983). A local angular unconformity separates the Terraba Formation from the overlying middle to late Miocene Curré Formation, which contains clastic lithologies recording a shoaling-upward trend from deep-sea to shallow-marine and locally terrestrial environments.

Western Colombian basin. Bowland (1993) interpreted seismic map units CB3 and CB4 as representing continuing deposition during a long period of turbiditic sedimentation in the Colombian basin. We correlate the upper section of clastic sedimentation (unit CB4) in the western Colombian basin with the deposition of the upper unit of the Tonosí Formation on the Azuero Peninsula and with deposition of the upper subunit GC1 in the Gulf of Chiriquí (Table 2).

Middle Miocene to late Miocene unconformity?

Fossils from outcrops around the Gulf of Chiriquí have produced no ages in the interval between the early Miocene (upper part of the Tonosí Formation) and the early Pliocene (base of the Charco Azul Formation on the Burica Peninsula) (Table 2). We correlate this interval with the prominent reflection surface that separates downlapping reflections of seismic map unit GC2 from seismic map unit GC1 (Fig. 3). It is possible that this surface represents an angular or "breakup" unconformity associated with the initiation of strike-slip faulting and pull-apart style rifting in the Gulf of Chiriquí. The presence of marine sedimentary rocks in the adjacent Terraba belt of Costa Rica (Terraba and Curré Formations) supports the idea of a local "breakup" unconformity related to local faulting rather than a widespread event.

Early Pliocene to Pleistocene rifting

The distribution, thickness, and geometry of seismic unit GC2 suggest a major rifting event in the Gulf of Chiriquí (Fig. 4). Field studies by ourselves and previous workers, including Terry (1956) and the United Nations team, in the land areas surrounding the eastern Gulf of Chiriquí have not identified any exposed half-graben structures containing syn-rift sediments that could correlate to seismic map unit GC2 in the Cébaco Basin complex (Fig. 5). The youngest sedimentary rocks identified to date on the Azuero and Soná Peninsulas and Coiba Island are early Miocene strata of the upper part of the Tonosí Formation, which we correlate to the upper part of seismic unit GC1. It is likely that the large vertical fault movements that created the half-grabens of the Cébaco basin complex and exposed basement rocks in land areas around the Gulf of Chiriquí acted to strip post–early Miocene strata off of older rocks.

We correlate seismic unit GC2 with the early Pliocene Peñita Member of the Charco Azul Formation on the Burica Peninsula (Fig. 6). Divergent reflections and wedge-shaped geometries of unit GC2 indicate fault control during deposition of the unit by normal and oblique-slip faults bounding the half-grabens of the Cébaco basin complex (Kolarsky and Mann, this volume). In the Cébaco basin complex, syn-rift fault control is also shown by the base of unit GC2 that onlaps the underlying seismic unit GC1 along a prominent unconformity surface on Texaco line 1 (Fig. 3).

Costa Rica and westernmost Panama. The Pliocene-Pleistocene interval on the Nicoya Peninsula is represented by a 120-m-thick, shallow-water section of Early Pliocene conglomerate and sandstone (Montezuma Formation of Lundberg, 1982, and McKee, 1985). In the Terraba belt, Pliocene-Pleistocene rocks are nonmarine, alluvial, and fluvial and rest with angular unconformity on underlying tilted and faulted marine rocks. The stratigraphy of Pliocene-Pleistocene rocks of the Burica Peninsula has been previously summarized.

Western Colombian basin. This interval is represented by seismic unit CB1 that Bowland (1993) interprets to be prograding submarine fan deposits related to increased erosion and volcanic activity in southern Central America to the west and south. We propose that the seismic unit GC2 that fills in half-grabens in the Gulf of Chiriquí correlates with the same period of increased erosion and volcanic activity.

Tectonic implications of this study

A question posed in the introduction of this chapter was: Can the stratigraphy of southwestern Panama help distinguish between two contrasting models for the origin of the crust beneath Panama: (1) The crust of southern Central America represents the western upturned edge of a Late Cretaceous Caribbean oceanic plateau inferred from deep sea drilling and seismic stratigraphic studies in the Colombian and Venezuelan basins (Bowland and Rosencrantz, 1988; Bowland, 1993). (2) The crust represents fragments of seamounts and other features accreted to the west- and southwest-facing Panama island arc during Late Cretaceous and early Tertiary time and is therefore distinct from the crust of the Colombian basin (DiMarco et al., this volume).

The similarities in stratigraphic history between southwestern Panama, Costa Rica, and the Colombian basin support the first hypothesis. However, pelagic and turbiditic rocks represent common facies that could have been deposited in widely separated areas. A test of these two hypotheses will require more detailed studies of the basement and sedimentary cover of Central America and deep sea drilling and sampling of the basement and cover of the Colombian basin.

Eustatic controls on stratigraphy

We are unable to make any correlations between the Exxon global sea-level curve (Haq et al., 1987) and stratigraphic events seen in southwestern Panama because of the lack of detail in our stratigraphic and age control (Fig. 19). In general, upper Cenozoic rocks of Panama and Costa Rica show a gradual shallowing-upward trend that is interpreted by most workers as tectonic uplift related to the formation of the present-day isthmus (e.g., Coates et al., 1992). It is possible that the Paleocene erosional event in southwestern Panama is in part related to an episode of low eustatic sea level during this interval (Fig. 19).

CONCLUSIONS

1. Interpretation of 3,100 km of seismic lines from the Gulf of Chiriquí reveals that the stratigraphy of the Gulf of Chiriquí consists of three major seismic map units: **(1) acoustic basement** that is correlated with outcrops of an unknown thickness of Jurassic?–Late Cretaceous basalt, chert, and pelagic rocks on the Azuero and Soná Peninsulas and Coiba and Cébaco Islands; **(2) lower subunit of seismic map unit GC1,** a sequence of discontinuous reflections that is correlated with outcrops of transgressive lower to middle Eocene conglomerate and reefal limestone that unconformably overlie basaltic basement on Coiba Island (~150 m), the Azuero Peninsula (~400 m), and the Burica Peninsula (<50 m); **upper subunit of seismic map unit GC1,** a sequence of semicontinuous, variable amplitude reflections that is correlated with outcrops of a ~500-m-thick section of mainly upper Eocene to early Miocene turbiditic sandstone and shale on the Azuero Peninsula; and **(3) seismic map unit GC2,** a 0- to 3-km-thick sequence of continuous, variable-amplitude reflections correlated with outcrops of a ~3,000-m-thick section of mainly Pliocene-Pleistocene turbiditic sandstone and shale on the Burica Peninsula.

2. Correlations of seismic map units in the Gulf of Chiriquí and the western Colombian basin with onland formations allow recognition of six stages in the stratigraphic development of the region (Fig. 19).

3. The first stage is recorded by basaltic basement rocks of the Jurassic–Late Cretaceous age interbedded and overlain by pelagic sedimentary rocks. Upper Cretaceous basaltic and pelagic rocks of southwestern Panama formed in an intraoceanic setting removed from terrigenous influence. It is possible that these rocks are underlain by older (Jurassic?) rocks of oceanic affinity.

4. A second stage is represented by a major stratigraphic hiatus inferred to represent a major erosional event that affected the southwestern margin of Panama in Paleocene time. We are unsure of the tectonic significance of this event beyond the obvious conclusion that it marks the end of submarine basaltic volcanism and the beginning of terrigenous clastic sedimentation in southwestern Panama.

5. A third stage is represented by a ~50- to 400-m-thick, basal transgressive section of coarse clastic and reefal carbonate rocks of early to middle Eocene age. This initiation of clastic sedimentation is a widespread stratigraphic event across southern Central America that appears to represent the initial uplift and erosion of land areas.

6. A fourth stage is represented by a 500- to 800-m-thick, deepening-upward section of mainly marine turbiditic rocks that appears to represent continued erosion of the land areas in southern Central America.

7. A fifth stage is represented by a stratigraphic hiatus in middle Miocene to late Miocene time that may represent the "breakup" unconformity associated with the initiation of strike-slip faulting and rifting in the Gulf of Chiriquí.

8. A sixth and final stage is represented by major rifting and syn-rift sedimentation in the Gulf of Chiriquí (Kolarsky and Mann, this volume). Three km of sedimentary infill of rift basins may reflect accelerated uplift of southern Central America and increased activity of the Middle America volcanic arc.

9. The similarities in stratigraphic events (except numbers five and six above) across large areas of southern Central America and the western Colombian basin supports the hypothesis that the crust of southern Central America represents the western upturned edge of a Late Cretaceous Caribbean oceanic plateau known from deep sea drilling and seismic stratigraphic studies in the Colombian and Venezuelan basins of the Caribbean Sea (Bowland and Rosencrantz, 1988; Bowland, 1993).

ACKNOWLEDGMENTS

This study would not have been possible without geophysical data from the Gulf of Chiriquí that were kindly provided by Texaco and Shell International. We thank Q. Nary at Texaco and A. J. Bol at Shell International for locating and allowing us to use these data. We also thank N. Villareal of the Dirección General de Hidrocarburos for providing us with unpublished reports on the geology of southwestern Panama and giving us permission to use these data in this publication. Special thanks to L. Campos, J. Corrigan, D. Heil, R. Miranda, and B. Shelton for field and logistical assistance in Panama. Reviews by C. Bowland, R. Buffler, J. Case, and N. Lundberg significantly improved this chapter. Funding for this project was provided by the Petroleum Research Fund of the American Chemical Society (grant number 2076-AC2-2). UTIG contribution no. 1076.

APPENDIX 1. SUMMARY OF PALEONTOLOGIC SAMPLING AND RESULTS TO ACCOMPANY "STRATIGRAPHIC DEVELOPMENT OF SOUTHWESTERN PANAMA AS DETERMINED FROM INTEGRATION OF MARINE SEISMIC DATA AND ONSHORE GEOLOGY"

Radim A. Kolarsky and Paul Mann

Introduction

The purpose of this appendix is to provide a summary and overview of the locations and ages of chert samples collected for radiolarians, shale samples collected for foraminifera and nannofossils, and coal samples collected for pollen.

Data sources and sample locations

Chert samples collected by P. Mann and J. Corrigan in 1987 and 1988 from the basement complex of the Azuero and Soná Peninsulas and Cébaco Island (App. 1 Fig. 1) were prepared and analyzed for their radiolarian content by D. Meyerhoff-Hull and E. Pessagno, Jr., at the University of Texas at Dallas. Their results are reported in Appendix 2 of this chapter. Mudstone and siltstone samples collected by R. Kolarsky, Mann, and Corrigan in 1987, 1988, and 1991 (App. 1 Fig. 2) were prepared and analyzed for their radiolarian content by S. Monechi of the University of Florence. Her results are reported in Appendix 3 of this chapter.

We have also included the results of two unpublished reports provided to us by the Dirección General de Hidrocarburos in Panama City. Two studies by Dirección General de Hidrocarburos dated 1974 and 1971 report on the content of foraminifera from the Tonosí basin area of the Azuero Peninsula. We have reproduced these results as Table 1 of Appendix 1 and show the locations of the samples on Appendix 1 Figure 3. A Dirección General de Hidrocarburos study by R. K. Jain, dated 1974, reports on the pollen content from coal seams in the Tonosí and Macaracas areas of the Azuero Peninsula. We have reproduced these results as Table 2 of Appendix 1 and show the locations of the samples on Appendix 1 Figure 4. We discuss the stratigraphic and tectonic significance of these fossil dates and paleoenvironments in the accompanying chapter.

APPENDIX 2. RADIOLARIAN BIOSTRATIGRAPHIC ANALYSIS FROM THE AZUERO PENINSULA, PANAMA

Donna Meyerhoff Hull and Emile A. Pessagno, Jr.

Introduction

Fourteen chert samples from the Azuero Peninsula, Soná Peninsula, and Cébaco Island, south of the Azuero-Soná fault zone, have been analyzed for Radiolaria (App. 1 Fig. 1). The cherts were collected from a mélange of sheared basalt and chert blocks that make up part of the basement rock in southwestern Panama (Kolarsky, 1992; Kolarsky et al., this volume). Although Early Jurassic through Late Cretaceous radiolarian assemblages have been described north of this area in Costa Rica (e.g., Santa Elena Peninsula and Nicoya Peninsula; Baumgartner, 1984; De Wever et al., 1985; Schmidt-Effing, 1980), no radiolarian faunas have previously been described and illustrated from southwestern Panama.

Of the fourteen chert samples analyzed (App. 1 Fig. 1) only one,

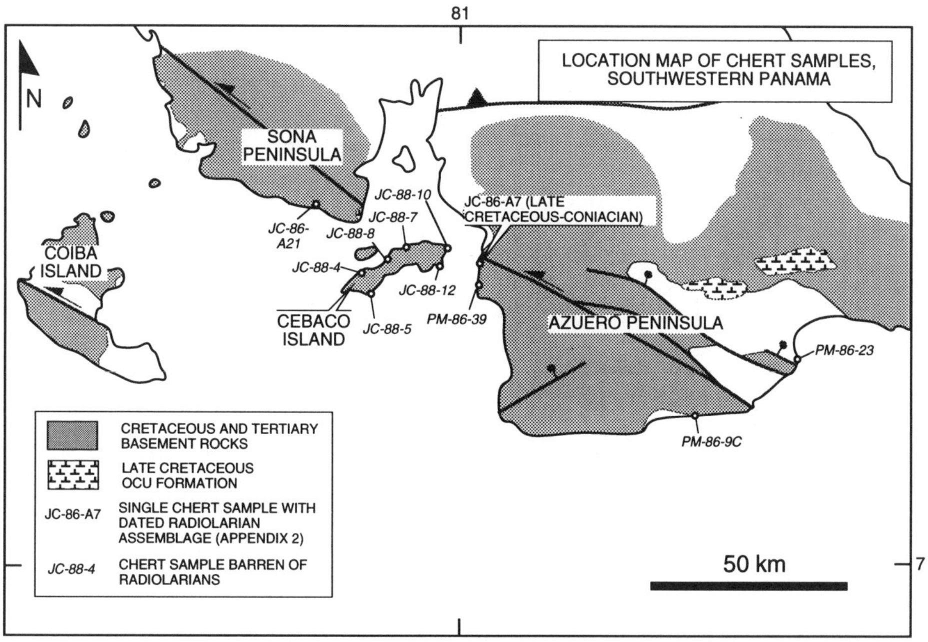

Appendix 1 Figure 1. Location map of chert samples, southwestern Panama.

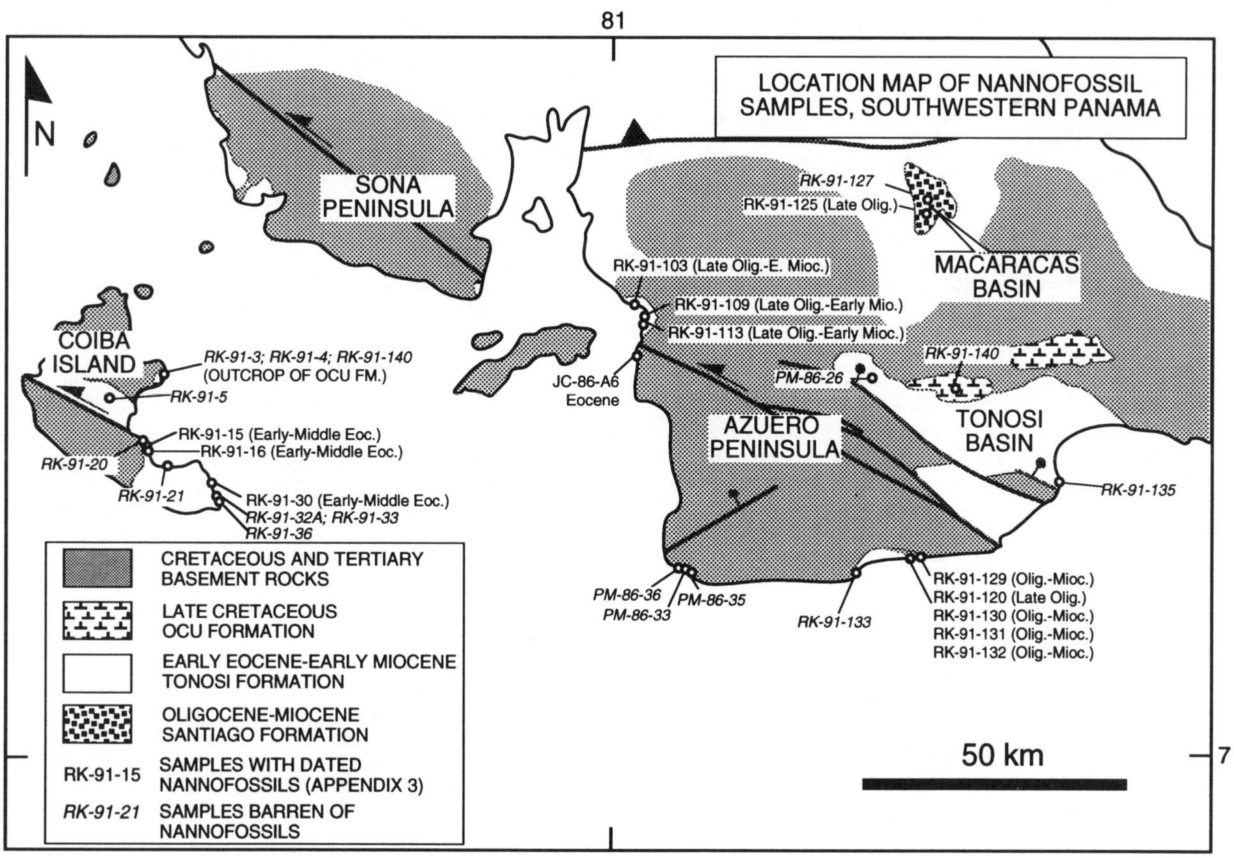

Appendix 1 Figure 2. Location map of nannofossil samples, southwestern Panama.

APPENDIX 1

TABLE 1. PLANKTIC FORAMINIFERA CONTENT, AGE, AND PALEOENVIRONMENT OF MUDSTONE AND SILTSTONE SAMPLES DESCRIBED IN UNPUBLISHED REPORTS BY DIRECCIÓN GENERAL DE HIDROCARBUROS, PANAMA CITY*

Sample 128 (Tonosi Basin, upper Tonosi Formation)

Abundant fauna, moderate preservation (Dirección General de Hidrocarburos, 1974).

Pliolepidina pustusoa tobleri
Lepidocyclina spp.
Operculinoides sp.
Red encrusting algae (Lithothaminum)

Paleoenvironment: Very shallow inner neritic, possible bank or near reefal conditions.

Age: Late Eocene.

Sample 134 (Tonosi Basin, upper Tonosi Formation)

Abundant fauna, moderate preservation (Dirección General de Hidrocarburos, 1974).

Pliolepidina pustuosa tobleri
Operculinoides cf. *moodybranchensis*
Globorotalia cf. *G. centralis*

Paleoenvironment: Inner to middle neritic.

Age: Late Eocene

Sample 135 (Tonosi Basin, upper Tonosi Formation)

Abundant fauna, moderate preservation (Dirección General de Hidrocarburos, 1974).

Pliolepidina pustuosa tobleri
Lepidocyclina spp.
Pelecypod fragments

Paleoenvironment: Shallow inner neritic.

Age: Late Eocene.

Sample 172 (Tonosi Basin, upper Tonosi Formation)

Very abundant fauna, good preservation (Dirección General de Hidrocarburos, 1971)

Opercullinoides bullbrooki
Lepidocyclina sp.
Miogypsina sp.

Paleoenvironment: Very shallow inner neritic, quiet, slightly hypersaline.

Age: Oligocene.

Sample 297 (Macaracas Basin, Santiago Formation)

Abundant fauna, moderate preservation (Dirección General de Hidrocarburos, 1974).

Globigerina venezuelana
Globigerinoides triloba immatura
Hastigerina cf. *H. siphonifera siphonifera*
Siphonegerina transversa
Cibicides trinitatensis
Globorotalia sp.

Paleoenvironment: Deeper inner to middle neritic.

Age: Miocene.

Sample 311 (Tonosi Basin, upper Tonosi Formation)

Abundant fauna, moderate preservation (Dirección General de Hidrocarburos, 1974)

Pliolepidina pustuosa toblieri

Paleoenvironment: Middle to outer neritic.

Age: Late Eocene.

*See Appendix 1, Figure 4 for sample locations.

JC-86-A7, contained radiolarians. Sample JC-86-A7 was collected along the west coast of the Azuero Peninsula, south of the town of Malena, Panama. The radiolarian assemblage is poorly diversified but well-preserved for a chert assemblage (App. 2 Plate I on p. 197). Biostratigraphic analysis of the radiolarian fauna indicates that the age of the assemblage is Coniacian (Late Cretaceous), providing the first age constraint on basement rocks south of the Azuero-Soná fault zone in Panama.

Methodology

Chert samples collected from Panama were processed in the laboratory at The University of Texas at Dallas, using the hydrofluoric acid technique of Pessagno and Newport (1972). Accordingly, cherts were mechanically broken down into 2- to 5-cm fragments and placed in a solution of 1 part hydrofluoric acid to 9 parts water for 24 hrs. The rock samples were then washed through a stack of 40-µm, 80-µm, and 230-µm mesh sieves; any remaining large fragments (>40 µm) were placed back into dilute hydrofluoric acid for a second 24-hr run. The residues of finer fractions (40- to 80-µm mesh and 80- to 230-µm mesh) were cleaned of iron-rich clays by placing the residues in hydrochloric acid and heating gently for 3 to 5 minutes. Cleaned residues were then washed and resieved, dried, and examined for Radiolaria. Initial study of Radiolaria was accomplished with a standard reflected light microscope at 100× magnification and higher. Subsequent studies and photographic documentation of the assemblage were accomplished with the JEOL JSM T-300 Scanning Electron Microscope. The basis for zonal age assignments and results of this study are presented in the following paragraphs.

Age determination

As noted above, only one of fourteen samples collected from southwestern Panama has yielded radiolarians. One sample collected from the Soná Peninsula (JC-86-A21) and six samples collected from Cébaco Island (JC-88-4, -5, -7, -8, -10, -12) were all barren of Radiolaria. Six of the seven samples collected from the western and southern coasts of the Azuero Peninsula were also barren (JC-86-9C, -23, -33, -34, -35, -36). The single radiolarian-bearing sample, JC-86-A7, was collected on the northwestern coast of the Azuero Peninsula, south of the town of Malena and south of the Azuero-Soná fault zone (App. 1 Fig. 1).

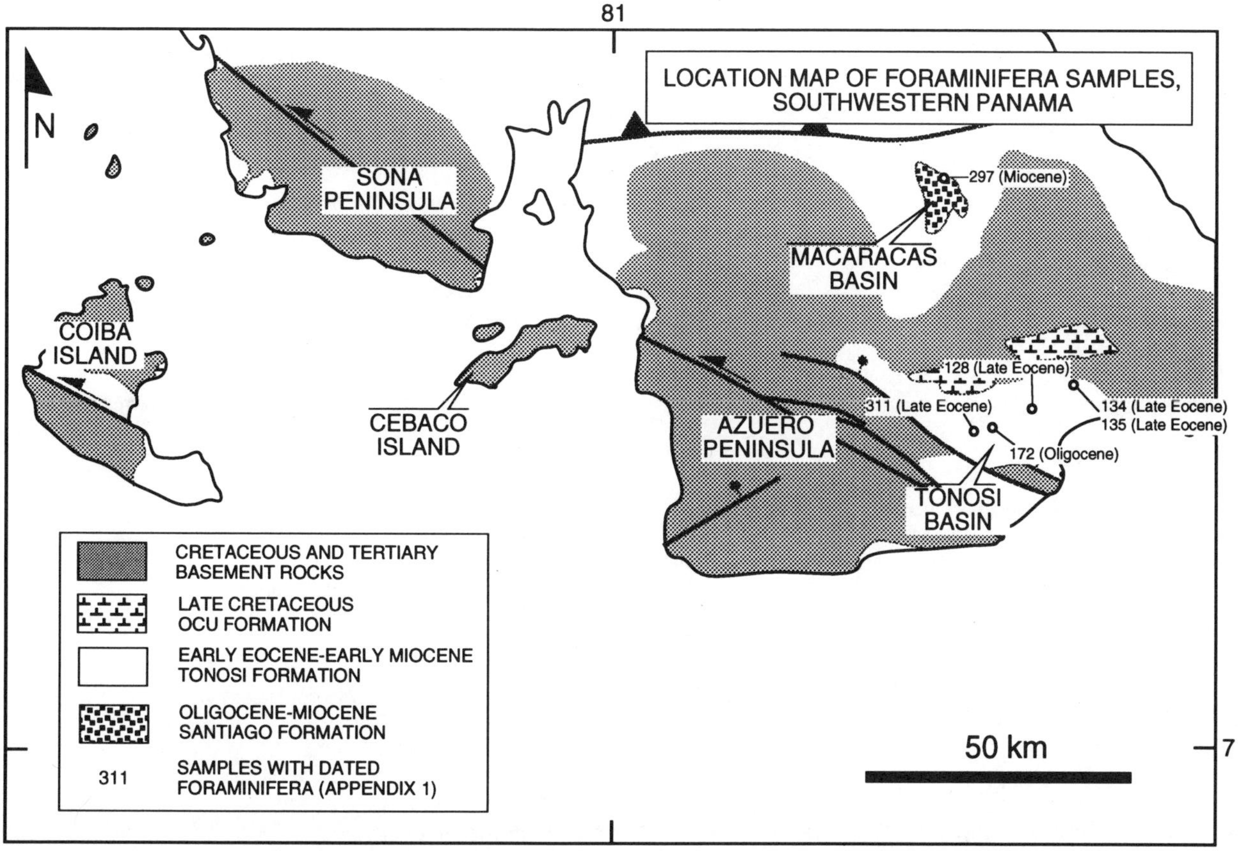

Appendix 1 Figure 3. Location map of foraminifera samples, southwe' ern Panama.

Species present in this sample include (App. 2 Plate I) *Acanthocircus* sp., *Alievium* sp. aff. *A. gallowayi* Pessagno, *Archaeodictyomitra* sp., *Dictyomitra formosa* Squinabol, *Dictyomitra* sp. cf. *D. formosa* Squinabol, *Crucella plana* Pessagno, *Hemicryptocapsa polyhedra* Dumitrica, *Praeconocaryomma universa* Pessagno, *Pseudoaulophacus lenticulatus* (White), and *Pseudoaulophacus venadoensis* Pessagno. Neosciadiocapsid fragments are common throughout the sample. In addition, numerous unnamed species of *Archeodictyomitra, Dictyomitra, Alievium,* and various dicyrtid and tricyrtid radiolarians are present. Overall, preservation of the detailed structure of radiolarian tests is good, but diversity of the fauna appears low. This low diversity may be due in part to limited diversity of the original assemblage, but destruction of the more fragile tests during diagenesis has almost certainly played a prominent role.

Biostratigraphic and chronostratigraphic determinations have been made utilizing the radiolarian zonation scheme of Pessagno (1976, 1977) for the Late Cretaceous (App. 2 Fig. 1). The presence of the primary marker taxa *Alievium praegallowayi* Pessagno and the absence of its evolutionary descendant, *Alievium gallowayi* Pessagno, results in the assignment of this assemblage to the *Alievium praegallowayi* Zone (Zone 12) of Pessagno (1976, 1977; App. 2 Fig. 1). This zone is assigned chronostratigraphically to the Coniacian. Pessagno (1976) further divided the *Alievium praegallowayi* Zone into two subzones: the *Archaeospongoprunum triplum* Subzone (early Coniacian) and the *Orbiculiforrma vacaensis* Subzone (late Coniacian). The absence of marker taxa for both of these subzones makes it difficult, if not impossible, to further constrain the age of this assemblage. The absence of *Dictyomitra densicostata* Pessagno, a relatively robust species usually preserved in chert assemblages, may suggest an early Coniacian age, but it should be noted that its absence may be due to other factors, including the low diversity and/or random sampling. Furthermore, forms such as *Alievium* sp. aff. *A. gallowayi* (App. 2 Plate I Fig. 5) possess characteristics approaching those of *Alievium gallowayi* (e.g., spines that are circular in cross section immediately above the base of the spine instead of triradiate, as in *A. praegallowayi*). This suggests an age close to the transition of the species *A. praegallowayi* to *A. gallowayi,* and thus a slightly younger age of late Coniacian is possible as well.

REFERENCES CITED IN APPENDIX 2

Baumgartner, P. O., 1984, El complejo ofiolítico de Nicoya (Costa Rica): Modelos estructurales analizados en función de las edades de los radiolarios (Calloviense a Santoniense), *in* Sprechmann, P., ed., Manual de geología de Costa Rica I: Estratigrafía: San Jose, Editorial Universidad Costa Rica, p. 114–122.

De Wever, P., Azéma, J., Tournon, J., and Desmet, A., 1985, Découverte de matériel océanique du Lias-Dogger inférieur dans la péninsule de Santa Elena (Costa Rica, Amérique Centrale): Paris, Compte Rendu de l'Academie des Sciences, t. 300, ser. II, no. 15, p. 759–764.

Kolarsky, R. A., 1992, Late-Cretaceous–Cenozoic tectonics and sedimentation in southern Central America [M.A. thesis]: Austin, The University of Texas, 326 p.

Pessagno, E. A., Jr., 1976, Radiolarian zonation and stratigraphy of the Upper

APPENDIX 1

TABLE 2. POLLEN CONTENT, AGE, AND PALEOENVIRONMENT OF COAL SAMPLES DESCRIBED IN UNPUBLISHED REPORT BY DIRECCIÓN GENERAL DE HIDROCARBUROS (1974), PANAMA CITY*

SAMPLE 72-2941 (TONOSI BASIN, LOWER SEAM 36 INCHES THICK, LOWER TONOSI FORMATION)

Abundant, poorly preserved pollen.

Proxapertites operculatus
Dictyophyllidites Sp. A
Polypodiaceoisporites retirugatus Muller
Punctatispores sp.
Verrucatosporites usmensis (Van der Hammen) Germeraad, Hoping, and Muller
Cyathidites australis Couper
Deltoidispora sp.
Cycadopites sp.
Trichotomocolpites sp.
Stephanoporites sp.
Psilatricolporites minutus Gonález Guzmán
Foveotricolporites voluminosus Gonález Guzmán
Malgocalporites sp. ef. *M. vanwijhei* Germeraad, Hoping, and Muller
Microthyriella sp. Dilcher

Paleoenvironment: Fresh-water forested swamp proximal to coast.

Age: Late-middle to early-late Eocene.

SAMPLE 72-2945 (TONOSI BASIN, UPPER SEAM SIX INCHES THICK, LOWER TONOSI FORMATION)

Abundant, poorly preserved pollen.

Proxapertites operculatus
Granamonoletes sp.
Laevigatosporites usmensis (Van der Hammen)
Verrucatosporites usmensis (Van der Hammen) Germeraad, Hoping, and Muller
Cyathidites australis Couper
Dictyophyllidites Sp. A
Trichotomocolpites sp.
Striatocolporites augustinus Gonález Guzmán
Polypodiaceoisporites retirugatus Muller

Paleoenvironment: Fresh-water forested swamp proximal to coast.

Age: Late-middle to early-late Eocene.

SAMPLE 72-29242 (MACARACAS BASIN, BOMBACHO COAL, SEAM 30 INCHES THICK).

Abundant, well-preserved pollen.

Psilatricolporites crassus Van der Hammen and Wymstra
Verrucatosporites usmensis (Van der Hammen) Germeraad, Hoping, and Muller.
Verrucatosporites sp. A
Verrucatosporites sp. B
Granamonoletes sp. B
Matonisporites sp. C
Matonisporites phlebopteroides Couper
Cyathidites sp.
Cyathidites minor Couper
Pistillipollenites sp.
Proxapetrites operculatus Van der Hammen
Duadosporites ellipsus Clarke
Callimothallus pertusus Dilcher
Microthyriella sp. cf. *M. Funguosa* Dilcher
Pluricellaesporites sp. A
Pluricellaesporites sp. B
Monoporisporonites globosus Clarke

Paleoenvironment: Lowland tropical swamp.

Age: Middle to early-late Eocene, possibly as young as late Oligocene-earliest Miocene.

*See Appendix 1, Figure 4 for sample locations.

Cretaceous portion of the Great Valley Sequence, California Coast Ranges: Micropaleontology Special Publication 2, 95 p.

Pessagno, E. A., Jr., 1977, Lower Cretaceous radiolarian biostratigraphy of the Great Valley Sequence and Franciscan Complex, California Coast Ranges: Cushman Foundation for Foraminiferal Research Special Publication 15, 87 p.

Pessagno, E. A., Jr., and Newport, R. L., 1972, A technique for extracting Radiolaria from radiolarian cherts: Micropaleontology, v. 18, p. 231–234.

Schmidt-Effing, R., 1980, Radiolarien der Mittel-Kreide aus dem Santa Elena-Massiv von Costa Rica: Neues Jahrbuch für Geologie und Paläontologie, Abhandlungen, v. 160, p. 241–257.

APPENDIX 3. CALCAREOUS NANNOFOSSIL ASSEMBLAGES AND BIOSTRATIGRAPHY OF TONOSÍ AND SANTIAGO FORMATIONS, SOUTHWEST PANAMA

Simonetta Monechi

Abstract

A biostratigraphic study based on calcareous nannofossils was performed on the basement and on some sedimentary formations outcropping in Panama. The nannofossil assemblages are shown in two distribution charts and in one plate.

Introduction and Methods

This study examines calcareous nannofossil assemblages of the sedimentary formations recognized in the Azuero and Soná Peninsulas of the Pacific margin of Panama in order to provide improved biostratigraphic control (App. 1 Fig. 2). Smear slides were made directly from the raw material. Semiquantitative analysis (300 species counted) was carried out, using a cross-polarized light microscope at a magnification of 1650×, to obtain consistent and reliable data. Unfortunately, half of the analyzed samples were barren of calcareous nannofossils, and the remaining ones showed few and poorly preserved nannofossil assemblages. Nevertheless, it has been possible to obtain some age data for the sedimentary formations of the Azuero Peninsula.

The nannofossil assemblages are shown in the distribution charts in Appendix 3 Figures 1 and 2. The reliability of the nannofossil age assignments depends on the abundance and preservation of the nannoflora so it has not been possible to assign each sample to a precise biostratigraphic zone but only to the time interval, as a result of the rarity (depleted assemblages) of the assemblages.

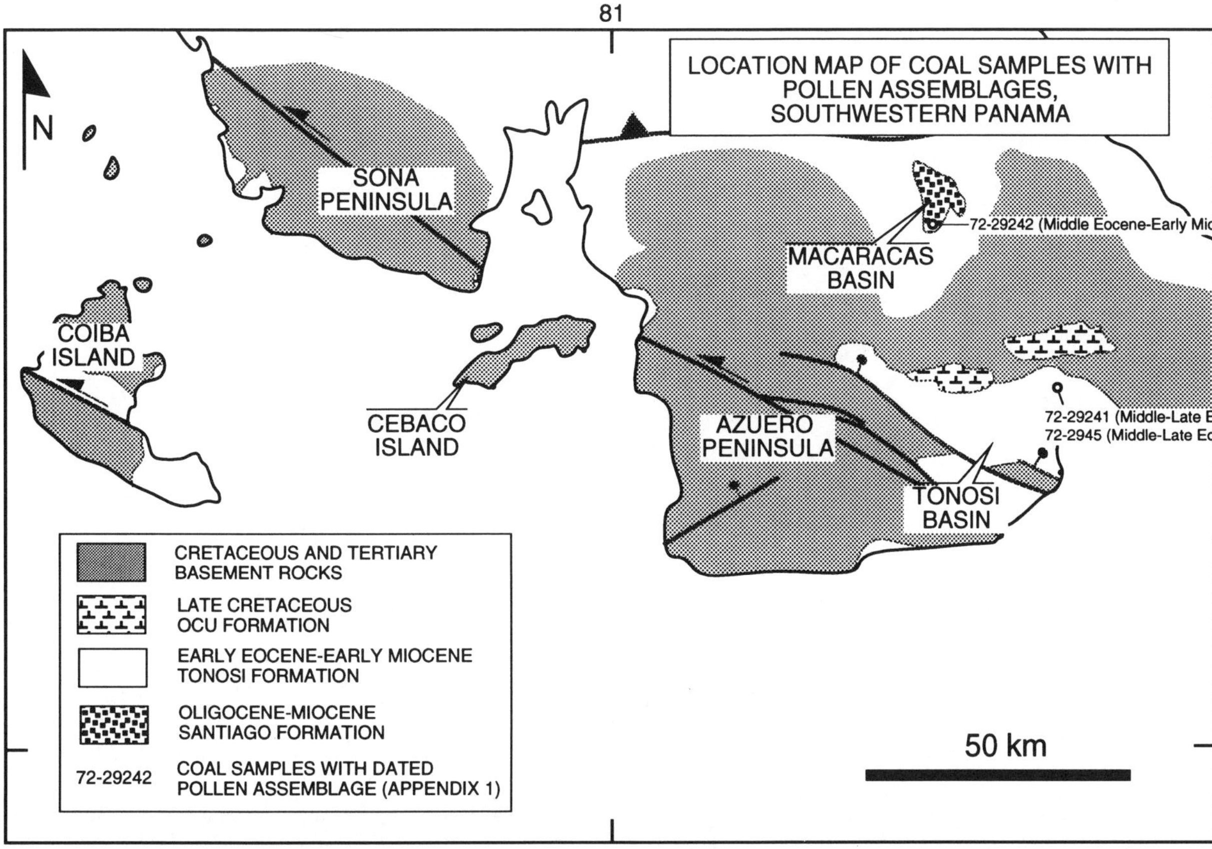

Appendix 1 Figure 4. Location map of coal samples with pollen assemblages, southwestern Panama.

STAGE	RADIOLARIAN ZONES			MARKER TAXA
SANTONIAN	13	Alievium gallowayi Zone		*Archaeospongoprunum bipartitum*; *Orbiculiforma quadratum*; *Alievium gallowayi*
CONIACIAN	12	Alievium praegallowayi Zone	Orbiculiforma vacaensis Subzone	*Orbiculiforma vacaensis*; *Dictyomitra densicostata*
			Archaeospongoprunum triplum Subzone	*Alievium praegallowayi*
TURONIAN	11	Alievium superbum Zone	Archaeospongoprunum venadoensis Subzone	*Archaeospongoprunum venadoensis*; *Crucella cachensis*
			Halesium sexangulum Subzone	*Alievium superbum*

Appendix 2 Figure 1. Radiolarian biostratigraphic zonation scheme for the Turonian through Santonian, modified from Pessagno (1976, 1977). Radiolarian zonal definitions used herein are the same as originally proposed by Pessagno (1976); numerical designations for these zones were proposed by Pessagno (1977). Sample JC-86-A7 from the Azuero Peninsula has been assigned to Zone 12, the *Alievium praegallowayi* Zone (Coniacian), based on the presence of *Alievium praegallowayi* Pessagno and the absence of its evolutionary descendant, *Alievium gallowayi* Pessagno.

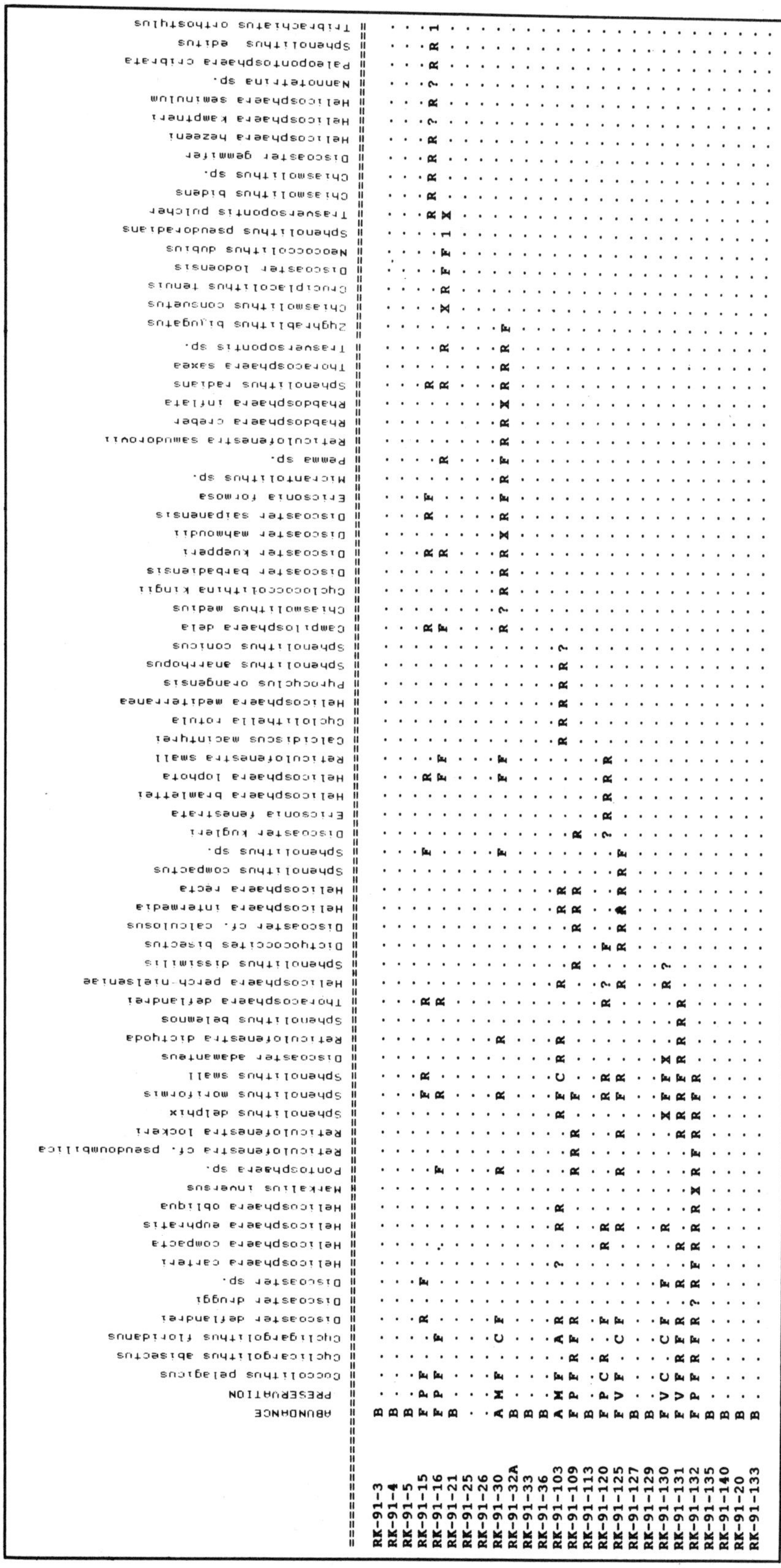

Appendix 3 Figure 1. Distribution chart of calcareous nannofossils of Coiba Island, West Azuero and South Azuero Peninsula. Abundance: B = Barren; A = Abundant; C = Common; F = Few; R = Rare; X = Very Rare. Preservation: M = Medium; P = Poor; X = Very Poor.

Sample	ABUNDANCE	PRESERVATION	Coccolithus pelagicus	Cyclicargolithus abisectus	Cyclicargolithus floridanus	Dictyococcites sp.	Discoaster sp.	Helicosphaera carteri	Pontosphaera sp.	Sphenolithus ciperoensis	Discoaster adamanteus	Discoaster cf. calculosus	Discoaster deflandrei	Discoaster gemmifer	Helicosphaera euphratis	Helicosphaera paleocarteri	Reticulofenestra sp.	Sphenolithus belemnos	Sphenolithus delphix	Sphenolithus moriformis	Cyclolitella rotula	Helicosphaera scissura	Sphenolithus conicus	Blackites sp.	Calcidiscus protoannula	Chiasmolithus solitus	Cribrocentrum reticulatum	Dictyococcites bisectus	Discoaster barbadiensis	Discoaster saipanensis	Discoaster tanii	Helicosphaera lophota	Helicosphaera seminulum	Reticulofenestra samudorovii	Reticulofenestra umbilica
PM86-2C	B	.	.	.	.	.	.	.	.	.	.	.	.	.	.	.	.	.	.	.	.	.	.	.	.	.	.	.	.	.	.	.	.	.	.
PM86-24	B	.	.	.	.	.	.	.	.	.	.	.	.	.	.	.	.	.	.	.	.	.	.	.	.	.	.	.	.	.	.	.	.	.	.
PM86-26	C	M	F	.	F	.	.	.	R	.	.	.	.	.	R	.	.	.	.	F	.	.	.	X	R	X	F	X	X	F	X	R	R	F	R
PM86-6C	F	P	F	.	C	.	R	R	.	.	.	.	R	.	.	.	R	.	R	F	R	R	R	.	.	.	.	.	.	.	.	.	.	.	.
PM86-4C	F	P	R	.	R	.	.	X	.	.	.	.	R	.	X	X	X	1	1	R	.	.	.	.	.	.	.	.	.	.	.	.	.	.	.
JC86-A6	R	V	X	.	.	.	F	.	.	.	R	X	F	R	.	.	.	.	.	.	.	.	.	.	.	.	.	.	.	.	.	.	.	.	.
PM86-53	R	V	R	R	F	R	R	X	R	1	.	.	.	.	.	.	.	.	.	.	.	.	.	.	.	.	.	.	.	.	.	.	.	.	.

Appendix 3 Figure 2. Distribution chart of calcareous nannofossils of Tonosí Formation (Azuero Peninsula).

The biostratigraphic results obtained for each investigated formation, species diversity, and some paleoecological remarks are reported in the next section.

Biostratigraphic data

The samples of the investigated basement are mostly barren in nannofossil content, and only one sample, JC86-A6 (App. 3 Fig. 2), collected in the northwest Azuero Peninsula (south of Azuero-Soná fault zone) yielded rare nannofossils. The assemblage consists only of discoasters and *Coccolithus pelagicus;* because discoasters are usually less prone to dissolution than associated coccoliths, I infer that this sample has probably undergone heavy dissolution. The discoasters recognized are Eocene in age.

No age assignment for the Ocú Formation was obtained because the samples (RK91-3, RK91-4, RK91-140) collected from Punta Damas section (northeast Coiba Island) and from Tonosí-Macaracas Road, North Tonosí basin (southeastern Azuero Peninsula) were barren of nannofossils.

The Tonosí Formation (lower and upper unit) exhibits facies changes in different areas across its outcrop area on the Azuero Peninsula; the samples collected at Punta Bucaró (RK91-135, PM86-24, PM86-2C) were barren of nannofossils. The samples of the upper Tonosí Formation collected at Punta Malena section (RK91-103, RK91-109, RK91-113) show a late Oligocene–early Miocene assemblage. The assemblage (App. 3 Fig. 1) is mainly characterized by the *Cyclicargolithus floridanus* group, helicosphaerids, and sphenoliths. Helicosphaerids are largely limited to low-latitude and shallow-water environments and are most diverse in the equatorial area. The presence of *C. abisectus, Helicosphaera recta, H. intermedia,* and *Sphenolithus dissimili* and the absence of *Dictyococcites bisectus* suggest an early Miocene age.

A few poorly preserved assemblages (App. 3 Figs. 1 and 2) have been recognized in the samples of the Punta Blanca section (RK91-120, RK91-120, RK91-129, RK91-130, RK91-131, RK91-132, RK91-132, RK91-133, PM86-4C, PM86-6C). All the samples investigated can be assigned to the Oligocene/Miocene boundary, except sample RK91-120, which showed an upper Oligocene assemblage.

The upper Tonosí Formation was also sampled: east of Camp Maria (RK91-15, RK91-16, RK91-20, RK91-21, RK91-30, RK91-32A, RK91-33, RK91-36), southeast Coiba Island, and Rio San Juan (RK91-5), central Coiba Island. Only three samples (RK91-15, RK91-16, RK91-30) yielded nannofloral assemblages. The recognized species suggest an early middle Eocene age. The assemblages are mainly characterized by discoasters, which are known to prefer warm waters (Bukry, 1973; Siesser and Haq, 1987). *Cyclicargolithus floridanus,* the genus *Campilosphera,* sphenoliths, and pentaliths (e.g. *Micrantholithus* and *Pemma*) are all specimens very common in middle Eocene nearshore and hemipelagic deposits (Perch-Nielsen, 1985).

The Santiago Formation was sampled (RK91-125, RK91-127) in the Macaracas basin, north-central Azuero Peninsula (App. 1 Fig. 2). Only sample RK91-125 yielded a few, poorly preserved nannofossil assemblage. The presence of *Helicosphaera perch-nielseniae, H. recta, H. intermedia,* and *D. bisectus* suggests a late Oligocene age.

Sample PM86-53, collected in the David basin area, showed a rare, poorly preserved nannofossil assemblage. The presence of *Helicosphaera paleocarteri* (App. 3 Plate I on pp. 198 and 199), *Cyclicargolithus abisectus,* and *Sphenolithus ciperoensis* suggests an Early Miocene age.

REFERENCES CITED IN APPENDIX 3

Bukry, D., 1973, Coccolith and silicoflagellate stratigraphy, Tasman Sea and Southeastern Pacific Ocean: Deep Sea Drilling Project Leg 21, *in* Burns, R. E., Andrew, J. E., eds., Initial reports of DSDP, vol. 21: Washington, D.C., U.S. Government Printing Office, p. 885–893.

Perch-Nielsen, K., 1985, Cenozoic calcareous nannofossils, *in* Bolli, H. M., Saunders, J. B., and Perch-Nielsen, K., eds., Plankton stratigraphy: Cambridge Earth Science Series, Cambridge, University Press, p. 427–554.

Siesser, W. G., and Haq, B. U., 1987, Calcareous nannoplankton, *in* Broadhead, T. W., ed., Fossil prokaryotes and protists: Note for a short course: University of Tennessee Department of Geological Sciences Stud. Geol. 18, p. 87–127.

APPENDIX 2 PLATE I EXPLANATION

All figures are scanning electron micrographs of Late Cretaceous Radiolaria from Sample JC-86-A7 from the northwest coast of the Azuero Peninsula, southwestern Panama. This sample has been assigned to the *Alievium praegallowayi* Zone (Zone 12) of Pessagno (1976, 1977). See Appendix 2 Figure 1 for explanation of the zonal scheme referred to in this plate and the text. Scale bar (upper right) is used as a reference for the magnification of all figures.

Figures 1, 7. *Hemicryptocapsa polyhedra* Dumitrica. Scale bar = 2.4 μm, 20.0 μm, respectively.

Figure 2. Neosciadiocapsid fragment, representing the latticed perimeter of the test of these spumellarians. Only these fragments remain; there are no wholly preserved species remaining of the Family Neosciadiocapsidae Pessagno. Scale bar = 43.5 μm.

Figure 3. *Crucella plana* Pessagno. Scale bar = 43.5 μm.

Figure 4. Nassellariina gen. et sp. indet. Scale bar = 20.0 μm.

Figure 5. *Alievium* sp. aff. *A. gallowayi* Pessagno. This form resembles *A. gallowayi* Pessagno by possessing spines that become circular in cross section just above the spine base. It differs from that species by possessing a more inflated test. Scale bar = 28.6 μm.

Figure 6. *Alievium praegallowayi* Pessagno. This species is the marker taxon or the base of Zone 12, also referred to as the *Alievium praegallowayi* Zone. Scale bar = 20.0 μm.

Figure 8. *Acanthocircus* sp. Scale bar = 24.4 μm.

Figure 9. *Dictyomitra* sp. cf. *D. formosa* Squinabol. This form differs from *Dictyomitra formosa* by possessing a more conical test and shallower strictures on the proximal portion of the test. Scale bar = 28.6 μm.

Figure 10. *Dictyomitra formosa* Squinabol. This species first appears at the base of the *Alievium praegallowayi* Zone (Coniacian), ranging into the lower Campanian. Scale bar = 24.4 μm.

Figure 11. *Archaeodictyomitra* sp. Although numerous unnamed species of this genus are present in Sample JC-86-A7, each individual form has very low abundance. Scale bar = 18.2 μm.

Figure 12. *Pseudoaulophacus venadoensis* Pessagno. First appearance of this species is at the base of the *Alievium praegallowayi* Zone; it extends to the top of the *Alievium gallowayi* Zone. Scale bar = 28.6 μm.

Figure 13. *Praeconocaryomma universa* Pessagno. *P. universa* first appears at the base of the *Alievium praegallowayi* Zone (Coniacian). Its full range extends into the middle Campanian. Scale bar = 13.3 μm.

Figure 14. *Pseudoaulophacus lenticulatus* (White). This species ranges from the base of the Coniacian (*Alievium praegallowayi* Zone) to the late Campanian. Scale bar = 28.6 μm.

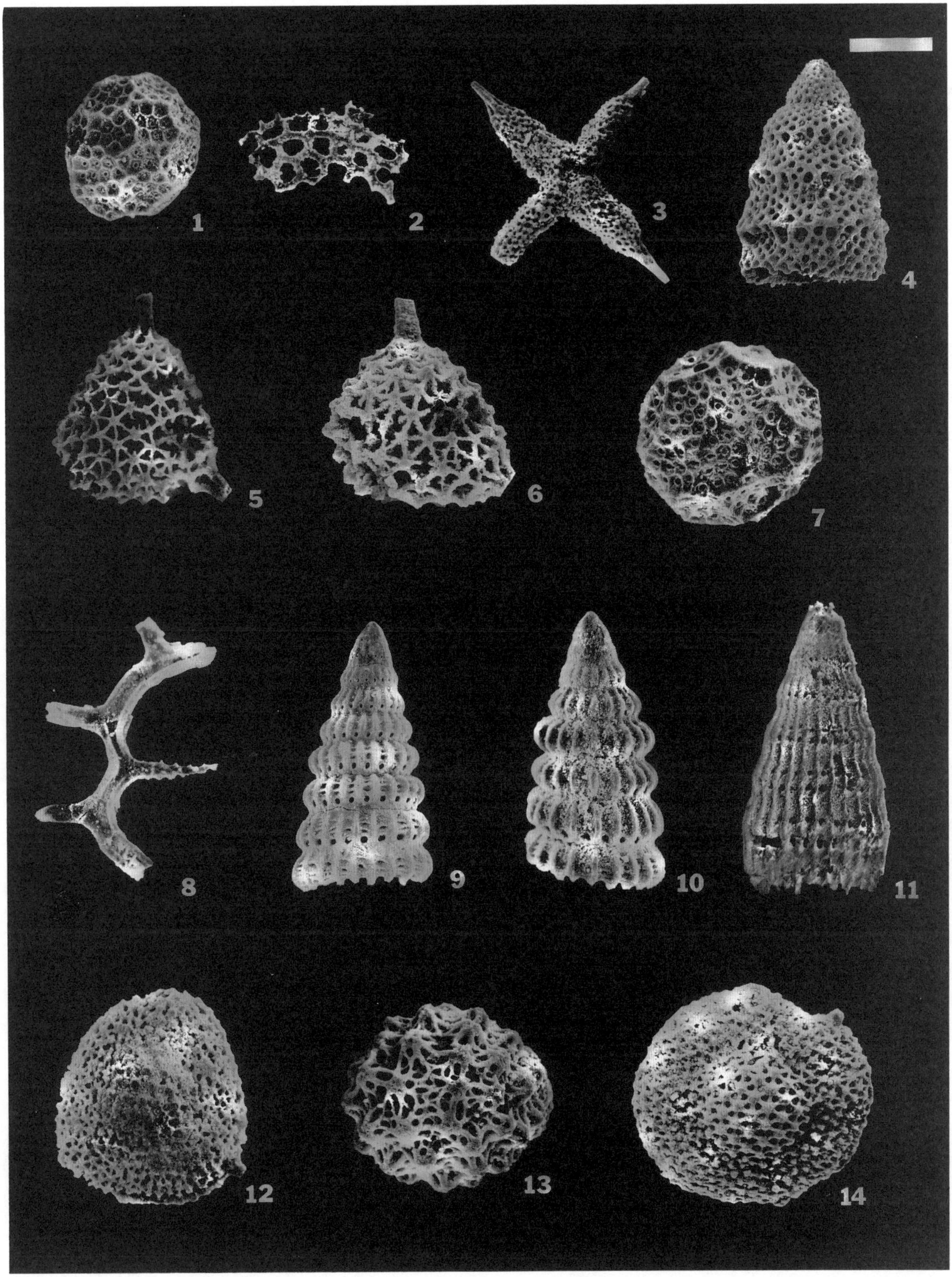
1
2
3
4
5
6
7
8
9
10
11
12
13
14

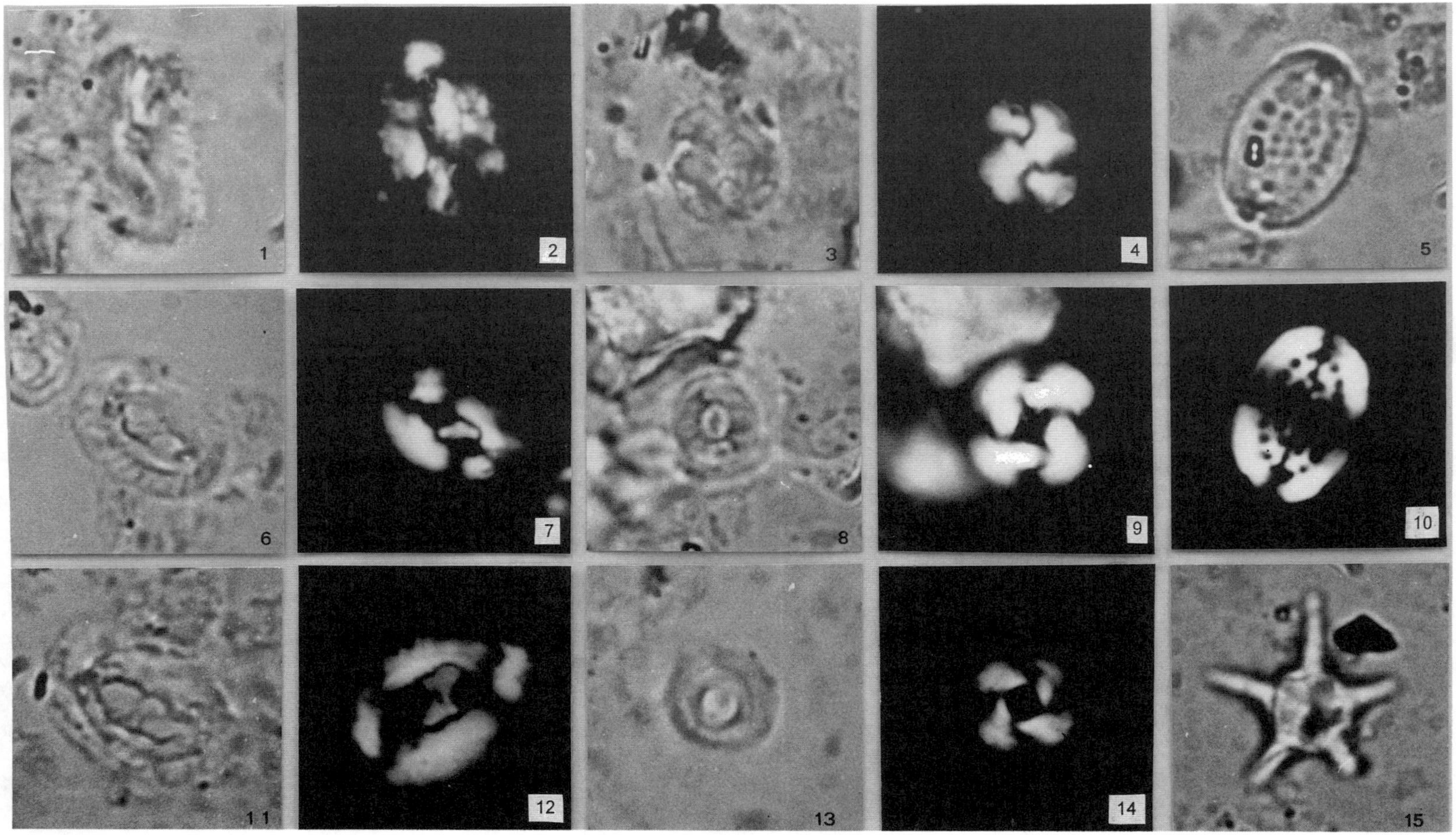

APPENDIX 3 PLATE I EXPLANATION (All pictures are at magnification 3,600× XN cross-polarized light.)

Figures 1-2. *Helicosphaera paleocarteri* Theodoridis 1984, sample PM86-4C, (1) PN, (2) XN.

Figure 3-4. *Dyctiococcites scrippsae* Bukry and Percival 1971, sample PM86-53, (3) PN, (4) XN.

Figure 5-10. *Pontosphaera multipora* (Kamptner, 1948) Roth 1970, sample PM86-26, (5) PN, (10) XN.

Figures 6-7. *Helicosphaera seminulum* Bramlette and Sullivan 1961, sample PM86-26, (6) PN, (7) XN.

Figures 8-9. *Reticulofenestra dictyoda* (Deflandre in Deflandre and Fert, 1954) Stradner in Stradner and Edwards, 1968, sample PM86-26, (8) PN, (9) XN.

Figures 11-12. *Helicosphaera bramlettei* Muller 1970, sample PM86-26, (12) XN.

Figures 13-14. *Cyclicargolithus cf. floridanus* (Roth and Hay in Hay et al., 1967) Bukry 1971, (14) XN.

Figure 15. *Discoaster mahmoudii* Perch-Nielsen 1981, sample PM86-26.

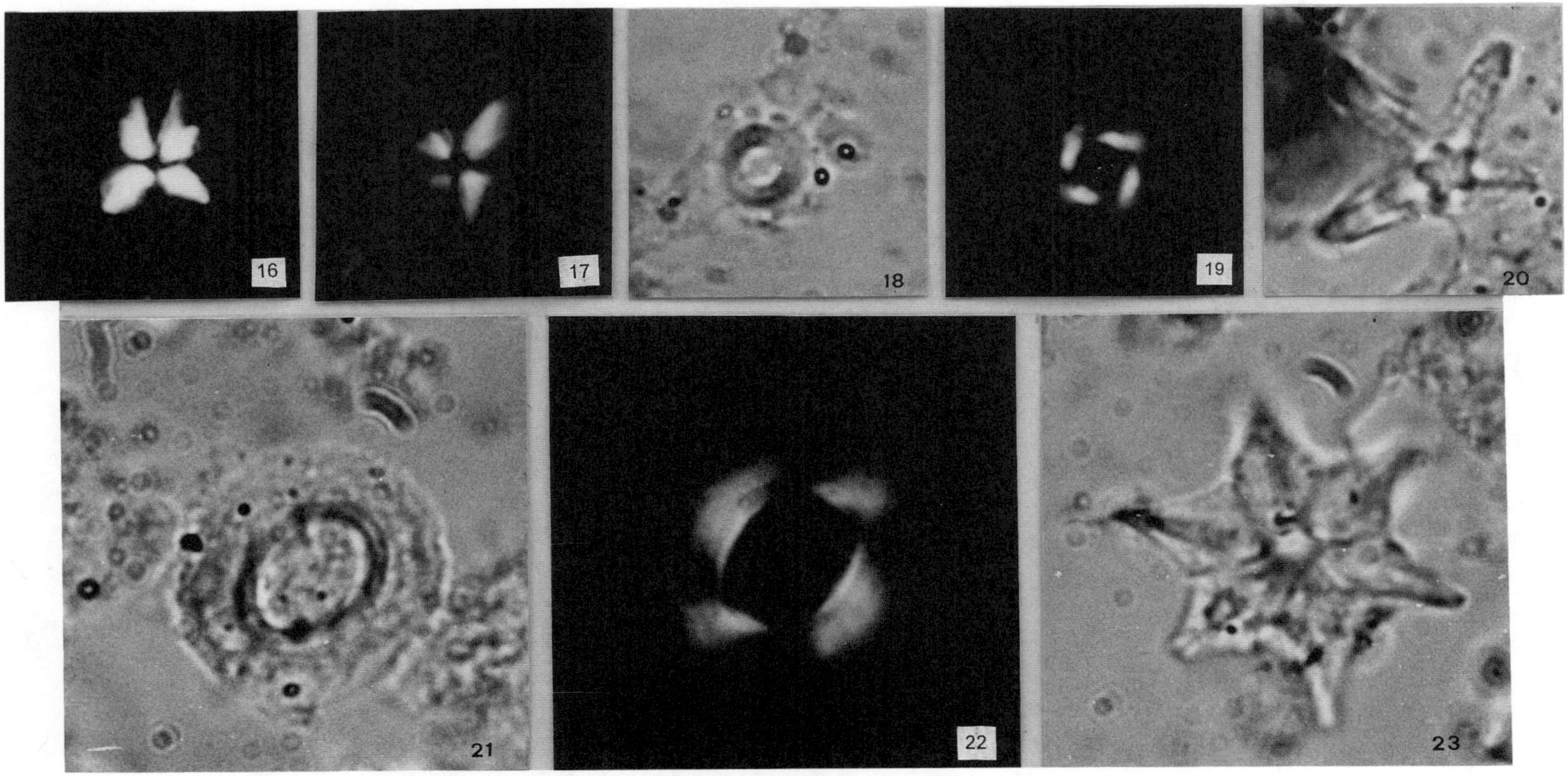

Figures 16-17. *Sphenolithus editus* Perch-Nielsen in Perch-Nielsen et al., 1978, (16) XN, (17) XN 45°.

Figures 18-19. *Calcidiscus protoannulus* (Gartner, 1971) Loeblich and Tappan, 1978, (19) XN.

Figure 20. *Discoaster cf. tanii* Bramlette and Riedel, 1954.

Figures 21-22. *Reticulofenestra umbilica* (Levin, 1965) Martini and Ritzkowski, 1968, (22) XN.

Figure 23. *Discoaster saipanensis* Bramlette and Riedel, 1954.

REFERENCES CITED

Bandy, O. L., and Casey, R. E., 1973, Reflector horizons and paleobathymetric history, eastern Panama: Geological Society of America Bulletin, v. 84, p. 3081–3086.

Berrangé, J., and Thorpe, R., 1988, The geology, geochemistry, and emplacement of the Cretaceous-Tertiary ophiolitic Nicoya Complex of the Osa Peninsula, southern Costa Rica: Tectonophysics, v. 147, p. 193–220.

Bourgois, J., and 10 others, 1982, Ages et structures des complexes basiques et ultrabasiques de la façade pacifique entre 3°N et 12°N (Colombie, Panama, et Costa-Rica): Geological Society of France Bulletin, v. 24, p. 545–554.

Bowland, C. L., 1993, Depositional history of the western Colombian basin, Caribbean Sea, revealed by seismic stratigraphy: Geological Society of America Bulletin, v. 105, p. 1321–1345.

Bowland, C. L., and Rosencrantz, E., 1988, Upper crustal structure of the western Colombian basin, Caribbean Sea: Geological Society of America Bulletin, v. 100, p. 534–546.

Case, J. E., 1974, Oceanic crust forms basement of eastern Panama: Geological Society of America Bulletin, v. 85, p. 645–652.

Coates, A. G., and seven others, 1992, Closure of the Isthmus of Panama: The near-shore marine record of Costa Rica and western Panama: Geological Society of America Bulletin, v. 104, p. 814–828.

Corrigan, J. C., Mann, P., and Ingle, J. C., Jr., 1990, Forearc response to subduction of the Cocos Ridge, Panama–Costa Rica: Geological Society of America Bulletin, v. 102, p. 628–652.

de Boer, J., Defant, M. J., Stewart, R. H., Restrepo, J. F., Clark, L. F., and Ramirez, A., 1988, Quaternary calc-alkaline volcanism in western Panama: Regional variation and implication for plate tectonic framework: Journal of South American Earth Sciences, v. 1, p. 275–293.

de Boer, J., Defant, M. J., Stewart, R. H., and Bellon, H., 1991, Evidence for active subduction below western Panama: Geology, v. 19, p. 649–652.

del Giudice, D., and Recchi, G., 1969, Geología del area del proyecto minero de Azuero: Administración de Recursos Minerales, Republica de Panama, Mapa 5, scale 1:250,000.

Gursky, H. J., Schmidt-Effing, R., Strebin, M., and Wildberg, H., 1982, The ophiolite sequence in northwestern Costa Rica (Nicoya Complex): Outlines of stratigraphical, geochemical, sedimentological, and technical data: 5th Latin American Geological Congress, v. 3, p. 607–619.

Haq, B. U., Hardenbol, J., and Vail, P. R., 1987, Chronology of fluctuating sea levels since the Triassic: Science, v. 235, p. 1156–1167.

Heil, D. J., 1988, Response of an accretionary prism to transform ridge collision south of Panama [M.S. thesis]: Santa Cruz, University of California, 88 p.

Kolarsky, R. A., 1992, Late Cretaceous–Cenozoic tectonics and sedimentation in southern Central America [M.A. thesis]: Austin, University of Texas, 351 p.

Lundberg, N., 1982, Evolution of the slope landward of the Middle America trench, Nicoya Peninsula, Costa Rica, *in* Leggett, J. K., ed., Trench-forearc geology: Geological Society of London Special Publication 10, p. 133–147.

Lundberg, N.,1991, Detrital record of the early Central American magmatic arc: Petrography of intraoceanic forearc sandstones, Nicoya Peninsula, Costa Rica: Geological Society of America Bulletin, v. 103, p. 905–915.

MacKay, M. E., 1988, Accretionary response to trench sediment variation along the south Panama margin [M.S. thesis]: Tulsa, Oklahoma, University of Tulsa, 44 p.

MacKay, M. E., and Moore, G. F., 1990, Variation in deformation of the South Panama accretionary prism: Response to oblique subduction and trench sediment variation: Tectonics, v. 9, p. 683–698.

Mann, P., and Corrigan, J., 1990, Model for late Neogene deformation in Panama: Geology, v. 18, p. 558–562.

McKee, W. H., 1985, Tertiary foraminifera of the Puntarenas Province, Costa Rica [M.S. thesis]: Baton Rouge, Louisiana State University, 279 p.

Metti, A., and Recchi, G., 1976, Geología de la peninsula de Soná e Isla de Coiba: Segundo Congreso Latinoamericano de Geología, v. 2, p. 541–553.

Metti, A., Recchi, G., and Esquivel, D., 1972, Mapa geologíco Soná–Isla de Coiba: Ministerio de Comecio y Industrias, Republica de Panama, scale 1:250,000.

Mutti, E., and Ricci Lucchi, R., 1978, Turbidites of the northern Appenines: Introduction to facies analysis (English translation by T. H. Nilsen): International Geological Review, v. 20, p. 125–166.

Okaya, D. A., and Ben-Avraham, Z., 1987, Structure of the continental margin of southwestern Panama: Geological Society of America Bulletin, v. 99, p. 792–802.

Phillips, P. J., 1983, Stratigraphy, sedimentology, and petrologic evolution of Tertiary sediments in southwestern Costa Rica [M.S. thesis]: Baton Rouge, Louisiana State University, 153 p.

Recchi, G., and Miranda, R., 1977, Calizas de los Planes-Guaniquito (Tonosí): Panama City, Dirección General de Recursos Minerales (unpublished report), 27 p.

Ross, D. A., and Shor, G. G., Jr., 1965, Reflection profiles across the Middle America trench: Journal of Geophysical Research, v. 70, p. 5551–5572.

Schmidt-Effing, R., Gursky, H. J., Strebin, M., and Wildberg, H., 1982, The ophiolites of southern Central America with special reference to the Nicoya Peninsula (Costa Rica): Transactions, Caribbean Geological Conference, 9th, Santo Domingo, Dominican Republic: Santo Domingo, Amigo del Hogar Publishers, p. 423–430.

Silver, E. A., Reed, D. L., Tagudin, J. E., and Heil, D. J., 1990, Implications of the North and South Panama thrust belts for the origin of the Panama orocline: Tectonics, v. 9, p. 261–281.

Terry, R. A., 1941, Notes on submarine valleys of the Panamanian coast: Geographical Review, v. 31, p. 377–384.

Terry, R. A., 1956, A geological reconnaissance of Panama: California Academy of Science Occasional Paper 23, 99 p.

Tournon, J., Triboulet, C., and Azema, J., 1989, Amphibolites from Panama: Anticlockwise *P-T* paths from a pre–Upper Cretaceous metamorphic basement in isthmian Central America: Journal of Metamorphic Geology, v. 7, p. 539–546.

Manuscript Accepted by the Society February 7, 1994

Printed in U.S.A.

Geological Society of America
Special Paper 295
1995

Fracture zone collision along the South Panama margin

G. F. Moore and K. L. Sender
Department of Geology and Geophysics, University of Hawaii, 2525 Correa Road, Honolulu, Hawaii 96822

ABSTRACT

Seismic reflection profiles and SeaMARC II imagery from the southwest Panama margin demonstrate that oblique convergence is presently occurring along what had previously been thought of as a transform margin. Our seismic profiles image landward-dipping thrust faults and seaward-verging folds at the toe of the slope. The frontal deformation zone as imaged on the SeaMARC II mosaic is 12 to 15 km wide with individual east-west–trending folds and thrusts that are laterally continuous for 5 to 10 km. Much of the terrigenous trench sediment is offscraped and accreted, forming an accretionary prism (South Panama deformed belt).

Three linear ridges (part of the Panama Fracture Zone complex) are being obliquely subducted along the southwest Panama margin. The oblique convergence causes the ridges to sweep eastward along the trench. The SeaMARC II mosaic shows that the regional structure of the South Panama deformed belt is dominated by east-west–trending trench segments that are separated by the north-south fracture zone ridges. The trench shallows where the ridges intersect the trench, and the deformation front is warped around the ridges. On the east side of each ridge the accretionary complex bends to a northwest-southeast trend, suggesting that the ridges are deforming the accretionary complex. As the accretionary prism rides up over each ridge, it thickens markedly. By the time the prism reaches the top of the ridge, its surface slope has been greatly oversteepened and large portions of accreted material slump into the trench. After passage of the ridge, the system returns to its "normal" state, and accretion resumes, adding the slumped material back into the accretionary prism. The accretionary prism is thus only temporarily disrupted by the subduction of the Panama Fracture Zone system ridges.

INTRODUCTION

The subduction of seafloor bathymetric features, such as seamounts, fracture zones, and aseismic ridges, is common along convergent margins. Much attention has recently been focused on the structural deformation associated with this process (e.g., Fryer and Hussong, 1985; von Huene, 1986; Fisher, 1986; Lallemand and Le Pichon, 1987; Lallemand et al., 1989; Yamazaki and Okamura, 1989; Corrigan et al., 1990; Masson et al., 1990). Most workers agree that the landward trench slope is strongly affected by such collisions and that the subducting feature may cause failure of the landward trench slope. In some areas, this slumped material is then subducted beneath the landward trench slope (e.g., von Huene and Lallemand, 1990). Whether or not the scars of the collisions remain long after the collisions is, however, still not clear.

Along the southwest margin of Panama, ridges and troughs of the Panama Fracture Zone system are being obliquely subducted to the northeast (Fig. 1). Because of the oblique nature of convergence, the topographic features of the Panama Fracture Zone system sweep eastward along the margin, and from east to west, the structures associated with the collision can be observed in the "before," "during," and "after" stages of collision. The purpose of this study is to document the structural development of the southwest Panama margin as a result of the oblique collision of the Panama Fracture Zone system.

Moore, G. F., and Sender, K. L., 1995, Fracture zone collision along the South Panama margin, *in* Mann, P., ed., Geologic and Tectonic Development of the Caribbean Plate Boundary in Southern Central America: Boulder, Colorado, Geological Society of America Special Paper 295.

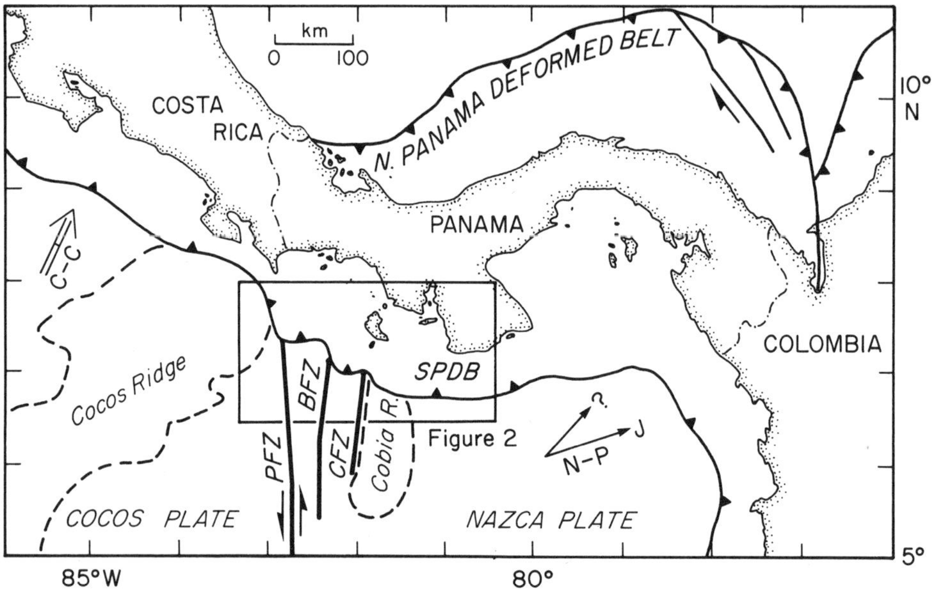

Figure 1. Tectonic setting of the South Panama deformed belt (SPDB). Tectonic elements after Silver and others (1990). The three fracture zones of the Panama fracture zone system are, from west to east: Panama Fracture Zone (PFZ), Balboa Fracture Zone (BFZ), and Coiba Fracture Zone (CFZ). Shapes of Coiba and Cocos Ridges are outlined by the 2,000-m contour. Nazca-Panama convergence vector (N-P) labeled J is from Jordan (1975); ? is hypothesized from our seismic data.

REGIONAL TECTONIC SETTING

The southwest Panama margin marks a diffuse zone of deformation associated with convergence between the Caribbean and Nazca plates (Fig. 1). Jordan (1975) originally proposed that this zone is a dominantly left-lateral strike-slip boundary, with the Nazca plate moving in an east-northeasterly (N70°E) direction at a velocity of 5.4 cm/yr. Okaya and Ben-Avraham (1987), Mann and Corrigan (1990), and Kolarsky and Mann (this volume) documented active and widespread left-lateral strike-slip tectonics both onshore in Panama and offshore in the Gulf of Chiriqui.

Other evidence, however, indicates that subduction has also been important in the recent development of the margin. Adamek and others (1988) interpret earthquake focal mechanisms under western Panama as indicating shallow underthrusting of the Cocos plate beneath the Caribbean plate or, more accurately, a microplate composed of the Panama isthmus (Silver and others, 1990). Regional plate boundaries, however, are not unambiguously constrained by seismicity studies alone (Adamek and others, 1988). Evidence for widespread Late Miocene to Quaternary volcanism in western Panama was used by De Boer and others (1988, 1991) to infer that subduction, not strike-slip tectonics, dominated the interaction of the Nazca plate and Panama. MacKay and Moore (1990) and Silver and others (1990) documented subduction deformation along the southwest margin of Panama and suggested an oblique-slip model in which plate motion is partitioned into offshore convergence and onshore strike-slip faulting. This model would imply that Nazca-Caribbean plate motion is more northerly than previously predicted, yielding convergence (although highly oblique) along the south Panama margin.

DATA ACQUISITION AND PROCESSING

We conducted a survey of the western segment of the South Panama deformed belt abroad the R/V *Moana Wave* in 1985 (Fig. 2), which acquired both SeaMARC II side-scan sonar and digital seismic reflection data. The ship's track was oriented obliquely to regional structural trends in order to optimize the imaging of the side-scan data. This orientation, however, was less desirable for the seismic reflection data.

The seismic reflection system consisted of two airguns (2 and 3.3l) and a single channel streamer. The seismic signals were recorded digitally on 9-track tape and processed using the following sequence: trace edit, deconvolution, band-pass filter (8-15-55-65 Hz), and finite difference migration. Although the oblique orientation of the seismic lines hampered migration (two-dimensional migration was not completely successful at imaging reflections), the structure of the deformed belt is sufficiently clear for interpretation. Because no velocity information was obtained during our survey, we used an approximate velocity function derived by Heil (1988) to construct dips of the descending oceanic crust.

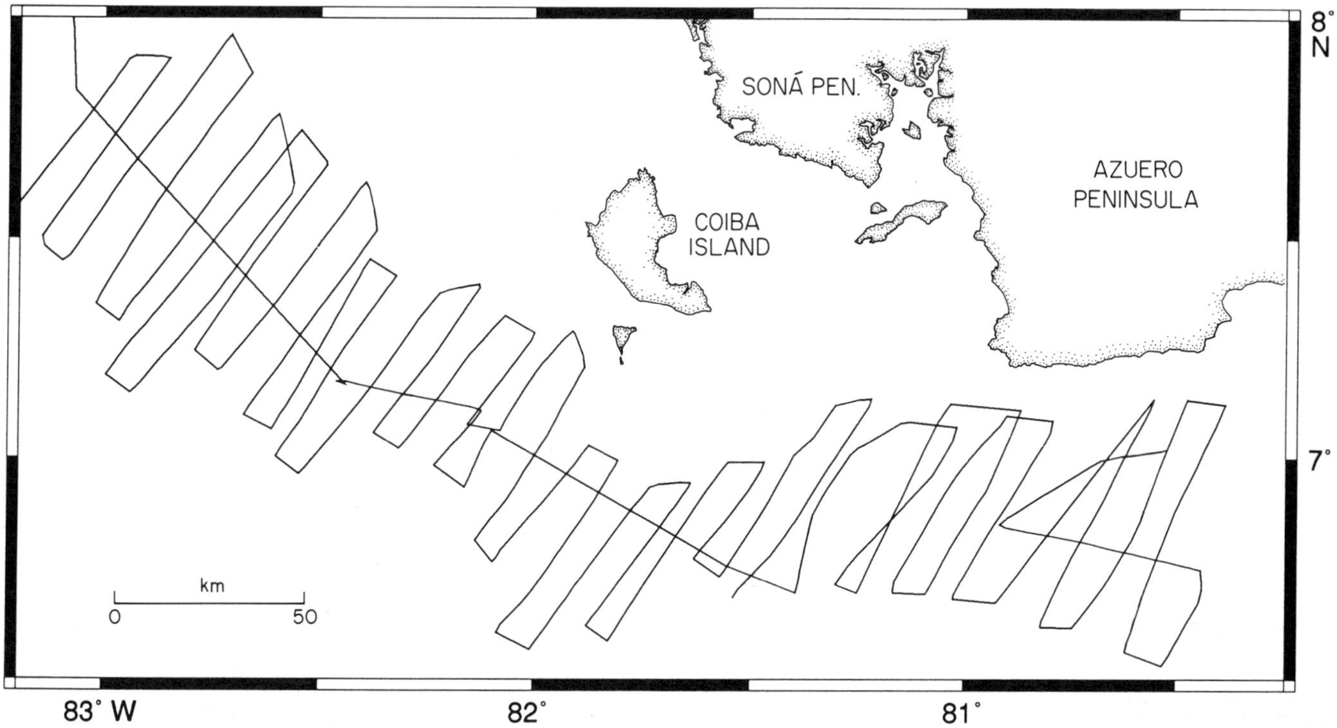

Figure 2. Track chart for SeaMARC II side-scan and seismic reflection survey of the south Panama deformed belt.

SeaMARC II side-scan and bathymetry data were recorded on 9-track tape. Postprocessing of these data was initially done at sea, and preliminary side-scan and bathymetric mosaics were produced shortly after the cruise. Bathymetric data were initially plotted along the ship tracks and hand contoured by Heil (1988). Advancements in data-processing techniques for both side-scan and bathymetry have allowed reprocessing of these data. Artifacts in the side-scan data that degraded the starboard swath in the original processing were greatly diminished by applying more sophisticated across-track gain adjustments than were available with the original processing. Along-track gain discontinuities were also removed by a gain ramping technique that minimized high and low data value clipping. The resulting side-scan data were processed to produce a "structural" image (Sender and others, 1989) by applying a histogram equalization scheme, which used statistics gathered and applied on a 15-minute time interval. The final side-scan data and improved navigation were merged to produce a digitally gridded image that was displayed on a Raytheon thermal gray-scale recorder (Plate 5). The improved detail of this enhanced image allowed for greater control of navigation, which in turn produced a more accurate bathymetric map (Plate 6). A combination of the SeaMARC structural mosaic, the bathymetric map, and the seismic reflection lines was used to construct a structural interpretation map (Fig. 3).

STRUCTURE OF THE SOUTH PANAMA MARGIN

Subducting Nazca plate

The Nazca plate south of the Panama margin contains three seismically active, north-trending fracture zones of the Panama Fracture Zone system: the Panama, Balboa, and Coiba Fracture Zones (Fig. 5; Heil, 1988). These fracture zones border prominent ridges characterized by variable but rough topography (Plate 6). The topography of the fracture zones is important because it controls the development of structures where the fracture zones impinge upon the accretionary prism. Coiba Fracture Zone has the most topographic relief, with a crest at approximately 1,000 m and isolated peaks shallower than 700 m. Twenty km to the west of Coiba Fracture Zone, the floor of the basin is greater than 3,000 m deep. Coiba Fracture Zone forms the western border of Coiba Ridge, an east-tilted piece of seafloor (Lonsdale and Klitgord, 1978; Fig. 1). East of Coiba Fracture Zone, depths of 3,000 m are not reached for 85 km because of Coiba Ridge's topographic asymmetry.

The Balboa Fracture Zone is narrower than the Coiba Fracture Zone, and the crest of Balboa Fracture Zone, at approximately 2,200 m, is much deeper than that of Coiba Fracture Zone. Like the Coiba Fracture Zone, the topography

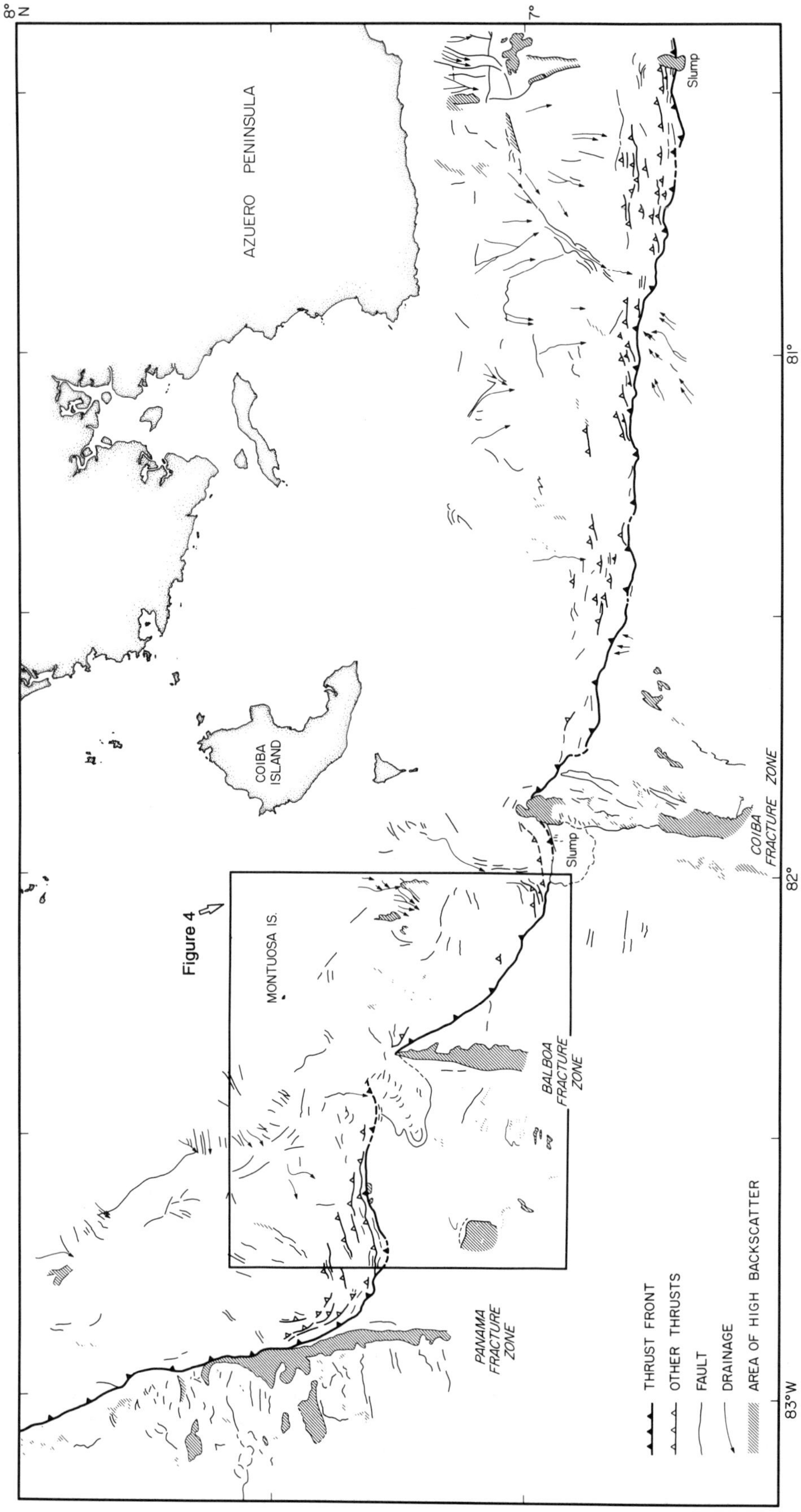

Figure 3. Interpretation of SeaMARC II side-scan mosaic of the South Panama deformed belt based on seismic reflection profiles and SeaMARC II side-scan imagery and bathymetry.

of Balboa Fracture Zone is somewhat asymmetric, with the steep flank on the west.

The Panama Fracture Zone forms the eastern border of the broad Cocos Ridge as well as the boundary between the Nazca and Cocos plates (Fig. 1). As opposed to Coiba Fracture Zone and Balboa Fracture Zone, Panama Fracture Zone is steepest on its east flank. The depth of the trench east of Panama Fracture Zone is 3,300 m, with the flank of Cocos Ridge reaching 2,000 m, approximately 4 km to the west.

The ridges associated with the fracture zones generally exhibit high side-scan backscatter (Fig. 4, Plate 5) that indicates a general lack of sediment cover. Seismic reflection profiles across the ridges confirm the absence of a thick sedimentary cover on the ridge tops (Fig. 5). Sediment on the ridge tops, if present at all, is less than 100 m thick. Isolated small (<5 km wide) ponds contain up to 500 m of sediment. The morphologic asymmetry of Balboa Fracture Zone and Coiba Fracture Zone is accented by thicker sediments onlapping their eastern flanks than their western flanks.

South Panama deformed belt

"Normal" sections of accretionary prism. Subduction-related deformation has built up an accretionary prism along the south Panama margin. Following Kolarsky and Mann (this volume), we refer to this region as the South Panama deformed belt. The South Panama deformed belt is a "typical" accretionary prism east of the Coiba Fracture Zone, where the trench is oriented approximately east-west (MacKay and Moore, 1990; Plates 5, 6). In the South Panama deformed belt the orientation of the thrust front ranges from east-west to N15°W, with east-west–trending linear ridges of accreted trench strata. The thrust ridges are highly reflective, and individual ridges are laterally continuous for up to 10 km. The lower portion of the accretionary prism as defined by MacKay and Moore (1990) is 12 to 15 km wide at the eastern end of our survey area and narrows to 7 to 10 km on the east flank of Coiba Ridge (Fig. 3). The surface slope of the lower accretionary prism shows a corresponding increase from 2 to 3° at the eastern edge of our survey area to 4 to 5° on the flank of Coiba Ridge. See MacKay and Moore (1990) for a more detailed discussion of the accretionary prism east of Coiba Ridge and Heil (1988) and Silver et al. (1990) for discussion of the prism west of the Coiba Ridge.

Similar typical accretionary features characterize segments of the margin between the fracture zones. For example, thrust ridges oriented parallel to the thrust front are imaged on the SeaMARC mosaic between Coiba Fracture Zone and Balboa Fracture Zone (approx. 6°58′N, 82°05′W on Fig. 4) and west of Balboa Fracture Zone (approx. 7°02′N, 82°40′W on Fig. 4). The seismic sections confirm that the ridges are bounded by landward-dipping thrusts (e.g., Fig. 5, seismic lines PAN-13 and PAN-16). The frontal thrust between Balboa Fracture Zone and Panama Fracture Zone trends northeast. Two nearly circular, highly reflective features, which we interpret as mud volcanoes, occur just seaward of the thrust front just east of Panama Fracture Zone (approx. 7°02′N, 82°40'W on Fig. 4).

Fracture zone intersection with South Panama deformed belt. Structural reentrants form where the ridges of the Panama Fracture Zone system intersect the south Panama trench (Fig. 3). The trend of the bathymetric trench bends around from east-west to northwest-southeast (Plate 6). The thrust front similarly changes orientation. The orientations of folds and other thrusts behind the thrust front also change but are only subparallel to the thrust front (Fig. 4, Plate 5). The thrust front is deflected approximately 12 km northward around Coiba Fracture Zone, 25 km northward around Balboa Fracture Zone, and 20 km northward around Panama Fracture Zone. The width of the lowermost accretionary prism decreases from 15 to 20 km to 5 to 10 km around each fracture zone (see, for example, Fig. 8 of MacKay and Moore, 1990), the distance between thrust faults decreases, and the surface slope of the prism increases (Figs. 4, 5).

Seismic reflection profiles across the eastern flanks of Coiba and Balboa Fracture Zones show that the internal structure of the accretionary prism also changes as the prism is underthrust by the fracture zone ridges (Fig. 5). The prism thickens, the surface slope steepens, and the angle between basal décollement and surface slope (critical taper, Davis and others, 1983) increases. For example, in the "normal" accretionary prism between Coiba and Balboa Fracture Zones, the dip of the décollement is about 3° and the surface slope is 4 to 5° (critical taper is about 7 to 8°). As the prism begins to ride up the ridge (e.g., Fig. 5, seismic line PAN-13), the décollement dips about 6 to 8° and the surface slope is 5 to 6° (taper is 11 to 14°); as the prism approaches the top of the ridge (e.g., Fig. 5, seismic line PAN-14), the décollement dips 13 to 15° and the surface slope increases to 9 to 10° (critical taper increases to approximately 22 to 25°).

In the "pre-collision" accretionary prism, the décollement is about 200 to 300 msec above the descending oceanic crust (MacKay and Moore, 1990), whereas near the intersection with Balboa Fracture Zone, the décollement deepens to within 50 to 100 msec of the top of the oceanic crust (Fig. 5, PAN-13 and PAN-14). A similar drop in the level of the décollement also occurs near the intersections of the prism with the Coiba and Panama Fracture Zones. On seismic line PAN-16 (Fig. 5), about 20 km west of the Balboa Fracture zone collision, the décollement is very close to the surface, implying accretion of only a small fraction of the incoming sediments.

Just west of the intersections of Balboa Fracture Zone and Coiba Fracture Zone with the accretionary prism, the orientation of the trench as well as folds and thrusts within the prism is again approximately east-west, but there is no easily identified frontal thrust (Fig. 4, Plate 5). Instead, seismic lines show that the seafloor is hummocky and the internal character just seaward of the prism is chaotic. The SeaMARC II image west

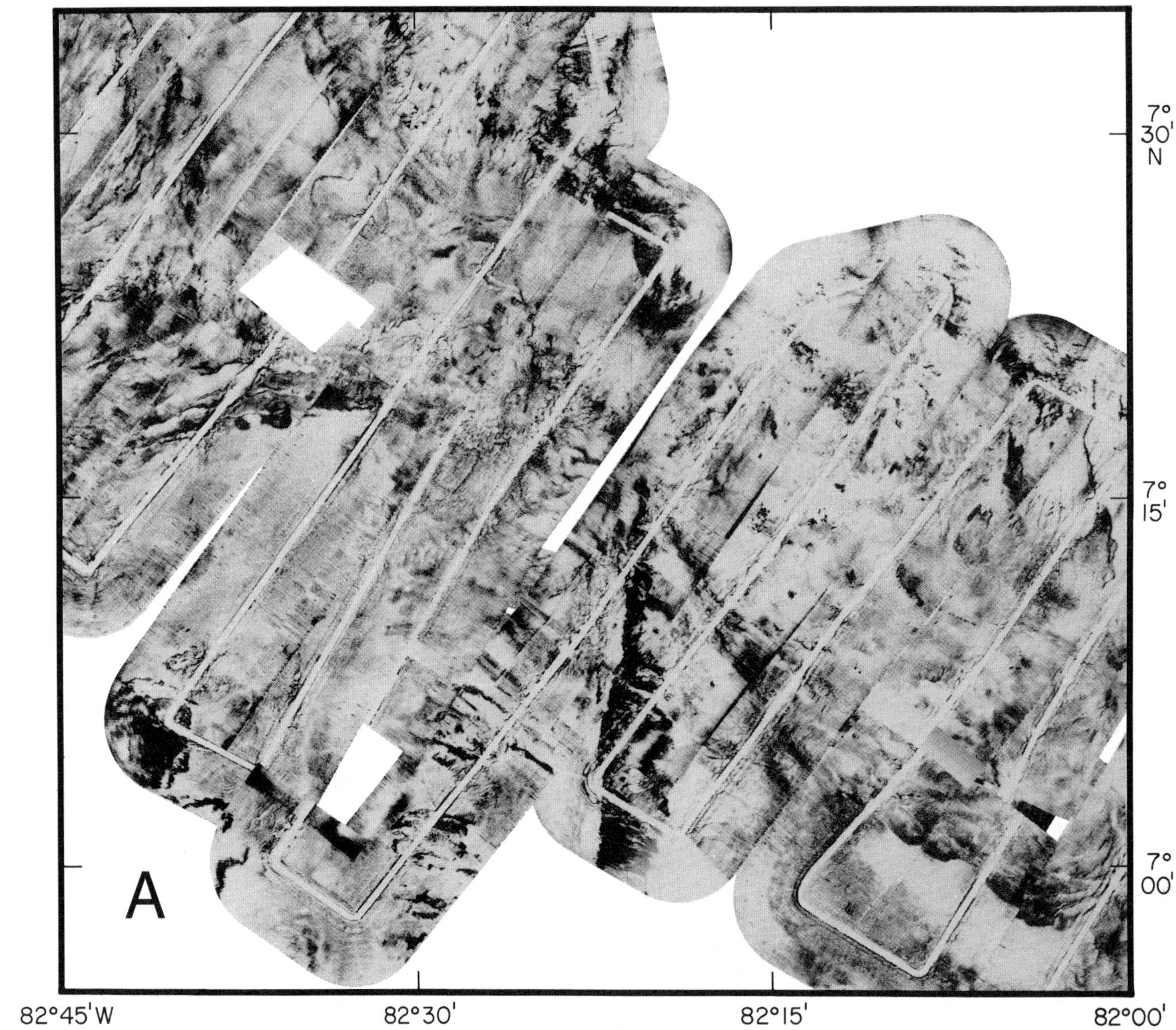

Figure 4. A, Detail of SeaMARC II side-scan mosaic across the Balboa Fracture Zone showing curvature of the thrust front around the fracture zone (see Fig. 3 for location). B, Interpretation of SeaMARC II side-scan and seismic reflection data (see Fig. 5).

of Balboa Fracture Zone shows a lobe-shaped feature at the surface ("slump mass" in Fig. 4B) that extends at least 15 km seaward of the base of the slope. On seismic line PAN-15, the slump thins from 0.06 sec (approx. 50 m) at the base of the slope to 0.02 sec (approx. 15 m) at its seaward edge (Fig. 5). We interpret this area to be composed of debris slumped from the accretionary prism as it climbs over the Balboa Fracture Zone. A similar slump has formed just west of Coiba Fracture Zone (Plate 6).

MODEL FOR OBLIQUE SUBDUCTION OF FRACTURE ZONES

As the ridges of the Panama Fracture Zone system are obliquely subducted under the Caribbean Plate at the south Panama margin, the accretionary prism is deformed. Because the ridges are being subducted obliquely, they sweep eastward along the margin. Thus, cross sections taken spatially along the margin reflect the temporal development of the margin.

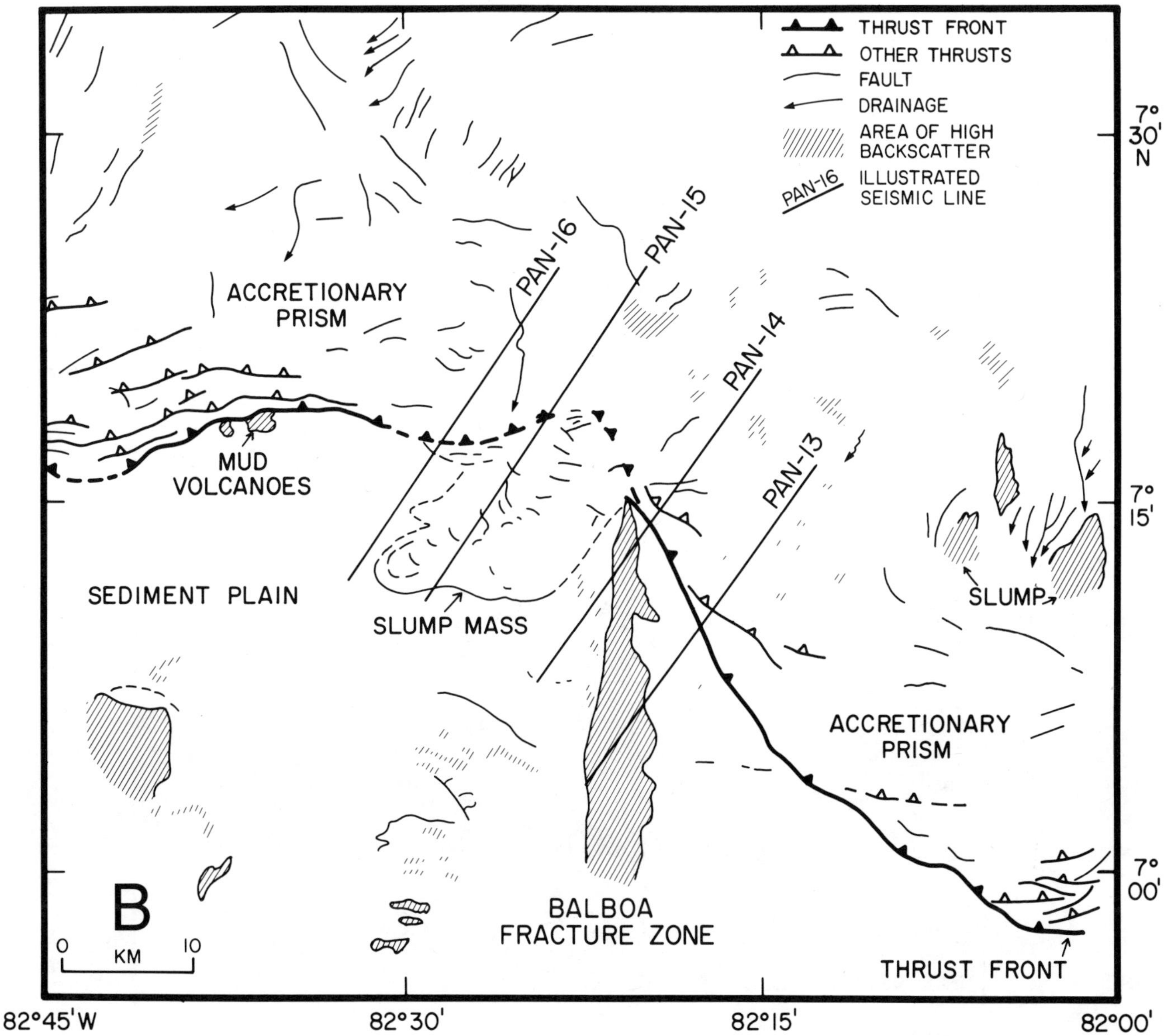

The temporal model presented below is based on the spatial development of structures imaged along the margin. A similar style of deformation has been documented where basement ridges are obliquely subducted beneath the Barbados Ridge Complex (Westbrook, 1982).

East of the Panama Fracture Zone system, the accretionary wedge has not yet been subjected to ridge collision (Fig. 6, "pre-collision"). Where a ridge intersects the margin, the thrust front as well as other thrusts and folds near the base of the South Panama deformed belt wraps around the ridges (Fig. 6, "collision"). During collision, the accretionary prism rides up the east flank of the ridge. The dip of the descending Nazca plate at the trench increases as a result of the presence of the ridge. The dip of the basal décollement increases as well, and the décollement deepens (Fig. 6, "collision," time 2). As the basal slope increases, the surface slope also increases, thus thickening the prism and increasing the critical wedge taper. Critical taper theory suggests that if both the dip of the basal décollement and the surface slope increase, either the coefficient of internal friction of the wedge should decrease or the basal friction should increase (Davis and others, 1983; Dahlen, 1984). The physical properties of the accretionary prism could be drastically changed during the collision process; however, we believe that it is more likely that the coefficient of friction along the basal décollement increases as it deepens during the collision. If the coefficient of basal friction

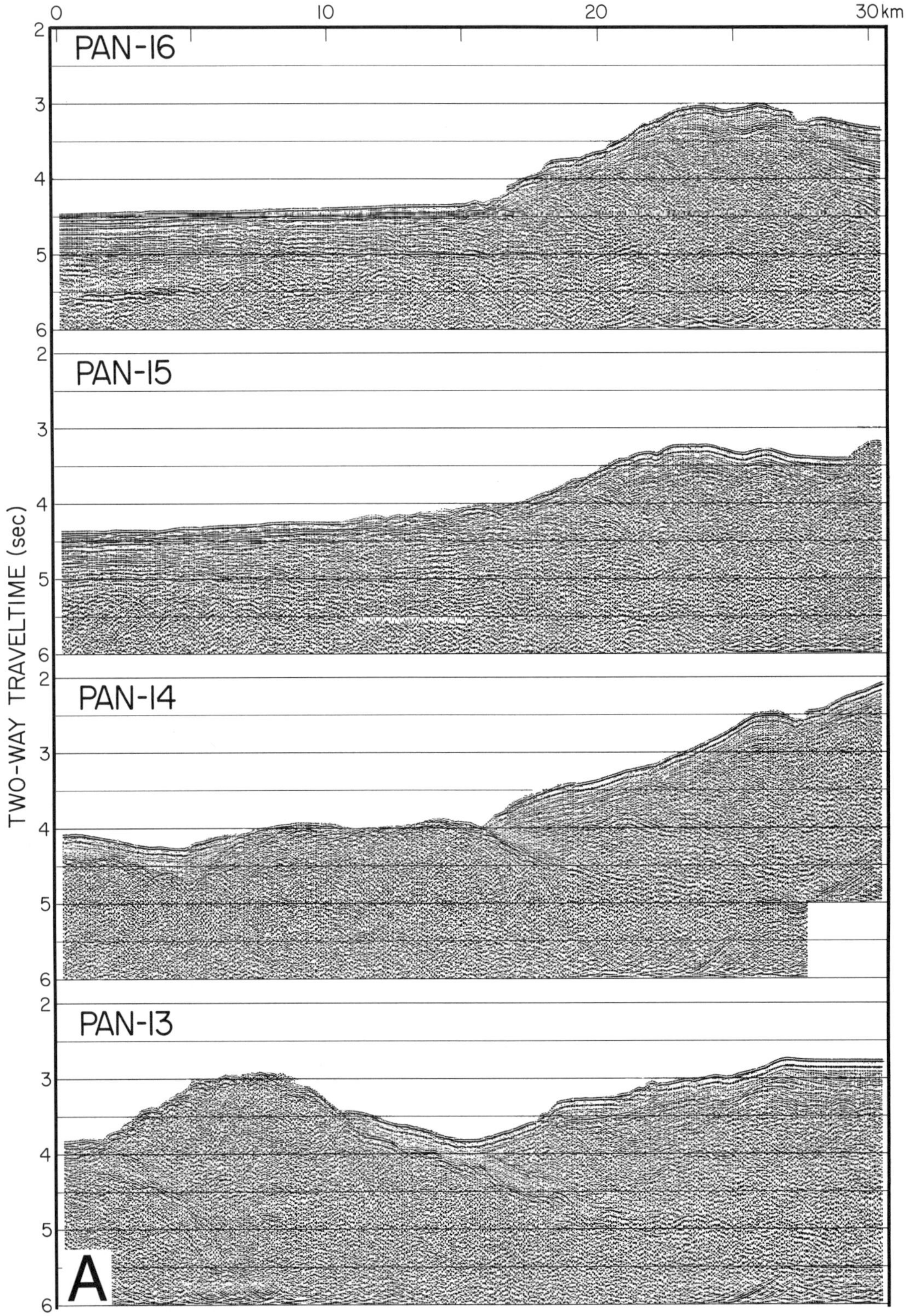

Figure 5. Seismic reflection profiles across the toe of the South Panama deformed belt at its intersection with the Balboa Fracture Zone. Locations of lines indicated on Figure 4B. A, Uninterpreted profiles. B, Interpreted profiles showing location of frontal thrust, approximate location of décollement, and base of sedimentary section on subducting ridges.

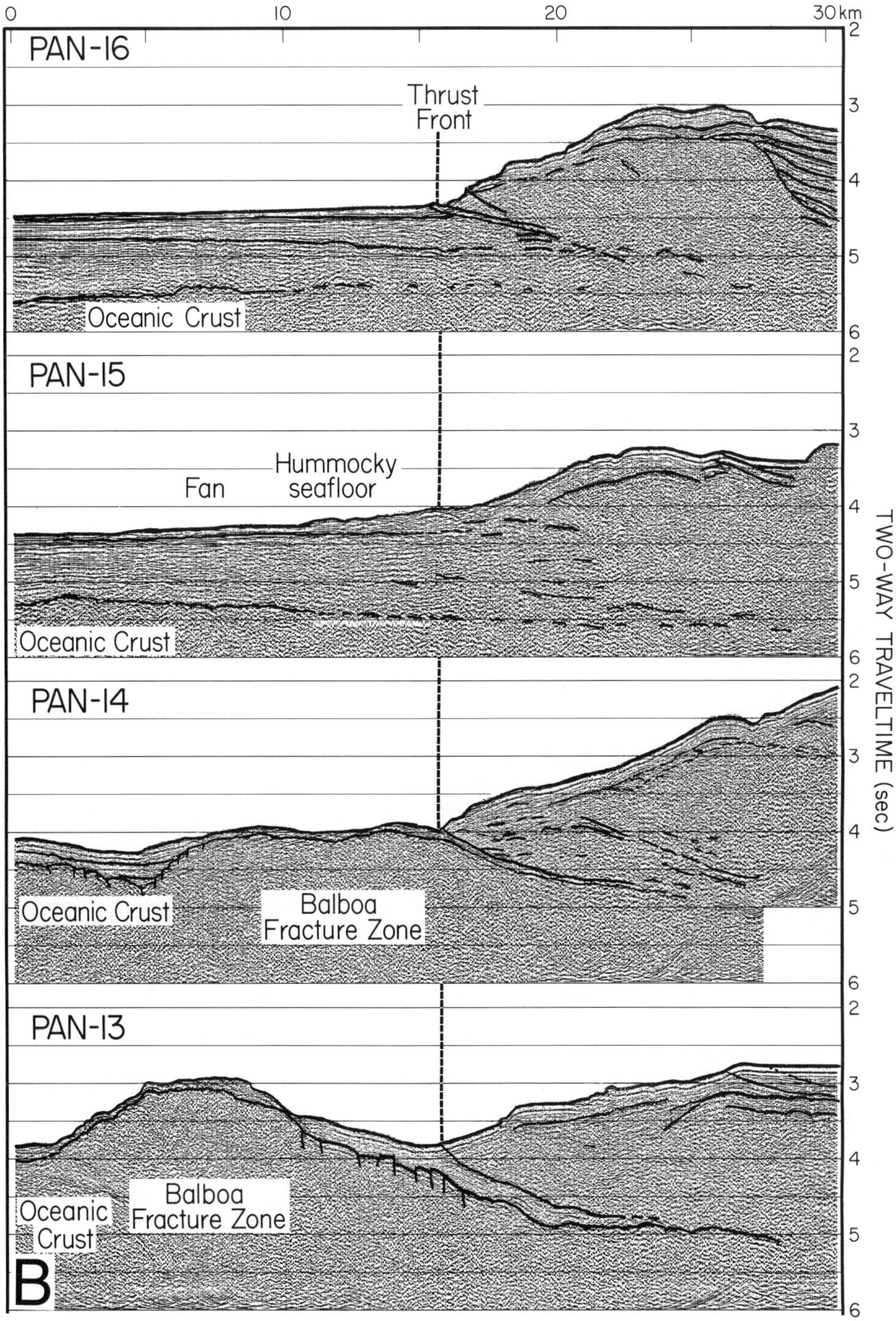
0
10
20
30 km
PAN-16
Thrust
Front
Oceanic Crust
PAN-15
Hummocky
seafloor
Fan
Oceanic Crust
PAN-14
Oceanic Crust
Balboa
Fracture Zone
PAN-13
Oceanic
Crust
Balboa
Fracture Zone
B
2
3
4
5
6
TWO-WAY TRAVELTIME (sec)

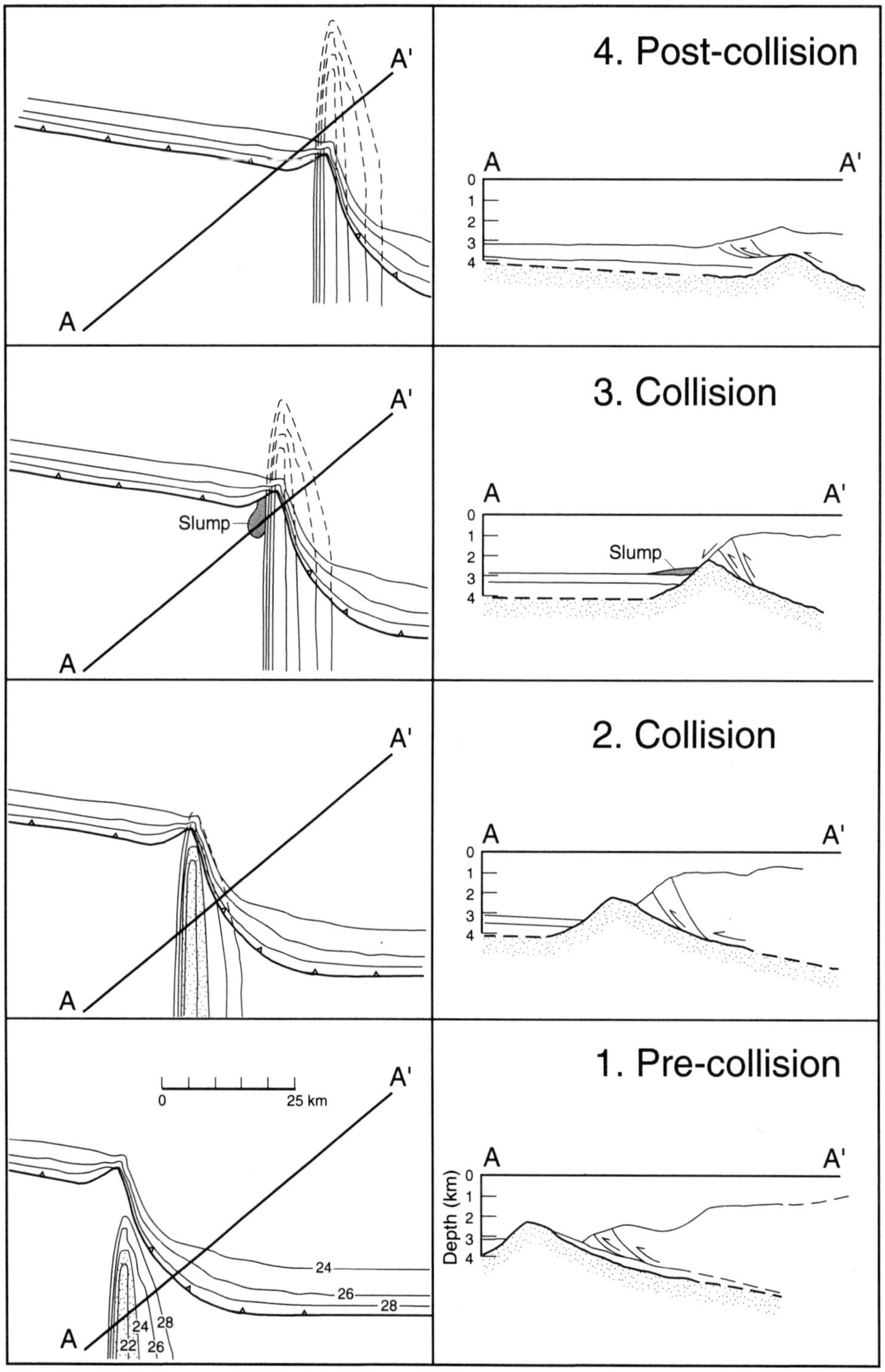

Figure 6. Evolutionary model for modification of the south Panama accretionary prism by oblique subduction ridges of the Panama Fracture Zone complex indicates sequential times. Bathymetric contours (in hundreds of meters) show orientation of regional slopes.

increases as the dip of the basal décollement increases during ridge subduction, the wedge taper increases, thus causing the wedge to become subcritical. The wedge will thicken internally to achieve a new critical taper with an increased surface slope. This increase in basal friction is likely to occur as the décollement drops down from the base of the sandy trench sediments east of the ridge collisions into the underlying pelagic sediments where the ridges are being subducted. We cannot determine whether the basal friction increases because of a difference in the physical properties of the accreting sediments or in response to decreased fluid pressures as the thinner sediment section is subducted.

As the prism slides over the top of the ridge and down the other flank, the basal slope reverses, causing the accretionary wedge to become unstable. Slope failure occurs, and much of the prism slumps down the western flanks of the ridges (Fig. 6, "collision," time 3). Slope failures of accretionary prisms following subduction of topographic features is very common, occurring during subduction of a seamount in the Japan Trench (Lallemand and Le Pichon, 1987), during oblique subduction of the d'Entrecasteaux Ridge in the New Hebrides Trench (Collot and Fisher, 1991), and during subduction of seamounts in the Middle America Trench (von Huene and others, this volume). In each case, the surface slope of the accretionary prism steepens during subduction of the anomalous topography, causing instability and slumping as the trailing slope of the feature passes beneath the prism.

A few kilometers west of the Coiba Fracture Zone and Balboa Fracture Zone, east-west–trending structures again dominate the slope, and a frontal thrust is imaged. This indicates that, following passage of the ridge, the tectonic regime quickly returns to one dominated by frontal accretion (Fig. 6, post-collision, time 4). This interpretation differs from that proposed for the subduction of seamounts in the Japan Trench, where the slumped material is believed to be subducted rather than reaccreted (Lallemand and Le Pichon, 1987; von Huene and Lallemand, 1990).

CONCLUSIONS

The structure of the South Panama deformed belt is affected by oblique subduction of the linear ridges of the Panama Fracture Zone system. The ridges are subducted beneath the accretionary prism, sweeping from west to east along the margin. As the prism rides up and over the ridges, it thickens vertically and narrows horizontally; in map view, the structures wrap around the ridges. As the ridges subduct further, the prism slides down the west sides of the ridges and the basal slope reverses, causing the prism to become unstable and slump into the trench. Within a short distance (and time), the subduction zone returns to a "normal" accretionary mode, and the slumped material is reaccreted. Therefore, the oblique subduction of the Panama Fracture Zone system ridges causes only temporary disruption of the margin.

ACKNOWLEDGMENTS

We thank Tom Shipley and Mark Wiederspahn for providing the digital seismic acquisition system for this cruise. We thank Don Reed, Eli Silver, and Darla Heil, who participated in the cruise and discussed Panama tectonics with us. Darla Heil also performed the original bathymetric contouring. Reviews by Reed, Silver, Paul Mann, Mary MacKay, and Dan Orange helped clarify the text. This work was funded by grants from the Division of Ocean Sciences, National Science Foundation. School of Ocean and Earth Science and Technology Contribution 3539.

REFERENCES CITED

Adamek, S. H., Frohlich, C., and Pennington, W., 1988, Seismicity of the Caribbean-Nazca boundary: Constraints on microplate tectonics of the Panama region: Journal of Geophysical Research, v. 93, p. 2053–2075.

Collot, J-Y., and Fisher, M. A., 1991, The collision zone between the north d'Entrecasteaux ridge and the New Hebrides island arc. 1: Sea beam morphology and shallow structure: Journal of Geophysical Research, v. 96, p. 4457–4478.

Corrigan, J. D., Mann, P., and Ingle, J. C., Jr., 1990, Forearc response to subduction of the Cocos ridge, Panama–Costa Rica: Geological Society of America Bulletin, v. 102, p. 628–652.

Dahlen, F. A., 1984, Noncohesive critical Coulomb wedges: An exact solution: Journal of Geophysical Research, v. 89, p. 10125–10133.

Davis, D., Suppe, J., and Dahlen, F. A., 1983, Mechanics of fold-and-thrust belts and accretionary wedges: Journal of Geophysical Research, v. 88, p. 1153–1172.

De Boer, J. Z., Defant, M. J., Stewart, R. H., Restrepo, J. F., Clark, L. F., and Ramirez, A. H., 1988, Quaternary calc-alkaline volcanism in western Panama: Regional variation and implications for the plate tectonic framework: Journal of South American Earth Sciences, v. 1, p. 275–293.

De Boer, J. Z., Defant, M. J., Stewart, R. H., and Bellon, H., 1991, Evidence for active subduction below western Panama: Geology, v. 19, p. 649–652.

Fisher, M. A, 1986, Tectonic processes at the collision of the d'Entrecasteaux zone and the New Hebrides island arc: Journal of Geophysical Research, v. 91, p. 10470–10486.

Fryer, P., and Hussong, D. M., 1985, SeaMARC II studies of subducting seamounts, *in* Nasu, N., et al., eds., Formation of active ocean margins, Tokyo, Terra Scientific Publishing Company, p. 291–306.

Heil, D. J., 1988, Response of an accretionary prism to transform ridge collision south of Panama [M.S. thesis]: Santa Cruz, University of California, 88 p.

Jordan, T. H., 1975, The present-day motions of the Caribbean plate: Journal of Geophysical Research, v. 80, p. 4433–4449.

Lallemand, S., and Le Pichon, X., 1987, Coulomb wedge model applied to subduction of seamounts in the Japan trench: Geology, v. 15, p. 1065–1069.

Lallemand, S., von Huene, R., and Cullota, R., 1989, Subduction of the Daiichi Kashima seamount in the Japan Trench: Tectonophysics, v. 160, p. 231–247.

Lonsdale, P., and Klitgord, K. D., 1978, Structure and tectonic history of the eastern Panama basin: Geological Society of America Bulletin, v. 89, p. 981–999.

MacKay, M. E., and Moore, G. F., 1990, Variation in deformation of the South Panama accretionary prism: Response to oblique subduction and trench sediment response: Tectonics, v. 9, p. 683–698.

Mann, P., and Corrigan, J. D., 1990, Model for late Neogene deformation in Panama: Geology, v. 18, p. 558–562.

Masson, D. G., and six others, 1990, Subduction of seamounts at the Java

Trench: A view with long-range sidescan sonar: Tectonophysics, v. 185, p. 51–65.

Okaya, D. A., and Ben-Avraham, Z., 1987, Structure of the continental margin of southwestern Panama: Geological Society of America Bulletin, v. 99, p. 792–802.

Sender, K. L., Shor, A., and Hagen, R., 1989, SeaMARC II side-scan processing techniques [abs.]: Eos (American Geophysical Union Transactions), v. 70, p. 1304.

Silver, E., Reed, D. L., Tagudin, J. E., and Heil, D. J., 1990, Implications of the North and South Panama thrust belts for the origin of the Panama orocline: Tectonics, v. 9, p. 261–281.

von Huene, R., 1986, To accrete or not accrete, that is the question: Geologische Rundschau, v. 75, p. 1–15.

von Huene, R., and Lallemand, S., 1990, Tectonic erosion along the Japan and Peru convergent margins: Geological Society of America Bulletin, v. 102, p. 704–720.

Westbrook, G. K., 1982, The Barbados Ridge Complex: Tectonics of a mature forearc system: Geological Society of London Special Publication 10, p. 275–290.

Yamazaki, T., and Okamura, Y., 1989, Subducting seamounts and deformation of overriding forearc wedges around Japan: Tectonophysics, v. 160, p. 207–229.

MANUSCRIPT ACCEPTED BY THE SOCIETY FEBRUARY 7, 1994

Printed in U.S.A.

Geological Society of America
Special Paper 295
1995

Sediment dispersal and accretionary growth of the North Panama deformed belt

Donald L. Reed
Department of Geology, San Jose State University, San Jose, California 95192-0102
Eli A. Silver
Earth Sciences Department and Institute of Marine Sciences, University of California, Santa Cruz, California 95064

ABSTRACT

This study examines the interplay between sediment dispersal along the Panama margin and the evolution of the North Panama deformed belt (NPDB) using SeaMARC II swath-mapping data and migrated seismic profiles. The NPDB is composed of folded and thrusted strata of the Colombian basin that rise to within 800 m of the sea surface along the San Blas ridge. The ridge extends for more than 200 km along strike, thereby acting as an important barrier to sediment dispersal in the eastern region of the NPDB and the Colombian basin. Sediments derived from eastern Panama are trapped in the San Blas basin, located along the landward margin of the ridge, and deflected westward before traversing the lower slope along the San Blas canyon in the central region of the NPDB.

Sediment entrapment within the San Blas basin and the westward deflection of transport paths by the San Blas ridge have contributed to a recent decrease in sediment thickness along the thrust front in the eastern region of the NPDB. Sediment thickness on the underthrusting basement of the Colombian basin also decreases in the region as a result of the blockage of Magdalena fan deposits by a basement ridge located less than 20 km from the thrust front. The decrease in sediment thickness is accompanied by a decrease in the width of the NPDB in this region, suggesting a concomitant decrease in the rate of frontal accretion as compared to other regions of the NPDB. A marked decrease in the thickness and fold wavelength of accreted strata from the interior of the belt to the thrust front may result from the diversion of terrigenous sediments away from the eastern region of the belt. Convergence rates along the frontal thrust are also significantly less in this region than elsewhere in the belt. Decreased frontal accretion along the eastern region has, however, been partially offset by the incorporation of thick deposits of the San Blas basin into the rear of the belt along backthrusts. Recent accretionary growth is therefore two-sided but may have varied considerably in the past in response to shifting patterns of sediment dispersal.

INTRODUCTION

Young orogenic belts rise above sea level and are eroded, resulting in the transport and deposition of large volumes of detritus in offshore basins. These deposits provide valuable insights into the history of orogenesis (Lundberg and Dorsey, 1990; Cochran, 1990). Orogenic deposits may be reincorporated into the active orogen along frontal thrusts in trench and foreland environments and at convergent structures marking the rear of orogenic belts and accretionary wedges (Dickinson and Seely, 1979; Reed et al., 1986; Silver and Reed, 1988; Torrini and Speed, 1989; Unruh et al., 1991). Here we exam-

Reed, D. L., and Silver, E. A., 1995, Sediment dispersal and accretionary growth of the North Panama Deformed Belt, *in* Mann, P., ed., Geologic and Tectonic Development of the Caribbean Plate Boundary in Southern Central America: Boulder, Colorado, Geological Society of America Special Paper 295.

ine the dispersal of deep-water sediments along the northern margin of Panama near the collision zone between the Panama volcanic arc and the South American continent. This region is composed of folded and thrusted strata of the Colombian basin that form the North Panama deformed belt (NPDB) (Fig. 1). Specifically, this study examines the effect of sediment dispersal in an offshore sedimentary basin, located at the rear of the NPDB, on the development of recent accretionary structures.

Using SeaMARC II swath-mapping and a grid of migrated seismic reflection profiles, we have mapped the geometry of offshore sedimentary basins and paths of sediment dispersal at water depths between 1,000 and 3,500 m along northeastern Panama. Uplift of the rear of the eastern NPDB along backthrusts has blocked the downslope transport of sediment, producing a westward shift of terrigenous sedimentation along the thrust front. We compare recent accretionary structures in the eastern NPDB with variations in sediment thickness and convergence rate along frontal thrusts. Silver et al. (1990) showed that the rate of convergence along the thrust front varies abruptly across a region of right-lateral strike-slip faulting, perhaps contributing to a decrease in frontal accretion in the eastern NPDB. Large volumes of sediment have, however, been incorporated into the rear of the belt along backthrusts, a situation similar to that of many submarine accretionary wedges (Silver and Reed, 1988), including the Barbados ridge complex (Westbrook, 1982; Speed et al., 1989).

GEOLOGIC SETTING OF NORTH PANAMA DEFORMED BELT

The Panama Isthmus is composed of a deformed Miocene-to-recent volcanic arc built on oceanic crust of Late Cretaceous to early Cenozoic age (Case, 1974; Weyl, 1980; Case et al., 1984; deBoer and others, 1988; Bowland and Rosencrantz, 1988). Both margins of the isthmus are marked by lithospheric convergence (Fig. 1). The Nazca and Cocos plates are subducting obliquely beneath the South Panama deformed belt (Silver and others, 1990; MacKay and Moore, 1990; Moore and Sender, this volume), whereas the North Panama deformed belt has formed by convergence between the Panama arc and Caribbean plate (Edgar et al., 1971; Lu and McMillen, 1982; Ademak et al., 1988). The contorted shaped of the isthmus has been attributed either to oroclinal bending (Wadge and Burke, 1983; Silver et al., 1990) or strike-slip faulting (Mann and Corrigan, 1990) during collision with South America beginning in the middle to late Miocene. Uplift of the arc, presumably associated with the collision with South America, closed the Caribbean-Pacific seaway by 1.8 Ma (Keller et al., 1989).

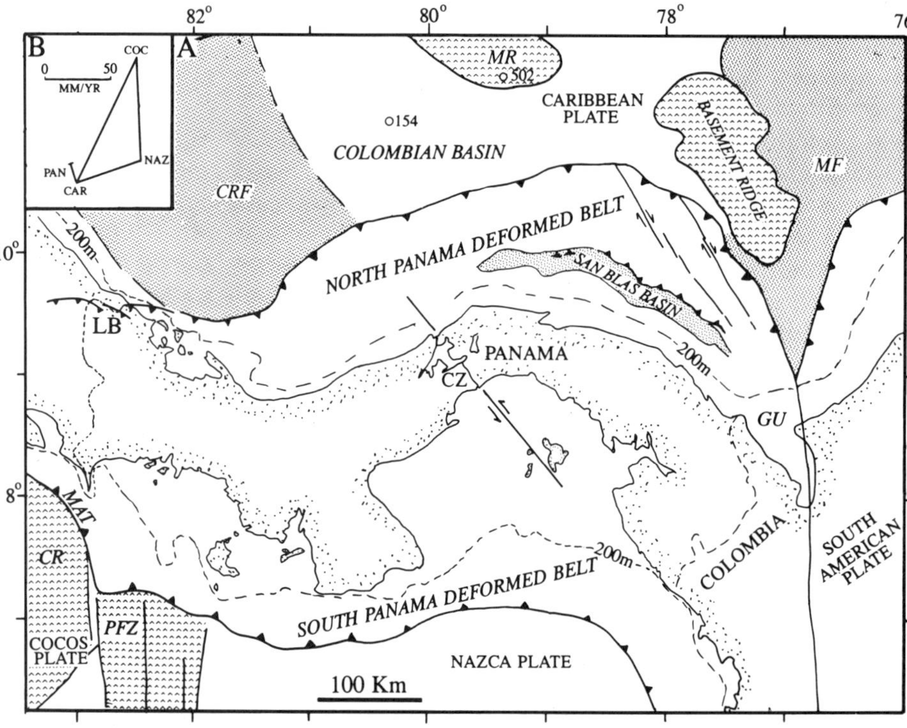

Figure 1. A, General tectonic map of offshore Panama (modified from Case and Holcombe, 1980; Silver et al., 1990). CR—Cocos Ridge, CRF—Costa Rica Fans, (CZ—Canal Zone) GU—Gulf of Uraba, LB—Limon Basin, MAT—Middle America Trench, MF—Magdalena Fan, MR—Mono Rise, PFZ—Panama Fracture Zone. B, Vector diagram of relative plate motions between the Caribbean (CAR), Nazca (NAZ), and Cocos (COC) plates and the Panama arc (PAN) (modified from Silver et al., 1990).

The North Panama deformed belt forms a large submarine fold-thrust belt, or accretionary wedge, consisting of deformed strata of the Colombian basin (Lu and McMillen, 1982; Silver et al., 1990). The belt occupies virtually the entire lower continental slope along northern Panama, extending from the Gulf of Uraba to the northern margin of Costa Rica (Fig. 1) (Case et al., 1984). A basement ridge, located southeast of the Mono Rise, separates sediments of the Magdalena fan from the thrust front in the eastern NPDB. In the west, the NPDB extends across the Costa Rica continental slope and on land along the Limon Basin (Silver et al., this volume). Silver et al. (1990) used SeaMARC II imagery to determine the structure of the NPDB, from which they inferred north-northwest movement of the Panama arc, with respect to the Caribbean plate, in response to collision with South America. Breen and others (1988) used these data to suggest that anticlinal ridges in the NPDB are cored by mud diapirs. Reed et al. (1990) noted, however, that the vast majority of mud volcanoes are located within a lower-slope depocenter and suggested the importance of methane and methane hydrate in the generation of these features.

DATABASE AND METHODS

This study is based on SeaMARC II swath-mapping and digital single-channel seismic reflection data acquired aboard the R/V Moana Wave in June 1985. The survey provides nearly 100% side-scan sonar and bathymetric coverage of the thrust belt (Fig. 2). The survey area is crossed by migrated single-fold seismic reflection profiles at spacing of approximately 8 km, which provide structural information to a depth of several kilometers beneath the seafloor. Digital processing of the reflection data included predictive deconvolution, band-pass filtering, finite-difference migration, automatic gain control, and display. Descriptions of seven core samples in the region were obtained from the National Geophysical Data Center, in order to correlate patterns of sonar backscattering with sediment type. Locations of the cores are shown in Figure 2. In general, cores composed of fine-grained hemipelagic sediments correspond to regions of low backscatter, whereas samples of volcanic ash or terrigenous sediment were collected within regions of moderate to high backscatter. These relations were used to trace transport paths of terrigenous sediment derived from northeastern Panama.

STRUCTURE OF THE NPDB

The North Panama deformed belt reaches a maximum width of 90 km northeast of the Panama Canal but decreases in width near Costa Rica in the west and the Gulf of Uraba in the east (Fig. 3). The NPDB can be divided into three regions (Figs. 2 and 3): (1) an eastern region where thin packages of accreted strata form short wavelength, seaward-verging folds; (2) a central region characterized by thick offscraped strata in mostly landward-verging folds; and (3) a western region in which the width of the belt narrows considerably and seaward- and landward-verging folds are observed. The latter region is the focus of a study by Silver and others (this volume).

Fold axes and fault trends in the central region, including the thrust front, exhibit a remarkably uniform east-northeast trend (Fig. 3). This region exhibits the longest wavelength folds in the NPDB, and thrust spacing is large, commonly

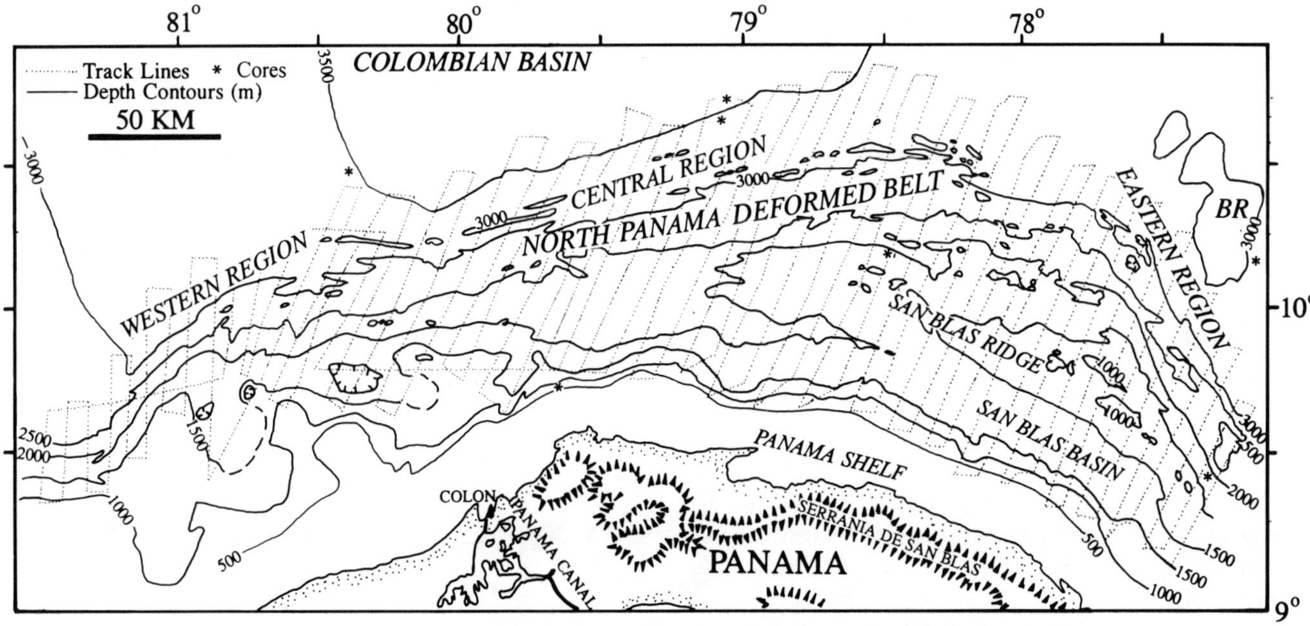

Figure 2. Ship tracks of SeaMARC II and seismic reflection survey superimposed on bathymetry map of the North Panama deformed belt. BR—Basement Ridge. Contour interval is 500 m.

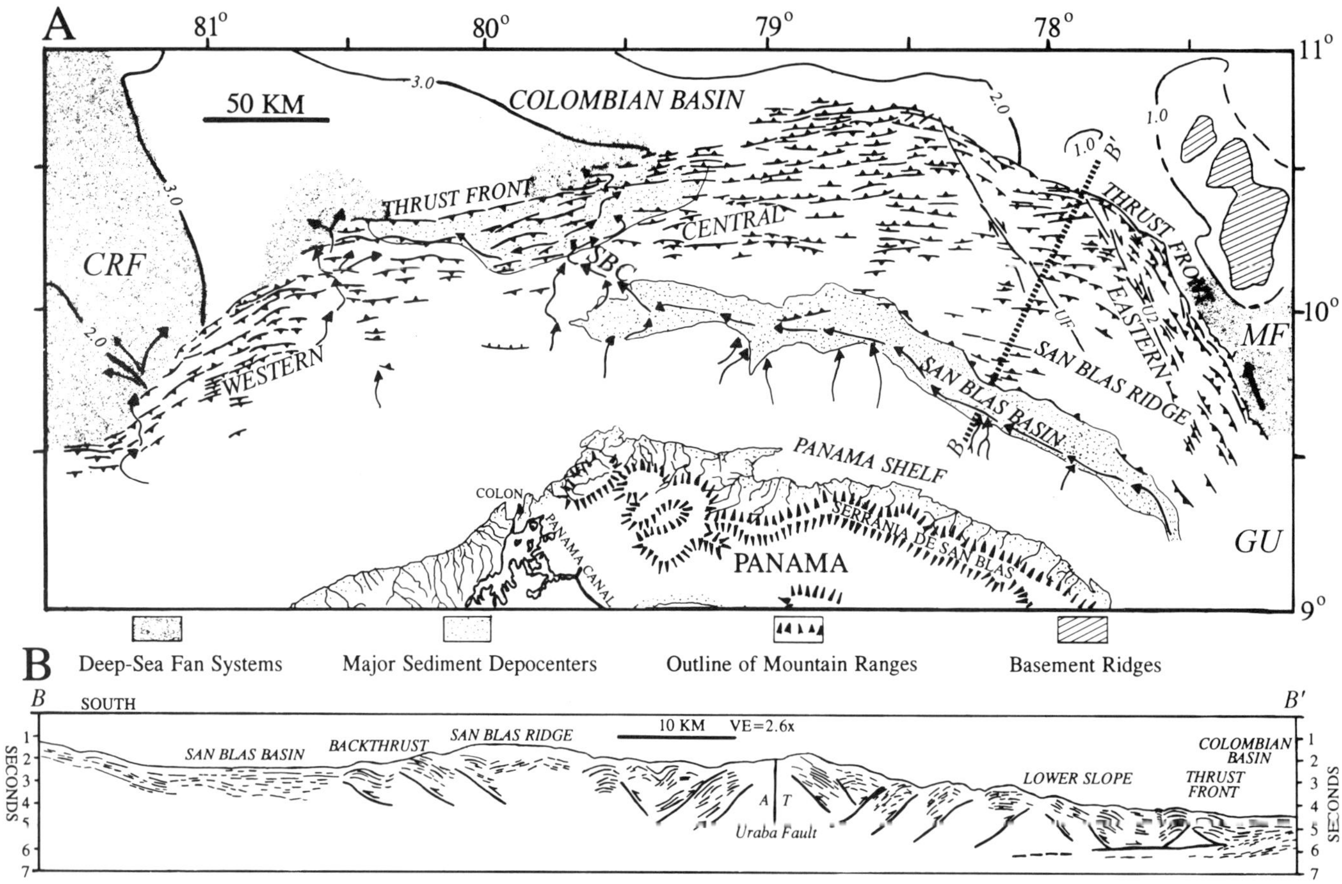

Figure 3. A, Structural map of North Panama deformed belt, modified from Reed et al. (1990) and Silver et al. (1990). Contour lines in the Colombian basin denote sediment thickness to basement in two-way travel-time, based on this study and Bowland (1993). Arrows indicate major routes of sediment dispersal. Teeth on upthrown side of thrust faults. UF—Uraba fault, UZ—UZ fault, GU—Gulf of Uraba, MF—Magdalena Fan, CRF—Costa Rica fan, SBC—San Blas Canyon. B, Line drawing of seismic Line 44 showing style of folding and faulting in deformed belt. Relative motions across right-lateral strike-slip denoted by A (away) and T (toward).

reaching distances of 6 to 10 km (Tagudin, 1988; Silver et al., 1990). The boundary between the central and eastern regions of the belt is marked by the northwest-trending Uraba fault, across which the thrust front bends to the east-southeast. Farther east, the frontal thrust is crossed by the U2 fault, and the trend of the thrust front changes to the southeast. Silver et al. (1990) associated the Uraba and U2 faults with right-lateral strike-slip displacement and suggested that convergence between the Panama arc and the Caribbean plate is partitioned into a combination of thrust and strike-slip faulting in the eastern region.

A large portion of the eastern region of the NPDB consists of the west-northwest–trending San Blas ridge. The ridge is composed of 2- to 3-sec-thick packages of accreted sediment that are separated by thrust faults spaced at an interval of 4 to 5 km (Figs. 3 and 4). The structural geometry of accreted strata composing the ridge is remarkably similar to that observed along the thrust front in the central region, where sediments reach a maximum thickness in the region of frontal accretion. In marked contrast, accreted strata along the thrust front in the eastern region are significantly thinner (<2 sec), and thrust spacing is smaller, although variable.

The San Blas ridge is more than 200 km long and rises to within 800 m of the sea surface in the eastern region. The water depth along the crest of the ridge increases westward and lies at approximately 2,000 m in the central region (Fig. 5). The southern, or landward, slope of the ridge exhibits south-directed thrusts or backthrusts that are especially well developed in the eastern region where the ridge rises to its highest elevation (Fig. 3) (Silver and Reed, 1988).

STRATIGRAPHY AT THE THRUST FRONT

Lu and McMillen (1982) characterized sediments seaward of the western and eastern regions as fan-type deposits that include the Costa Rica fan and Magdalena fan, respec-

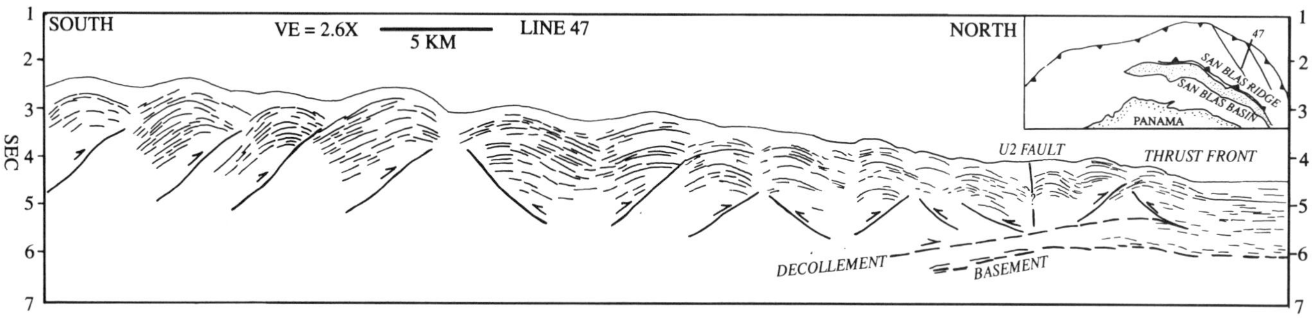

Figure 4. Line drawing of seismic line 47 showing variation in structure from the San Blas ridge to the thrust front.

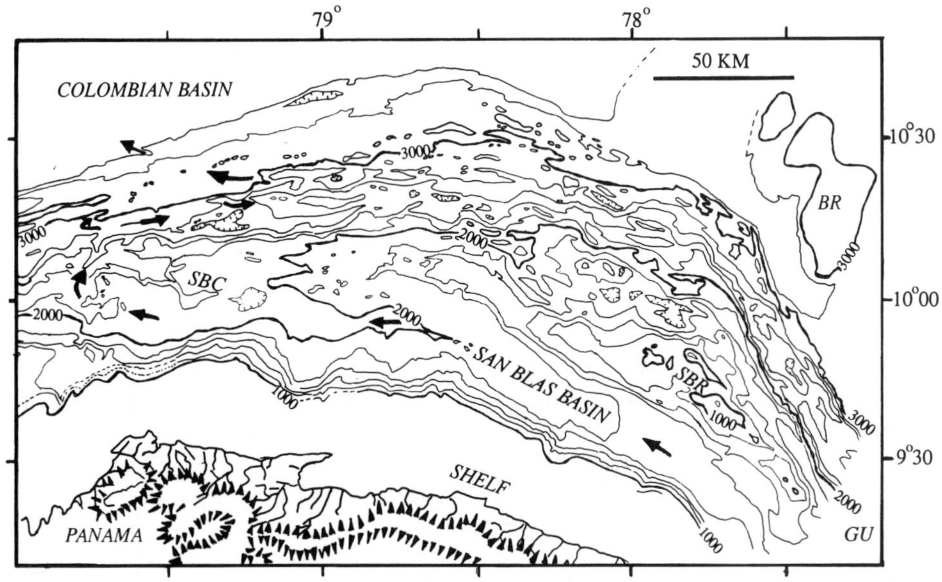

Figure 5. Bathymetric map showing the connection between the San Blas basin and San Blas canyon (modified from Silver et al., 1990). Contour interval is 200 m. BR—Basement Ridge, SBC—San Blas Canyon, SBR—San Blas Ridge.

tively. Submarine canyons traverse the western region of the belt, leading to the development of deep-sea fans in the Colombian basin seaward of the thrust front. Sedimentary deposits located seaward of the central region were classified as basin-floor type. Using a regional grid of seismic profiles and drilling results at DSDP Sites 154 and 502, Bowland (1993) divided the sedimentary section in the western Colombian basin into five major depositional units. Upper Cretaceous to Eogene stratigraphic sequences overlie basement rocks and are correlated with dominantly pelagic strata of the central Venezuelan basin. Clastic and hemipelagic sedimentary units are probably of Eocene age, followed by the deposition of extensive turbidite fans in the late Miocene. The latter units were attributed to increased volcanism and uplift along southern Central America, including uplift associated with the collision of the Panama arc with South America.

Sediment thickness along the thrust front decreases from more than 3.0 sec near the Costa Rica fan to less than 1 sec seaward of the eastern region of the NPDB, largely as a result of the thinning of turbidite fan deposits of late Miocene to recent age (Bowland, 1984; this study). Upper Cretaceous to Eocene units are also thinner in this region where they lap onto a basement ridge located southeast of the Mono Rise (Figs. 1 and 3) (Tagudin, 1988; Bowland, 1993). This ridge may be an extension of the Mono Rise, which is an uplifted basement block of probable Late Cretaceous age (Bowland and Rosencrantz, 1988).

SAN BLAS BASIN

The San Blas basin covers an area of approximately 4,500 km^2, forming one of the most significant physiographic features along northern Panama (Fig. 3). The basin is located between the Panama shelf and the San Blas ridge; the latter serves as a barrier to the down-slope transport of terrigenous detritus derived from the mountains of the Serrania de San Blas and Serrania del Darien along eastern Panama. These sediments are either trapped within the San Blas basin or transported westward to the San Blas canyon, which wraps around the western edge of the ridge in the central region of the NPDB (Fig. 5).

The landward margin of the San Blas basin is underlain by a sequence of slope sediments that reaches a thickness in excess of 2.0 sec (two-way travel time). These sediments overlie a north-dipping basement reflector that probably delineates the top of Late Cretaceous through Eocene volcanic and plutonic rocks exposed onshore along the Serrania de San Blas (Weyl, 1980; Bowland and Rosencrantz, 1988) (Figs. 3 and 6). The basement surface is generally smooth and continuous, although it is offset along a south-dipping normal fault in one area, producing a half-graben. The slope sequence changes in reflective character and thickness across the fault, indicating that it existed prior to the deposition of the slope sequence. Displacement along the fault has, however, continued throughout the deposition of these strata, as suggested by the divergence of reflectors toward the fault within the half-graben. Thrust loading of the arc slope by the rear of the NPDB may contribute to flexure in this region, although normal faulting is not widespread.

The slope sequence is cut by numerous submarine canyons that provide conduits for sediment transport from the Panama shelf to the San Blas basin. The canyons terminate in deep-sea fans that coalesce northward in the form of flat-lying basinal turbidites (Fig. 6). These deposits reach a thickness of more than 3 sec in the eastern region of the basin but progressively decrease in thickness westward along the length of the San Blas basin.

The slope sequence and the underlying basement reflector dip gently northward beneath flat-lying deposits of the San Blas basin (Fig. 6). Along the northern margin of the basin, basinal deposits and the underlying slope strata are uplifted and tilted northward along backthrusts marking the southern slope of the San Blas ridge. The backthrusts are overlain by anticlinal ridges that exhibit growth folding. A bottom-simulating reflector (BSR), associated with the base of a gas hydrate layer, is commonly observed in hinge zones of anticlines along the margin of the ridge and can be traced southward into the strata of the San Blas basin (Reed et al., 1990; Silver et al., 1990). Sedimentary basins, lying between the anticlinal ridges, contain up to 1.5 sec of sediment and are perched hundreds of meters above the San Blas basin. These deposits are isolated from modern channels and canyons extending from Panama. Reflectors from within the San Blas basin can be tracked northward beneath these basins, indicating that the basins along the San Blas ridge are underlain by uplifted portions of the San Blas basin. Recent sedimentation in the basins may have resulted from (1) submarine erosion of the crest of the San Blas ridge and (2) turbidity flows that cross the San Blas basin and climb the landward slope of the ridge, similar to up-

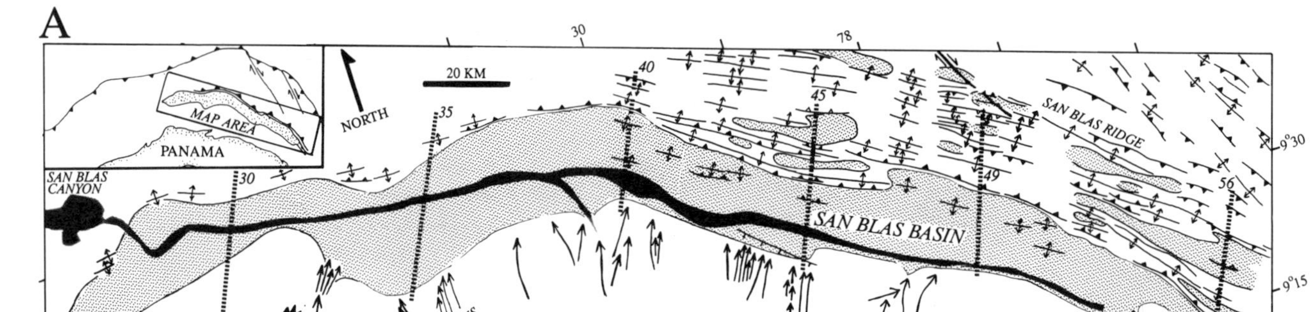

Figure 6. A, Structural map of the San Blas basin. Basin deposits in dotted pattern; axial channel of the San Blas basin shown in gray. GU—Gulf of Uraba. B, Migrated seismic profile along line 45 showing stratigraphy of the San Blas basin and backthrusting at boundary with the San Blas ridge. BSR denotes the bottom-simulating reflector.

slope turbidity flows proposed along the Tiburon Rise in the west-central Atlantic (Dolan et al., 1989) and from physical modeling studies (Muck and Underwood, 1990). Thick deposits in the eastern region of the San Blas basin, near the Gulf of Uraba, have been tilted southward but do not appear to be offset along shallow thrusts. The structure in this region is similar to backthrust deformation that has been described along the rear of accretionary wedges elsewhere in the world (Silver and Reed, 1988; Westbrook, 1982). The term "tectonic wedging" has been used to describe this process, whereby a wedge of accreted sediment is thrust beneath the strata of the forearc basin, resulting in uplift and tilting of latter (Torrini and Speed, 1989; Speed et al., 1989; Unruh et al., 1991).

PATHS OF SEDIMENT DISPERSAL

The North Panama deformed belt exhibits three major transport systems of terrigenous sediment. The two located in the western region transport sediment directly to the thrust front; a third is composed of the San Blas basin and canyon system that extends across the eastern and central regions (Fig. 3). Terrigenous sediment is transported to the thrust front in the western region along the two relatively straight canyons, thereby allowing detritus to bypass the lower slope. The foot of the each canyon is marked by a deep-sea fan that extends northward into the Colombian basin. Lu and McMillen (1982) included these deposits in their fan-type sequence. In contrast, sediments in the central and eastern regions are transported through a chain of basins prior to reaching the thrust front and the Colombian basin (Reed et al., 1990). These basins are linked together along a 300-km-long corridor that begins with a system of small canyons extending from the northeastern Panama shelf. The canyons act as tributaries directing detritus to the San Blas basin (Figs. 6 and 7). Within the basin, sediments are transported westward from a water depth of 1,600 to 1,700 m near the Gulf of Uraba to 2,200 m at the head of the San Blas canyon (Fig. 5). A 1- to 3-km-wide channel marks the main transport path that runs along the entire length of the basin. The axial channel receives sediment from fan valleys and channels that extend northward from submarine canyons (Fig. 7). The channel is relatively straight and has cut more than 75 m into strata of the San Blas basin (Fig. 8), in indicating recent erosion and incision. Other areas of the channel show such evidence of deposition as the development of small fan and levee deposits.

Near the western edge of the San Blas basin, the axial channel extends into the head of the San Blas canyon, which in turn wraps around the San Blas ridge before continuing downslope (Fig. 6). The canyon provides the only transport corridor to the lower slope in the eastern and central regions. Several channels extend outward from the foot of the canyon, forming a major sediment depocenter (Reed et al., 1990). The channels meander around anticlinal ridges, resulting in the deposition of sediments within slope basins (Fig. 3). Sediment thicknesses exceed 1.0 sec in several of the basins. The main channel in this system steps down through several slope basins and ultimately breaches the frontal thrust, where it terminates in a deep-sea fan in the Colombian basin. The fan complex contributes to sedimentation along the thrust front in the central region of the NPDB. Therefore a large proportion of terrigenous detritus derived from the mountains of northeastern Panama is either trapped in one of several depocenters within the NPDB or diverted approximately 300 km westward, effectively reducing terrigenous sedimentation along the thrust front in the eastern region. In contrast this system increases sedimentation along the rear of the belt and at the thrust front in the central region.

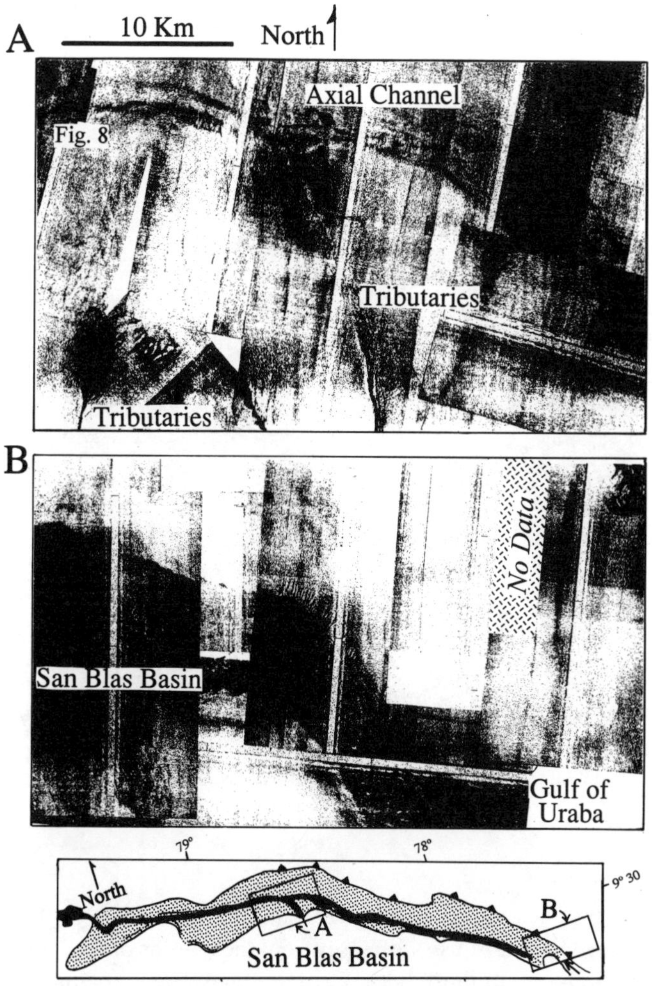

Figure 7. A, SeaMARC II image showing tributaries to axial channel in the San Blas basin. Dark regions represent areas of high acoustic backscatter, whereas low backscatter areas are marked by light-colored regions. B, SeaMARC II image showing canyon in the Gulf of Uraba entering the eastern edge of the San Blas basin.

IMPLICATIONS FOR ACCRETIONARY GROWTH

The eastern region exhibits three features that set it apart from other regions of the NPDB. First, a change in the orientation of the thrust front to the southeast is accompanied by a

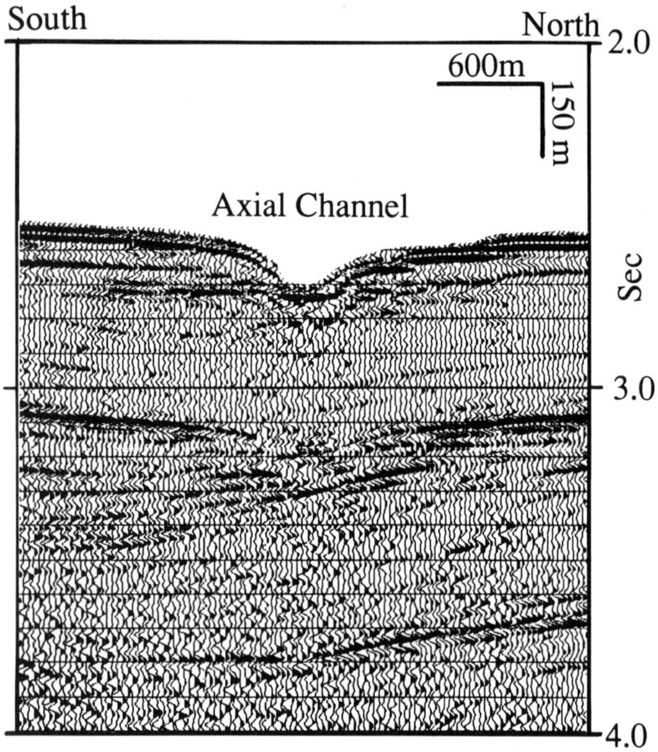

Figure 8. Portion of migrated reflection profile showing incision of axial channel in strata of the San Blas basin. Location shown in Figure 7A.

decrease in width of the belt. Second, the wavelength of folds and spacing of thrusts decreases from the interior of the belt to the thrust front; third, northwest-trending strike-slip faults cut across the wedge (Figs. 3 and 4). The decreased width of the belt and change in structural style may be indicative of a recent decrease in the rate of sediment accretion along the thrust front in the eastern region. We first describe the conditions that are responsible for the decrease in thickness of sedimentary sequences along the thrust front in this region, then discuss the role of strike-slip faulting in accretionary processes.

Two factors have combined to reduce the thickness of sedimentary sequences along the thrust front in this region; this situation has in turn affected the recent development of structures along the thrust front. First, thick deposits of the Magdalena fan are prevented from reaching the thrust front in the eastern region by the basement ridge located southeast of the Mono Rise (Bowland, 1993; Kolla et al., 1984). The ridge has yet to intersect the thrust front, although thin deposits along its southwestern slope are currently being offscraped at the toe of the eastern region. Second, sediments derived from the subaerial portion of the Panama orogen do not contribute to sediment accumulation along the thrust front in the eastern region. The San Blas ridge confines sediment transport along the San Blas basin, which acts as a funnel diverting sediment away from the eastern region of the NPDB and focusing sedimentation in the central region. Westward transport of sediment within the basin provides a source of detritus to the head of the San Blas canyon. Indeed, the canyon provides the only path for terrigenous sediment to reach the thrust front in the eastern and central regions of the NPDB. Sediment thicknesses are therefore significantly reduced seaward of the eastern region as compared to the western and central regions (Bowland, 1984).

The decrease in sediment thickness along the thrust front is accompanied by a change in structural style. The spacing of the thrust faults and thickness of accreted strata generally decrease from the central to eastern region (Tagudin, 1988). A similar change in structural geometry is observed from the interior of the eastern region to the thrust front (Figs. 3 and 4). The San Blas ridge is composed of 2- to 3-sec-thick packages of accreted strata in the eastern region, whereas significantly thinner sequences are offscraped along the thrust front. This variation is accompanied by a decrease in the spacing of thrust faults toward the Colombian basin. The change in structural style suggests that sediment thicknesses along the thrust front, and therefore rates of sediment accretion, in the eastern region may have been greater in the past. Accretion of thick Colombian basin strata, either by offscraping or underplating, may have been a critical factor leading to the formation of the San Blas ridge. In addition, backthrusting along the rear of the belt has contributed to the uplift of the ridge, at least in recent times. Uplift of the ridge subsequently diverted sediment transport away from the thrust front, leading to a recent decrease in sediment accretion in the region. The opposite effect may have occurred in the central region, where terrigenous sedimentation has increased along the thrust front owing to the development of a deep-sea fan deposit at the foot of the San Blas canyon; accretion of these sedimentary sequences may have contributed to the great width of NPDB in this region.

The development of strike-slip faulting has profoundly influenced accretionary structures in the eastern region. The Uraba and U2 faults cut across anticlines and thrust faults, indicating relatively recent movement along the faults (Fig. 3). Silver et al. (1990) noted the location of recent earthquakes near the faults and proposed that Panama–Caribbean plate motion is partitioned into very slow convergence, normal to the thrust front, in the eastern region and right-lateral strike-slip displacement along the Uraba and U2 faults. The rate of convergence along the frontal thrust calculated from this model is on the order of 1 to 2 mm/yr, significantly less than the 6 to 10 mm/yr rate of convergence calculated between Panama and the Caribbean plate (Fig. 1). Consequently, the convergence rate along the frontal thrust at present may decrease by at least a factor of three from the central to the eastern region of the belt. Moreover, the onset of strike-slip faulting may have acted to decrease convergence rates along the thrust front in the eastern region. The timing of the initiation of strike-slip faulting is, however, poorly constrained. Assuming slip rates of 1 to 5 mm/yr along the Uraba and U2 faults (Silver et al., 1990), we might expect to find discernable offsets of features across the fault, especially along the San Blas ridge, if the faults are more than a few million years old. We have not identified offsets of this magnitude, although such offsets are very difficult to recognize within the interior of the belt.

EVOLUTION OF THE SAN BLAS RIDGE

The eastern region of the NPDB has continued to grow, albeit slowly, by the incorporation of thick San Blas basin strata into the rear of the belt along backthrusts. The region of backthrusting makes up nearly 20% of the surface area of the eastern NPDB (Fig. 6) and has reduced the width of the San Blas basin near the Gulf of Uraba to less than 10 km (Fig. 7).

Sediment dispersal in the eastern and central regions of the NPDB is controlled by the San Blas ridge, although its role as a sediment barrier may be a relatively recent phenomenon. The crest of the ridge exhibits several perched basins, each containing a significant accumulation of turbidites (Fig. 9). One of the basins contains a 1.5-sec-thick sequence of flat-lying strata, even though it lies along the crest of the ridge and presently has no apparent means of receiving terrigenous sediments. Our interpretation is that this basin, and others like it, have been uplifted during the formation of the San Blas ridge, resulting in their isolation from sediment transport routes. Sedimentary sequences lying within the uplifted basins extend southward beneath the recent basin-fill deposits of the San Blas basin. Consequently, strata occupying the slope basins are older than the basinal turbidites of the San Blas basin and mark paleo-depocenters of detritus derived from eastern Panama, prior to the formation of the San Blas ridge. Reflectors in the uplifted basins also extend northward over the crest of the San Blas ridge, providing evidence that the ridge was not a significant barrier to sediment dispersal during the early history of the NDPD.

Silver et al. (1990) suggested that uplift in the interior of the belt along the San Blas ridge is propagating westward. Figure 9 contains a series of line drawings of seismic profiles arranged from west to east, or from youthful to mature stages in the basin history, using the concept of time-space equivalence for a westward-propagating collision. The development of backthrusting correlates well with the increased relief of the ridge, which in turn traps thick deposits in the San Blas basin. These deposits are either incorporated into the rear of the NPDB or thrust northward over a wedge of deformed sediments of the NPDB. Uplift of these strata has resulted in the southward migration of basin depocenter with time.

The position of the San Blas canyon is controlled by the western extent of the San Blas ridge. Silver et al. (1990) therefore proposed that the westward propagation of uplift along the San Blas ridge may have resulted in a westward migration of the San Blas canyon, thus altering sediment distribution along the thrust front throughout the history of the NPDB. This situation would have resulted in the westward migration of sedimentation, away from the thrust front in the eastern re-

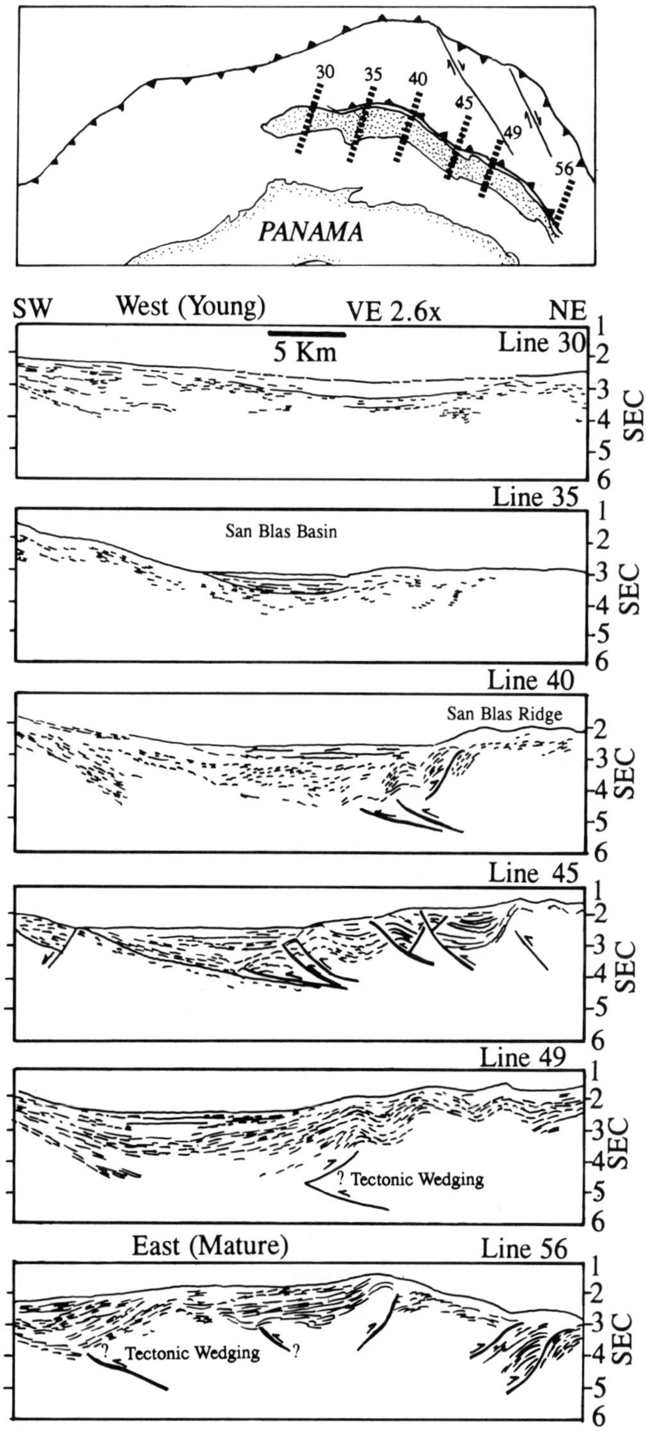

Figure 9. Serial cross sections of the San Blas basin from west to east showing the distribution of sedimentary deposits and development of the basin. Line locations also shown in Figure 6A.

gion, and may be responsible for a decrease in frontal accretion in this area.

SUMMARY

Sediment transport and the deposition of terrigenous detritus derived from eastern and central Panama are closely linked to the structural evolution of the North Panama deformed belt. Currently, sediment transport is deflected away from the thrust front in the eastern region of the belt, resulting in a recent decrease in frontal accretion. A basement ridge, located seaward of the eastern region, also prevents sediments of the Magdalena fan from reaching the thrust front. Terrigenous sediments instead accumulate as thick turbidite deposits along the rear of the belt or are diverted to the thrust front in the central region. The development of strike-slip faulting in the eastern region has reduced convergence rates along the frontal thrust, perhaps compounding the effects of decreased sediment thickness on frontal accretion. The timing of decreased sedimentation along the thrust front is poorly known; however, it may be related to the diversion of sediment transport following the formation of the San Blas ridge. The change in structural style in the eastern region from thick accreted strata along the San Blas ridge to thin sheets of accreted sediment along the thrust front may be one result of the diversion of sediment transport. The eastern region, however, has continued to grow, albeit probably at a slower rate, as backthrusting has incorporated thick deposits of the San Blas basin into the rear of the belt.

ACKNOWLEDGMENTS

This study was supported by grant OCE84-15714 from the National Science Foundation. We thank the captain and crew of the R/V Moana Wave, Don Hussong, and Tom Shipley for their role in data acquisition. Jill Tagudin and Greg Moore processed the seismic reflection data at the University of Tulsa, and Karen Sender of the University of Hawaii was responsible for the enhancement of the SeaMARC data. Jim Dolan, Ken McMillen, and Dan Davis provided helpful reviews that improved the final manuscript. We greatly appreciate the constructive criticism of Paul Mann in the preparation of the manuscript.

REFERENCES CITED

Ademak, S. H., Frohlich, C., and Pennington, W., 1988, Seismicity of the Caribbean-Nazca boundary: Constraints on microplate tectonics of the Panama region: Journal of Geophysics, v. 93, p. 2053–2075.

Bowland, C. L., 1984, Seismic stratigraphy and structure of the Colombian basin, Caribbean Sea [M.A. thesis]: Austin, University of Texas, 247 p.

Bowland, C. L., 1993, Depositional history of the western Colombian basin, Caribbean Sea, revealed by seismic-stratigraphic analysis: Geological Society of America Bulletin, v. 105, p. 1321–1345.

Bowland, C. L., and Rosencrantz, E., 1988, Upper crustal structure of the western Colombian Basin, Caribbean Sea: Geological Society of America Bulletin, v. 100, p. 534–546.

Breen, N. A., Tagudin, J. E., Reed, D. L., and Silver, E. A., 1988, Mud-cored parallel folds and possible melange development in the north Panama thrust belt: Geology, v. 16, p. 207–210.

Case, J. E., 1974, Oceanic crust forms basement of eastern Panama: Geological Society of America Bulletin, v. 85, p. 645–652.

Case, J. E., and Holcombe, T. L., 1980, Geologic-tectonic map of the Caribbean region: U.S. Geological Survey Miscellaneous Investigations Map I-1100 (scale 1:2,500,000).

Case, J. E., Holcombe, T. L., and Martin, R. G., 1984, Map of geologic provinces in the Caribbean region, *in* Bonini, W. E., Hargraves, R. B., and Shagam, R., eds., The Caribbean–South American plate boundary and regional tectonics: Geological Society of America Memoir 162, p. 1–30.

Cochran, J. R., 1990, Himalayan uplift, sea level, and the record of Bengal sedimentation at the ODP Leg 116 sites, *in* Cochran, J. R., Stow, D.A.V., eds., Proceedings of the Ocean Drilling Program, Scientific Results, 116: College Station, Texas, Ocean Drilling Program, p. 397–414.

deBoer, J., Defant, M. J., Stewart, R. H., Restrepo, J. F., Clark, L. F., and Ramirez, A., 1988, Quaternary calc-alkaline volcanism in western Panama: Regional variation and implication for plate tectonic framework: Journal of South American Earth Science, v. 1, p. 275–293.

Dickinson, W. R., and Seely, D. R., 1979, Structure and stratigraphy of forearc regions: American Association of Petroleum Geologists Bulletin, v. 63, p. 2–31.

Dolan, J. F., Beck, C., and Ogawa, Y., 1989, Upslope deposition of extremely distal turbidites: An example from the Tiburon Rise, west-central Atlantic: Geology, v. 17, p. 990–994.

Edgar, N. T., Ewing, J. T., and Hennion, J., 1971, Seismic refraction and reflection in the Caribbean Sea: American Association of Petroleum Geologists Bulletin, v. 55, p. 833–870.

Keller, G., Zenker, C., and Stone, S., 1989, Late Neogene history of the Pacific-Caribbean gateway: Journal of South American Earth Sciences, v. 2, p. 73–108.

Kolla, V., Buffler, R. T., and Ladd, J. W., 1984, Seismic stratigraphy and sedimentation of the Magdalena fan, southern Colombian basin, Caribbean Sea: American Association of Petroleum Geologists Bulletin, v. 68, p. 316–332.

Lu, R. S., and McMillen, K. J., 1982, Multichannel seismic survey of the Colombian Basin and adjacent margins, *in* Watkins, J. S., and Drake, C. L., eds., Studies in continental margin geology: American Association of Petroleum Geologists Memoir 34, p. 395–410.

Lundberg, N., and Dorsey, R. J., 1990, Rapid Quaternary emergence, uplift and denudation of the Coastal Range, eastern Taiwan: Geology, v. 18, p. 638–641.

MacKay, M. E., and Moore, G. F., 1990, Variation in deformation of the south Panama accretionary prism: Response to oblique subduction and trench sediment variation: Tectonics, v. 9, p. 683–698.

Mann, P., and Corrigan, J., 1990, Model for late Neogene deformation in Panama: Geology, v. 18, p. 558–561.

Muck, M. M., and Underwood, M. B., 1990, Upslope flow of turbidity currents: A comparison between empirical data, theory, laboratory models: Geology, v. 18, p. 54–57.

Reed, D. L., Silver, E. A., Prasetyo, H., and Meyer, A. W., 1986, Deformation and sedimentation along a developing terrane suture: Geology, v. 14, p. 1000–1003.

Reed, D. L., Silver, E. A., Tagudin, J. E., Shipley, T. H., and Vrolijk, P., 1990, Relations between mud volcanoes, thrust deformation slope sedimentation and gas hydrate, offshore north Panama: Marine and Petroleum Geology, v. 7, p. 44–54.

Silver, E. A., and Reed, D. L., 1988, Backthrusting in accretionary wedges: Journal of Geophysical Research, v. 93, p. 3116–3126.

Silver, E. A., Reed, D. L., Tagudin, J. E., and Heil, D. J., 1990, Implications of the North and South Panama thrust belts for the origin of the Panama orocline: Tectonics, v. 9, p. 261–281.

Speed, R. C., Torrini, R., Jr., and Smith, P. L., 1989, Tectonic evolution of the Tobago Trough forearc basin: Journal of Geophysical Research, v. 94, p. 2913–2936.

Tagudin, J. E., 1988, A correlation between fault vergence, fault spacing, and sediment type in the north Panama thrust belt [M.S. thesis]: Santa Cruz, University of California, 73 p.

Torrini, R., and Speed, R. C., 1989, Tectonic wedging in the forearc basin–accretionary prism transition, Lesser Antilles forearc: Journal of Geophysical Research, v. 94, p. 10549–10584.

Unruh, J. R., Ramirez, V. R., Phipps, S. P., and Moores, E. M., 1991, Tectonic wedging beneath fore-arc basins: Ancient and modern examples from California and the Lesser Antilles: GSA Today, v. 1, p. 185–190.

Wadge, G., and Burke, K., 1983, Neogene Caribbean plate rotation and associated Central American tectonic evolution; Tectonics, v. 2, p. 633–643.

Westbrook, G. K., 1982, The Barbados Ridge complex: Tectonics of a mature forearc system, *in* Leggett, J. K., ed., Trench-forearc geology, sedimentation and tectonics on modern and ancient active plate margins: Geological Society of London Special Publication 10, p. 275–290.

Weyl, R., 1980, Geology of Central America: Berlin, Gebruder Borntraeger, 371 p.

Manuscript Accepted by the Society February 7, 1994

Printed in U.S.A.

Geological Society of America
Special Paper 295
1995

Variation in structure, style, and driving mechanism of adjoining segments of the North Panama deformed belt

Eli A. Silver, Joseph Galewsky, and Kirk D. McIntosh
Earth Sciences Department, University of California, Santa Cruz, California 95064

ABSTRACT

The western part of the North Panama deformed belt (NPDB) undergoes a number of changes in the orientation of the frontal thrusts, in marked contrast to the eastern part of the NPDB, which shows very little change over distances of 150 km. In the western part of the belt, vergence and structural style of the frontal thrusts vary rapidly along the belt, again in contrast to the eastern part of the belt. The northeast-trending part of the western NPDB appears to be constrained in orientation by the slope of the Panama arc, which lies parallel to the thrust front. Farther west the belt trends northwest, and the thrust front migrates across the slope to the shelf area, where it has been identified by studies of the April 22, 1991, earthquake. Seismic data offshore from the surface location of this earthquake show no signs of crustal deformation, further constraining the location of the thrust front. The northwest orientation of the thrust belt on- and offshore of Costa Rica is consistent with a kinematic mechanism related to collision of Costa Rica by the Cocos Ridge. A proposal to explain the fact that the thrust front is located on the shelf, rather than at the base of the slope in its westernmost position, is impedance of thrusting as a result of the heavy load of sediment deposited by the Costa Rica fan. Thus several kinematic mechanisms are operating around the NPDB, producing very different directions of thrust belt orientation. In addition, several more mechanisms, such as slope stress, sediment loading, and possibly the structure of the lower plate, are operating to modify the orientation of the frontal thrusts.

INTRODUCTION

The North Panama deformed belt (NPDB) is an accretionary wedge that bounds the northern margin of Panama and the northeast coast of Costa Rica (Fig. 1). It is part of a series of thrust systems that make up the southern tectonic boundary of the Caribbean plate. These thrust systems are very irregular in trend, as they tend to follow the orientation of the irregular southern margin of the Caribbean. Lack of Benioff zone seismicity along much of the southern Caribbean, except for that in western Colombia, is evidence against significant subduction along this system, bringing forth the question of what drives these thrust systems. Can any one driving mechanism explain the highly irregular nature of these thrust systems, or are multiple mechanisms required? The western end of the NPDB lies onshore in eastern Costa Rica. North of about 10°N latitude, crustal extension prevails in Costa Rica and Nicaragua (e.g., Plafker, 1976). South of 10°N horizontal shortening dominates in Costa Rica.

In order to evaluate the variation in driving mechanism for the NPDB, we have reprocessed a set of digital seismic profiles across the western part of the deformed belt. We reinterpreted the existing side-scan data and incorporated the results of studies on the April 22, 1991, earthquake in eastern Costa Rica. These studies support multiple mechanisms for directing the orientation of the deformed belt. We compare these results with others from zones of backarc thrusting and note some commonality among them. Specifically, the orientations

Silver, E. A., Galewsky, J., and McIntosh, K. D., 1995, Variation in structure, style, and driving mechanism of adjoining segments of the North Panama deformed belt, *in* Mann, P., ed., Geologic and Tectonic Development of the Caribbean Plate Boundary in Southern Central America: Boulder, Colorado, Geological Society of America Special Paper 295.

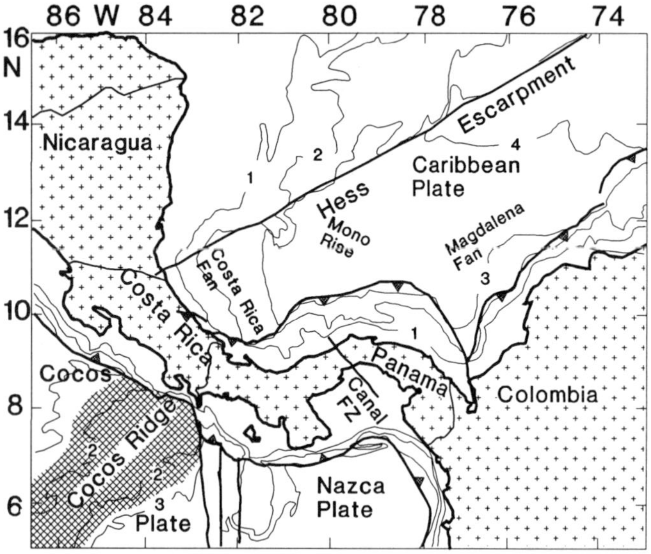

Figure 1. Tectonic elements of southern Central America and offshore regions. Note relationship between the Cocos Ridge (hachured) and the east Costa Rica segment of the backarc thrust system. Submarine contour interval 1 km. Major thrusts shown with triangular teeth on upper plate. The Chorotega block of Dengo (1985) is defined as Costa Rica and western Panama and their margins, whereas the Choco block is defined as eastern Panama and its margins. The blocks are separated by the Canal fault zone, shown on this figure.

of the thrusts represents the sum of regional and slope stresses. Slope stresses decrease with distance from the arc slope, and regional stresses vary as the changing effects of collision and sedimentation along the arc produce differing structural responses. For the NPDB, the combination of collision of Panama with South America, collision of Costa Rica with the Cocos Ridge, variation of sediment thickness, and variability of slope stresses provide workable explanations for the observed thrust orientations.

FACTORS GOVERNING THRUST FRONT ORIENTATION

In a homogeneous medium, thrust faults develop perpendicular to the local direction of the maximum horizontal principal stress axis, σ_1. Among a variety of factors, preexisting fabric in the rock or preexisting faults cutting the sediment pile can act to modify the effects of a homogeneous medium. However, such inhomogeneities are generally minor in a pile of newly deposited strata about to be deformed in a thrust belt. Even where such faults exist, the strength of the strata is generally low enough to be easily overcome by the regional stress. Where older, competent strata or rock are accreted, inhomogeneities will play a more important role in determining fault orientations.

The local direction of σ_1 may be determined by several factors. One is the local direction of convergence between the two structural blocks. A second factor is the slope stress, which takes on more importance when the developing thrust plane lies at the base of a significant slope, especially the base of an arc, seamount, or continental slope (e.g., Lambeck and Nakiboglu, 1980; Dalmayrac and Molnar, 1981; Silver et al., 1983; Breen et al., 1989). The closer the new fault surface is to the base of such a slope, the larger will be the component of the slope stress acting to drive and direct the thrust surface.

To identify the local directions of movement we use the orientation of the frontal one or two fold and thrusts. We do not suggest that these are the only important structures acting to distribute strain across the deformed region, but as strain accumulates on aging features they may rotate more closely into parallelism with the gross bathymetric structures of the margin (e.g., Lewis et al., 1988). Thus, provided the structures do not lie close to the base of the slope, the outer one or two structures have the best chance of indicating the local direction of tectonic stress.

PRIOR INTERPRETATIONS OF THE NORTH PANAMA DEFORMED BELT

Interpretations of all or parts of the NPDB have been made by Case and Holcomb (1980), Lu and McMillen (1982), Wadge and Burke (1983), Bowland (1984), Winslow (1988), Reed et al. (1990), Silver et al. (1990), Mann and Corrigan (1990), Plafker and Ward (1992), Kolarsky et al. (this volume), Goes et al. (1992), Astorga et al. (1991), and Tajima an Kikuchi (this volume). Suggested driving mechanisms include (1) collision between the Panama arc and South America (Wadge and Burke,1983; Silver et al., 1990); (2) cross-isthmus faulting along left slip faults to form crustal slices moving northwest relative to the Colombian Basin (Mann and Corrigan, 1990) and (3) Collision of Costa Rica by the Cocos Ridge (Plafker and Ward, 1992; Kolarsky et al., this volume; Goes et al., 1992; Tajima and Kikuchi, this volume). In addition to the primary driving mechanisms, thrust orientation may be affected by the slope stress along steep slopes (Silver et al., 1990), by thick sedimentary deposits on the slope or abyssal plain (Breen, 1989), or by variations in the crustal thickness (and surface roughness) on the lower plate (Bowland and Rosencrantz, 1988).

Wadge and Burke (1983) were the first to point out that nearly end-on collision between the Panama arc and South America would produce a bend similar to the shape of the Panama orocline. Silver et al. (1990) carried out a simple model of the expected kinematic directions of movement along the northern edge of the orocline and found a predominant north-northeast direction of slip between Panama and the Colombian Basin all along its north margin. That model predicted a nearly uniform orientation of frontal thrusting along the NPDB, based on the kinematic driving mechanism and deformation of a homogeneous medium. They then carried out a complete side-scan (SeaMARC II) imaging program of the

NPDB and found that in fact a large part of the belt does indeed have a very uniform orientation and that it is normal to the expected direction of convergence based on the model.

Mann and Corrigan (1990) carried out a careful study of the fault pattern onshore in Panama and discovered a predominance of northwest-trending, left-slip faults, consistent with but improving on the base maps of Wing (1971) and Case and Holcomb (1980). Mann and Corrigan (1990) proposed to explain the NPDB by movements on these faults, producing a thrust belt striking dominantly northeast. Silver et al. (1990) compared the expected thrust orientations based on the Mann and Corrigan and the Silver et al. models with the observed thrust orientations for the NPDB north of Panama but not Costa Rica. They concluded that their model best explained the observations for the main central part of the NPDB, whereas the Mann and Corrigan model better explained the western part of the NPDB.

Bowland (1984), in contrast to either of the above models, proposed to explain the orientation of the NPDB as an escape phenomenon, squeezing the deformed belt between thick crust in the Colombian Basin on either side of the belt. Evidence for the existence of such thickened crust was presented by Bowland and Rosencrantz (1988).

Deformation along the Caribbean coast of Costa Rica was noted by Case and Holcomb (1980) and studied onshore by Winslow (1988). Industry studies of the deformed belt both onshore and offshore were summarized by Astorga et al. (1991), in which they showed that thrusting characterizes the region south of Limon, extension to the north. Sheehan et al. (1990) made similar observations. Plafker and Ward (1992) constrained the location of a major thrust surface at the northern end of the belt by studying the uplift patterns and tsunami timing after the April 22, 1991, Limon earthquake. Their proposed thrust front is shown on Figure 2. Their thrust orientation is similar to those of fault mechanism solutions determined by teleseismic studies (Goes et al., 1992; Tajima and Kikuchi, this volume). Each of the studies of the April 22, 1991, earthquake concluded that the driving mechanism for the earthquake was collision of the Cocos Ridge through subduction of the Cocos plate.

Several geological studies have concluded that the Cocos Ridge collision has played a significant role in the regional strain field of southeast Costa Rica. Heywood (1984) built on the earlier study of Mora (1979), which showed that thrust deformation characterizes the forearc strata of the Terraba basin, just to the northeast of the Osa Peninsula. Gardner et al. (1992) examined stratigraphy of the Pacific coast of Costa Rica and Collins et al. (this volume) studied the Limon and Bocas del Toro Basins farther east, in order to determine the history of vertical motions in the region of collision of the Cocos Ridge. Gardner et al. found significantly higher uplift rates on the Burica and Osa Peninsulas than to the northwest, forming a spatial pattern similar to that expected from passage of the Cocos Ridge beneath the coastal region. Collins et al. found that uplift of the Burica Peninsula began at 3.6 Ma and ceased at 1.6 Ma, whereas emergence of the Bocas del Toro and Limon Basins began at 1.6 Ma. They proposed that the Cocos Ridge would have taken about 2 m.y. to traverse the isthmus from Burica to Bocas del Toro at the present rate of convergence.

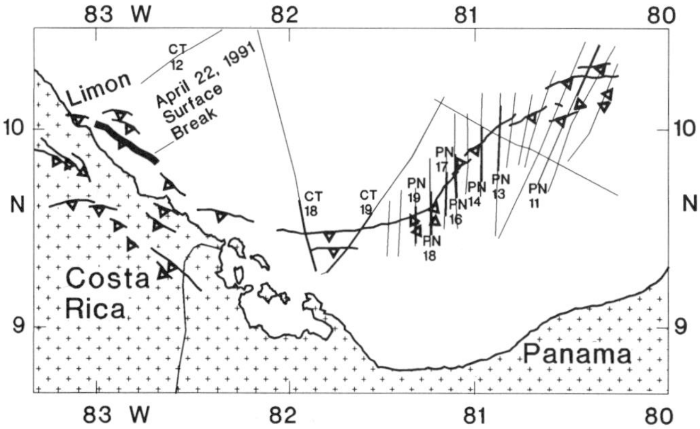

Figure 2. Structural map of the west Panama and east Costa Rica segments of the North Panama thrust belt. Also shown are tracks and locations of seismic lines shown in Figures 3 through 11. North Panama seismic lines taken on the *Moana Wave* are prefaced with PN, whereas University of Texas multichannel seismic data are prefaced with CT. Locations of thrust faults are shown with triangular teeth on the upper plate. Offshore of Limon one thrust segment is shown as abnormally thick. This segment was determined by modeling uplift after the April 22, 1991, earthquake by Plafker and Ward (1992).

Kolarsky et al. (this volume) have compiled further evidence of the tectonic effects of Cocos Ridge collision, including a volcanic gap, shallowing dip of the Benioff zone, forearc indentation, structural inversion of the Terraba and Limon Basins, a radial stress pattern around the Cocos Ridge collision, and arching of acoustic basement north of Limon. The latter effect is different from the interpretation of Bowland and Rosencrantz (1988), who demonstrated the shallowing of basement approaching the coast and viewed its cause as a thickening of the Colombian Basin crust by a middle Cretaceous episode of mid-plate volcanism. Although we cannot resolve these differing viewpoints, we note that the arching here may be of significantly greater wavelength than other features with demonstrably thickened crust in the Colombian Basin (Bowland and Rosencrantz, 1988), thus favoring somewhat the hypotheses of Collins et al. and Kolarsky et al.

INTERPRETATION OF SEISMIC REFLECTION DATA

Data processing procedures

We have reprocessed all of the seismic lines along the western part of the NPDB (Fig. 2), which included parts of two data sets. One was from the 1985 cruise of the R/V *Moana Wave,* in which digital single-channel seismic reflection records were collected simultaneously with SeaMARC II side-scan and bathymetry data. The second set was from the University of Texas (discussed by Lu and McMillen, 1982; Bowland, 1984; Bowland and Rosencrantz, 1988) (Fig. 2). The *Moana Wave* data were originally processed through migration, but off northwestern Panama the original migration was incomplete and the deconvolution was insufficient to bring out much of the detail, especially in the uppermost part of the sections. The University of Texas data were stacked and deconvolved, but were not migrated (Lu and McMillen, 1982). For the present study we have reprocessed the *Moana Wave* data from the original field tapes, using better deconvolution and migration. We have had success on some sections using F-X deconvolution (Gulunay, 1986; Hornbostel, 1991) but did not rely on this technique for most of the records. Despite the fact that the earlier migration utilized finite-difference methods, we have produced better results with this single-channel data using a constant (water) velocity F-K migration (Stolt, 1978). The better success with F-K here probably results from careful quantity control to remove bad traces, the superior ability of this technique to migrate steep dips, and the fact that a velocity model is not required. We have also successfully utilized residual migration, which employs initial F-K migration followed by residual finite-difference migration. The University of Texas data were reasonably well processed through stack, and our only additional effort involved application of residual migration with a velocity model based on stacking velocities. These worked quite well on UT seismic lines CT1-12, 18, and 19 (Fig. 2).

Interpretation

We have interpreted the seismic lines that lie west of the main part of the NPDB. The latter has been discussed by Silver et al. (1990) and by Reed et al. (1990), and our interpretation of that part of the thrust front remains essentially unchanged. However, the kinematic mechanism suggested by Silver et al. (1990) does not explain the region to the west, nor does it deal with the deformation along the northeastern margin of Costa Rica. Utilizing the improved seismic data and the results of the April 22, 1991, earthquake and other geological studies, we are now able to better evaluate this enigmatic region. We will discuss the interpretation of seismic lines PN11, PN13, PN14, PN16, PN17, PN18, PN19, CT-19, and CT-18, which progress westward from the western edge of the main segment of the NPDB.

Seismic line PN11 crosses the westernmost part of the main segment of the NPDB. Because of the very low surface slope, minor internal deformation, and clear stratigraphy, we were able to produce a geometrically consistent interpretation of the outer two fault-bend folds of seismic line PN11 (Fig. 3). The uppermost 0.5 sec of this section show pinchout against the folds, indicating deposition during the growth of the folds. The rest of the section appears to be of essentially uniform thickness. Ramp heights and displacements are similar for

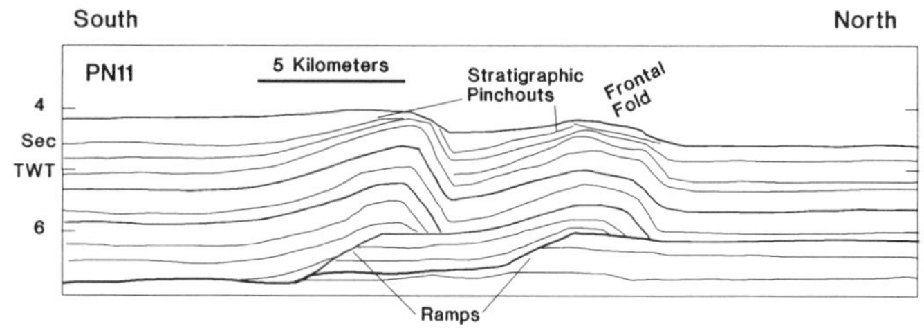

Figure 3. Interpretation of seismic profile PN-11, showing the structure of the frontal fold. Vertical scale in seconds of two-way travel time. Located on Figure 2.

each fold, suggesting a similar mechanical constraint on their development (e.g., similar fluid pressures along the décollement). The basin south of the second fold is approximately 10 km wide, indicating either very strong turbidites or (more likely) a very weak basal décollement beneath this basin. With such a weak décollement, very low surface slope, and wide, undeformed basin separating the outer folds from the rest of the deformed sediments, it is likely that the dominant factor determining the local $\sigma 1$ direction is simply the kinematic direction of motion. Thus the orientation of the frontal fold or thrust appears to be a good indicator of the relative movement direction between Panama and the Colombian Basin in this area. Silver et al. (1990) has discussed the evidence for a relatively consistent direction of motion along the thrust front for 150 km to the east of this location. For the rest of this section we will discuss the structure of the thrust belt to the southwest.

Seismic line PN13 (Fig. 4) appears different from PN11. The incoming turbidite sequence is essentially the same thickness in the two profiles, but in PN13 the frontal thrust lies at the base of a slope, and no undeformed basin strata separate any thrust package. The frontal thrust package is a pop-up block, with a south-dipping frontal thrust and a north-dipping secondary thrust. The frontal thrust is complex near the surface, simpler at depth. In our interpretation it is cut off at depth by the deeper secondary thrust. Upslope from the frontal pop-up structure is a series of overlapping thrusts, with complex internal structure. The basal décollement that underlies the frontal pop-up structure is very flat (in the time section) and undoubtedly steepens somewhat as the slope steepens.

Seismic line PN14 (Fig. 5) has a structure broadly similar to that of PN13, except that both of the frontal thrusts dip southward, forming a frontal pop-up between two south-dipping thrusts. In this case it appears likely that the frontal thrust cuts off the secondary thrust at depth and probably continues into the main décollement, which is not clearly displayed on this line. The slope is steeper on PN14 than on the previous line, underlain by a complex array of thrust packages. Acoustic penetration is good on this section.

Seismic line PN16 (Fig. 6) is bounded by a north-dipping frontal thrust, though all of the structures south of the front appear to be governed by south-dipping thrust packages. The secondary thrust projects to near the top of the frontal thrust,

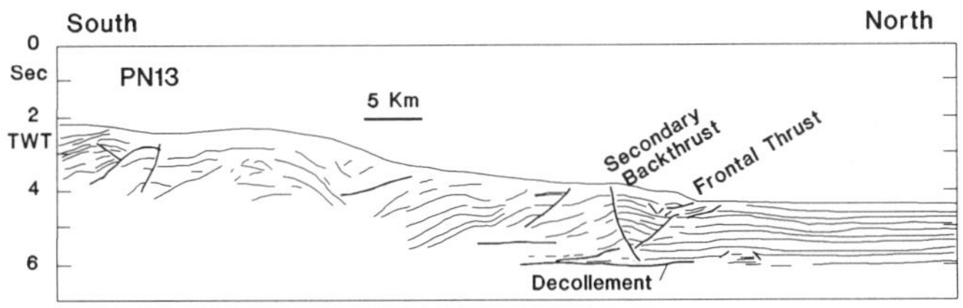

Figure 4. Interpretation of seismic profile PN-13, showing the structure of the frontal thrust and secondary backthrust. Vertical scale in seconds of two-way travel time. Located on Figure 2.

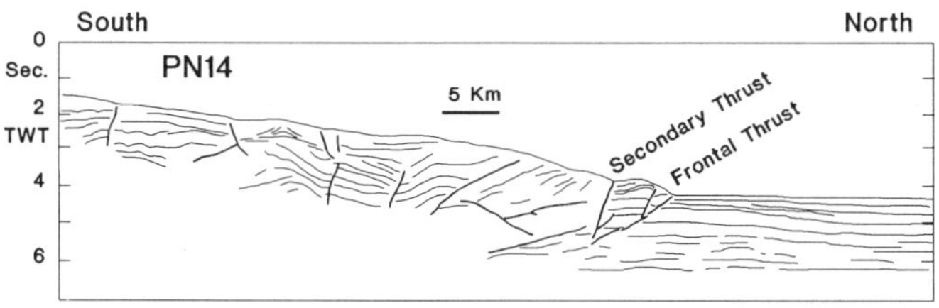

Figure 5. Interpretation of seismic profile PN-14, showing the structure of the lower slope and frontal thrusts. Vertical scale in seconds of two-way travel time. Located on Figure 2.

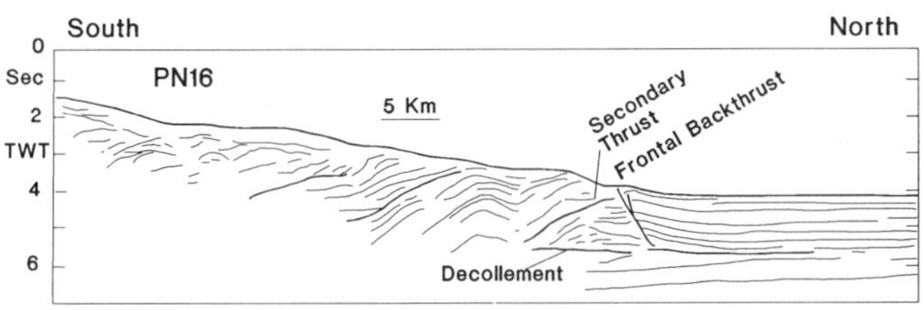

Figure 6. Interpretation of seismic profile PN-16, showing a frontal backthrust. Vertical scale in seconds of two-way travel time. Located on Figure 2.

forming a push-down rather than pop-up structure at the front of the wedge. The décollement here, similar to that of PN13, is very flat in the time section. The surface slope is slightly greater than those of the previous two lines and much greater than that of line PN11.

The frontal thrust in line PN17 (Fig. 7) is high angle and south dipping, rooting into a very flat décollement. The slope may be somewhat less than that of PN16. Not recorded is the possibility of an incipient backthrust forming a frontal pop-up structure here. We do not see displaced reflectors, but there is a north-dipping zone of folding that could be interpreted as a developing backthrust. As in PN16, all other structures to the south appear to have formed as a result of south-dipping thrusts. Also well displayed in this section is a south-dipping zone of reflectors that extends from 0.5 sec beneath the surface to nearly the décollement. There is some hint of develop-

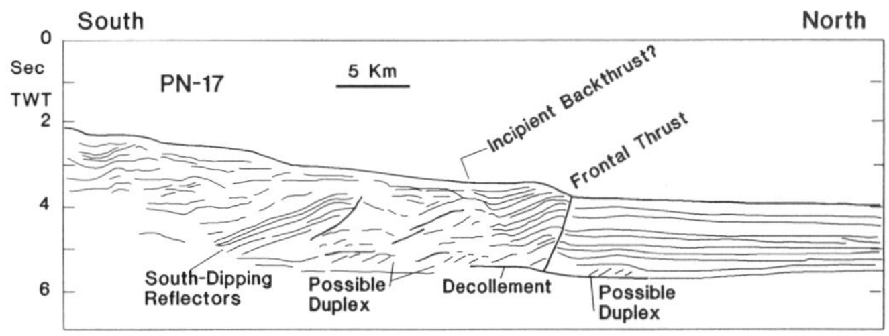

Figure 7. Interpretation of seismic profile PN-17, showing a well-developed frontal thrust, an incipient backthrust, and possible duplex. Vertical scale in seconds of two-way travel time. Located on Figure 2.

ing duplexes at depth in this section, including one beneath a very gentle bulge in the strata north of the frontal thrust.

Most of the mapped thrusts in seismic line PN18 (Fig. 8) dip southward, with the exception of a possible antithetic fault on the footwall of the frontal thrust. In this section there is a wide zone of folded strata behind the frontal thrust that forms a high ridge near the toe of the slope. North of the frontal thrust the strata are bowed up and apparently truncated near the surface. A zone of disturbed bedding at depth beneath this upbowed sequence may indicate incipient duplexing, as we inferred for seismic line PN17.

Seismic line PN19 (Fig. 9) contains a frontal pop-up structure, similar to that seen in line PN13 (Fig. 4). In this case both faults appear to meet just at the décollement, which appears to dip to the north on this time section as a result of velocity pull-up. South of the pop-up structure all the fault packages appear to dip southward. North of the frontal thrust we have noted several very low angle displacement features within the incoming sedimentary package. These may be very low angle normal faults.

Seismic lines CT-19 and CT-18 (Fig. 10) cross the thrust belt where it is east-west oriented. Each of these lines documents south-dipping thrust packages, with no indication of the backthrusting that seemed so common on most of the lines discussed above from the northeast-oriented thrust segment. Line CT-19 has a large frontal fold that is clearly formed above a major ramp in the décollement surface. Line CT-18 has developed a much thinner frontal fold set, but it too has several ramps. Here the basal décollement is not entirely clear, but it appears to dip at a low angle beneath the frontal deformed strata. North of the frontal fold is a fold or upwarp in the incoming strata that does not affect the material beneath the seaward projection of the decollement. This upwarp may represent a local diapir, or else it is another example of incipient duplexing. The latter seems less likely in this case than in the previous profiles because the structure lies on an apparently flat portion of the décollement with no nearby ramps visible.

The strata underlying line CT-18 (Fig. 10) are significantly thinner than those underlying the previous lines, located to the east. This thinning coincides with the region where Bowland and Rosencrantz (1988) have interpreted thicker crust, forming a high, shoaling toward the coast. It is also the region where Collins et al. (this volume) and Kolarsky et al. (this volume) have inferred uplift due to the shallow underthrusting of the Cocos Ridge. If the sedimentary section on the Colombian Basin ranges from Cretaceous to Present in age, the fact that the entire section seems to be thinned on line CT-18 relative to other lines to the east does not seem to be consistent with a Quaternary development of the structural high. The latter should have produced stratigraphic sections equally thick on both CT lines, at least.

We note, however, that seismic line CT-12 (Fig. 11) has a very thick sedimentary section on it, significantly thicker than that on line CT-19 (Fig. 10). Unfortunately this region is sparsely covered by seismic data, but it does not appear that the sediment is regionally thin on the Costa Rica fan. Instead, the sediment there is thicker than in the surrounding regions. Bowland and Rosencrantz (1988) point out that the basement appears to dip to the northeast away from the coast, and we suggest on the basis of the stratigraphy and thickness on line CT-12 that the structure is consistent with the hypothesis of a young uplift of the basement resulting from subduction of the Cocos Ridge.

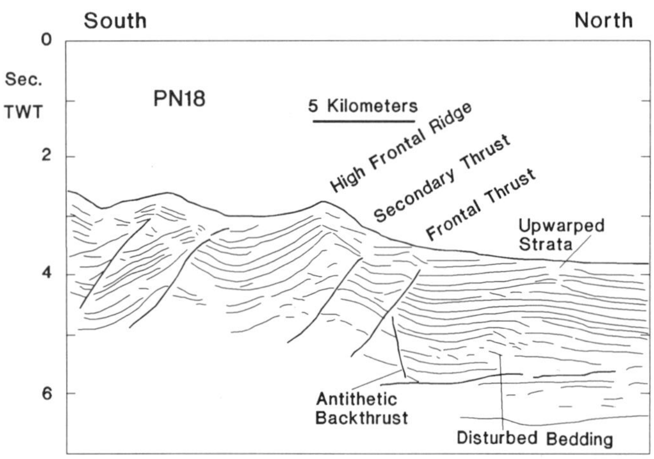

Figure 8. Interpretation of seismic profile PN-18, showing the structure of the lower slope and upwarped strata north of the frontal thrust. Vertical scale in seconds of two-way travel time. Located on Figure 2.

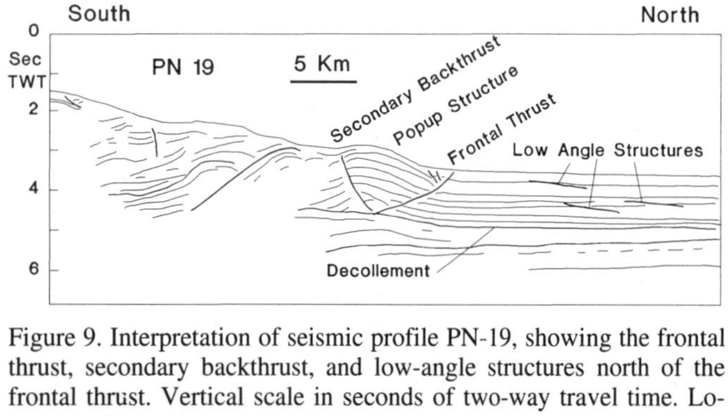

Figure 9. Interpretation of seismic profile PN-19, showing the frontal thrust, secondary backthrust, and low-angle structures north of the frontal thrust. Vertical scale in seconds of two-way travel time. Located on Figure 2.

Interpretation of the SeaMARC II and regional bathymetry

The SeaMARC II side-scan sonar data used in this work were acquired along tracks with 8-km spacing for nearly continuous or overlapping coverage. However, bathymetry determined from the side-scan data (by a phase-shift method [Blackinton et al., 1983]) generally occurs in a slightly narrower swath. In this case the bathymetry data covered a little more than half the survey area because of the narrower swath and some noise problems. Nonetheless, the data were of sufficient quality to construct a map at 100-m contour intervals (Fig. 12) by following the continuity of backscattered reflec-

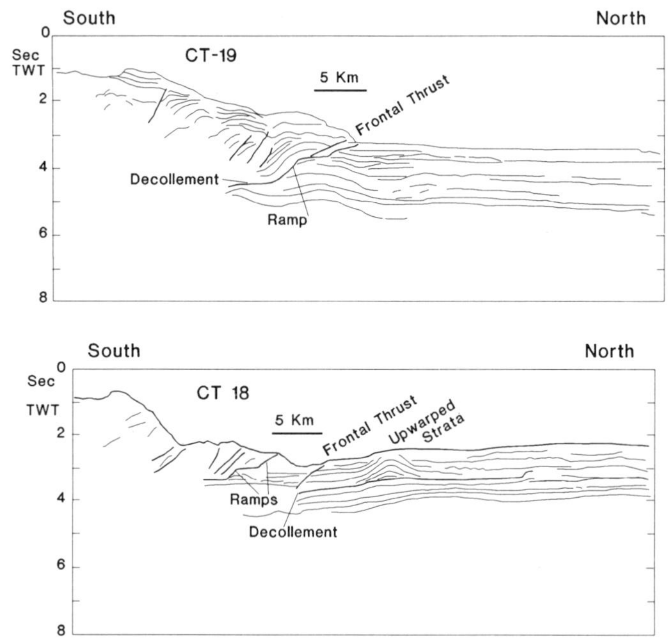

Figure 10. Interpretation of seismic profiles CT-19 and CT-18, showing the structure of the lower slope. Note upwarped strata in CT-18, north of the frontal thrust. Vertical scale in seconds of two-way travel time. Located on Figure 2.

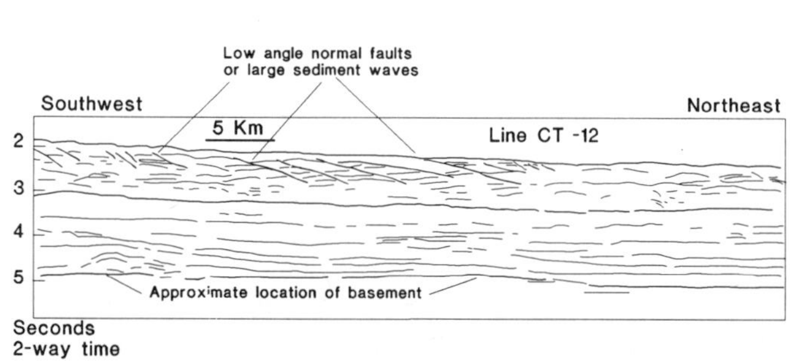

Figure 11. Interpretation of seismic profile CT-12, located on Figure 2. Lines drawn as low-angle normal faults might alternatively be interpreted as parts of large sediment waves (Bowland, 1984).

tions on the side-scan amplitude plots between the swaths of computed bathymetry (see discussion in Silver et al., 1990).

The front of the NPDB between 78°30′ and 80°30′W has a nearly uniform orientation of 075° (Fig. 12). We refer to this region as the main segment of the NPDB. The front changes to more nearly east-west in the western end of this zone, west of which it changes trend sharply to about 065°, between 80°40′ and 81°00′W. From 81°00′ to 81°20′W the front trends 045°. In the main segment of the NPDB the orientation of the front (though highly regular) shows no relationship to that of the arc slope to the south. Note that the strike of the arc slope in this zone changes trend from north of east to south of east (going from west to east). Linear escarpments in the bathymetric data show a gradual fanning from nearly east-west in the central part of the segment to 075° at the thrust front. Thrust vergence along the main segment is northward between 79°45′ and 80°35′W, whereas it is southward between 79°45′ and 78°45′W.

From about 80°30′ to 81°20′W the bathymetry is more complex. The 065° segment is irregular in structure and shows no clear relation to the orientation of the arc slope. The 045° segment from 81°00′ to 81°20′W shows a very clear parallelism between the thrust front and the 1,000 m and 1,500 m contours, suggesting control of their orientations by the arc slope. This segment has the steepest frontal slope found anywhere along the NPDB. West of 81°20′W the slope changes to a nearly east-west orientation, again following the changing trend of the arc slope. In contrast to the long stretches of similar structural style of the main segment of the NPDB, the northeast-trending segment shows significant changes in style of the zone of frontal thrusting along strike. Most of the profiles show northward (or northwestward) vergence, though several (lines 15—not shown—and 16) show south vergence. In addition, lines 13 and 19 image pop-up structure at the front, and in line 13 the south-verging thrust is dominant and has the most displacement. Line 14, between lines 13 and 15, shows dominance of northward vergence.

Along the east-west–trending segment of the front, at the Panama–Costa Rica border region, the two seismic lines (CT-19 and CT-18) both show northward vergence, and both show several ramps near the front of the wedge as the décollement rapidly steps down to deeper levels, especially in CT-19. This structure contrasts with a number of profiles to the northeast in which the décollement stays reasonably flat beneath the frontal thrusts. Line CT-12 does not cross the frontal thrust, but modeling and the mechanism of the April 22, 1991, earthquake each indicate a landward-dipping thrust plane, consistent with the sense of vergence shown in lines CT-19 and CT-18. Similar vergence is suggested from thrusts along the eastern Costa Rica margin in the industry data presented by A. Astorga et al. (1991).

DISCUSSION

What is responsible for the rapid change in orientation of the NPDB? We assume (based on the study of Silver et al., 1990) that the main segment of the thrust front reflects the dominance of kinematics, regardless of the mechanism by which the kinematics occur. If so, then the northeast-trending segment either responds to a separate kinematic driving mechanism or is governed by nonkinematic factors. If by a separate kinematic mechanism, there must be a separately moving block in western Panama responsible for this segment of the thrust. Some reconstructions of Panama (e.g., Dengo, 1985) imply that the Chocos and Chorotega blocks (see caption of Fig. 1) were once quite separate but were later juxtaposed. Weyl (1980) proposed that the two blocks were not greatly separated but moved

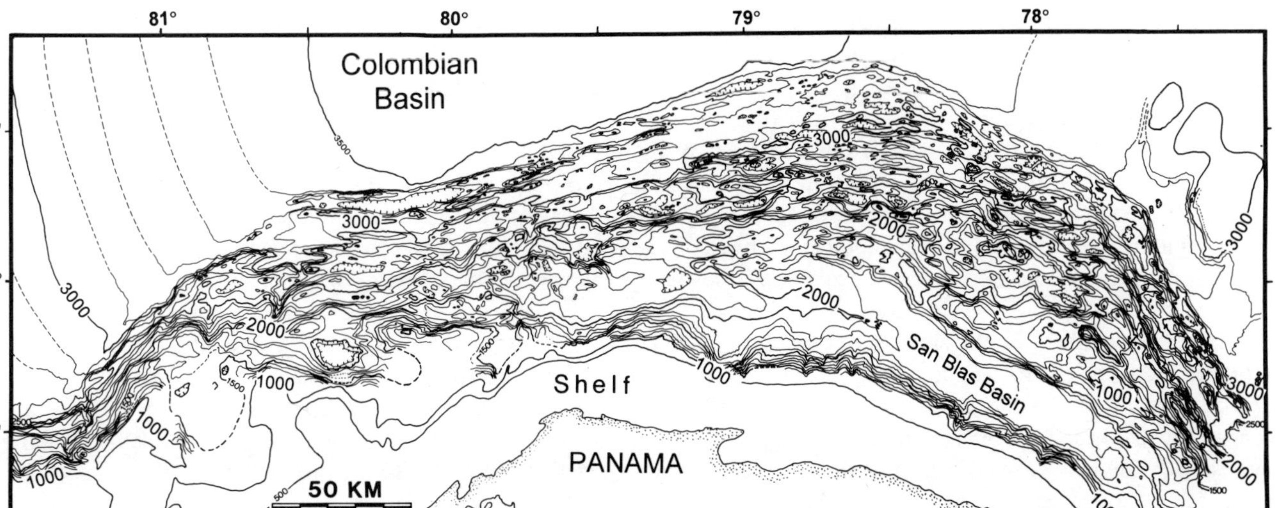

Figure 12. Bathymetric map of the northern Panama margin, constructed from SeaMARC II color bathymetry. Contour interval 100 m. From Silver et al. (1990).

along a canal fault zone in the Eocene. We prefer the reconstruction of Wadge and Burke (1983), which produces the orocline as a result of late Cenozoic Collision by the leading edge of the Choco block.

Although we cannot rule out an independent motion of west Panama, we feel it is most likely that the orientation of the thrust front west of 80°30′W is governed by the slope orientation. The 045° segment is parallel to the 2,000-m and 1,500-m contours, and its bend to the west follows the slope orientation. Furthermore, the front is located at the base of a relatively steep slope.

A similar phenomenon occurs in eastern Indonesia. Breen et al. (1989) showed that the Wetar thrust is oriented partway between that of the arc slope and normal to the kinematic direction of motion. The thrust front is about 20 km from the arc slope. Breen et al. (1989) observed that the farther the thrust front is from the arc, the less will be the effect of the slope stress.

The westward continuation of the thrust belt into Costa Rica is associated with another major change in orientation of the thrust front. Though only poorly controlled by seismic data in this zone, the orientation of the zone of frontal thrusting between 82° and 83°W becomes decidedly northwest trending. This orientation is constrained by the April 22, 1991, earthquake (Plafker and Ward, 1992), by the map of Case and Holcomb (1980), and by the sparse reports of industry data in this region (e.g. Astorga et al., 1991). This northwest-trending orientation of thrusts cannot be explained by the oroclinal bending proposed by Silver et al. (1990) or by the suggestion of slip on a series of blocks bounded by northwest-trending left-slip faults (Mann and Corrigan, 1990). Neither can it be explained simply by the effect of a slope stress, because the thrust front migrates shoreward to the northwest. The orientation is best explained as a result of backarc thrusting of eastern Costa Rica as a result of the collision of the Cocos Ridge with Costa Rica (Gardner et al., 1992; Plafker and Ward, 1992; Collins et al., this volume). In that case the NPDB results from at least two kinematic mechanisms and one or more modifying mechanisms.

The apparent continuity of the thrust front from the base of the slope to the coastline raises the question of why the thrust front migrates across the continental slope onto the continent in the westernmost segment of the thrust belt. A possible explanation may be the effect of sediment loading by the large fan of sediment offshore of Limon. Breen (1989) illustrated this effect on the Magdalena fan off Colombia. The load of rapidly deposited sediment on the slowly converging thrust zone stabilized the taper and forced the zone of thrusting to migrate upslope. Away from the thickest part of the fan the thrusts then stepped down to the base of the slope. We suggest that the northwest-trending segment is behaving similarly to the Magdalena fan off Colombia. Seismic line CT1-12 (Fig. 11) shows that thick sediments occur offshore of Limon, and the fan sediments thin toward the east. The bathymetric map of Case and Holcomb (1980) supports the idea of a sediment fan offshore of Limon. The thrust does not step back downslope to the north for two reasons. One is that in northern Costa Rica the backarc crust becomes very thick, owing to the presence of the Nicaragua Rise (Fig. 1) and possibly to thickened crust in the westernmost Colombian Basin (Bowland and Rosencrantz, 1988). The second is that the driving mechanism for thrusting (the Cocos Ridge collision) dissipates in that direction.

CONCLUSIONS

Interpretation of reprocessed digital seismic data along the western part of the North Panama thrust belt has documented wide variation in orientation and structure of the frontal fold and fault system. Rapidly changing thrust vergence and structural style occurs along and across the northeast-trending segment of the thrust belt. Incoming sediment north of the thrust belt increases westward with proximity to the Costa Rica fan deposits, and the water depth at the frontal thrust decreases westward as well. A local deviation from this trend is seen along one seismic line (CT-18, Fig. 10), which is explained by local variations in crustal thickness of the Colombian Basin (Bowland and Rosencrantz, 1988). Marked changes in orientation of the thrust belt occur on line with the Canal fault zone and near the Costa Rica–Panama border. The thrust front crosses the continental slope off Costa Rica and approaches the coast near Limon, the site of the April 22, 1991, earthquake. One hypothesis to explain the onshore migration of the thrust front is the loading of sediment by the Costa Rica fan, which could inhibit thrusting near the lower slope.

The great variability in structure and orientation of the thrust belt precludes direct extrapolation of the well-studied areas to the northernmost tip of the belt, which broke in the large (Mw = 7.7) earthquake on April 22, 1991. The northwest orientation of the segment of the thrust belt off Costa Rica is most easily explained as a result of collision of Costa Rica by the Cocos Ridge (Fig. 1). Thrusts in this segment of the thrust belt lie nearly perpendicular to the front of the ridge and also to the direction of convergence here. In contrast, such an explanation does not help our understanding of the rest of the thrust belt off Panama. Although much of that structure may be related to the collision and oroclinal bending of Panama against South America (Silver et al., 1990), the evidence available suggests an important contribution of slope stresses to the northeast-trending segment. This modification has been shown to be important in other backarc thrusting environments.

The North Panama deformed belt may serve as a model for other thrust belts that are not driven by self-sustaining subduction. Here the driving mechanism for the thrust belt's existence and orientation varies along the belt. Oroclinal bending of the main segment of the Panama arc is responsible for the eastern part of the deformed belt, and over much of this region the thrust belt orientation lies perpendicular to the direction of local convergence. The northeastern-trending segment of the deformed belt may be best explained by the orientation of the slope in this location, and this explanation provides a consistent but not unique interpretation of the orientation of the belt and its difference from structures to the east. Off the Caribbean coast of Costa Rica, the deformed belt is oriented northwest, perpendicular to the collision direction of the Cocos Ridge. The fact that the thrust front seems to migrate across the slope to the continental shelf in the northwestern part of the belt may be related to sediment loading on the belt, stabilizing the taper of the thrust system.

ACKNOWLEDGMENTS

We thank Fumiko Tajima, Saskia Goes, and Steve Ward for giving us unpublished versions of their papers on the Costa Rica earthquake of April 22,1991, and Patty Ganey-Curry for her help in obtaining the data tapes for this study. The paper was improved significantly as a result of careful reviews by Chris Bowland, Ken McMillen, Eric Rosencrantz, Tom Shipley, and Paul Mann. This work was supported by National Science Foundation grants OCE-8512068 and OCE-8917351 to Silver.

REFERENCES CITED

Astorga, A., and six others, 1991, Cuencas sedimentarias de Costa Rica: Evolucion geodinamica y potencial de hidrocarburo: [Sedimentary basins of Costa Rica: Geodynamics evolution and hydrocarbon potential] Revista Geologica de America Central, v. 13, p. 25–59.

Blackinton, J. G., Hussong, D. M., and Kosalos, J. G., 1983, First results from a combination side-scan sonar and seafloor mapping system (SeaMARC II), *in* Proceedings, Offshore Technology Conference, Houston, OTC 4478, v. 1: Society of Petroleum Engineers, Dallas, Texas, v. 1, p. 307–314.

Bowland, C. L., 1984, Seismic stratigraphy and structure of the western Colombian Basin, Caribbean Sea [M.A. thesis]: Austin, The University of Texas at Austin, 238 pp.

Bowland, C. L., and Rosencrantz, E., 1988, Upper crustal structure of the western Colombian basin, Caribbean Sea: Geological Society of America Bulletin, vol. 100, p. 534–536.

Breen, N. A., 1989, Structural effect of Magdalena fan deposition on the northern Colombia convergent margin: Geology, vol. 17, p. 34–37.

Breen, N. A., Silver, E. A., and Roof, S., 1989, The Wetar back arc thrust belt, eastern Indonesia: The effect of accretion against an irregularly shaped arc: Tectonics, vol. 8, p. 85–98.

Case, J. E., and Holcomb, T. L., 1980, Geologic-tectonic map of the Caribbean region: U.S. Geological Survey Miscellaneous Investigations Map I-1100, scale 1:2,500,000.

Dalmayrac, B., and Molnar, P., 1981, Parallel thrust and normal faulting in Peru and constraints on the state of stress, Earth and Planetary Science Letters, v. 55, p. 473–481.

Dengo, G., 1985, Mid America: Tectonic setting for the Pacific margin from southern Mexico to northwestern Colombia, *in* Nairn, A.E.M., Stehli, F. G., and Uyeda, S., The ocean basins and margins, vol. 7A: New York, Plenum Press, p. 123–180.

Gardner, T. W., and six others, 1992, Quaternary uplift astride the aseismic Cocos Ridge, Pacific coast, Costa Rica: Geological Society of America Bulletin, v. 104, p. 219–232.

Goes, S.D.B., Velasco, A. A., Schwartz, S. Y., and Lay, T., 1992, The April 22, 1991 Valle de la Estrella, Costa Rica (Mw=7.7) earthquake and its tectonic implications: A broadband seismic study: Journal of Geophysical Research, v. 98, p. 8127–8142.

Gulunay, N., 1986, F-X Decon and the complex Weiner prediction filter for random noise reduction on stacked data: Presented at the 56th Annual International Meeting of the Society for Exploration Geophysicists, Houston, Texas.

Heywood, C. E., 1984, Forearc deformation in southern Costa Rica: a consequence of the collision of the aseismic Cocos Ridge [M.S. thesis]: Santa Cruz, University of California, 104 p.

Hornbostel, S., 1991, Spatial prediction filtering in the t-x and f-x domains: Geophysics, v. 56, p. 2019–2026.

Lambeck, K., and Nakiboglu, S. M., 1980, Seamount loading and stress in the ocean lithosphere: Journal of Geophysical Research, v. 85, p. 6403–6418.

Lewis, S. D., Ladd, J. W., and Bruns, T. R., 1988, Structural development of an accretionary prism by thrust and strike-slip faulting: Shumagin region, Aleutian Trench: Geological Society of America Bulletin, v. 100, p. 767–782.

Lu, R. S., and McMillen, K. J., 1982, Multichannel seismic survey of the Colombia basin and adjacent margins, *in* Watkins, J. S., and Drake, C. L., eds., Studies in continental margin geology: American Association of Petroleum Geologists Memoir 34, p. 395–410.

Mann, P., and Corrigan, J., 1990, Model for late Neogene deformation in Panama: Geology, vol. 18, p. 558–562.

Mora, S., 1979, Estudio geologico de una parte de region sureste del Valle del General, Provincia Puntarenas, Costa Rica [undergraduate senior thesis]: San Jose, University of Costa Rica, 185 p.

Plafker, G., 1976, Tectonic aspects of the Guatemala earthquake of 4 February, 1976: Science, vol. 193, p. 490–509.

Plafker, G., and Ward, S. N., 1992, Backarc thrust faulting and tectonic uplift along the Caribbean sea coast during the April 22, 1991 Costa Rica earthquake: Tectonics, v. 11, p. 709–718.

Reed, D. L., Silver, E. A., Tagudin, J. E., Shipley, T. H., and Vrolijk, P., 1990, Relations between mud volcanoes, thrust deformation, slope sedimentation, and gas hydrate, offshore northern Panama: Marine and Petroleum Geology, vol. 7, p. 44–54.

Sheehan, C. A., Penfield, G. T., and Morales, E., 1990, Costa Rica basins lure wildcatters: Oil and Gas Journal (April 30), p. 73–79.

Silver, E. A., Reed, D. L., McCaffrey, R., and Joyodiwiryo, Y., 1983, Back arc thrusting in the eastern Sunda arc, Indonesia: A consequence of arc-continent collision: Journal of Geophysical Research, vol. 88, p. 7429–7448.

Silver, E. A., Reed, D. L., Tagudin, J. E., and Heil, D. J., 1990, Implications of the North and South Panama thrust belts for the origin of the Panama orocline: Tectonics, vol. 9, p. 261–281.

Stolt, R. H., 1978, Migration by Fourier transform: Geophysics, v. 43, p. 23–48.

Wadge, G., and Burke, K., 1983, Neogene Caribbean plate rotation and associated Central American tectonic evolution: Tectonics, v. 2, p. 633–643.

Weyl, R., 1980, Geology of Central America: Berlin, Borntraeger, 371 p.

Wing, R. S., 1971, Structural analysis from radar imagery of the eastern Panama Isthmus, Part II: Modern Geology, v. 2, p. 75–127.

Winslow, M., 1988, Tectonic evolution of an actively deforming back arc basin: The Limon Basin of Costa Rica: Transactions of the 11th Caribbean Geological Conference, Barbados, p. 19:1–19:17.

Manuscript Accepted by the Society February 7, 1994

Printed in U.S.A.

Geological Society of America
Special Paper 295
1995

Island arc response to shallow subduction of the Cocos Ridge, Costa Rica

Radim A. Kolarsky*
Department of Geological Sciences, University of Texas at Austin, Austin, Texas 78712
Paul Mann
Institute for Geophysics, University of Texas at Austin, 8701 Mopac Boulevard, Austin, Texas 78759
Walter Montero
Centro de Investigaciones Geofisicas, Escuela Centroamericana de Geología, Universidad de Costa Rica, San José, Costa Rica

ABSTRACT

Shallow subduction of the Cocos Ridge beneath the Costa Rican island arc results in six major tectonic effects. These effects include a volcanic gap in the Costa Rican volcanic arc chain, a shallowing of the dip of the subducted Cocos plate beneath Costa Rica, forearc indentation of the Pacific margin of Costa Rica, structural inversion of forearc (Terraba) and backarc (Limon) basins, arching of on- and offshore acoustic basement in a direction parallel to plate convergence between Costa Rica and the Cocos plate, and a radial stress pattern around the underthrust area of the Cocos Ridge as inferred from earthquake and geologic indicators.

Structures formed in forearc basin sedimentary and volcanic rocks of the Térraba belt above the subducted Cocos Ridge include major reverse faults that consistently place older lithologic units over younger lithologic units. One of these faults, the Ballena-Celmira fault zone, forms a prominent linear contact between Quaternary alluvium of the Pacific coastal plain and the Térraba belt. Bedding plane and fault data in the Térraba belt constrain a maximum shortening direction of N30–34°E for the central and eastern Térraba belt. This direction of maximum shortening corresponds closely to the N35°E direction of maximum shortening of Corrigan et al. (1990) from Plio-Pleistocene rocks of the outer forearc in the Burica/Osa area to the south and southeast of the Térraba belt. Assuming that the predicted plate convergence direction (N32°E) and the direction of maximum shortening in the forearc subparallel, thrusting and tilting in the forearc of westernmost Panama and eastern and central Costa Rica is interpreted as the result of regional northeast-southwest–oriented maximum compressive stresses exerted by post-Miocene shallow subduction of the Cocos Ridge.

INTRODUCTION

Four forces are generated at the line of contact between a subducting plate and the overriding island arc (Fig. 1). These forces include: (1) a horizontal force applied to the island arc by the subducting plate, (2) a horizontal reaction force applied to the subducting plate by the island arc, (3) a vertical topographic load applied to the subducting plate by the island arc, and (4) a vertical reaction force applied to the island arc by the subducting plate (Fig. 1). The purpose of this paper is to investigate what types of structures are present in island arc rocks above a zone of shallow subduction of thicker-than-average oceanic lithosphere and what forces or combinations of forces

*Present address: Eastern E&P Region, Texaco USA, P.O. Box 60252, New Orleans, Louisiana 70160.

Kolarsky, R. A., Mann, P., and Montero, W., 1995, Island arc response to shallow subduction of the Cocos Ridge, Costa Rica, *in* Mann, P., ed., Geologic and Tectonic Development of the Caribbean Plate Boundary in Southern Central America: Boulder, Colorado, Geological Society of America Special Paper 295.

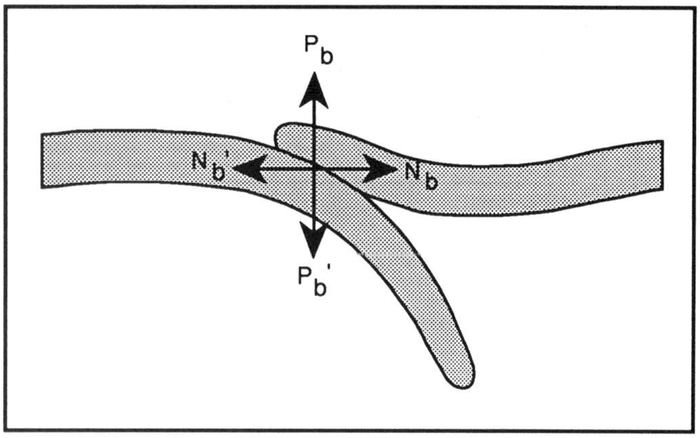

Figure 1. Schematic representation of forces acting on a subducting slab and an overriding island arc (from Chung and Kanamori, 1978). N_b is the horizontal force applied to the island arc by the subducting plate, N_b' is the reaction force applied to the subducting plate by the island arc, P_b' is the vertical force applied to the subducting slab by the island arc, and P_b is the reaction force applied to the island arc by the subducting plate.

are responsible for producing these structures. In the case of shallow subduction, one might expect the following differences from the case of subduction of normal oceanic crust: (1) larger horizontal forces associated with a larger region of contact between the island arc and underriding plate, (2) larger horizontal forces associated with resistance to subduction by the thicker crust, and (3) larger vertical forces associated with larger buoyant forces of the thicker-than-average oceanic lithosphere (Fig. 1).

The response of active island arcs to shallow subduction of anomalously thick areas of seafloor formed in seamounts, oceanic plateaus, and aseismic ridges has been studied worldwide over the past ten years using seismic reflection data, sidescan sonar, and onland geologic mapping. Literature review reveals three basic suites of mapped structures produced in island arcs by shallow subduction: (1) *convergent structures striking perpendicular to the convergence direction* (that is, fold and thrust faults interpreted as the result of horizontal forces applied to the island arc by shallow subduction—Fig. 2C); (2) *divergent structures striking parallel to the convergence direction* between the island arc and downgoing plate (that is, normal faults and conjugate strike-slip faults interpreted as a result either of horizontal forces produced by shortening of the island arc—Fig. 2A—or of vertical forces produced by isostatic uplift of the subducted plate and subsequent arching of the forearc—Fig. 2B); (3) *gravitational slump structures sliding parallel to the convergence direction* and usually in a seaward direction. Oversteepened slopes producing the slumps are thought to be an effect of convergent structures described above and presented in Figure 2C.

Folding, normal, reverse, and strike-slip faults have been mapped in both arc-parallel and arc-perpendicular orientations in areas of shallow subduction that include the New Hebrides (Vanuatu), Java, Tonga, Lesser Antilles, Peru, and southern Middle America (see Table 1 for a summary of key references). In their previous study of Plio-Pleistocene forearc rocks in the shallow subduction area of the Cocos Ridge of southern Middle America, Corrigan et al. (1990) were unable to establish the relative effects of horizontal and vertical forces on structure of the outer forearc area. Corrigan et al. (1990) acknowledged that both vertical (Fig. 2B) and horizontal (Fig. 2A, 2C) forces may be important in forming mapped structures in the outer forearc area.

In this study we report on continued investigation of the segment of the Costa Rican forearc that is overthrusting the Cocos Ridge. We carried out structural mapping to better establish the types and orientations of structures present and the relative roles of horizontal and vertical forces. The Middle America island arc in Costa Rica is an excellent area for detailed studies of island arc structures because, unlike other circum-Pacific island arcs, extensive and deeply incised land areas are present both within and outside of the zone of shallow subduction. In this study, we incorporate the results of the Corrigan et al. (1990) study from the outer forearc of Costa Rica and Panama but also include data from a much larger area of the island arc (inner forearc, volcanic arc, and backarc) to establish regional effects of shallow subduction.

ORGANIZATION OF THIS CHAPTER AND RELATION TO PREVIOUS STUDIES

First, we describe the tectonic setting of the area of shallow subduction of the Cocos Ridge in southern Central America (Costa Rica and Panama). Second, we review geological and geophysical maps showing evidence for six regional effects of shallow subduction on the island arc structure of Costa Rica and Panama and present a regional cross section showing the relation between the subducted Cocos Ridge and the overriding island arc of Costa Rica. Data used to compile both maps and the regional cross section are drawn from previous work by others and from our own observations. Third, we present new structural field data from Eocene to Pliocene sedimentary and igneous rocks of the deformed Costa Rican inner forearc basin (Térraba belt) that directly overlies the area of Cocos Ridge shallow subduction. We include structural mapping of the Térraba belt that formed part of a "Licenciatura" thesis by Dr. Sergio Mora; the thesis was supervised by Professor Walter Montero at the University of Costa Rica in San José. Biostratigraphic and lithologic correlation of Cenozoic sedimentary rocks of the Térraba belt is derived from two master's theses (Phillips, 1983; Lowery, 1982) and one doctoral dissertation (Yuan, 1984), all supervised by Dr. Don Lowe at Louisiana State University. Finally, we integrate structural and stratigraphic data with the structural and stratigraphic field study of the outer forearc basin of Panama and Costa Rica by Corrigan et al. (1990) and with results of DSDP

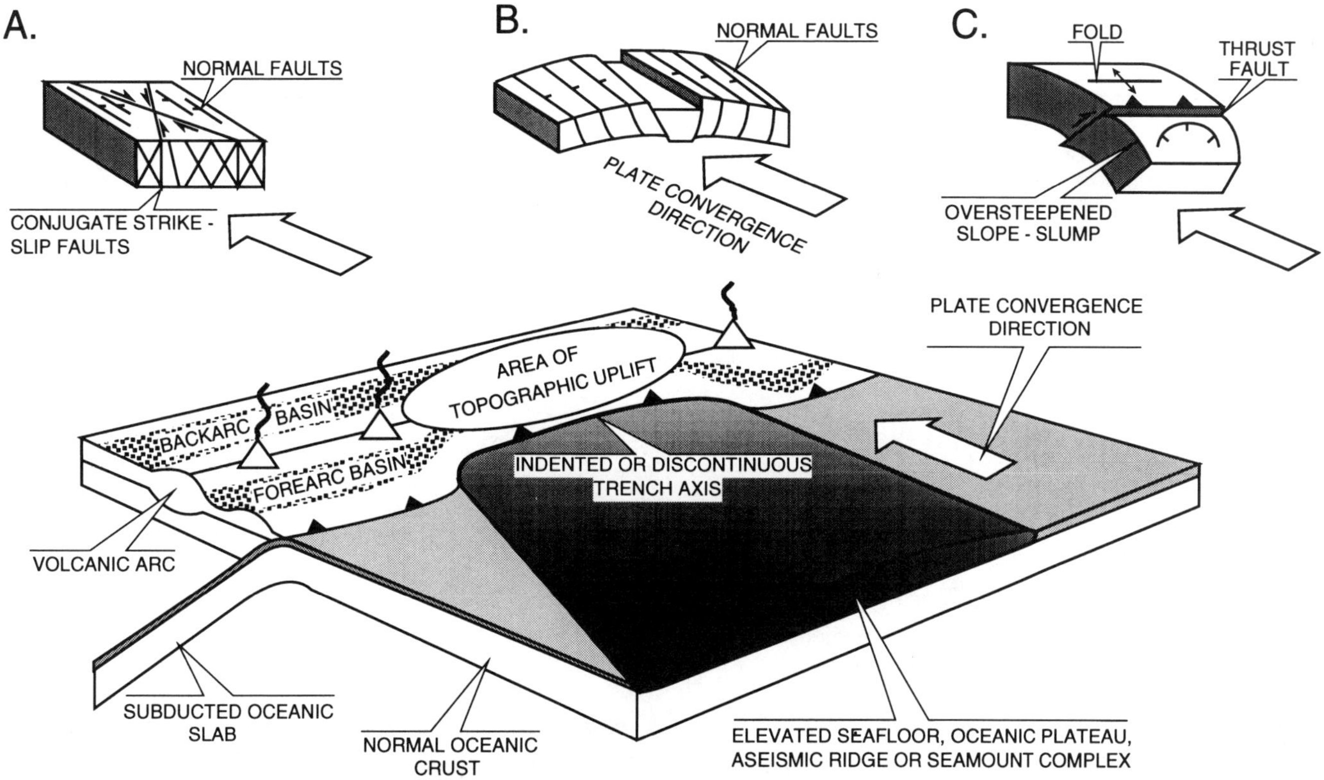

Figure 2. Summary of mapped structures produced by shallow subduction in active island arc settings. Shallow subduction is produced by subduction of thicker-than-normal oceanic crust (oceanic plateau, aseismic ridge, seamount complex) and produces an indented or discontinuous trench axis and an area of topographic uplift in the island arc. Margin indentation and topographic uplift may shorten and invert forearc and backarc basins. Insets show three basic suites of mapped structures recognized in areas of shallow subduction: A, divergent structures (normal faults) striking parallel and conjugate strike-slip faults striking subparallel to the convergence direction between the island arc and the subducting slab, B, divergent structures (normal faults) striking parallel to the convergence direction between the island arc and the subducting slab, and C, convergent structures (folds, thrust faults) striking perpendicular to the convergence direction; note formation of oversteepened slopes on fold flanks and occurrence of gravitational slumps.

leg 84 along the offshore forearc areas of Guatemala and Costa Rica (von Huene et al., 1980).

TECTONIC SETTING OF THE SOUTHERN MIDDLE AMERICA TRENCH AND COCOS RIDGE

Island arc structure

The southern Middle America Trench and Cocos Ridge are located near the triple junction between the Caribbean, Nazca, and Cocos plate (Fig. 3). The Nazca plate converges obliquely in an east-northeasterly direction on the southwestern corner of the Caribbean plate at a predicted velocity of about 3.7 cm/yr (DeMets et al., 1990). Convergence between the Cocos and Caribbean plates in Costa Rica is predicted to be in a northeasterly direction (032). Calculated convergence rate increases southeastward along the strike of the trench from about 7 cm/yr off southeastern Mexico and Guatemala to more than 9 cm/yr in the area of Cocos Ridge subduction beneath southeastern Costa Rica (DeMets et al., 1990) (Fig. 3). The Cocos and Nazca plates are separated by the Panama fracture zone, a transtensional, right-lateral transform fault (Adamek and Tajima, 1987; von Huene et al., this volume).

Active island arc features produced by subduction of the Cocos plate beneath the Caribbean plate in Central America include a trench axis, forearc basin, volcanic arc, and backarc basin (Fig. 3). Drilling by the Deep Sea Drilling Project during Leg 84 off the Pacific margin of Guatemala (von Huene et al., 1980) and oil exploration wells drilled by industry (Seely, 1979) indicate an early Tertiary age for the development of the forearc basin. The Quaternary volcanic arc of Central America extends as a continuous 1,100-km-long line of 40 calc-alkaline volcanic centers from the Mexico-Guatemala border to central Costa Rica. The close spacing of one major stratovolcano oc-

TABLE 1. SELECTION OF PROPOSED ISLAND ARC RESPONSES TO SHALLOW SUBDUCTION OF ASEISMIC RIDGES AND SEAMOUNTS

1. Convergent structures striking perpendicular to the convergence direction; structures produced by sub-horizontal island arc shortening.	A. New Hebrides (Collot et al., 1985; Collot and Fisher, 1991) B. Southern Middle America trench in Costa Rica and Panama (this chapter) C. Java trench (Masson et al., 1990) D. Various examples (McCann and Habermann, 1989) E. Various examples (Green and Wong, 1989)
2. Apparent lack of convergent structures produced by sub-horizontal island arc shortening.	A. New Hebrides (Fisher, 1986)
3. Isostatic uplift of the island arc possibly accompanied by divergent structures striking parallel to the convergence direction.	A. Tonga trench (Dupont and Herzer, 1985) B. New Hebrides trench (Moretti and Ngokwey, 1985) C. Lesser Antilles (Moretti and Ngokwey, 1985) D. New Hebrides (Chung and Kanamori, 1978) E. Mariana and Izu-Bonin trenches (Fryer and Smoot, 1985)
4. Gravitational slump structures striking perpendicular to the convergence direction; structures are related to oversteepened slopes in the forearc	A. Java trench (Masson et al., 1990) B. Japan trench (Lallemand and Le Pichon, 1987) C. Peru trench (von Huene et al., 1989)
5. Both convergent and divergent structures related to isostatic uplift and sub-horizontal island arc shortening.	A. Southern Middle America trench in Costa Rica and Panama (Corrigan et al., 1990)

curring approximately every 26 km is one of the highest densities of active volcanic centers along any convergent plate margin (Carr and Stoiber, 1990; de Boer et al., this volume). An elongate Quaternary backarc and intra-arc basin extends approximately 600 km from El Salvador to the Caribbean Sea in Costa Rica. The backarc/intra-arc basin is most prominent in Nicaragua, where it forms a large Quaternary graben occupied by two large lakes (Fig. 3).

Island arc and strike-slip features produced by oblique subduction of the Nazca plate beneath the Caribbean plate in western Panama include an accretionary wedge (MacKay and Moore, 1990; Moore and Sender, this volume), forearc basin cut by strike-slip faults (Okaya and Ben-Avraham, 1987; Kolarsky and Mann, this volume), and a 300-km-long chain of historically active arc volcanoes that has only been recognized in the past several years (de Boer et al., 1988; Defant et al., 1991).

Cocos Ridge

The Cocos Ridge is a 200- to 300-km-wide, northeast-trending area of elevated, thickened oceanic crust that rises 2 to 2.5 km above the surrounding oceanic floor of the Cocos and Nazca plates (Fig. 3). The highest part of the ridge is to the east adjacent to the Panama fracture zone. To the west, the Cocos Ridge deepens to an area of scattered seamounts described by von Huene et al. (this volume). The high area of the Cocos Ridge and the area of seamounts is collectively referred to as the Cocos Ridge and Seamount Domain (von Huene et al., this volume) (Fig. 3).

The age of oceanic crust being subducted in the vicinity of the Cocos Ridge is early to middle Miocene (15 to 20 Ma) (Hey, 1977). Marine refraction studies indicate a crustal thickness of approximately 15 km beneath the Cocos Ridge that is significantly thicker than surrounding normal oceanic crust of the Nazca and Cocos plates (Bentley, 1974). Hey (1977) interpreted the Cocos Ridge as a hot-spot trace formed on the Cocos plate as it passed over the Galapagos hot spot during Miocene time. In a regional study of magnetic anomalies of the Cocos and Nazca plates. Lonsdale and Klitgord (1978) concluded that the northeastern end of the Cocos Ridge encountered the southern Middle America Trench about 1 Ma.

REGIONAL EFFECTS OF COCOS RIDGE SHALLOW SUBDUCTION IN COSTA RICA AND PANAMA

Based on data compiled from previous studies by other workers, we identify six major effects apparently related to the shallow subduction of the Cocos Ridge. We have compiled data supporting these effects in six maps that are described below (Figs. 4A–F). These effects are: (1) Volcanic arc gap and topographic uplift, (2) decrease of the dip of the subducting slab beneath Costa Rica relative to southern Nicaragua, (3) forearc indentation and uplift, (4) inversion of forearc and backarc basins of Costa Rica, (5) basement arching,and (6) regional stress field modification.

Volcanic arc gap and topographic uplift

In Costa Rica, there is a 175-km-wide gap in the chain of normally closely spaced Central America stratovolcanoes (Fig. 4A). The gap extends from the presently active Irazú-Turrialba complex in central Costa Rica to the historically active Barú complex of western Panama. Weyl (1980), de Boer et al. (1988), and de Boer et al. (this volume) have demonstrated the widespread occurrence of small volumes of radiometrically dated Neogene volcanic products in the gap of large Costa Rican stratovolcanoes. The active volcanic arc extends 300 km farther east from Barú volcano near the Costa Rica–Panama boundary across western and central Panama as an east-trending chain of 10 historically active stratovolcanoes (de Boer et al., 1988; Defant et al., 1991).

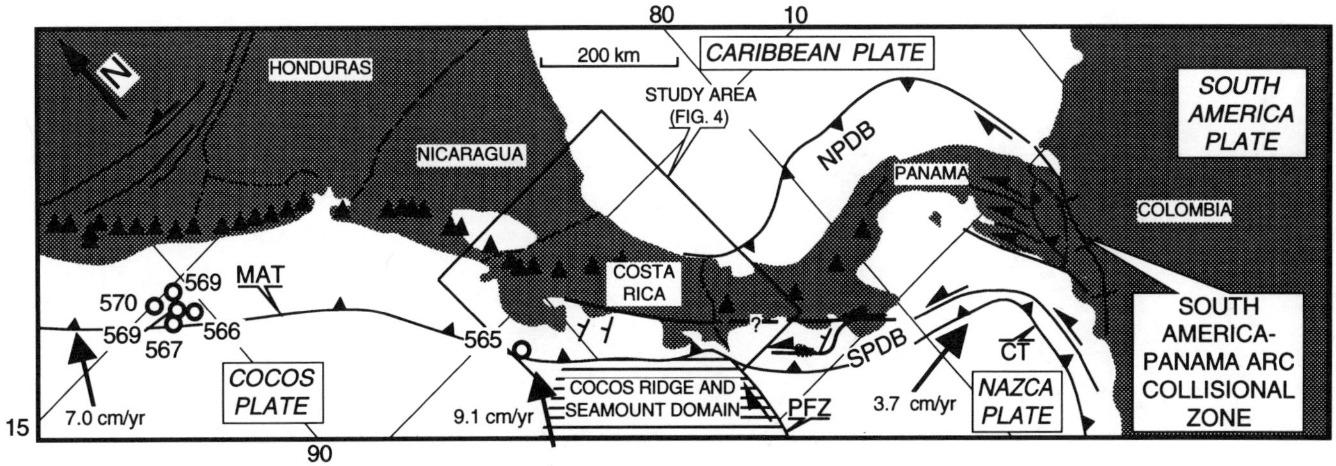

Figure 3. Tectonic setting of volcanic arc of Central America, southern Middle America Trench (MAT), and Panama fracture zone (PFZ). Black triangles represent Quaternary calc-alkaline volcanoes; arrows show plate convergence directions with numbers indicating rates of relative plate motion in cm/yr; open circles represent Deep Sea Drilling Project (DSDP) sites. Study area (box) encompasses area of shallow subduction of the Cocos Ridge beneath the Costa Rica forearc (Fig. 4).

The width of the gap in stratovolcanoes corresponds precisely to the only area of southern Central America presently above 2,000 m in topographic elevation (Cordillera Talamanca) (Fig. 4A). The maximum elevation of the Cordillera Talamanca is 3,800 m. The width of the volcanic gap and the Cordillera Talamanca correlates approximately with the area of shallowest bathymetry on the crest of the Cocos Ridge. The shallowest areas range from approximately 1,000 to 2,000 m in water depth and are named here the eastern and western elevation of the Cocos Ridge.

McGeary et al. (1985) have compared the gap in the Costa Rican volcanic arc with 23 areas worldwide exhibiting major spatial gaps (>200 km) in volcanic arcs. They concluded on the basis of their survey that the Costa Rican gap along with 19 other gaps worldwide was associated with shallow subduction of a large aseismic plateau or ridge. They suggested that in areas of shallow subduction no asthenospheric wedge exists to produce source material for the volcanic arc.

Dip of subducting slab beneath southeastern Costa Rica

The dip of the subducted Cocos plate decreases significantly beneath central Costa Rica. Shallowing in subduction angle for the depth interval of 25 to 250 km, along with an abrupt decrease in seismicity below 75 km, has been recognized from previous earthquake studies (Burbach et al., 1984; Guendel, 1986). Guendel (1986) indicates an eastward deflection in the 200- and 250-km isodepth contours to the top of the Benioff zone (Fig. 4B).

We have compiled seismic interval velocity values from marine multichannel seismic reflection profiles shot perpendicular to the Middle America Trench in the area of the Cocos Ridge (Crowe and Buffler, 1983; Shipley and Moore, 1986). These lines image the top of the subducted oceanic crust of the Cocos plate for the depth interval 0 to 14 km. Immediately landward of the Middle America Trench off the Nicoya Peninsula, the subduction angle is approximately 7° (Shipley and Moore, 1986), decreasing to approximately 3° off the Osa Peninsula (Fig. 4B). The depth contours based on seismic reflection data parallel the eastward deflection of the deeper contours based on earthquake data.

Forearc indentation and uplift

Crowe and Buffler (1985) produced an isopach map of the sedimentary shelf-slope section overlying acoustic basement along the Pacific margin of Costa Rica (Fig. 4C). We have improved the eastern isopachs of this map using line P-1600 provided by Shell Internationale Petroleum Mij. B.V. (Fig. 5). On line P-1600, the shelf-slope section is interpreted as a minimum of seven stratigraphic sequences. The age of the upper shelf-slope section is latest Miocene to Quaternary, based on microfauna recovered from DSDP hole 565 drilled into the shelf-slope section rocks approximately 40 km southwest of the Nicoya Peninsula (von Huene et al., 1985). The base of the slope could be much older, as hole 565 did not penetrate the base of the section. Where relatively undeformed, the shelf-slope section thins downslope and downlaps the underlying acoustic basement (Fig. 5). The basement below the shelf-slope section composes a thick wedge that is up to 10 km thick on line P-1600 and is believed to be composed of either accreted sedimentary material or upper Cretaceous–Eocene oceanic material of the Nicoya complex that has been studied in outcrops on the Nicoya Peninsula (Dengo, 1962; Lundberg, 1982, 1983; Gursky et al., 1982),

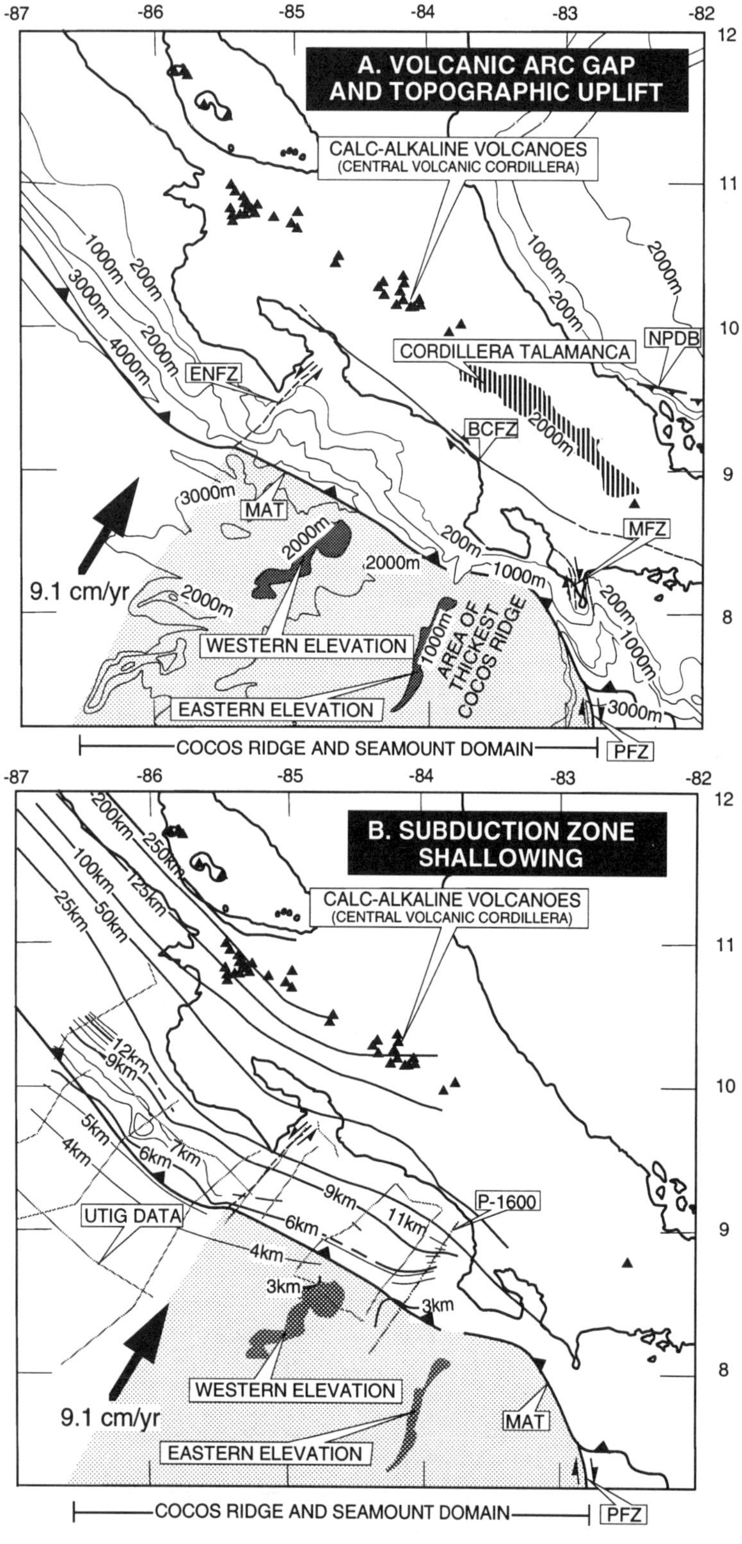

Figure 4 (on this and following two pages). A, Gap in Middle America volcanic arc and localized topographic uplift spatially associated with subduction of the bathymetrically highest areas of the Cocos Ridge and Seamount Domain (shaded in gray). Shallowest parts of the underthrusting Cocos Ridge are named here Eastern and Western Elevation (shaded in darker gray). Note spatial correlation of bathymetrically highest parts of the Cocos Ridge (dark gray) with topographically highest area of southern Central America (Cordillera Talamanca). Key to abbreviations of fault zones and deformed areas: Medial (MFZ); Ballena-Celmira (BCFZ); East Nicoya (ENFZ); NPDB (North Panama deformed belt); MAT (Middle America Trench). B, Shallowing of subduction zone spatially associated with subduction of the bathymetrically highest areas of the Cocos Ridge and Seamount Domain (shaded in gray). Contours in kilometers are on top of subducting Cocos plate. Contours from 4 to 14 km are based on seismic lines near the Middle America Trench; lines were depth converted using seismic interval velocities from Crowe and Buffler (1985). Shell line P-1600 is shown in Figure 5. Contours from 25 to 250 km are earthquake studies by Guendel (1986). Note the clustering of calc-alkaline volcanoes (black triangles) in areas where the Cocos plate is subducted to depths of 125 to 200 km and the absence of calc-alkaline volcanoes above the eastern part of the subducted Cocos Ridge and Seamount Domain.

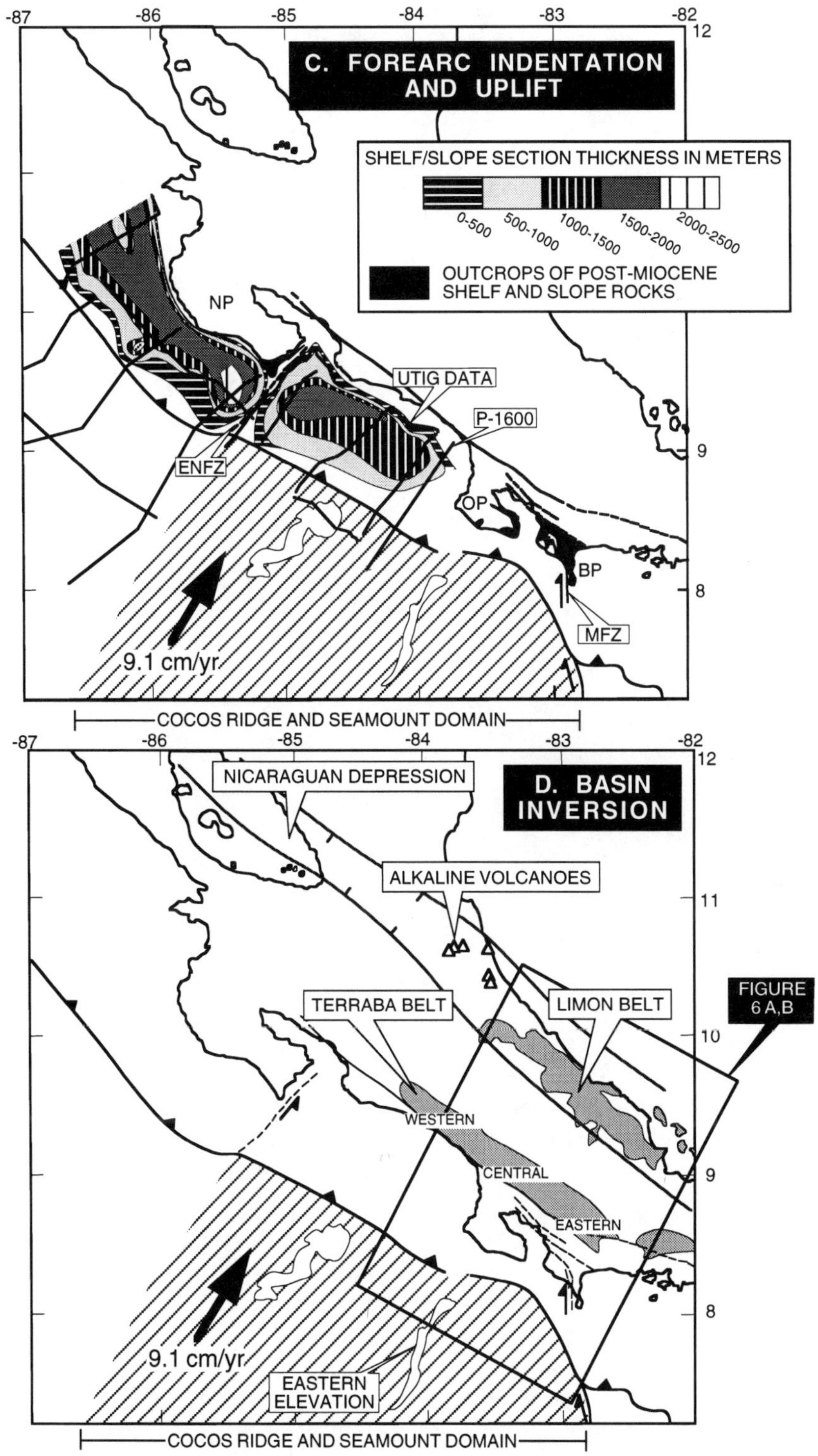

C, Indentation, thinning, and subaerial exposure of post-Miocene outer forearc marine sedimentary rocks spatially associated with the thickest part of the Cocos Ridge. Isopachs in meters are from Crowe and Buffler, 1985), and limits of onshore outcrops are from Corrigan et al. (1990). Lines indicate UTIG and Shell regional seismic lines used to constrain isopachs. Shell line P-1600 is shown in Figure 5. Note landward indentation of the trench and the post-Miocene forearc depocenter at the East Nicoya fault zone (ENFZ) and the Medial fault zone (MFZ). D, Basin inversion spatially associated with the bathymetrically highest part of the Cocos Ridge and Seamount Domain (eastern elevation). Térraba belt represents the inverted inner forearc basin of the Costa Rica arc; Limón belt represents the inverted backarc basin of the Costa Rica arc. In northwestern Costa Rica and Nicaragua, the backarc basin is not uplifted and forms a topographically low backarc basin filled by Quaternary lake and lahar deposits and alkaline volcanoes related to backarc rifting (Nicaraguan Depression).

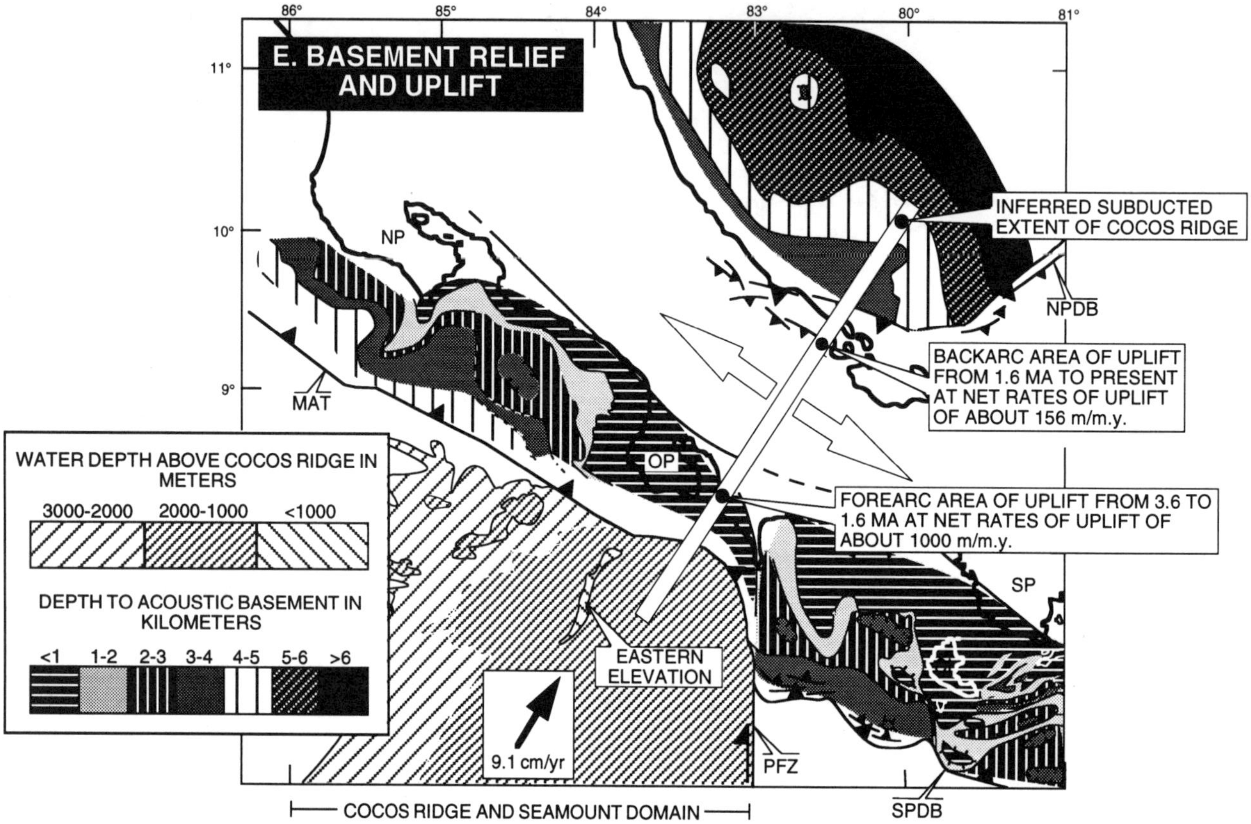

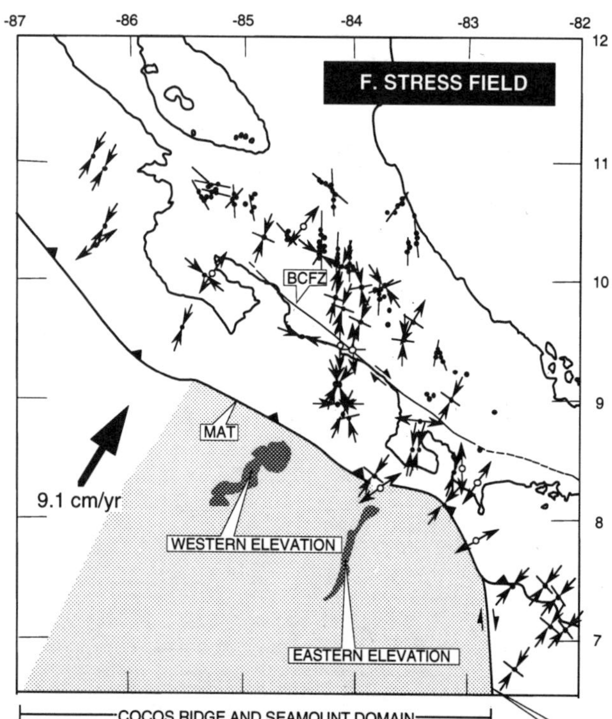

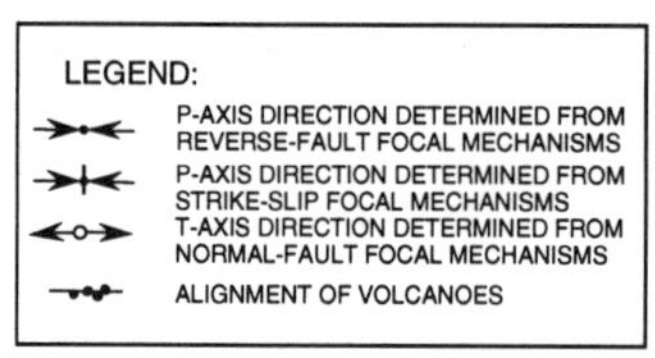

E, Depth to acoustic basement calculated from seismic interval velocities in the Costa Rica forearc (this study and Crowe and Buffler, 1985), the Colombian basin (Bowland, 1984), and the Gulf of Chiriquí of southwestern Panama (Kolarsky and Mann, this volume). Note the spatial association of the bathymetrically highest area of the Cocos Ridge and Seamount Domain at the eastern elevation and the area of shallow basement in the forearc of southeastern Costa Rica and the Colombian basin. The convergence vector of the subducting Cocos Ridge and Seamount Domain connects the areas of shallow basement in the forearc with the area of shallow basement in the Colombian basin. F, Late Quaternary stress field based on earthquakes and aligned late Quaternary volcanoes in Costa Rica and western Panama, modified from Montero and Morales (1990). Note the suggestion of a radial pattern in maximum compressive stresses in front of the Cocos Ridge and Seamount Domain.

Osa Peninsula (Obando, 1986; Berrangé and Thorpe, 1988), and Burica Peninsula (Obando, 1986; Corrigan, 1986; Corrigan et al., 1990) and in DSDP leg 82 drill holes along the Guatemala margin (Bourgois et al., 1984). The younger part of the shelf-slope section onlaps basement along a prominent unconformity near the landward edges of seismic lines near the Osa Peninsula.

Near the eastern edge of the Nicoya Peninsula, the slope depocenter axis is truncated in a left-lateral sense by the East Nicoya fault zone (ENFZ), a down-to-the-northwest high-angle fault that strikes perpendicular to the Middle America Trench (Crowe and Buffler, 1983; Corrigan et al., 1990). Apparent left-lateral offset of the shelf-slope section isopachs by the ENFZ is approximately 20 km. The ENFZ is located at the approximate northeastern margin of the Cocos Ridge and Seamount Province (von Huene et al., this volume).

Corrigan et al. (1990) correlated the offshore shelf-slope section deposits with outcrops of post-Miocene shelf and slope rocks exposed on the Nicoya, Osa, and Burica Peninsulas (Fig. 4C). Based on the distribution and thickness variations in these rock units, there are several important changes in slope structure and morphology in the Middle America forearc adjacent to the subducting Cocos Ridge: (1) Northwest of the ENFZ along the margins of northern Costa Rica and Guatemala, the trench slope is relatively smooth whereas in the vicinity of and southeast of the ENFZ, the bathymetry of the lower slope is very irregular and is characterized by large, fault-bounded basement blocks (Crowe and Buffler, 1983, 1985; Bourgois et al., 1984); (2) Northwest of the ENFZ the forearc region is characterized by a much wider shelf (>70 km width) than the shelf to the southeast of the ENFZ (~10 to 50 km); (3) Onshore outcrops of post-Miocene shelf and slope rocks are exposed in the area of the ENFZ on the Nicoya Peninsula and on the Osa and Burica Peninsulas but are not found in other areas of the Middle America Trench to the northwest of the Nicoya Peninsula or southeast of the Burica Peninsula (Corrigan et al., 1990); (4) Uplift rates based on the age and paleobathymetry of marine microfauna in post-Miocene shelf and slope rocks increase from the Nicoya Peninsula (0.05 mm/yr) to the Osa Peninsula (0.2 mm/yr) to the Burica Peninsula (0.94 mm/yr) (Corrigan et al., 1990); a similar southeastward increase in relative uplift rates is observed from geomorphologic studies on the Costa Rican mainland (Wells et al., 1988); (5) The Middle America Trench axis loses its well-defined morphologic expression east of the ENFZ and shoals from a depth of greater than 4 km to depths of less than 1,600 m near the Osa and Burica Peninsula; (6) The inferred axis of maximum shelf-slope section thickness narrows and shifts landward east of the ENFZ (Fig. 4C). All of these observations support the model of Corrigan et al. (1990) for landward indentation and uplift of both submarine and onshore shelf-slope section deposits by shallow subduction of the Cocos Ridge.

Inversion of forearc and backarc basins of Costa Rica

"Structural inversion" of sedimentary basins occurs when basin-controlling divergent faults reverse their movement because of convergent tectonics and the basin is turned inside out to form a present-day mountain range. In Costa Rica, there are two extensive outcrops of mainly marine sedimentary rocks deposited in near-arc basins formed by Cenozoic divergence or transtension within the Costa Rican arc. These rocks were subsequently folded, thrust faulted, and uplifted to form the present-day Térraba belt in the Fila Costeña (Coastal Range) and Limón belt in rolling hills along the Caribbean coastal plain (Fig. 4D).

Phillips (1983), Yuan (1984), and Lowery (1982) interpreted the Térraba belt as a pre-Pliocene forearc basin because of its position between the Pacific peninsulas of Costa Rica and western Panama and the volcanic arc of Costa Rica. Based on its position with respect to the Middle America Trench and the Costa Rican volcanic arc, we propose that the Térraba belt is the inverted southeastward extension of the undeformed submerged forearc basin along the Pacific margin of Nicaragua (Sandino basin), Guatemala, and southern Mexico (Seely, 1979) (Fig. 3). Winslow (1988) interpreted the Limón belt as a pre–late Miocene backarc basin adjacent to a coeval volcanic arc presently exposed in the Cordillera Talamanca. Based on its position with respect to the Middle America Trench and the Costa Rican volcanic arc, we propose that the Limón belt is the inverted southeastward extension of the undeformed intra-arc and backarc basin of Nicaragua and El Salvador (Fig. 3). The presence of Quaternary alkaline volcanoes and thick Quaternary sediments in a backarc setting immediately to the northeast of the Limón belt suggests that basin inversion is limited to the area of the present-day outcrop area of the Cenozoic sediments (Fig. 4D). Because both the Térraba and Limón areas are in exact alignment with the bathymetrically highest part of the Cocos Ridge at its eastern elevation, we attribute the cause of basin inversion to shallow subduction of the Cocos Ridge.

Basement structural relief

We have compiled the results of previous seismic stratigraphic studies of the depth to the top of acoustic basement along the Pacific margin of Costa Rica (Crowe and Buffler, 1985), the Pacific margin of western Panama in the Gulf of Chiriquí (Kolarsky and Mann, this volume), and the southwestern Colombian basin in the Caribbean Sea (Bowland, 1984; Bowland and Rosencrantz, 1988) (Fig. 4E). There are three areas of prominent basement structural highs: (1) in the Colombian basin, the top of acoustic basement dips strongly to the northeast below the Caribbean Sea; a semicircular basement high that is approximately 100 km wide and exhibits 3 km of structural relief is present along the eastern coast of Costa Rica; (2) along the Pacific margin of Costa Rica, the top

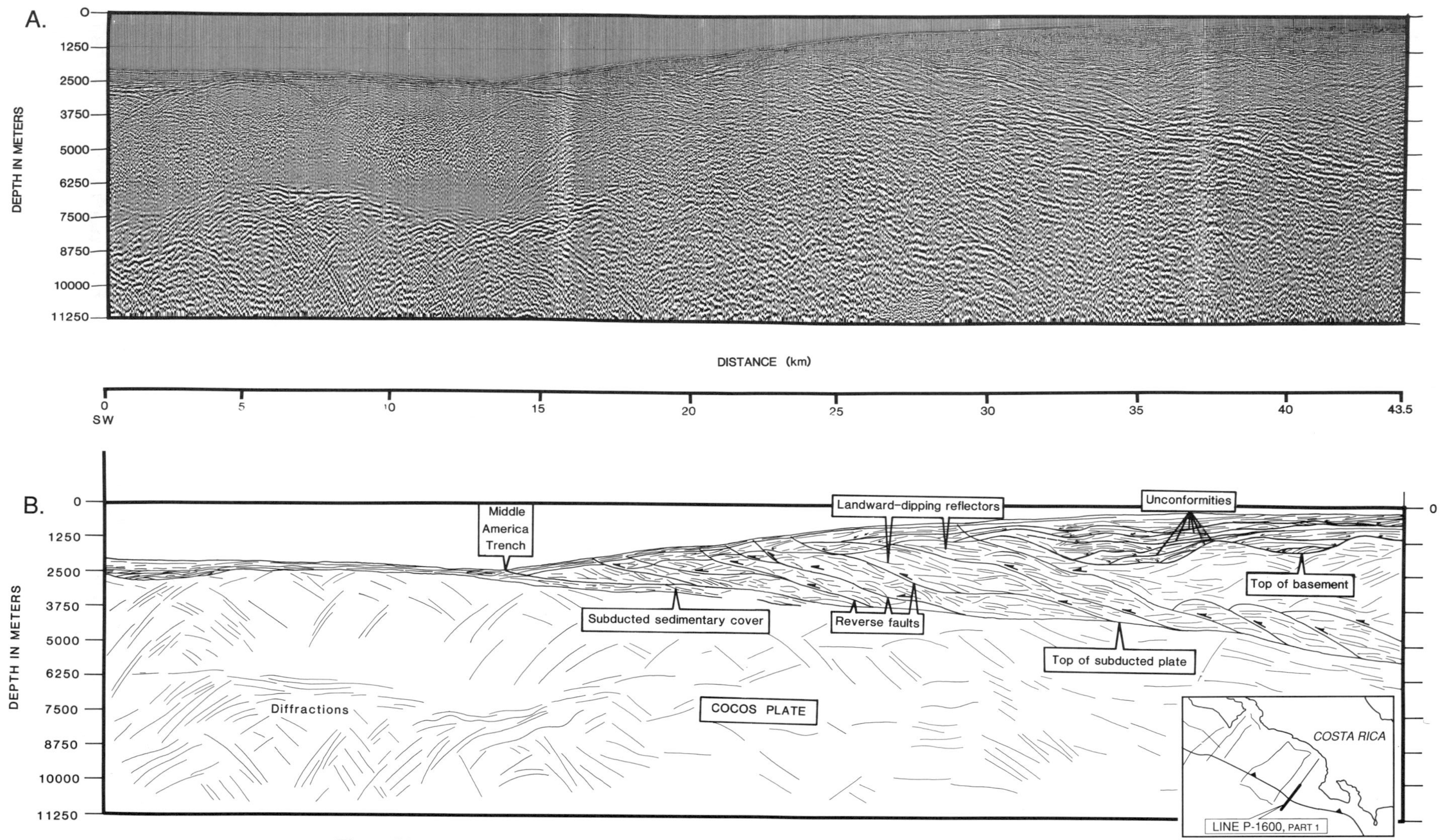

Figure 5A, B, C, and D. Interpretation of depth-converted multichannel seismic reflection profile P-1600 from the Cocos Ridge, Middle America Trench, and Costa Rica forearc (original seismic data provided courtesy of Shell Internationale Petroleum Mij. B.V.). The data were recorded in 1973 using a 1,200-cubic-inch air gun array and a 48-channel seismic streamer. The line shown is a line-drawing of a depth converted unmigrated 24-fold CDP-stack. Note that the top of the subducted Cocos plate can be followed for at least 66 km beneath the Costa Rica forearc.

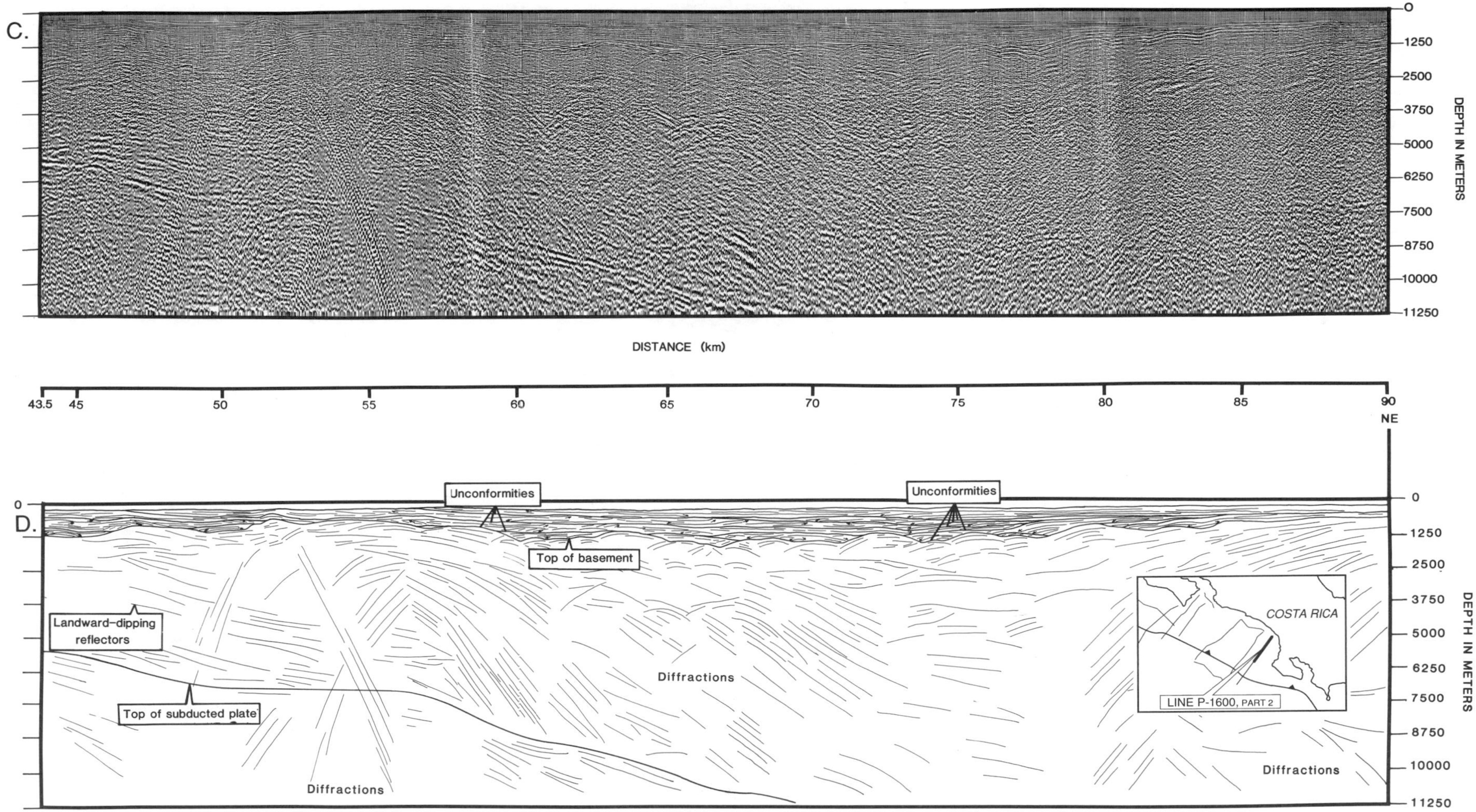
C.
D.
DEPTH IN METERS
0
1250
2500
3750
5000
6250
7500
8750
10000
11250
DISTANCE (km)
43.5
45
50
55
60
65
70
75
80
85
90
NE
Unconformities
Top of basement
Landward-dipping reflectors
Top of subducted plate
Diffractions
COSTA RICA
LINE P-1600, PART 2

of acoustic basement dips to the east and southwest; a rectangular to semicircular basement high that is approximately 100 to 130 km wide is entered on the Osa and Burica Peninsulas; and (3) an elongate, 25- to 50-km-wide basement high with 3 km of structural relief extends parallel to the ENFZ along the southeastern margin of the Nicoya Peninsula. The top of acoustic basement in the Gulf of Chiriquí is heavily faulted but exhibits less structural relief than basement on the Caribbean and Pacific margins of Costa Rica.

We attribute the Colombian basin and Osa-Burica highs to shallow subduction of the Cocos Ridge because both highs are aligned with the area of the Cocos Ridge that is bathymetrically shallowest and are aligned along the Caribbean-Cocos plate convergence vector. The basement high parallel to the ENFZ that is also aligned with the plate convergence vector may have formed in response to indentation and left-lateral shearing of the Costa Rican forearc at the northwestern edge of the Cocos Ridge and Seamount Domain (von Huene et al., this volume). The tectonic origin of arches in the Gulf of Chiriquí is discussed in Kolarsky and Mann (this volume).

Regional stress field

Regional stress fields can be inferred from P and T axes of earthquake focal mechanisms (Zoback and Zoback, 1980) and the alignment of dikes and cinder cones of Quaternary stratovolcanoes (Nakamura, 1977). Using these techniques, Montero and Morales (1990) compiled earthquake and volcanic information published in Costa Rica over the past 10 years. We have reproduced their map in Fig. 4F. P axes derived from reverse and strike-slip focal mechanisms and T axes derived from normal fault focal mechanisms are thought to provide an average value of the direction of maximum compression. This assumption is consistent with the observation that both types of axes are subparallel to the Cocos-Caribbean convergence direction (032) calculated by DeMets et al. (1990). Anomalous east-trending P and T axes are observed near the vicinity of the Ballena-Celmira fault zone, a possibly active right-lateral strike-slip or reverse fault that formed as a result of shallow subduction of the Cocos Ridge (Corrigan et al., 1990). Alignments of Quaternary dikes and cinder cones in volcanic arc and backarc volcanoes of central and western Costa Rica suggest a more north-south direction of maximum compression, similar to that observed in Nicaragua, Honduras, El Salvador, and Guatemala (Montero and Morales, 1990; Mann et al., 1990). P and T axes describe a radial pattern about the subducting Cocos Ridge, particularly in the area of eastern Costa Rica and the Gulf of Chiriquí.

OVERVIEW OF COSTA RICAN GEOLOGY IN A REGIONAL TRANSECT FROM COCOS RIDGE TO CARIBBEAN SEA

The five tectonic elements in a transect across Costa Rica are: (1) Cocos Ridge on subducting Cocos plate; (2) outer Middle America forearc defined by Corrigan et al. (1990) as the forearc region between the Middle America Trench and the Ballena-Celmira fault zone; in Costa Rica the outer forearc includes the trench, uplifted upper Cretaceous–Eocene basaltic basement and upper Miocene–Quaternary shelf-slope section rocks on the Osa Peninsula, and submerged shelf-slope section rocks in the Golfo Dulce; (3) inner Middle America forearc area defined by Corrigan et al. (1990) as deformed Eocene to Pliocene forearc basin sedimentary and volcanic rocks between the Ballena-Celmira fault zone and the uplifted volcanic arc of the Cordillera Talamanca (Térraba belt); (4) upper Cretaceous to Neogene volcanic, plutonic, and sedimentary rocks of the Middle America volcanic arc of Cordillera Talmanca; and (5) deformed Cenozoic sedimentary and volcanic rocks ofthe Middle America backarc basin (Limón belt). The age relationships and locations of the four onshore tectonic elements of the Middle America island arc are shown in Figure 6A and 6B. We have constructed a regional cross section, shown in Figure 7, to illustrate the relationships at depth between the Cocos Ridge and the tectonic elements described above.

Map relationships of major tectonic elements in Costa Rica

A map showing the ages of rocks mapped in eastern Costa Rica was modified from Weyl (1980) and is shown in Figure 6B. Superimposed on this map is the northeast-trending axis of the basement high recognized from our compilation of depth to acoustic basement in offshore areas in the Caribbean Sea and Pacific margin (Fig. 4E). The basement high is approximately aligned along the Cocos-Caribbean plate convergence vector. Ages of rock units form a roughly domal pattern, with the oldest rocks generally in the center of the dome. A variation in the domal pattern is the presence of upper Miocene plutonic rocks in the uplifted volcanic arc in the Cordillera Talamanca. These rocks are typically quartz monzonitic rocks of calc-alkaline affinity (Weyl, 1980; de Boer et al., this volume) and form one of the youngest exposed plutonic belts in the world. The Ballena-Celmira fault zone juxtaposes rocks of Paleocene-Oligocene age of the inner forearc (Térraba belt) against Quaternary alluvium and rocks of Jurassic-Cretaceous age of the outer forearc (Obando, 1986). Rocks to the northeast of the Ballena-Celmira fault zone are tilted to the northeast and expose younger Miocene-Pliocene sections in the Térraba belt. Rocks of the Limón belt are tilted to the southwest and young in that direction.

The structural dome is bisected by the northeast-trending axis of the basement high identified using offshore data. The confluence of the structural dome, the offshore basement highs, and their approximate alignment with the Cocos-Caribbean plate convergence vector suggests that basement doming is produced by shallow subduction of the Cocos Ridge beneath the Costa Rican island arc.

Regional transect of tectonic elements in Costa Rica

The cross section in Figure 7 is modified from the section presented by Corrigan et al. (1990) across the outer and inner

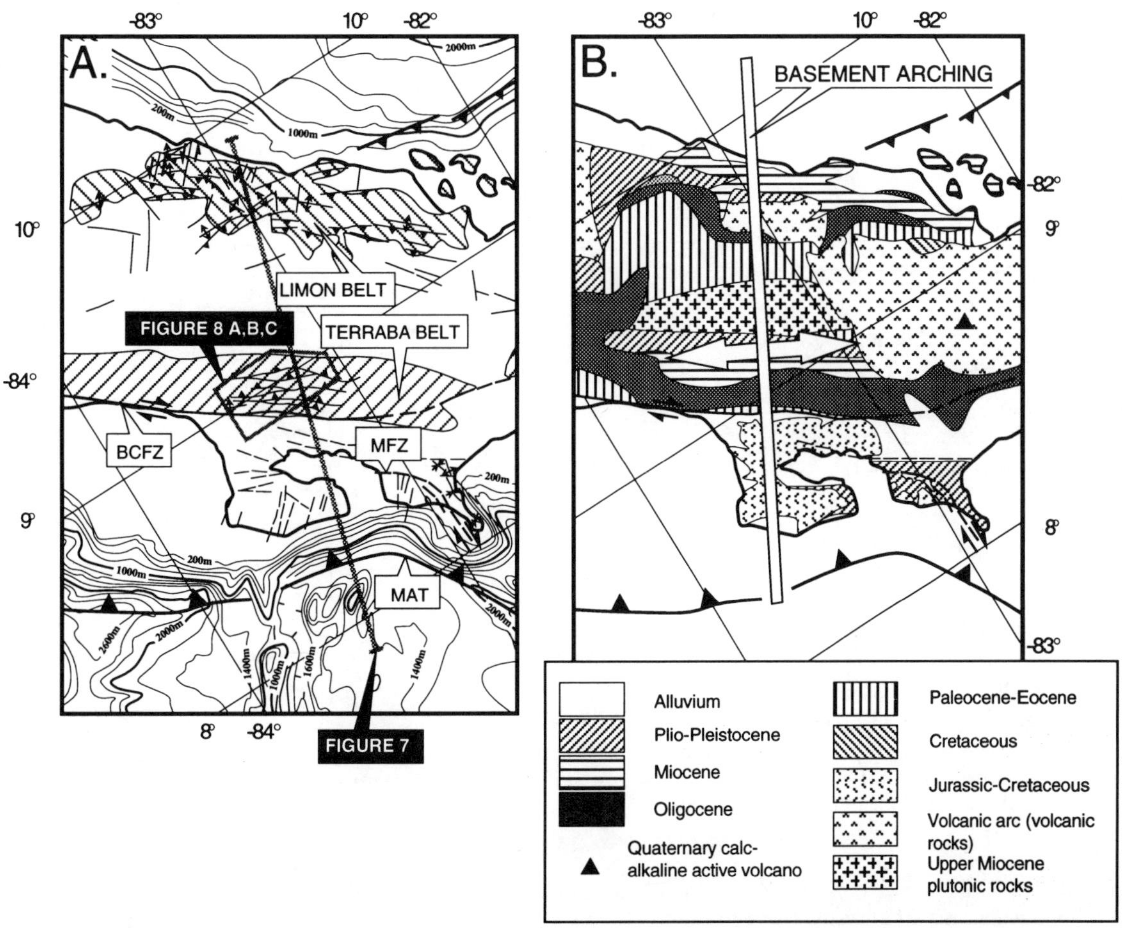

Figure 6. A, Structural map of eastern Costa Rica and western Panama showing onshore faulting related to the Cocos Ridge collision and inverted inner forearc (Térraba) and backarc (Limón) basins (ruled areas). Faults are compiled from this study (Térraba belt); Corrigan et al. (1990) (outer forearc area); and Winslow (1988) (Limón belt). Regional cross section across area affected by Cocos Ridge subduction is shown in Figure 7. Area of detailed geologic map of Térraba belt is shown in Figure 8. B, Generalized geologic map of eastern Costa Rica and western Panama showing ages of onshore rock units (from Weyl, 1980). Note extensive outcrops of upper Miocene plutonic rocks as well as Jurassic-Cretaceous ophiolites in the area affected by Cocos Ridge subduction. White arrow indicates suggested sense of basement arching.

forearc areas and the longer cross section by Rivier (1985) across the area from the Middle America Trench to the Caribbean Sea. We base our reinterpretation of the dip of the subducted Cocos Ridge on the offshore Shell seismic line P-1600 (Fig. 5) along with earthquake data of Guendel (1986) (Fig. 4B).

The top of the Cocos Ridge on seismic line P-1600 is defined by a high-amplitude reflector. Beneath this reflector is at least a 9-km-thick section characterized by diffraction hyperbolae. These hyperbolae probably originate on basalt flows and sills forming the thickened oceanic crust of the Cocos Ridge, known to be up to 15 km thick from seismic refraction experiments (Bentley, 1974). Overlying the top of the igneous body of the Cocos Ridge is a 200- to 500-m-thick section of parallel and wavy reflectors that represent Miocene to Recent pelagic and terrigenous hemipelagic sediments deposited on the top of the Cocos Ridge (Shipley and Moore, 1986). Landward of the trench the top of the igneous body of the Cocos Ridge can be traced beneath the forearc for a horizontal distance of 55 km and to a depth of 10 km.

The average dip of the top of the Cocos Ridge is 3° and significantly less than the 23° dip of Corrigan et al. (1990) that was based on the assumption that hypocenters of earthquakes below the outer forearc area occur along a Benioff zone between the top of the Cocos Ridge and the base of the outer forearc (Adamek, 1986; Adamek and Tajima, 1987). The lower dip of the subducted slab indicated by seismic profiles suggests that some of the earthquakes relocated by Adamek (1986) and Adamek and Tajima (1987) may have occurred within the subducted Cocos Ridge and perhaps reflect internal deformation of the Cocos Ridge during shallow subduction.

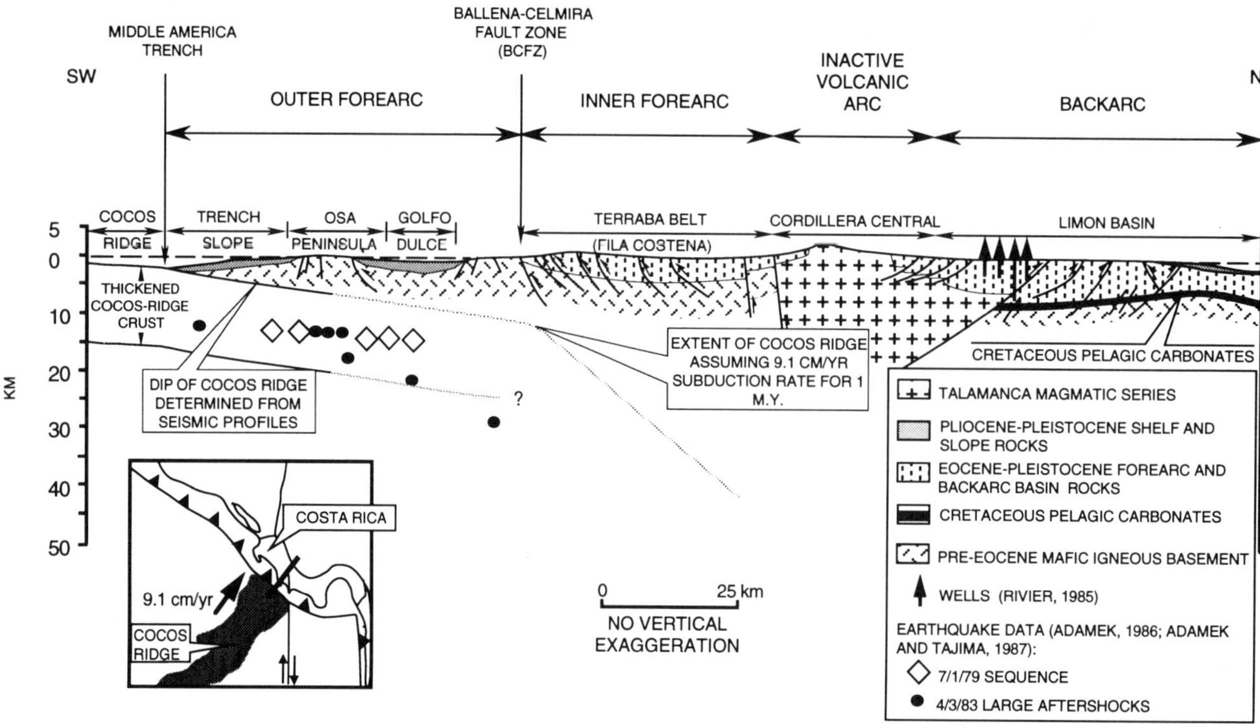

Figure 7. Regional cross section across southeastern Costa Rica. Forearc structure is modified from Corrigan et al., 1990, and seismic line shown in Figure 5; backarc structure and well data are from Rivier (1985); earthquake data are from Adamek and Tajima (1987).

We have extended the area of the shallowing dipping Cocos Ridge 40 km beyond the limit of control by seismic line P-1600, based on the assumption that the Cocos Ridge arrived at the southern middle America Trench approximately 1 Ma (Lonsdale and Klitgord, 1978; Corrigan et al., 1990) and was subducted at the present-day convergence rate of 9.1 cm/yr (DeMets et al., 1990).

In cross section, the outer forearc forms a large anticline with the core corresponding to exposures of the Nicoya complex on the Osa Peninsula (Berrangé and Thorpe, 1988) and the flanks corresponding to the upper Miocene–Quaternary shelf-slope section, around the coastline of the Osa Peninsula and in the Golfo Dulce (Corrigan et al., 1990). The Ballena-Celmira fault zone abruptly truncates this large-scale anticline in igneous and overlying sedimentary rocks of the outer forearc area and juxtaposes it with a northeast-dipping monocline of mainly sedimentary rocks in the inner forearc area of the Térraba belt. Mora (1979) suggested that folds and thrust faults in the Térraba belt verge to the southwest.

Poorly studied high-angle faults separate upper Miocene plutonic rocks of the Cordillera Talamanca with rocks of the Térraba belt. To the north of the Cordillera Talamanca, mapping (Winslow, 1988) and oil exploration wells (Rivier, 1985) in mainly sedimentary rocks of the Limón belt both show northeast-verging folds and thrust faults. The entire regional transect reveals opposed sense of structural vergence on either side of the Cordillera Talamanca: folds and thrust faults in the Térraba belt verge to the southwest, whereas folds and thrusts in the Limón belt verge to the northeast.

To better establish the types of convergent structures reported in the Térraba belt by Mora (1979) and to compare them with structures previously mapped in both the outer forearc area (Corrigan et al., 1990) and the backarc area (Winslow, 1988), we carried out detailed mapping in the forearc area of the central Térraba belt. In the next sections, we describe the stratigraphy and structure of the Térraba belt and relate these to shallow subduction of the Cocos Ridge.

Geologic map of the central Térraba belt

Deformed rocks of the Térraba belt are exposed over an along-strike distance of 220 km and an across-strike distance of 20 km in the Fila Costeña (Fig. 4D). The most accessible part of the Térraba belt is in the valley of the Rio Térraba that drains the Valle General and Cordillera Talamanca to the north (Fig. 8A). The Térraba river valley obliquely crosses erosionally resistant ridges formed by northeastwardly dipping strata within the Térraba belt and provides a natural cross section through the belt (Fig. 8B). Térraba belt rocks are well exposed along the banks of the river and its main tributaries as well as in numerous road cuts along the paved Interamerican Highway that follows the northwest river bank through the

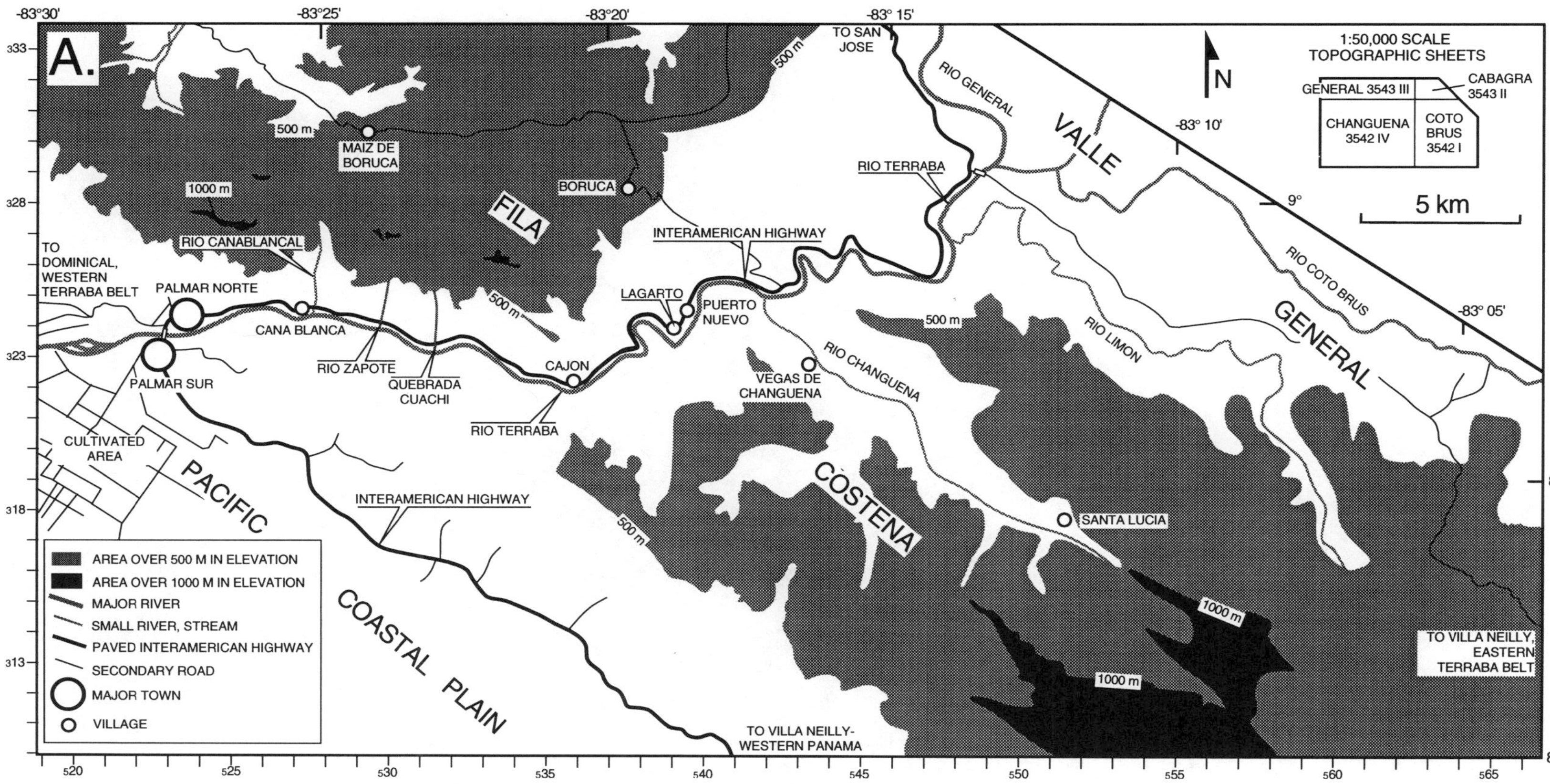

Figure 8 (on this and following two pages). A, Topographic and locality map of the central Térraba belt. Areas above 1,000 m in elevation are shaded black. Access to the study area is along the paved Interamerican Highway, which crosses the Térraba belt (Fila Costeña, or "Coastal Range") in the valley of the Rio Térraba. Inset shows key to 1:50,000 scale U.S. Defense Mapping Agency topographic maps used as base maps for field mapping in the Térraba belt (available from Costa Rican cartographic office in San José, Costa Rica). B, Geologic map and legend of the central Térraba fold and thrust belt along the Rio Térraba valley. Inset shows source map for geologic data compiled on the map. Major reverse faults superimpose older stratigraphic units over younger units. Map uses standard geologic symbols for bedding attitude, faults, and folds. Dip symbols with arrow show photointerpreted strikes and dip senses. Numbers identify regional reverse faults discussed in text.

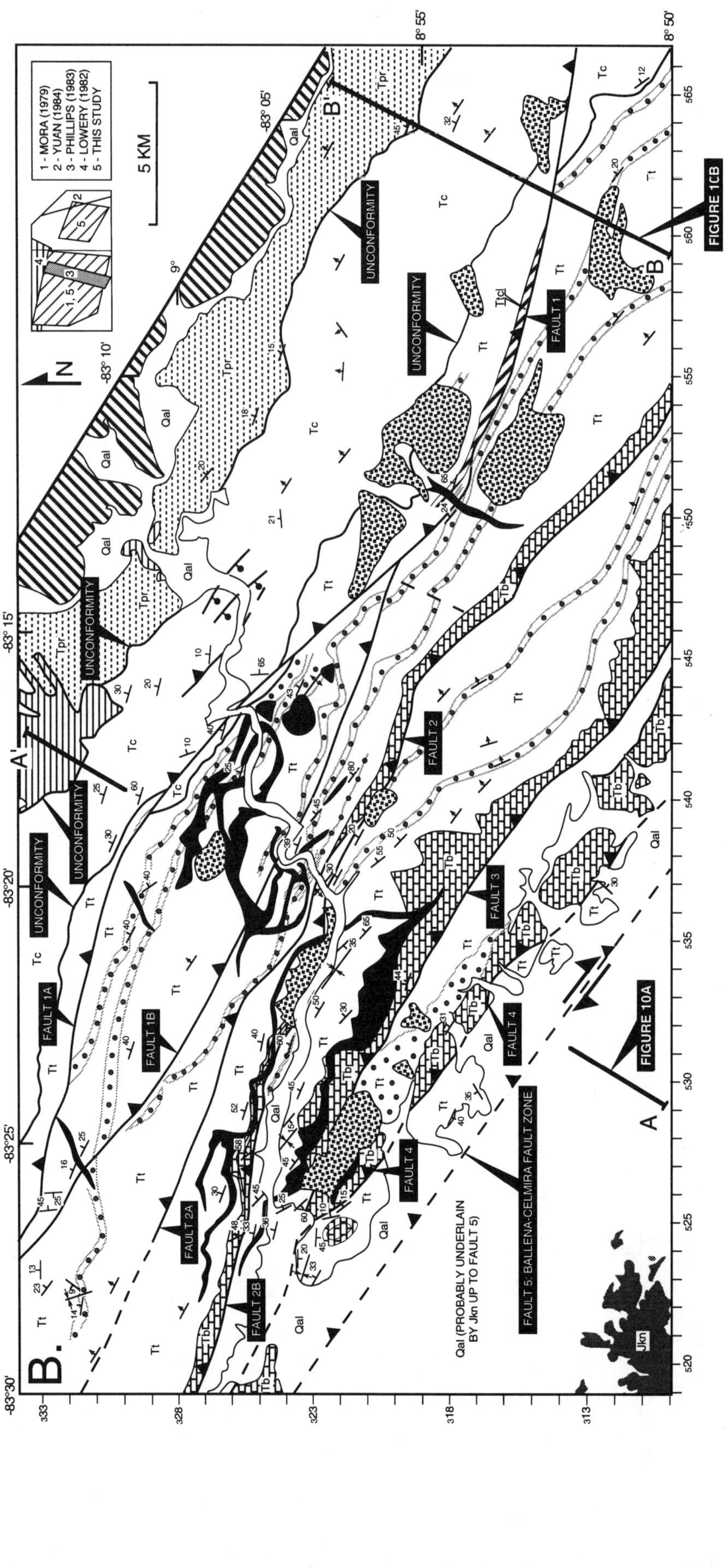
B.
1 - MORA (1979)
2 - YUAN (1984)
3 - PHILLIPS (1983)
4 - LOWERY (1982)
5 - THIS STUDY
5 KM
N
UNCONFORMITY
FAULT 1
FAULT 1A
FAULT 1B
FAULT 2
FAULT 2A
FAULT 2B
FAULT 3
FAULT 4
FAULT 5: BALLENA-CELMIRA FAULT ZONE
Qal (PROBABLY UNDERLAIN BY Jkn UP TO FAULT 5)
FIGURE 10A
FIGURE 1CB
A
A'
B
B'
-83°30'
-83°25'
-83°20'
-83° 15'
-83° 10'
-83° 05'
8° 55'
8° 50'
9°

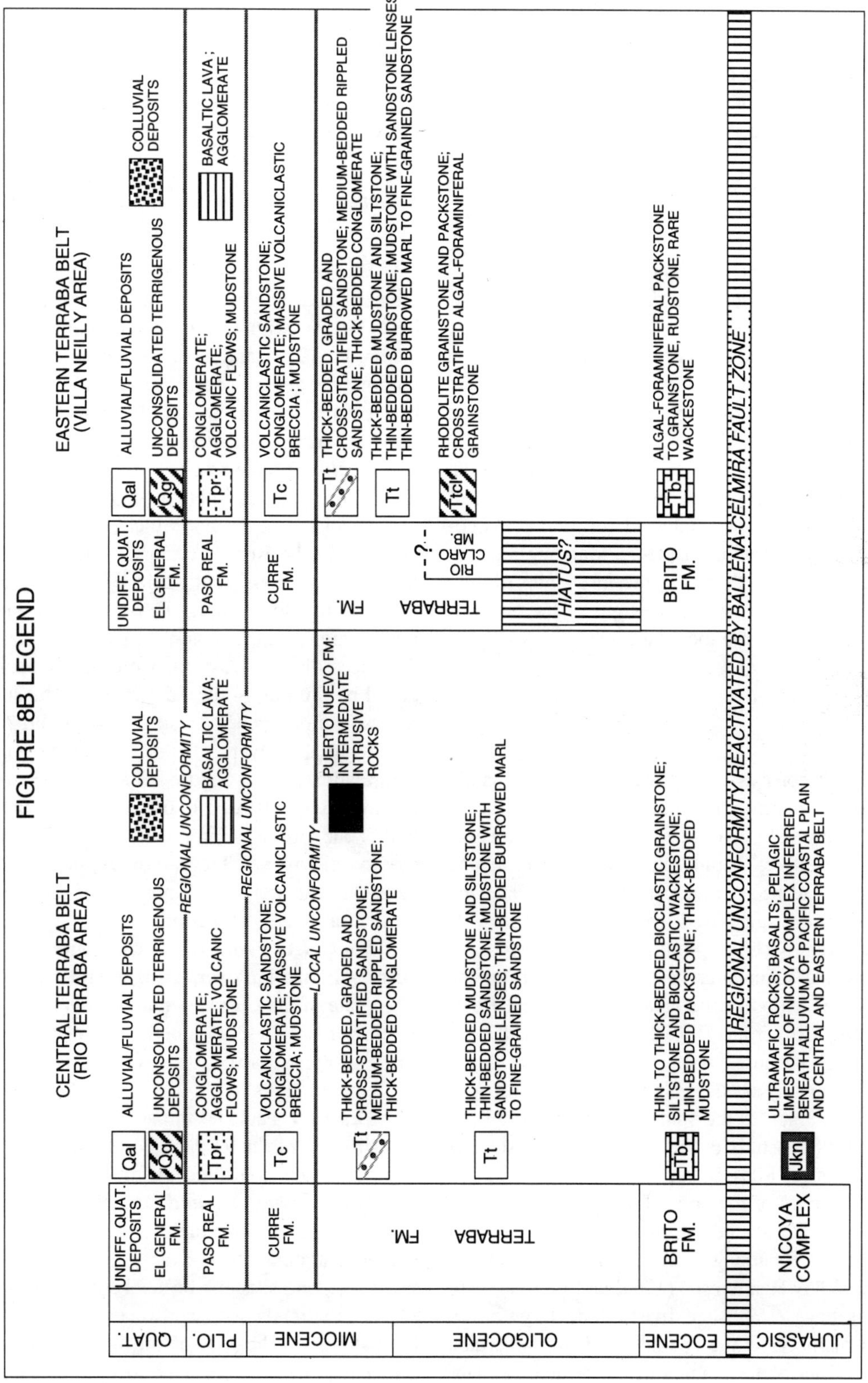
FIGURE 8B LEGEND
CENTRAL TERRABA BELT (RIO TERRABA AREA)
EASTERN TERRABA BELT (VILLA NEILLY AREA)
QUAT.
PLIO.
MIOCENE
OLIGOCENE
EOCENE
JURASSIC
UNDIFF. QUAT. DEPOSITS
EL GENERAL FM.
PASO REAL FM.
CURRE FM.
TERRABA FM.
BRITO FM.
NICOYA COMPLEX
Qal ALLUVIAL/FLUVIAL DEPOSITS
COLLUVIAL DEPOSITS
Qg UNCONSOLIDATED TERRIGENOUS DEPOSITS
REGIONAL UNCONFORMITY
Tpr CONGLOMERATE; AGGLOMERATE; VOLCANIC FLOWS; MUDSTONE
BASALTIC LAVA; AGGLOMERATE
REGIONAL UNCONFORMITY
Tc VOLCANICLASTIC SANDSTONE; CONGLOMERATE; MASSIVE VOLCANICLASTIC BRECCIA; MUDSTONE
LOCAL UNCONFORMITY
Tt THICK-BEDDED, GRADED AND CROSS-STRATIFIED SANDSTONE; MEDIUM-BEDDED RIPPLED SANDSTONE; THICK-BEDDED CONGLOMERATE
PUERTO NUEVO FM: INTERMEDIATE INTRUSIVE ROCKS
Tt THICK-BEDDED MUDSTONE AND SILTSTONE; THIN-BEDDED SANDSTONE; MUDSTONE WITH SANDSTONE LENSES; THIN-BEDDED BURROWED MARL TO FINE-GRAINED SANDSTONE
Tb THIN- TO THICK-BEDDED BIOCLASTIC GRAINSTONE; SILTSTONE AND BIOCLASTIC WACKESTONE; THIN-BEDDED PACKSTONE; THICK-BEDDED MUDSTONE
REGIONAL UNCONFORMITY REACTIVATED BY BALLENA-CELMIRA FAULT ZONE
Jkn ULTRAMAFIC ROCKS; BASALTS; PELAGIC LIMESTONE OF NICOYA COMPLEX INFERRED BENEATH ALLUVIUM OF PACIFIC COASTAL PLAIN AND CENTRAL AND EASTERN TERRABA BELT
Qal ALLUVIAL/FLUVIAL DEPOSITS
COLLUVIAL DEPOSITS
Qg UNCONSOLIDATED TERRIGENOUS DEPOSITS
Tpr CONGLOMERATE; AGGLOMERATE; VOLCANIC FLOWS; MUDSTONE
BASALTIC LAVA; AGGLOMERATE
Tc VOLCANICLASTIC SANDSTONE; CONGLOMERATE; MASSIVE VOLCANICLASTIC BRECCIA; MUDSTONE
Tt THICK-BEDDED, GRADED AND CROSS-STRATIFIED SANDSTONE; MEDIUM-BEDDED RIPPLED SANDSTONE; THICK-BEDDED CONGLOMERATE
Tt THICK-BEDDED MUDSTONE AND SILTSTONE; THIN-BEDDED SANDSTONE; MUDSTONE WITH SANDSTONE LENSES; THIN-BEDDED BURROWED MARL TO FINE-GRAINED SANDSTONE
RIO CLARO MB.
?
Ttcl RHODOLITE GRAINSTONE AND PACKSTONE; CROSS STRATIFIED ALGAL-FORAMINIFERAL GRAINSTONE
HIATUS?
Tb ALGAL-FORAMINIFERAL PACKSTONE TO GRAINSTONE, RUDSTONE, RARE WACKESTONE

Fila Costeña. Because of the monoclinal northeastwardly dip, rocks of the Térraba belt generally young from southwest to northeast. The northeastern edge of the Térraba belt is defined by the onlap of Quaternary alluvium on northeastward-dipping strata of the youngest section of the belt. The southwestern edge of the Térraba belt is defined by the Ballena-Celmira fault zone that abruptly truncates the oldest section of the belt and separates it from Quaternary alluvium of the Pacific coastal plain.

We spent six man weeks in January and February of 1990 mapping approximately 30% of the most accessible central part of the Térraba belt in the valley of the Rio Térraba. Knowledge of key outcrops from previous mapping by Mora (1979), Montero (unpublished field maps), and other unpublished thesis projects from the University of Costa Rica in San José (e.g., Alvarado, 1979) expedited field mapping. Mapping was done on 1:25,000 scale enlargements of U.S. Defense Mapping Agency 1:50,000 topographic sheets available through the Instituto Geografico in San José.

The Universal Transverse Mercator (UTM) grid location system is used throughout this chapter to locate outcrops mentioned in the text. Mapping was conducted between UTM coordinates 320–348, 513–529 on the Defense Mapping Agency 1:50,000 topographic sheets shown in the inset map of Fig. 8A. For brevity, we use only the last six digits of the grid reference (G. R.) to specify outcrop localities.

Regional stratigraphic correlation

The Térraba belt consists of at least 3,500 m of mainly Eocene to Pleistocene marine, fluvial, and alluvial fan facies sedimentary rocks intruded by volumetrically minor Miocene-Pliocene (?) sills and dikes. Stratigraphy of the central and eastern Térraba belt had been previously studied in detail by three graduate students supervised by Dr. Don Lowe at Louisiana State University (LSU). Phillips (1983) carried out a sedimentary facies and sedimentary petrography study of the Eocene to Oligocene section in the area of Dominical, Pilas, and the valley of the Rio Térraba (Fig. 8A). Phillips (1983) refined the stratigraphy proposed in an earlier study by Mora (1979) that focused mainly on the valley of the Rio Térraba. Lowery (1982) complemented the study of Phillips (1983) by completing a similar type study of the Miocene to Pleistocene section of the Dominical-Pilas-Térraba area (Fig. 8A). Yuan (1984) completed a study of the entire Eocene to Pleistocene section of the eastern Térraba belt in the vicinity of the town of Villa Neilly, Costa Rica (Fig. 8A). Yuan's study refined stratigraphy proposed by Henningsen (1965) and Weyl (1980).

These studies have established five sedimentary formations and several members in the central and eastern Térraba belt. We recognized all of these units in our structural mapping of the area and show them on our geologic map (Fig. 8B). Because ages and lateral relationships of stratigraphic units are critical for structural interpretation of folding and faulting in the Térraba belt, we summarize lithologic and age information from each of these units. A summary of the stratigraphy of the Térraba belt at five localities as derived from the LSU master's theses and dissertation is presented in Figure 9. These columns show that the five main formations of the Térraba belt (Brito, Térraba, Curré, Paso Real, El General) are correlatable over distances up to 200 km.

Structural sections of the central and eastern Térraba belt

We have integrated the stratigraphic correlation of sedimentary and igneous units in the central and eastern Térraba belt (Fig. 9) with structural data collected by Mora (1979), Alvarado (1979), and us in the same area (Fig. 8B). Special attention was given to confirming the sense of movement along major faults both from bedding offsets and from fault striations and other small-scale features.

Central Térraba belt. In the Rio Térraba valley, the central Térraba belt is imbricated by five southwest-verging reverse faults. The faults, which were first mapped by Mora (1979), cut the Brito, Térraba, Puerto Nuevo, and Curré Formations and consistently place older lithologic units over younger lithologic units. Crystalline basement of the inner forearc area of the Osa and Burica Peninsulas or of the Middle America volcanic arc of the Cordillera Talamanca is never exposed by faulting in the Térraba belt. The surface trace of the reverse faults is generally straight or slightly curving. We have numbered the reverse faults 1 through 5 on the geological compilation map of Figure 8B and on the structural cross sections of Figure 10A. One of these faults, the Ballena-Celmira fault zone, forms a prominent linear contact between Quaternary alluvium of the Pacific coastal plain and the Térraba belt. Unfortunately, no exposures of this fault were found in the central or eastern Térraba belt.

Fault 1A. Fault 1A thrusts Miocene sandstone of the Térraba Formation over younger Miocene sandstone and conglomerate of the Curré Formation. Because rocks of the Curré Formation are present on both sides of Fault 1A in areas along strike of the cross section, reverse offset on Fault 1A is relatively minor (Fig. 8B).

Fault 1B. Fault 1B thrusts Miocene mudstone, sandstone, and siltstone of the Térraba Formation over younger Miocene mudstone, sandstone, and siltstone of the Térraba Formation.

Fault Zones 2A and 2B. Faults 2A and 2B are two closely spaced reverse faults that enclose a southwest-verging, faulted anticline. Fault zone 2A thrusts Miocene mudstone, sandstone, and siltstone of the Térraba Formation over younger Miocene mudstone, sandstone, and siltstone of the Térraba Formation. To the east of the intersection of Faults 2A and 2B, Fault zone 2 thrusts Eocene limestone of the Brito Formation over Oligocene sandstone and siltstone of the Térraba Formation.

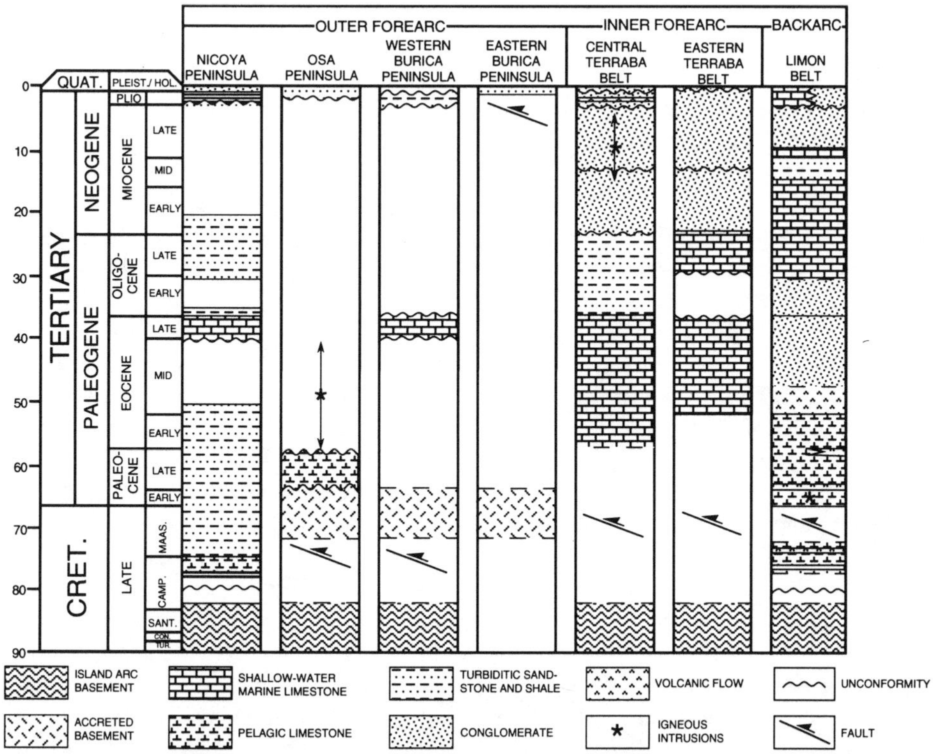

Figure 9. Stratigraphic sections from outer forearc (Nicoya and Osa Peninsulas of Costa Rica; Burica Peninsula of Panama–Costa Rica), inner forearc (Térraba belt), and backarc (Limón belt) of Costa Rica. Data from Lundberg (1982, 1983) (Nicoya Peninsula); Lew (1983) (Osa Peninsula); Corrigan et al. (1990) (East and West Burica Peninsula); Phillips (1983), Lowery (1982), and Yuan (1984) (East and West Térraba Belt); and Winslow (1988) (Limón belt).

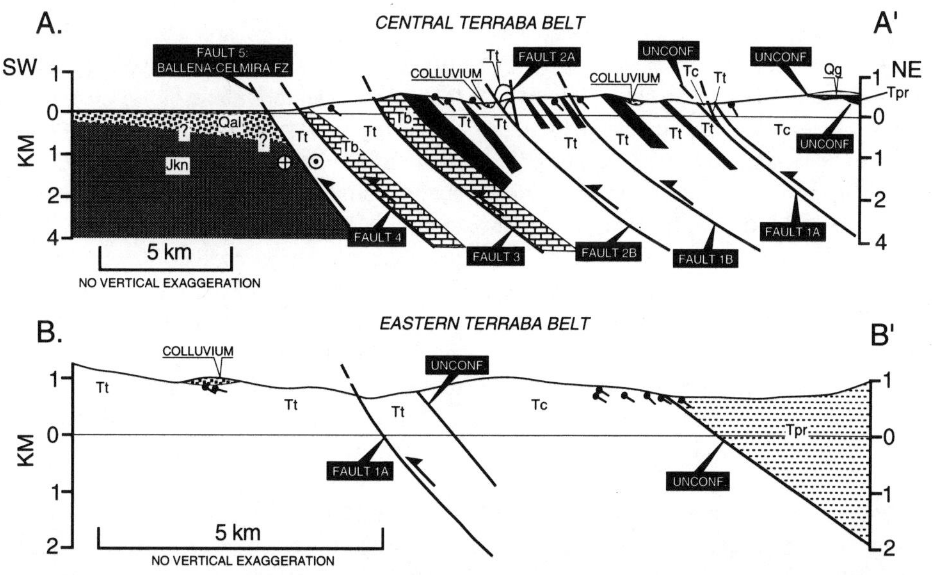

Figure 10. True scale cross sections across the central Térraba belt in the area of the Rio Térraba valley (A) and the eastern Térraba belt in the area of the town of Villa Neilly near the Panama–Costa Rica border (B). Symbols on Ballena-Celmira fault zone indicate right-lateral and reverse motion. See map legend of Figure 8B for key to stratigraphic units. Numbers identify regional reverse faults discussed in text.

Fault 3. Fault 3 thrusts Eocene limestone of the Brito Formation over undifferentiated Oligocene mudstone, siltstone, and sandstone of the Térraba Formation.

Fault 4. Fault 4 thrusts Eocene limestone of the Cañablancal Member of the Brito Formation over undifferentiated Oligocene mudstone, siltstone, and sandstone of the Térraba Formation.

Eastern Térraba belt. The eastern Térraba belt is cut by one northeast-dipping reverse fault that represents the merged eastward continuation of Faults 1A and 1B described in the central Térraba belt. The fault, which was first mapped by Yuan (1984), places Oligocene carbonate lithologies of the Rio Claro Member of the Térraba Formation over Oligocene mudstone, siltstone, and sandstone of the Térraba Formation. We assume that this is a reverse fault because of its continuity with reverse faults 1A and 1B from the central Térraba belt. However, there is no clear stratigraphic repetition along this fault because different members of the Térraba Formation are not recognizable on either side of the fault. As in the central Térraba belt, crystalline basement of the inner forearc area of the Osa and Burica Peninsulas or of the volcanic arc area of the Cordillera Talamanca is never exposed by faulting. The surface trace of the reverse faults is generally straight or slightly curving. The Ballena-Celmira fault zone forms a prominent linear contact between Quaternary alluvium of the Pacific coastal plain and the eastern Térraba belt. Unfortunately, no exposures of this fault were found in the central or eastern Térraba belt.

Structural analysis of the central and eastern Térraba belt

In order to plot the orientations of bedding, faults, and folds, the map area in the central and eastern Térraba belt was divided into six structural domains (Fig. 11, inset). The structural domains are named A through F and are bounded by Faults 1 through 5 described above. Generally, domains A and B in the southwest include older rocks of the Brito and lower Térraba Formations, and domains C, D, and E include rocks of the upper Térraba Formation. Domain F includes mostly rocks of the Curré and Paso Real Formations. Most of the bedding data are concentrated in the central Térraba belt, where we combined our own bedding measurements with those of Mora (1979).

Poles to bedding. Lower hemisphere equal area stereoplots of poles to 292 measurements of bedding indicate northeastward tilting by an average of 30° (Fig. 11). Southwest dips, particularly in domains C, D, and E, are consistent with the presence of reverse fault-bounded anticlines, such as the one described between Faults 2A and 2B of the central Térraba belt. The similarity in dip values for domains A through E suggests a single deformational event postdating the deposition of marine Miocene rocks of the upper Térraba belt in domain E. There is a distinct shallowing of dip between domains E and F, which reflects the presence of a major angular unconformity between the upper part of the marine Térraba Formation and the lower part of the Pliocene alluvial Paso Real Formation.

Minor reverse faults and fault striations. Minor, outcrop-scale reverse faults were identified in domains C, D, E, and F using both apparent offsets in bedding planes and fault plane striations. Sense of slip on all fault planes was determined from various types of fault striations using the kinematic criteria of Petit (1987). The orientation of reverse faults yields an average strike of 154° and an average dip of 53°E and generally shows up-to-the-northwest movement consistent with larger-scale movement on Faults 1 through 5 in the central Térraba belt. Striations indicate an average pitch of 66°, indicative of primarily dip slip reverse movement. Domain D exhibits some anomalous north-northwest– to northeast–striking reverse fault planes.

Minor normal faults and fault striations. Minor, outcrop-scale normal faults were identified in domains A, B, C, D, and E using both apparent offsets in bedding planes and fault plane striations. The orientation of normal faults yields an average strike of 163° and an average dip of 43° and generally shows down-to-the-northeast movement. Striations indicate an average pitch of 52°, indicative of oblique dip-slip normal movement. Domains D and E exhibit some anomalous northwest- to northeast-striking normal fault planes. Because of the parallelism of most of the normal faults with both large- and small-scale northwest-dipping reverse faults, we interpret the normal faults as late-stage reversals of movement along preexisting reverse fault planes. Normal reactivation of reverse faults suggests the possibility of limited topographic collapse of the Fila Costeña.

Minor strike-slip faults and fault striations. Four minor, outcrop-scale strike-slip faults were identified in domains D and E using both apparent offsets in bedding planes and fault plane striations. The four strike-slip faults show no preferred fault plane orientation and both right and left-lateral offsets.

Poles to axial planes of minor folds. Ten minor, outcrop-scale folds were identified in domains A, C, and D. Fold shapes range from chevron to isoclinal even over short distances in a single outcrop (Fig. 12). The best exposed minor folds are present in incised streams or road cuts through indurated limestone of the Brito Formation. Axial planes of minor folds dip to the northeast in all examples and support the interpretation of northeast-side-up movement on reverse faults 1 through 5 that bound the structural domains.

Age of deformation. We observed right-lateral offset of the dikes and sills of the Puerto Nuevo Formation (4.5 to 14.8 Ma) along minor faults at locality 538, 324 in the Rio Térraba Valley and locality 522, 331 in the Rio Reventazon Valley (Fig. 8B). We interpret the development of major reverse faults and earliest thrusting as post–middle Miocene. Uplift, in addition to incipient thrusting, is the probable cause of the unconformity between the Curré and Paso Real Formations. Gardner et al. (1992) suggest reactivation of the major faults as late as Pleistocene time.

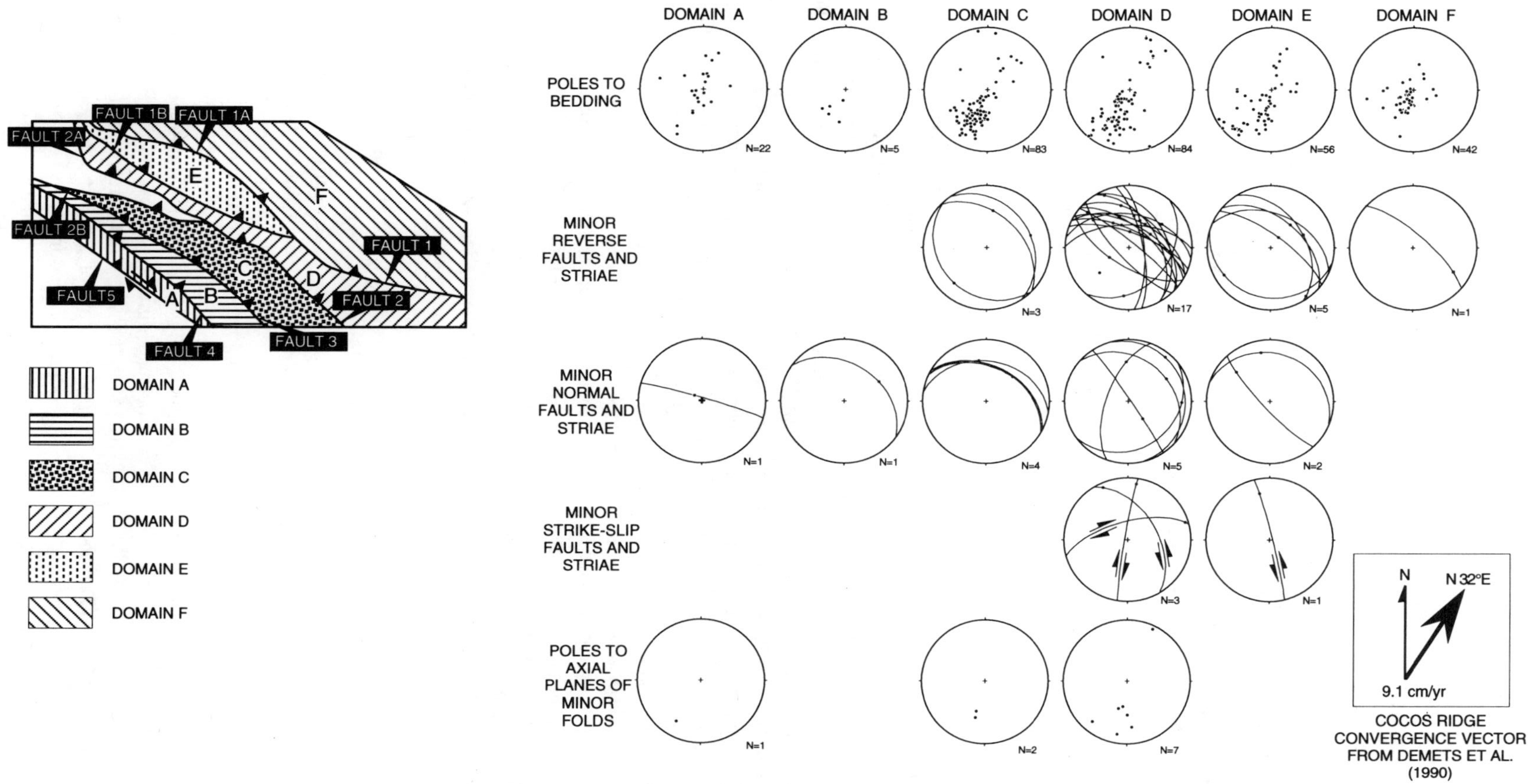

Figure 11. Key to structural domains A to F bounded by faults 1 to 5 in the central Térraba belt and lower hemisphere stereoplots of minor fault planes and poles to bedding planes and minor fold axes. Note that average shortening from bedding and fault attitudes (N30–34°E) is subparallel to the N32°E convergence direction of Cocos Ridge and Caribbean plate that was calculated by DeMets et al. (1990).

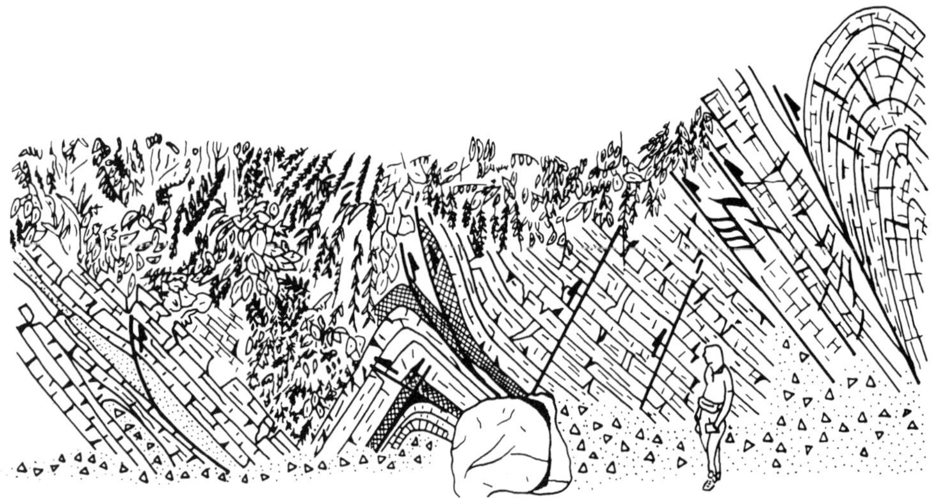

Figure 12. Outcrop drawing of folded and faulted Eocene bioclastic limestone of the Brito formation in the Rio Cañablancal valley of the central Térraba belt. Rio Cañablanca valley shown on Figure 8A (UTM grid 525, 327.5). Left side of drawing is southwest, right side is northeast.

DISCUSSION

Shortening of forearc rocks in the Térraba belt in response to shallow subduction of the Cocos Ridge

Literature review reveals three basic suites of mapped structures produced in island arcs by shallow subduction: (1) convergent structures striking perpendicular to the convergence direction, (2) divergent structures striking parallel to the convergence direction, and (3) gravitational slump structures sliding parallel to the convergence direction (Fig. 2; Table 1).

In the Térraba belt we relate folds and reverse faults of the first category (Fig. 10) to shallow subduction of the Cocos Ridge. A causal relationship is inferred between shallow subduction of the Cocos Ridge and convergent structures in the Térraba belt because the outcrop of the Térraba belt disappears to the east and west of the projected downdip extension of the Cocos Ridge (Fig. 4D). Structural mapping constrained a maximum shortening direction of N30°–34°E for the central and eastern Térraba belt (Fig. 11). This direction of maximum shortening corresponds closely to the N35°E direction of maximum shortening of Corrigan et al. (1990) from Plio-Pleistocene rocks of the outer forearc in the Burica/Osa area to the south and southeast of the Térraba belt. Assuming that the predicted plate convergence direction (N32°E) (DeMets et al., 1990) and the direction of maximum shortening in the forearc are subparallel, we propose that thrusting and tilting in the forearc of westernmost Panama and eastern and central Costa Rica resulted from regional northeast-southwest–oriented maximum compressive stresses exerted by shallow subduction of the Cocos Ridge.

Uplift and arcward tilting of the forearc basin (Térraba belt)

Uplift and progressive arcward tilting of the rocks of the Térraba belt are manifested by two major angular unconformities that developed on the Miocene-Pliocene boundary and in early to middle Pleistocene (Fig. 13). Northeast-dipping deep-marine sedimentary rocks of the Brito, Térraba, and Curré Formations of Eocene to Miocene age are onlapped by less steeply dipping beds of mainly terrestrial deposits of the Pliocene Paso Real Formation and the Quaternary Formation (Fig. 10A). Gently tilted deposits of the Quaternary El General Formation indicate that a significant portion of the uplift and tilting has occurred within the past 20,000 years (Kesel, 1983). Thrusting and folding in the inner forearc is post–middle Miocene, although some of the reverse faults appear to have been reactivated as normal faults as recently as the Pleistocene (Gardner et al., 1992). Normal faulting may reflect topographic collapse of rocks in the Térraba belt following their shortening and uplift along reverse fault planes (Fig. 11).

Timing of Cocos Ridge–related deformation in Costa Rica and westernmost Panama

Corrigan et al. (1990) compared the pattern of Pleistocene deformation of the outer forearc area to the longer-term, late Cenozoic history of the entire southern Middle America arc-trench system. They concluded that during the early Eocene to Pleistocene the inner forearc area of the Térraba belt was subsiding and receiving sediment while the outer forearc block of the Burica, Osa, and Nicoya Peninsulas was uplifting and eroding (Fig. 9). A major hiatus between Eocene and lower Pliocene sedimentary rocks on the Burica and Osa Peninsulas suggests that the outer forearc region was a topographic high during the time of greatest subsidence of the Térraba trough. Paleocurrent measurements from the Térraba belt indicate belt-parallel sediment dispersal, suggestive of the presence of an outer forearc high that restricted trenchward transport of the

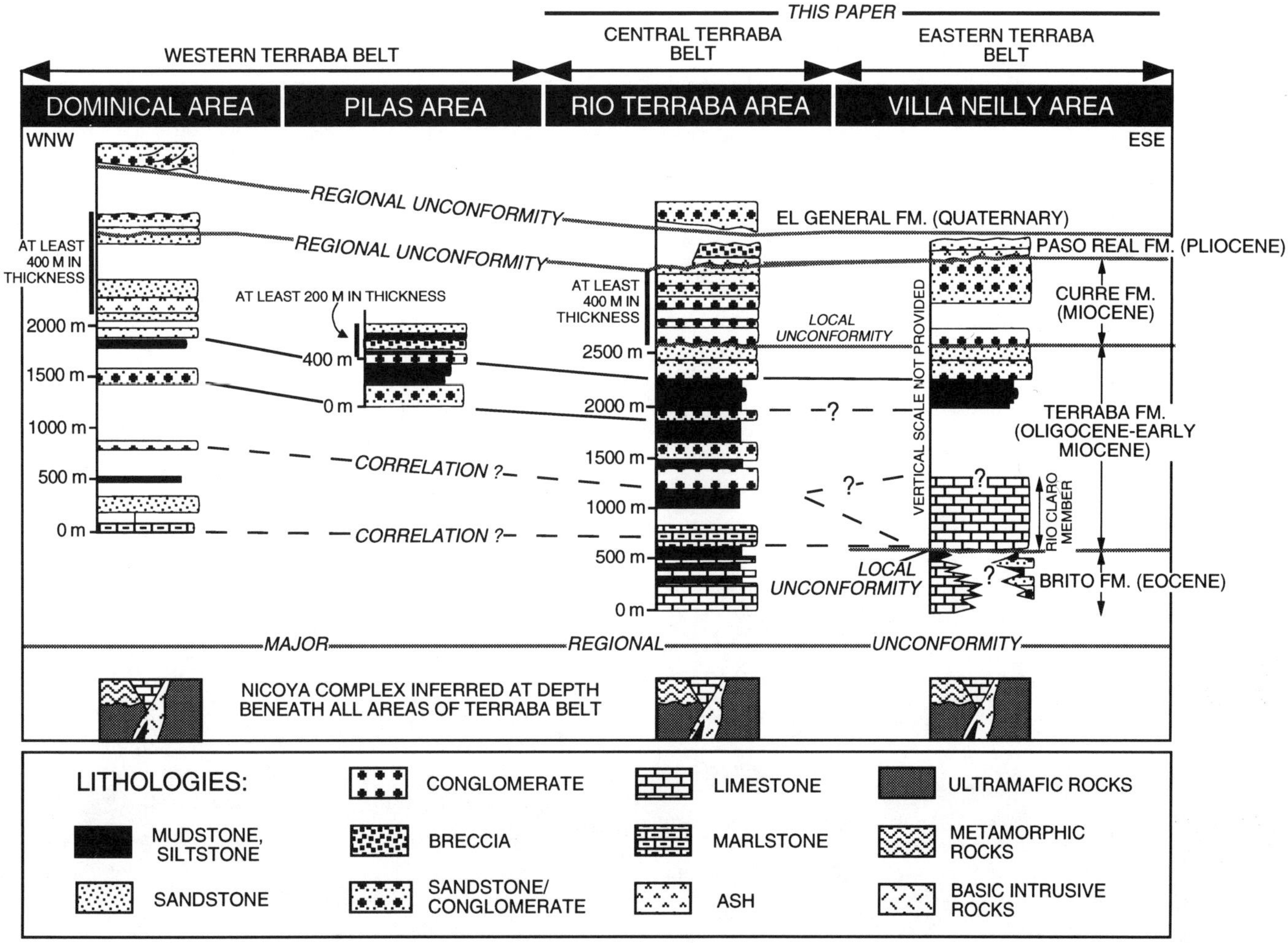

Figure 13. Regional correlation of Eocene to Holocene sedimentary lithologic formations in the western and central (Lowery, 1982; Phillips, 1983) and eastern (Yuan, 1984) Térraba belt.

arc-derived volcaniclastic sediment into the outer forearc area of the Burica, Osa, and Nicoya Peninsulas (Lowery, 1982; Phillips, 1983; Yuan, 1984). The Costa Rica backarc basin (Limón belt) subsided like the Térraba trough throughout Cenozoic time, although it was isolated from the Térraba trough by volcanic highlands in central Costa Rica (Winslow, 1988; Cassell and Sen Gupta, 1989a, b) (Fig. 14).

A reversal in the subsidence pattern of the inner and outer forearc areas of Costa Rica near the Miocene-Pliocene boundary about 5 Ma may reflect either early shallow subduction of the Cocos Ridge or the proximity of an unstable Cocos-Nazca-Caribbean triple junction presently located at the northern end of the Panama fracture zone (Fig. 3). Shallow-marine, fluvial, and lacustrine depositional environments in the Paso Real and El General Formations of the Térraba belt indicate complete filling of the Térraba trough by the late Miocene (Lowery, 1982) (Fig. 14). Rapid, outer arc subsidence of the Burica area began about 5 Ma and continued until the early Pleistocene (Corrigan et al., 1990).

To establish whether the rapid outer forearc subsidence in the area of eastern Costa Rica and westernmost Panama was a regional or local event, we compiled relative sea-level curves from DSDP site 565 of Leg 84 off the Nicoya Peninsula of northwestern Costa Rica and seven DSDP sites from the same leg off of Guatemala. All of these wells indicate a gradual shallowing-upward trend from abyssal or lower bathyal depth zones to middle and upper bathyal zones. Rapid Pleistocene shallowing to middle and upper bathyal depths is observed at sites 565 and 566. Because the paleobathymetry of these wells does not suggest an abrupt shallowing event similar in age or magnitude to that seen in the Burica-Osa area, it appears that the Burica shallowing event is a localized tectonic event that may be related to either early Cocos Ridge shallow subduction or the presence of an unstable triple junction in this sector of

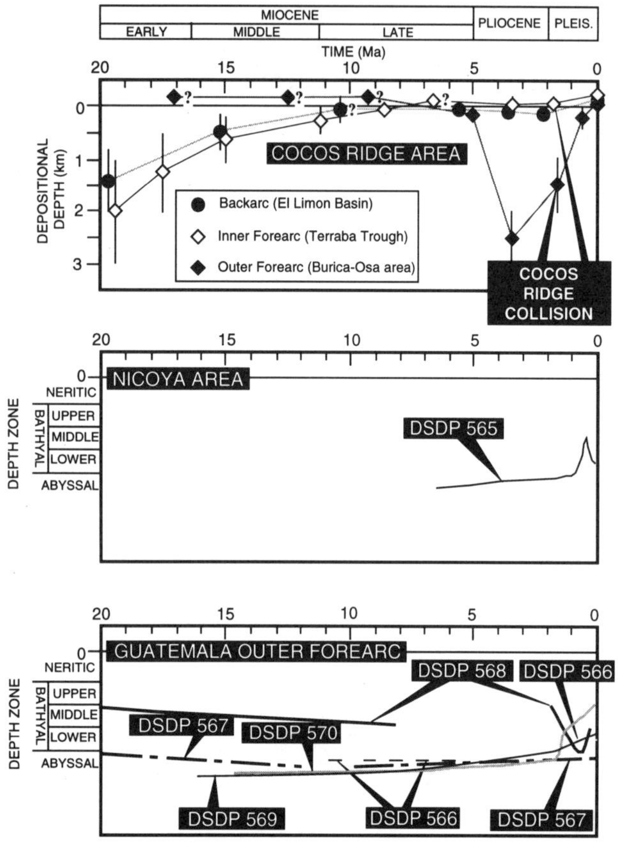

Figure 14. Paleobathymetry from submarine and subaerially exposed outer forearc areas of the Cocos Ridge area (Corrigan et al., 1990), the Nicoya Peninsula of Costa Rica (DSDP site 565; Shipboard Scientific Party, 1982a), and the Guatemala outer forearc (DSDP sites 566, 567, 568, 569, 570; Shipboard Scientific Party, 1985a, b, c, d, e, f). For comparative purposes with data from the Cocos Ridge area, curves from DSDP sites have been rescaled assuming constant depositional rates during individual geologic epochs. Note the general shallowing trends at all sites and the anomalous Plio-Pleistocene deepening and subsequent shallowing in the Cocos Ridge area.

the island arc. If Cocos Ridge shallow subduction is responsible for the abrupt and localized early Pliocene shallowing of the Burica-Osa areas, then the initiation of shallow subduction was well before the 1 Ma age for arrival of the Cocos Ridge at the Middle America Trench that was predicted by Lonsdale and Klitgord (1978) using magnetic anomalies from the eastern Pacific.

Extent of Cocos Ridge–related deformation in Costa Rica and westernmost Panama

Maps showing the volcanic arc gap and topographic uplift (Fig. 4A), shallowing of the Benioff zone (Fig. 4B), forearc indentation and uplift (Fig. 4C), forearc and backarc basin inversion (Fig. 4D), basement relief (Fig. 4E), and regional stress field (Fig. 4F) suggest that shallow subduction of the Cocos Ridge may affect up to 45,000 km^2 of southern Central America. Scale models of continental deformation also support our proposed genetic link between shallow subduction and the above large-scale tectonic features of southern Central America. For example, shallow subduction of bathymetric features such as the Cocos Ridge can be approximated experimentally by models of continental indentation (Cobbold and Davy, 1988; Davy and Cobbold, 1988). Assuming that the bathymetric high of the Cocos Ridge behaves as an undeformed, rigid indenter, we can predict subhorizontal compressive stresses and shear stresses in front of the indenter (Fig.15). This predicted stress field can be compared to the stress field observed from earthquakes, aligned Quaternary volcanoes, and fault studies in Costa Rica (Fig. 4F).

Theoretically, shear stress value increases from zero at the long axis of the indenter to its maximum at the indenter edges. The maximum values are numerically equal unless the stress vectors have opposite orientations. In the case of the Cocos Ridge, shear stresses concentrated at the edges of the ridge would result in the development of left- and right-lateral fault zones as observed at the sites of the East Nicoya and Medial fault zones (Fig. 4C). Such fault zones would act to accommodate most of the margin indentation by laterally confining the zone of thrusting along the Térraba belt and Ballena-Celmira fault zone.

Subhorizontally oriented compressive stress in front of the Cocos Ridge causes shortening of the forearc, demonstrated by folding and reverse faulting (Fig. 15). The deformation is most intense in front of the colliding ridge and gradually decays with distance.

SUMMARY AND CONCLUSIONS

1. Based on compiled data we identify six major tectonic effects produced by shallow subduction of the Cocos Ridge. These effects include: (1) a 175-km-wide volcanic gap and anomalous topographic uplift are found in the normally closely spaced Central American stratovolcanoes; (2) the dip of the subducted Cocos plate decreases significantly beneath central Costa Rica; (3) the upper Miocene to Quaternary shelf-slope forearc section overlying acoustic basement is locally uplifted and indented or displaced arcward in the contact area of the Cocos Ridge; (4) the forearc (Térraba) and backarc (Limón) basins of Costa Rica are structurally inverted and topographically uplifted in the contact area of the Cocos Ridge; (5) relief of acoustic basement in the western Colombian basin and in the area of the Osa, Burica, and Nicoya Peninsulas is arched in a direction parallel to the direction of plate convergence (032) and overlies the downdip projection of the Cocos Ridge; (6) the regional stress field of Costa Rica that was inferred from P and T axes of earthquake focal mechanisms and the alignment of dikes and cinder cones of Quaternary stratovolcanoes describe a radial pattern about the subducting Cocos Ridge, particularly in the area of eastern Costa Rica and westernmost Panama.

2. Previous work has defined five tectonic elements in a

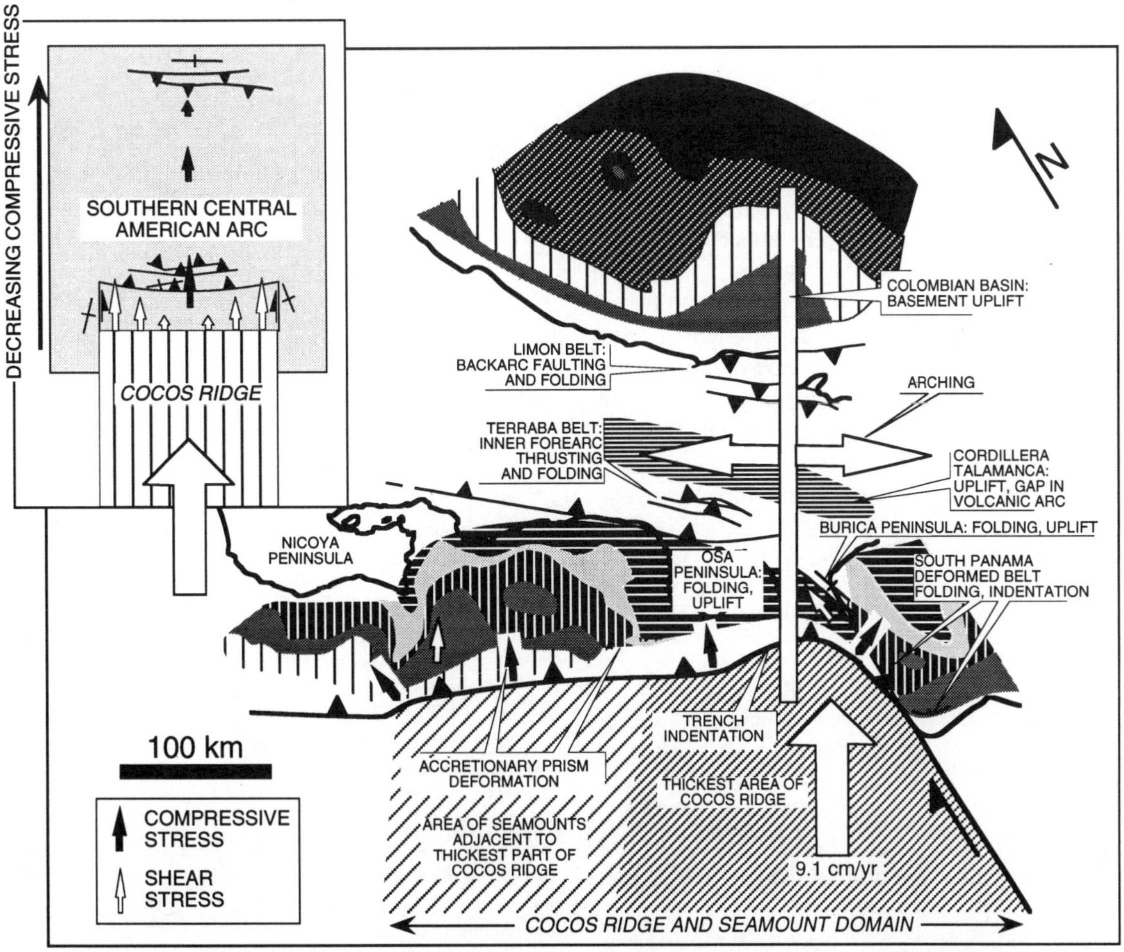

Figure 15. Indentor model for the major Neogene structures observed in Cocos Ridge subduction area. Shading indicates depth of basement (see Fig. 4E for detailed map of basement structure and surface structures). Inset shows predicted shear (open arrows) and compressive (black arrows) stresses associated with the Cocos Ridge–induced indentation of the forearc (modified from Davey and Cobbold, 1988).

transect across Costa Rica: (1) the Cocos Ridge on the subducting Cocos plate, (2) the outer Middle America forearc, (3) the inner Middle America forearc, (4) the Middle America volcanic arc, and (5) the Middle America backarc basin.

3. A northeast-trending basement high within these five tectonic elements is defined by depth to acoustic basement in offshore areas in the Caribbean Sea and Pacific margin and is approximately aligned along the Cocos–Caribbean plate convergence vector. Ages of rock units of onshore tectonic elements form a roughly domal pattern with the oldest rocks generally in the center of the dome.

4. The top of the subducted Cocos Ridge can be traced on a depth-converted Shell multichannel seismic line for a horizontal distance of 55 km and a depth of 10 km beneath the Costa Rican forearc. The average dip of the top of the Cocos Ridge based on this line is 3°.

5. Structures formed in forearc basin sedimentary and volcanic rocks of the Térraba belt above the subducted Cocos Ridge include five major reverse faults that consistently place older lithologic units over younger lithologic units. One of these faults, the Ballena-Celmira fault zone, forms a prominent linear contact between Quaternary alluvium of the Pacific coastal plain and the Térraba belt. The surface traces of reverse faults are generally straight or curving and can be traced for approximately 75 km along strike.

6. Poles to 292 bedding planes of Eocene to Pleistocene sedimentary rocks of the Térraba belt indicate northeastward tilting by an average of 30°. The similarity in dip values for Eocene to Miocene rocks of the Térraba belt suggests a single deformational event postdating the deposition of Miocene rocks. A distinct shallowing of dip between Miocene and Pliocene rocks reflects the presence of a major angular unconformity between the upper part of the marine Térraba Formation and the lower part of the alluvial Paso Real Formation.

7. Minor reverse faults in the central Térraba belt yield an average strike of 154° and an average dip of 53°E and gen-

erally show up-to-the- northwest movement consistent with larger-scale reverse movement on major faults. Fault striations indicate an average pitch of 66°, indicative of dip-slip reverse movement.

8. Reverse offsets of sills and dikes dated from 4.5 to 14.8 Ma in the central Térraba belt and the presence of an angular unconformity between Miocene and Pliocene sedimentary rocks suggest thrusting was active by the Miocene-Pliocene boundary.

9. Bedding plane and fault data constrain a maximum shortening direction of N30–34°E for the central and eastern Térraba belt. This direction of maximum shortening corresponds closely to the N35°E direction of maximum shortening of Corrigan et al. (1990) from Plio-Pleistocene rocks of the outer forearc in the Burica/Osa area to the south and southeast of the Térraba belt. Assuming that the predicted plate convergence direction (N32°E) and the direction of maximum shortening in the forearc are subparallel, thrusting and tilting in the forearc of westernmost Panama and eastern and central Costa Rica is interpreted as the result of regional northeast-southwest–oriented maximum compressive stresses exerted by shallow subduction of the Cocos Ridge.

10. Rapid marine transgression of the outer forearc of Costa Rica near the Miocene-Pliocene boundary about 5 Ma may be controlled by either eustatic variations or subsidence, or both. If the transgression is not due solely to eustacy, it may reflect either early shallow subduction of the Cocos Ridge or the proximity of an unstable Cocos-Nazca-Caribbean triple junction presently located at the northern end of the Panama fracture zone.

11. Shallow subduction of the Cocos Ridge can be approximated experimentally by models of continental indentation. Assuming that the Cocos Ridge behaves as an undeformed, rigid indenter, indentor models predict subhorizontal compressive stresses responsible for forearc shortening, concentrating in front of the indentor. Shear stresses oriented parallel to the axis of the indentor cause indentation of the forearc and the accretionary prism along trench-perpendicular strike-slip faults. This predicted stress field is similar to the stress field observed from earthquakes, aligned Quaternary volcanoes, and fault studies in Costa Rica.

ACKNOWLEDGMENTS

We thank A. J. Bol of Shell International for providing Shell line 1600 for use in this study. We thank Richard Buffler, Sergio Mora, Alan Lopez, and Jorge Obando for discussions and reviews of early drafts of this chapter and the Escuela Centroamericana de Geologia for logistic support. Special thanks to Eduardo Vega for assistance in the field. Reviews by J. Corrigan, J. de Boer, and K. McIntosh significantly improved this chapter. Funding for this project was provided by the Petroleum Research Fund of the American Chemical Society (grant number 20676-AC2-2). UTIG contribution no. 1077.

REFERENCES CITED

Adamek, S., 1986, Earthquake studies in the Panama–Costa Rica region [M.A. thesis]: Austin, University of Texas at Austin, 211 p.

Adamek, S., and Tajima, F., 1987, Seismic rupture associated with the subduction of the Cocos ridge: Tectonics, v. 6, p. 757–774.

Alvarado, R., 1979, Estudio geologico de los rios Cañablancal, Retinto y Alrededores, Canton de Osa, Palmar, Provincia Puntarenas [undergraduate senior thesis]: San José, University of Costa Rica School of Geology, 41 p.

Bentley, L., 1974, Crustal structure of the Carnegie ridge, Panama basin, and Cocos ridge [M.S. thesis]: Honolulu, University of Hawaii, 49 p.

Berrangé, J., and Thorpe, R., 1988, The geology, geochemistry, and emplacement of the Cretaceous-Tertiary ophiolitic Nicoya complex of the Osa Peninsula, southern Costa Rica: Tectonophysics, v. 147, p. 193–200.

Bourgois, J., Azéma, J., Baumgartner, P., Tournon, J., Desmet, A., and Auboin, J., 1984, The geologic history of the Caribbean-Cocos plate boundary with special reference to the Nicoya ophiolite complex (Costa Rica) and DSDP results (Legs 67 and 84 off Guatemala): A synthesis: Tectonophysics, v. 108, p. 1–32.

Bowland, C. L., 1984, Seismic stratigraphy and structure of the western Colombian basin, Caribbean sea [M.A. thesis]: Austin, University of Texas at Austin, 238 p.

Bowland, C. L., and Rosencrantz, E., 1988, Upper crustal structure of the western Colombian basin, Caribbean Sea: Geological Society of America Bulletin, v. 100, p. 534–546.

Burbach, G. V., Frohlich, C., Pennington, W. D., and Matsumoto, T., 1984, Seismicity and tectonics of the subducted Cocos plate: Journal of Geophysical Research, v. 89, p. 7719–7735.

Carr, M. J., and Stoiber, R. E., 1990, Volcanism, *in* Dengo, G., and Case, J. E., eds., The Caribbean Region: Boulder, Colorado, Geological Society of America, The Geology of North America, v. H, p. 375–391.

Cassell, D. T., and Sen Gupta, B. K., 1989a, Pliocene foraminifera and environments, Limón basin, Costa Rica: Journal of Paleontology, v. 63, p. 146–157.

Cassell, D. T., and Sen Gupta, B. K., 1989b, Foraminiferal stratigraphy and paleoenvironments of the Tertiary Uscari formation, Limón basin, Costa Rica: Journal of Foraminiferal Research, v. 19, p. 52–71.

Chung, W., and Kanamori, H., 1978, A mechanical model for plate deformation associated with aseismic ridge subduction in the New Hebrides arc: Tectonophysics, v. 50, p. 29–40.

Cobbold, P. R., and Davy, P., 1988, Indentation tectonics in nature and experiment. 2: Central Asia: Bulletin of the Geological Institute of Uppsala (Sweden), v. 14, p. 143–162.

Collot, J.-Y., and Fisher, M. A., 1991, The collision zone between the North d'Entrecasteaux ridge and the New Hebrides island arc. 1: Seabeam morphology and shallow structure: Journal of Geophysical Research, v. 96, p. 4432–4479.

Collot, J.-Y., Daniel, J. and Burne, R. V., 1985, Recent tectonics associated with the subduction/collision of the d'Entrecasteaux zone in the central New Hebrides arc: Tectonophysics, v. 112, p. 325–356.

Corrigan, J. D., 1986, Geology of the Burica peninsula, Panama–Costa Rica: Neotectonic implications for the southern Middle America convergent margin [M.A. thesis]: Austin, University of Texas at Austin, 152 p.

Corrigan, J. D., Mann, P., and Ingle, J. C., Jr., 1990, Forearc response to subduction of the Cocos Ridge, Panama–Costa Rica: Geological Society of America Bulletin, v. 102, p. 628–652.

Crowe, J. C., and Buffler, R. T., 1983, Regional seismic reflection profiles across the Middle America Trench and convergent margin of Costa Rica, *in* Bally, A. W., ed., Seismic expansion of structural styles: AAPG Studies in Geology 15, v. 3, p. 3.4.2.-147–3.4.2-162.

Crowe, J. C., and Buffler, R. T., 1985, Top of acoustic basement and thickness of sediments, Costa Rica margin, *in* Land, J. W., and Buffler, R. T., eds., Middle America trench of western central America: Atlas 7, Ocean Margin Drilling Program, Regional Atlas Series: Woods Hole, Massachusetts, Marine Margin Sciences International, sheet 18.

Davy, P., and Cobbold, P. R., 1988, Indentation tectonics in nature and experiment. 1: Experiments scaled for gravity: Bulletin of the Geological Institute of Uppsala (Sweden), v. 14, p. 129–141.

de Boer, J., Defant, M., Stewart, R., Restrepo, J., Clark, L., and Ramirez, A., 1988, Quaternary calc-alkaline volcanism in western Panama: Regional variation and implications for plate tectonic framework: Journal of South American Earth Sciences, v. 1, p. 275–293.

Defant, M., and eight others, 1991, Andesite and dacite genesis via contrasting processes: The geology and geochemistry of the El Valle volcano, Panama: Contributions to Mineralogy and Petrology, v. 106, p. 309–324.

DeMets, C., Gordon, R. G., Argus, D. F., and Stein, S., 1990, Current plate motions: Geophysical Journal International, v. 101, p. 425–478.

Dengo, G., 1962, Tectonic-igneous sequence in Costa Rica, *in* Engel, A.E.J., James, H. L., and Leonard, B. F., eds., Petrologic studies: A volume in honor of A. F. Buddington: Boulder, Colorado, Geological Society of America, p. 133–161.

Dupont, J., and Herzer, R. H., 1985, Effect of subduction of the Louisville Ridge on the structure and morphology of the Tonga arc, *in* Scholl, D. W., and Vallier, T. L., eds., Geology and offshore resources of Pacific islands arcs—Tonga region: Houston, Texas, Circum-Pacific Council for Energy and Mineral Resources Earth Science Series 2, p. 323–332.

Fisher, M. A., 1986, Tectonic process at the collision of the D'Entrecasteaux zone and the New Hebrides island arc: Journal of Geophysical Research, v. 91, p. 10470–10486.

Fryer, P., and Smoot, N. C., 1985, Processes of seamount subduction in the Mariana and Izu-Bonin trenches: Marine Geology, v. 64, p. 77–90.

Gardner, T. W., Pinter, N. M., Bullard, T., and Slingerland, R., 1992, Quaternary uplift astride the aseismic Cocos Ridge, Pacific coast, Costa Rica: Geological Society of America Bulletin, v. 104, p. 219–232.

Green, H. G., and Wong, F. L., 1989, Ridge collisions along the plate margins of South America compared with those in the southwest Pacific, *in* Ericksen, G. E., Cañas Pinochet, M. T., and Reinemund, J. A., eds., Geology of the Andes and its relation to hydrocarbon and mineral resources: Houston, Texas, Circum-Pacific Council for Energy and Mineral Resources Earth Science Series, v. 11, p. 39–57.

Guendel, F. D., 1986, Seismotectonics of Costa Rica: An analytical view of the southern terminus of the Middle America trench [Ph.D. thesis]: Santa Cruz, University of California, 156 p.

Gursky, H. J., Schmidt-Effing, R., Strebin, M., and Wildberg, H., 1982, The ophiolite sequence in northwestern Costa Rica (Nicoya complex): Outlines of stratigraphical, geochemical, sedimentological and tectonical data: Quinto Congreso Latinoamericano de Geologia, v. 3, p. 607–619.

Henningsen, D. W., 1965, Stratigraphy and paleogeography of upper Cretaceous and Tertiary sediments in southern Costa Rica, *in* Saunders, John B., ed., Fourth Caribbean Geological Conference, Port of Spain, Trinidad, p. 353–356.

Hey, R., 1977, Tectonic evolution of the Cocos-Nazca spreading center: Geological Society of America Bulletin, v. 88, p. 1404–1420.

Kesel, R. H., 1983, Quaternary history of the Rio General valley, Costa Rica: National Geographic Society Research Reports, v. 15, p. 339–358.

Lallemand, S., and Le Pichon, X., 1987, Coulomb wedge model applied to the subduction of seamounts in the Japan trench: Geology, v. 15, p. 1065–1069.

Lew, L. R., 1983, The geology of the Osa Peninsula, Costa Rica: Observations and speculations about the evolution of part of the outer arc of the southern Central American orogen [M.A. thesis]: College Park, Pennsylvania State University, 128 p.

Lonsdale, P., and Klitgord, K. D., 1978, Structure and tectonic history of the eastern Panama basin: Geological Society of America Bulletin, v. 89, p. 981–999.

Lowery, B. J., 1982, Sedimentology and tectonic implications of the Middle to Upper Miocene Curré formation [M.S. thesis]: Baton Rouge, Louisiana State University, 100 p.

Lundberg, N. S., 1982, Evolution of the forearc landward of the Middle America Trench, Nicoya peninsula, Costa Rica and southern Mexico [Ph.D. thesis]: Santa Cruz, University of California, 256 p.

Lundberg, N. S., 1983, Development of forearcs of intraoceanic subduction zones: Tectonics, v. 2, p. 51–61.

MacKay, M., and Moore, G., 1990, Variation in deformation of the South Panama accretionary prism: Response to oblique subduction and trench sediment variation: Tectonics, v. 9, p. 683–698.

Mann, P., Schubert, C., and Burke, K., 1990, Review of Caribbean neotectonics, *in* Dengo, G., and Case, J. E., eds., The Caribbean Region: Geological Society of America, The Geology of North America, v. H, p. 307–338.

Masson, D. G., and six others, 1990, Subduction of seamounts at the Java trench: A view with long-range sidescan sonar: Tectonophysics, v. 185, p. 51–65.

McCann, W. R., and Habermann, R. E., 1989, Morphologic and geologic effects of the subduction of bathymetric highs: Pure and Applied Geophysics, v. 129, p. 41–69.

McGeary, S., Nur, A., and Ben-Avraham, Z., 1985, Spatial gaps in arc volcanism: The effect of collision or subduction of oceanic plateaus: Tectonophysics, v. 119, p. 195–221.

Montero, W., and Morales, L., 1990, Deformacion y esfuerzos neotectonicos en Costa Rica: Revista Geologica de America Central, v. 11, p. 69–87.

Mora, S., 1979, Estudio geológico de una parte de la región sureste del Valle del General, Provincia Puntarenas, Costa Rica [undergraduate senior thesis]: San José, University of Costa Rica, 185 p.

Moretti, I., and Ngokwey, D., 1985, Aseismic ridge subduction and vertical motion of the overriding plate, *in* Proceedings, Symposium Geodynamique des Caribes, Paris, France, p. 245–254.

Nakamura, K., 1977, Volcanoes as possible indicators of tectonic stress orientations—Principle and proposal: Journal of Volcanology and Geothermal Research, v. 2, p. 1–16.

Obando, J., 1986, Sedimentología y tectónica del Cretácico y Paleógeno de la region de Golfito, Península de Burica y Península de Osa, Provincia de Puntarenas, Costa Rica [undergraduate senior thesis]: San José, University of Costa Rica, 211 p.

Okaya, D. A., and Ben-Avraham, Z., 1987, Structure of the continental margin of southwestern Panama: Geological Society of America Bulletin, v. 99, p. 792–802.

Petit, J., 1987, Criteria for the sense of movement on fault surfaces in brittle rocks: Journal of Structural Geology, v. 9, p. 597–608.

Phillips, P. J., 1983, Stratigraphy, sedimentology, and petrologic evolution of Tertiary sediments in southwestern Costa Rica [M.S. thesis]: Baton Rouge, Louisiana State University, 153 p.

Rivier, F., 1985, Sección geologíca del Pacifico al Atlantico á través de Costa Rica: Revista Geologica de America Central (San José, Costa Rica), v. 2, p. 23–32.

Seely, D., 1979, The evolution of structural highs bordering major forearc basins, *in* Watkins, J., et al., eds., Geological and geophysical investigation of continental margins: American Association of Petroleum Geologists Memoir, v. 29, p. 245–260.

Shipboard Scientific Party, 1985a, Site 565, Initial reports of the Deep Sea Drilling Project, v. 84, p. 21–78.

Shipboard Scientific Party, 1985b, Site 566, Initial reports of the Deep Sea Drilling Project, v. 84, p. 79–110.

Shipboard Scientific Party, 1985c, Site 567, Initial reports of the Deep Sea Drilling Project, v. 84, p. 111–166.

Shipboard Scientific Party, 1985d, Site 568, Initial reports of the Deep Sea Drilling Project, v. 84, p. 167–232.

Shipboard Scientific Party, 1985e, Site 569, Initial reports of the Deep Sea Drilling Project, v. 84, p. 233–282.

Shipboard Scientific Party, 1985f, Site 570, Initial reports of the Deep Sea Drilling Project, v. 84, p. 283–336.

Shipley, T., and Moore, G., 1986, Sediment accretion, subduction, and dewatering at the base of the slope off Costa Rica: A seismic reflection view of the decollement: Journal of Geophysical Research, v. 91, p. 2019–2028.

von Huene, R., and 17 others, 1980, The Deep Sea Drilling Project in the mid-America trench transect off Guatemala: Geological Society of America Bulletin, v. 91, p. 412–432

von Huene, R., and 13 others, 1985, Site 565, *in* von Huene, R., Aubouin, J., and others, eds., Initial reports of the Deep Sea Drilling Project, vol. 84: Washington, D.C., U.S. Government Printing Office, p. 21–77.

von Huene, R., Bourgois, J., Miller, J., and Pautot, G., 1989, A large tsunamogenic landslide and debris flow deposit along the Peru trench: Journal of Geophysical Research, v. 94, p. 1703–1714.

Wells, S., and seven others, 1988, Regional variations in tectonic geomorphology along a segmented convergent plate boundary, Pacific coast of Costa Rica: Geomorphology, v. 1, p. 239–265.

Weyl, R., 1980, Geology of Central America: Berlin, Gebruder Borntraeger, 371 p.

Winslow, M., 1988, Tectonic evolution of an actively deforming backarc basin: The Limón basin of Costa Rica: Transactions of the 11th Caribbean Geological Conference, Barbados, p. 19:1–19:17.

Yuan, P., 1984, Stratigraphy, sedimentology and geologic evolution of the eastern Térraba trough, southeastern Costa Rica [Ph.D. thesis]: Baton Rouge, Louisiana State University, 110 p.

Zoback, M. L., and Zoback, M., 1980, State of stress in the conterminous United States: Journal of Geophysical Research, v. 85, p. 6113–6156.

MANUSCRIPT ACCEPTED BY THE SOCIETY FEBRUARY 7, 1994

Printed in U.S.A.

Geological Society of America
Special Paper 295
1995

Timing and rates of emergence of the Limón and Bocas del Toro basins: Caribbean effects of Cocos Ridge subduction?

Laurel S. Collins
Museum of Paleontology, University of Michigan, Ann Arbor, Michigan 48109-1079
Anthony G. Coates, Jeremy B.C. Jackson, and Jorge A. Obando
Smithsonian Tropical Research Institute, Unit 0948, APO AA 34002-0948

ABSTRACT

Paleobathymetries of consecutively dated stratigraphic sections within the Limón belt, Costa Rica, and Bocas del Toro belt, Panama, are used to calculate rates of emergence for the Caribbean coast of southern Central America over the past 6 m.y. We present estimates of paleoenvironment, paleobathymetry, and rates of emergence for a series of sedimentary units precisely dated as 6.5 to 1.6 Ma. Paleodepths and paleoenvironments are interpreted primarily from preferred habitats of common benthic foraminifera, combined with lithofacies and neotectonics.

The Limón and Bocas del Toro belts lie on either side of the northeast-southwest–trending axis of a structural arch that formed from subduction of the Cocos Ridge beneath the Costa Rica–Panama arc complex. Uplift caused by Cocos Ridge subduction began on the southern margin of the arc complex at Burica Peninsula ~3.6 Ma, and relatively little uplift occurred after ~1.6 Ma; net rates of emergence from 3.6 to 1.6 Ma were about 1,000 m/m.y. and near 0 m/m.y. thereafter. Effects of Cocos Ridge subduction were not manifest on the northern margin of the arc complex in the Limón and Bocas del Toro basins until after 1.6 Ma. Although sedimentation patterns and coeval, isobathyal benthic foraminiferal faunas were distinctly different in these basins, indicating at least partial isolation, their histories of uplift are similar. Average emergence rates from ~6 to 3.5 Ma for the Bocas del Toro basin varied from 24 to 90 m/m.y., which could be accounted for by sea-level fall, and from 3.5 to 1.6 Ma there was no measurable emergence in either basin. However, after 1.6 Ma, rates of emergence were as high as 156 m/m.y. on the Caribbean margin, which we attribute to the penetration of the subducting Cocos Ridge. Thus, uplift of the Caribbean margin in the vicinity of the Cocos Ridge northeast extension began 2 m.y. later and was an order of magnitude slower than uplift of the Pacific margin. Timing for the arrival of the Cocos Ridge at the Caribbean margin is supported by its current subduction rate: If the ridge had traveled northeast from the southern Burica Peninsula at 9 cm/yr, in 2 m.y. it would have crossed 180 km, the approximate distance to the Limón basin.

INTRODUCTION

The southern Central American isthmus, from Costa Rica to the Atrato Valley of northwestern Colombia, forms the thrust-bounded, independent Panama microplate (Bowin, 1976; Case in Mattson, 1984; Adamek et al., 1988; Mackay and Moore, 1990) at the trailing margin of the Caribbean plate (Fig. 1). Along the southwestern microplate margin lies the convergent Middle America Trench, intersected near the Burica Peninsula by the Panama Fracture Zone (PFZ) transform fault. The southeastern margin of the microplate, along the northern edge of the Nazca plate, is bounded by the South

Collins, L. S., Coates, A. G., Jackson, J.B.C., and Obando, J. A., 1995, Timing and rates of emergence of the Limón and Bocas del Toro basins: Caribbean effects of Cocos Ridge subduction? *in* Mann, P., ed., Geologic and Tectonic Development of the Caribbean Plate Boundary in Southern Central America: Boulder, Colorado, Geological Society of America Special Paper 295.

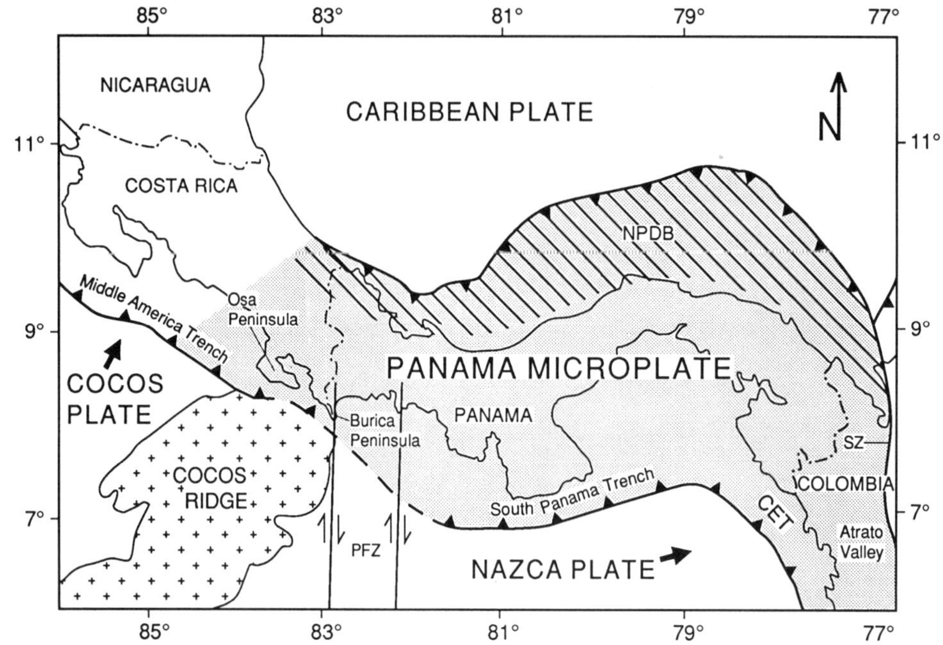

Figure 1. Tectonic boundaries of the Panama microplate (shaded). Arrows show directions of plate convergence. CET = Colombia–Ecuador Trench, NPDB = North Panama deformed belt (diagonal lines), PFZ = Panama Fracture Zone, SZ = Panama–South America Suture Zone formed by the Uramita Fault Zone. Adapted from Case et al. (1984), Vergara Muñoz (1988), and Mackay and Moore (1990).

Panama Trench and thrust belt (Mackay and Moore, 1990; Silver et al., 1990), which curves sharply south to become the Colombia-Ecuador Trench (Pennington, 1981). The microplate is bounded to the north by the North Panama deformed belt (NPDB), an accretionary wedge of sediments overthrusting the Caribbean plate (Dengo, 1973; Bowin, 1976; Case and Holcombe, 1980; Lu and McMillen, 1982; Silver et al., 1990; Silver et al., this volume; and Tajima, this volume). The NPDB thrust curves southward through the Atrato Valley to form the Uramita Fault, the suture with the South American plate (Dento, 1983; Duque-Caro, 1990).

Two major events affecting the evolution of the Panama microplate are the shallow subduction of the Cocos Ridge in the region of the Osa and Burica Peninsulas and collision of the microplate with South America. The Cocos Ridge, an aseismic, anomalously thickened area of oceanic crust, moved northeastward with the Cocos plate to encounter the Middle America Trench (Hey, 1977), where it has been partially subducted beneath the volcanic arc system of Costa Rica and western Panama (Lonsdale and Klitgord, 1978; Gardner et al., 1987, 1992). The primary tectonic effect of Cocos Ridge subduction (Corrigan et al., 1990; Kolarsky et al., this volume) has been major uplift of the basement and sedimentary basins along a zone that follows the projected axis of the Cocos Ridge beneath the Costa Rica–Panama arc complex (Fig. 2, arrow). Collision of the Panama microplate with South America caused the oroclinal bending of Panama (Wadge and Burke, 1983) and tectonic escape of the northern microplate margin, expressed as convergence between the microplate and Caribbean plate (Adamek et al., 1988), strike-slip faulting in northeastern and southwestern Panama (Mann and Corrigan, 1990), and thrust faulting within the NPDB (Dengo, 1973; Bowin, 1976; Lu and McMillen, 1982). The NPDB thrusts extend westward as far as the backarc region of the Limón belt (Fig. 2; Plafker and Ward, 1992; Silver et al., 1990, this volume; Tajima, this volume).

In this chapter we focus on differences in timing and rates of emergence of the Caribbean versus Pacific margins of the Panama microplate in the vicinity of the Cocos Ridge and its northeastward extension. We have calculated rates of emergence for the Caribbean basins of Limón and Bocas del Toro on the basis of chronologic, paleobathymetric changes as interpreted from assemblages of benthic foraminifera. Emergence rates for the Burica Peninsula on the Pacific margin are based on new,well-constrained biochronologic ages (Coates et al., 1992) and previously documented paleobathymetries (Corrigan et al., 1990). We suggest that the history of Caribbean and Pacific uplift is explained most effectively by subduction

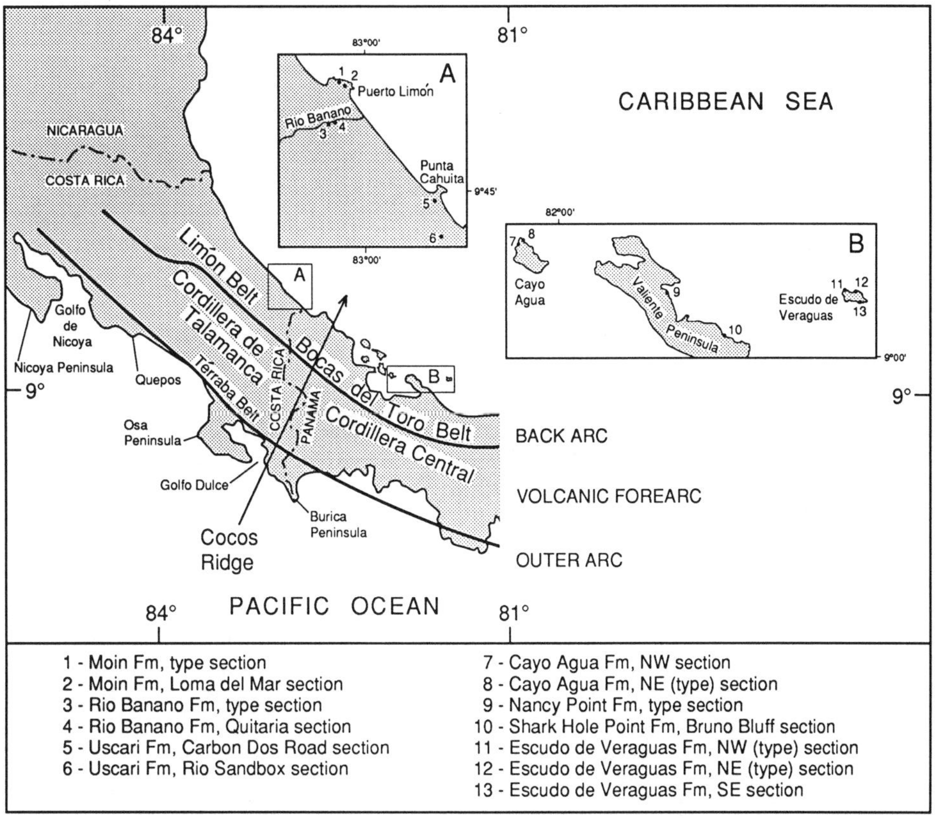

Figure 2. Structural and tectonic units of the southern Central American isthmus (shaded). Arrow across isthmus is convergence vector of subducting Cocos Ridge (Kolarsky et al., this volume). Adapted from Dengo (1962) and Escalante (1990). Insets A and B: locality numbers of geologic sections (listed at bottom) in the Limón (A) and Bocas del Toro (B) belts.

of the buoyant Cocos Ridge, which began about 3.6 Ma, or 2 to 3 m.y. earlier than proposed by Lonsdale and Klitgord (1978), Corrigan et al. (1990), and Gardner et al. (1992).

RATES AND TIMING OF EMERGENCE, CARIBBEAN MARGIN

Use of benthic foraminifera

This study uses assemblages of benthic foraminifera from the Limón and Bocas del Toro Groups of Caribbean Panama and Costa Rica (Coates et al., 1992) to evaluate sedimentary relationships between the Bocas del Toro and Limón basins (Fig. 2), and to calculate the timing of latest Miocene to Quaternary emergence in this region. Benthic foraminiferal species are bathymetrically zoned (Natland, 1933; Bandy, 1953). Because about 90% of the common species found in these deposits are still extant in Caribbean to Gulf of Mexico waters, we can apply their modern water-depth ranges, combined with lithofacies, to interpretations of marine paleobathymetries and paleoenvironments. Relatively continuous, dated, stratigraphic sections are used to calculate rates of change in water depth, or emergence. Geologic sections were biochronologically dated with planktic foraminifera and calcareous nannofossils (Table 1), using the geologic time scale of Berggren et al. (1985).

Stratigraphic sections

Late Miocene to latest Pliocene sediments of the Bocas del Toro and Limón belts were deposited in back-arc basins, north of the volcanic arc that forms the Cordillera Central in Panama and the Cordillera de Talamanca in Costa Rica (Fig. 2). Stratigraphic sequences (Fig. 3) that are relatively continuous (Coates et al., 1992) occur in Bocas del Toro Province, Panama, on the Valiente Peninsula, and the islands of Cayo Agua and Escudo de Veraguas (Table 1; Fig. 2, inset B). Collins (1993; Table 2) described the foraminiferal assemblages and paleoenvironments from these sections (localities 7 through 13, Fig. 2) of the Bocas del Toro Group.

TABLE 1. BIOCHRONOLOGIC DATES FOR GEOLOGIC SECTIONS*

No.†	Stratigraphic Formation	Section	Dates (Ma)
1	Moín	Type	~3.0 to 1.6–1.9§
2	Moín	Loma del Mar	1.7–1.9**
3	Rio Banano	Type	2.4–2.5
4	Rio Banano	Quitaria	3.5–3.6
5	Uscari	Carbon Dos	5 – 6
6	Uscari	Rio Sandbox	5 – 6
7	Cayo Agua	NW	3.5–3.6
8	Cayo Agua	NE (type)	4.6–5.0
9	Nancy Point	Type	5.6–6.5
10	Shark Hole Point	Bruno Bluff	3.5–3.6
11	Escudo de Veraguas	NW (type)	1.8–1.9
12	Escudo de Veraguas	NE (type)	3.5–3.6
13	Escudo de Veraguas	SE	3.5–3.6

*From Coates et al, 1992, and Collins, 1993, except as noted.
†Section numbers of Figure 2.
§1.6–1.9 based on cooccurrence of planktic foraminifera *Globorotalia truncatulinoides* (1.9 Ma-R) and *Globigerinoides fistulosus* (3.0–1.6 Ma).
**Based on presence of *G. truncatulinoides* (1.9 Ma-R) and absence of large nannofossil *Gephyrocapsa* spp. (1.7 Ma-R).

Costa Rican localities (Table 1; Fig. 2, inset A) in the Limón belt (Castillo M., 1984; Sprechmann, 1984; Winslow et al., 1988; Coates et al., 1992) include relatively continuous stratigraphic sequences (Fig. 2; localities 1, 3, 4) used to determine chronologic changes in water depth as well as sections with single dates (localities 2, 5, 6) that serve to distinguish facies changes in coeval deposits. Benthic foraminifera from type sections of the Moín Formation (locality 1), Rio Banano Formation (locality 3), and Uscari Formation are well described by Taylor (1975), Cassell (1986), and Cassell and Sen Gupta (1989a, b). In this chapter, we interpret paleoenvironments of additional facies of these units (localities 2, 4 through 6). Locality 2, the Loma del Mar section of the Moín Formation, was exposed at a construction site excavation within the city of Limón and consisted of a sequence of silty clays to sands, overlain by coral patch reefs in a clastic matrix. Sediment samples of Loma del Mar stratigraphically span 6 m, from above to below the patch reefs. Locality 4, the Quitaria section of the Rio Banano Formation, is an exposure on the south bank of the Banano River, ~100 m from a Dole banana-packing plant near the town of Quitaria, 2 km west of the type Rio Banano locality at Bomba. Samples from the Quitaria site stratigraphically span 12 m.

Two sections of the Uscari Formation (localities 5 and 6, Table 1; Fig. 2, inset A) were designated "post-Uscari beds" by Coates et al. (1992) because they are younger than previously described Uscari sections (Cassell and Sen Gupta, 1989b). We now include them in the Uscari Formation because they were mapped as such by Castillo (1984) and their dark shale lithology corresponds to that of the Uscari. In this general area south of Limón (Fig. 2), deposition of the Uscari continued from late Oligocene (Cassell and Sen Gupta, 1989b) to latest Miocene time, after which the younger Rio Banano Formation was thinly deposited to absent. West of Limón, the Uscari was replaced as early as late Miocene by a thick (maximum 1,500 m), deltaic wedge of the Rio Banano Formation (G. Gonzalez, personal communication, 1992; Bowland, 1993), that in its upper part is coeval with the deeper-water Moín type section (locality 1). The late Miocene Uscari sections of this study (localities 5, 6) are coeval with the Nancy Point Formation (locality 9) of the Bocas del Toro Group and the Gatun Formation of central Panama (Whittaker and Hodgkinson in Woodring, 1982; Coates et al., 1992). The uppermost Uscari differs from the shallow neritic, sandy Gatun in its deeper-water facies and from the Nancy Point (blue-gray clays to fine sands) in its lithology and benthic foraminiferal assemblages (described below).

Methods

Sediment samples were collected in the Bocas del Toro and Limón belts from exposures of relatively fresh, compacted, mostly unindurated sediments occurring in short sections along coastlines, rivers and streams, construction sites, and road cuts. Sixty-seven samples from 13 sections (Fig. 2, App. 1) were selected for paleobathymetric analysis on the basis of biochronologic age constraint; samples from sections restricted to the shortest intervals of time (Table 1) were analyzed.

Samples were processed using the method described by Collins (1993). In nearly all cases, benthic foraminifera were picked from sample splits containing 400 or more individuals and sorted by species onto faunal slides, for which the U.S. National Museum of Natural History, Washington, D.C., is the designated repository.

Paleobathymetric interpretations of foraminiferal assemblages were based primarily on Recent distributions of the same species from the Caribbean and Gulf of Mexico regions. Paleodepth interpretations utilize deepest-dwelling, abundant benthic foraminifera, as downslope redeposition of sediments may be common near the continental margin (Ingle, 1980).

Percentages per sample of common taxa were subjected to hierarchical cluster analyses to assess differences in species assemblages within and among sections. This served two purposes: (1) to demonstrate that geologic sections represent discrete paleoenvironments and paleobathymetries and (2) to determine the similarity between coeval, isobathyal deposits from the Limón and Bocas del Toro belts. Cluster analysis places individual samples into groups (clusters) according to their degree of similarity in species' abundances. A hierarchical algorithm merges the two most similar samples to form a cluster, then successively connects the next two most similar samples or clusters until one cluster remains (Everitt, 1980; Davis, 1986). Canonical variate or discriminant analysis (Davis, 1986), an alternative approach, is impractical for whole tropical assemblages because it requires more samples than species, and the enormous diversity of tropical foraminifera virtually prohibits this unless the data set is considerably reduced.

Counts of 94 species with ≥2.0% abundance in any sample and occurring in three or more samples (Apps. 1, 2) were transformed from percentages (P) using the transformation: 2 arcsin $\sqrt{P}$ (Buzas, 1970). Counts per sample within species (variables) were standardized by calculation of z values (Everitt, 1980), a procedure that gives equal weight to all species in calculating the similarity of samples. Samples were clustered hierarchically by the Euclidean distance between species counts using the complete linkage method (Sokal and Michener, 1958), although results from five other hierarchical methods (Ward's, average linkage, single linkage, median linkage, and centroid) were also compared in the following section. The statistical software SYSTAT (Wilkinson, 1989) was used in all calculations and production of dendrograms.

Paleobathymetric data

The cluster analyses demonstrate that geologic sections consist of distinct assemblages of taxa, allowing assignment of a discrete paleobathymetric range to each section. Samples cluster primarily with other samples from the same geologic section (App. 1), and sections cluster hierarchically by general paleoenvironment. Figure 4 is a dendrogram produced by the complete linkage algorithm, but results from average linkage and Ward's methods are quite similar. Other clustering algorithms were less satisfactory; single linkage produced long, stringy clusters primarily reflecting geologic sections but having little paleoenvironmental meaning, whereas median linkage and centroid linkage produced relatively few clusters.

Paleobathymetries of Bocas del Toro sections are explained by Collins (1993) and summarized in Figures 3 and 4 and in Table 2. Bocas del Toro samples clustered according to their sections (Fig. 4), except for the following four samples. With all clustering methods, sample 86-38-1 (Cayo Agua NE section) clustered with more shoal- and reef-influenced Cayo Agua NW samples, primarily due to its abundant *Amphistegina gibbosa, Cassidulina* sp. A, *Nodobaculariella cassis, Quinqueloculina compta,* and *Q. lamarckiana.* Most cluster methods did not group sample 87-9-1 with the other samples

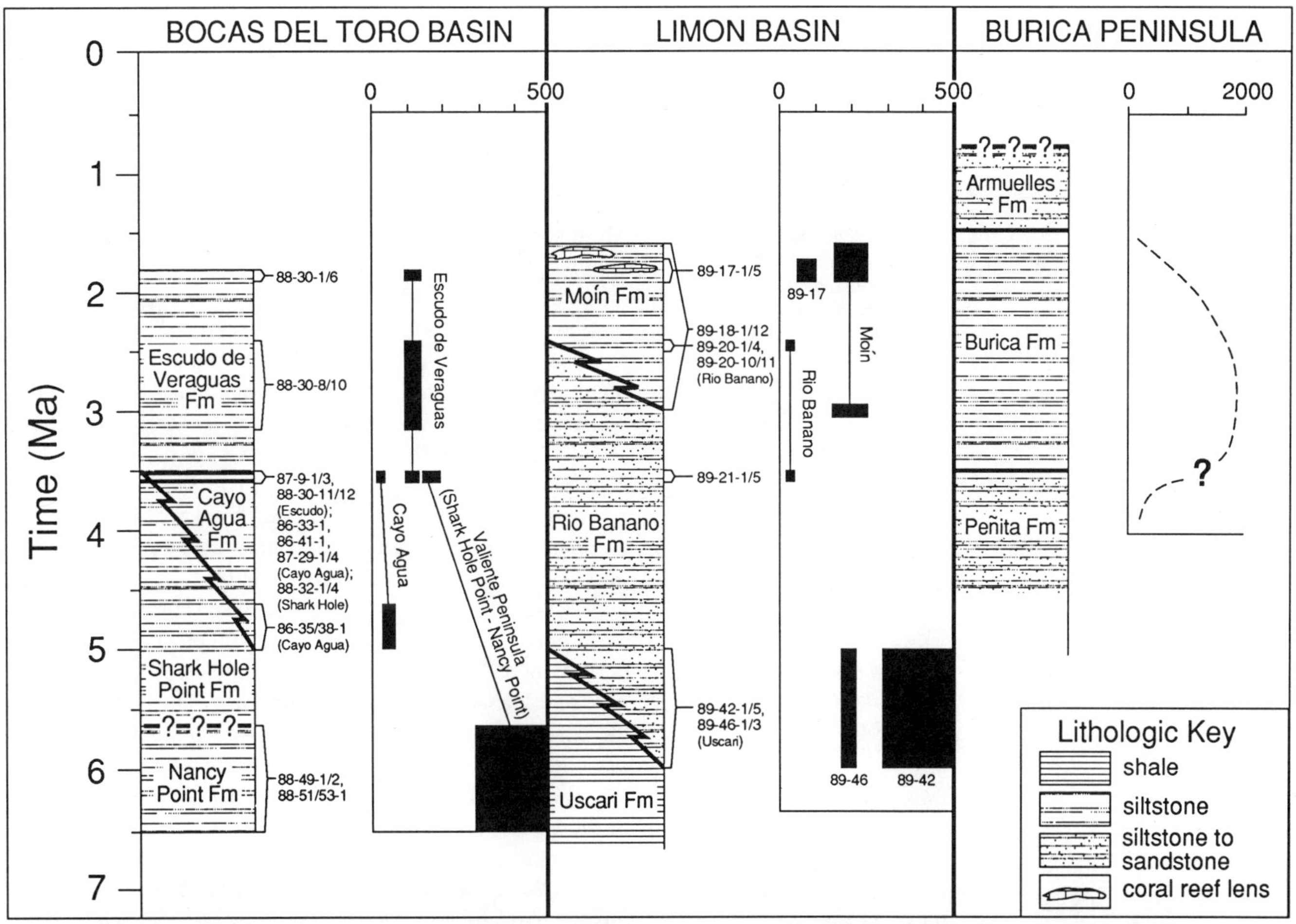

Figure 3. Predominant lithologies, stratigraphic sampling levels, and paleobathymetry (m) of the Bocas del Toro and Limón belts, Caribbean margin, and Burica Peninsula, Pacific margin. Stratigraphy is from Coates et al. (1992). Paleobathymetry of the Caribbean margin is from Collins (1993) and this chapter; black boxes represent uncertainty in age and water depth estimates, and lines connect sections within continuous stratigraphic sequences. Paleobathymetry for Burica is from Corrigan et al. (1990), fit to stratigraphic boundary dates of Coates et al. (1992).

from its Escudo de Veraguas (EdV) SE section, largely because of its greater abundances of *Cymbaloporetta atlantica* and *Peneroplis carinatus,* suggesting a stronger reef-to-shoal influence. Although samples 88-30-12 (EdV NE section) and 88-30-6 (EdV NW section) are not grouped with other samples from their sections in Figure 4, other clustering methods commonly include them in their section clusters.

For paleodepth interpretations of the Limón Group given below, citations in brackets are references for species' modern environmental distributions rather than taxonomic authorship. Paleobathymetries for the type section of the Moín Formation (locality 1), Limón Group, were previously determined by Cassell (1986) to be 150 to 200 m. Our paleodepth interpretations largely agree, although age assignments differ. Our 12 samples cover approximately half of Cassell's geographic range; his 13 samples are from banks of an unnamed creek flowing through the community of El Cangrejo, and we collected up the creek for about half the distance covered by his samples. Type Moín samples contain a shelf-edge assemblage (150 to 250 m) with transported shallower-water species. Shelf-edge indicators (i.e., those exhibiting a transitional distribution from outer neritic to upper bathyal depths) include *Cassidulina minuta* [Ingle, 1980], *Gyroidina regularis* [Pflum and Frerichs, 1976], *Planulina foveolata* [van Morkhoven et al., 1986], and *Trifarina eximia* [McLaughlin, 1989]. Abundant *Bulimina marginata* [Collins, 1988] suggest an upper depth limit of 150 m; abundant *B. aculeata* [Collins, 1988] and *Planulina foveolata* [Parker, 1954] indicate a lower depth limit of about 250 m. Cassell (1986) assigned a Pleistocene age to this type section of the Moín Formation, but our samples range in age from about 3.0 Ma to 1.6 to 1.9 Ma (Table 1). Although Cassell based his age on the presence of *G. truncatulinoides,* he did not identify *Sphaeroidinellopsis,* which, together with its cooccurrence with *Sphaeroidinella dehiscens,* indicates an age of ~3.0 Ma for the base of the Moín type section (Coates et al., 1992). Smaller clusters within the type Moín section cluster (Fig. 4) do not correspond to chronological groupings.

TABLE 2. PALEOBATHYMETRIES AND ABUNDANT INDICATOR SPECIES, BOCAS DEL TORO GROUP*

Section	Depth (m)	Diagnostic Taxa
Cayo Agua NW	20–40	*Elphidium discoidale* *Eponides antillarum* *Fursenkoina pontoni* *Hanzawaia concentrica* *Nodobaculariella cassis* *Nonionella atlantica* *Quinqueloculina compta* *Quinqueloculina lamarckiana*
Cayo Agua NE	40–80	*Cassidulina subglobosa* *Elphidium discoidale* *Eponides antillarum* *Fursenkoina pontoni* *Hanzawaia concentrica* *Nonionella atlantica* *Planulina ariminensis* *Reussella spinulosa* *Sigmoilina tenuis*
Escudo NE, NW, S	100–150	*Bolivina paula* *Bulimina marginata* *Cassidulina curvata* *Cassidulina minuta* *Gyroidina regularis* *Hanzawaia concentrica* *Melonis barleeanum* *Reussella spinulosa* *Sigmoilina tenuis* *Siphonina pulchra* *Uvigerina laevis* *Uvigerina peregrina*
Bruno Bluff	150–200	*Bulimina marginata* *Cassidulina curvata* *Cassidulina minuta* *Cassidulina norcrossi australis* *Trifarina eximia* *Uvigerina peregrina*
Nancy Point	300–500	*Bolivina imporcata* *Bolivina mexicana* *Bulimina inflata mexicana* *Hoeglundina elegans* *Siphogenerina lamellata* *Sphaeroidina bulloides*

*From Collins, 1993.

The Loma del Mar section (locality 2) of the Moín Formation is a middle neritic, reefal facies coeval with the youngest type Moín (Table 1). An upper depth limit of 50 m is based on abundant *Cassidulina curvata* [Manning, 1985], *Sigmoilina tenuis,* and *Spirillina vivipara* [Drooger and Kaasschieter, 1958]. A lower, 100-m depth limit is suggested by abundant *Elphidium discoidale* [Parker, 1954] and *Planulina ariminensis* var. *exorna* [Phleger, 1951] and by a well-developed, in situ hermatypic coral reef, which must have grown within the photic zone to maintain obligate photosynthetic, symbiotic algae (Wells, 1956). The cluster analyses commonly grouped one Loma del Mar sample, 89-17-4, with other middle to outer neritic, more clastically influenced samples of the Cayo Agua NW and Escudo de Veraguas sections (Fig. 4), primarily on the basis of its high abundance of *Buccella hannai, Elphidium discoidale, Eponides antillarum,* and *Reussella spinulosa.*

Cassell and Sen Gupta (1989b) determined that sediments of the type Rio Banano Formation (locality 3) were deposited on the inner to middle continental shelf, about 20 m or slightly deeper. We agree and place an upper depth limit of about 20 m on the basis of abundant *Pararotalia magdalenensis* [Golik and Phleger, 1977, as *Ammonia decorata;* Cassell and Sen Gupta, 1989b, as *Pararotalia sarmientoi*] and *Rotorbinella umbonata* [Sellier de Civrieux, 1977], and a lower, 40-m depth limit on the basis of *Sigmoilina tenuis* [Drooger and

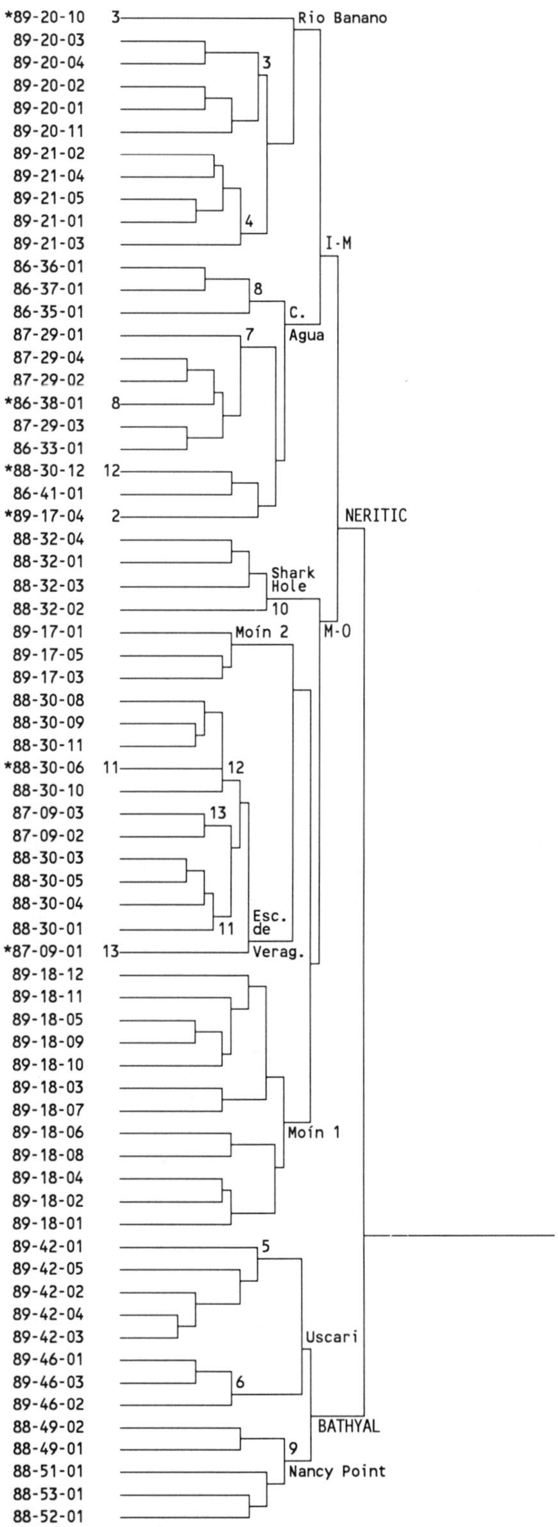

Figure 4. Cluster analysis of benthic foraminiferal assemblages by sample number (at left), complete linkage method. Paleobathymetric clusters of bathyal, neritic, inner to middle (I-M) neritic, and middle to outer (M-O) neritic form main clusters. Names and numbers of stratigraphic sections (Fig. 2) are placed at main cluster branches. * = sample excluded from its section's cluster.

Kaasschieter, 1958] and *R. umbonata.* The type section, composed primarily of sand, has a strong carbonate shoal to reef influence with abundant species of *Quinqueloculina, Amphistegina gibbosa, Hauerina fragillissima,* and *Nodobaculariella cassis.* The presence of burrowing sand dollars and callianassid burrow zones suggests a high-energy shoreface to inner neritic zone, so we infer the paleodepth to be at the shallow end of the foraminiferal depth range, i.e., about 20 m deep. Sample 89-20-10 from the type section fell within the Rio Banano cluster (Fig. 4) but not within its section cluster, primarily as a result of high abundances of *Sorites marginalis, Peneroplis carinatus,* and *Q. costata,* which indicate an even stronger reefal influence at this point.

The Quitaria section of the Rio Banano Formation (Fig. 2, locality 4) is distinguished from the type section by the cluster analysis (Fig. 4), but similar depth indicators (App. 1) also place it in the range of 20 to 40 m. Cassell and Sen Gupta (1989b) placed the middle part of the type Rio Banano section in biostratigraphic plankton zone N18 at about 5.1 Ma, based on the cooccurrence of the ostracod *Radimella ovata* (N17–lower N18) and a Pliocene foraminiferal fauna. However, recent work by Coates et al. (1992) uses a global calcareous nannofossil zonation, which indicates an age for the type section of 2.4 to 2.5 Ma, based on cooccurring *Discoaster pentaradiatus* (last occurrence 2.4 Ma) and small *Gephyrocapsa* sp. (first occurrence ~2.5 to 3.0 Ma), and an age for the Quitaria section of 3.5 to 3.6 Ma from cooccurring *Pseudoemiliana lacunosa* (3.6 to 0.5 Ma) and *Sphenolithus abies* (last occurrence 3.5 Ma).

The Carbon Dos and Rio Sandbox sections (localities 5, 6) of the Uscari Formation, southern Limón belt, are coeval with the Nancy Point Formation (locality 9) of the Bocas del Toro Group (Coates et al., 1992), and Carbon Dos and Nancy Point sections have similar paleobathymetries. Rio Sandbox contains shelf-edge assemblages from around 200 m characterized by abundant *Bolivina imporcata* [Renz, 1948; Ingle, 1980; Kohl, 1985], *B. lowmani* [Phleger, 1951; Drooger and Kaaschieter, 1958], *B. mexicana* [Pflum and Frerichs, 1976], and *Cassidulina minuta.* Few specimens of predominantly bathyal species such as *Hoeglundina elegans* [Pflum and Frerichs, 1976] and *Sphaeroidina bulloides* [van Morkhoven et al., 1986] are found at Rio Sandbox, although common in the Carbon Dos section. Compared to Carbon Dos, Rio Sandbox has more abundant neritic species such as *Nonionella atlantica* [Drooger and Kaasschieter, 1958], *Pararotalia magdalenensis,* and *Rotorbinella umbonata.* Carbon Dos contains upper bathyal assemblages of about 300- to 500-m water depth. The upper depth limit of 300 m is placed on the basis of abundant *Sphaeroidina bulloides* [van Morkhoven et al., 1986], *B. barbata,* and *Pullenia bulloides* [Pflum and Frerichs, 1976]. The lower depth limit of about 500 m is based on *B. mexicana* and *Eponides turgidus* [Pflum and Frerichs, 1976].

Paleobathymetries for the 13 sections are plotted against time in Figure 3. Lines connect portions of the five continuous, sequentially dated exposures of the Valiente Peninsula (Nancy

Point to Bruno Bluff sections), Escudo de Veraguas (NE to NW sections), Cayo Agua (NE to NW sections), type Moín section, and Rio Banano (Quitaria to type sections). Slopes of the lines between sections graphically depict rates of emergence. Calculated rates are presented in Table 3; Quaternary rates use present-day altitudes of the youngest sections of latest Pliocene to earliest Pleistocene age.

BASINAL HISTORIES OF THE CARIBBEAN AND PACIFIC MARGINS

Limón and Bocas del Toro basins, Caribbean margin

Formation names such as Uscari and Gatun have been used to refer to stratigraphic units from both the Limón and Bocas del Toro areas (Olsson, 1922; Terry, 1956), and they have been frequently regarded as a single past structural and depositional basin (Case et al., 1984; Escalante, 1990). However, late Neogene faunal and sedimentation patterns in the Limón and Bocas del Toro belts serve to differentiate their histories.

Different faunal patterns, indicating different depositional environments for the Bocas del Toro and Limón regions, are suggested by the cluster analyses of benthic foraminiferal assemblages (Fig. 4), which illustrate dissimilarities between coeval deposits with comparable paleobathymetries. Complete linkage, average linkage, and Ward's methods produce two main clusters (Fig. 4): a latest Miocene bathyal cluster of the coeval Uscari and Nancy Point sections and a Pliocene neritic cluster. Within the bathyal cluster, Limón (localities 5, 6) and Bocas del Toro (locality 9) sections are separated, even though Carbon Dos (locality 5) and Nancy Point (locality 9) have comparable paleodepths. Compared with Nancy Point, the Carbon Dos section has relatively more *Bolivina lowmani, Epistominella vitrea, Eponides turgidus,* and *Pullenia* cf. *P. bulloides* and relatively fewer *B. barbata, Cibicides floridanus, Hanzawaia isidroensa,* and *Siphogenerina lamellata* (App. 1). Within the neritic cluster, sections are grouped into an inner- to middle-neritic (I-M) cluster composed of Cayo Agua and Rio Banano sections and a middle- to outer-neritic (M-O) cluster containing Moín and Escudo de Veraguas sections. Again, coeval, isobathyal sections in the Limón belt (Rio Banano, Quitaria section, locality 4) and the Bocas del Toro belt (Cayo Agua NW section, locality 7) fall into different clusters, although at these shallower depths there is more potential for geographic variability. Sediments of the Cayo Agua NW section have a stronger shoal or reef influence than those of the Rio Banano Quitaria section, based on relatively greater abundances of *Amphistegina gibbosa, Bigenerina irregularis,* and *Eponides antillarum* at Cayo Agua NW. Rio Banano Quitaria sediments show a stronger clastic facies, indicated by relatively higher proportions of *Ammonia beccarii, Buliminella elegantissima,* and *Pararotalia magdalenensis.*

Large-scale sedimentation patterns also differed between basins. From Late Miocene through Pliocene time, a broad, thick (up to 1,500 m) deltaic wedge of shallow-water, volcaniclastic sediments composing the Rio Banano Formation in the region west of Limón was derived from erosion of the Cordillera de Talamanca (G. Gonzalez, personal communication, 1992). The land-based Rio Banano Formation is correlative in part with sediments of the thick, Miocene to Recent Costa Rica fan system extending off the coast of Caribbean Costa Rica–southern Nicaragua (Sheehan et al., 1990; Bowland, 1993). In contrast to the Limón belt, coeval deposits of the Bocas del Toro belt have no such facies and thus, presumably, no analogously large, uplifted sediment source.

Although faunal and sedimentation patterns differed between the Limón and Bocas del Toro basins, patterns of emer-

TABLE 3. RATES OF EMERGENCE (M/M.Y.): RANGES FOR CARIBBEAN SECTIONS, INTERVAL AVERAGES FOR CARIBBEAN, AND VALUES FOR BURICA PENINSULA

Caribbean Section	Basin	Time Interval	Midpoints*	Maximum†	Minimum§	Caribbean** (Interval Average)	Pacific (Burica Peninsula)
Moin (type)	Limón	Quaternary	114	156	79		
Escudo de Veraguas	Bocas del Toro	Quaternary	68	83	53	91	0
Rio Banano	Limón	Late Pliocene-Quaternary	40	50	32		
Rio Banano	Limón	Late Pliocene	0	20	-17		
Escudo de Veraguas	Bocas del Toro	Late Pliocene	0	31	-28	0	1,000
Moin (type)	Limón	Late Pliocene	0	91	-71		
Cayo Agua	Bocas del Toro	Early Pliocene	24	60	0	24	-333‡
Valiente	Bocas del Toro	Late Miocene-Early Pliocene	90	175	33		

*Calculated with midpoints of ranges in age and paleobathymetry.
†Calculated with shortest time interval and largest possible depth change.
§Calculated with longest time interval and smallest possible depth change.
**Averages of midpoint values for time intervals in third column.
‡Subsidence based on Peñita Formation with base at 5 Ma (Corrigan et al., 1990), top at 3.5 Ma (Coates et al., 1992), and minimal depth change of 500 m (Corrigan et al., 1990).

gence, including tectonic uplift, were similar. Figure 5 summarizes emergence rates for the Caribbean margin, which depend on tectonic uplift, basinal subsidence, and sea-level change. In the early Pliocene, the Caribbean margin experienced moderate emergence within a range that could be attributable to estimated sea-level fall (Haq et al., 1987; Prentice and Matthews, 1988; Matthews, 1988), so a tectonic component is not necessary to explain it. No measurable shoaling occurred in either basin in late Pliocene time during the three intervals of 3.6 to 2.4 Ma (Rio Banano), 3.6 to 1.8 Ma (Escudo de Veraguas), and 3.0 to 1.6 (type Moín), indicating little to no tectonic uplift or subsidence. However, relatively rapid uplift during Quaternary time affected both areas. Assuming sea level 1.6 to 1.9 Ma was between 20 m above to 100 m below the present level (Haq et al., 1987; Prentice and Matthews, 1988), an average Quaternary emergence rate of 91 m/m.y. (Table 3) requires a tectonic component. If type Moín sediments, deposited at ~200 m 1.6 to 1.9 Ma and at a present altitude of 20 m, are used to calculate uplift, uplift rates would be 105 to 200 m/m.y. The Caribbean margin continues to undergo tectonic uplift today, as evidenced in the April 1991 Limón earthquake, which raised 135 km of coastline from Limón to the Panama border up to 1.57 m (Plafker and Ward, 1992).

The timing of emergence of the Limón and Bocas del Toro areas supports an interpretation of Cocos Ridge subduction as an uplift mechanism. The history of the Limón Basin is associated with the evolution of the Cordillera de Talamanca,

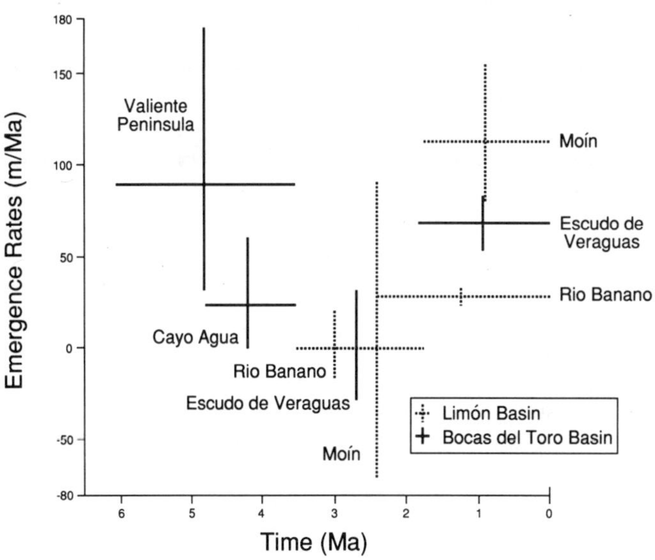

Figure 5. Rates of emergence over time, Caribbean margin. Lines reflect uncertainties in age and emergence rate determinations. Horizontal time lines are drawn between midpoints of age determinations; vertical lines show ranges in emergence rates. Minimum rates reflect the least possible change in water depth over the longest possible time; maximum rates reflect the largest possible change in water depth over the shortest possible time. Time lines intersect vertical lines at rates calculated with midpoints of age and paleobathymetry ranges. Quaternary rates are based on altitudes of outcrops.

the Térraba belt, and the Burica Peninsula terranes to the south (Fig. 2). Kolarsky et al. (this volume) relate the alignment of these features to their correlated geological history, inferring them to be a result of shallow subduction of the Cocos Ridge. Extension of the convergence vector of the subducting Cocos Ridge to the Caribbean coast (Fig. 2) explains the timing of emergence of the Limón and Bocas del Toro basins. Negligible emergence between about 3.6 Ma and 1.6 Ma (youngest age for the Moín Formation), followed by subsequent tectonic uplift in both basins, suggests a response to the Quaternary arrival of the shallowly subducted Cocos Ridge.

The interpretation of uplift from the April 22, 1991, Limón earthquake by Plafker and Ward (1992) might suggest that back-arc thrusting associated with the western end of the North Panama deformed belt near Limón (NPDB, Fig. 1) was the mechanism of uplift. However, as modeled by McGeary et al. (1985) and proposed by Tajima (this volume), such back-arc thrusting, rupture, and uplift are predicted effects of ridge-trench intersections in which a buoyant ridge subducted beneath an arc causes the arc to ride up over the plate on the back-arc side to form such features as the NPDB. Our Caribbean data do not extend eastward to more thrusted, deformed sediments of the central to eastern NPDB, which may reflect the collision and oroclinal bending of Panama against South America (Mann and Corrigan, 1990; Silver et al., 1990) and some convergence between the Panama microplate and the Caribbean plate (Mackay and Moore, 1990). Caribbean effects of Cocos Ridge subduction are also suggested by Silver et al. (this volume), who explain the northwest orientation of the western NPDB as a result of the collision of Costa Rica by the Cocos Ridge. Hence, uplift attributable to movement within the western NPDB may be primarily related to the effect of buoyant Cocos Ridge subduction.

Burica Peninsula, Pacific margin

Fitting our new biochronologic dates (Coates et al., 1992) to the paleobathymetric data of Corrigan et al. (1990) for Burica Peninsula, we calculate the timing of tectonic uplift. The Burica Formation consists of about 2,600 m of trench-apron turbidites, dated as 3.5 to 3.6 Ma at the base and as 1.5 to 1.7 Ma at the top, where it is in contact with the Armuelles Formation (Fig. 3). Turbidites reflecting deepest water depths around 2,000 m occur near the base of the Burica Formation (Corrigan et al., 1990), and the Armuelles Formation includes very shallow water deposits (<10 m) close to or slightly above sea level today. This yields a rapid late Pliocene emergence rate of about 1,000 m/m.y. (2,000 m/2 m.y.) that would have outpaced sea-level change and requires rapid tectonic uplift, since sea level is estimated to have fallen no more than 120 m (and possibly even risen) during this time (Haq et al., 1987; Prentice and Matthews, 1988). To calculate a Quaternary uplift rate based on Armuelles deposits, estimates of sea level from 1.5 to 1.7 Ma range from 20 m above to 100 m below the present level (Haq et al., 1987; Prentice and Matthews, 1988). A past sea level from 1.5 to 1.7 Ma of +20 m would require Quaternary subsidence, whereas an estimate of –100 m would yield a maximum Quaternary uplift of 67 m/m.y., far slower than uplift during the late Pliocene. Thus, rapid late Pliocene uplift of the Burica Peninsula was followed by minor Quaternary uplift.

Corrigan et al. (1990) suggested that rapid shallowing between upper Burica and Armuelles times recorded the start of subduction of the Cocos Ridge, an estimate in agreement with that of Lonsdale and Klitgord (1978), who used marine magnetic anomalies to calculate initial subduction at about 1 Ma. However, the calculation by Corrigan et al. (1990) assumed that most uplift had occurred within Quaternary time. Because net uplift since deposition of the Armuelles Formation ~1.5 to 1.7 Ma has been relatively minor, we suggest that initial subduction of the Cocos Ridge had commenced by about 3.5 Ma.

Our evidence for very limited uplift during Pleistocene time for both the Burica and Osa Peninsulas (Fig. 2) is based upon exposures of the Armuelles Formation close to sea level in the rivers Claro (Osa), El Higo (west side of Burica), and Rabo de Puerco (east side of Burica) and along the beach near the river Manzanillo on Burica's east coast. Only some slight uplift in this region is confirmed by Fischer (1980), whose bioerosional data for approximately the last 600 years show the most uplift on the Nicoya Peninsula and very little to no uplift from Quepos south to the Osa Peninsula. Fischer's data also indicate zones of subsidence through the Golfo de Nicoya and Golfo Dulce, a finding that agrees with repeated leveling surveys from 1948 to 1969 by Miyamura (1975) that show subsidence of 0 to 87 mm at 15 sites on the mainland side of the Golfo de Nicoya. However, these uplift patterns conflict strongly with those of Gardner et al. (1992). Using radiometrically dated Pacific marine terraces and beach ridges, they posit maximum uplift rates over the last 36,000 years of 6.5 m/ka for Osa, decreasing on either side to 2.5 m/ka for Nicoya and 4.7 m/ka for Burica. Their elastic flexure model for Cocos Ridge uplift is optimally fit to an initial subduction time of 0.5 Ma, penetration to about 50 km, and truncation of the ridge by the Panama Fracture Zone (Fig. 1). Our data suggest a much older time for initial subduction. Clearly, locally varied rates of tectonic uplift and subsidence are also caused by differential neotectonic movements along faults cutting the Osa and Burica Peninsulas perpendicular to the trench and others in the Golfo Dulce parallel to the trench (Fischer, 1980; Corrigan et al., 1990). We conclude that our long-term data effectively time-average the uplift trends expected from isostatic Cocos Ridge effects.

Comparison of Caribbean and Pacific margins

Differences in the average rates and timing of emergence for the two margins during Pliocene to Quaternary time are summarized in Table 3 (right two columns). In the early Pliocene, the Caribbean margin experienced a moderate emergence rate that could have been due to sea-level fall (Haq et al., 1987; Prentice and Matthews, 1988; Matthews, 1988), so tectonic uplift is not invoked. On the Pacific margin, the Burica region rapidly subsided during the early Pliocene, perhaps as a result of the proximity of the unstable Cocos-Nazca-Caribbean triple junction at the northern end of the PFZ (Corrigan et al., 1990). The Pacific-Caribbean contrast is large in the late Pliocene: whereas Burica experienced rapid uplift, no measurable change occurred in the Bocas del Toro or Limón basins, suggesting little to no tectonic uplift or subsidence. An opposite pattern holds for Quaternary time: Although the Pacific margin recorded little emergence and a relatively low rate of uplift, the Caribbean margin experienced its fastest rates, which require a tectonic component in addition to a eustatic one. Thus, the late Pliocene to Recent pattern of emergence reflects a similar pattern of uplift. The late Pliocene emergence of Burica was over an order of magnitude faster than Quaternary emergence on the Caribbean margin, reflecting a decrease in deformational effects away from the Pacific zone of active subduction.

Opposite intermarginal patterns of emergence for late Pliocene and Quaternary intervals may reflect the history of subduction for the Cocos Ridge. Figure 6 shows our reconstructed sequence for the timing of Cocos Ridge subduction and summarizes the changing distribution of depositional environments. We suggest that initial subduction at the southern margin of the Panama microplate was recorded by the rapid emergence of the Burica Peninsula from about 3.6 to 1.6 Ma. If the Cocos Ridge began subduction at 3.6 Ma and traveled beneath the microplate at about 9 cm/year—the present rate of plate convergence (Minster and Jordan, 1978)—over a trans-isthmian distance of about 180 km from the southern tip of the Burica Peninsula to the eastern Limón belt, the ridge margin would have arrived at the northern microplate margin about 2 m.y. later (180 km/(90 km/m.y.) = 2 m.y.), around 1.6 Ma. Increased rates of emergence on the Caribbean margin at this time may reflect the arrival of the edge of the Cocos Ridge beneath the Limón and Bocas del Toro sedimentary basins. Although net Quaternary emergence rates for Limón and Bocas del Toro are comparable (Table 3), differential patterns of uplift and subsidence due to the April 22, 1991, earthquake indicate short-term differential uplift between the Caribbean basins in Recent time due to rupturing of the NPDB fault system that intersects the coast north of Limón (Fig. 1; Plafker and Ward, 1992).

CONCLUSIONS

1. Deposits of the Bocas del Toro and Limón belts on the northern margin of the Panama microplate recorded distinctly different coeval, bathyal paleoenvironments and large-scale patterns of sedimentation, indicating at least partially isolated sedimentary basins, although their uplift histories are similar.

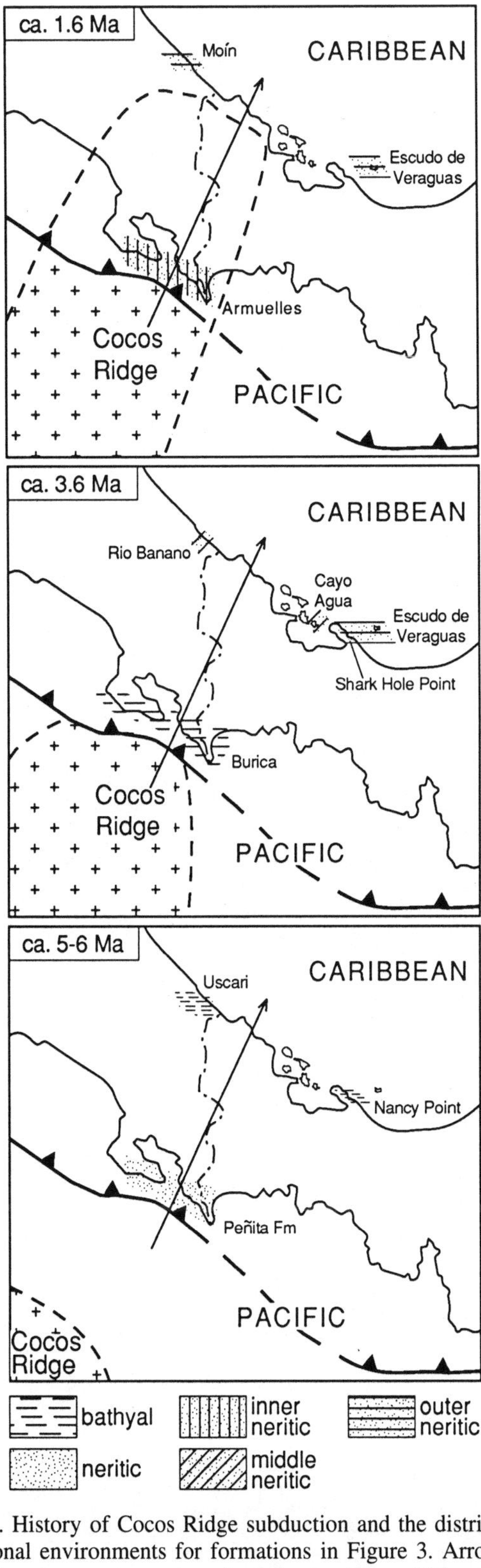

Figure 6. History of Cocos Ridge subduction and the distribution of depositional environments for formations in Figure 3. Arrow across isthmus is convergence vector of subducting Cocos Ridge (Kolarsky et al., this volume).

2. The fossil data for the northern, Caribbean margin of the Panama microplate constrain a pattern of little to no Pliocene uplift, followed by relatively rapid Quaternary uplift. Moderate emergence from late Miocene through early Pliocene time could be attributable to sea-level fall and requires no uplift component. No bathymetric changes are detectable in three sections spanning late Pliocene time, implying little to no uplift or subsidence. Rapid emergence rates for Quaternary time outpace estimated eustatic changes and require a tectonic component.

3. The pattern of emergence for the northern, Caribbean margin contrasts with that of the Burica Peninsula on the southern, Pacific margin, where rapid uplift in the late Pliocene was followed by relatively little uplift during Quaternary time. The differential transisthmian uplift pattern is explained most effectively by the arrival of the subducting, buoyant Cocos Ridge at the southern margin of the microplate about 3.6 Ma and at the northern margin about 1.6 Ma.

ACKNOWLEDGMENTS

Laurie Anderson, John-Mark Coates, Tim Collins, Mathew Cotton, Tom Cronin, Dana Geary, Peter Jung, René Panchaud, and Tom Stemann were valuable members of the various field expeditions to Bocas del Toro and Limón. In Costa Rica, generous logistical support was provided by Lic. José F. Castro Muñoz, Director de Geología y Minas, and Lic. Maria Elena Rodríguez A., Sección Ambiental, both of the Dirección de Geología y Minas, Ministerio de Recursos Naturales, Energía y Minas, San José; and by Lic. César Laurito Mora, Encargado de la Sección de Geología y Paleontología, Departamento de Historia Natural, Museo Nacional de Costa Rica. Advice from professors Enrique Malavassi, Universidad de Costa Rica, and Peter Baumgartner, Université de Lausanne, greatly facilitated fieldwork near Limón. Mercedes Arosemena, Argelis Ruiz de Guevara, Elena Lombardo, and David West, Smithsonian Tropical Research Institute (STRI), Panama, provided invaluable logistical planning and support. Helena Fortunato and Magnolia Calderon, STRI, processed and prepared foraminiferal samples. Laurel Bybell, Tom Cronin, and Harry Dowsett, U.S. Geological Survey, contributed a biochronologic date in addition to those previously published. Martin Buzas, U.S. National Museum, provided access to foraminiferal collections and helpful discussions of taxonomy. Bonnie Miljour, Museum of Paleontology, University of Michigan, prepared the figures. Reviews by Paul Mann, Thomas Gardner, Martin Lagoe, and Gerta Keller improved the quality of the paper.

Fieldwork was supported by two grants from the National Geographic Society to Coates and Jackson, by grants from Sigma Xi and the Roger Tory Peterson Institute to Collins, and by STRI. Additional support for Collins came from an STRI postdoctoral fellowship, NSF grant RII-9002977, and NSF grant BSR-9006523 to Coates, Jackson, and Collins.

APPENDIX 1. PERCENTAGES OF TOTAL BENTHIC FORAMINIFERA PER SAMPLE FOR SPECIES MORE THAN TWO PERCENT IN MORE THAN THREE SAMPLES*

	Cayo Agua NW					Cayo Agua NE				C. Solarte
	86-33-1	87-29-1	87-29-2	87-29-3	87-29-4	86-35-1	86-36-1	86-37-1	86-38-1	86-41-1
Ammonia beccarii	0.3									
Amphistegina gibbosa	1.5	11.7	14.6	4.5	12.3	1.8	0.4	1.4	14.4	6.4
Bigenerina irregularis	0.9	1.5	1.6		1.1	1.0		1.4		
Bolivina alata							3.4	1.0		
Bolivina barbata							0.2			
Bolivina imporcata										
Bolivina lowmani										
Bolivina mexicana							0.5	0.3		
Bolivina paula		0.5								0.2
Bolivina pozonensis										
Bolivina sp. *A*						0.3	1.2	0.3		0.5
Bolivina striatula	0.9	0.7		0.5		0.3	0.7	0.3	0.4	0.2
Bolivina subexcavata	1.9			0.8		0.3	1.1	0.5	0.4	0.5
Bolivina subspinescens						0.3		0.3		
Buccella hannai	1.2	0.5	0.6	0.3	1.5	6.5	0.9	0.9	1.0	
Bulimina aculeata										
Bulimina inflata										
Bulimina marginata							0.2			
Buliminella curta										
Buliminella elegantissima										
Cancris sagra	0.9		0.4	0.8	0.3	2.8	1.1	1.2	0.6	0.9
Cassidulina curvata								0.4		
Cassidulina laevigata	6.3	6.6	5.5	4.0	7.7	0.7	0.2	1.3	1.2	0.9
Cassidulina minuta							0.7	0.5		
Cassidulina norcrossi australis						0.6		1.3		
Cassidulina sp. *A*	9.7	2.2	11.4	16.1	7.0	2.4	3.7	4.5	9.2	3.5
Cassidulina subglobosa	2.0	1.2		1.0		2.5	12.3	18.9		5.9
Cibicides floridanus										
Cibicides pachyderma	8.1	2.0	4.5	7.5	7.4	1.7	2.3	3.1	3.8	12.6
Cornuspira planorbis										
Cymbaloporetta atlantica	1.5			2.1	0.7		2.1	0.3	1.4	0.4
Discorbis bulbosa	0.7	1.2					0.3	0.5		0.7
Discorbina patelliformis	0.3	0.5		0.3		0.3	0.2		0.8	
Dyocibicides biserialis		1.7				0.4	0.7	0.4	0.4	1.7
Elphidium discoidale	3.2	0.7	2.6	4.6	3.9	7.6	4.3	2.1	1.2	4.0
Elphidium excavatum	0.3			0.6	0.3	2.3	0.4		0.4	0.2
Elphidium mexicanum							0.5	0.1	0.8	
Epistominella vitrea					0.3		1.9	0.7		
Eponides antillarum	3.1	4.2	7.9	8.0	7.7	5.7	0.7	1.6	6.0	1.1
Eponides turgidus										
Fissurina laevigata						0.3	0.7	0.6		
Fursenkoina pontoni	3.1	2.4	2.8	2.9	2.0	3.7	3.8	0.6	3.6	1.2
Gyroidina regularis							0.2			0.4
Gyroidina sp. *B*										
Gyroidina umbonata							0.4	0.2		
Hanzawaia concentrica	7.0	17.3	14.0	12.7	12.6	10.2	2.5	6.7	12.0	4.1
Hanzawaia isidroensa								0.1		
Hauerina fragillissima										0.4
Haynesina depressula	2.4	0.7		0.6		1.1	0.9	1.0		1.0
Hoeglundina elegans										
Lenticulina calcar		0.5				0.3				
Lenticulina sp. *A*						0.3	0.4	0.2		
Marginulinopsis marginulinoides								0.1		
Melonis barleeanum	0.3						0.5	0.5		1.8
Miliolinella californica										
Nodobaculariella cassis	1.7	4.6	2.4	0.5	6.1				3.6	
Nonionella atlantica	4.2	4.2	3.2	4.1	0.8	10.3	6.3	3.3	4.2	3.0
Nonionella basiloba										

APPENDIX 1. PERCENTAGES OF TOTAL BENTHIC FORAMINIFERA PER SAMPLE FOR SPECIES MORE THAN TWO PERCENT IN MORE THAN THREE SAMPLES* (continued, p. 2)

	Cayo Agua NW					Cayo Agua NE				C. Solarte
	86-33-1	87-29-1	87-29-2	87-29-3	87-29-4	86-35-1	86-36-1	86-37-1	86-38-1	86-41-1
Nonionella incisa										
Oridorsalis umbonatus										
Pararotalia magdalenensis										
Peneroplis carinatus										0.2
Planorbulina acervalis									1.0	0.7
Planulina ariminensis		1.7	0.6	0.3	0.3	2.7	3.6	5.8		13.3
Planulina foveolata										
Pullenia cf. *P. bulloides*										
Quinqueloculina compta	1.9	3.9	3.2	2.4	4.7		0.8		2.2	4.0
Quinqueloculina costata										
Quinqueloculina lamarckiana	1.9	4.4	6.3	1.6	5.2				4.8	6.2
Quinqueloculina sp. *A*	1.9			0.5						1.0
Quinqueloculina sp. *B*					0.5					
Reussella spinulosa	4.1	2.9	0.8	0.8	1.1	10.7	3.1	4.8	2.0	3.1
Rosalina concinna	0.3			0.5			0.6		0.8	0.5
Rosalina globularis	4.9	0.7		2.5		1.7	1.6	0.8	1.0	0.7
Rosalina sp. *A*										
Rosalina subaraucana	1.7	0.7	0.4	4.6	1.0	3.8	1.3	0.1	2.0	
Rotalia garveyensis	2.4	3.2	2.2	0.3	1.5	8.9	5.3	2.8	3.6	1.8
Rotorbinella umbonata										0.4
Scutuloris sp. *B*		1.0								0.5
Sigmoilopsis schlumbergeri	1.5		0.8	4.1	0.3	0.9	12.1	14.0	0.6	0.7
Sigmoilina tenuis							2.4	2.1		0.2
Siphogenerina lamellata										
Siphonina pulchra		0.7	1.2	0.3	0.5	0.7	0.3	1.6		2.2
Sorites marginalis										
Sphaeroidina bulloides										
Spirillina vivipara	0.3									
Textularia schencki					0.3		0.2			
Trifarina eximia						0.3	1.0	2.1		
Trifarina illingi										
Trifarina occidentalis	4.6		1.0	1.9	4.6	0.6	3.4	2.3	0.6	2.4
Uvigerina laevis						0.3				
Uvigerina peregrina						0.9	1.4	0.8		
Valvulineria palmerae	0.9	0.5	2.2	0.5				0.1	0.4	
Valvulineria sp. *A*		0.5					0.7	0.7		

APPENDIX 1. PERCENTAGES OF TOTAL BENTHIC FORAMINIFERA PER SAMPLE FOR SPECIES MORE THAN TWO PERCENT IN MORE THAN THREE SAMPLES* (continued - p.3)

	Escudo S			Escudo NW					Escudo NE		
	87-9-1	87-9-2	87-9-3	88-30-1	88-30-3	88-30-4	88-30-5	88-30-6	88-30-8	88-30-9	88-30-10
Ammonia beccarii		0.3				0.2			0.5	1.0	
Amphistegina gibbosa		0.5		0.2	0.4	0.4	0.2		2.3	1.0	1.9
Bigenerina irregularis											
Bolivina alata	1.1				0.6	0.7	0.5	0.9			
Bolivina barbata											
Bolivina imporcata											
Bolivina lowmani			0.4							0.3	
Bolivina mexicana		0.6	0.3					1.4	2.0		
Bolivina paula	2.6	2.7	2.7	3.7	6.3	5.1	9.2	4.5	1.7	3.5	1.2
Bolivina pozonensis											
Bolivina sp. *A*	0.8	2.0	1.5	1.6	2.8	2.8	2.7	1.1	0.6	1.4	1.0
Bolivina striatuia	1.9	0.5	0.9	1.7	0.6	0.6	1.1	0.8	1.7	2.0	0.4
Bolivina subexcavata	2.1	2.4	0.5	3.0	3.0	1.9	1.6	0.9	0.7	1.6	1.2
Bolivina subspinescens		0.4		1.6	0.7	1.3	0.2	0.5		0.1	
Buccelia hannai		0.2							1.0	0.2	
Bulimina aculeata											
Bulimina inflata											
Bulimina marginata	2.1	0.8	2.2	1.1		1.8		1.1	1.5	2.1	2.9
Buliminella curta		0.3									
Buliminella elegantissima		0.2		0.7		0.3					
Cancris sagra	0.5		0.8	0.3	0.2	0.2	0.2	0.5	0.4		0.4
Cassidulina curvata		0.4	1.2	0.5	1.2	0.9	1.1	1.5	1.4	1.4	2.1
Cassidulina laevigata	4.0	4.5	7.4	7.6	10.6	15.3	13.5	9.8	7.0	11.5	6.9
Cassidulina minuta	1.1	1.5	2.4	1.5	3.0	3.5	2.3	1.7	1.6	1.2	1.2
Cassidulina norcrossi australis	1.1	1.4	0.6		0.4	0.7	0.6	1.8		0.4	0.8
Cassidulina sp. *A*	1.1	0.5	1.0	0.8	0.2	0.5	2.3	0.9	3.7	3.5	2.3
Cassidulina subglobosa	8.2	15.1	17.7	19.1	14.4	12.0	7.0	5.5	4.3	13.4	6.7
Cibicides floridanus											
Cibicides pachyderma	4.8	3.9	5.8	2.0	3.3	1.9	4.3	4.0	11.3	6.9	6.9
Cornuspira planorbis		0.3	0.6	0.9			0.2				
Cymbaloporetta atlantica	5.0	3.0	2.2	1.0	4.0		3.8	5.2	1.7	1.2	4.2
Discorbis bulbosa				0.3		0.2		0.3	0.6	0.3	
Disvorbina petelliformis	7.1	4.1	1.0	3.7	1.1	0.5	0.3		0.8		1.5
Dyocibicides biserialis		0.2						0.6	0.5		
Elphidium discoidale		0.4	0.4	0.6		1.0	0.2	1.5	2.1	1.1	0.4
Elphidium excavatum						0.3			0.4	0.1	
Elphidium mexicanum			0.6				0.3			0.4	
Epistominella vitrea		1.0		1.0	0.2	1.7	1.1				
Eponides antillarum										0.1	
Eponides turgidus	2.1	1.7	0.6		0.7	1.1		0.3		0.3	0.8
Fissurina laevigata		0.5	0.4								
Fursenkoina pontoni		0.4	0.5	0.4	0.2	0.3	0.3	0.3	0.2	0.1	0.4
Gyroidina regularis	2.9	2.7	6.5	2.2	6.4	2.1	6.1	5.8	2.3	2.2	3.3
Gyrodina sp *B*											
Gyrodina umbonata	0.5	0.4	0.4	0.5	1.0	0.3	0.2	0.8	0.7	0.1	0.4
Hanzawaia concentrica	1.9	0.3	0.8	0.5	1.9	1.9	3.8	1.2	4.0	4.1	2.5
Hanzawaia isidroensa											
Hauerina fragillissima						0.5	0.2		1.6	0.9	
Haynesina depressula	2.4	1.9	0.8	1.7	1.1	1.0	1.7	1.1	1.2	1.9	0.6
Hoeglundina elegans											
Lenticulina calcar										0.1	
Lenticulina sp. *A*											
Marginulinopsis marginulinoides											
Melonis barleeanum	2.6	2.2	1.4	1.3	0.2	1.5	1.1	1.7	1.2	0.3	1.9
Miliolinella californica		1.8	1.7		1.5	0.4	0.2			0.4	1.2
Nodobaculariella cassis	1.9								0.6	0.1	
Nonioneila atlantica	0.5	0.6	0.6		0.2	0.6		0.3	1.0	1.0	0.4
Nonionella basiloba		0.2									

APPENDIX 1. PERCENTAGES OF TOTAL BENTHIC FORAMINIFERA PER SAMPLE FOR SPECIES MORE THAN TWO PERCENT IN MORE THAN THREE SAMPLES* (continued, p. 4)

	Escudo S			Escudo NW					Escudo NE		
	87-9-1	87-9-2	87-9-3	88-30-1	88-30-3	88-30-4	88-30-5	88-30-6	88-30-8	88-30-9	88-30-10
Nonionella incisa											
Oridorsalis umbonatus											
Pararotalia magdalenensis						0.1	0.5		0.5	0.3	
Peneroplis carinatus	1.1			0.7	0.4	0.2	0.5				
Planorbulina acervalis	0.5					0.5		0.8	0.9		0.8
Planulina ariminensis	1.9	1.7	3.3	1.0	0.7	1.3	0.7	4.6	3.7	1.8	2.9
Planulina foveolata											
Pullenia cf. *P. bulloides*											
Quinqueloculina compta			0.3			0.4		0.3	1.9	1.4	
Quinqueloculina costata	1.3	2.3	1.0		0.6				0.8	0.4	
Quinqueloculina lamarckiana		0.4	0.5	0.4		0.9	0.8	0.6	1.4	1.4	1.9
Quinqueloculina sp. *A*		0.5	0.8		0.4	0.6					
Quinqueloculina sp. *B*	0.5	0.2		0.3		0.1					0.6
Reussella spinulosa				2.0	1.2	2.6	1.0	3.4	2.3	2.1	1.9
Rosalina concinna	1.1		0.5	0.4		0.1			0.5	0.7	
Rosalina globularis		0.4	0.4	2.0	1.3	3.1	1.4	1.1	4.7	2.4	2.5
Rosalina sp. *A*											
Rosalina subaraucana											
Rotalia garveyensis	0.8	0.6	1.2	0.7	1.0	2.1	1.6	3.1	0.8	1.1	1.9
Rotorbinella umbonata		0.4	0.3						0.6	0.3	
Scutuloris sp. *B*			0.6			0.1				0.1	0.4
Sigmoilopsis schlumbergeri				0.4	0.2	1.0	0.8	2.0	1.0	1.1	1.5
Sigmoilina tenuis	1.1	1.3	1.4	1.0	1.6	1.1	0.8	1.5	1.4	2.8	
Siphogenerina lamellata											
Siphonina pulchra	1.1	1.5	1.9	0.6	0.8	1.4	1.6	4.3	2.2	1.5	2.9
Sorites marginalis											
Sphaeroidina bulloides											
Spirillina vivipara	1.1	0.6	0.6	0.5	0.4				0.2	0.3	1.0
Textularia schencki				0.3							
Trifarina eximia		0.2		1.0	0.1	0.5	1.9	2.2	0.5	0.2	4.1
Trifarina illingi											
Trifarina occidentalis	2.4	2.2	0.4	1.4	1.7	0.9	2.7	2.3	1.2	1.1	
Uvigerina laevis	1.3	2.0	1.4	0.9	5.2	2.6	1.0	3.1	0.9	0.6	
Uvigerina peregrina	2.6	4.0	2.8	3.8	1.3	1.8	1.6	3.7	1.3	1.2	1.9
Valvulineria palmerae		0.9	1.0	1.7	0.8	0.7	0.8	0.5	0.6	1.1	0.4
Valvulineria sp. *A*		1.1	1.7	2.1	1.5	2.2	1.7	0.8	0.4		

APPENDIX 1. PERCENTAGES OF TOTAL BENTHIC FORAMINIFERA PER SAMPLE FOR SPECIES MORE THAN TWO PERCENT IN MORE THAN THREE SAMPLES* (continued - p.5)

	Escudo NE		Bruno Bluff				Nancy Point				
	88-30-11	88-30-12	88-32-1	88-32-2	88-32-3	88-32-4	88-49-1	88-49-2	88-51-1	88-52-1	88-53-1
Ammonia beccarii	0.7			0.5							
Amphistegina gibbosa	1.2	3.1	0.7		0.5				0.5		
Bigenerina irregularis					0.8						
Bolivina alata	0.2			1.9	0.8	0.7					
Bolivina barbata			1.9			1.5	6.1	4.8	6.1	2.2	2.6
Bolivina imporcata							2.6	7.8	4.4	7.0	10.2
Bolivina lowmani	0.7					0.7	2.0	2.1	2.6	2.8	0.7
Bolivina mexicana	0.3		0.7			1.5	0.6	15.7	3.6	7.6	3.4
Bolivina paula	2.7	1.0	1.2	2.6							
Bolivina pozonensis											
Bolivina sp. *A*	2.3	0.6		1.2			2.8	2.4	1.8		0.7
Bolivina striatuia	0.8	0.8						1.2	0.6	1.7	
Bolivina subexcavata	1.2	2.5							4.0		
Bolivina subspinescens		0.4		1.0							1.0
Buccelia hannai	0.2	1.4			0.5						
Bulimina aculeata									0.3		0.4
Bulimina inflata									0.3	2.6	1.8
Bulimina marginata	3.3	0.6	4.9		1.0						
Buliminella curta								2.6		2.4	0.9
Buliminella elegantissima		0.4									
Cancris sagra		0.6			0.8						0.2
Cassidulina curvata	2.3		2.8	10.5	11.4	1.1	1.6		0.5	1.1	0.4
Cassidulina laevigata	13.2	11.2	3.1	1.7	0.5	5.5	5.9		0.5		0.9
Cassidulina minuta	1.8		2.8	3.1	0.5	1.1	4.1	3.1	9.3	5.4	3.9
Cassidulina norcrossi australis	1.3		7.3	1.7	1.8	1.8	0.4	0.5	1.2		2.8
Cassidulina sp. *A*	1.7	2.5	0.7	1.9	1.8						
Cassidulina subglobosa	8.5	6.3	6.1	7.9	1.8	5.5	3.5	0.7	6.9	1.7	1.7
Cibicides floridanus							3.4	9.5	3.6	2.8	2.7
Cibicides pachyderma	6.4	8.6	15.8	1.4	5.2	8.8	1.6		0.8		0.5
Cornuspira planorbis	0.3										
Cymbaloporetta atlantica	3.0	2.5				0.7	0.4				
Discorbis bulbosa	0.4	0.4				1.5					
Disvorbina petelliformis	0.2	2.4									
Dyocibicides biserialis		0.1				0.7					
Elphidium discoidale	0.7	0.8	0.7		4.2	2.6					0.2
Elphidium excavatum	0.1										
Elphidium mexicanum	0.2		1.2	0.5	2.1	2.9			0.2		
Epistominelia vitrea							9.1	1.0	0.6	1.5	
Eponides antillarum		0.6			1.3	0.7					
Eponides turgidus	0.7			2.9					0.8	0.4	1.1
Fissurina laevigata	0.4		0.9	1.0		2.2			1.0	0.4	0.5
Fursenkoina pontoni						1.8	0.4		0.7	0.9	0.4
Gyroidina regularis	3.8	0.6	0.7	1.0			2.0	1.7	2.3	1.1	2.0
Gyroidina sp B											
Gyroidina umbonata		0.4	0.7	2.4		1.8	0.4		0.2	0.4	0.5
Hanzawaia concentrica	2.6	1.4	2.4	1.9	1.3	1.8	1.4	0.7	1.0		1.1
Hanzawaia isidroensa			2.8	1.0	0.5	1.1	6.3	2.4	4.1	3.3	4.9
Hauerina fragillissima	0.8										
Haynesina depressula	2.8	0.6		1.0			0.4		0.5		0.2
Hoeglundina elegans			0.5			1.8	3.0	2.1		0.7	2.1
Lenticulina calcar	0.4		2.4	1.2	0.8	1.5	1.4	1.4	0.7	1.7	1.0
Lenticulina sp. *A*									0.2	2.2	0.4
Marginulinopsis marginulinoides				0.5	2.3	0.7					0.6
Melonis barlleeanum	1.7	0.6	0.5								
Miliolinella californica											
Nodobaculariella cassis	0.2										
Nonioneila atlantica	0.6	1.2	1.4	1.9	10.7	11.7		0.7	0.2		0.4
Nonionella basiloba										0.4	

APPENDIX 1. PERCENTAGES OF TOTAL BENTHIC FORAMINIFERA PER SAMPLE FOR SPECIES MORE THAN TWO PERCENT IN MORE THAN THREE SAMPLES* (continued, p. 6)

	Escudo NE		Bruno Bluff				Nancy Point				
	88-30-11	88-30-12	88-32-1	88-32-2	88-32-3	88-32-4	88-49-1	88-49-2	88-51-1	88-52-1	88-53-1
Nonionella incisa								1.0		1.5	0.2
Oridorsalis umbonatus								0.7			0.7
Pararotalia magdalenensis	0.3										
Peneroplis carinatus											
Planorbulina acervalis	0.2	1.0									
Planulina ariminensis	2.5	7.6	0.9	4.3	15.1	2.9			0.2		
Planulina foveolata											
Pullenia cf. *P. bulloides*											
Quinqueloculina compta	0.2	0.6									
Quinqueloculina costata	0.2	0.4									
Quinqueloculina lamarckiana	2.3	0.4									
Quinqueloculina sp. *A*	0.5										
Quinqueloculina sp. *B*											
Reussella spinulosa	0.4	2.0		1.0	0.8						
Rosalina concinna		2.7							0.2		
Rosalina globularis	1.3	1.4									
Rosalina sp. *A*											
Rosalina subaraucana			1.2		0.8	2.2				0.4	
Rotalia garveyensis	0.6	1.2	1.2	1.0	0.5	2.2	2.0	6.7	4.7	5.9	4.4
Rotorbinella umbonata	0.2										
Scutuloris sp. *B*	0.2										
Sigmoilopsis schlumbergeri	4.0		6.6	2.4	9.1	3.3	0.8		1.1	0.4	1.1
Sigmoilina tenuis	0.5		2.1	1.2	1.6		0.8		0.2		1.5
Siphogenerina lamellata							2.2	1.9	0.7	1.7	0.7
Siphonina pulchra	2.7	1.4	3.3	0.7	1.0	2.6	4.5		2.1		
Sorites marginalis											
Sphaeroidina bulloides				1.7					1.9		0.9
Spirillina vivipara	0.2	0.8									
Textularia schencki		0.6									0.2
Trifarina eximia	1.6	1.6	0.5	6.4		6.2	3.5		5.2		
Trifarina illingi									2.6	0.4	
Trifarina occidentalis	0.9	3.1	1.4	1.4	0.5		1.0		1.3	0.4	0.4
Uvigerina laevis	0.4			9.1	0.8		1.6	1.4	0.8		3.7
Uvigerina peregrina	2.3	0.4	3.8	3.6	8.3	3.7	10.6	18.1	5.6	21.7	26.6
Valvulineria palmerae	0.2			2.6			0.6		0.5	0.4	1.0
Valvulineria sp. *A*	0.3	2.5	1.7	3.3	0.8	2.6			0.9	1.3	

APPENDIX 1. PERCENTAGES OF TOTAL BENTHIC FORAMINIFERA PER SAMPLE FOR SPECIES MORE THAN TWO PERCENT IN MORE THAN THREE SAMPLES* (continued, p. 7)

	Loma del Mar				Moin, type section						
	89-17-1	89-17-3	89-17-4	89-17-5	89-18-1	89-18-2	89-18-3	89-18-4	89-18-5	89-18-6	89-18-7
Ammonia beccarii		0.9	0.7	0.5			0.3		0.2		
Amphistegina gibbosa	5.5	0.8	15.7								
Bigenerina irregularis	0.5	0.7	0.5	0.7			1.5				4.0
Bolivina alata			0.5								
Bolivina barbata					1.8	0.4				0.4	
Bolivina imporcata			0.5								
Bolivina lowmani	0.3			0.5	5.3	5.9	2.5	2.0	1.6	5.3	0.6
Bolivina mexicana	0.5		0.5		6.4	19.0	0.3	1.6			3.1
Bolivina paula	1.2	0.2	1.2				0.6	0.2			
Bolivina pozonensis					1.6	1.1	2.4	3.9	6.9		
Bolivina sp. *A*	0.5	0.7		2.8	5.4	0.4		1.9			
Bolivina striatula		0.2									
Bolivina subexcavata	0.8		0.5	0.7	1.6	0.8	1.0	2.0	1.6	2.6	
Bolivina subspinescens					1.3	2.5		0.7			
Buccella hannai	0.3	0.3	1.6								
Bulimina aculeata					0.5			0.4	1.3		
Bulimina inflata											
Bulimina marginata	0.6	0.2	0.5	0.5	3.3	5.2	1.3	0.2	1.6		0.9
Buliminella curta					0.7						
Buliminella elegantissima		0.2								0.6	
Cancris sagra	0.6		0.7								0.6
Cassidulina curvata	1.8	3.1		0.9		5.1	4.4	0.4	7.1	1.4	2.1
Cassidulina laevigata	4.4	1.4	2.1	6.7	22.0	14.3	12.3	22.6	9.3	4.1	8.5
Cassidulina minuta						2.1	1.2	7.1	3.0	2.8	9.8
Cassidulina norcrossi australis					3.3	7.5	3.2	2.5	6.3	4.3	1.2
Cassidulina sp. *A*	2.1	1.7		2.6		0.5			0.5		0.6
Cassidulina subglobosa	6.4	6.3	2.3	9.3	1.3	3.1	3.1	4.7	2.9	3.5	3.1
Cibicides floridanus											
Cibicides pachyderma	5.6	10.9	9.6	6.7	0.5	3.4	14.8	4.9	9.6	2.6	6.7
Cornuspira planorbis	0.9			0.9	0.7		0.3	0.8	0.7	3.1	
Cymbaloporetta atlantica	0.8	0.2			1.3	1.1	2.4	3.0	1.8	2.0	3.1
Discorbis bulbosa		0.7		2.3					0.6		
Discorbina patelliformis	1.2				0.5		0.4	1.7	0.4	1.8	0.6
Dyocibicides biserialis	1.8	0.2	3.7	0.5			0.5				
Elphidium discoidale	1.1	3.8	6.1	1.6				0.4	0.2		
Elphidium excavatum									0.7		
Elphidium mexicanum											
Epistominella vitrea	0.8				2.5	0.3		0.2		0.8	
Eponides antillarum	0.3	0.5	2.8								0.6
Eponides turgidus					0.3	0.3	0.2	2.7			
Fissurina laevigata				0.5			0.4				
Fursenkoina pontoni	1.1	0.5	0.7	1.4							
Gyroidina regularis					0.7	0.3	0.2	1.6	2.2	2.4	
Gyroidina sp. *B*											
Gyroidina umbonata	0.3						0.4	0.3	0.4	0.6	
Hanzawaia concentrica	1.8	5.5	3.7	3.0					0.4		
Hanzawaia isidroensa											
Hauerina fragillissima										0.4	0.6
Haynesina depressula	0.8	0.4		1.4		0.5	0.4	2.5		3.1	
Hoeglundina elegans							0.6				
Lenticulina calcar					0.3	0.8	1.3				
Lenticulina sp. *A*											
Marginulinopsis marginulinoides									0.2		
Melonis barleeanum	0.9	2.3	0.9	1.6		0.3	6.3	1.4	4.5	0.4	4.3
Miliolinella californica	3.0	1.1		4.2			1.0			2.0	
Nodobaculariella cassis	0.6										
Nonionella atlantica	1.2	0.3	1.9	0.7	0.8	1.1	0.2	0.3	0.4		
Nonionella basiloba					2.8						

APPENDIX 1. PERCENTAGES OF TOTAL BENTHIC FORAMINIFERA PER SAMPLE FOR SPECIES MORE THAN TWO PERCENT IN MORE THAN THREE SAMPLES* (continued, p. 8)

	Loma del Mar				Moin, type section						
	89-17-1	89-17-3	89-17-4	89-17-5	89-18-1	89-18-2	89-18-3	89-18-4	89-18-5	89-18-6	89-18-7
Nonionella incisa											
Oridorsalis umbonatus						0.3			0.2		
Pararotalia magdalenensis	0.6	0.4									
Peneroplis carinatus	0.5										
Planorbulina acervalis	0.6	1.7		2.1						0.4	
Planulina ariminensis	5.3	24.0	13.1	2.3							
Planulina foveolata						0.3	0.5		5.3	1.2	1.5
Pullenia cf. *P. bulloides*											
Quinqueloculina compta		1.2		0.5							1.2
Quinqueloculina costata								0.2			
Quinqueloculina lamarckiana	1.2	0.3					0.9				
Quinqueloculina sp. *A*	0.9	0.8		1.4			0.3			2.4	
Quinqueloculina sp. *B*	0.5	0.3		2.3			0.3				
Reussella spinulosa	2.3	1.8	7.0	3.5							0.6
Rosalina concinna					0.3					1.2	
Rosalina globularis	2.7	1.1	0.5	1.2	0.7		0.4	1.2	0.5	1.4	0.6
Rosalina sp. *A*											
Rosalina subaraucana											
Rotalia garveyensis		2.2		1.4							
Rotorbinella umbonata		0.7				0.3	0.3		0.2	0.8	
Scutuloris sp. *B*					2.5	1.2	0.5	0.9			
Sigmoilopsis schlumbergeri	0.3			0.5	0.5	0.3	2.5	0.4	0.7	1.0	4.6
Sigmoilina tenuis	0.8	2.7		0.9	0.5	1.2	0.8	0.4	6.3	2.0	
Siphogenerina lamellata											
Siphonina pulchra	4.1	1.2	8.4	0.9		0.4	5.2	0.8	1.8	0.8	3.4
Sorites marginalis											
Sphaeroidina bulloides											
Spirillina vivipara	2.6			2.6	1.6	0.3	0.5	0.2		1.2	
Textularia schencki										0.4	
Trifarina eximia							2.2		0.6	0.6	3.1
Trifarina illingi		0.4	1.4	0.5			2.2		0.6	0.6	3.1
Trifarina occidentalis	1.1	0.5	3.0	2.1	0.3	0.5	2.1	1.1	0.6	2.4	0.6
Uvigerina laevis					1.6	0.3	4.2	4.3	4.8	2.6	1.2
Uvigerina peregrina	0.5	2.4	0.5		4.1	10.4	7.4	3.9	4.6	0.4	11.6
Valvulineria palmerae	0.3	0.7		1.4	1.3	0.7		0.2	0.7	1.6	
Valvulineria sp. *A*	1.1	0.5		3.0	2.3	0.8	0.2	1.6		1.6	1.2

APPENDIX 1. PERCENTAGES OF TOTAL BENTHIC FORAMINIFERA PER SAMPLE FOR SPECIES MORE THAN TWO PERCENT IN MORE THAN THREE SAMPLES* (continued, p. 9)

	Moin, type section					Rio Banano, type section					
	89-18-8	89-18-9	89-18-10	89-18-11	89-18-12	89-20-1	89-20-2	89-20-3	89-20-4	89-20-10	89-20-11
Ammonia beccarii	0.2	0.5						1.4	0.7	2.2	
Amphistegina gibbosa						4.9	2.0	0.5		0.6	4.7
Bigenerina irregularis											
Bolivina alata											
Bolivina barbata					0.8						
Bolivina imporcata											
Bolivina lowmani	5.6	2.1	4.5	7.6	2.2		0.5	5.6	9.6	0.6	0.4
Bolivina mexicana	0.2		1.9	0.4							
Bolivina paula			0.4				0.3				
Bolivina pozonensis	1.1	6.5	2.7	2.1	6.2			0.5		0.4	
Bolivina sp. *A*	0.7	0.5	0.4		0.2	0.5				0.4	
Bolivina striatula					0.4					1.2	0.4
Bolivina subexcavata	2.2	1.7	1.0	1.5	0.2	1.3	1.2		1.6	0.8	1.3
Bolivina subspinescens											
Buccella hannai										0.6	1.3
Bulimina aculeata		0.6	0.9	2.3	0.2						
Bulimina inflata											
Bulimina marginata	0.6	0.9			9.8	0.5					0.4
Buliminella curta											
Buliminella elegantissima	0.2	0.3	0.3						1.1	0.8	
Cancris sagra						0.8					0.9
Cassidulina curvata	4.7	1.8	2.7	4.3	0.5						
Cassidulina laevigata	12.6	8.2	12.2	6.2	3.5	13.3	11.4	20.1	28.0	22.4	14.2
Cassidulina minuta	2.3	1.7	1.8	2.6	0.4						
Cassidulina norcrossi australis	5.5	4.7	22.1	4.7	3.3					0.4	
Cassidulina sp. *A*						5.7	1.2	1.2	0.9	1.2	8.3
Cassidulina subglobosa	3.9	1.2	4.0	7.9	13.4	1.3	0.8	0.9	1.1	2.4	1.5
Cibicides floridanus											
Cibicides pachyderma	2.2	8.2	2.4	13.2	9.9	9.4	6.6	0.7	1.4	1.8	9.1
Cornuspira planorbis			0.9		0.2						
Cymbaloporetta atlantica	1.1	1.1	1.6	3.0	1.1	2.0	0.5				0.9
Discorbis bulbosa	0.5					0.8	0.5				0.6
Discorbina patelliformis	1.7	0.6	0.5					0.5			
Dyocibicides biserialis	0.2		0.5			1.3	2.5	0.7	0.5		0.9
Elphidium discoidale					0.2	1.3	1.3	0.9		2.8	1.9
Elphidium excavatum											0.4
Elphidium mexicanum										1.2	
Epistominella vitrea	0.2					0.3			0.5		0.6
Eponides antillarum							0.3			0.6	0.6
Eponides turgidus			0.8	1.1	2.2						
Fissurina laevigata	0.2							0.5			
Fursenkoina pontoni						0.5		1.4		0.6	0.9
Gyroidina regularis	1.3	5.2	0.3	0.9	4.2						0.9
Gyroidina sp. *B*											
Gyroidina umbonata	1.1	0.5	0.4		2.4				0.5		0.4
Hanzawaia concentrica		0.5		0.6		2.9	5.5	2.6	1.8	2.6	2.6
Hanzawaia isidroensa											
Hauerina fragillissima						1.7	4.0	0.5	2.5	2.8	1.7
Haynesina depressula	1.0	0.9		1.3		1.0	0.5		0.7	0.8	
Hoeglundina elegans				1.1							
Lenticulina calcar		0.6	0.5	0.4	0.5						
Lenticulina sp. *A*											
Marginulinopsis marginulinoides											
Melonis barleeanum	2.4	4.9	1.9	1.3	5.5	0.3					
Miliolinella californica	1.9							1.4	0.9	2.0	
Nodobaculariella cassis							1.2			0.6	
Nonionella atlantica	0.7	0.8	0.4	0.6		0.5	1.0	1.9	1.6	7.7	2.1
Nonionella basiloba	1.5					0.3					

APPENDIX 1. PERCENTAGES OF TOTAL BENTHIC FORAMINIFERA PER SAMPLE FOR SPECIES MORE THAN TWO PERCENT IN MORE THAN THREE SAMPLES* (continued, p. 10)

	Moin, type section					Rio Banano, type section					
	89-18-8	89-18-9	89-18-10	89-18-11	89-18-12	89-20-1	89-20-2	89-20-3	89-20-4	89-20-10	89-20-11
Nonionella incisa											
Oridorsalis umbonatus		0.3		2.1	1.5						
Pararotalia magdalenensis								1.2	0.5		
Peneroplis carinatus										2.9	
Planorbulina acervalis							0.5				
Planulina ariminensis						0.3				0.4	
Planulina foveolata	1.0	5.0	6.1	4.0							
Pullenia cf. *P. bulloides*											
Quinqueloculina compta						0.8	1.0	0.9		0.8	
Quinqueloculina costata										1.0	
Quinqueloculina lamarckiana	0.2	0.3	0.9		1.4		1.2	1.2	0.5	2.2	0.4
Quinqueloculina sp. *A*			0.5			0.7	1.2	3.3	1.6	2.4	
Quinqueloculina sp. *B*							0.5	0.5	0.7	0.4	
Reussella spinulosa	0.2					3.2	2.5	0.7		1.0	5.1
Rosalina concinna	1.0					3.2	2.0				1.5
Rosalina globularis	1.8	0.6	1.5	0.6		4.0	6.1	3.7	1.8	3.1	4.3
Rosalina sp. *A*						3.7	3.8	3.7	3.6	1.2	0.6
Rosalina subaraucana							0.5	0.7	1.6	3.3	
Rotalia garveyensis	0.4				0.7	1.0	0.5			0.4	1.5
Rotorbinella umbonata						2.9	6.9	15.6	14.1	0.8	2.1
Scutuloris sp. *B*	0.6	0.3	1.1	0.4	0.2						
Sigmoilopsis schlumbergeri	0.5	2.9	0.5	1.5	3.5						
Sigmoilina tenuis	1.6	4.1	3.7	2.3	1.7	0.8	0.3			0.4	0.4
Siphogenerina lamellata											
Siphonina pulchra	0.6	3.2	1.1	4.0	4.1	0.3					
Sorites marginalis								0.5		4.1	
Sphaeroidina bulloides			0.3								
Spirillina vivipara	0.5		0.4	0.4							
Textularia schencki						2.2	7.3	3.7	3.0	0.4	0.4
Trifarina eximia	0.2	0.9	0.9	0.4	0.2						
Trifarina illingi											
Trifarina occidentalis	1.2	0.6	0.5		1.3	5.0	5.5	1.9	0.7	2.0	8.9
Uvigerina laevis	6.3	6.1	3.1	4.9	4.4						
Uvigerina peregrina	0.9	7.6	2.2	4.9	5.9	0.5	0.5			0.8	2.1
Valvulineria palmerae	0.4	1.1	1.0		0.6						0.4
Valvulineria sp. *A*	9.1	0.6	0.3	1.5	1.1					0.6	1.3

APPENDIX 1. PERCENTAGES OF TOTAL BENTHIC FORAMINIFERA PER SAMPLE FOR SPECIES MORE THAN TWO PERCENT IN MORE THAN THREE SAMPLES* (continued, p. 11)

	Rio Banano, Quitaria					Uscari, Carbon Dos				
	89-21-1	89-21-2	89-21-3	89-21-4	89-21-5	89-42-1	89-42-2	89-42-3	89-42-4	89-42-5
Ammonia beccarii	0.3	3.4	1.6	1.7	1.6					
Amphistegina gibbosa	1.2	0.2	0.7	0.5	0.3					
Bigenerina irregularis										
Bolivina alata										
Bolivina barbata						0.8	0.6	0.6	1.3	2.2
Bolivina imporcata						10.0	0.3			1.1
Bolivina lowmani	0.9	6.9		4.9	3.9	7.9	11.3	21.0	16.6	3.3
Bolivina mexicana						11.1	1.1	0.5	0.6	3.7
Bolivina paula			0.5	0.5	0.7					
Bolivina pozonensis		0.5								
Bolivina sp. *A*	0.2	0.6		1.3	0.4	1.0	2.2	2.1	1.2	
Bolivina striatula		0.7		0.3						
Bolivina subexcavata	1.2	1.9	0.9	1.7	2.4		0.9	1.1	1.1	
Bolivina subspinescens						0.3		0.2	0.2	
Buccella hannai	0.9	0.3	0.4		0.4				0.2	
Bulimina aculeata										
Bulimina inflata										
Bulimina marginata										
Buliminella curta						0.8	0.4	0.2	0.7	2.1
Buliminella elegantissima	0.3	1.5		1.0	0.9	0.3	0.6	1.0	1.2	
Cancris sagra	0.3	0.3			0.4					
Cassidulina curvata						2.1	0.4	0.5	0.4	
Cassidulina laevigata	5.7	4.8	4.5	6.7	6.1					
Cassidulina minuta		0.5				11.6	9.6	12.1	15.7	3.7
Cassidulina norcrossi australis						1.5	2.1	1.9	1.1	5.5
Cassidulina sp. *A*	1.9	0.3	0.4	2.0	1.3			0.3		
Cassidulina subglobosa	1.7	7.3	2.1	2.9	3.4	1.4	5.7	7.5	6.5	6.5
Cibicides floridanus						2.2	2.2	0.6	0.6	0.6
Cibicides pachyderma	1.6	1.1	3.0	0.5	1.8	0.4	0.5	2.1	0.4	
Cornuspira planorbis	0.7				0.3					
Cymbaloporetta atlantica	1.3	0.3	1.3	1.8	0.2					
Discorbis bulbosa	0.2	0.5		0.5	0.4			0.2		
Discorbina patelliformis					0.3				0.2	
Dyocibicides biserialis	1.1	0.2	1.8		0.2					
Elphidium discoidale	6.8	2.7	3.9	2.9	4.4	1.5				
Elphidium excavatum	0.2	0.5		0.3	0.8					
Elphidium mexicanum		0.4					0.2	0.2	0.6	
Epistominella vitrea					0.8	3.1	8.7	6.6	6.1	9.8
Eponides antillarum	1.0		0.4							
Eponides turgidus						0.8	15.4	5.9	9.5	12.7
Fissurina laevigata		0.5			0.6			0.2	0.6	
Fursenkoina pontoni	0.9	2.0	5.9	1.5	2.8	0.7		0.6	0.6	0.4
Gyroidina regularis	0.2	0.6				3.6	3.2	2.0	2.0	0.7
Gyroidina sp. *B*						4.1	0.2		0.2	1.2
Gyroidina umbonata						1.1	2.0	1.1	0.8	1.9
Hanzawaia concentrica	6.0	2.1	2.1	1.2	0.8	0.7	0.9	0.6	0.7	
Hanzawaia isidroensa						0.7				
Hauerina fragillissima	0.9		3.0	0.5	0.8					
Haynesina depressula	2.1	2.1	1.3	4.2	2.7	1.5	1.9	1.3	2.2	
Hoeglundina elegans						0.4	1.2	0.2	0.4	0.3
Lenticulina calcar						0.8	0.4	1.2	0.4	0.4
Lenticulina sp. *A*						0.4				
Marginulinopsis marginulinoides										
Melonis barleeanum		0.3			0.2					
Miliolinella californica	1.0			0.8	3.9					
Nodobaculariella cassis	6.3	0.3	11.9	0.5	2.4					
Nonionella atlantica	8.2	14.9	7.7	11.9	7.2	1.1	0.6	0.7	0.2	0.8
Nonionella basiloba				1.7		0.3	0.5	0.8	1.2	0.7

APPENDIX 1. PERCENTAGES OF TOTAL BENTHIC FORAMINIFERA PER SAMPLE FOR SPECIES MORE THAN TWO PERCENT IN MORE THAN THREE SAMPLES* (continued, p.12)

	Rio Banano, Quitaria					Uscari, Carbon Dos				
	89-21-1	89-21-2	89-21-3	89-21-4	89-21-5	89-42-1	89-42-2	89-42-3	89-42-4	89-42-5
Nonionella incisa										
Oridorsalis umbonatus										
Pararotalia magdalenensis	1.5	5.8	6.2	10.6	6.4	0.4				
Peneroplis carinatus										
Planorbulina acervalis	0.5	0.4		0.3	0.3					
Planulina ariminensis	0.2					0.3	0.8		0.5	
Planulina foveolata										
Pullenia cf. *P. bulloides*							5.7	1.9	2.6	11.6
Quinqueloculina compta	2.3	1.5	3.2	1.0	0.7					
Quinqueloculina costata	0.4		0.4							
Quinqueloculina lamarckiana	9.4	0.5	6.4	4.2	2.5					
Quinqueloculina sp. *A*	3.2	1.5	2.1	2.2	2.4					
Quinqueloculina sp. *B*	0.7			0.3	1.0					
Reussella spinulosa	0.8	0.5	0.5	0.7	1.3					
Rosalina concinna	0.8	0.4	0.5	1.2	0.9			0.2		
Rosalina globularis	1.1	3.7	2.0	1.7	2.3			1.0	0.6	
Rosalina sp. *A*										
Rosalina subaraucana	1.6	1.8	1.6	1.3	0.9	1.2	0.4	0.7	0.4	
Rotalia garveyensis	0.7	1.3			1.5	5.6	1.2	0.9	0.7	2.5
Rotorbinella umbonata	1.5	2.0	0.9	3.0	4.9					
Scutuloris sp. *B*			0.5							
Sigmoilopsis schlumbergeri			0.4							
Sigmoilina tenuis	0.2	0.4	1.8	0.7	0.5		0.2	0.6		
Siphogenerina lamellata						0.3			0.2	
Siphonina pulchra				0.3						
Sorites marginalis			0.5	0.3	0.2					
Sphaeroidina bulloides						0.7		0.9	1.2	2.6
Spirillina vivipara	0.4				0.4					
Textularia schencki					0.3					
Trifarina eximia										
Trifarina illingi										
Trifarina occidentalis	1.3	0.8	0.7	2.4	0.6			0.5	0.4	
Uvigerina laevis						1.4	0.2	0.8	0.3	
Uvigerina peregrina		0.2				7.8	2.6	2.8	1.2	7.6
Valvulineria palmerae	0.2	0.2			0.5	0.8	0.2			2.9
Valvulineria sp. *A*	0.6	1.0		2.0	0.8		0.4	1.2	1.3	

APPENDIX 1. PERCENTAGES OF TOTAL BENTHIC FORAMINIFERA PER SAMPLE FOR SPECIES MORE THAN TWO PERCENT IN MORE THAN THREE SAMPLES* (continued - p.13)

	Uscari, Sandbox		
	89-46-1	89-46-2	89-46-3
Ammonia beccarii	0.8	0.4	0.1
Amphistegina gibbosa	0.3		
Bigenerina irregularis			
Bolivina alata			0.1
Bolivina barbata		1,2	
Bolivina imporcata	0.5	1.9	0.6
Bolivina lowmani	10.6	8.2	10.8
Bolivina mexicana	4.3	7.6	3.0
Bolivina paula			
Bolivina pozonensis			0.1
Bolivina sp. *A*	0.6	1.2	0.7
Bolivina striatuia	0.3	2.2	1.0
Bolivina subexcavata	3.0	0.6	5.4
Bolivina subspinescens		0.3	0.1
Buccella hannai			0.4
Bulimina aculeata			
Bulimina inflata			
Bulimina marginata			
Buliminella curta		0.9	0.4
Buliminella elegantissima	0.9	1.0	3.7
Cancris sagra			0.3
Cassidulina curvata			0.2
Cassidulina laevigata	1.1	0.3	0.2
Cassidulina minuta	3.6	6.0	3.8
Cassidulina norcrossi australis		1.5	0.7
Cassidulina sp. *A*			0.2
Cassidulina subglobosa	3.5	3.8	2.6
Cibicides floridanus		1.3	
Cibicides pachyderma	2.3	0.6	2.5
Cornuspira planorbis			
Cymbaloporetta atlantica			
Discorbis bulbosa	0.3		0.8
Disvorbina petelliformis			
Dyocibicides biserialis			
Elphidium discoidale			0.2
Elphidium excavatum			
Elphidium mexicanum	0.3		0.1
Epistominelia vitrea	4.6	15.2	9.6
Eponides antillarum			
Eponides turgidus		0.4	
Fissurina laevigata	0.5		0.4
Fursenkoina pontoni	0.6	0.4	0.5
Gyroidina regularis	0.8	0.8	0.8
Gyrodina sp B			
Gyrodina umbonata	0.3		
Hanzawaia concentrica	0.8		0.8
Hanzawaia isidroensa		0.1	

	Uscari, Sanbox		
	89-46-1	89-46-2	89-46-3
Hauerina fragillissima			
Haynesina depressula	2.9	1.4	5.6
Hoeglundina elegans		0.3	
Lenticulina calcar		0.5	
Lenticulina sp. *A*		1.0	1.0
Marginulinopsis marginulinoides			
Melonis barleeanum			
Miliolinella californica			
Nodobaculariella cassis			
Nonionella atlantica	2.6	2.9	4.8
Nonionella basiloba	0.3	0.3	0.1
Nonionella incisa	0.9	2.7	0.3
Oridorsalis umbonatus			
Pararotalia magdalenensis	29.3	4.0	17.8
Peneroplis carinatus			
Planorbulina acervalis			
Planulina ariminensis			
Planulina foveolata			
Pullenia cf. *P. bulloides*			
Quinqueloculina compta			
Quinqueloculina costata			
Quinqueloculina lamarckiana			
Quinqueloculina sp. *A*		0.3	
Quinqueloculina sp. *B*			
Reussella spinlosa	0.3		
Rosalina concinna			0.4
Rosalina globularis	0.6		
Rosalina sp. *A*			
Rosalina subaraucana			1.0
Rotalia garveyensis	0.9	2.8	1.0
Rotorbinella umbonata	1.8		1.7
Scutuloris sp. *B*			
Sigmoilopis schlumbergeri			
Sigmoilina tenuis		0.3	0.1
Siphogenerina lamellata		0.3	
Siphonina pulchra			
Sorites marginalis			
Sphaeroidina bulloides		0.3	
Spirillina vivipara			
Textularia schencki			
Trifarina eximia			0.2
Trifarina illingi			
Trifarina occidentalis	0.5	0.4	1.0
Uvigerina laevis			
Uvigerina peregrina	4.6	14.5	3.8
Valvulineria palmerae	0.9	0.9	0.4
Valvulineria sp A	0.5	1.5	0.6

*Samples numbered in hierarchial format XX-YY-ZZ; XX = final digits of year collected, YY = section number, ZZ = site or horizon.

APPENDIX 2. FULL TAXONOMIC NAMES OF SPECIES*

Ammonia beccarii (Linne) [*Rotalia*]
Amphistegina gibbosa d'Orbigny
Bigenerina irregularis Phleger and Parker
Bolivina alata (Seguenza) [*Vulvulina*]
Bolivina barbata Phleger and Parker
Bolivina imporcata Cushman [*B. floridana* Cushman var. *imporcata*]
Bolivina lowmani Phleger and Parker
Bolivina mexicana Cushman [*B. subaenariensis* Cushman var. *mexicana* Cushman]
Bolivina paula Cushman and Cahill
Bolivina pozonensis Cushman and Renz
Bolivina sp. A
Bolivina striatula Cushman
Bolivina subexcavata Cushman and Wickenden
Bolivina subspinescens Cushman
Buccella hannai (Parker and Phleger) [*Eponides*]
Bulimina aculeata d'Orbigny
Bulimina inflata Seguenza var. *mexicana* Cushman
Bulimina marginata d'Orbigny
Buliminella curta Cushman
Buliminella elegantissima (d'Orbigny) [*Bulimina*]
Cancris sagra (d'Orbigny) [*Rotalina (Rotalina)*]
Cassidulina curvata Phleger and Parker
Cassidulina laevigata d'Orbigny
Cassidulina minuta Cushman
Cassidulina norcrossi Cushman var. *australis* Phleger and Parker
Cassidulina sp. A = *C.* sp. B of Todd and Bronnimann, 1957
Cassidulina subglobosa Brady
Cibicides floridanus (Cushman) [*Truncatulina*]
Cibicides pachyderma (Rzehak) [*Truncatulina*]
Cornuspira planorbis Schulze
Cymbaloporetta atlantica (Cushman) [*Tretomphalus*]
Discorbis bulbosa Parker
Discorbina patelliformis (Brady) [*Discorbis*]
Dyocibicides biserialis Cushman and Valentine
Elphidium discoidale (d'Orbigny) [*Polystomella*]
Elphidium excavatum (Terquem) [*Polystomella*]
Elphidium mexicanum Kornfeld [*E. incertum* (Williamson) var. *mexicanum*
Epistominella vitrea Parker
Eponides antillarum (d'Orbigny) [*Rotalina*]
Eponides turgidus Phleger and Parker
Fissurina laevigata Reuss
Fursenkoina pontoni (Cushman and Ponton) [*Virgulina*]
Gyroidina regularis Phleger and Parker
Gyroidina sp. B
Gyroidina umbonata Silvestri
Hanzawaia concentrica (Cushman) [*Truncatulina*]
Hanzawaia isidroensa (Cushman and Renz) [*Cibicides*]
Hauerina fragillissima (Brady) [Spiroloculina]
Haynesina depressula (Walker and Jacob) [*Nautilus*]
Hoeglundina elegans (d'Orbigny) [*Rotalia (Turbinulina)*]
Lenticulina calcar (Lamarck) [*Cristellaria*]
Lenticulina sp. A
Marginulinopsis marginulinoides (Göes) [*Cristellaria aculeata* d'Orbigny var. *marginulinoides*]
Melonis barleeanum (Williamson) [*Nonionina*]
Miliolinella californica Rhumbler
Nodobaculariella cassis (d'Orbigny) [*Vertebralina*]
Nonionella atlantica Cushman
Nonionella basiloba Cushman and McCulloch
Nonionella incisa (Cushman) [*Nonionina*]
Oridorsalis umbonatus (Reuss) [*Rotalina*]
Pararotalia magdalenensis Lankford
Peneroplis carinatus d'Orbigny
Planorbulina acervalis Brady
Planulina ariminensis (Brady) [*Anomalina*]
Planulina foveolata (Brady) [*Anomalina*]
Pullenia cf. *P. bulloides* (d'Orbigny) [*Nonionina*]
Quinqueloculina compta Cushman
Quinqueloculina costata d'Orbigny
Quinqueloculina lamarckiana d'Orbigny
Quinqueloculina sp. A
Quinqueloculina sp. B
Reussella spinulosa Cushman
Rosalina concinna (Brady) [*Discorbina*]
Rosalina globularis d'Orbigny
Rosalina sp. A
Rosalina subaraucana (Cushman) *Discorbis*]
Rotalia garveyensis Natland
Rotorbinella umbonata Sellier de Civrieux
Scutuloris sp. B
Sigmoilopsis schlumbergeri (Silvestri) [*Sigmoilina*]
Sigmoilina tenuis (Czjzek) [*Quinqueloculina*]
Siphogenerina lamellata Cushman
Siphonina pulchra Cushman
Sorites marginalis (Lamarck) [*Orbulites*]
Sphaeroidina bulloides d'Orbigny
Spirillina vivipara Ehrenberg
Textularia schencki Cushman and Valentine
Trifarina eximia (Cushman and Jarvis) [*Angulogerina*]
Trifarina illingi (Cushman and Renz) [*Angulogerina*]
Trifarina occidentalis (Cushman) [*Uvigerina*]
Uvigerina laevis Göes [*U. auberiana* d'Orbigny var. *laevis*]
Uvigerina peregrina Cushman
Valvulineria palmerae Cushman and Todd
Valvulineria sp. A

*Original names in brackets where subsequently modified.

REFERENCES CITED

Adamek, S., Frohlich, C., and Pennington, W. D., 1988, Seismicity of the Caribbean-Nazca boundary: Constraints on microplate tectonics of the Panama region: Journal of Geophysical Research, v. 93, p. 2053–2075.

Bandy, O. L., 1953, Ecology and paleoecology of some California Foraminifera. Pt. I: The frequency distribution of Recent Foraminifera off California: Journal of Paleontology, v. 27, p. 161–182.

Berggren, W. A., Kent, D. V., Flynn, J. J., and Van Couvering, J. A., 1985, Cenozoic geochronology: Geological Society of America Bulletin, v. 96, p. 1407–1418.

Bowin, C., 1976, Caribbean gravity field and plate tectonics: Geological Society of America Special Paper 169, p. V–79.

Bowland, C. L., 1993, Depositional history of the western Colombian Basin, Caribbean Sea, revealed by seismic stratigraphy: Geological Society of America Bulletin, v. 105, p. 1321–1345.

Buzas, M. A., 1970, On the quantification of biofacies: Proceedings, North American Paleontological Convention, Chicago, September 1969: Lawrence, Kansas, Allen Press, p. 101–116.

Case, J. E., and Holcombe, T. L., 1980, Geologic-tectonic map of the Caribbean region: U.S. Geological Survey Miscellaneous Investigations Series Map 1-1100, scale 1:2,500,000.

Case, J. E., Holcombe, T. L., and Martin, R. G., 1984, Map of geologic provinces in the Caribbean region, *in* Bonini, W. E., Hargraves, R. B., and Shagam, R., eds., The Caribbean–South American plate boundary and regional tectonics: Geological Society of America Memoir 162, p. 1–30.

Cassell, D. T., 1986, Neogene foraminifera of the Limon Basin of Costa Rica [Ph.D. thesis]: Baton Rouge, Louisiana State University, 322 p.

Cassell, D. T., and Sen Gupta, B. K., 1989a, Pliocene foraminifera and environments, Limon Basin of Costa Rica: Journal of Paleontology, v. 63, p. 146–157.

Cassell, D. T., and Sen Gupta, B. K., 1989b, Foraminiferal stratigraphy and paleoenvironments of the Tertiary Uscari Formation, Limon Basin, Costa Rica: Journal of Foraminiferal Research, v. 19, p. 52–71.

Castillo, M., R., 1984, Geología de Costa Rica, una sinopsis: San José, Universidad de Costa Rica, 182 p.

Coates, A. G., and seven others, 1992, Closure of the Isthmus of Panama: The near-shore marine record in Costa Rica and western Panama: Geological Society of America Bulletin, v. 104, p. 814–828.

Collins, L. S., 1988, Environmental gradients and morphologic variation in *Bulimina aculeata* and *Bulimina marginata* [Ph. D. thesis]: New Haven, Connecticut, Yale University, 235 p.

Collins, L. S., 1993, Neogene paleoenvironments of the Bocas del Toro Basin, Panama: Journal of Paleontology, v. 67, p. 699–710.

Corrigan, J., Mann, P., and Ingle, J. C., 1990, Forearc response to subduction of the Cocos Ridge, Panama–Costa Rica: Geological Society of America Bulletin, v. 102, p. 628–652.

Davis, J. C., 1986, Statistics and data analysis in geology (second edition): New York, John Wiley and Sons, 646 p.

Dengo, G., 1962, Tectonic-igneous sequence in Costa Rica, *in* Engel, A.E.J., James, H. L., and Leonard, B. F., eds., Petrologic studies: A volume to honor A. F. Buddington: Boulder, Colorado, Geological Society of America, p. 133–161.

Dengo, G., 1973, Estructura geológica, historia tectónica y morfología de America Central (second edition): Guatemala City, Instituto Centroamericano de Investigacion Tecnología Industrial (ICAITI), p. 1–52.

Dengo, G., 1983, Mid-America: Tectonic setting for the Pacific margin from southern Mexico to northern Colombia: Guatemala City, Centro de Estudios Geológicos de America Central, 56 p.

Drooger, C. W., and Kaasschieter, J.P.H., 1958, Foraminifera of the Orinoco-Trinidad-Paria shelf, Vol. IV of Reports of the Orinoco shelf expedition: Verhandelingen der Koninklijke Nederlandse Akademie van Wetenschappen, afd. Natuurkunde, Eerste Reeks 22, 108 p.

Duque-Caro, H., 1990, The Choco Block in the northwestern corner of South America: Structural, tectonostratigraphic, and paleogeographic implications: Journal of South American Earth Sciences, v. 3, p. 71–84.

Escalante, G., 1990, The geology of southern Central America and western Colombia, *in* Dengo, G., and Case, J. E., eds., The Caribbean Region: Geological Society of America, The Geology of North America, v. H, p. 201–230.

Everitt, B., 1980, Cluster analysis: London, Heinemann Educational Books, 122 p.

Fischer, R., 1980, Recent tectonic movements of the Costa Rican Pacific coast: Tectonophysics, v. 70, p. T25–T33.

Gardner, T. W., and 11 others, 1987, Central America and the Caribbean, *in* Graf, W. L., ed., Geomorphic systems of North America: Geological Society of America, Centennial Special Volume 2, p. 343–402.

Gardner, T. W., and six others, 1992, Quaternary uplift astride the aseismic Cocos Ridge, Pacific coast, Costa Rica: Geological Society of America Bulletin, v. 104, p. 219–232.

Golik, A., and Phleger, F. B., 1977, Benthonic foraminifera from the Gulf of Panama: Journal of Foraminiferal Research, v. 7, p. 83–99.

Haq, B. U., Hardenbol, J., and Vail, P. R., 1987, Chronology of fluctuating sea levels since the Triassic: Science, v. 235, p. 1156–1167.

Hey, R., 1977, Tectonic evolution of the Cocos-Nazca spreading center: Geological Society of America Bulletin, v. 88, p. 1404–1420.

Ingle, J. C., 1980, Cenozoic paleobathymetry and depositional history of selected sequences within the southern California continental borderland, *in* Sliter, W. V., ed., Studies in marine micropaleontology and paleoecology, a memorial volume to Orville L. Bandy: Cushman Foundation for Foraminiferal Research Special Publication 19, p. 163–195.

Kohl, B., 1985, Early Pliocene benthic foraminifers from the Salina Basin, southeastern Mexico: Bulletins of American Paleontology, v. 88, 173 p.

Lonsdale, P., and Klitgord, K. D., 1978, Structure and tectonic history of the eastern Panama Basin: Geological Society of America Bulletin, v. 89, p. 981–999.

Lu, R. S., and McMillen, K. J., 1982, Multichannel seismic survey of the Colombia Basin and adjacent margins, *in* Watkins, J. S., and Drake, C. L., eds., Studies in continental margin geology: American Association of Petroleum Geologists Memoir 34, p. 395–410.

Mackay, M. E., and Moore, G. F., 1990, Variation in deformation of the south Panama accretionary prism: Response to oblique subduction and trench sediment variation: Tectonics, v. 9, p. 683–698.

Mann, P., and Corrigan, J., 1990, Model for late Neogene deformation in Panama: Geology, v. 18, p. 558–562.

Manning, E. M., 1985, Ecology of Recent Foraminifera and ostracods of the continental shelf of southeastern Nicaragua [M.S. thesis]: Baton Rouge, Louisiana State University, 451 p.

Matthews, R. K., 1988, Comment *on* 'Sea level history': Science, v. 241, p. 597–599.

Mattson, P. H., 1984, Caribbean structural breaks and plate movements, *in* Bonini, W. E., Hargraves, R. B., and Shagam, R., eds., The Caribbean–South American plate boundary and regional tectonics: Geological Society of America Memoir 162, p. 131–149.

McGeary, S., Nur, A., and Ben-Avraham, Z., 1985, Spatial gaps in arc volcanism: The effect of collision or subduction of oceanic plateaus: Tectonophysics, v. 119, p. 195–221.

McLaughlin, P., 1989, Neogene basin evolution in the southwestern Dominican Republic: A foraminiferal study [Ph.D. thesis]: Baton Rouge, Louisiana State University, 318 p.

Minster, J. B., and Jordan, T. H., 1978, Present-day plate motions: Journal of Geophysical Research, v. 83, p. 5331–5354.

Miyamura, S., 1975, Recent crustal movements in Costa Rica disclosed by relevelling surveys: Tectonophysics, v. 29, p. 191–198.

Natland, M. L., 1933, The temperature and depth-distribution of some Recent and fossil Foraminifera in the southern California region: Bulletin of the Scripps Institution of Oceanography, Technical Series, v. 3, p. 225–230.

Olsson, A. A., 1922, The Miocene of northern Costa Rica with notes on its general stratigraphic relations: Bulletins of American Paleontology, v. 9, p. 181–192.

Parker, F. L., 1954, Distribution of the Foraminifera in the northeastern Gulf of Mexico: Bulletin of the Museum of Comparative Zoology at Harvard College, v. 111, 588 p., 13 pl.

Pennington, W. D., 1981, Subduction of the eastern Panama Basin and seismotectonics of northwestern South America: Journal of Geophysical Research, v. 86, p. 10753–10770.

Pflum, C. E., and Frerichs, W. E., 1976, Gulf of Mexico deep-water foraminifers: Cushman Foundation for Foraminiferal Research Special Publication 14, 125 p.

Phleger, F. B., 1951, Ecology of Foraminifera, northwest Gulf of Mexico. Pt. I: Foraminifera distribution: Geological Society of America Memoir 46, 88 p.

Plafker, G., and Ward, S. N., 1992, Backarc thrust faulting and tectonic uplift along the Caribbean Sea coast during the April 22, 1991, Costa Rica earthquake: Tectonics, v. 11, p. 709–718.

Prentice, M. L., and Matthews, R. K., 1988, Cenozoic ice-volume history: Development of a composite oxygen isotope record: Geology, v. 16, p. 963–966.

Renz, H. H., 1948, Stratigraphy and fauna of the Agua Salada Group, State of Falcón, Venezuela: Geological Society of America Memoir 32, 219 p.

Sellier de Civrieux, J. M., 1977, Las Discorbidae del Mar Caribe, Frente a Venezuela: Cumana, Venezuela, Universidad de Oriente Cuadernos Oceanograficos, v. 6, 44 p., 29 pl.

Sheehan, C. A., Penfield, G. T., and Morales, E., 1990, Costa Rica geologic basins lure wildcatters: Oil and Gas Journal, v. 88, p. 73–79.

Silver, E. A., Reed, D. L., Tagudin, J. E., and Heil, D. J., 1990, Implications of the north and south Panama thrust belts for the origin of the Panama orocline: Tectonics, v. 9, p. 261–281.

Sokal, R. R., and Michener, C. D., 1958, A statistical method for evaluating systematic relationships: University of Kansas Science Bulletin, v. 38, p. 1409–1439.

Sprechmann, P., ed., 1984, Manual de geología de Costa Rica. Vol. 1: Estratigrafía: San José, Universidad de Costa Rica, 320 p.

Taylor, G., 1975, The geology of the Limon area of Costa Rica [Ph.D. thesis]: Baton Rouge, Louisiana State University, 116 p.

Terry, R A., 1956, A geological reconnaissance of Panama: Occasional Papers of the California Academy of Sciences, v. 23, 91 p.

Todd, R., and Bronnimann, P., 1957, Recent Foraminifera and Thecamoebina from the eastern Gulf of Paria: Cushman Foundation for Foraminiferal Research Special Publication 3, 43 p.

van Morkhoven, F.P.C.M., Berggren, W. A., and Edwards, A. S., 1986, Cenozoic cosmopolitan deep-water benthic foraminifera: Bulletin du Centre de Recherches de Pau Mémoire 11, 433 p.

Vergara Muñoz, A., 1988, Tectonic patterns of the Panama Block deduced from seismicity, gravitational data and earthquake mechanisms: Implications to the seismic hazard: Tectonophysics, v. 154, p. 253–267.

Wadge, G., and Burke, K., 1983, Neogene Caribbean plate rotation and associated Central American tectonic evolution: Tectonics, v. 2, p. 633–643.

Wells, J. W., 1956, Scleractinia, *in* Moore, R. C., ed., Treatise on invertebrate paleontology. Part F: Coelenterata: Boulder, Colorado, Geological Society of America (and University of Kansas Press), p. F328–F444, text figures 222–239.

Wilkinson, L., 1989, SYSTAT: The system for statistics: Evanston, Illinois, SYSTAT, Inc., 822 p.

Winslow, M. A., Vega, L. A., and Hall, R. C., 1988, Tectonic evolution of an actively deforming back-arc basin: The Limón basin of Costa Rica: Transactions, Caribbean Geological Conference, 11th, Barbados, p. 19:1–19:17 [published incorrectly as Wilson et al.].

Woodring, W. P., 1982, Geology and paleontology of Canal Zone and adjoining parts of Panama, description of Tertiary mollusks (pelecypods: Propeamussiidae to Cuspidariidae; additions to families covered in P 306-E; additions to gastropods; cephalopods): U.S. Geological Survey Professional Paper 306-F, p. 542–845.

MANUSCRIPT ACCEPTED BY THE SOCIETY FEBRUARY 7, 1994

Printed in U.S.A.

Geological Society of America
Special Paper 295
1995

Morphotectonics of the Pacific convergent margin of Costa Rica

R. von Huene, J. Bialas, E. Flueh, B. Cropp, T. Csernok, E. Fabel, and J. Hoffmann
GEOMAR, Research Center for Marine Geosciences, Division of Oceanic Geodynamics, Christian-Albrechts-University, Wischhofstrasse 1-3, 24148 Kiel, Germany
K. Emeis, P. Holler, and G. Jeschke
Geologisch-Paläontologisches Institut, Christian-Albrechts-University, Wischhofstrasse 1-3, 24148 Kiel, Germany
C. Leandro M., I. Peréz Fernandéz, and J. Chavarria S.
Instituto Costarricense de Electricidad, San Jose, Costa Rica
A. Florez H. and D. Escobedo Z.
Universidad Nacional Autonoma de Mexico, Mexico D.F., Mexico
R. León
Escuela Centraoamericana de Geologia, Universidad de Costa Rica, San Jose, Costa Rica
O. Barrios L.
Universidad Autonoma de Nuevo Leon, Facultad de Ciencas de la Tierra, Linares, Mexico

ABSTRACT

Offshore of the Pacific side of Costa Rica, the Caribbean plate converges with the subducting Cocos plate along the Middle America Trench. The tectonics of both plates, from the Cocos Ridge to the Nicoya Peninsula, were studied with swathmapping, magnetic anomalies, and samples. Three morphological domains on the Cocos plate were defined by mapping. The broadly arched Cocos Ridge forms the southeastern domain. Adjacent to the northwest flank of Cocos Ridge is a domain where seamounts and their aprons cover about 40% of the ocean floor. Farther northwest, a sharp juncture in the oceanic crust separates the seamount domain from a deep sea plain. These three contrasting oceanic seafloor morphologies are mimicked in the morphology of the Pacific continental margin of Costa Rica. Opposite the subducting Cocos Ridge are a broad continental shelf and Osa Peninsula, which are attributed to large-scale domal uplift. Where the seamount domain has been subducted, a rugged continental slope has developed, including 55-km-long furrows trending parallel to the Cocos-Caribbean interplate convergence direction. We propose that the furrows represent paths of disruption produced by subducting seamounts. Where the smooth deep sea plain has been subducted, a well-organized accretionary prism covered by slope deposits forms a relatively smooth morphology. The Costa Rican margin illustrates the effects of subducting seafloor morphology on the continental margin structure and morphology.

INTRODUCTION

During the 76th cruise (SO-76) of the German Research Vessel *Sonne* off Costa Rica (December 1991 and January 1992), a new Hydrosweep swathmapping system was used to record more than 8,500 track-line-kilometers of bathymetric data along with magnetic anomaly data. The grid of lines was navigated using GPS (Global Positioning System) positions. Wide-angle reflection and refraction measurements with OBH (Ocean Bottom Hydrophone) and land recordings were also accomplished and are reported elsewhere (Flueh et al., 1992). Samples of older rock from the continental massif and sediment cores were recovered. These data and samples were acquired for study by Costa Rican and German scientists in a joint continuing pro-

von Huene, R., Bialas, J., Flueh, E., Cropp, B., Csernok, T., Fabel, E., Hoffmann, J., Emeis, K., Holler, P., Jeschke, G., Leandro M., C., Peréz Fernandéz, I., Chavarria S., J., Florez, H., A., Escobedo Z., D., León, R., and Barrios L., O., 1995, Morphotectonics of the Pacific convergent margin of Costa Rica, *in* Mann, P., ed., Geologic and Tectonic Development of the Caribbean Plate Boundary in Southern Central America: Boulder, Colorado, Geological Society of America Special Paper 295.

ject (see acknowledgments). The project, named PACOMAR (PAcific COntinental MARgins), concerns the understanding of natural hazards, particularly earthquakes and tsunamis, and basic knowledge of plate boundary processes. Major tectonic boundaries on the oceanic plate not obvious from previous data are detailed with high-resolution morphology, and the continental slope shows several features uncommon in accretionary prisms. Our maps of morphology, which were constructed onboard the *Sonne,* the magnetic anomalies, and sampling data are the principal subjects of this report.

TECTONIC SETTING

Middle America Trench

The Middle America Trench (MAT) is the morphological expression of a convergent plate boundary produced by subduction of the oceanic Cocos plate beneath the continental crust of Central America. The Cocos plate off Osa Peninsula includes Cocos Ridge (Fig. 1), the trace of the Galapagos hot spot, which near the MAT is of middle Miocene age (Hey, 1977; van Andel et al., 1973). At the crest of the Cocos Ridge the MAT is about 1,400 m deep, and the depth of the axis increases northwestward down the ridge flank to about 3,700 m. Farther northwest off the Nicoya Peninsula the axis is 4,000 m deep. Little trench-filling sediment is observed in the MAT axis, and most of the sediment reaching the trench must soon be subducted. Efficient sediment subduction has previously been recorded in seismic records (Crowe and Buffler, 1983; Shipley and Moore, 1986; Shipley et al., 1992).

Oceanic crustal domains

Seaward of the trench axis a rough and a smooth morphologic domain have been noted in studies of the ocean crust (Fisher, 1961; Hey, 1977; Case and Holcombe, 1980). Oppo-

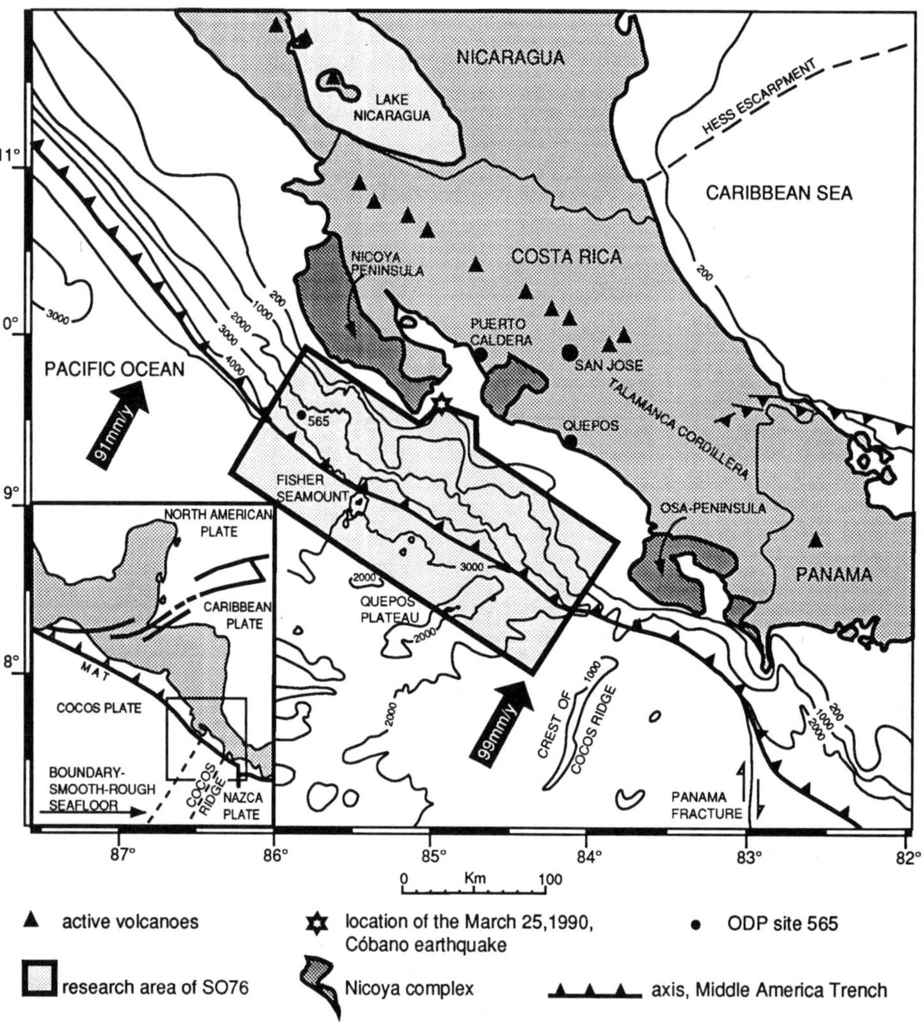

Figure 1. Location map showing the area of study during SO-76 off Costa Rica. Onshore, outcrops of the Nicoya Ophiolitic complex, which forms the framework of this margin, are indicated. Bathymetry shown by 1,000-m contours is after Case and Holcombe (1980) and this chapter. MAT = Middle America Trench.

site Nicoya Peninsula, the ocean-basin morphology is smoothed by sediment 500 m thick covering igneous crust that was formed along the East Pacific Rise (Buffler, 1982; Lonsdale and Klitgord, 1978; Klitgord and Mammerickx, 1982). Crust formed at the east-west–trending Galapagos Rift has a complex magnetic pattern that is incompletely mapped (Hey, 1977). The Galapagos hot spot is thought to have constructed the Cocos Ridge on this crust. Near the MAT the ridge is of upper Miocene age (van Andel et al., 1973). The rough crust from Galapagos Rift and the crustal domain with contrasting smooth crust from the East Pacific Rise compose the Cocos plate, which is subducted along the MAT. The direction of plate convergence is nearly normal to the trench axis at a rate between 91 and 99 mm/yr (DeMets et al., 1990) (Fig. 1).

On the Panamanian side of Cocos Ridge, southeast of Osa Peninsula (Fig. 1), the Panama fracture zone cuts diagonally across the ridge, offsetting it right-laterally. It is argued from plate tectonic history (Lonsdale and Klitgord, 1978; Gardner et al., 1992) that the Malpelo and Cocos Ridges were a single feature that was offset beginning about 9 Ma (Fig. 2). According to the plate tectonic reconstructions, a segment 100 km long between the two offset parts of the ridge is now subducted. Interestingly, the northwest flank of Malpelo Ridge has seamounts like the corresponding flank of Cocos Ridge (Lonsdale and Klitgord, 1978). Thus both exposed ridge segments on either side of the subducted segment have seamounts, indicating that a seamount domain has been subducted beneath central Costa Rica.

Continental slope

The continental slope opposite to the smooth morphology of the Cocos plate is similarly without much local relief. Here some of the most extensive seismic reflection work across a convergent margin has been done (Crowe and Buffler, 1983; Shipley and Moore, 1986; Stoffa et al., 1991; Shipley et al., 1990, 1992; Hinz et al., 1992). The seismic data include a three-dimensional survey indicating an upper plate consisting of a typical wedge-shaped accretionary prism covered by continental margin sediment. The base of the continental margin sediment has a minimum age of 5 Ma at Site 565 where the section was partially penetrated by DSDP drilling (von Huene et al., 1985). The boundary between the contemporary accretionary prism and the Cretaceous Nicoya Complex exposed on the Nicoya Peninsula lies in an undetermined position offshore.

The continental slope morphology to the southeast, where rough oceanic crust is subducted (between the Nicoya and Osa Peninsulas), is marked by a highly irregular topography (Lonsdale and Klitgord, 1978; Case and Holcombe, 1980). Seismic reflection records (Buffler, 1982; Crowe and Buffler, 1983) indicate that the irregular morphology is associated with structure differing from that in the better-known forearc sector facing the area of smooth Cocos plate. Uplifted basement blocks, slumping, normal faults, deformed slope sediment, and ac-

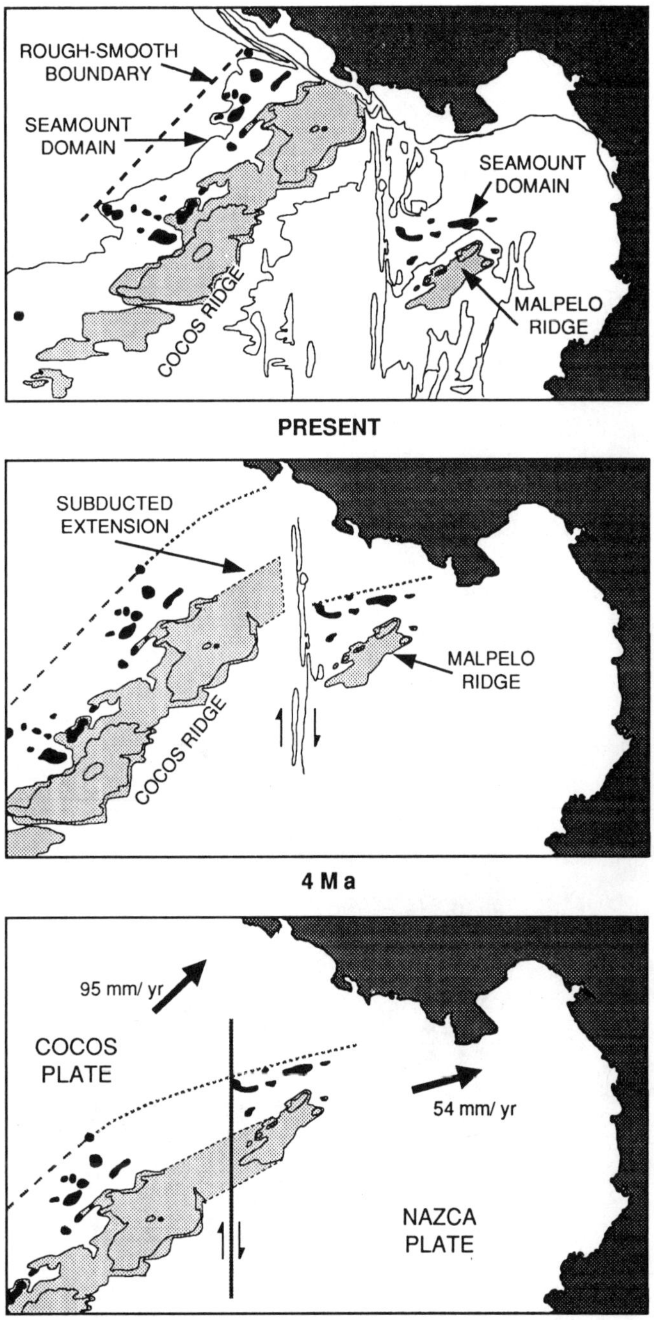

Figure 2. Plate tectonic reconstruction showing displacement of Cocos and Malpelo Ridges and the adjacent seamount domain. Ridges are shaded and seamounts are blackened above 2,000 m. Extension of Cocos Ridge to Malpelo Ridge is based on the bathymetry of Lonsdale and Klitgord (1978); plate tectonic reconstruction is after those authors and Gardner et al. (1992).

creted masses are inferred features that provide a far greater structural variability than observed to the northwest.

The continental slope off Osa Peninsula rises abruptly from the crest of the Cocos Ridge. Contemporary vertical tectonism measured on the Osa Peninsula is greater than that observed on the Nicoya Peninsula (Corrigan et al., 1990; Gardner et al., 1992). This accelerated rate of vertical tectonism and the associated morphology are inferred to result from uplift accompanying subduction of Cocos Ridge. Rates of contemporary uplift appear greater opposite the crest of the ridge than opposite its flanks (Gardner et al., 1992).

Variability in the seafloor morphology appears related to differences in structure of the oceanic crust. Differences in the thickness, density, and thermal state of the lithosphere affect the relative sinking or buoyancy of crustal segments as they are subducted. Such crustal differences are given as an explanation for a steeply dipping Benioff Zone beneath the Nicoya Peninsula relative to that beneath the Osa Peninsula (Protti et al., this volume). The shallow dip of the Benioff Zone opposite the crest of the Cocos Ridge as been noted by several investigators (Liaw, 1981; Kolarsky et al., this volume; Protti et al., this volume).

GENERAL MORPHOLOGY FROM SWATHMAPPING WITH THE HYDROSWEEP MULTIBEAM ECHOSOUNDER AND MAGNETIC ANOMALY OBSERVATIONS

Data acquisition

In the fall of 1991, an Atlas Hydrosweep multibeam echosounder system was installed during the rebuilding of the RV *Sonne*. The system's onboard contour-map plotting capability consists of the Caris mapping software interfaced to the Hydrosweep instrument through an Atlas Hydromap (HMS) software shell. The SO-76 cruise was the first field use of the system for extensive swathmapping. Preliminary maps were produced onboard (Fig. 3 and Plates 7 and 8) and are not yet succeeded by postcruise maps for lack of a comparable data reduction system onshore.

Simultaneously with the Hydrosweep mapping, total field magnetic values were measured with a Geometrics G866 proton procession magnetometer. These data were reduced after the cruise by removing the diurnal variation determined with measurements at the Instituto Costarricense de Electricidad Geomagnetic Observatory at Tilaran, Costa Rica (80 km from Puntarenas). After subtracting the International Geomagnetic Reference Field, the data were plotted and contoured with general purpose software. The initial results were locally distorted by systematic shifts between lines that occurred because the sensor was not towed sufficiently far from the magnetic field of the ship. After a systematic correction, the shifts disappeared (Fig. 4).

Morphotectonic features resolved

We begin a general description of the surveyed area's morphology by portraying the axial profile of the MAT (Fig. 5) and then describe the ocean basin crustal features on one side and the continental margin on the other. The axial profile illustrates a section across features of the ocean-basin crust that are about to be subducted. The profile contains a steep slope to the southeast defining the Cocos Ridge and its thickened crust. The ridge crest stands 2 km above the adjacent ocean-basin floor. Along the trench axis the northwest ridge flank is 100 km wide or about 60% of the width of the average profile constructed by Gardner et al. (1992). Profiles of the ridge averaged by those authors are 50 km to 250 km seaward of the trench axis, and the difference suggests either a greater ridge width in a seaward direction or inclusion and averaging of many seamounts into their profile. One seamount of the northwest ridge flank has arrived in the trench axis and constricts it. This seamount is a long northeast-trending plateau, informally called Quepos Plateau (Western Elevation of Kolarsky et al., this volume; Figs. 1, 3, #2).

Where Cocos Ridge is subducted, the trench axis has a narrow V-shaped profile consistent with its relatively steep axial gradient and lack of sediment. Normal faulting, which commonly characterizes the seaward slope of trenches, is absent along the thickened crust of the Cocos Ridge between Quepos Plateau and the ridge crest. In fact, the Cocos Ridge morphology has significantly fewer local positive features relative to the adjacent northwestern seafloor (Case and Holcombe, 1980).

Across the MAT axis on the landward side of the trench, the slope rises steeply to a broad shelf containing the Osa Peninsula (Fig. 1; Case and Holcombe, 1980). Northwest, the broad shelf is bounded by a transverse escarpment in the middle and upper slope directly opposite the Quepos Plateau (Fig. 3, #3). The escarpment does not continue across the lower slope on one side or across the shelf on the other.

The axial gradient over the 90 km in the segment adjacent to Cocos Ridge drops only 200 m and includes several sediment-filled basins (Fig. 5). Individual basins are separated by diagonal east-west–trending scarps formed where the ocean crust is flexed into the trench axis (Fig. 3). The scarps represent normal faults (Crowe and Buffler, 1983) breaking the oceanic crust parallel to its structural grain or the trend of magnetic anomalies (Fig. 4).

Figure 3. Map of Hydrosweep data from the PACOMAR project. Soundings were binned at 100- to 120-m grid point intervals for contouring at a 50-m interval. Conventional bathymetry in areas without Hydrosweep data is recognized from smooth contours near the edges of the map. Features referred to in the text are numbered as follows: 1. Fisher Seamount and ridge; 2. Quepos Plateau; 3. transverse escarpment, Osa shelf; 4, 5, 6. furrows and associated uplifts; and 7, 8. features along the ENFZ trend.

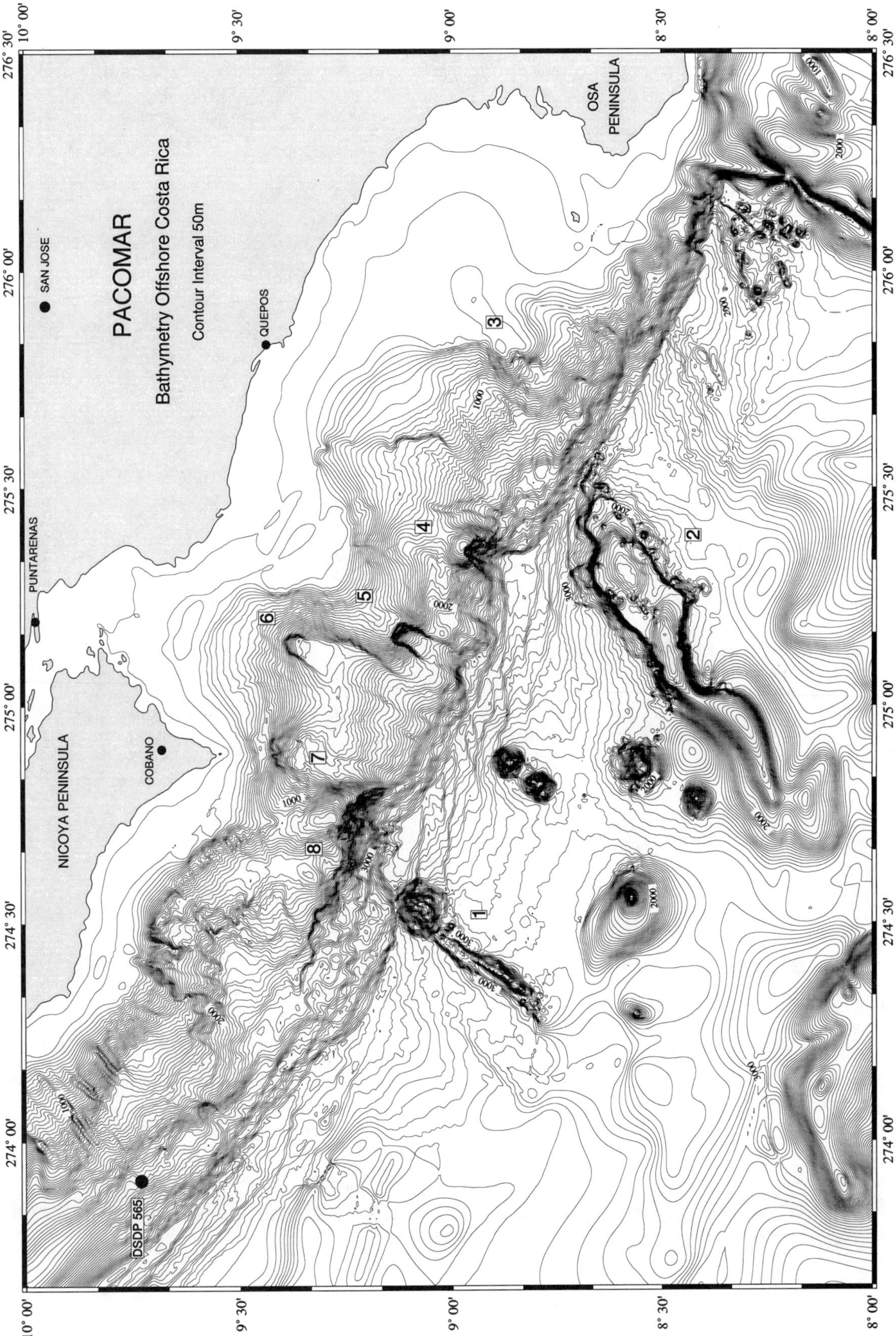
PACOMAR
Bathymetry Offshore Costa Rica
Contour Interval 50m
SAN JOSE
QUEPOS
PUNTARENAS
NICOYA PENINSULA
COBANO
OSA PENINSULA
DSDP 565
1
2
3
4
5
6
7
8
10° 00'
9° 30'
9° 00'
8° 30'
8° 00'
274° 00'
274° 30'
275° 00'
275° 30'
276° 00'
276° 30'

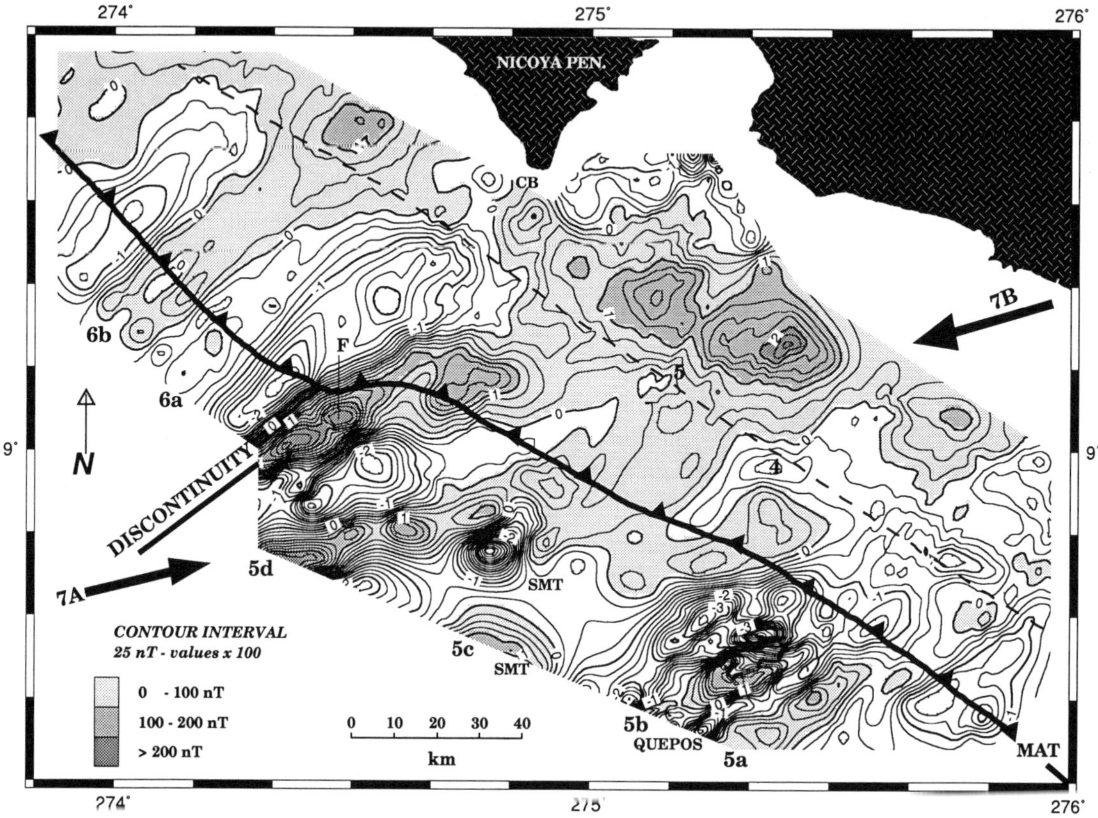

Figure 4. Map of magnetic anomalies recorded simultaneously with the Hydrosweep measurements. Contour interval is 25 nT, and bold lines are 100-nT contours annotated in hundreds. Positive values are shaded. Arrows show direction of view in Figure 7. Magnetic anomalies are numbered along the southwest parameter of the surveyed area. The age gap along the discontinuity is between anomaly 5d and 6a—18 Ma and 21 Ma, respectively. F = Fisher seamount; SMT = seamount; 4 and 5, inferred buried seamounts keyed to Figure 2; MAT = Middle America Trench, with trench axis indicated by barbed line. Dashed line is the limit of visible magnetic anomalies from the subducted ocean crust.

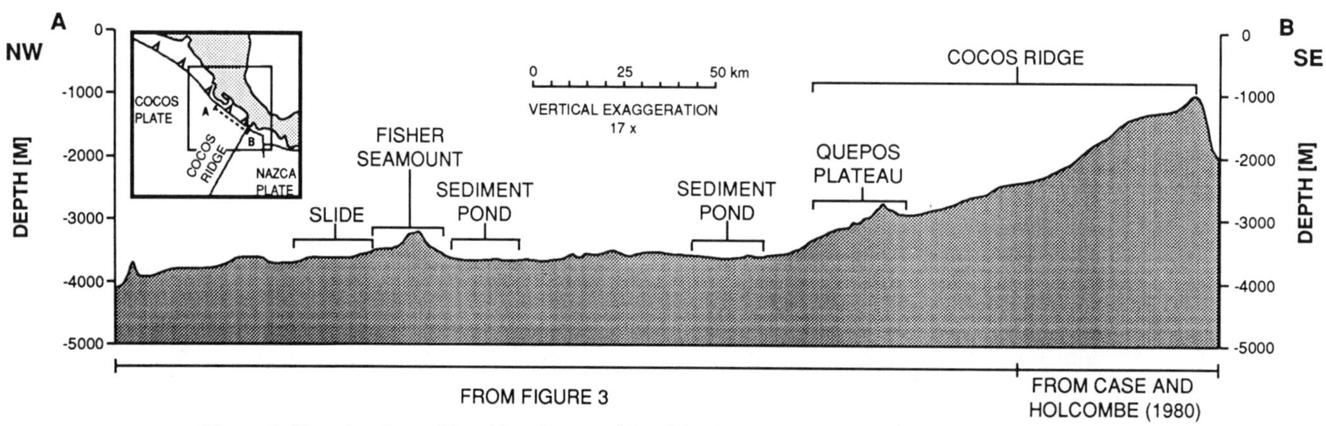

Figure 5. Trench axis profile with points outside of the SO-76 area from Case and Holcombe, 1980.

The northwestern constriction of the trench axis seen in the axial profile (Fig. 3, #1, and Fig. 5) is caused by Fisher Seamount. Extending southwest on the Cocos plate from the seamount is a 35-km-long and 5- to 7-km-wide volcanic ridge: across the ridge, the seafloor is vertically displaced 250 to 300 m. A similar morphologic step in the seafloor but without the volcanic ridge occurs 40 km seaward along strike, across a normal fault (Crowe and Buffler, 1983). The Cocos plate between the Quepos Plateau and the Fisher Seamount lineament is characterized by many seamounts, as recognized commonly since the mapping of Fisher (1961). Despite considerable conventional soundings in previous compilations, the three new seamounts discovered in the small portion of the seamount domain surveyed during SO-76 suggest many others may be undiscovered. The morphological contrast between Quepos Plateau and the conical seamounts is clear from the SO-76 swathmapping.

Opposite the seamount domain the continental slope between Nicoya and Osa Peninsulas has developed a morphology disrupted more than that of any previously swathmapped part of the MAT (cf. Aubouin et al., 1982; Shipley and Moore, 1986). Striking features of the margin here are furrows across the continental slope, some with a sharp morphology and others that are more obscured by later sedimentation and erosion (Fig. 3, #4–#7). At the landward heads of some furrows, 50- to 150-m-deep closed lows not yet obscured by sedimentation indicate a recent origin (Fig. 3, #4–#7). Some of these furrows (Fig. 3, #4 and #5) end at a low raised mound. In the middle slope a conical mound (Fig. 3, #4) trails behind it a short furrow that breaches the lower slope (Fig. 6A). Upslope from that mound are curved canyons in the morphology conforming to the shape of the mound. The adjacent furrow (Fig. 3, #5) reaches to the middle slope, and it too has a raised conical mound at its head (Fig. 6B) that is lower and broader than the first. A third furrow (Fig. 6C) is without a mound and extends 55 km (Fig. 3, #6). Each of the furrows strikes parallel to the direction of relative plate motion (Figs. 1 and 2; McNally and Minster, 1981; DeMets et al., 1990).

Fisher Seamount and its trailing long ridge mark the northwest boundary of the rough seafloor and the disrupted slope topography. The trend of Fisher Seamount and ridge is repeated on the landward side of the MAT by two broad transverse-oriented features crossing the continental slope. Opposite Fisher Seamount, the slope rises steeply from the trench as a semiarcuate unstable cliff. The top is a north-trending bank aligned with Cabo Blanco on the upper slope, the southernmost point of the Nicoya Peninsula, and the linear southeast coastline of Nicoya Peninsula, after which the trend ends. These transverse features separate distinct morphologies to the northwest and southeast on both the seaward and landward sides of the trench axis. The trend has been proposed as a major tectonic feature (Crowe and Buffler, 1983; Corrigan et al., 1990), who called it the ENFZ (East Nicoya Fracture Zone). The swathmap morphology shows that in the upper plate the trend is composed of multiple aligned morphotectonic elements.

The trench axis profile northwest of Fisher Seamount has a gradient similar to that between the Quepos and Fisher constrictions (Fig. 5). The ocean-basin seafloor is smooth, and swathmapping, particularly that of Shipley and Moore (1986), differs little from the conventional bathymetry (Case and Holcombe, 1980). The flexure of ocean crust into the trench axis produces normal faults paralleling the trend of the trench axis (see Fig. 1 of Shipley and Moore, 1986). Thus extensional tectonism from flexural bending produces normal faults paralleling the axis of flexure and across the magnetic anomalies that trend perpendicular to the trench axis (Fig. 4).

In accordance with the smooth morphology of the subducting plate, the continental slope morphology off northwestern Nicoya Peninsula includes a lower continental slope with little local relief (Shipley et al., 1990, 1992). The upper slope has numerous canyons that indicate sufficient tectonic stability for seafloor processes to proceed uninterrupted. North of Cabo Blanco is a broad embayment in the upper slope rilled with short canyons, and the mid-slope has a similarly mature morphology with canyons and flats. The lower slope west-northwest of Fisher Seamount has the morphology of a debris slide emanating from a low escarpment. Since the front of the slide fills the trench axis and projects about 4 km seaward, and the debris is not modified by subsequent subduction, the slide appears to be a recent feature. The slide has flowed against Fisher Seamount, forming a major constriction in the trench axis there. Additionally, the deep lower slope east-southeast of Fisher Seamount is in failure and has retreated landward. The apparent lateral offset of the trench axis noted by previous investigators (cf. Crowe and Buffler, 1983; Corrigan et al., 1990; Kolarsky et al., this volume) would disappear if the advancing slide and retreating slope on either side of Fisher Seamount were restored to their original positions.

Magnetic anomaly measurements

Magnetic anomalies (Fig. 4) show principally the remnant magnetism of the igneous oceanic crust upon which is superimposed the strong local effect of seamounts on the oceanic plate. Off the Nicoya Peninsula the anomalies trend about N50°E and can be followed from the ocean basin to the upper slope. A change in contour patterns at the upper slope could indicate a subtle effect from the Nicoya Complex that continues to where these rocks crop out on land. The linear boundary between the two magnetic anomaly patterns is best observed on perspective presentations (Fig. 7). Near the ENFZ the trend of the oceanic anomalies shifts from N50° to N80°E, and the boundary between two ocean crustal domains can be followed from the ocean basin into the mid-slope. The boundary is marked by strong anomalies on the ocean plate, but these rapidly decrease in amplitude beneath the middle slope. Fisher Seamount and its adjacent ridge trend parallel to the ENFZ.

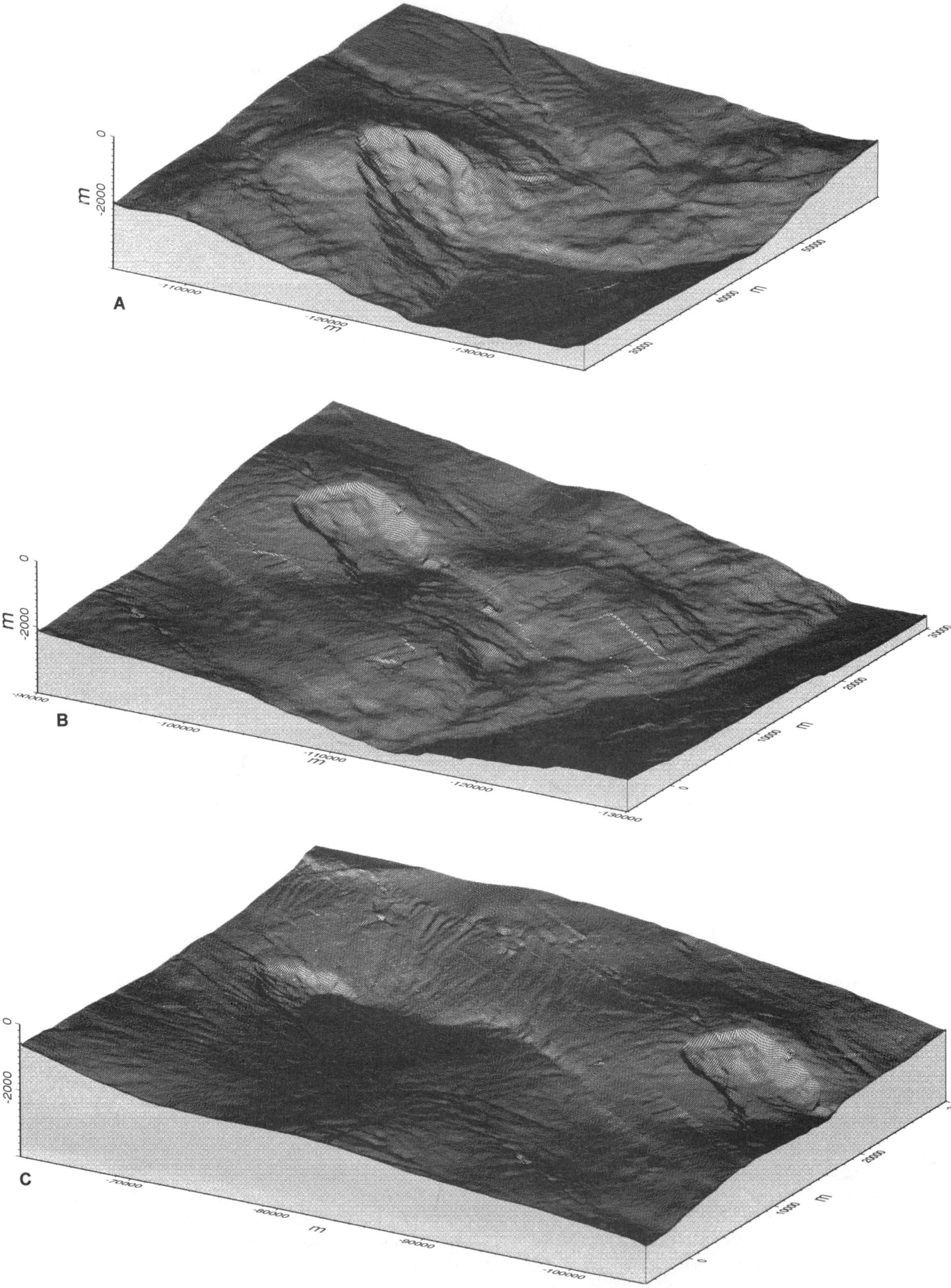
0
-2000
m
-110000
-120000
m
-130000
30000
40000
m
50000
A
0
-2000
m
-90000
-100000
-110000
m
-120000
-130000
0
10000
m
20000
30000
B
0
-2000
m
-70000
-80000
m
-90000
-100000
0
10000
m
20000
30000
C

Quepos Plateau is associated with a complex magnetic field, but its topography and thus its magnetic field also reflect a N50°E directional grain. The shallowest underthrust seamount on the lower slope (Fig. 3, #4) is marked by a 75 to 100 gamma magnetic low (Fig. 4), but deeper underthrust seamounts are without clearly distinguishable magnetic anomalies. The largest magnetically positive area of the map is along the indented shelf edge between the Nicoya and Osa Peninsulas. These positive anomalies mask anomalies associated with the oceanic crust. The landward positive magnetic anomaly pattern is explained as an effect from the Nicoya Ophiolitic Complex, which becomes shallow (1 km below sea level) beneath the upper slope and shelf, whereas the ocean crust is subducted considerably deeper (10 km below sea level).

Based on a comparison with a standard anomaly sequence, the oceanic anomalies were assigned numbers (Fig. 4) along a profile across the southwestern parameter of the surveyed area. That comparison indicates a missing segment between magnetic anomalies 5d and 6a representing a temporal gap of about 3 m.y. The regional magnetic anomaly coverage is insufficient to reconstruct the lateral displacement causing this juxtaposition of different-aged crust.

EFFECTS OF LOWER PLATE MORPHOLOGY ON THE UPPER PLATE

Mapping during SO-76 resolved in detail a continental slope morphology inferred to result from the subduction of differing seafloor morphologies. Even from conventional onland and offshore mapping, previous authors recognized in a generic way the morphological similarity between the Cocos plate and the continental slope. The seamount domain is opposite the disrupted slope morphology between the Nicoya and Osa Peninsulas, Cocos Ridge is opposite the broad Osa shelf, and the smooth oceanic crust is opposite a relatively orderly Nicoya slope morphology. But the boundaries between morphological provinces were poorly defined by incomplete coverage and imprecise navigation. The SO-76 data indicate sharp boundaries and distinct morphological domaines. With new insights from swathmapping, we discuss the apparent consequences resulting from the subduction of the three ocean-basin morphologies beneath the frontal wedge of the upper plate offshore Costa Rica.

Figure 6. Mesh-net perspective diagrams showing furrows and low mounds. A shows the shortest furrow, uplift of the sides of the furrow, and subsequent collapse sending slump blocks from the side into the center. Landward of the uplifted mound are short ridges and a canyon that conforms with the uplift. B shows next-longer furrow where the breached lower slope has been reconstructed along the trench axis. C shows the longest furrow (55 km), which has an older subdued topography at its foot and a relatively low slump scar at its head compared with that in the right corner, which is also shown in B.

Models of ridge subduction

Uplift along the coast and its relation with subduction of the Cocos Ridge have been the subject of several recent studies to quantify uplift (cf. Corrigan et al., 1990; Gardner et al., 1992; Kolarsky et al., this volume). Because Cocos Ridge trends perpendicular to the margin and in the plate convergence direction (Fig. 1), about 100 km of the ridge has subducted essentially in one place during the past 1 to 3 m.y. (Fig. 2). Opposite the crest of Cocos Ridge, the Osa Peninsula has experienced accelerated rates of uplift, whereas opposite the northwest flank of Cocos Ridge rates of uplift are lower (Gardner et al., 1992). Along the Nicoya Peninsula where normal Cocos plate subducts, uplift is assumed to reflect a background rate for the general vertical tectonics that accompanies plate convergence. Arched uplifted coastal terraces have been observed opposite the subducting Nazca Ridge in Peru (Hsu, 1988; Macharé and Ortlieb, 1992) where broad domal uplift over the past 125,000 yrs occurred at about one-tenth the rate proposed by some at the Osa Peninsula (Gardner et al., 1992).

The effects from the subducted ridge on the upper plate can be simplistically modeled as a vertical uplift beneath an elastic sheet (Gardner et al., 1992). Such modeling demonstrates how the domal uplift of the surface becomes broader and lower as elastic sheet thickens (i.e., the ridge becomes more deeply buried).

Cocos Ridge–Osa Shelf

Cocos Ridge with its thickened crust (Bentley, 1974) is associated with a shallowing of the subducting plate under the forearc of Costa Rica (Buffler, 1982; Adamek and Tajima, 1987; Corrigan et al., 1990; Kolarsky et al., this volume; Protti et al., this volume). The steep gradient of the trench axis where the crestal ridge subducts is not associated with a horizontal shift in its position as along the Louisville Ridge (Ballance et al., 1989). Unlike Louisville Ridge, Cocos Ridge has been subducted essentially in one location rather than migrating along the margin and leaving a trailing wake of local erosion. The generally low local relief of the ridge-crest morphology is broken by Quepos Plateau, which differs morphologically from other seamounts mapped. Its nearly flat top is covered with sediment (Buffler, 1982), and many small conical miniseamounts. Other seamounts of this type may be present in a similar position on the ridge seaward of Quepos Plateau (Case and Holcombe, 1980).

Adjacent to the crest of the ridge on the upper plate is the Osa shelf, characterized by steeply dipping flanks. An uplifting area opposite the ridge crest is consistent with the uplift mechanisms as proposed by Corrigan et al. (1990) and Gardner et al. (1992). The northwest part of the shelf mapped during SO-76 includes the middle- to upper-slope transverse escarpment offshore of Quepos (Fig. 3, #3). The scarp is near the Quepos constriction and the imminent collision of the

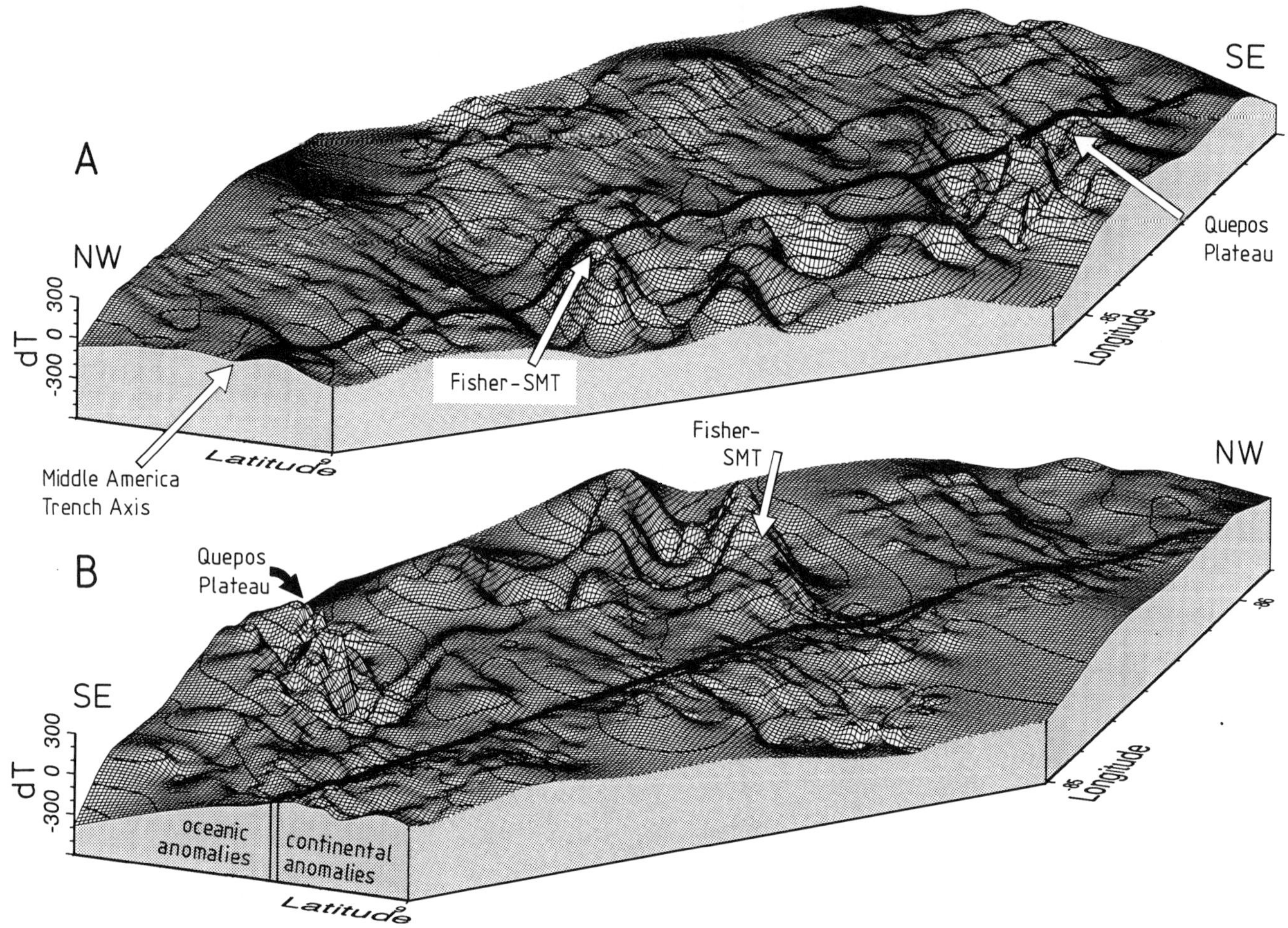

Figure 7. Perspective diagrams of magnetic anomalies. A, looking NE from the ocean basin (see Fig. 4 for direction). B, looking in the reciprocal direction from the continent. The approximate trace of the trench axis and the change from oceanic to continental magnetic anomalies are sketched across the diagrams.

Quepos Plateau. Alignment of the escarpment with the direction of convergence opposite Quepos Plateau suggests possible erosion by subduction of a positive feature of similar size during the past 1 m.y.

Models of seamount subduction

We focus here on seamount subduction because the swathmapped morphology and older geophysical data (Buffler, 1982) related to furrows are explained as tracks from seamounts that were thrust beneath the continental margin. Four principal morphological observations that are most simply explained by this hypothesis are: (1) The furrows only occur opposite the seamount domain of the subducting Cocos plate. (2) Uplift at the head of a furrow from a conical feature at depth has dimensions similar to those of many seamounts on the adjacent Cocos plate. (3) The furrows are linear and parallel the direction of relative plate convergence. (4) Furrow morphology includes a straight trace, locally raised edges that collapsed after uplift, a younger morphology at the head, and a mature distal morphology modified by common seafloor processes (Fig. 8). Our explanation is discussed more fully below.

An analog to the morphology of Costa Rica is found in morphological features observed where seamounts are subducted along the convergent Japan (Lallemand et al., 1989), Tonga (Ballance et al., 1989), and New Hebrides (Collot and Fisher, 1989) margins. Positive features on the subducting plate uplift and disrupt the upper plate where it is thin and weak. A general model of this process (Fig. 9) follows a model of seamount subduction along the Japan Trench (von Huene and Lallemand, 1990) modified to better fit Costa Rica. Initially (Fig. 9a) the seamount enters the trench and constricts it. Fisher seamount and Quepos Plateau constrict the trench off Costa Rica (Fig. 3, #1; Fig. 5). When the front of the seamount initially impacts the accretionary wedge it oversteepens the

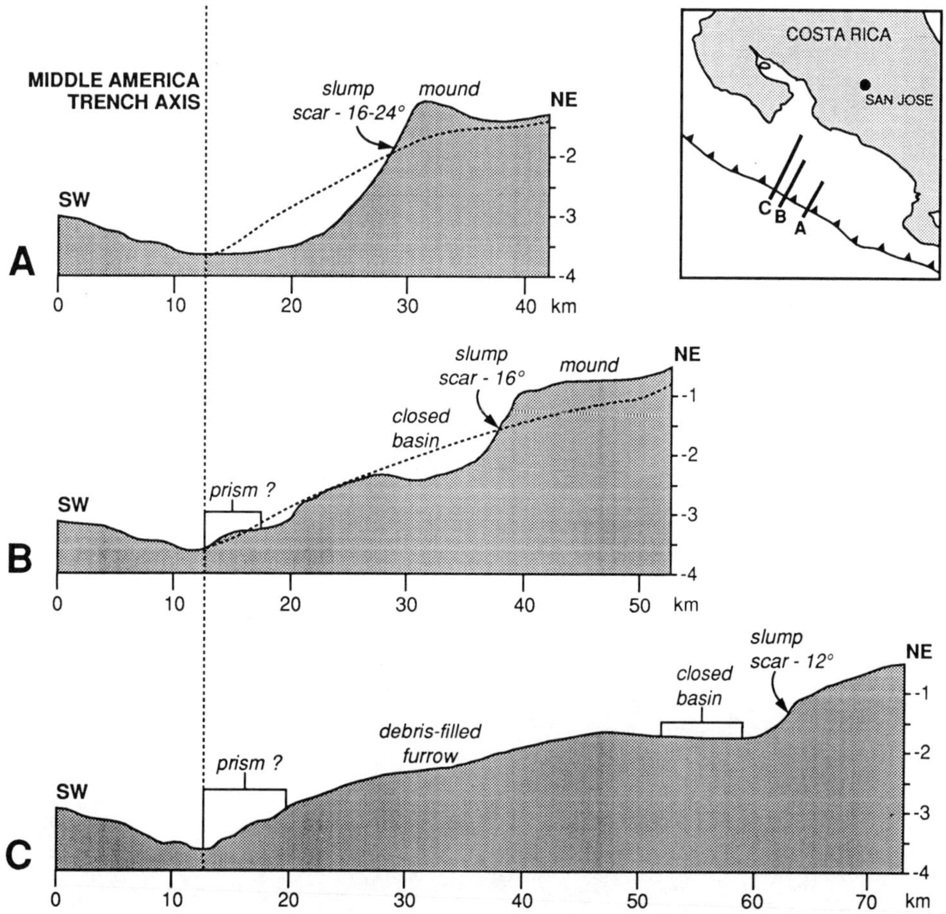

Figure 8. Annotated profiles of morphology along the axis of three furrows keyed to Figure 3 as follows: A runs through #4, B through #5, and C through #6, Figure 3. Dashed lines indicate general morphological profile to the side of a furrow, showing the initial configuration prior to seamount subduction. Note the decrease in height of the slump scars as lower plate is subducted deeper.

slope, a condition that produces slides (Fig. 9B and C). These stages are well known from the studies of Daiichi Kashima Seamount (Lallemand et al., 1989). The leading flank of Daiichi Kashima Seamount induces fracturing and contractional faulting as it uplifts the accretionary wedge (Lallemand and LePichon, 1987). When the seamount has been completely underthrust beneath the accretionary prism and the upper plate has ridden up and over its crest (Fig. 9C), the upper plate above the trailing flank of the seamount is oversteepened, causing it to fail in mass wasting and sending debris into the trench. This stage (Fig. 9C) is represented off Costa Rica by the seamount at the foot of the slope (Fig. 6A). As the trailing flank continues to subduct, the uplifted upper plate increases in thickness until it retains continuity (Fig. 9D and E). Soon afterward the foot of the margin is reconstructed by accretion. Only at the foot of the margin is accretion active, and the scars from seamount collision remain visible longest on the middle slope.

Seamount domain—disrupted continental margin

Within the mapped area many seamounts on the oceanic plate are of conical shape and are roughly 15 km in diameter and 1.5 km high. Despite the lack of much seismic data as this is written (on *Sonne* cruise 81 seismic reflection lines were shot across much of the Costa Rican margin [Hinz and others, 1992]; however, processing of the data was not begun until spring 1993), we interpret certain features of the disrupted Costa Rican morphology in accordance with the general model of Figure 9. The three adjacent parallel furrows (Fig. 3, #4–#6) illustrate a progression of seamount subduction. The lower slope circular mound and short furrow is the latest such feature (Fig. 3, #4; Figs. 6A and 8A). In map view the furrow is a broad V-shaped embayment adjacent to the trench axis, where the accretionary wedge is probably thinner than the height of most seamounts. Here much of the wedge material must flow around a subducting seamount, thereby leaving a

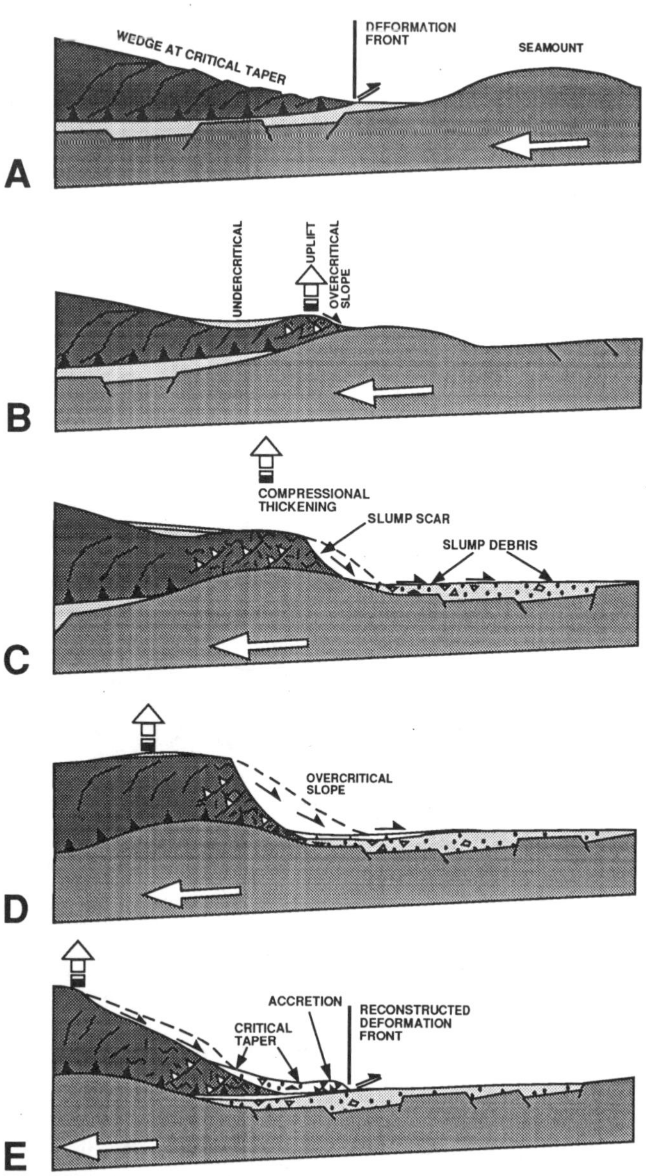

Figure 9. Diagrams showing a general model of the subduction of a seamount (after von Huene and Lallemand, 1990). A shows initial approach of the seamount and the incipient change in critical taper. B shows initial subduction of the seamount, which not only changes the critical taper but also introduces a nonsedimented surface in the subduction zone. C shows the tunneling stage in which the disturbed upper plate wedge overtops the seamount crest and collapses down the trailing flank as it is oversteepened above the negative slope. D shows th stage in which the upper plate remains coherent in its lower part and only the surface is oversteepened. Not all of the debris flows reach the trench and begin to partially fill in the seamount track. In E the deformation front has closed the breach made by the seamount track and begins to accrete the accumulated debris and trench sediment.

broad embayment. The deformation in the direction of seamount movement is clearly reflected in the morphology by troughs 8 to 10 km from the crest and around the leading flank of the seamount. A 1,000-m-thick mass of incoherently reflective rock (Crowe and Buffler, 1983) appears to be uplifted over the seamount, which overtops the crest and slumps down its trailing flank. Trailing the crest and below a 16° to 24° slump scarp the morphology indicates debris from mass wasting that partially fills the furrow. Along segments of the furrow its sides were raised and subsequently slumped. These failed features would be found in the wacke of relief on the subducting plate as it plowed across the margin.

Cores from a sediment pond in the trench axis opposite the furrow recovered sediments containing several intervals of debris-flow origin intercalated with laminated and graded, fine-grained olive-gray hemipelagic sediments (Fig. 10). Varicolored pebble- to sand-sized indurated rock fragments in the debris flow are composed of massive, dark greenish-gray siltstones. Major components in the pebbles are angular quartz grains and volcanic glass (15% each), diatoms and radiolaria (10% each), and sponge spicules (5%).

Above the seamount's crest, sediment 1,000 m thick has been uplifted about 600 to 800 m. The lower plate on which the seamount stands has descended about 500 m below its elevation at the trench axis.

The second positive topographic feature (Fig. 3, #5) has traversed 33 km, and in the absence of any seismic image, its dimensions are unknown. From neighboring seismic records (Buffler, 1982), the lower plate at its base is 1 to 2 km below basement elevation in the trench axis. The accretionary wedge has thickened to 5 km, and the slope sediment is 1.6 km thick. Here the surface mound is no more than 200 m high, but its seaward flank is still oversteepened beyond the angle of slope sediment stability (16°). The debris no longer reaches the trench, and it stops at a low constriction in the furrow that produces a closed low. The broad V-shaped embayment in the slope is bridged at the trench axis. This bridge is interpreted as an accretionary ridge 300 m high and 5 km wide (Fig. 6b and 8b). Slump debris partially fills the furrow landward of the accreted ridge. This example is very similar to that of the seamount subducted landward of Bougainville Seamount off the New Hebrides (Collot and Fisher, 1989).

The longest intact furrow projects from the lower slope 55 km landward and shows no clear morphologic mound at its head (Figs. 6C and 8C). At present plate convergence rates, a track of this length is produced by a subducting seamount after 0.5 Ma. Its 12° steep failure scarp at the head of the furrow is considerably lower and less angular in form than those of its neighbors. It has numerous erosional rills, but the upper part of the furrow still has a closed depression. At the head of the furrow the oceanic crust is about 13 km deep (~9 km below basement in the trench axis) as measured a few tens of kilometers away with refraction techniques (Liaw, 1982). The first 8 km across the lower slope are bridged by a presumed accre-

tionary mass, and the lowermost part of the furrow is sufficiently modified by seafloor processes to obscure all but the apex of the V-shaped embayment adjacent to the trench axis.

Assuming the seamount responsible for this scar is of similar size to the larger conical ones on the Cocos plate (20 km diameter, 2.5 km height), such a conical seamount can apparently disrupt the overlying wedge to a thickness of roughly four times its height. At that point the pushed-up mound over the seamount has become so low and broad that it no longer leaves a morphological trace from failure of surface materials by gravity sliding. Here the furrow ends, and this mound was no longer detectable with our swathmapping. Accretion restores thc lower slope (a zone 8 km wide) in less than 0.5 m.y., in which time the furrow is modified by surface processes or by another colliding positive feature.

East Nicoya Fracture Zone (ENFZ)

From the swathmapping it can be seen that the strong linear trend noted along the east end of the Nicoya Peninsula in reconnaissance mapping is instead several individual morphological features. On the oceanic plate, Fisher Seamount, its trailing elongate magmatic ridge, and the associated magnetic anomaly define a major lower plate fracture. This ocean plate fracture is a loci for the construction of volcanic features that impact the upper plate as the lower plate is subducted.

On the upper plate, a distinct north-trending morphological ridge occupies the middle and lower slopes along the ENFZ (Fig. 3 and Plates 7 and 8). The two older seismic lines that cross it (Buffler, 1982) show a core of acoustic basement approximately 2.5 km high and 28 km wide across which a 500- to 600-m-thick deformed sediment blanket is draped. The steep seaward-facing failure slope exposes this sediment blanket from which igneous and sedimentary rock was dredged (Fig. 11).

Dredging yielded rock samples of three lithologies (Emeis et al., 1992): a volcanic breccia of as yet unknown chemical affinities and age (Fig. 11A); a turbiditic sandstone/mudstone (Fig. 11B) devoid of siliceous microfossils and thus of presumed pre-Miocene age (in analogy with other nearby areas); and indurated, massive siliceous- and glass-rich siltstones of inferred post-Miocene age (Fig. 11C). The dredged specimens were either freshly broken or rounded, partly bored blocks that

Figure 10. X-ray radiograph of a core section (Core 16971-3; interval 275 cm to 300 cm below seafloor) recovered from a sediment pond at 08°48,2′N and 84°39,1′W (3,592 m water depth) at the mouth of a lower slope furrow. Finely laminated hemipelagic muds overlie distorted and slumped muds over a coarse-grained debris flow (pebble- to sand-sized). Clasts in the debris flow are indurated mudstones and siltstones containing quartz, ash, and opaline fossils as dominant components.

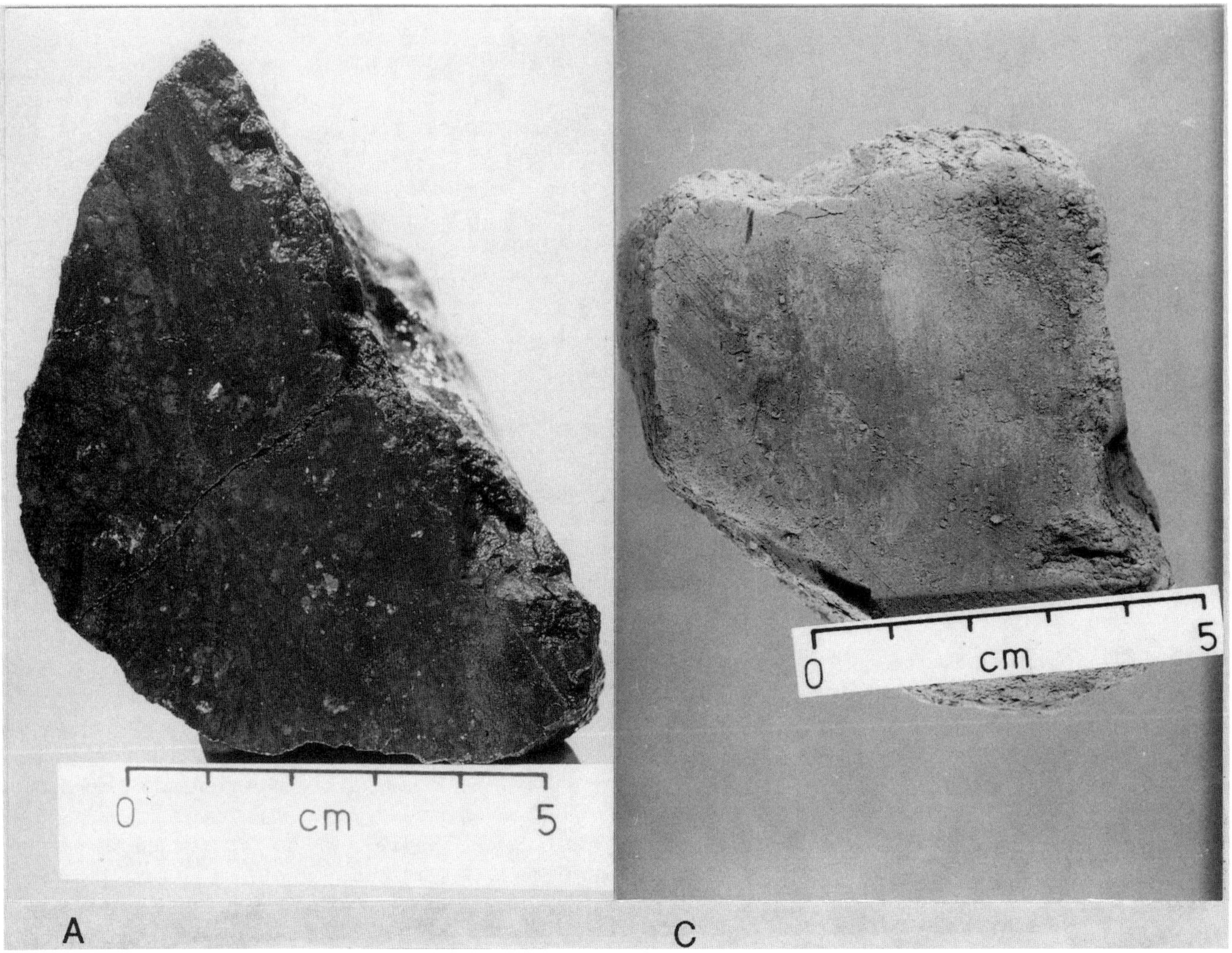

Figure 11. A, specimen of basalt breccia recovered by dredging from 09°12N and 85°,17′W to 09°14′N and 85°16,4′W (2,583 to 1,503 m water depth). The primary structure and mineral content of the breccia are strongly altered; a millimeters-thick oxide crust covering some specimens indicates exposure at the seafloor. In the dredge hauls, the volcanic breccia was associated with turbiditic sandstones (shown in B) similar to those in the Nicoya Complex. C, siltstone of inferred post-Miocene age.

show oxidation rims and oxide coatings. They were thus partly exposed at the seafloor.

Individual fragments of the volcanic breccia (Fig. 11) are moderately vesicular. Considerable alteration of the primary mineral inventory (formation of smectite, zeolites, and carbonates) has only left clinopyroxenes (diopside) and magnetite as primary minerals.

Specimens of the turbiditic sandstone/mudstone were devoid of microfossils. They are cemented by quartz and dissected by quartz-filled veins. In analogy with land exposures on the south coast of Nicoya, we consider these specimens to represent Eocene sedimentary rock.

The lithologies recovered at the steep scarp on the lower slope south of the Nicoya Peninsula are similar to lithologies exposed on land, suggesting continuation of that section at least to the middle slope. We speculate that a basement high on the oceanic plate collided with the headland between two embayments. The basement mass pushed up and exposed at the seafloor a slope section that has similarities with that exposed along the coast. The ENFZ may thus be a line of positive volcanic features on a lower plate fracture that are thrust into the front of the margin, much like the individual seamounts. Although no historic earthquake epicenters have been recorded from its seaward continuation (Protti et al., this volume), the fracture zone might be reactivated beneath the margin during subduction.

Smooth oceanic crust—orderly Nicoya slope

The ocean-basin floor opposite the northwestern Nicoya Peninsula has only two small seamounts within a few hundred kilometers (Case and Holcombe, 1980; Buffler, 1982) and is covered with about 500 m of regularly layered sediment. The adjacent continental margin consists of a regular wedge-shaped accretionary prism covered by a mildly faulted blanket of slope sediment (cf. Shipley et al., 1990, 1992). SO-76 mapping across the upper slope disclosed numerous linear canyons trending down dip from the edge of the shelf. This area was inferred by Corrigan et al. (1990) to represent the continental margin of Costa Rica prior to collision of the Cocos Ridge. A steady-state tectonism of the area since upper Miocene is supported by uniform rates of sedimentation in the upper 330 m of the slope section recovered at DSDP Site 565, the absence of strong tectonic deformation in those cores, and seismic records (von Huene et al., 1985; Shipley et al., 1992).

The three-dimensional seismic data obtained since DSDP drilling reveal structural diversity and intraprism faulting across the whole prism at a scale of tens to hundred of meters, which were undetected in earlier two-dimensional seismic records. The architecture of this prism is considered consistent with steady growth and steady-state geologic conditions since upper Miocene time (Shipley et al., 1992).

Relation to earthquake seismicity

If positive features on the subducting plate disrupt roughly 10 km of the upper plate above them, one would expect to see some seismological effects associated with asperities at the plate boundary. Locating individual asperities is beyond the scope of this study; however, the number of earthquakes increases where the seamount domain is subducted (Protti et al., this volume). The seismogenic zone begins where refraction measurements indicate that the plate boundary is roughly 15 km deep, which is shallow compared to the depth of other seismogenic zones (Zhang and Schwartz, 1992). The seamounts appear to remain on the subducting plate into the seismogenic zone because there is no evidence of sheared-off seamounts along the continental slope. Possibly the increased friction along the plate boundary in the seismogenic zone is sufficient to shear seamounts off (Cloos, 1992). Considering the diameter of most seamounts mapped in the seamount domain, present techniques of hypocentral location are sufficiently accurate to test this speculation.

The mouth of the Gulf of Nicoya has been a historic site for epicenters of large earthquakes. The history of magnitude 7 earthquakes suggests a recurrence interval of 22 ± 11 years (Montero et al., 1991). On 25 March 1990 such an earthquake occurred. A Ms 6.4 earthquake was followed a few minutes later by a Ms 7.0; both were felt throughout Costa Rica. The hypocenter was just above the plate boundary, and the rupture covered an area of about 690 km^2 (Montero et al., 1991). The initial epicenter and the area of aftershock hypocenters are along a projection of the ENFZ. A previous magnitude 7 earthquake in 1952 was located at the same position.

CONCLUSIONS

The Costa Rican margin provides a unique opportunity to study the tectonics in the upper plate that are caused by the subduction of large positive features on the lower plate. Four morphologies on the subducting plate can be associated with distinctly structured segments of the continental margin: (1) a smooth sediment-covered igneous ocean crust that subducts an orderly structured accretionary prism; (2) a seamount domain that subducts a highly disrupted continental margin; (3) the broad Cocos Ridge crest, which subducts a wide uplifted shelf; and (4) an elongate volcanic trend (Fisher Seamount and adjacent features) that subducts opposite a series of features aligned across the margin. The tectonics of the margin is sufficiently influenced by the subducted morphology, with a relief of 1 to 2 km, to be observed with swathmapping at the seafloor. Broad positive features such as Cocos Ridge are opposite a broad uplifted continental margin; sharp positive features such as the inferred seamounts are associated with local relief; a featureless plain is opposite an orderly accretionary prism.

It can be argued that the unique transverse morphology paralleling the direction of plate convergence off Costa Rica is caused by subduction of seamounts. Strong transverse features occur only opposite the oceanic plate with many seamounts; furrows have lateral flanks that were uplifted and then collapsed; and where the upper plate is less than 10 km thick, furrows end at a conical uplift. The unusual length of the furrows compared with other areas of seamount subduction may result from a relatively shallow dip of the subducting plate. As the lower plate becomes deeper, the scarp at the surface above a seamount becomes lower. When the diminished height of the scarp provides insufficient potential energy for a slide to reach the trench axis, the furrow fills and builds the closed depression near its head. Active accretion can then begin to bridge the broad scar at the trench axis. When a seamount first breaches the lower slope, the broad V-shaped morphological scar may remain for about 100,000 yrs before restoration by accretion begins. About 200,000 yrs after the trailing flank of a seamount passed the foot of the MAT slope, a 300-m-high and 5-km-wide accretionary mass has developed. Upslope the seamount tracks end at a water depth of about 1 km, presumably where seamounts become so deeply buried that they can no longer cause slope failure at the seafloor. The result is a distinctive linear morphological track in the middle slope parallel to the direction of plate convergence. After 0.5 m.y. this linear track is significantly modified by erosion and sedimentation. The morphology that clearly distinguishes these features probably remains recognizable for only a short period of geological time.

We interpret the ENFZ morphology as the result of subducted relief on the lower plate. On the oceanic plate, extrusive rock has erupted through a fracture zone and built the seamount and volcanic ridge that today enter and constrict the trench axis. Along a continuation of the boundary beneath the middle slope is a 2.5-km-high basement block. Seismicity beneath the margin and along the ENFZ trend may indicate reactivation in the subduction zone.

A zone of increased numbers of clustered earthquakes appears where the seamount domain is subducted (Protti et al., this volume). This observation contributes to the arguments concerning the role of seamounts on the subducted oceanic crust as a type of asperity for subduction-zone earthquakes.

ACKNOWLEDGMENTS

The PACOMAR (from PAcific COntinental MARgins) project is a cooperative study between German and Costa Rican institutions. The PACOMAR study concerns the understanding of natural hazards. Its principal support is from the German Ministry of Research and Technology (BMFT) through a grant to the GEOMAR Research Center for Marine Geosciences and the Geological and Paleontological Institute (GPI) at the Christian-Albrechts-University in Kiel as well as the Federal Ministry for Geology and Resources (BGR) in Hannover. Cooperating and supporting institutions in Costa Rica are the Instituto Costarricense de Electricidad (ICE); the Escuela Centroamericana de Geologia, Universidad de Costa Rica (UCR); and the Refinadora Costarricense de Petroleo (RECOPE). We thank the Bundesamt fuer Seeschiffrahrt und Hydrographie (BSH), Hamburg, for the loan of their magnetometer, and visiting scientist Lu Wenzheng for his help in reducing the magnetic data.

REFERENCES CITED

Adamek, S., and Tajima, F., 1987, Seismic rupture associated with subduction of the Cocos Ridge: Tectonics, v. 6, p. 757–774.

Auboin, J., and 15 others, 1982, Initial reports of the Deep Sea Drilling Project (Volume 67): Washington, D.C., U.S. Government Printing Office, 799 p.

Ballance, P. F., Scholl, D. W., Vallier, T. L., Stevenson, A. J., Ryan, H., and Herzer, R. H., 1989, Subduction of a Late Cretaceous seamount of the Louisville Ridge at the Tonga Trench: A model of normal and accelerated tectonic erosion: Tectonics, v. 8, p. 952–962.

Bentley, L. R., 1974, Crustal structure of the Carnegie Ridge, Panama Basin and Cocos Ridge [M.S. thesis]: Honolulu, University of Hawaii, 49 p.

Buffler, R. T., 1982, Geologic structure of the forearc region off the west coast of Costa Rica in the vicinity of the Nicoya Peninsula—Results of a multifold seismic reflection survey: University of Texas Technical Bulletin, 56 p.

Case, J. E., and Holcombe, T. L., 1980, Geologic-tectonic map of the Caribbean region: U.S. Geological Survey Miscellaneous Investigations Map I-1100, scale 1:250,000.

Cloos, Mark, 1992, Thrust-type subduction-zone earthquakes and seamount asperities: A physical model for seismic rupture: Geology, v. 20, p. 601–604.

Collot, J. Y., and Fisher, R. L., 1989, Formation of forearc basins by collision between seamounts and accretionary wedges: An example from the New Hebrides subduction zone: Geology, v. 17, p. 930–933.

Corrigan, J., Mann, P., and Ingle, J. C., 1990, Forearc response to subduction of the Cocos Ridge, Panama–Costa Rica: Geological Society of America Bulletin, v. 102, p. 628–652.

Crowe, J. C., and Buffler, R. T., 1983, Regional seismic reflection profiles across the Middle America Trench and convergent margin of Costa Rica, *in* Bally, A. W., ed., Seismic expression of structural styles: American Association of Petroleum Geological Studies in Geology 15, v. 3, p. 3.42-147–3.4.2-162.

DeMets, C., Gordon, R. G., Argus, D. F., and Stein, S., 1990, Current plate motions: Geophysical Journal International, v. 101, p. 425–478.

Emeis, K., Holler, P., and Jeschke, G., 1992, *Fahrtbericht SO-76, Meeresgeologische Arbeitsgruppe:* Geologisch-Paläontologisches Institut und Museum; Christian-Albrechts-Universität Kiel, Deutschland, Internal Cruise Report, p. 51.

Fisher, R. L., 1961, Middle America Trench: Topography and structure: Geological Society of America Bulletin, v. 72, p. 703–720.

Flueh, E. R., Ye, S., Stavenhagen, A., Leandro, G., and Bialas, J., 1992, Pacomar, *Sonne* 76, Compilation of seismic data: Kiel, GEOMAR Data Report, 23 p.

Gardner, T. W., and six others, 1992, Quaternary uplift astride the aseismic Cocos Ridge, Pacific coast, Costa Rica: Geological Society of America Bulletin, v. 104, p. 219–232.

Hey, R. N., 1977, A new class of pseudofaults and their bearing on plate tectonics: A propagating rift model: Earth and Planetary Science Letters, v. 37, p. 321–325.

Hinz, K., et al., 1992, *Geoscientific Investigations off Costa Rica—"PAKOMAR 2," Sonne cruise SO81/1&2, 18.08.-15.09.1992:* Bericht Bundesanstalt für Geowissenschaften und Rohstoffe, Archiv-Nr. 110.148, Nov. 192, 134 p.

Hsu, J. T., 1988, Emerged Quaternary marine terraces in southern Peru: Sea level changes and continental margin tectonics over the subducting Nazca Ridge: [PH.D. thesis]: Ithaca, New York, Cornell University, 309 p.

Klitgord, K. D., and Mammerickx, J., 1982, East Pacific Rise: Magnetic anomaly and bathymetric framework: Journal of Geophysical Research, v. 87, p. 6725–6750.

Lallemand, S., and LePichon, X., 1987, Coulomb wedge model applied to subduction of seamounts in the Japan Trench: Geology, v. 15, p. 1065–1069.

Lallemand, S., von Huene, R., and Culotta, R., 1989, Subduction of the Daiichi Kashima seamount in the Japan trench: Tectonophysics, v. 160, p. 231–247.

Liaw, H. B., 1981, Seismic velocity modeling from an ensemble of earthquakes [Ph.D. thesis]: Dallas, University of Texas at Dallas, 135 p.

Lonsdale, P., and Klitgord, K. D., 1978, Structure and tectonic history of the eastern Panama Basin: Geological Society of America Bulletin, v. 89, p. 981–999.

Macharé, J., and Ortlieb, L., 1992, Plio-Quaternary vertical motions and the subduction of the Nazca Ridge, central coast of Peru: Tectonophysics, v. 205, p. 97–108.

McNally, K. D., and Minster, J. B., 1981, Nonuniform seismic slip rates along the Middle America Trench: Journal of Geophysical Research, v. 86, p. 4949–4959.

Montero, W., Boschini, I., and Rojas, W., 1991, El terremoto de Cobano, Costa Rica, 25 de Marzo de 1990, *in* La crisis sismica del Golfo de Nicoya y eventos sismicos relacionados Costa Rica, 1990: San Jose, Costa Rica, Instituto Costarricense de Electricidad, 126 p.

Shipley, T. H., and Moore, G. F., 1986, Sediment accretion, subduction and dewatering at the base of slope off Costa Rica: A seismic reflection view of the decollement: Journal of Geophysical Research, v. 91, p. 2019–2028.

Shipley, T. H., Stoffa, P. L., and Dean, D., 1990, Underthrust sediments, fluid migration paths and mud volcanoes associated with the accretionary wedge off Costa Rica: Middle American Trench: Journal of Geophysical Research, v. 95, p. 8743–8752.

Shipley, T. H., McIntosh, K. D., Silver, E. A., and Stoffa, P. L., 1992, Three dimensional seismic imaging of the Costa Rica accretionary prism: Structural diversity in a small volume of the lower slope: Journal of Geophysical Research, v. 97, B4, p. 4439–4459.

Stoffa, P. L., and seven others, 1991, Three-dimensional seismic imaging of the Costa Rica accretionary prism: Field program and migration examples: Journal of Geophysical Research, v. 96, p. 21693–21712.

van Andel, T. H., and others, 1973, Initial reports of the Deep Sea Drilling Project (Volume 16): Washington, D.C., U.S. Government Printing Office, p. 151–229.

von Huene, R., and Lallemand, S., 1990, Tectonic erosion along the Japan and Peru convergent margins: Geological Society of America Bulletin, v. 102, p. 704–720.

von Huene, R., and 14 others, 1985, Site 565, *in* von Huene, R., and others, Initial reports of the Deep Sea Drilling Project (Volume 84): Washington, D.C., U.S. Government Printing Office, p. 21–77.

Zhang, Zhi, and Schwartz, S. Y., 1992, Depth distribution of moment release in underthrusting earthquakes at subduction zones: Journal of Geophysical Research, v. 97, p. 537–548.

MANUSCRIPT ACCEPTED BY THE SOCIETY FEBRUARY 7, 1994

Printed in U.S.A.

Geological Society of America
Special Paper 295
1995

Correlation between the age of the subducting Cocos plate and the geometry of the Wadati-Benioff zone under Nicaragua and Costa Rica

Marino Protti and Federico Güendel
Observatorio Vulcanológico y Sismológico de Costa Rica, Universidad Nacional, Apartado 86-3000, Heredia, Costa Rica
Karen McNally
Charles F. Richter Seismological Laboratory, Institute of Tectonics, University of California, Santa Cruz, California 95064

ABSTRACT

High-resolution seismicity data from Costa Rica and southern Nicaragua have been used to image the geometry of the Wadati-Benioff zone under southern Central America. The data include 9,514 events with computed horizontal and vertical errors smaller than 4 and 5 km respectively.

We found that under the Nicaragua–Costa Rica border the Wadati-Benioff zone contorts (from steep to shallow dip angles, northwest to southeast) but does not show evidence of a brittle tear, as postulated by others. Further to the southeast—northeast from Puerto Caldera—the Wadati-Benioff zone does, however, show a tear (the Quesada Sharp Contortion) at intermediate depths (h > 70 km). Overall the dip angle of the Wadati-Benioff zone decreases from 84° under Nicaragua to 60° under central Costa Rica. The maximum depth of intraplate Wadati-Benioff zone earthquakes also decreases from northwest (~200 km under Nicaragua) to southeast (~125 km under central Costa Rica). In southern Costa Rica, east of 83°55′W, we find no evidence of the Wadati-Benioff zone below 50 km.

This geometry of the Wadati-Benioff zone and other tectonic features related to the subduction of the Cocos plate beneath the Caribbean plate have been integrated into a model that correlates them with along-trench variations in age of the subducted Cocos plate. These tectonic features are (1) the shallowing of the Middle America Trench bathymetry from northwest to southeast, (2) differences in coupling between Cocos and Caribbean plates, (3) the termination of the Central America Volcanic Chain in central Costa Rica, and (4) distinct stress field variations on the overriding Caribbean plate.

The subduction of the Cocos Ridge under southern Costa Rica is partially responsible for some of these features (shallowing of the Middle America Trench in southern Costa Rica) and for the high uplift rates of the outer arc. However, since the presence of the Panama Fracture Zone limits the subducted extension of the Cocos Ridge to less than 100 km from the trench, we propose that the overall geometry of the Wadati-Benioff zone is controlled by abrupt along-trench changes in the age of the subducted Cocos plate.

Protti, M., Güendel, F., and McNally, K., 1995, Correlation between the age of the subducting Cocos plate and the geometry of the Wadati-Benioff zone under Nicaragua and Costa Rica, *in* Mann, P., ed., Geologic and Tectonic Development of the Caribbean Plate Boundary in Southern Central America: Boulder, Colorado, Geological Society of America Special Paper 295.

INTRODUCTION

This work presents new data that refine previously published geometries of the Wadati-Benioff zone along the southern terminus of the Middle America Trench, between Nicaragua and Costa Rica, where the Cocos plate subducts beneath the Caribbean plate (Fig. 1). This unique earthquake data set, collected by a local seismographic network, also allows the definition of new features of this subduction zone. The local network has been jointly operated since 1984 by the Observatorio Vulcanologico y Sismologico de Costa Rica, Universidad Nacional (OVSICORI-UNA), and the Charles F. Richter Seismological Laboratory, University of California at Santa Cruz (CFRSL-UCSC)

A new high-resolution three-dimensional plot of the top of the Wadati-Benioff zone from northern Nicaragua to southern Costa Rica is the main result presented in this chapter. This new definition of the Wadati-Benioff zone is combined with previously known regional tectonic features to construct a model that correlates them with abrupt age changes in the subducted Cocos plate. The amount and quality of located earthquakes and the existence in this region of along-strike abrupt age changes in a relatively young lithosphere (less than ~40 m.y. old) made correlations on a small scale possible.

Correlations between the age of the subducting plate and the geometry (dip angle, length, and maximum depth) of Wadati-Benioff zones have been made, on a large scale (subduction segments larger than 500 km), for subduction zones with very different tectonic settings and evolutions (e.g., Sugi and Uyeda, 1984; Yamaoka et al., 1986; Jarrard, 1986). In general, these works indicate a direct correlation between the age of the plate, or a combination of age and convergence rate, with the Wadati-Benioff zone dip angle and its length. Such a correlation has been explained by differences in the thickness, density, and thermal state of the subducting lithosphere that affect the sinking or buoyancy of the slab and its thermal assimilation by the aesthenosphere.

Here we attempt to correlate the geometry of the Wadati-Benioff zone along a single subduction zone on a smaller scale (segments as small as 150 km) and with age differences of less than ~20 m.y. Such a detailed study has the potential for revealing clues about the processes and dynamics of subduction zones.

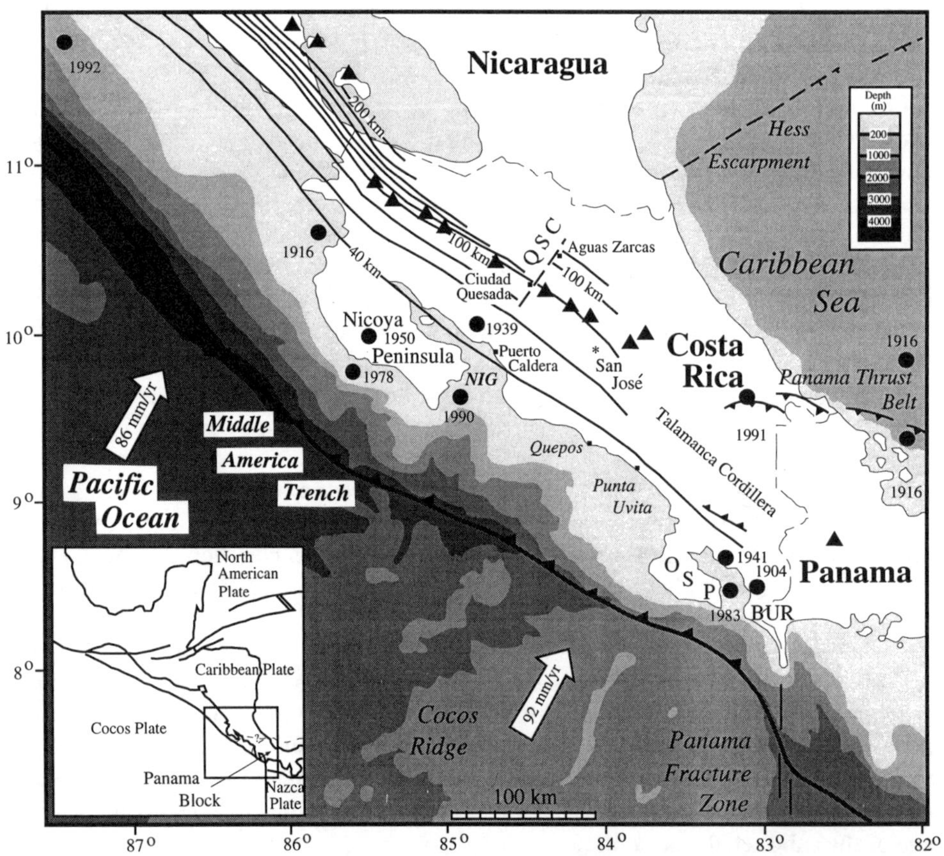

Figure 1. Tectonic setting of southern Central America and isodepth contours (in km) of the top of the Wadati-Benioff zone obtained in this study. Convergence velocities between Cocos and Caribbean plates are after DeMets et al., 1990. Closed triangles are active volcanoes, and solid circles show the location of large (Ms ≥ 7.0) interplate earthquakes that occurred this century. NIG: Nicoya Gulf; OSP: Osa Peninsula; BUR: Burica Peninsula; QSC: Quesada Sharp Contortion. The locations of other places mentioned in the text are also given.

TECTONIC SETTING

Costa Rica, as part of Central America, is located on the western margin of the Caribbean plate (Fig. 1). Here, the Cocos plate subducts under it along the Middle America Trench. The direction of convergence of these two plates is N 25 to 30° E, and their relative velocity varies from 69 ± 3 mm/year offshore of northern Guatemala to 95 ± 4 mm/year offshore of southern Costa Rica (De Mets et al., 1990).

South of the border between Costa Rica and Panama is the Panama Fracture Zone. This right-lateral transform fault is one of the plate boundaries between the Cocos and Nazca plates. West of the Cocos-Nazca-Caribbean triple junction and southwest of the Osa Peninsula, the Cocos Ridge is being subducted beneath southern Costa Rica as part of the Cocos plate. The Cocos Ridge is a trace of the Galapagos hot spot (Hey, 1977) and has an orientation N 45° E, which indicates the direction of movement of the Cocos plate with respect to the underlying mantle. Cocos plate generated at the East Pacific Rise subducts under Nicaragua and northern Costa Rica, whereas Cocos plate generated at the Galapagos Spreading Center subducts under central and southern Costa Rica.

In Costa Rica, off the southern tip of the Nicoya Peninsula, the Middle America Trench axis shows a prominent left bend from northwest to southeast, and its depth rapidly decreases southeast of this point. East of Costa Rica is the Panama Block. The northern boundary of the Panama Block with the Caribbean plate is a convergent margin, the Panama Thrust Belt, which extends from the Caribbean coast of Colombia to south of Limon, inland within Costa Rica (Silver et al., 1990). The western limit of the Panama Block with the Caribbean plate, in Costa Rica, could be the northwest extension of the April 22, 1991, Valle de la Estrella earthquake (Fig. 1) rupture zone.

PREVIOUS WORK

The Wadati-Benioff zone geometry in southern Central America has been deduced from both surface tectonic features and direct seismic observations. Stoiber and Carr (1973), based on differences in the strikes and position of volcanic lineaments, proposed tear faults in the subducted plate under central Nicaragua, under the Nicaragua–Costa Rica border, and under Irazu-Turrialba volcanoes. Burbach et al. (1984) and Burbach and Frohlich (1986) combined ISC (International Seismological Center) relocated events with locally recorded events and proposed a tear under the Nicoya Peninsula. Bevis and Isacks (1984) used hypocentral trend surface analysis for teleseismically recorded events and found no evidence for pronounced discontinuities in the region. Wolters (1986), using PDE (Preliminary Determination of Epicenters) data, mapped limits of seismic sections ("not necessarily breaks") under central Nicaragua and under the Nicoya Peninsula. Yamaoka et al. (1986), using ISC locations to fit an inextensible spherical shell, concluded that no tearing of the slab was needed to fit the data. "Güendel (1986), using mainly local data, mapped the mean depth of the Wadati-Benioff zone for the same region covered in this chapter and suggested that smooth contortions of the slab fit the hypocentral distribution of intermediate-depth earthquakes in the region, without requiring tear faults.

The data set used by Güendel (1986) has been updated in this chapter, with almost three times the number of earthquakes, to map the top of the Wadati-Benioff zone. We refined the smooth contortion under the Nicaragua–Costa Rica border and defined a tear under central Costa Rica.

DATA ANALYSIS

Data used and location errors

Both teleseismic and local data sets have been used for studies conducted in Nicaragua and Costa Rica (Table 1). Data from four local seismographic networks are used in this chapter (Fig. 2): Arenal (1978), Nicaragua (1978), OVSICORI-UNA (1984–1990), and a temporal network (October 1987–July 1988) funded by the National Science Foundation (NSF) as part of a joint research project between OVSICORI-UNA AND CFRSL-UCSC. Three of these networks overlap in northern Costa Rica.

To study the geometry of the Wadati-Benioff zone, data sets DS04 to DS09 were used, for a total of 9,514 events (Fig. 3). This analysis does not include DS01 to DS03, used by Güendel (1986), because it is deemed desirable to maintain consistency by using only earthquakes located by local networks. The majority of these data sets (88%) was recorded by OVSICORI-UNA's network, and we have complete location parameters for them (e.g., 77% of the events were located with 11 or more readings including P and S; 50% are within an azimuthal gap less than 225°; 63% are at an epicentral distance of less than 40 km to the nearest station; 85% have root mean square (RMS) equal to or smaller than 0.30 seconds; 93% have horizontal errors equal to or smaller than 6 km; and 84% have vertical errors equal to or smaller than 6 km). For Costa Rica we only used earthquakes located with horizontal and vertical errors smaller than 4 and 5 km respectively. For Nicaragua no values of horizontal and vertical errors exist in the data base; however, the data correspond to a time period (1978) when the network was fully operational (David Harlow, personal communication, 1985).

Local versus global earthquake locations

Local networks are capable of locating earthquakes of magnitudes smaller than those global networks are capable of recording. Several works have pointed out a tendency for local networks to yield a steeper dip of the Wadati-Benioff zone (e.g., Adams and Ware, 1977; McLaren and Frohlich, 1985); however that tendency is not significant for the region of our study. McLaren and Frohlich (1985) showed that the maximum errors occur for earthquakes deeper than 200 km and where the local network does not extend above the deepest part of the seismic slab. In the Wadati-Benioff zone under

TABLE 1. SEISMIC DATA SETS FOR NICARAGUA AND COSTA RICA MENTIONED IN THIS CHAPTER

Name*	From (DD-MM-YY)	To (DD-MM-YY)	Mag. Minimum	Mag. Maximum	Events	Region†	Remarks§
DS01	05-10-50	11-11-50	6.0	7.7	7	NCR	a
DS02	06-05-51	23-12-72	5.0	7.3	73	N	b
DS03	16-10-65	25-09-85	5.0	6.3	75	CR	c
DS04	01-01-78	31-12-78	0.6	3.3	568	N, NCR	d
DS05	01-05-78	15-10-78	1.7	7.0	421	NCR	e
DS06	01-04-84	22-10-86	1.0	5.2	2,601	SN, CR	f
DS07	23-10-86	01-04-90	1.7	6.8	5,748	SN, CR	f
DS08	15-10-87	05-07-88	1.6	4.7	176	SN, NCR	g
DS09	02-04-84	30-12-87	2.2	4.9	416	SN, CR	h

*Names used in this chapter for simplicity.
†N = Nicaragua; CR = Costa Rica; SN = southern Nicaragua; NCR = northern Costa Rica.
§a = Relocations of the Nicoya 1950 sequence by Güendel, 1986; b = Relocation of WWSSN data by Dewey and Algermissen, 1974; c = Relocation of WWSSN data by Güendel, 1986; d = Events located in 1978 by the Nicaraguan network (Güendel, 1986); e = Relocations by Güendel, 1986, of events recorded by the Arenal network; f = Events reported by OVSICORI-UNA (1984–1990); g = Events recorded in a joint project by OVSICORI-UNA and CFRSL-UCSC with financial support of NSF (Protti, 1991); h = Relocated events from DS06 and DS07 (Protti, 1991).

Nicaragua (where the coverage of the network is not as good as in Costa Rica), from 0 to 100 km in depth there is no difference between the local results and the angle of subduction obtained with relocations of globally recorded earthquakes (Dewey and Algermissen, 1974). Deeper than 100 km the difference in dip angle is only 11°, being steeper for the local network, resulting in less than 30 km of horizontal discrepancy at a maximum depth of 210 km. A qualitative description of this difference for Nicaragua, using the same data sets, was made by Burbach et al. (1984). In Costa Rica, where the local network is denser and covers the entire seismic slab and where the Wadati-Benioff zone is not as deep as in Nicaragua, this difference becomes less significant. The relative lack of moderate- to large-magnitude earthquakes within the slab under Costa Rica makes both quantitative and qualitative comparisons difficult.

Wadati-Benioff zone

To obtain a three-dimensional view of the top of the Wadati-Benioff zone, two procedures were used. First, maps of seismicity with depth intervals of 20 km were plotted to obtain the general trend of seismicity as a function of depth. Second, cross sections of seismicity were constructed perpendicular to the general trend obtained in the first step, producing sections almost perpendicular to the Middle America Trench (Fig. 4). We also constructed cross sections parallel to the Middle America Trench (Fig. 5). Given the amount of data used, no overlap in the cross sections was needed. Cross sections for Nicaragua are wider, since the Nicaragua data base is not as large or as complete as the Costa Rica data base. Therefore, resolution in the geometry obtained for the slab under Nicaragua is not as good as it is for Costa Rica. Another advantage of the Costa Rica data set resides in the fact that it has been collected by networks that extend closer to the trench, overlying most of the seismic slab. In Nicaragua most of the seismic slab lies outside the network coverage.

The top of the Wadati-Benioff zone was drawn on each cross section and projected to the surface in intervals of 20 km, starting at the 40-km depth contour. The slices of seismicity from the first step were also used for the interpolation of the projected depths and for the construction of the isodepth contours.

DISCUSSION

Proposed subduction segments

We have divided the subduction zone from Nicaragua to Costa Rica into four segments: Nicaragua and northern, central, and southern Costa Rica (Fig. 6). This subdivision is based on the geometry of the Wadati-Benioff zone and on other tectonic features (Table 2) that differ from segment to segment (Protti, 1991).

Comparison to previous results

Although Güendel (1986) studied the geometry of the subducted slab for the same region presented in this chapter (from Nicaragua to Costa Rica), the two studies differ: In Güendel (1986) the mean depth of the seismic slab was projected to the surface, whereas in the present study the top of the Wadati-Benioff zone has been used to define the geometry. Also, for this chapter almost three times the number of earthquakes, not available for the first study, have been included in the data base, mostly from southern Nicaragua and all of Costa Rica.

For Nicaragua and northern Costa Rica the general shape of the seismic slab obtained in this work does not differ substantially from that obtained by Güendel (1986). This is because almost the same data set was used in both works for Nicaragua. The major difference in the results of the two studies in this region is the relative location of the isodepths and the maximum depth of the seismic slab. This difference arises because in this study we mapped the top of the Wadati-Benioff zone. The general shape of the smooth contortion in the Nicaragua–Costa Rica border remains almost the same (Fig. 1). A more detailed sampling of the Wadati-Benioff zone under the border was obtained by using data base DS08, collected specifically for that purpose.

Quesada Sharp Contortion

Given the total amount of data used, we were able to identify a tear in the slab under central Costa Rica (Fig. 7), hereafter called the Quesada Sharp Contortion because it lies under Ciudad Quesada (Fig. 1). For the final analyses we adjusted the zones covered by profiles I-I′ and J-J′ (Fig. 3) in such a way that the boundary between the zones included in these cross sections was coincident with the location obtained for the Quesada Sharp Contortion. In this way we avoided the overlapping of the two Wadati-Benioff zone segments and ob-

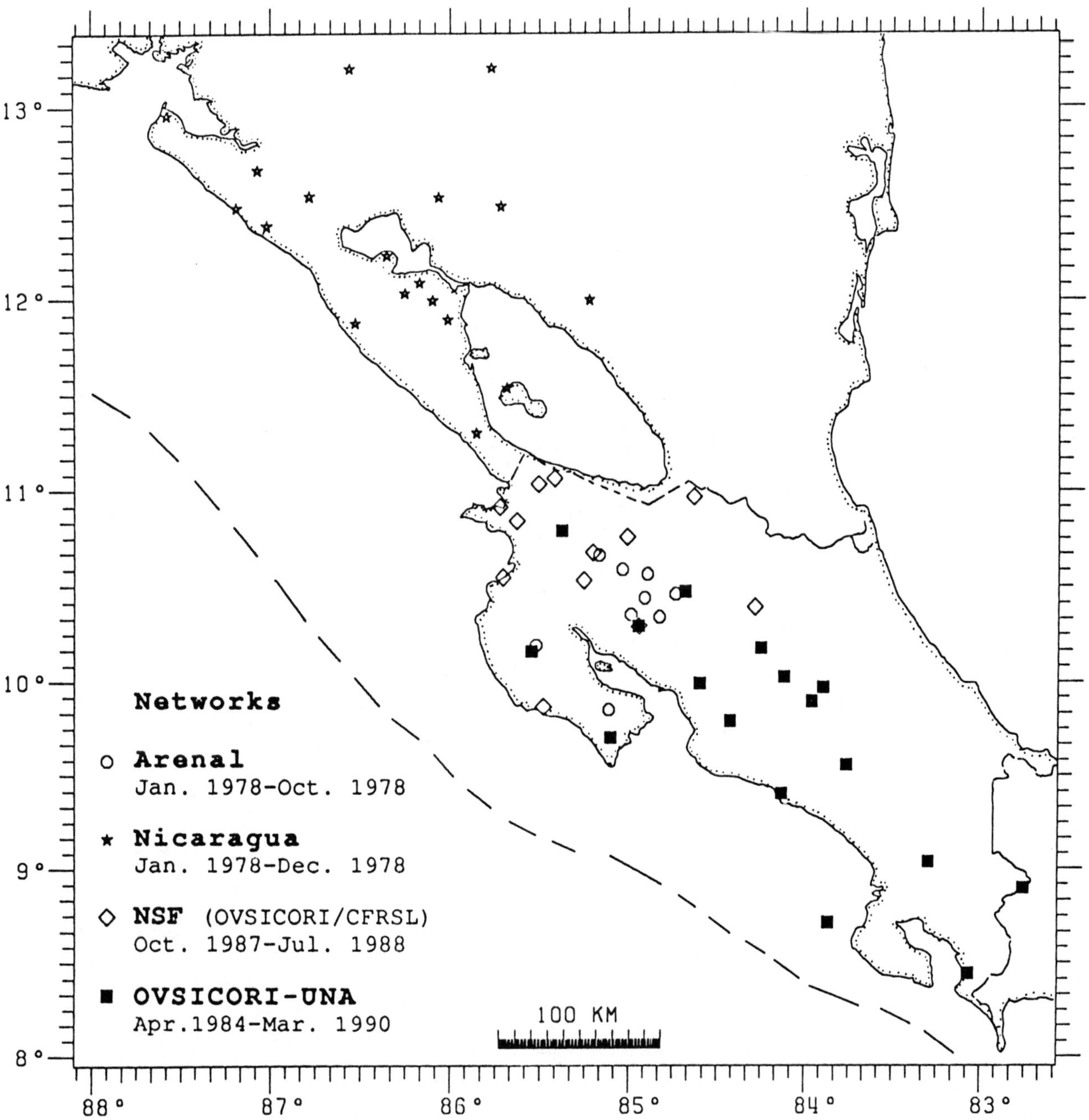

Figure 2. Map showing the location of seismographic stations and period of operation of local networks from which data was used.

tained a more realistic picture in cross section. The Quesada Sharp Contortion is recognizable by means of seismicity only below depths of 70 km. It extends N 34° E from northeast of Puerto Caldera to Aguas Zarcas de San Carlos (Fig. 1). The horizontal offset of the isodepths along this feature is on the order of 15 km, higher than the statistical errors inherent in the earthquake locations selected for this work (ERH ≤ 4 km, ERZ ≤ 5 km). To the northwest of this contortion a deeper portion of the seismic slab dips about 80° and reaches maximum depths ranging from 200 km near the border between Nicaragua and Costa Rica to 135 km under Ciudad Quesada. To the southeast of the Quesada Sharp Contortion the Wadati-Benioff zone dips about 60°, and the seismicity does not extend below depths ranging from 125 km, behind the volcanic chain, to 50 km, northeast of Quepos. On both sides of the Quesada Sharp Contortion the maximum depth of earthquakes in the Wadati-Benioff zone shallows up from northwest to southeast at a rate of 7 km per each 10 km of horizontal distance parallel to the trench axis (Fig. 5).

SE termination of the seismic slab

Another remarkable feature of the Wadati-Benioff zone under Costa Rica is its abrupt termination at 83°55'W (Fig. 8), coincident with the southeast end of the Central America ac-

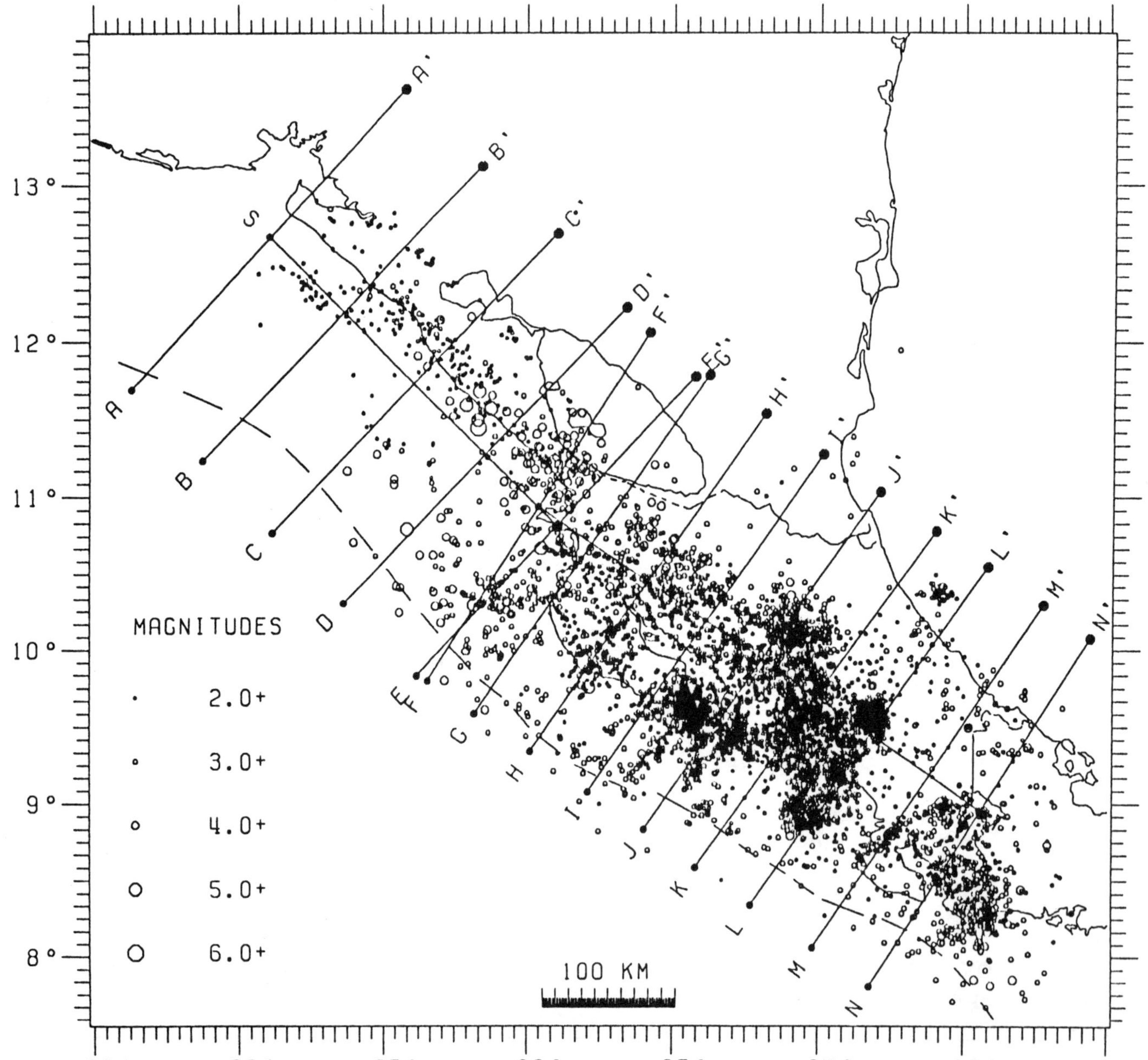

Figure 3. Map showing the epicentral distribution of all events used and the location of cross sections perpendicular (A through N) and parallel (S-T) to the Middle America Trench. The cross sections are shown in Figures 4 and 5.

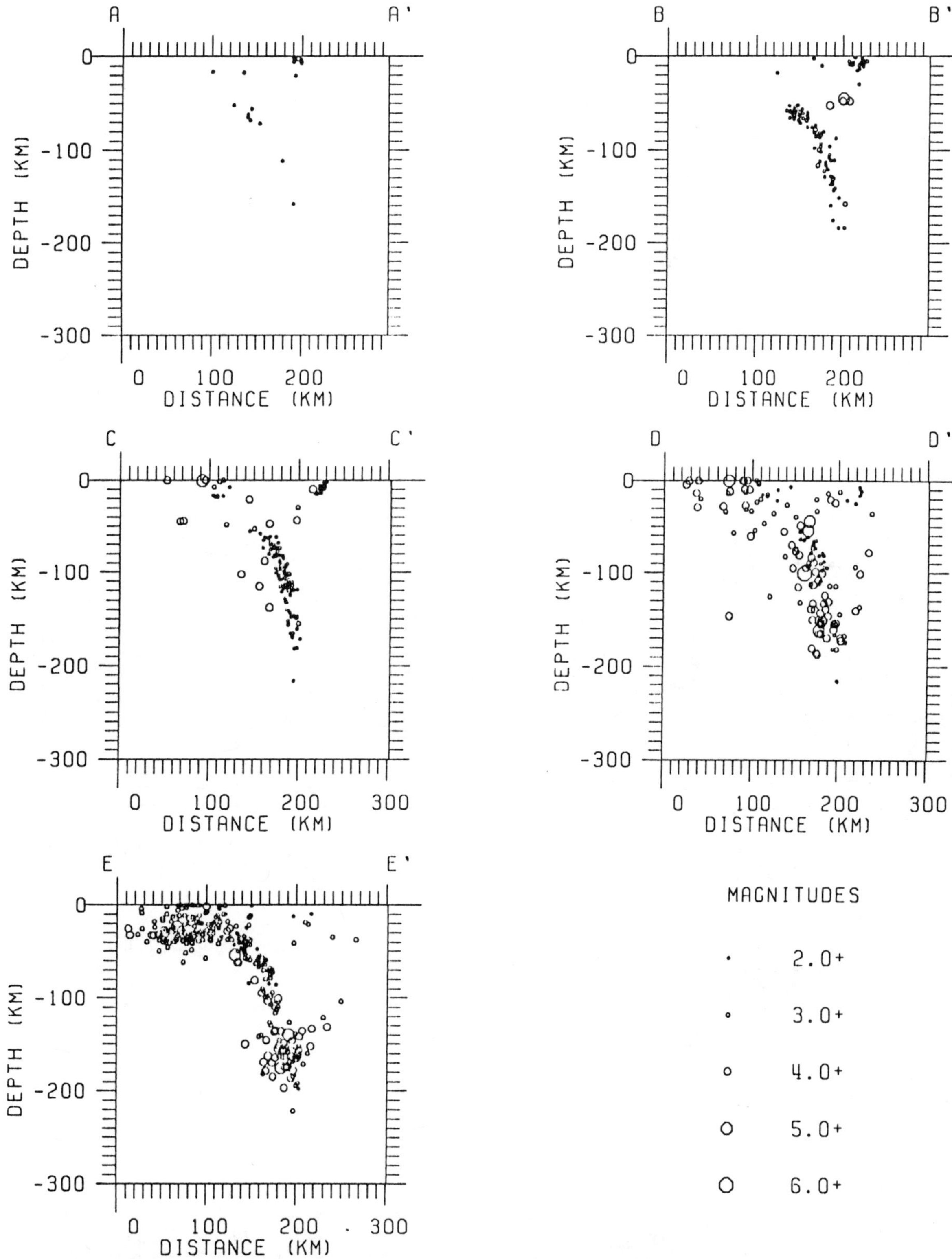

Figure 4 (on this and following two pages). Depth distribution of earthquakes on cross sections perpendicular to the Middle America Trench. See Figure 3 for location of the cross sections. For Nicaragua, earthquakes within 37.5 km on both sides of the profile line were plotted on the cross sections, whereas for Costa Rica that distance was only 25 km on each side.

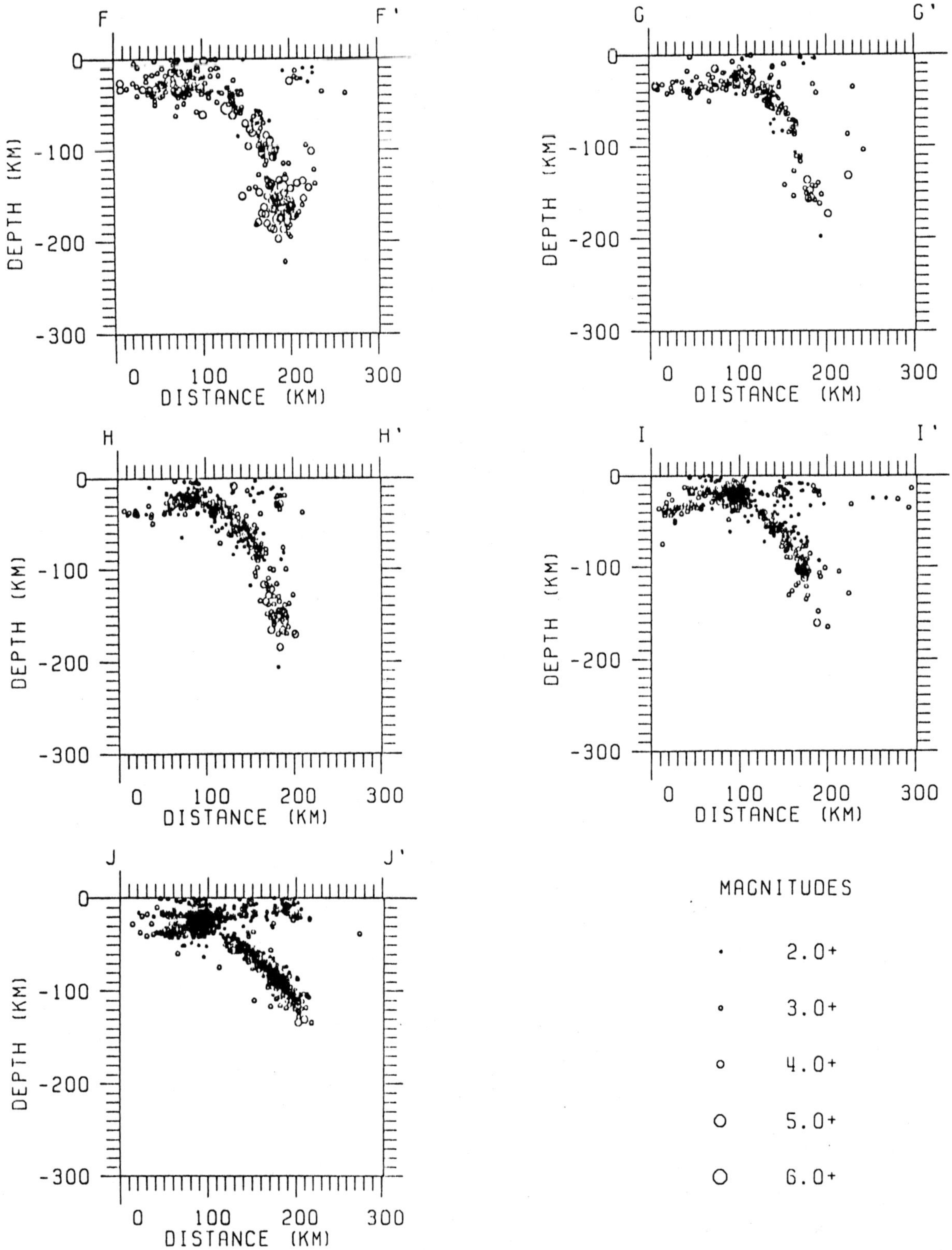
F
F'
G
G'
H
H'
I
I'
J
J'
DEPTH (KM)
0
-100
-200
-300
0
100
200
300
DISTANCE (KM)
MAGNITUDES
2.0+
3.0+
4.0+
5.0+
6.0+

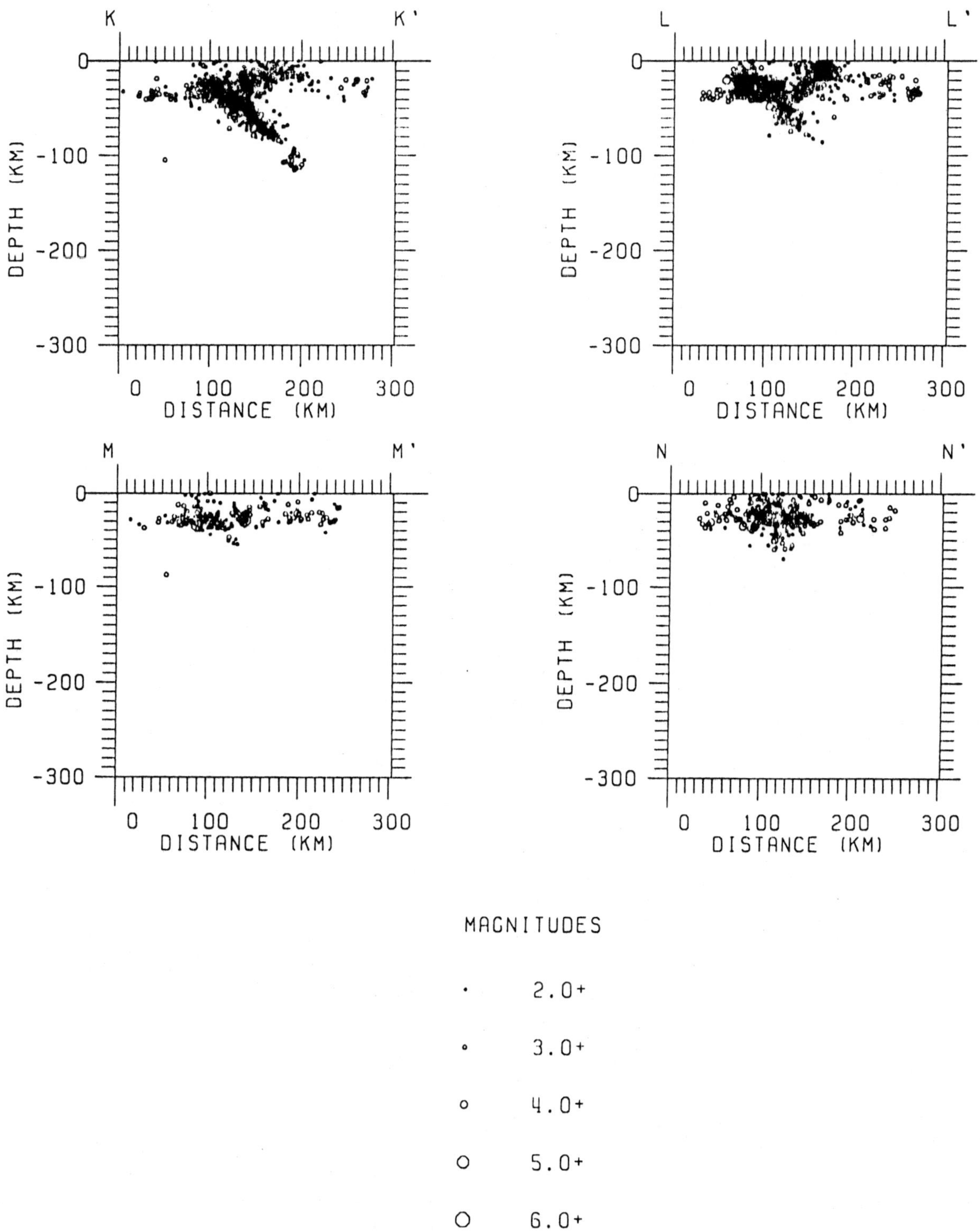
K
K'
L
L'
M
M'
N
N'
DEPTH (KM)
0
-100
-200
-300
0 100 200 300
DISTANCE (KM)
MAGNITUDES
2.0+
3.0+
4.0+
5.0+
6.0+

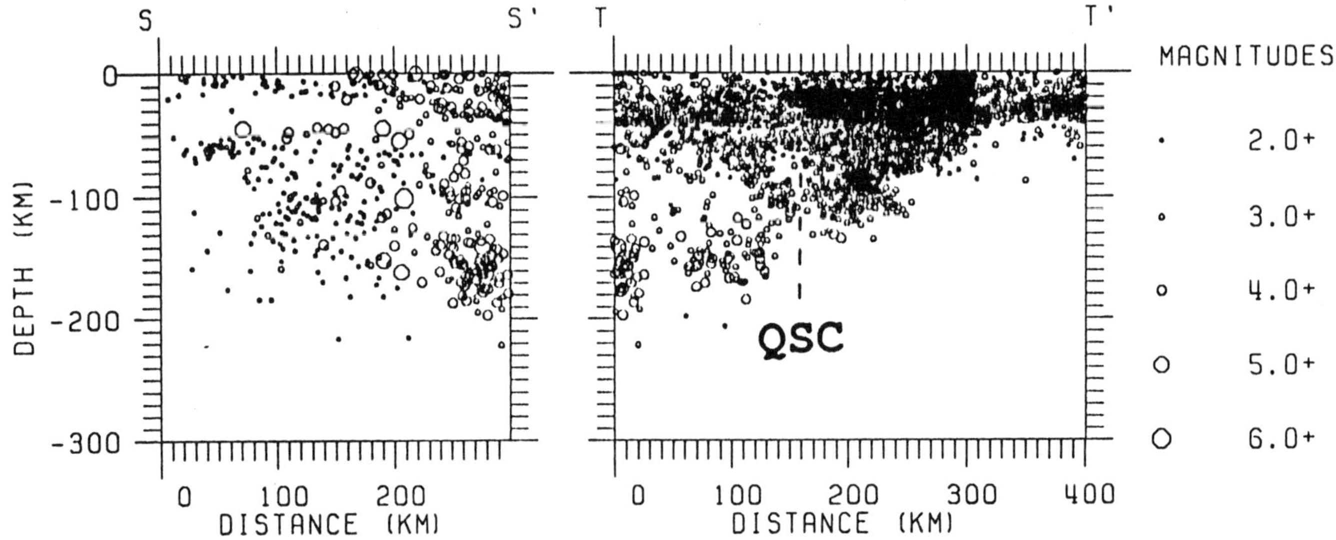

Figure 5. Depth distribution of earthquakes on cross sections parallel to the Middle America Trench. Also shown is the depth projection of the Quesada Sharp Contortion (QSC) discussed in the text. See Figure 3 for location of the cross sections.

tive volcanic chain. No evidence of Wadati-Benioff zone seismicity deeper than 50 km is found southeast of Punta Uvita (cross sections M-M' and N-N' on Figs. 3 and 4).

Relation of Wadati-Benioff zone geometry to volcanic arc

It is noteworthy that even though there is an abrupt displacement of the volcanic chain from Nicaragua to Costa Rica (Fig. 1), no major changes in the geometry of the seismic slab were found. In addition, no appreciable changes in the orientation of the volcanic chain axis exist across the Quesada Sharp Contortion. Nevertheless, and even though there are no changes in the orientation of the volcanic chain axis across the Quesada Sharp Contortion, Malavassi (1991) found important changes in the geochemical composition of Tertiary and Quaternary lavas across the contortion. In addition, he also found important geochemical signatures in the Aguas Zarcas cones (near Aguas Zarcas, Fig. 1), beneath which the Quesada Sharp Contortion has its maximum morphological expression. It is important to mention that by means of focal mechanisms (for events deeper than 50 km), no evidence has been found of either tearing along the Quesada Sharp Contortion or changes in the general internal deformation of the slab (Protti, 1991).

Variations in the shallow dip of the subduction zone

Because our earthquake location procedure uses a flat-layer velocity model (Matumoto et al., 1977), without lateral variations, the depth resolution for earthquakes offshore is not as good as for those events within the network. Therefore, the dip of the shallow part of the Wadati-Benioff zone is assumed constant from the trench axis to the shallower part of the resolved seismic slab (35 to 40 km) (i.e., the deepest extension of the brittle coupling zone as defined by Protti, 1991). Given the bend of the trench axis off the southern tip of the Nicoya Peninsula, the subduction angle seems to change from 23° in northern Costa Rica to 30° in southern Costa Rica. Whether or not this feature is real cannot be ascertained until better depth resolution is obtained for events offshore (i.e., through a better crustal model).

MODEL FOR ALONG-STRIKE CHANGES IN THE WADATI-BENIOFF ZONE GEOMETRY AS A FUNCTION OF CHANGES IN AGE OF THE SUBDUCTED PLATE

Age and geometry changes along the trench

The updip projection of the Quesada Sharp Contortion coincides with an abrupt change in the bathymetry of the Cocos plate along a northeast alignment. This alignment extends from the Cocos-Nazca-Pacific triple junction, on the East Pacific Rise, to where the sharpest bend in the Middle America Trench occurs offshore of the southern tip of the Nicoya Peninsula (Fig. 9); here the Fisher seamounts (von Huene et al., this volume) are being subducted as part of the Cocos plate. This bathymetric change (the northern "rough-smooth boundary" of Hey, 1977), mainly marked by the 3,000-m depth contour, reflects an age and, therefore, a density contrast on the Cocos plate on either side of the alignment (younger and lighter lithosphere on the southeast side) (Fig. 9). The ages shown in Figures 9 and 10, near the East Pacific Rise and the Galapagos Spreading Center, were obtained directly from the magnetic anomalies reported in the plate tectonic map by the Circumpacific Council for Energy and Mineral Resources (southeast quadrant) (1981). Farther from

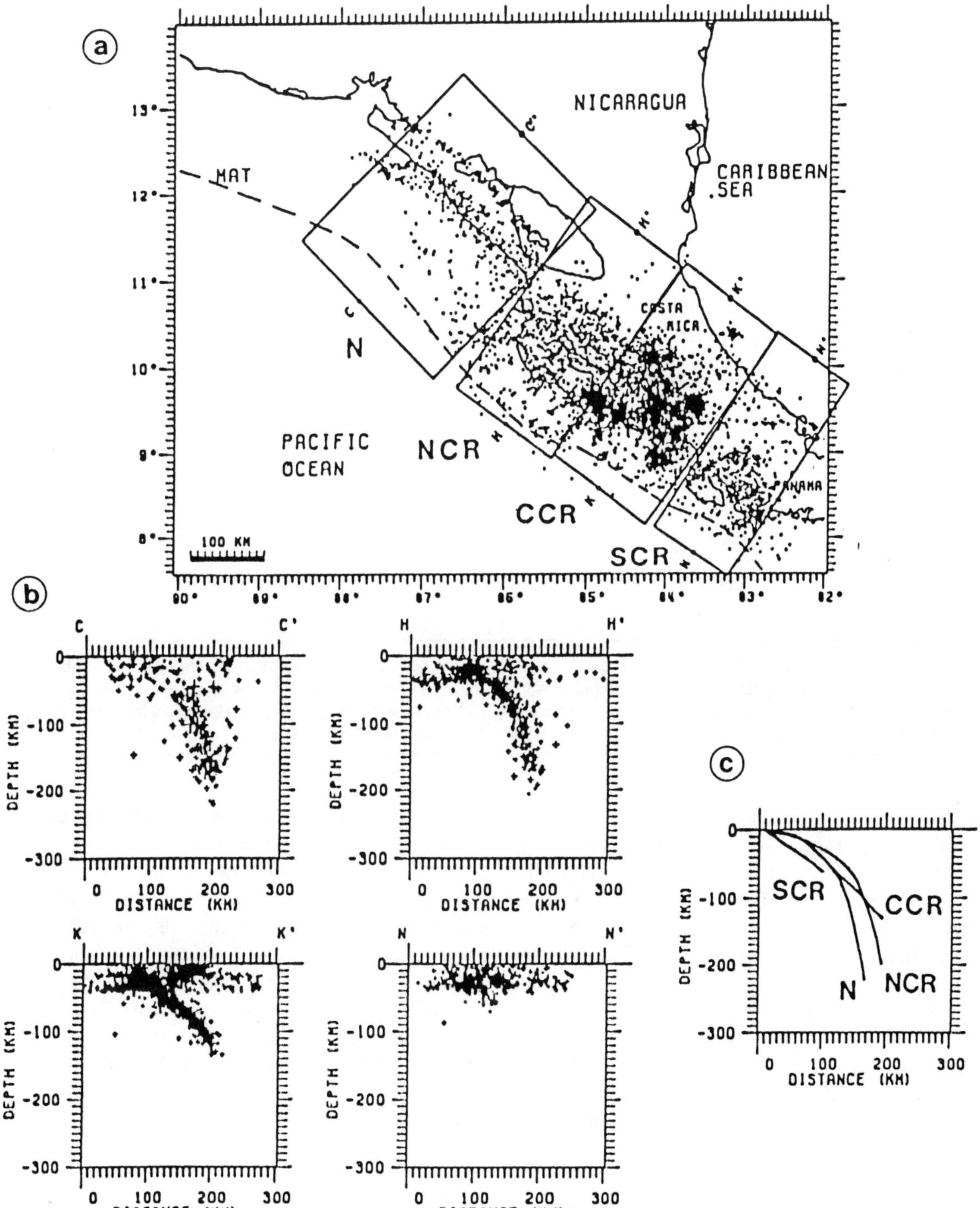

Figure 6. Subduction segments described in the text: N: Nicaragua; NCR: northern Costa Rica; CRR: central Costa Rica; SCR: southern Costa Rica. a, Map showing the area and seismicity covered by each segment; b, depth distribution of seismicity for each segment; c, overlapping of all segments with the trench axis as common point.

TABLE 2. PARAMETERS FOR EACH SUBDUCTION SEGMENT DESCRIBED IN THIS CHAPTER*

Subduction Parameters	Nicaragua	Northern Costa Rica	Central Costa Rica	Southern Costa Rica
Length along trench (km)	~250	~150	~160	~120
Trench depth (m)	NW-5,000 SE-4,500	NW-4,500 SE-3,400	NW-3,400 SE-2,000	NW-2,000 SE-2,000
Trench azimuth	N 40° W (320°)	N 50° W (310°)	N 65° W (295°)	N 65° W (296°)
Arc-trench gap (km)	185	160	160	125
Highest peak on inner arc (m)	1,745	2,028	2,919	3,819
Age at trench (m.y.)	25–30	25–30	20–22	13–17
Age at tip (m.y.)	30–35	30–35	20–22	17–19
Convergence rate (mm ± 4/y)	NW-81 SE-86	NW-86 SE-89	NW-89 SE-92	NW-92 SE-95
WBZ length (km)	NW-260 SE-310	NW-310 SE-260	NW-240 SE-170	NW-100
WBZ horizontal extent (km)	NW-180 SE-180	NW-180 SE-170	NW-210 SE-110	NW-110
Dip (0–60 km)	25°	23°	24°	30°(?)
Dip (0–100 km)	34°	32°	30°	
Dip (>100 km)	84°	80°	60°	
Maximum depth (km)	NW-160 SE-220	NW-200 SE-135	NW-125 SE-50	NW-45
Strain class	3	4	5	6-7
Slab pull force (*10^{12} N/m)	11.0	11.0	8.1	3.7
Largest earthquake (Ms)	7.2	7.7	7.0	7.6

*Shown on map view and on cross section in Figure 6. The NW boundary of the Nicaragua segment was arbitrarily set at the border between Nicaragua and Honduras. Ages of the Cocos plate are projections of spreading rates (see text and Figs. 9 and 10). Convergence rates are after De Mets et al., 1990; the dip ranges for the Wadati-Benioff zone (WBZ) are those used by Jarrard, 1986; and the strain classification is also that suggested by Jarrard, 1986. Values for strain class and slab pull force are from Protti, 1991.

the spreading centers (i.e., near the trench) the age was obtained following Hey (1977), Lonsdale and Klitgord (1978), Klitgord and Mammerickx (1982), and Mammerickx and Klitgord (1982) and from projections of spreading rates and spreading patterns of both rifts. The age obtained for the oceanic plate subducting under the Nicoya Peninsula is a conservative value (it could be over 40 m.y. old according to Klitgord and Mammerickx, 1982). The emphasis of this chapter is on relative age differences (or age contrasts across boundaries) rather than absolute ages, not available yet for this region.

The rough-smooth boundary marks the boundary between Cocos lithosphere created at the East Pacific Rise, to the north, and that created at the Galapagos Rift System, to the south. The northeast section of the rough-smooth boundary, which turns into the Quesada Sharp Contortion, is a relic of the fracture along which the Farallon plate broke into Cocos and Nazca plates between 25 m.y. (Hey, 1977) and 28 m.y. ago (Lonsdale and Klitgord, 1978). As a consequence of this opening, the age of the subducted plate under the Nicoya Peninsula is older and, in the direction of subduction, increases more rapidly than under the central Costa Rica. To clarify this, we have unbent the Wadati-Benioff zone, and with it our estimated ages, to a dip of 0° (Fig. 10). This age distribution correlates not only to the steeper angle in the northwest segment and the shallower angle found in the southeast segment but also with an almost constant Wadati-Benioff zone dip angle in central Costa Rica, in contrast to the downdip increase in the angle under Nicaragua and northern Costa Rica. According to our age estimates the age contrast at the tip of the Wadati-Benioff zone (across the Quesada Sharp Contortion) is on the order of 5 to 10 m.y. (Figs. 9 and 10) and could be even larger. The age at the tip, shown in Figure 10, is the present age based on projections of spreading rates and not the effective age when the tip was at the trench as used by Jarrard (1986).

Age variations of this order are effective only in changing the Wadati-Benioff zone geometry for subduction of young, rapidly cooling lithosphere (less than 40 to 50 m.y. old) as in this case. Older lithosphere is almost thermally stable and does not experience major density and thickness changes (Leeds, 1975; Parsons and Sclater, 1977; Wortel, 1980; Sacks, 1983). Also, these changes in age have to occur abruptly across fracture zones, transform faults, ridges, or boundaries between plates generated at different spreading centers like the rough-smooth boundary.

Maximum depth of the seismic slab

The proposed age contrast also correlates with the deeper Wadati-Benioff zone under Nicaragua and with the shallower one under central Costa Rica. Such a correlation is to be expected since the deepest seismicity, which reflects the transition from brittle to ductile deformation in the slab as a response to gradual heating (Isacks et al., 1968; McKenzie, 1969), is also a function of the slab age (Vlaar and Wortel, 1976; Wortel and Vlaar, 1978). The slab length is also a function of convergence rate (e.g., Jarrard, 1986), but in our case differences in convergence rates are not significant (~5 mm/yr according to De Mets et al., 1990), and the faster rate occurs where the Wadati-Benioff zone is the shortest, contrary to what the model predicts. Thus, the slab age is the main parameter controlling the length and maximum depth of the Wadati-Benioff zone at the southern terminus of the Middle America Trench. The shallowing of the maximum depth of the Wadati-Benioff zone from Nicaragua to the Quesada Sharp Contortion, along lithosphere of the same age, may be accounted for

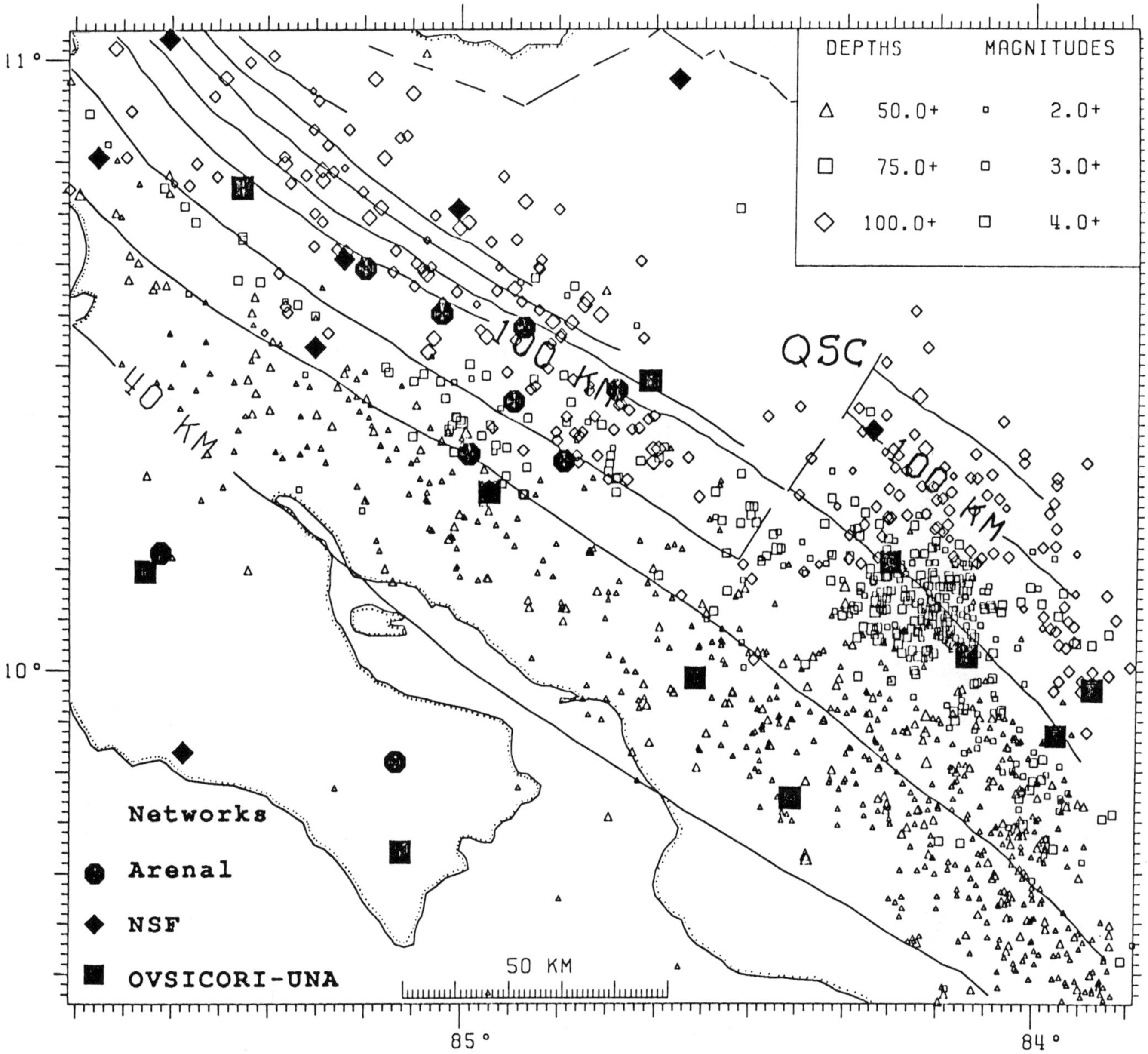

Figure 7. Detailed map of northern and central Costa Rica showing seismicity, station distribution, and the geometry of the top of the Wadati-Benioff zone obtained in this work. The Quesada Sharp Contortion (QSC) represents the boundary between two segments of the subducted Cocos plate with different ages and geometries. The different-sized squares that indicate magnitudes refer to the size for all of the symbols.

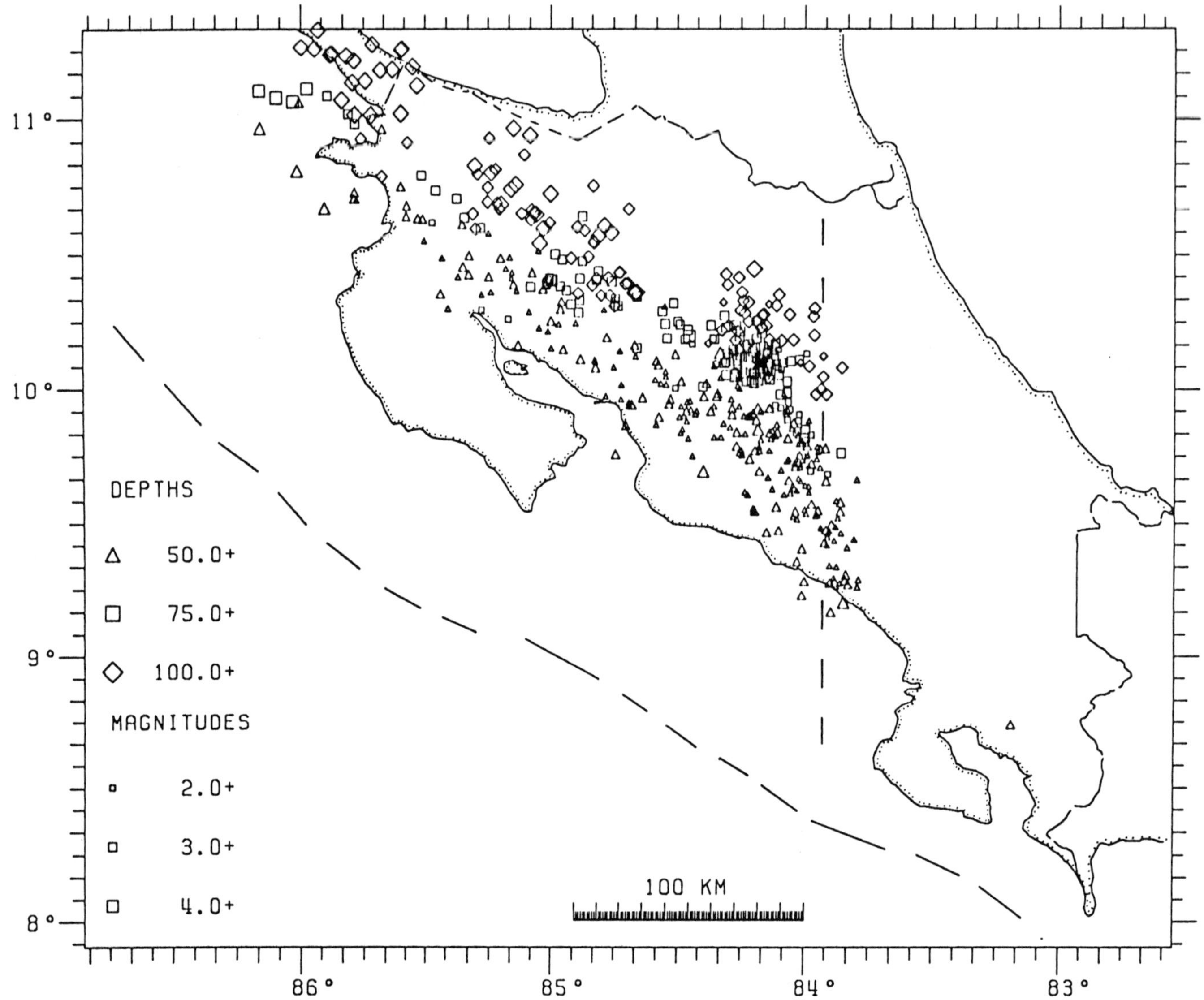

Figure 8. Epicentral distribution of earthquakes with focal depth greater than 50 km (DS09 of Table 1). Note the sudden termination of the seismicity within the slab east of the 83°55′W meridian. The different-sized squares that indicate magnitudes refer to the size for all of the symbols.

by lateral thermal assimilation, since the slab is being exposed to mantle heat flow from three directions: top, bottom, and laterally along the tear of the Quesada Sharp Contortion.

Southeast end of the seismic slab

The abrupt termination of the Wadati-Benioff zone at 83°55′W (Figs. 1 and 8) correlates with and is parallel to another along-trench age change (~5 m.y.) on the Cocos plate across a north-south transform fault that originated at the Cocos-Nazca spreading center. This fault marks the west limit of the Costa Rica rift and the east limit of the Ecuador rift (CRR and ECR on Fig. 9). East of the location where this fracture is being subducted, the oceanic plate is very young (~15 m.y.); its thermal structure is probably beyond the brittle-plastic limit, and it therefore cannot deform elastically (i.e., lack of earthquakes beyond 50-km depth). We predict that this segment of the Cocos plate has to be subducting with a shallower angle than its western counterpart and probably bends upward, continuing an almost horizontal path like that found under Peru (Isacks and Barazangi, 1977) and southern Mexico (Suarez et al., 1990). In such horizontal paths, the subducted plate underrides the base of the upper plate—in this case the roots of the plutonic Talamanca Cordillera—and thus does not allow for an aesthenospheric wedge between the plates. This driving mechanism precludes the oceanic plate from reaching the depth needed for the dehydration of the slab, thus inhibiting the subduction-related volcanic activity in southeastern Costa Rica. The underriding of the Cocos plate in southern

Costa Rica may also be responsible for the high compressive stress regime on the upper plate, which probably contributed to the genesis of the April 22, 1991 (Mw = 7.6), Valle de la Estrella earthquake (Fig. 1).

Cocos-Caribbean coupling and seismic risk

The differences in coupling between the Cocos and Caribbean plates, for Nicaragua and Costa Rica, can also be correlated with the characteristics of the subducted ocean floor. Based on historical and instrumentally recorded seismicity in Costa Rica, three main segments with different coupling are recognized, corresponding to the subduction segments described in Table 2: northwestern Costa Rica (from Papagayo Gulf to the southeastern end of the Nicoya Peninsula), central Costa Rica (from the entrance of the Nicoya Gulf to northwestern Osa Peninsula), and southeastern Costa Rica (from Osa Peninsula to the inland projection of the Panama Fracture Zone). Historically, large (Ms > 7.0) underthrust earthquakes have occurred along the northwestern and southeastern segments but not within the central segment (Table 2).

The boundary between the northwestern and the central segment is the updip projection of the Quesada Sharp Contortion (i.e., the down dip projection of the rough-smooth boundary), marking a difference in the ocean-floor relief. Under the northwestern segment, relatively smooth plate is being subducted, facilitating a larger area of contact with the overriding plate and therefore increasing the potential for large earthquakes. Under the central segment, the subducted plate consists of isolated seamounts (von Huene et al., this volume), which reduces the coupling area to a set of smaller patches that can break by moderate ($M_s \leq 7.0$) earthquakes. Von Huene et al. (this volume) suggest that these seamounts act as asperities for large subduction earthquakes. We suggest that the small size of these asperities inhibits the nucleation of large subduction earthquakes and therefore reduces the seismic potential of the central segment. Based on this interpretation, the central segment can be classified as Kurile type according to the asperity model of Lay et al. (1982). The March 25, 1990, M_s = 7.0, earthquake at the entrance of the Nicoya Gulf (Protti and McNally, 1990) (Fig. 1) occurred at

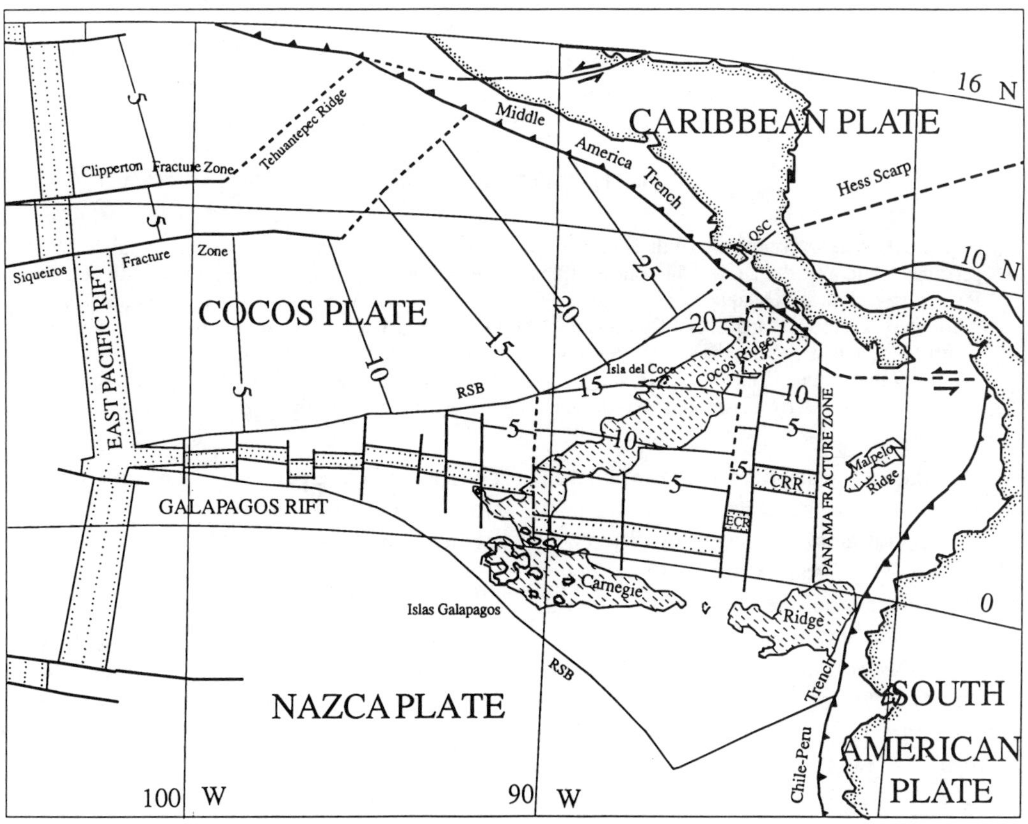

Figure 9. Distribution of lithospheric age, in m.y., of the Cocos plate. Base map is a segment of the plate-tectonic map of the Circum Pacific region, southeast quadrant (Circumpacific Council for Energy and Mineral Resources, 1981). The isochrons were drawn following the pattern of magnetic anomalies reported in that map and from projections of spreading rates of the East Pacific Rise and the Galapagos Rift. Dotted bands represent oceanic lithosphere created during the last million years. Cocos, Malpelo, and Carnegie Ridges are shown by the 2,000-m depth contour. The two RSB are the Rough-Smooth boundaries of Hey, 1977. The northern RSB divides lithosphere of the Cocos plate created along the Galapagos Rift from that created along the East Pacific Rise. The southern RSB is the analogous boundary for the Nazca plate. QSC is the Quesada Sharp Contortion; CRR and ECR are the Costa Rica and Ecuador rifts, respectively.

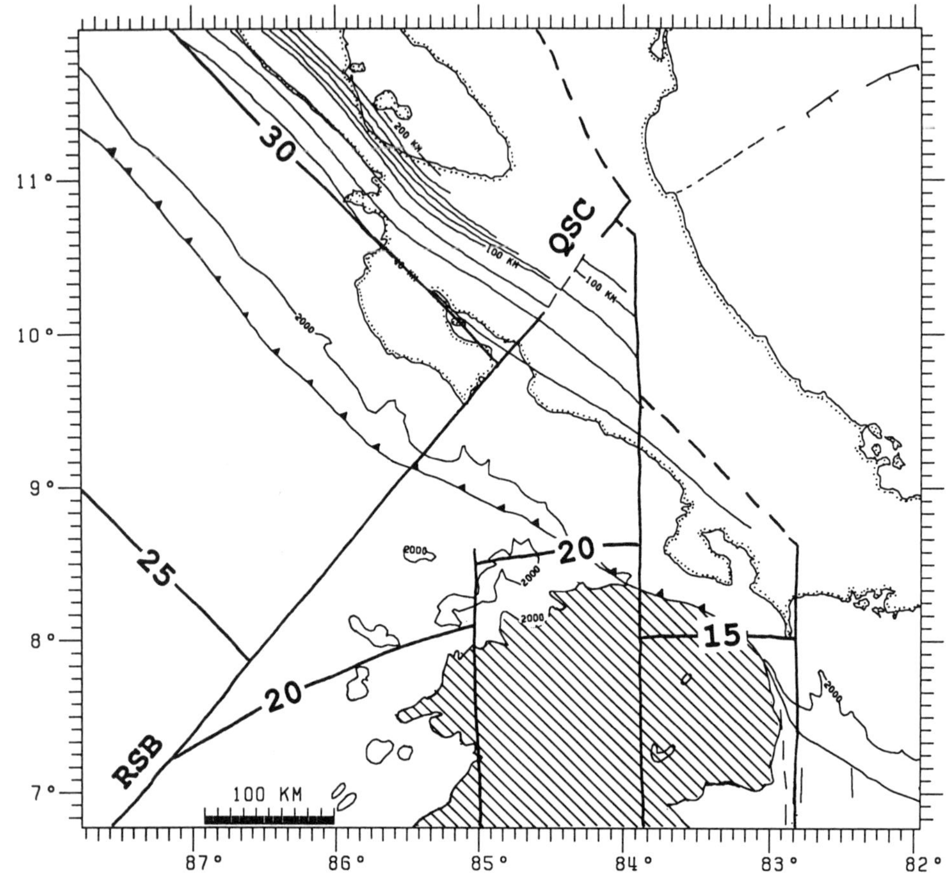

Figure 10. Map with projected lithospheric ages, in m.y., of the Cocos plate after unbending the seismic slab to a 0° dip angle. These projections of age are based on the spreading rates of the East Pacific Rise and the Galapagos Rift. Dashed line is the maximum extension of the seismic slab. Northwest of the Quesada Sharp Contortion (QSC) the subducted Cocos plate is older than under central Costa Rica. Also shown for reference are the isodepth contours (in kilometers) of the Wadati-Benioff zone. The Cocos Ridge is represented by the area enclosed by the 2,000-m depth contour; RSB is the rough-smooth boundary. Compare this figure with Figure 8.

the northwestern end of the central segment and may represent the largest earthquake that can be generated in that segment. That earthquake is a good indicator of the coupling differences between the northwestern and central segments: Even though it occurred right at the boundary between the segments, no aftershocks occurred toward the northwest (Protti and McNally, 1990) where a much stronger coupling exists (a locked gap) (Nishenko, 1989). Rather, all aftershocks occurred southeast of where the fracture initiated. Teleseismic analyses of the rupture propagation of that earthquake (Pacheco and Protti, 1990) agree with this interpretation.

The southeastern segment once again has a stronger coupling, relative to the central segment, as a result of the subduction of the Cocos Ridge, which is not only more buoyant but is also a much higher and more constant bathymetric anomaly. The expected existence of deeper roots beneath the Nicoya and Osa Peninsulas, practically absent in the central segment, may also contribute to the stronger coupling of the northwestern and southeastern segments with respect to the central one. The roots of these peninsulas may not only act as bumps at the base of the overriding plate but also increase the vertical load on the coupling interface.

Analogous regions

The main correlation we are making in this chapter is of Wadati-Benioff zone geometries with abrupt changes in age of the oceanic plate. These abrupt changes occur across fracture zones, transform faults, ridges, or boundaries between plates generated at different spreading centers like the rough-smooth boundary.

Our model could be applied to the isodepth contours of the Wadati-Benioff zone under Ecuador, which are almost perpendicular to the trench (Bevis and Isacks, 1984). There, the southern counterpart of the rough-smooth boundary that controls the Quesada Sharp Contortion, with abrupt changes of age, is on the same order as in Costa Rica.

Another possible analogy for the Costa Rica case may be represented by the differences in the subduction geometry found on either side of the Tehuantepec Ridge (Bevis and

Isacks, 1984). Across this ridge the plate shows a strong difference in bathymetry that may correspond with an abrupt change in age.

Southern Chile is another region where this model can be tested. We believe that a sharp contortion or tear might exist under southern Chile, where an abrupt jump in age on the order of 10 m.y. occurs across the Valdivia Fracture Zone on relatively young (20 to 30 m.y.) oceanic lithosphere of the subducted Nazca plate.

Another promising region in looking for the same Wadati-Benioff zone geometry variations discussed in this chapter is the South Sandwich subduction zone. A major change in trench depth across a fracture zone separating crust of very different ages is found there (Jarrard, 1986).

Furthermore, an analogous correlation to the one presented in this chapter was found by Farrar and Lowe (1978) for Peru and Chile. In that subduction zone conditions similar to those in Costa Rica are present, that is, the same convergence rates but abrupt contrasts in slab ages and slab lengths.

Subduction of the Cocos Ridge

The Cocos Ridge, here considered as the bathymetric high on the Cocos plate enclosed by the 2,000-m contour, subducts under southern Costa Rica. This ridge is one of the traces of the Galapagos hot spot (Hey, 1977) and therefore represent a thermal and density anomaly within the Cocos plate. Its buoyancy and bathymetric characteristics are responsible for the fast uplift rates of Osa and Burica Peninsulas (e.g., Corrigan et al., 1990; Gardner et al., 1992). Another effect that the subduction of the Cocos Ridge has in southern Costa Rica, as mentioned above, is an increase in the coupling between oceanic and upper plate.

Several dates have been suggested for the initiation of the subduction of the Cocos Ridge under southern Costa Rica: 0.5 Ma from onland work on the Pacific coast of Costa Rica and elastic deformation modeling (Gardner et al., 1992), 1 Ma from ocean-floor magnetic anomalies and plate reconstructions (Lonsdale and Klitgord, 1978), and 3.6 Ma from onland work on the Caribbean coast of Costa Rica and Panama (Collins et al., this volume). We believe the younger ages are correct since they better fit physical models of crustal deformation. Uplift of the Caribbean basins of Costa Rica and Panama can be accounted for by shortening and underthrusting along the Panama Thrust belt as observed after the April 24 1991, Valle de la Estrella earthquake. With 1 m.y. of subduction of the Cocos Ridge, its subducted portion would not be longer than 100 km. The presence of the Panama Fracture Zone, which truncated the Cocos Ridge around 9 m.y. ago (Lonsdale and Klitgord, 1978), also limits the northeastern extension of the subducted Cocos Ridge to ~100 km. This leads us to suggest that it is the subduction of young lithosphere across the transform fault that offsets the Costa Rica and Ecuador rifts that produces the shallow subduction under southern Costa Rica and not the subduction of the Cocos Ridge as postulated by several workers, including ourselves (e.g., Adamek et al., 1987; Güendel et al., 1989; Corrigan et al., 1990; Kolarsky et al., this volume). Geochemical signals of igneous rocks outcropping in the Talamanca Cordillera indicate that the subduction of young lithosphere started in southern Costa Rica approximately 6 or 7 m.y. ago (de Boer et al., this volume).

CONCLUSIONS

A new model, based on age variations of the subducted Cocos plate along the Middle America Trench, correlates well with the complex geometry of the Wadati-Benioff zone under Nicaragua and Costa Rica (the Quesada Sharp Contortion, changes in dip angle, length and maximum depth of the seismic slab) as well as with regional tectonic features on the overriding Caribbean plate, such as distinct stress field variations and the termination of the Central America Volcanic Chain. This model better correlates the changes in the Wadati-Benioff zone geometry than previous models based on the subduction of the Cocos Ridge.

The difference in coupling between the Cocos and Caribbean plates along the southern terminus of the Middle America trench can be associated with bathymetric features on the subducted Cocos plate. These bathymetric features are also related to age and genetic differences within the Cocos Plate.

All of the above conclusions make Nicaragua–Costa Rica a potential area in which to study subduction processes and effects in a small area with variations from mildly tensional- to compressional-type subductions. This subduction zone represents almost the entire spectrum from Marianas- to Chilean-type subductions of Wilson and Burke (1972) and Uyeda and Kanamori (1979).

ACKNOWLEDGMENTS

We offer special thanks to the personnel of the Costa Rica Volcanological and Seismological Observatory, Universidad Nacional (OVSICORI-UNA), for the excellence of their work collecting and processing the best seismic data set that has ever existed in Costa Rica. Universidad Nacional in Heredia, Costa Rica, and Fulbright Foundation through the Latin American Scholarship Program of American Universities (LASPAU) provided financial and institutional support for M. Protti to complete this work. Robert Garrison, Casey Moore, Elise Knitle, and Eli Silver reviewed an earlier version of this chapter and made excellent suggestions to improve it. We thank reviewers Jim Dewey, Cliff Frohlich, Wayne Pennington, Fumiko Tajima, and especially the editor of this volume, Paul Mann, for their constructive suggestions. The final version was proofread by Anne McCaffrey. NSF grant number INT-8604157 made possible the collection of key data for imaging of the Wadati-Benioff zone under northern Costa Rica. W. M. Keck Foundation provided partial support for instrumentation and computer facilities. This work represents OVSICORI's scientific contribution no. 150 and contribution no. 191 of the Institute of Tectonics, University of California, Santa Cruz.

REFERENCES CITED

Adamek, S., Tajima, F., and Wiens, D. A., 1987, Seismic rupture associated with subduction of the Cocos Ridge: Tectonics, v. 6, p. 757–774.

Adams, R. D., and Ware, D. E., 1977, Subcrustal earthquakes beneath New Zealand: Locations determined with a laterally inhomogeneous velocity model: New Zealand Journal of Geology and Geophysics, v. 20, p. 59–83.

Bevis, M., and Isacks, B. L., 1984, Hypocentral trace surface analysis: Probing the geometry of the Benioff zones: Journal of Geophysical Research, v. 89, p. 6153–6170.

Burbach, G. V., and Frohlich, C., 1986, Intermediate and deep seismicity, and lateral structure of the subducted lithosphere in the Circum-Pacific region: Reviews of Geophysics, v. 20, p. 833–874.

Burbach, G. V., Frohlich, C., Pennington, W. D., and Matumoto, T., 1984, Seismicity and tectonics of the subducted Cocos plate: Journal of Geophysical Research, v. 89, p. 7719–7735.

Circumpacific Council for Energy and Mineral Resources, 1981, Plate-tectonic map of the Circum-Pacific region, southeast quadrant: American Association of Petroleum Geology, scale 1:10,000,000.

Corrigan, J., Mann, P., and Ingle, J. C., 1990, Forearc response to subduction of the Cocos Ridge, Panama–Costa Rica: Geological Society of America Bulletin, v. 102, p. 628–652.

De Mets, C., Gordon, R. G., Argus, D. F., and Stein, S., 1990, Current plate motions: Geophysical Journal International, v. 101, p. 425–478.

Dewey, J. W., and Algermissen, S. T., 1974, Seismicity of the Middle America arc-trench system near Managua, Nicaragua: Bulletin of the Seismological Society of America, v. 64, p. 1033–1048.

Farrar, E., and Lowe, R. M., 1978, Age-length dependence of inclined seismic zones: Nature, v. 273, p. 292–293.

Gardner, T. W., and six others, 1992, Quaternary uplift astride the aseismic Cocos Ridge, Pacific coast, Costa Rica: Geological Society of America Bulletin, v. 104, p. 219–232.

Güendel, F., 1986, Seismotectonics of Costa Rica: An analytical view of the southern terminus of the Middle America Trench [Ph.D. thesis]: Santa Cruz, University of California, Santa Cruz, 157 p.

Güendel, F., and 14 others, 1989, First results from a new seismographic network in Costa Rica, Central America: Bulletin of the Seismological Society of America, v. 79, p. 205–210.

Hey, R. N., 1977, Tectonic evolution of the Cocos-Nazca spreading center: Geological Society of America Bulletin, v. 88, p. 1404–1420.

Isacks, B. L., and Barazangi, M., 1977, Geometry of Benioff zones: Lateral segmentation and downwards bending of the subducted lithosphere, *in* Talwani, M., and Pitman, W. C., III, eds., Island arcs, deep sea trenches, and back-arc basins: Maurice Ewing ser., vol. 1, Washington, D.C., American Geophysical Union, p. 99–114.

Isacks, B. L., Oliver, J., and Sykes, L. R., 1968, Seismology and the new global tectonics: Journal of Geophysical Research, v. 73, p. 5855–5899.

Jarrard, R. D., 1986, Relations among subduction parameters: Reviews of Geophysics, v. 24, p. 217–284.

Klitgord, K. D., and Mammerickx, J., 1982, Northern East Pacific Rise: Magnetic anomaly and bathymetric framework: Journal of Geophysical Research, v. 87, p. 6725–6750.

Lay, T., Kanamori, H., and Ruff, L., 1982, The asperity model and the nature of large subduction zone earthquakes: Earthquake Prediction Research, v. 1, p. 3–71.

Leeds, A. R.,1975, Lithospheric thickness in the western Pacific: Physics of the Earth and Planetary Interiors, v. 11, p. 61–64.

Lonsdale, P. F., and Klitgord, K. D., 1978, Structure and tectonic history of the eastern Panama basin: Geological Society of America Bulletin, v. 89, p. 981–999.

Malavassi, E., 1991, Magma sources and crustal processes at the southern terminus of the Central American volcanic arc [Ph.D. thesis]: Santa Cruz, University of California, Santa Cruz, 148 p.

Mammerickx, J., and Klitgord, K. D., 1982, Northern East Pacific Rise: Evolution from 25 m.y. B.P. to the present: Journal of Geophysical Research, v. 87, B8, p. 6751–6759.

Matumoto, T., Othake, M., Latham, G., and Umaña, J., 1977, Crustal structure of southern Central America: Bulletin of the Seismological Society of America, v. 67, p. 121–134.

McKenzie, D. P., 1969, Speculations on the consequences and causes of plate motions: Geophysical Journal of the Royal Astronomical Society, v. 18, p. 1–32.

McLaren, J., and Frohlich, C., 1985, Model calculations of regional network locations for earthquakes in subduction zones: Bulletin of the Seismological Society of America, v. 75, p. 397–413.

Nishenko, S. P., 1989, Circumpacific seismic potential 1989–1999: U.S. Geological Survey, National Earthquake Information Center open file report 89-86, 168 p.

Pacheco, J., and Protti, M., 1990, The March 25, 1990, earthquake sequence at the entrance of the Nicoya Gulf, Costa Rica: Teleseismic source parameters and rupture history: EOS (American Geophysical Union Transactions), v. 71, p. 1438.

Parsons, B., and Sclater, J. G., 1977, An analysis of the variation of ocean floor bathymetry and heat flow with age: Journal of Geophysical Research, v. 82, p. 803–827.

Protti, M., 1991, Correlation between the age of the subducting Cocos Plate and the geometry of the Wadati-Benioff zone under Nicaragua and Costa Rica [M.Sc. thesis]: Santa Cruz, University of California, Santa Cruz, 66 p.

Protti, M., and McNally, K., 1990, The March 25, 1990 (ML=6.8, Ms=7.0) earthquake sequence at the entrance of the Nicoya Gulf, Costa Rica: A near field study: EOS (American Geophysical Union Transactions), v. 71, p. 1438.

Sacks, I. S., 1983, The subduction of young lithosphere: Journal of Geophysical Research, v. 88, p. 3355–3366.

Silver, E. A., Reed, D. L., Tagudin, J. E., and Heil, D. J., 1990, Implications of the north and south Panama thrust belt for the origin of the Panama orocline: Tectonics, v. 9, p. 261–281.

Stoiber, R. E., and Carr, M., 1973, Quaternary volcanic and tectonic segmentation of Central America: Bulletin of Volcanology, v. 37, p. 304–325.

Suarez, G., Monfret, T., Wittlinger, G., and David, C., 1990, Geometry of subduction and depth of the seismogenic zone in the Guerrero gap, Mexico: Nature, v. 345, p. 336–338.

Sugi, N., and Uyeda, S., 1984, Subduction of young oceanic plates without deep focus earthquakes: Bulletin de la Societe Gologique de France, v. 26, p. 245–254.

Uyeda, S., and Kanamori, H., 1979, Back-arc opening and the mode of subduction: Journal of Geophysical Research, v. 84, p. 1049–1061.

Vlaar, N. J., and Wortel, M.J.R., 1976, Lithospheric aging, instability, and subduction: Tectonophysics, v. 32, p. 331–351.

Wilson, J. T., and Burke, K., 1972, Two types of mountain building: Nature, v. 239, p. 448–449.

Wolters, B., 1986, Seismicity and tectonics of southern Central America and adjacent regions with special attention to the surroundings of Panama: Tectonophysics, v. 128, p. 21–46.

Wortel, R., 1980, Age-dependent subduction of oceanic lithosphere [Ph.D. thesis]: Utrecht, The Netherlands, State University of Utrecht, 147 p.

Wortel, M.J.R., and Vlaar, N. J., 1978, Age-dependent subduction of oceanic lithosphere beneath western South America: Physics of the Earth and Planetary Interiors, v. 17, p. 201–208.

Yamaoka, K., Fukao, Y., and Kumazawa, M., 1986, Spherical shell tectonics: Effects of sphericity and inextensibility on the geometry of the descending lithosphere: Reviews of Geophysics, v. 24, p. 27–53.

Manuscript Accepted by the Society February 7, 1994

Printed in U.S.A.

Geological Society of America
Special Paper 295
1995

Tectonic implications of the seismic ruptures associated with the 1983 and 1991 Costa Rica earthquakes

Fumiko Tajima
Institute for Geophysics, The University of Texas at Austin, 8701 North Mopac Boulevard, Austin, Texas 78759-8397
Masayuki Kikuchi
Department of Physics, Yokohama City University, 22-2 Seto, Kanazawa-ku, Yokohama, 236, Japan

ABSTRACT

Two large earthquakes occurred in Costa Rica in 1983 and 1991. The source areas are in the vicinity of the triple junction of the Cocos, Nazca, and Caribbean plates, where a large bathymetric feature, the Cocos Ridge, is colliding/subducting beneath the Caribbean plate. The 1983 event occurred in the subduction zone of the Cocos Ridge near the southeastern terminus of the Middle America trench just west of the Panama Fracture zone. The 1991 earthquake took place about 100 km north of the 1983 epicenter and is considered to be a back-thrusting event representing the overthrusting of the North Panama deformed belt over the Caribbean plate. In this region the subduction of the buoyant Cocos Ridge should play an important role in the occurrence of unique seismological and tectonic features, for example, a decrease in seismicity, a cessation of volcanic activity, and a shallowing of the trench.

This chapter summarizes several different seismological studies for the rupture processes and associated seismicity patterns of the 1983 and 1991 Costa Rica earthquakes, attempting to constrain the tectonic structure in this region. The results provide a clearer picture of the mechanical interaction and geometry between the subducting ridge and overriding Caribbean plate than previous hypothetical models schematically illustrated for ridge-trench intersections.

INTRODUCTION

Two large earthquakes (Ms = 7.3, depth ≅ 30 km on April 3, 1983; and Ms = 7.5, depth ≅ 12 km on April 22, 1991) occurred in Costa Rica near the triple junction of the Cocos, Nazca, and Caribbean plates. In this area a large bathymetric feature, the Cocos Ridge, is colliding/subducting beneath the Caribbean plate (Fig. 1). The 1983 earthquake took place in the subduction zone, which had been the site of two previous events with similar magnitudes—Ms 7.6 in 1904 and Ms = 7.5 in 1941—and therefore had been identified as a seismic gap prior to the recent event (Astiz and Kanamori, 1984). The 1991 earthquake took place about 100 km north of the 1983 epicenter, with noticeable vertical deformation of the coast, and is considered to be a back-thrusting event (thrusting on the side away from the subduction zone). This event represents the thrusting of the North Panama deformed belt (NPDB) over the Caribbean plate (Goes et al., 1993; Plafker and Ward, 1992; Suárez et al., unpublished manuscript; Tajima and Kikuchi, 1991). The occurrence of large back-thrusting events are rare in general.

The intersection of the Cocos Ridge and the southeastern terminus of the Middle America trench (MAT) is just west of the Panama Fracture Zone (PFZ), where the two Costa Rica earthquakes took place. The Cocos Ridge is a track of the Galapagos hot spot and is genetically related to the Carnegie Ridge of the Nazca plate (Hey, 1977). Numerous papers have reported the unique seismological and tectonic features observed in this area, for example, a decrease in seismicity, a cessation of volcanic arc activity, and a shallowing of the

Tajima, F., and Kikuchi, M., 1995, Tectonic implications of the seismic ruptures associated with the 1983 and 1991 Costa Rica earthquakes, *in* Mann, P., ed., Geologic and Tectonic Development of the Caribbean Plate Boundary in Southern Central America: Boulder, Colorado, Geological Society of America Special Paper 295.

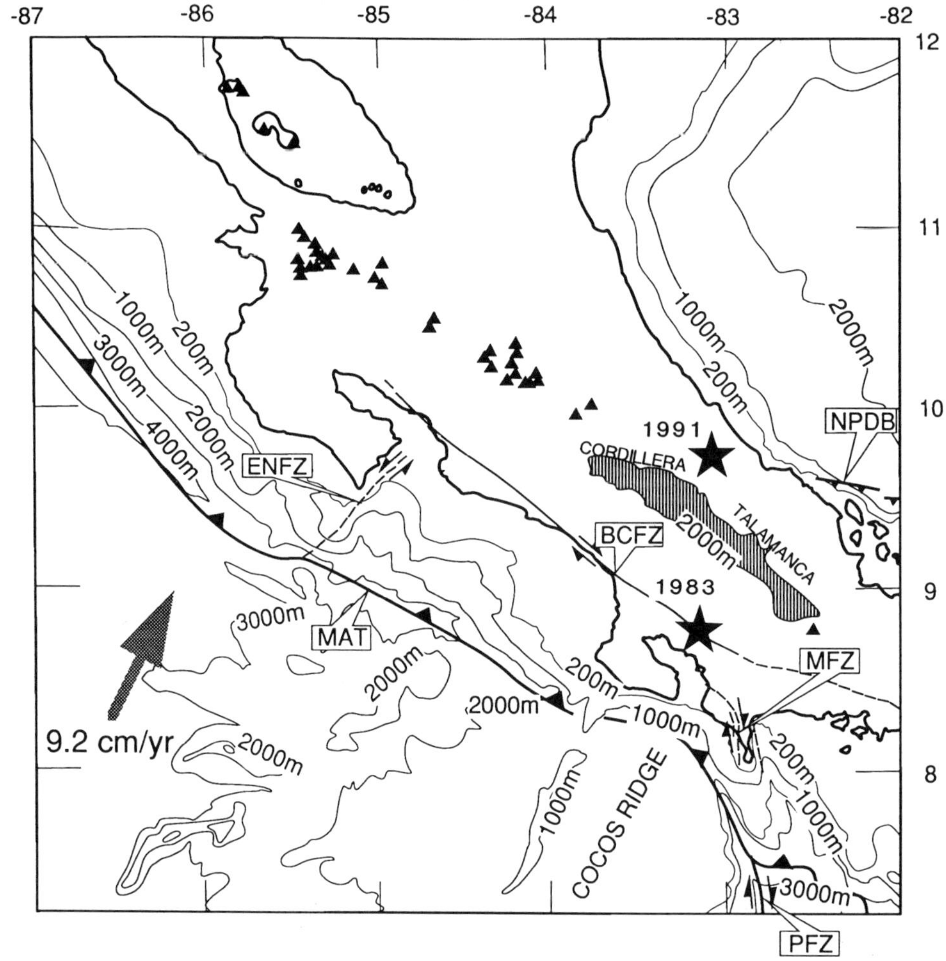

Figure 1. Tectonic map of the Cocos Ridge–Middle America Trench (MAT) intersection area, including the epicenters of the 1983 and 1991 Costa Rica earthquakes. Subduction zone shallowing (with equidepth lines) and cessation of volcanic activity are shown. The shaded area between the 1983 and 1991 epicenters illustrates the topographic uplift (modified from Kolarsky et al., this volume). The plate convergence rate 9.2 cm/yr determined by DeMets et al. (1990) is indicated. MAT: Middle America Trench; PFZ: Panama Fracture Zone; NPDB: North Panama deformed belt; BCFZ: Ballena-Celmira fault zone; ENFZ: East Nicoya fault zone; MFZ: Medial fault zone.

trench as compared to other parts of the MAT. These observations imply disturbance of the plate convergence, although the Cocos-Caribbean convergence rate is estimated to be about 92 mm/year (DeMets et al., 1990).

The uplifted area between the two earthquakes (see the shaded area in Fig. 1; also Miyamura, 1975; Rivier, 1985) is in line with the Cocos Ridge and is the topographic high of the Costa Rican isthmus. Similar topography is observed worldwide at (aseismic) ridge-trench intersections, revealing the buoyant nature of these features (e.g., Vogt et al., 1976; Kelleher and McCann, 1976; Chung and Kanamori, 1978a, b; McCann and Sykes, 1984; Eissler and Kanamori, 1982; Astiz and Kanamori, 1984; LeFevre and McNally, 1985; McGeary et al., 1985; Kolarsky et al., this volume), which should also have an important role in the stress transfer between the earthquake source area. McGeary et al. (1985) summarized the unique seismo-tectonic features at ridge-trench intersections worldwide, focusing on the spatial gaps of arc volcanism, and presented several schematic models for the interaction between a subducting ridge and the overriding plate.

A conventional approach to constrain hypothetical tectonic models of a seismogenic area is to use properties obtained from seismological analyses, such as moment tensor fault mechanism solutions, or seismicity patterns. As mentioned above, back-thrusting events are rare, and the geometry of the southern terminus of the MAT is still ambiguous, based on seismicity studies. At present several papers are available that examine seismological properties associated with the two recent Costa Rica earthquakes. Some of the studies described in these papers use modern digital seismological techniques to analyze the seismic rupture of the 1991 Costa Rica earthquake. The results obtained are somewhat complementary to each other, but none of them has succeeded in presenting a comprehensive model by itself. In addition, they are still in review (or in press). Therefore, in this chapter we summarize the recent source rupture studies presented for the 1983 and 1991 Costa Rica earthquakes, including Goes et al. (1993), and associated seismicity (Fan et al., 1993; Protti et al., this volume; and Suárez et al., unpublished manuscript) and then attempt to constrain the model of interaction between the descending Cocos Ridge and the overriding Caribbean plate.

SEISMOLOGICAL STUDIES OF THE COSTA RICA EARTHQUAKES

Overview of seismicity

Figure 2 summarizes seismicity patterns (for events with magnitudes ≥ 4.7) obtained from the International Seismological Center (ISC) and National Earthquake Information Center (NEIC) catalogs for the vicinity of the two Costa Rica earthquakes during the period between January 1964 and September 1991: a, for the entire period; b, from 1964 to 1978; c, from 1979 to March 1983; d, from April to September 1983; e, from April 1983 to March 1991; f, from April to September 1991. Figures 2b and c show the activity before the 1983 event, 2d the five months of aftershock activity following the 1983 event, 2e the period between the 1983 and 1991 events including the 1983 aftershocks, and 2f the aftershock activity of the 1991 event. For some of the events (M ≥ 5.5), Centroid Moment Tensor (CMT) solutions determined by the Harvard group are shown. The rupture zones of the 1983 and 1991 main events that were inferred from our study are indicated with contours.

In the future rupture zone of the 1983 event there were two moderate events in 1979: one a normal event with substantial aftershocks near the southeastern boundary (July 1, Ms = 6.5, depth = 12 km) and the other a thrust event near the northwestern boundary (August 24, Ms = 6.4, depth = 19 km) (Fig. 2c). The normal faulting event is consistent with the present-day uplift of the Burica Peninsula, which may behave as a detached block from the Osa Peninsula. The aftershock area of the 1983 event expanded mostly to the northwest, parallel to the trench axis during the five months after the event (Fig. 2d). Adamek et al. (1987) noted aftershocks to the southeast of the epicenter as well. This aftershock expansion pattern indicates that there are no distinct boundaries that act as barriers to stress propagation beyond the main source rupture area along the trench axis. The influences of the detached blocks or the distribution of partial coupling are reflected in the aftershock expansion pattern and in their fault mechanisms, which show substantial variation (Fig. 2e).

Figures 2d and 2e show that the area between the 1983 rupture zone and the future 1991 rupture zone was very quiet during the period between the two earthquakes except for one event that occurred on March 11, 1988, with an Ms = 6 at about 50 km north of the 1983 epicenter. This implies that the stress created beyond the 1983 source area in the direction normal to the trench axis could have been transmitted toward the 1991 source area through the topographic uplift. In this circumstance the 1988 event is a peculiar and interesting event in terms of location (PDE [Preliminary Determinations of Epicenters] depth = 23 km and CMT depth = 37 km) and fault mechanism (oblique strike-slip) and may offer clues to understanding the stress propagation from the 1983 source area to the 1991 rupture zone.

The aftershock distribution of the 1991 event widened in the direction northwest of the main rupture zone and merged with the activity near the trench. Since the 1991 event, the area between the two rupture zones has remained seismically quiescent. The stress propagation through the topographic uplift seems to be aseismic. The strike-slip mechanisms northwest of the 1983 rupture zone (Fig. 2e) and the 1991 rupture zone (Fig. 2f) indicate segmentation of the Caribbean plate boundary in that area.

Subduction zone geometry

Many researchers pointed out that the dip of the Benioff zone along the MAT becomes shallow where the Cocos Ridge is subducting (Molnar and Sykes, 1969; Dean and Drake, 1976; Liaw, 1981; Burbach et al., 1984). A steeply dipping Benioff zone is well defined up to Nicaragua, and the transition of the Benioff zone seismicity is clearly observed from Nicaragua to Costa Rica (Dewey and Algermissen, 1974). The hypocentral depths listed in the ISC catalog indicate little or no deep (h > 70 km) activity from southeast of the Nicoya Peninsula to the southeastern terminus of the MAT.

Liaw (1981) used data from a local network and showed that the Benioff zone of northwest Costa Rica is less well defined than beneath Nicaragua and becomes distinctly shallower, with little activity below 100 km near the Nicoya Peninsula. Burbach et al. (1984) incorporated this observation with teleseismic data and proposed that the change of the slab dip is due to a tear fault, which separates the subducting lithosphere beneath the Nicoya Peninsula. They also noted that deeper seismic activity (h > 70 km) ceases southeast of the Nicoya Peninsula, and accordingly the dip angle of the slab is undetermined. Other researchers correlated the slab segmentation with the cessation of the volcanic chain in central Costa Rica (Stoiber and Carr, 1973; Carr and Stoiber, 1977).

Protti et al. (this volume) reevaluated the seismicity using high-resolution location data of 9,514 events with computed horizontal and vertical errors smaller than 4 and 5 km, respectively. They imaged the geometry of the Wadati-Benioff zone under southern central America and summarized the transition of the slab geometry from a steep dip angle to a shallow one focusing on three areas: Nicaragua, Nicoya Peninsula, and the Cocos Ridge collision zone. They found that the Wadati-Benioff zone under the Nicaragua–Costa Rica border contorts

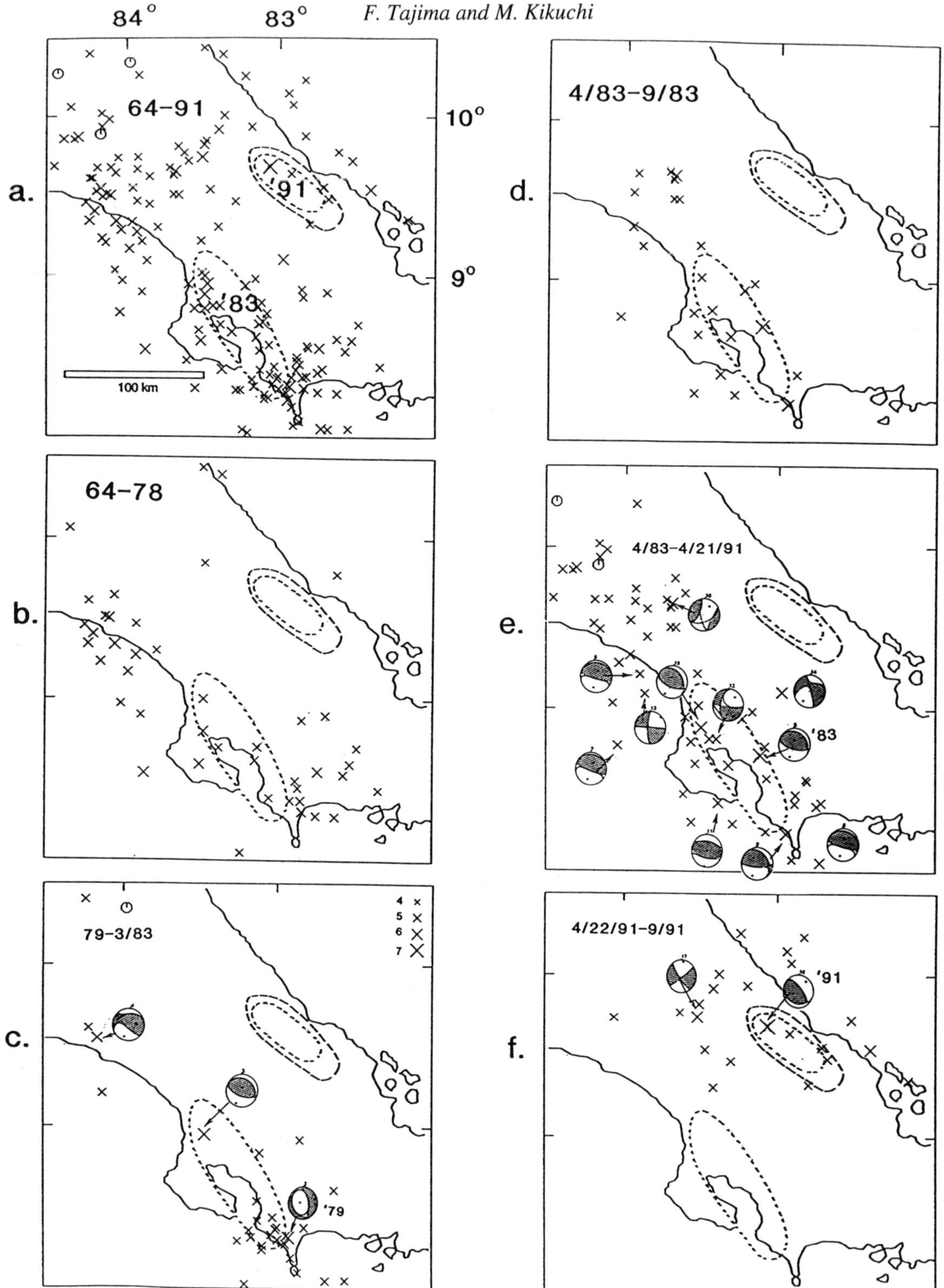

Figure 2. Epicenter plots of events (M > 4.7) that occurred in the region of the Cocos Ridge–Middle America trench intersection using the International Seismological Center and National Earthquake Information Center catalogs and some Centroid Moment Tensor solutions by the Harvard group during different periods: a, between 1964 and 1991; b, from 1964 to 1978; c, from 1979 to March 1983 (note that there was a moderate event with a normal fault mechanism [Ms = 6.4, July 1] near the southeastern boundary of the future source area and a thrust event near the northwestern boundary in 1979 [Adamek et al., 1987]); d, from April to September 1983; e, from April 1983 to March 1991 (note that the aftershock activity of the 1983 event expanded along the fault strike but not widthwise; there were also some strike slip mechanisms to the northwest of the source areas); f, from April to September 1991. The future 1991 earthquake source area and its vicinity were quiet before the event in 1991 (b, c, d and e). There was some expansion of the aftershock activity of the 1991 event toward the 1983 source area.

(from steep to shallow dip angles, northwest to southeast), but it does not show evidence of a brittle tear, as postulated by others (Stoiber and Carr, 1973; Burbach et al., 1984; Burbach and Frohlich, 1986). Based on age variations of the subducted Cocos plate along the Middle America Trench, Protti et al. (this volume) propose a new model that correlates well with the complex geometry of the Wadati-Benioff zone under Nicaragua and Costa Rica (the Quesada Sharp Contortion, changes in dip angle, length and maximum depth of the seismic slab) as well as with regional tectonic features on the overriding Caribbean plate.

April 3, 1983, earthquake and its aftershocks

Preliminary solutions for the source mechanism of the 1983 event indicate that this event is a thrust event along the MAT (Dziewonski et al., 1983). The body waveforms show a complex time history of moment release as compared with the waveforms of comparable events along the MAT in southern Mexico (Tajima, 1984). If the time history of moment release is complex, it also implies that the source rupture may include source mechanism changes due to the changing stress condition during the rupture process. Adamek et al. (1987) studied the rupture process of this main event using a one-dimensional inversion method (Kikuchi and Kanamori, 1982) with long-period body waveforms and a fixed focal mechanism (ϕ = 295°, δ = 32°, λ = 90°). The source-time function obtained shows a relatively long lasting source process with some complexity (see the upper box in Fig. 3a). From a directivity analysis, Adamek et al. determined the locations of strong moment release that indicate the rupture propagation to the northwest along the fault strike (Fig. 3a). The aftershock activity is moderate and sparse (shown with crosses and small solid circles in Figs. 3a and b). The rupture zone based on the one-day aftershock distribution 70 × 110 km^2 (illustrated with a contour in Fig. 3a) was somewhat overestimated, and the stress drop (Δ ~ 2 to 3 bars) and the seismic slip (D ~ 58 cm) calculated for the rupture area were underestimated according to our analysis. The first motion focal mechanisms determined for major aftershocks show oblique strike slip faults northwest and southeast beyond the main rupture zone (Fig. 3b; modified from Adamek et al., 1987).

Adamek et al. (1987) also compared the waveforms of the 1941 large earthquake (Ms = 7.5) that occurred near the present

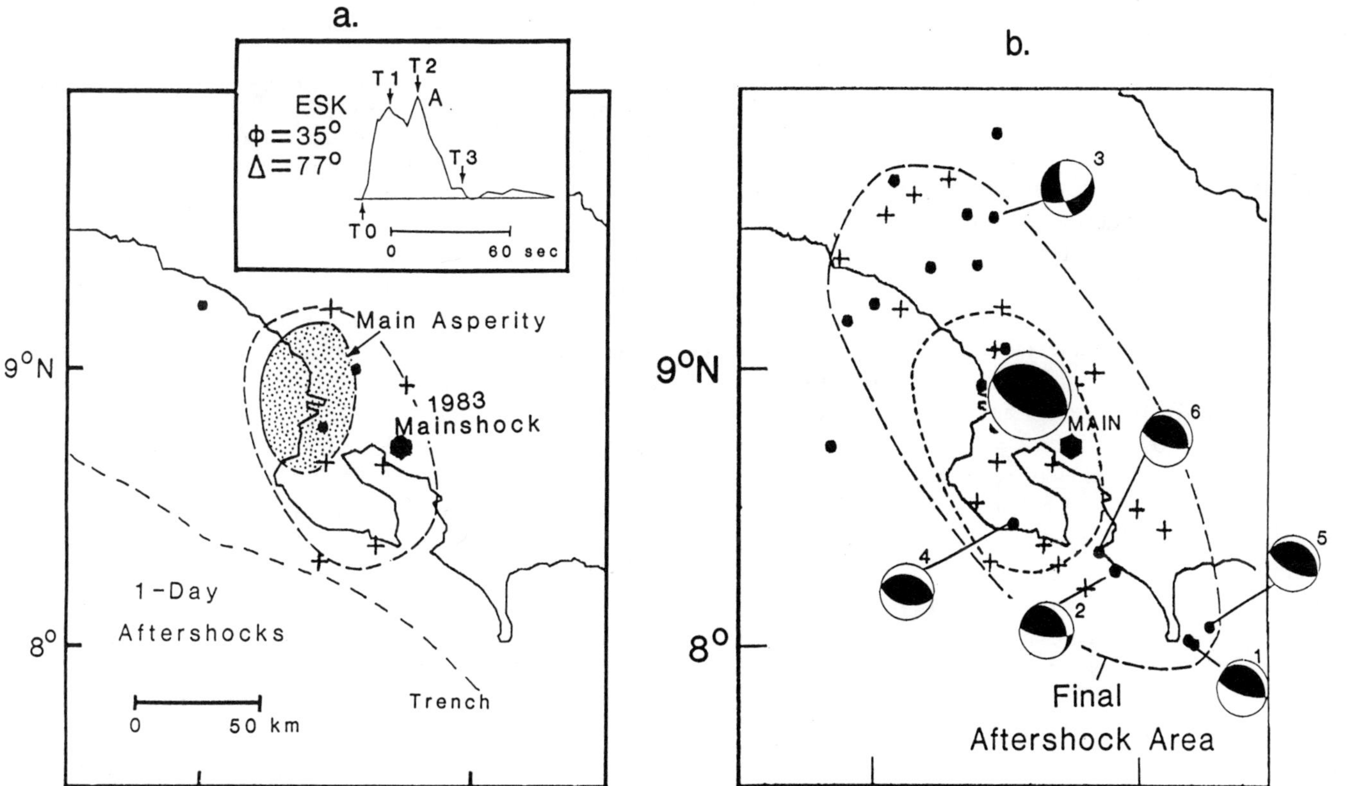

Figure 3. a, Illustration for the location of moment release inferred for the 1983 main event by Adamek et al. (1987). The crosses and small solid circles (relocated by Adamek et al. [1987]) show the epicenters of aftershocks. The source time function shown in the upper box indicates two subevents during the main rupture. b, Aftershock distribution and some focal mechanism solutions. Events (1) 04/07/83, (2) 05/09/83, (3) 07/03/83, (4) 09/23/83, (5) 10/12/83, and (6) 04/09/84 (modified from Adamek et al., 1987).

source area to those of the 1983 event and suggested a similar source process. The P wave directivity and aftershock distribution of the 1983 event as well as the source process similarity of the two large events suggest that the major zone of coupling along this section of the MAT is located just west of the epicenter of the 1983 event. This is the locally highest-stressed zone of contact between the Cocos ridge and the overriding Caribbean plate. The gradual expansion of the aftershocks southeast and trenchward of the epicenter (east Osa and Burica area) suggests that the coupling beneath the outer forearc region is relatively weak and heterogeneous. However, as mentioned before, this main rupture process could also include fault mechanism changes as a result of the changing stress condition during the process. If this is the case, the assumption of a constant fault mechanism in the inversion may not be adequate.

We were interested in the spatial and temporal proximity of the 1983 and 1991 earthquakes and analyzed the source rupture processes of these events in detail, using an iterative inversion method (Kikuchi and Kanamori, 1991) with long-period body waves. The P and SH waveforms used in the inversion were obtained from the Global Seismographic Network (GSN) and World Wide Standardized Seismographic Network (WWSSN) long-period records. The inversion method models the entire source rupture process with a series of subevents and determines the moment-tensors $[M_{ij}]$ of subevents as well as the spatial extent of subevents and the time-function of moment release. Each of the subevents is determined separately and can have a different mechanism from others; therefore, fault mechanism changes during the rupture can be accounted for. All the source parameters are allocated at assigned grids on a fault plane, and, accordingly, the fault geometry and source area are well determined.

Unlike the preliminary solutions mentioned above, we modeled the entire rupture process with three subevents of varying mechanisms for the main event (denoted with s1, s2, and s3 in Fig. 4). The primary subevent shows a thrust mechanism ($\phi = 357°$, $\delta = 46°$, and $\lambda = 103°$) with a source depth of 32 km; it was succeeded by two comparable subevents, one located at 45 km southeast of the epicenter with a strike-slip mechanism ($\phi = 126°$, $\delta = 78°$, and $\lambda = 173°$) and a depth of 21 km in 13 s and the other one at 60 km northwest of the epicenter with a thrust mechanism ($\phi = 298°$, $\delta = 40°$, and $\lambda = 98°$) and a depth of 37 km in 23 s (see the fault mechanisms and locations of the subevents in Fig. 4). The rupture area defined by the two thrust subevents is about 30 × 75 km^2, but with the strike-slip subevent it is about 60 × 110 km^2.

The rupture pattern determined by our analysis is remarkably consistent with the result by Adamek et al. (1987), who used a one-dimensional inversion method (Kikuchi and Kanamori, 1982) with a fixed-thrust fault mechanism. According to their analysis, the rupture front propagated westward 63 km from the epicenter to where the major moment release was truncated. The aftershock distribution, which also expanded southeastward, indicates a possibility of bilateral rupture propagation, although the inversion with a fixed mechanism could not resolve this. With the new approach of varying fault mechanism, the rupture propagation to the southeast was also retrieved. A few major aftershocks ($5 \leq M \leq 6.1$; see Adamek et al. [1987] for the mechanisms) are located around the second and third subevents identified in our study (Fig. 3b).

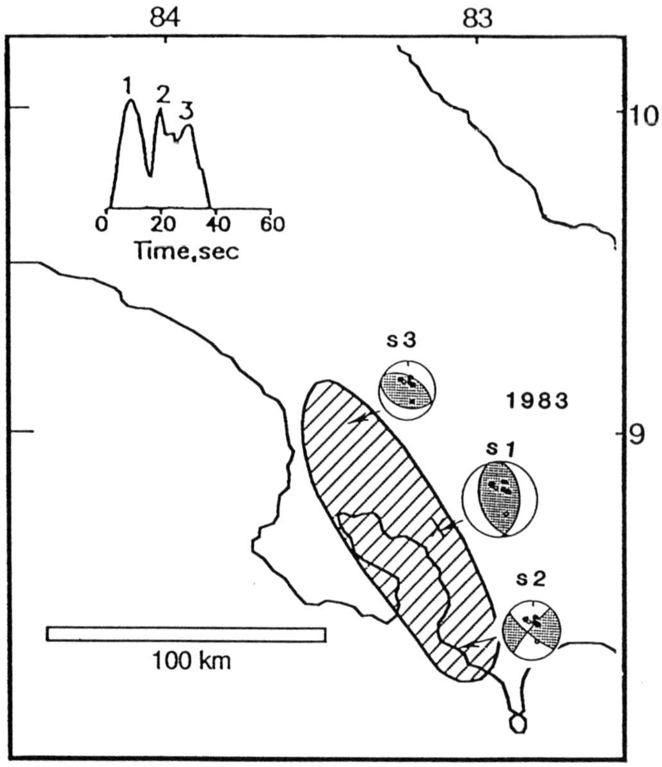

Figure 4. Results obtained for the 1983 main event source process by our study. The entire rupture was modeled with three subevents of varying focal mechanisms (denoted with s1, s2, and s3 in the order of occurrence). The source-time function is also shown to distinguish the three subevents. The rupture zone is indicated with a hatched area.

The total duration of the rupture is about 40 s from the source-time function, and the estimated rupture velocity ranges between 2.8 and 3.5 km/s. The moments estimated for the three subevents are 0.54×10^{27}, 0.32×10^{27}, and 0.33×10^{27} dyne-cm, respectively. The total moment release is 1.2×10^{27}, which is similar to the moment 1.35×10^{27} dyne-cm estimated from long-period surface waves at 256 s (Tajima, 1985) but between the estimate from the CMT solution, 1.8×10^{27} dyne-cm (Dziewonski et al., 1983) and that listed in the 1983 PDE catalog by the United States Geological Survey. The mechanism obtained by summing up the moment tensors of the subevents is similar to the CMT solution, including the nondouble couple component (14%). A description of the entire rupture process is that it started with a thrust mechanism primary rupture, propagated bilaterally, and triggered two subevents, one with a thrust mechanism northwest of the epicenter and the other with a strike-slip mechanism southeast of the epicenter. Interestingly, the strike of the primary thrust subevent is parallel to the strike of the Panama fracture zone, and the mechanism of the third subevent is parallel to the strike of the Panama fracture zone, and the mechanism of the third subevent is between the CMT solution (ϕ = 310°, δ = 25°, λ = 110°) and the first motion solution (ϕ = 295°, δ = 32°, λ = 90°) by Adamek et al. (1987), except with a steeper dip angle.

Though a change of the subduction angle from 23° in northern Costa Rica to 30° in southern Costa Rica is suggested, the subduction geometry (dip angle and depth extent) near the source area is ambiguous even from the local seismicity study due to the lack of an adequate crustal velocity model (Protti et al., this volume). A comparison of the properties inferred for the subduction zone by our study and Protti et al. (this volume) is given in Table 1.

April 22, 1991, main event

A special session to discuss the 1991 Costa Rica earthquake was held at the 1991 American Geophysical Union fall meeting. Papers presented at the session pointed out that the source consists of multiple subevents (Goes and Schwartz, 1991; Montero et al., 1991; Tajima and Kikuchi, 1991), and the major moment release occurred in the up-dip direction with a steeper dip angle (Velasco et al., 1991; Güendel et al., 1991). Plafker and Ward (1992) combined geodetic data and focal mechanism solutions and suggested that the main rupture of the 1991 event dips landward beneath Costa Rica and northern Panama at an angle of about 30°, striking between 105° and 120° and is approximately 40 km wide and 80 km long.

Recently Goes et al. (1993) studied the 1991 earthquake using teleseismic broad-band body waveforms and long-period (157 to 288 s) surface (Love and Rayleigh) waves obtained from the Global Seismographic Network (GSN), International Deployment of Accelerometers (IDA), and GEOSCOPE. Their preferred solution from the surface waves has a major double couple with strike (ϕ) = 107 ± 5°, dip (δ) = 21 ± 10°, rake (λ) = 56 ± 11°, centroid depth (h) = 22 ± 8 km, and duration of source time function (τ) = 40 ± 6 s, with an 8% non–double couple component. They noted that shallow-dipping, shallow-depth earthquakes do not strongly excite surface wave moment tensor terms M_{XZ} and M_{YZ} (dip-slip component) and have large uncertainties in dip and moment estimates.

From the broad band body waveform inversion analysis, Goes et al. (1993) identified six subevents in two groups, the first one at approximately 15 km down-dip of the hypocenter and the second in an area 15 to 30 km up-dip. Results suggest oblique thrust motion (λ = 50 to 80°) on a shallow-dipping (δ = 15 to 20°), southeast-striking (ϕ = 90 to 120°) fault plane. The resolvable rupture area is no larger than 60 × 80 km^2, and most of the moment release occurred within an area of about 45 × 45 km^2 (see Fig. 5). They noted that the extent of the determined rupture is quite small for an event with M_w = 7.7.

After examining source models for various combinations of parameters including variable fault mechanisms, Goes et al. (1993) preferred a fixed fault mechanism source model, which was determined consistently with the body waveform and surface waveform modeling. This model suggests that the event occurred on a shallow southwest-dipping rupture plane, that the source process is relatively simple, and that most energy is released up-dip of the hypocentral location. Their preferred model has a focal mechanism with strike 102 ± 10°, dip 17 ± 14°, and rake 63 ± 17°; a seismic moment of 3.8 ± 1.5 × 10^{20} Nm (1.2 to 2.0 × 10^{20} Nm from body waveforms); and a total rupture duration of 40 ± 6 s. They also pointed out that whether the NPDB is a full-fledged or developing plate boundary is not clear.

We studied the source rupture process of the 1991 main event along with the 1983 earthquake, as described above. They used the iterative inversion method of Kikuchi and Kanamori (1991) with broad-band body waveforms from the GSN and GEOSCOPE broad-band/intermediate-band and long-period records. The broad- and intermediate-band wave-

TABLE 1. SUBDUCTION ZONE FOR THE 1983 EARTHQUAKE*

	Authors	
	Tajima and Kikuchi	Protti et al.
Objectives	Main source rupture process.	Subduction geometry.
Method and data	Teleseismic body waveform inversion for the main rupture.	Seismicity patterns using local and regional data.
Geometry of the subduction coupling.	Dip angle between 40° and 46°; the depth extent from 22 to 37 km (inferred from the mechanism of the main rupture).	30° (?) with a depth extent to as much as 70 km.

*From Protti et al., this volume, and Tajima and Kikuchi, present study.

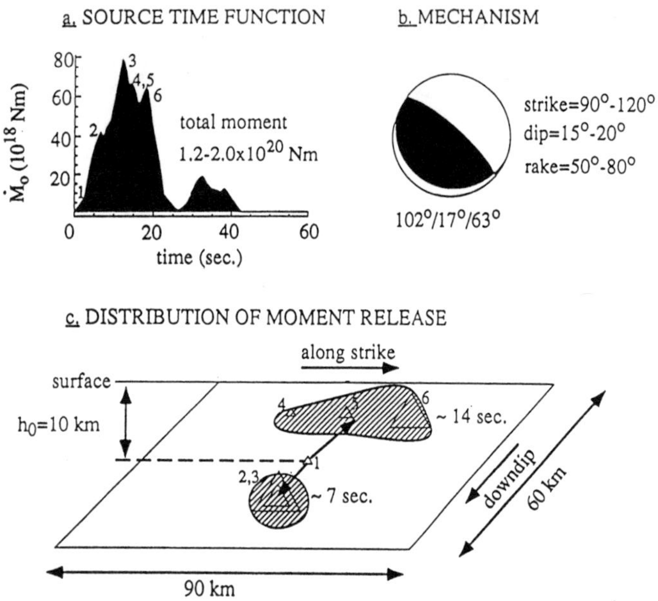

Figure 5. Results for the preferred model of the 1991 main source rupture obtained using teleseismic body waveforms by Goes et al. (1993). a, Source time function and six subevents identified. b, Source mechanism and its allowable range for the preferred model. c, Interpretation of the rupture propagation. The hatched regions depict the areas where most of the moment was released.

forms of the 1991 event show a somewhat complex time history of moment release, whereas such a feature is not obvious from the long-period waveforms. This rupture process was modeled as three thrust subevents followed by a minor strike-slip subevent. The primary moment release has been carried out by the thrust fault mechanism in a time span of about 20 s (Fig. 6a).

The subevent locations and onset times suggest that the rupture started at the hypocenter (depth = 12 km) and the major moment release occurred in the up-dip direction. The dip angles of the first and second subevents are 36° and 39°, steeper than the CMT solution (δ = 25°). The dip angle of the third subevent located at a depth of 17 km is 16°, less steep than the CMT solution. The varying dip angles of the subevents imply a rupture zone that flattens with depth and may reflect the local tectonic structure. In comparison, the study by Goes et al. (1993) preferred a rather simple source model (e.g., with no mechanism change) in which a range of the dip angle is 17 ± 14°. However, the mechanism obtained from the summation of the moment tensors of the four subevents is a thrust mechanism (ϕ = 123°, δ = 32°, and λ = 89°) comparable to those determined by CMT or Goes et al. (1993). Figure 6b illustrates the bending rupture zone with a steeper dip angle (36 to 39°) at a shallow depth (12 to 7 km) and a smaller dip angle (~16°) at a depth of 17 km, inferred from the body waveform inversion.

1991 aftershocks

The 1991 Costa Rica earthquake produced a number of aftershocks. Fan et al. (1993) studied waveform data recorded by a network of three Program for Array Seismic Studies of the Continental Lithosphere (PASSCAL)–type, portable instruments that was deployed to monitor the aftershock activity in southern Costa Rica two to six weeks after the main shock. They determined source parameters of 15 relatively small aftershocks with magnitudes between 3.2 and 4.4, using a regional moment tensor inversion method (Fan and Wallace, 1991). In this analysis they admitted that the focal depths are not well constrained owing to the uncertainty of the local velocity structure model, but the relative source depths can be compared.

The focal mechanisms obtained show a variety of faulting behavior: Thrust faulting, strike-slip faulting, and normal faulting are all present. From the distribution of the focal mechanisms, Fan et al. (1993) noted a clear spatial segmentation of focal mechanism types. Most aftershocks near or southeast of the main shock were thrusting events, with focal mechanisms similar to the main shock. On the other hand, a cluster of aftershocks northwest of the main shock showed dominantly left-lateral, strike-slip motion on a northeasterly striking nodal plane. Figure 7 shows a block diagram modified from Isacks et al. (1969) to illustrate the nature of faulting for the main shock and aftershocks of the 1991 earthquake.

Suárez et al. (unpublished manuscript) studied aftershock locations using data recorded by local network instruments, including nine portable seismographs installed in the aftershock zone. Several hundred aftershocks of a duration magnitude (Lee et al., 1972) ranging from 2.4 to 4.0 were recorded during the field campaign from 29 April through 4 May, 1991. The epicenters of the aftershocks determined from the local network data are located in the back arc of Costa Rica, to the northeast of the Talamanca Cordillera (see Fig. 1). When they compared the teleseismically determined epicenters to those based on local arrival times, the teleseismic locations appear to be displaced 15 to 25 km to the northwest. The aftershock zone defines an ellipse of approximately 45 × 85 km^2. The long axis of the inferred rupture zone is parallel to the orientation of the Caribbean coast (see Fig. 8). The cross-section projections show that the average focal depth of the aftershocks increases from about 15 km near the coastline to an average of 20 to 30 km beneath the continent. Suárez et al. (unpublished manuscript) pointed out that the cross sections show an almost horizontal distribution of seismicity; the distribution of the aftershocks along cross sections BB′ and CC′ (Fig. 8) suggest that nucleation of the source rupture began at a depth of about 20 km. The aftershock distribution agrees with the hypocentral location (depth = 24 km) of the main event determined with the local data. Suárez et al. (unpublished manuscript) concluded that the actual rupture probably took place on a thrust sheet that is subhorizontal beneath the continent and steepens

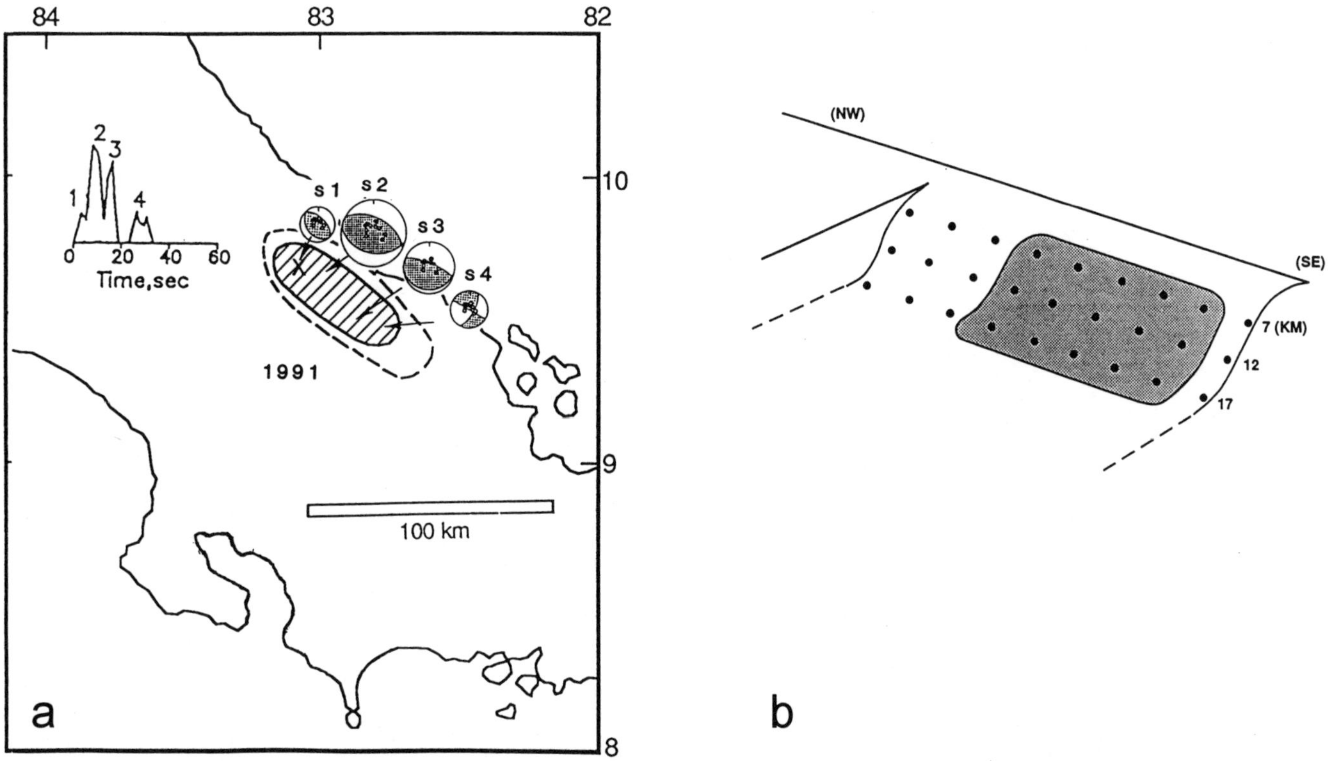

Figure 6. Results obtained for the 1991 main event by our study. The notations are the same as in Figure 4. a, The entire rupture was modeled with subevents S1, S2, S3, and S4. The primary moment release has been carried out with three thrust subevents that occurred in the first 20 sec. b, A schematic illustration of the bending rupture zone inferred from the inversion result.

toward the east, reaching the surface offshore at a high angle (see Fig. 9). This rupture pattern, which suggests a bending rupture zone, is similar to our result.

The seismological properties obtained for the 1991 Costa Rica earthquake by the present study, Goes et al. (1993), Suárez et al. (unpublished manuscript), and Fan et al. (1993) are summarized in Table 2.

DISCUSSION

Implications from seismological approaches

Although there are some disagreements of results or interpretations among the seismological studies above, we can extract some common or complementary factors to describe the seismotectonic process in the Costa Rica region and the role of the subducting buoyant ridge. The seismicity patterns in Figure 2 illustrate the stress accumulation and release associated with the two large Costa Rica earthquakes and other major events in this region during the period between 1964 and 1991. The 1983 source area had some precursory events including a normal faulting event (Ms = 6.5) with its aftershocks near the southeastern boundary and a thrust event (Ms = 6.4) near the northwestern boundary. The distribution of the subevents suggests that the rupture propagated bilaterally with its primary rupture in the Cocos Ridge subduction zone and a reactivated subevent in the loose fault zone southeast of the epicenter. The source duration time of about 40 s indicates a relatively slow rupture process. As was suggested by Bonafede et al. (1983), the effect of viscoelastic coupling near the rupture zone may produce less-brittle failure, resulting in a slow rupture process as well as a depletion of high-frequency seismic waves. This depletion has been observed for the source spectra of short period records from the 1983 event (Houston and Kanamori, 1986).

The aftershock area expansion pattern supports the idea of bilateral rupture propagation of the 1983 main event. A few major aftershocks ($5 \leq M \leq 6.1$; see Adamek et al. [1987] for the mechanisms) are located around the second and third subevents identified in the inversion of the present study. The aftershock expansion pattern of the 1983 event indicates that the fault coupling northwest and southeast of the main rupture zone is weak (Adamek et al., 1987). The observations northwest of the rupture zone correspond to the diffuse, transcurrent deformation zone northwest of the 1991 main rupture area pointed out by our study, Goes et al. (1993), Fan et al. (1993), and Suárez et al. (unpublished manuscript).

The future 1991 source area had no teleseismically observable activity (for events with magnitudes ≥ 4.7) before the

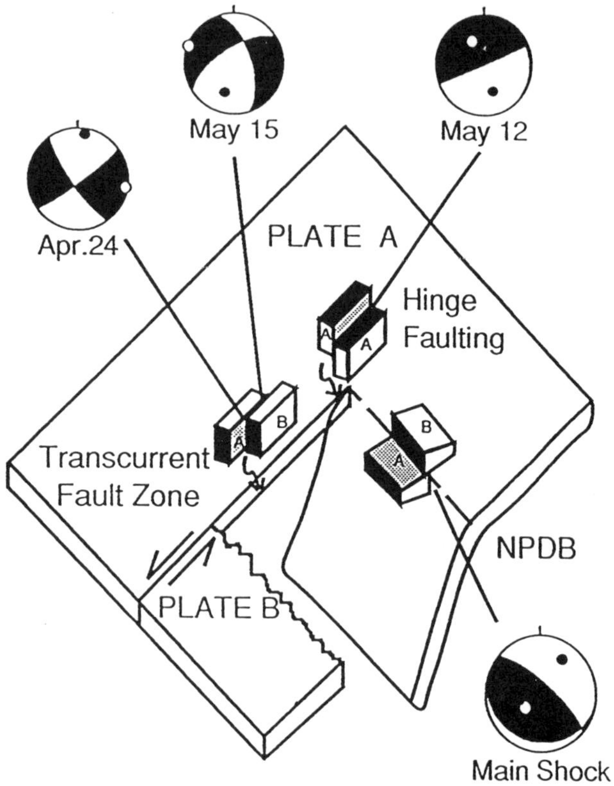

Figure 7. A model proposed by Fan et al. (1993) showing the faulting characteristics associated with the 1991 main event and its aftershocks.

main event. The region between the 1983 and 1991 source areas was quiet as well. This implies that the coupling of the 1991 fault zone is strong; the stress produced by the 1983 event normal to the strike was not released seismically but possibly transmitted through the buoyant zone to the impending 1991 source area where the stress accumulation was high. The presence of the Talamanca range in Costa Rica and its geologic structure, together with the occurrence of the 1991 back-thrusting event, reflects an increase in the compressive stress transmitted to the upper plate by the subduction of the Cocos Ridge (Kolarsky et al., this volume). The concentration of major moment release in a relatively small rupture zone for the 1991 source process as analyzed by our study and Goes et al. (1993) supports this hypothesis.

The varying dip angles of the thrust subevents at different depths (39° at 7 km and 16° at 17 km) suggest a slightly bending rupture zone for the 1991 back-thrusting event (e.g., steeper angle at a shallower depth and flattening at a deeper depth). This rupture pattern supports the model proposed by Suárez et al. (unpublished manuscript) based on the aftershock locations (see the illustration in Fig. 9). The strong ground motion data recorded at 68 km southwest of the epicenter suggest a multiple and complex rupture, and at least five subevents can be identified; the fifth subevent is the largest of the sequence, occurring at 21.2 s after the first subevent.

Suárez et al. (unpublished manuscript) also noted that the epicenter of the main event reported by the NEIC lies about 10 km to the northeast of the location determined using local data. The focal depth of 24 km determined from local data agrees better with the aftershock distribution (see Fig. 8) than the focal depth of about 10 km determined teleseismically by Goes et al. (1993) (note that the centroid depth determined with a long-period surface wave inversion by Goes et al. [1993] is 22 ± 8 km). The aftershock distribution along cross sections CC′ and DD′ near the epicenter suggests a relatively flat fault at a depth of approximately 20 km (Fig. 8).

A clear spatial segmentation was observed from the focal mechanism types of the 1991 aftershocks (Fan et al., 1993). A similar spatial segmentation occurred after the 1983 earthquake (Adamek et al., 1987). The two earthquakes with the aftershocks taken together define a diffuse deformation zone in central Costa Rica, characterized by left-lateral strike-slip motion (Goes et al., 1993; Fan et al., 1993). The diffuse, transcurrent deformation zone coincides with the termination of the volcanic chain in central Costa Rica and a change of the Wadati-Benioff zone geometry along the southwestern coast of Costa Rica (Fan et al., 1993).

The rupture patterns and seismicity associated with the two large Costa Rica earthquakes provide us a clearer picture of geometry for the interaction between the subducting ridge and overriding Caribbean plate than the schematically illustrated models at ridge-trench intersections (McGeary et al., 1985). The models by McGeary et al. (1985) as well as others (Kelleher and McCann, 1976; Chung and Kanamori, 1978a, b) attempted to explain the mechanisms of resisting forces to subduction, which affect both the overriding and underlying lithosphere and eventually deform the margin of the overriding plate. But the consequent results are described rather schematically, reflecting the simplified models, and do not provide quantitative constraints for the mechanical interaction between the forearc and backarc source areas.

The mechanical interaction between a subducting buoyant ridge and the overriding plate may be investigated further by assessing the associated crustal deformation for coseismic and postseismic displacement and interseismic strain accumulation, using the source parameters of large earthquakes. The gravitational stress due to the height of the buoyant ridge should also be accounted for. When a buoyant ridge is thrust beneath an "island arc," the island arc rides up over the plate (backthrusting), which replaces the asthenospheric wedge. The topographic uplift illustrated in Figure 1 is evidence of this tectonic situation.

Plate convergence and seismic slip ratio

Earthquake source characteristics and occurrence cycles at various convergent plate boundaries have been studied by many researchers. Assuming that the source mechanisms of the historic events were similar to those of the 1983 event, the

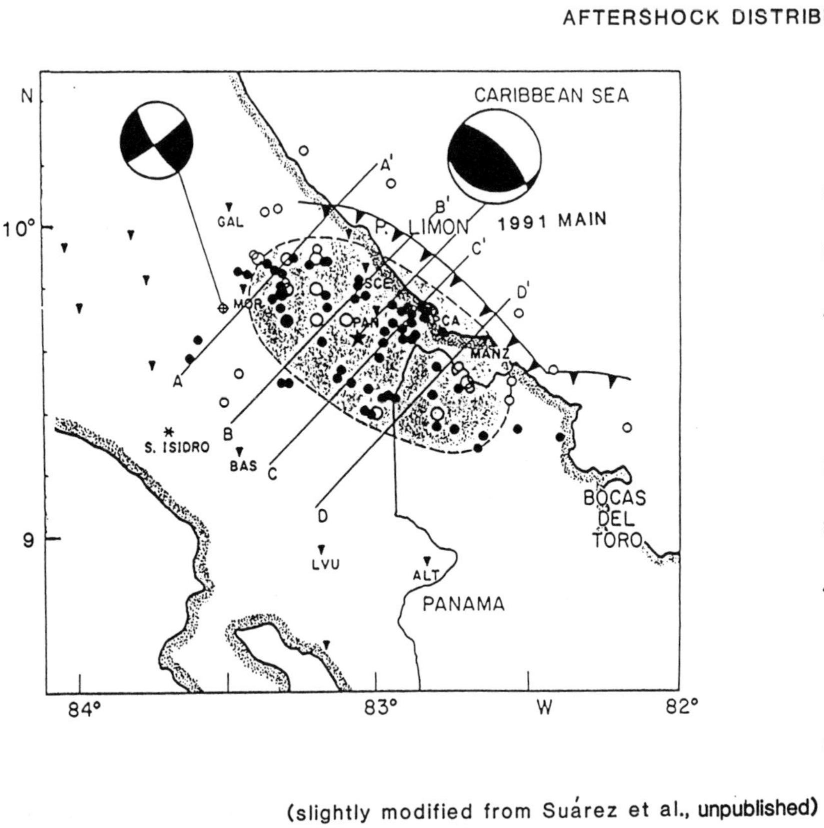

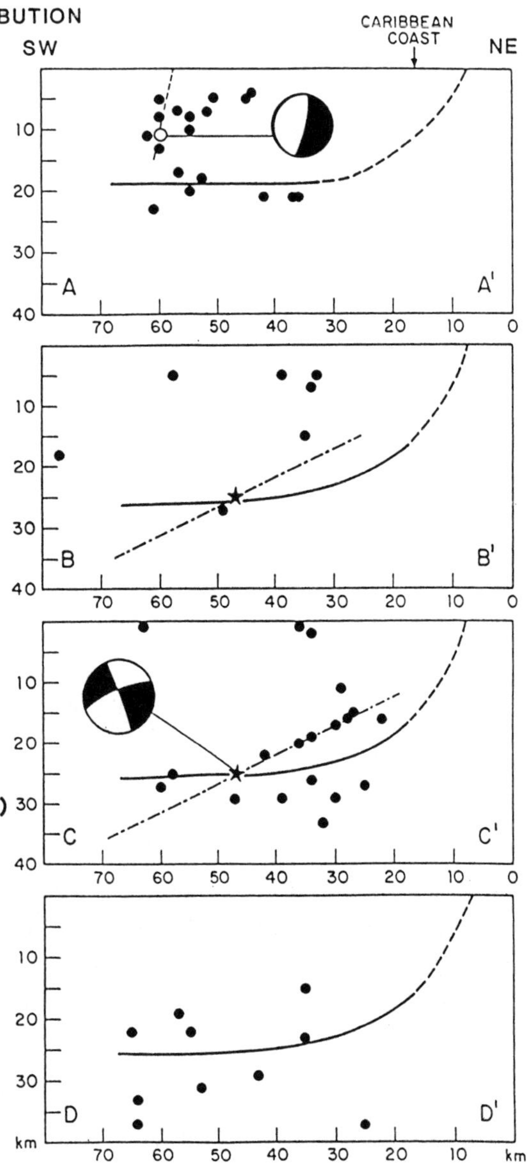

Figure 8. Aftershock distributions adopted and modified from Suárez et al. (unpublished manuscript). Left, Epicentral locations of aftershocks recorded by the temporarily installed network (solid circles) and those reported by NEIC (open circles) for the first seven days. The star indicates the main shock. (Right, Cross sections of the aftershocks shown on the left. The dashed lines on BB′ and CC′ show the average dip of the fault used to model the coastal uplift. The solid lines represent the geometry of the fault plane inferred from the aftershocks.

recurrence time of a large event in this subduction zone was estimated to be approximately 40 years (Astiz and Kanamori, 1984). If the plate convergence rate V, average slip of a characteristic large earthquake D, and the ratio of seismic slip to total plate motion η are known, the recurrence time τ can be calculated by

$$\tau = D/\eta V \tag{1}$$

Although V and D are known or can be determined reasonably well, a large uncertainty is involved in η (Kanamori and Astiz, 1985). When the seismic slip ratio is not certain because of the lack of reliable historical data or the variable rupture length of large individual earthquakes from time to time, an estimate of recurrence interval is not certain, either. The source rupture process of the 1983 event (Adamek et al., 1987; present study) implies a small seismic slip ratio in the Cocos Ridge subduction zone and also disturbance in the recurrence interval due to the interaction of the two source areas (see also Mann et al., 1990).

Plafker and Ward (1992) estimated that if the full 92 mm/y (N29°) (DeMets et al., 1990) of Cocos-Caribbean convergence is partitioned between forearc (Pacific side) and backarc (Caribbean side) components and backarc convergence falls between 2 and 11 mm/y at the Cocos Ridge subduction zone, then the recurrence time of a similar back-thrusting event is estimated to be from 200 to 1,100 years. Given the effects of crustal deformation associated with the back-thrusting event in 1991, the recurrence interval, estimated at approximately 40 years in the subduction zone (1983 source area), may be disturbed as well because of the Cocos Ridge involvement.

Resolution and limitation of seismological analyses

We reviewed the source ruptures and seismicity patterns associated with the 1983 and 1991 Costa Rica earthquakes with the goal of applying seismological approaches to con-

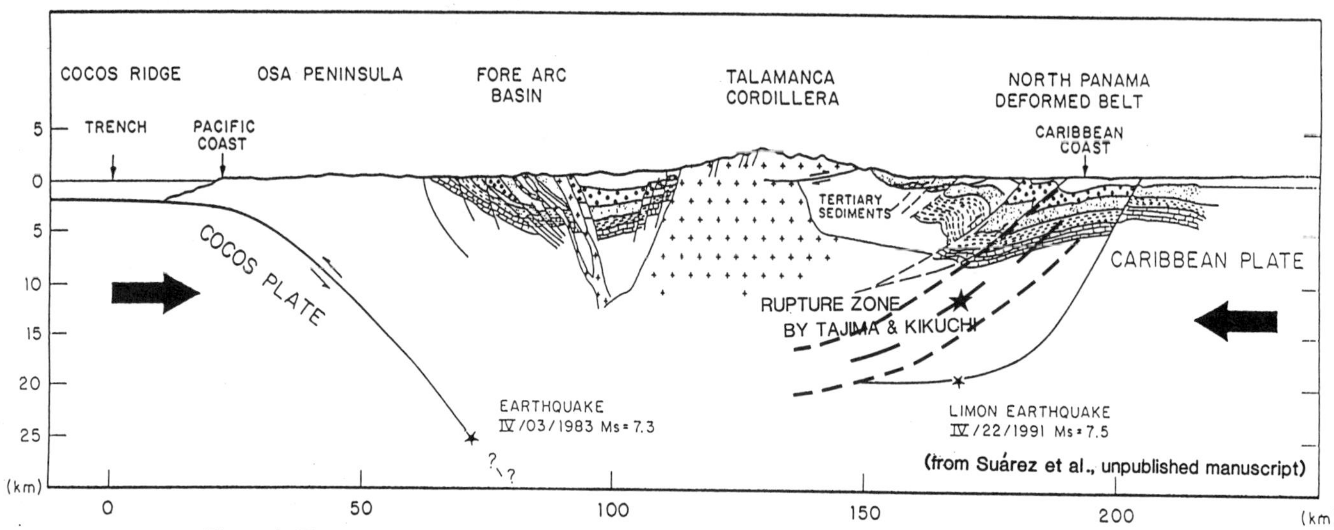

Figure 9. The rupture zone model for the 1991 source area obtained by our study is illustrated with thick dashed lines on a schematic cross section adopted from Suárez et al. (unpublished manuscript). The surface geology from the Osa Peninsula on the Pacific side to Puerto Limón on the Caribbean side was taken from Rivier (1985).

TABLE 2. COMPARISON OF SEISMOLOGICAL STUDIES FOR THE 1991 EARTHQUAKE*

	Authors			
	Tajima and Kikuchi	Goes et al.	Suárez et al.	Fan et al
Determined quantities	Main rupture process.	Main rupture process.	Locations of aftershocks.	Moment tensors of 15 aftershocks (3.2 to 4.4).
Used data and method	Inversion modeling using teleseismic body waveforms (BWV) (IRIS and GEOSCOPE broad band, and DWWSSN mid-period).	Inversion with teleseismic BWV and surface waves (SWV).	Location determination using travel times from locally recorded data by temporarily installed seismometers.	Inversion using locally recorded body waveforms by three PASSCAL instruments.
Main source duration time.	29 s	40 s (20s[†]) from BWV. 40 ± 6 s from SWV.	N.A.	N.A.
Source complexity of main rupture	Four subevents (three thrust subevents).	Single event from SWV. Six subevents from BWV.	N.A.	N.A.
Mechanism: Strike = ϕ Dip = δ Rake = λ	ϕ = 123° (91°-138°)[§] δ = 32° (16°-39°)[§] λ = 89° (69°-96°)[§]	ϕ_S = 107 ± 5°, δ_S = 21 ± 10°, λ_S = 56 ± 11° from SWV. ϕ_B = 90°-120°, δ_B = 15° - 20°, λ_B = 50° - 80° from BWV.	N.A.	(Varying mechanisms also with time.)
Rupture area	40 x 60 km^2, depth = 7 to 17 km.	60 x 80 km^2 (45 x 45 km^2[†]); depth = 15 to 30 km from BWV. Centroid depth = 22 ± 8 km from SWV.	45 x 85 km^2; depth ≈20 km.	Depth = 4 to 16 km

*From Fan et al., 1993; Goes et al., 1993; Suárez et al., unpublished; and Tajima and Kikuchi, present study.
[†]Most of the moment was released within this area and process time (Goes et al., 1993).
[§]The ranges are shown only for the three thrust subevents.

strain regional tectonics and structural geology. Using high-quality digital seismograms and robust inversion methods, we now have the means to determine complex earthquake rupture processes that provide useful clues for the involved tectonic structure. Nonuniqueness and instability of the solutions remain a problem, however, and these are commonly encountered in general.

One of the causes for such problems is the lack of knowledge of the near-source velocity structure that affects the source depth determination and waveform inversion. The only available crustal model in this region, that of Matumoto et al. (1977), was not adequate to model the observed body waveforms (present study; Goes et al., 1993). As described by Suárez et al. (unpublished manuscript), the locally determined hypocentral depth of the main event is 24 km, whereas the hypocentral depth determined teleseismically by Goes et al. (1993) is about 10 km. The uncertainty of the near-source velocity structure and hypocentral depth can cause the inversion results to be unstable. We and Goes et al. (1993) tested teleseismic body waveforms for an adequate near-source structure. Fan et al. (1993) tested regional waveform data of the aftershocks for the influence of the structure and admitted that the depth determinations are not stable within several kilometers but that the source mechanism parameters are well constrained. This situation leads the results to be less convincing even though the varying dip angle of the fault mechanism during the rupture provides useful information for the rupture zone and is consistent with the hypocentral distribution of the aftershocks obtained from local data (Suárez et al., unpublished manuscript). Under this circumstance a preferred model, which is a sort of averaged solution with a fixed fault mechanism, is given with the allowable ranges of the dip and rake angles for the subevents (Goes et al., 1993).

Currently high-quality global digital waveform data are accessible from worldwide networks shortly after a large event occurs. However, local information such as the near-source structure is not generally available immediately after an earthquake, even though it is crucial in the following seismological analysis. We are still in the process of improving our practical ability to obtain a stable solution of waveform modeling or inversion for events that take place in tectonically complex regions.

ACKNOWLEDGMENTS

The authors thank P. Mann for the opportunity to contribute a chapter to this special issue. We benefited from discussions with him, from his useful review comments for the manuscript, and from his patience throughout the preparation of this manuscript. The authors also would like to thank J. Dewey and L. Astiz for their constructive review comments and S. Schwartz, G. Suárez, and G. Plafker for sending preprints. K. McIntosh and P. Nyffenegger read and provided helpful comments for the final manuscript. This study was partially supported by NSF grants EAR-8816993, and EAR-9011845.

REFERENCES CITED

Adamek, S., Tajima, F., and Wiens, D. A., 1987, Seismic rupture associated with subduction of the Cocos Ridge: Tectonics, v. 6, p. 757–774.

Astiz, L., and Kanamori, H., 1984, An earthquake doublet in Ometepec, Guerrero, Mexico: Physics of the Earth and Planetary Interior, v. 34, p. 24–45.

Bonafede, M., Boschi, E., and Dragoni, M., 1983, Viscoelastic stress relaxation on deep fault sections as a possible source of very long period elastic waves: Journal of Geophysical Research, v. 88, p. 2251–2260.

Burbach, G. V., and Frohlich, C., 1986, Intermediate and deep seismicity and lateral structure of the subducted lithosphere in the Circum-Pacific region: Reviews of Geophysics, v. 20, p. 833–874.

Burbach, G. V., Frohlich, C., Pennington, W. D., and Matumoto, T., 1984, Seismicity and tectonics of the subducted Cocos plate: Journal of Geophysical Research, v. 89, p. 7719–7735.

Carr, M. J., and Stoiber, R. E., 1977, Geologic setting of some destructive earthquakes in Central America: Geological Society of America Bulletin, v. 88, p. 151–156.

Chung, W. Y., and Kanamori, H., 1978a, Subduction process of a fracture zone and oceanic plateau ridges—The focal mechanism and source characteristics of the New Hebrides earthquakes of 1969 January 19 and some related events: Geophysical Journal of Royal Astronomical Society, v. 54, p. 221–240.

Chung, W. Y., and Kanamori, H., 1978b, A mechanical model for plate deformation associated with aseismic ridge subduction in the New Hebrides arc: Tectonophysics, v. 50, p. 29–40.

Dean, B. W., and Drake, C. L., 1976, Focal mechanism solutions and tectonics of the Middle America arc: Journal of Geology, v. 86, p. 111–128.

DeMets, C., Gordon, R. G., Argus, D. F., and Stein, S. A., 1990, Current plate motions: Geophysical Journal International, v. 101, p. 425–478.

Dewey, J. W., and Algermissen, S. T., 1974, Seismicity of the Middle America arc-trench system near Managua, Nicaragua: Bulletin of Seismological Society of America, v. 64, p. 1033–1048.

Dziewonski, A. M., Franzen, J. E., and Woodhouse, J. H., 1983, Centroid-moment tensor solutions for April-June, 1983: Physics of the Earth and Planetary Interior, v. 33, p. 243–249.

Eissler, H., and Kanamori, H., 1982, A large normal-fault earthquake at the junction of the Tonga trench and the Louisville ridge: Physics of the Earth and Planetary Interior, v. 29, p. 161–172.

Fan, G.-W., and Wallace, T. C., 1991, The determination of source parameters for small earthquakes from a single very broadband seismic station: Geophysical Research Letters, v. 18, p. 1385–1388.

Fan, G.-W., Beck, S. L., and Wallace, T. C., 1993, The seismic source parameters of the 1991 Costa Rica aftershock sequence: Evidence for a transcurrent plate boundary: Journal of Geophysical Research, v. 98, p. 15759–15778.

Goes, S., and Schwartz, S. Y., 1991, Rupture process of the April 22, 1991 Valle de la Estrella, Costa Rica earthquake from teleseismic body waves: Eos (American Geophysical Union Transactions), v. 72, p. 301.

Goes, S.D.B., Velasco, A. A., Schwartz, S. Y., and Lay, T., 1993, The April 22, 1991 Valle de la Estrella, Costa Rica (Mw = 7.7) earthquake and its tectonic implications: A broadband seismic study: Journal of Geophysical Research, v. 98, p. 8127–8142.

Güendel, F., and 14 others, 1991, Mainshock-aftershock sequence associated with the Costa Rica earthquake of April 22, 1991: Eos (American Geophysical Union Transactions), v. 72, p. 300.

Hey, R., 1977, Tectonic evolution of the Cocos-Nazca spreading center: Bulletin of Geological Society of America, v. 88, p. 1404–1420.

Houston, H., and Kanamori, H., 1986, Source spectra of great earthquakes: Teleseismic constraints on rupture process and strong motion: Bulletin of Seismological Society of America, v. 76, p. 19–42.

Isacks, B., Sykes, L. R., and Oliver, J., 1969, Focal mechanisms of deep and shallow earthquakes in the Tonga-Kermadec region and the tectonics of island arcs: Geological Society of America Bulletin, v. 80, p. 1443–1470.

Kanamori, H., and Astiz, L., 1985, The 1983 Akita-Oki earthquakes (MW =

7.8) and its implication for systematics of subduction earthquakes: Earthquake Prediction Research, v. 3, p. 305–317.

Kelleher, J., and McCann, W., 1976, Bouyant zones, great earthquakes, and unstable boundaries of subduction: Journal of Geophysical Research, v. 81, p. 4885–4896.

Kikuchi, M., and Kanamori, H., 1982, Inversion of complex body waves: Bulletin of Seismological Society of America, v. 72, p. 491–506.

Kikuchi, M., and Kanamori, H., 1991, Inversion of complex body waves—III: Bulletin: Bulletin of Seismological Society of America, v. 81, p. 2335–2350.

Lee, W., Bennett, R., and Meagher, K., 1972, A method of estimating magnitude of local earthquakes from signal duration: U.S. Geological Survey Open-file Report, 28 p.

LeFevre, L. V., and McNally, K. C., 1985, Stress distribution and subduction of aseismic ridges in the Middle America subduction zone: Journal of Geophysical Research, v. 90, p. 4495–4510.

Liaw, H. B., 1981, Seismic velocity modeling from an ensemble of earthquakes [Ph.D. thesis]: Dallas, University of Texas at Dallas, 135 p.

Mann, P., Schubert, C., and Burke, K., 1990, Review of Caribbean neotectonics, in Dengo, G., and Case, J. E., eds., The Caribbean Region: Geological Society of America, The Geology of North America, v. H, p. 307–338.

Matumoto, T., Ohtake, M., Latham, G., and Umana, J., 1977, Crustal structure in southern central America: Bulletin of Seismological Society of America, v. 67, p. 121–134.

McCann, W. R., and Sykes, L. R., 1984, Subduction of aseismic ridges beneath the Caribbean plate: Implications for the tectonics and seismic potential of the northeastern Caribbean: Journal of Geophysical Research, v. 89, p. 4493–4519.

McGeary, S., Nur, A., and Ben-Avraham, Z., 1985, Spatial gaps in arc volcanism: The effect of collision or subduction of oceanic plateaus: Tectonophysics, v. 119, p. 195–221.

Miyamura, S., 1975, Recent crustal movements in Costa Rica disclosed by leveling surveys: Tectonophysics, v. 29, p. 191–198.

Molnar, P., and Sykes, L. R., 1969, Tectonics of the Caribbean and Middle America regions from focal mechanisms and seismicity: Geological Society of America Bulletin, v. 80, p. 1639–1684.

Montero, W., and seven others, 1991, The Limon, Costa Rica earthquake of April 22, 1991 (Ms = 7.5): Seismicity, focal mechanism and tectonic implication: Eos (American Geophysical Union Transactions), v. 72, p. 301.

Plafker, G., and Ward, S. N., 1992, Thrust faulting and tectonic uplift along the April 22, 1991 Costa Rica earthquake: Tectonics, v. 11, p. 709–718.

Rivier, F., 1985, Sección geologica del Pacífico al Atlántico a traves de Costa Rica, Revista: Geológica de America Central, v. 2, p. 22–32.

Stoiber, R. E., and Carr, M., 1973, Quaternary volcanic and tectonic segmentation of Central America: Bulletin of Volcanology, v. 37, p. 304–325.

Tajima, F., 1984, Study of source processes of the 1965, 1968, and 1978 Oaxaca earthquakes using teleseismic short-period records: Journal of Geophysical Research, v. 89, p. 1867–1873.

Tajima, F., 1985, The 1983 Costa Rica earthquake: A complex and long-lasting event: Earthquake Notes, v. 55, p. 14–15.

Tajima, F., and Kikuchi, M., 1991, The 1983 and 1991 Costa Rica earthquakes: Complex source processes with mechanism changes: Eos (American Geophysical Union Transactions), v. 72, p. 302.

Velasco, A. A., Lay, T., and Zhang, J., 1991, Long period source parameters of the April 22, 1991, Valle de la Estrella, Costa Rica, earthquake: Eos (American Geophysical Union Transactions), v. 72, p. 301.

Vogt, P. R., Lowrie, A., Bracey, D. R., and Hey, R. N., 1976, Subduction of aseismic oceanic ridges: Effects on shape, seismicity, and other characteristics of consuming plate boundaries: Geological Society of America Special Paper 172, 59 p.

Manuscript Accepted by the Society February 7, 1994

Printed in U.S.A.

Geological Society of America
Special Paper 295
1995

Selected publications of Richard Weyl on the geology of Central America

Paul Mann
Institute of Geophysics, University of Texas at Austin, 8701 Mopac Boulevard, Austin, Texas 78759

1. Weyl, R., 1953, Beiträge zur Geologie El Salvadors: I. Geologisch-morphologische Übersicht: Neues Jahrbuch für Geologie und Paläontologie, Monatshefte, p. 198–202.

2. Weyl, R., 1953, Beiträge zur Geologie El Salvadors: II. Lithogenetische Studien in den Mangroven der Pazifik-Küste: Neues Jahrbuch für Geologie und Paläontologie, Monatshefte, p. 202–218.

3. Weyl, R., 1953, Aktiver und erloschener Vulkanismus in El Salvador (Zentralamerika): Geologische Rundschau, v. 42, p. 146.

4. Weyl, R., 1954, Beiträge zur Geologie El Salvadors: IV. Die Bimsaschen in der Umgebung San Salvadors: Neues Jahrbuch für Geologie und Paläontologie, Monatshefte, p. 49–70.

5. Weyl, R., 1954, Beiträge zur Geologie El Salvadors: V. Die Schmelztuffe der Balsamkette: Neues Jahrbuch für Geologie und Paläontologie, Abhandlungen, v. 99, p. 1–32.

6. Weyl, R., 1954, Beiträge zur Geologie El Salvadors: VI. Die Lavae der jungen Vulkane: Neues Jahrbuch für Geologie und Paläontologie, Abhandlungen, v. 101, p. 12–38.

7. Weyl, R., 1954, Die Schmeltztuffe des zirkumpazifischen Vulkangürtels: Zeitschrift der Deutschen Geologischen Gesellschaft, v. 105, p. 207.

8. Weyl, R., 1955, Geologische Studien in der Cordillera de Talamanca von Costa Rica: Neues Jahrbuch für Geologie und Paläontologie, Monatshefte, p. 262–269.

9. Weyl, R., 1956, Spuren eiszeitlicher Vergletscherung in der Cordillera de Talamanca Costa Ricas (Mittelamerika): Neues Jahrbuch für Geologie und Paläontologie, Abhandlungen, v. 102, p. 283–294.

10. Weyl, R., 1956, Eiszeitliche Gletscherspuren in Costa Rica (Mittelamerika): Zeitschrift für Gletscherkunde und Glazialgeologie, v. 3, p. 318–325.

11. Weyl, R., 1956, Neue Erkenntnisse über den geologischen Aufbau der Ozeanböden: Nachrichten der Giessener Hochschulgesellschaft, v. 25, p. 60–73.

12. Weyl, R., 1957, Erdgeschichte und Bau der Cordillera de Talamanca in Costa Rica (Mittelamerika): Zeitschrift der Deutschen Geologischen Gesellschaft, v. 109, p. 663–664.

13. Weyl, R., 1957, Erdgeschichte und Bau der mittelamerikanishen Festlandsbrücke: Zeitschrift der Deutschen Geologischen Gesellschaft, v. 109, p. 644–646.

14. Weyl, R., 1957, Beiträge zur Geologie der Cordillera de Talamanca Costa Ricas (Mittelamerika): Neues Jahrbuch für Geologie und Paläontologie, Abhandlungen, v. 105, p. 123–204.

15. Weyl, R., 1957, Vulkanismus und Plutonismus in südlichen Mittelamerikas: Geologische Rundschau, v. 46, p. 220–228.

16. Weyl, R., 1960, The Cordillera de Talamanca of Costa Rica in the western rim of the Caribbean area: Mayaguez, Puerto Rico, Transactions of the Second Caribbean Geological Conference, p. 162.

17. Schulz, R., and Weyl, R., 1960, Erdbeben und Krustenaufbau im nördlichen Mittelamerika: Neues Jahrbuch für Geologie und Paläontologie, Monatshefte, p. 193–201.

18. Weyl, R., 1960, Die Orogene Mittelamerikas: Geologische Rundschau, v. 50, p. 605–619.

19. Weyl, R., 1961, Die Geologie Mittelamerikas: Berlin, Verlag Borntraeger, 221 p.

20. Weyl, R., 1961, Mittelamerikanische Ignimbrite: Neues Jahrbuch für Geologie und Paläontologie, Abhandlungen, v. 113, p. 23–46.

21. Weyl, R., 1961, Volcanes en Centroamérica: Alemania, v. 2, p. 39–43.

22. Weyl, R., 1964, Volcanoes and volcanic rocks in Central America and the West Indies: Port-of-Spain, Trinidad, Transactions of the 4th Caribbean Geological Conference, p. 357–360.

23. Weyl, R., 1965, Erdgeschichte und Landschaftsbild in Mittelamerika: Frankfurt am Main, Verlag W. Kramer, 175 p.

24. Weyl, R., 1966, Die Geologie der Antillen: Berlin, Verlag Borntraeger, 410 p.

25. Weyl, R., 1966, The paleogeographic development of Central America—West Indian region: Boletin Informativo de Asociación Venezolana Geología, Minería, y Petróleo (Caracas, Venezuela), v. 9, p. 99–120.

26. Weyl, R., 1966, Tektonik, Magmatismus und Krustenbau in Mittelamerika und Westindien: Geotektonische Forschungen, v. 23, p. 67–109.

27. Weyl, R., 1966, Ozeanische Kruste im südlichen Mittelamerika?: Neues Jahrbuch für Geologie und Paläontologie, Monatshefte, v. 5, p. 275–281.

28. Weyl, R., 1967, Krustenbau und sialischer Magmatismus: Geologische Rundschau, v. 56, p. 369–372.

29. Weyl, R., and Henningsen, D., 1967, Ozeanische Kruste im Nicoya-Komplex von Costa Rica (Mittelamerika): Geologische Rundschau, v. 57, p. 33–47.

30. Weyl, R., 1969, Magmatische Förderphasen und Gesteinschemismus in Costa Rica (Mittelamerika): Neues Jahrbuch für Geologie und Paläontologie, Monatshefte, p. 423–446.

31. Weyl, R., 1969, Magnetitsande der Küste Nicoyas (Costa Rica, Mittelamerika): Neues Jahrbuch für Geologie und Paläontologie, Monatshefte, p. 499–511.

32. Pichler, H., and Weyl, R., 1974, Petrochemical aspects of Central American magmatism: Geologische Rundschau, v. 62, p. 357–396.

33. Weyl, R., 1974, Die Paläogeographische Entwicklung Mittelamerikas: Zentralblatt für Geologie und Paläontologie, Teil 1, v. 5/6, p. 432–466.

Mann, P., 1995, Selected publications of Richard Weyl on the geology of Central America, *in* Mann, P., ed., Geologic and Tectonic Development of the Caribbean Plate Boundary in Southern Central America: Boulder, Colorado, Geological Society of America Special Paper 295.

Index

[Italic page numbers indicate major references]

H

I

J

K

L

M

T

U

V

W

X

Z

Typeset by WESType Publishing Services, Inc., Boulder, Colorado
Printed in U.S.A. by Johnson Printing, Boulder, Colorado